AF327338

Springer Series in
MATERIALS SCIENCE 103

A.S. Alexandrov (Ed.)

Polarons in Advanced Materials

With 223 Figures

 Springer

A.S. Alexandrov (Ed.)
Department of Physics
Loughborough University
Loughborough LE11 3TU
United Kingdom

Series Editors:

Professor Robert Hull
University of Virginia
Dept. of Materials Science and Engineering
Thornton Hall
Charlottesville, VA 22903-2442, USA

Professor R. M. Osgood, Jr.
Microelectronics Science Laboratory
Department of Electrical Engineering
Columbia University
Seeley W. Mudd Building
New York, NY 10027, USA

Professor Jürgen Parisi
Universität Oldenburg, Fachbereich Physik
Abt. Energie- und Halbleiterforschung
Carl-von-Ossietzky-Strasse 9–11
26129 Oldenburg, Germany

Professor Hans Warlimont
Institut für Festkörper-
und Werkstofforschung,
Helmholtzstrasse 20
01069 Dresden, Germany

A C.I.P. Catalogue record for this book is available from the Library of Congress

ISSN 0933-033x

ISBN 978-1-4020-6347-3 (HB)

ISBN 978-1-4020-6348-0 (e-book)

Published by Springer,
P.O. Box 17, 3300 AA Dordrecht, The Netherlands
In association with
Canopus Publishing Limited,
27 Queen Square, Bristol BS1 4ND, UK

www.springer.com and www.canopusbooks.com

Dedicated to Sir Nevill Francis Mott (1905-1996), whose research on metal-insulator transitions, polarons and amorphous semiconductors has had tremendous impact on our current understanding of strongly correlated quantum systems

Preface

Conducting electrons in inorganic and organic matter interact with vibrating ions. If the interaction is sufficiently strong, a local deformation of ions, created by an electron, transforms the electron into a new quasiparticle called a polaron as observed in a great number of conventional semiconductors and polymers. The polaron problem has been actively researched for a long time. The electron Bloch states and *bare* lattice vibrations (phonons) are well defined in insulating parent compounds of semiconductors including the advanced materials discussed in this book. However, microscopic separation of electrons and phonons might be rather complicated in doped insulators since the electron-phonon interaction (EPI) is strong and carriers are correlated. When EPI is strong, the electron Bloch states and phonons are affected.

If characteristic phonon frequencies are sufficiently low, local deformations of ions, caused by the electron itself, create a potential well, which binds the electron even in a perfect crystal lattice. This *self-trapping* phenomenon was predicted by Landau in 1933. It was studied in great detail by Pekar, Fröhlich, Feynman, Rashba, Devreese, Emin, Toyozawa and others in the effective mass approximation for the electron placed in a continuous polarizable (or deformable) medium, which leads to a so-called *large* or *continuum* polaron. Large polaron wave functions and corresponding lattice distortions spread over many lattice sites, which makes the lattice discreteness unimportant. The self-trapping is never complete in a perfect lattice. Since phonon frequencies are finite, ion polarizations can follow polaron motion if the motion is sufficiently slow. Hence, large polarons with a low kinetic energy propagate through the lattice as free electrons with an enhanced effective mass.

When the characteristic polaron binding energy E_p becomes comparable with or larger than the electron half-bandwidth, D, of the rigid lattice, all states in the Bloch band become "dressed" by phonons. In this strong-coupling regime, $\lambda = E_p/D > 1$, the finite bandwidth and lattice discreteness are important and polaronic carriers are called *small* or *lattice* polarons. In the last century many properties of small polarons were understood by Tyablikov, Yamashita and Kurosava, Sewell, Holstein and his school, Firsov, Lang and

Kudinov, Reik, Klinger, Eagles, Böttger and Bryksin and others. The characteristic fingerprints of small polarons are a band narrowing and multi-phonon features in their spectral function (the so-called phonon side-bands).

Interest in the role of EPI and polaron dynamics in contemporary materials has recently gone through a vigorous revival. There is overwhelming evidence for polaronic carriers in novel high-temperature superconductors, colossal-magnetoresistance (CMR) oxides, conducting polymers and molecular nanowires. Here we encounter novel multi-polaron physics, which is qualitatively different from conventional Fermi-liquids and conventional superconductors. The recent interest in polarons extends, of course, well beyond physical descriptions of advanced materials. No general solution to the polaron problem exists for intermediate λ in finite dimensions. It is the enormous differences between weak and strong coupling limits and adiabatic and nonadiabatic limits which make the polaron problem in the intermediate λ regime extremely difficult to study analytically and numerically. The field is a testing ground for modern analytical techniques, including the path integral approach, unitary transformations, diagrammatic expansions, and numerical techniques, such as exact numerical diagonalisations, advanced variational methods, and novel Quantum-Monte-Carlo (QMC) algorithms reviewed in this book.

Polarons in Advanced Materials is written in the form of self-consistent pedagogical reviews authored by well-established researchers actively working in the field. It will lead the reader from single-polaron problems to multi-polaron systems and finally to a description of many interesting phenomena in high-temperature superconductors, ferromagnetic oxides, and molecular nanowires.

The book naturally divides into four parts, following historical reminiscences by Rashba on the early days of polarons. Part I opens with a comprehensive overview by Devreese of the optical properties of continuum all-coupling polarons in all dimensions in the path-integral based theory. The second chapter by Firsov introduces small polarons, the Lang-Firsov canonical transformation and small polaron kinetics. Detailed analysis of magnetotransport and spin transport in the hopping regime of small polarons is presented by Böttger, Bryksin and Damker in chapter 3. Chapter 4 (Cataudella, De Filippis and Perroni) presents large and small polaron models from a unified variational point of view. The fifth chapter by Kornilovitch offers a comprehensive tutorial on the path-integral approach to all-coupling lattice polarons with any-range EPI including Jahn-Teller polarons based on novel continuous-time QMC (CTQMC). Part I closes with the path integral description of polarons by Zoli in the Su-Schrieffer-Heeger model of the EPI important in low-dimensional conjugated polymers and related systems.

Part II opens with the strong-coupling bipolaron theory of superconductivity and a discussion of small mobile bipolarons in cuprate superconductors by Alexandrov (chapter 7). Aubry analyses a small *adiabatic* polaron, an adiabatic bipolaron and multi-polaron adiabatic systems in chapter 8, where theorems for the adiabatic Holstein-Hubbard model are formulated and the role of quantum fluctuations for bipolaronic superconductivity is emphasised. Nano-scale phase separation and different mesoscopic structures in multi-polaron systems are described by Kabanov in chapter 9 with the realistic EPI and a long-range Coulomb repulsion.

Part III starts with a complete numerical solution of the Holstein polaron problem by exact diagonalization (ED) including bipolaron formation and relates the results to strongly-correlated polarons in high-temperature super-conductors, CMR oxides and other materials (Fehske and Trugman). Chapter 11 by Hohenadler and von der Linden describes the canonical transforma-tion based QMC and variational approaches to the Holstein-type models with any number of electrons. Strongly-correlated polarons in relation to high-temperature superconductors are further reviewed by Mishchenko and Na-gaosa in chapter 12, where basics of recently developed Diagrammatic Monte Carlo (DMC) method are discussed.

Part IV includes a comprehensive review by Mihailovic of photoinduced polaron signatures in conducting polymers, cuprates, manganites, and other related materials (chapter 13). Zhao reviews polaronic isotope effects and elec-tric transport in CMR oxides and high-temperature superconductors (chapter 14), which are further reviewed by Bussmann-Holder and Keller in chapter 15, where a two-component approach to cuprate superconductors with pola-ronic carriers is described. The final chapter by Bratkovsky presents a detailed description of electron transport in molecular scale devices including rectifica-tion, extrinsic switching, noise, and a theoretically proposed polaronic intrinsic switching of molecular quantum dots.

This contemporary encyclopedia of polarons is easy to follow for senior undergraduate and graduate students with a basic knowledge of quantum mechanics. The combination of viewpoints presented within the book can provide comprehensive understanding of strongly correlated electrons and phonons in solids. The book would be appropriate as supplementary read-ing for courses in Solid State Physics, Condensed Matter Theory, Theory of Superconductivity, Advanced Quantum Mechanics, and Many-Body Phenom-ena taught to final year undergraduate and postgraduate students in physics and math departments. The subject of the book is of direct relevance to the design of novel semiconducting, superconducting, and magnetic bulk and nano-materials. Their long term potential could be fully realised if an increase in fundamental understanding is achieved. The book will benefit researchers working in condensed matter, theoretical and experimental physics, quantum chemistry and nanotechnology.

It is a great pleasure and honor for the Editor to present these collected reviews. I thank our distinguished authors for sharing their insights and ex-pertise in polarons.

Loughborough University, August 2006   *Sasha Alexandrov*

Reminiscences of the Early Days of Polaron Theory

Emmanuel I. Rashba

Department of Physics, Harvard University, Cambridge, Massachusetts 02138, USA erashba@physics.harvard.edu

This volume, covering various aspects of modern polaron physics, its highlights and challenges, appears at a time that is quite remarkable in the history of polarons. Exactly 60 years before the compilation of this volume, the seminal paper by Solomon Pekar that initiated the theory of polarons was published [1, 2]. In it, a model of the large polaron was developed and the term *polaron* was proposed. The volume will appear early in 2007, close to the 90th anniversary of Pekar's birthday, March 16, 1917.

Referring to the famous Landau paper in which the possibility of electron self-trapping was first conceived [3], Pekar developed a macroscopic model that became a cornerstone of the theories that were to follow. The coupling of an electron to a polar lattice was expressed in terms of a dielectric continuum. The inertial part of its polarization supported the electron self-consistently in a self-trapped state. The coupling constant for this mechanism is the Pekar factor $\kappa^{-1} = \epsilon_\infty^{-1} - \epsilon_0^{-1}$, where ϵ_∞ and ϵ_0 are high and low frequency dielectric constants, respectively. For the strong (adiabatic) coupling limit, Pekar calculated the ground state energy of a polaron, proved that the energies of its optical and thermal dissociation differ by a factor of 3, and established exact relations between different contributions to the polaron energy. The title of the paper emphasizes local states in an ideal ionic crystal, but drift of the polaron in an electric field was also envisioned. The concept of a polaron as a charge carrier in ionic crystals was developed in a following paper [4], and was supported by calculating the polaron mass by Landau and Pekar [5]. In this way, the original concept of self-trapping as formation of crystal defects like F-centers in alkali-halides [3] evolved into the concept of polarons as free charge carriers in polar crystals.

The beginning of Pekar's scientific career was quite remarkable. On the eve of the Nazi invasion of the USSR in the spring of 1941, he was awarded the degree of a Doctor of Science (similar to Habilitation) for his PhD dissertation, which was an extraordinary event. Landau concluded Pekar's talk at his seminar with the comment: "The self-conception of theoretical physics has happened in Kiev." During the war Pekar worked on defense projects, and after returning to Kiev he established there a Theoretical Division at the Institute of Physics and a Chair in Theoretical Physics at the University. Pekar's former colleagues and friends, who came back from their service in the Army, became his graduate students and worked enthusiastically on different aspects of the polaron theory. In the framework of the strong coupling limit, it was shown that for electrons coupled to the lattice by a deformation potential there are no macroscopic self-trapped states in 3D [6], but they exist in 1D with 2D as a critical dimension [7], and the possibility of exciton self-trapping in polar crystals was also proven [8].

As an undergraduate student, I joined Pekar's group in the late 1940s and remain the last witness of those developments in Kiev where polaron theory was initiated, and of the loose contacts with the related work in the West that were possible only through published papers, under the conditions of the self-imposed political isolation of the USSR. Hence, the present brief note is restricted to this subject of which I have primary knowledge, and is not intended to cover the different aspects of polaron theory that are reflected in the vast review literature.

The next step was generalizing the semiclassical approach of [1] and [5] to a consistent quantum theory. Pekar drafted the first version of such a theory [9] by introducing a zero-mode related to the motion of the polaron center. Its three degrees of freedom come from the phonon system whose energy is automatically reduced by $3\omega/2$, ω being the phonon frequency. The scattering of carriers by phonons is reduced because the dominant part of electron–phonon coupling is included through the polaron energy. The field-theoretical aspects of polaron theory attracted the attention of Bogoliubov who was working in Kiev at that time, and in collaboration with Tyablikov he developed another version of the quantum dynamics of adiabatic polarons [10, 11]. Because, after dressing an electron by a phonon cloud, the effective Hamiltonian is quadratic in phonon amplitudes, it allows the finding of the renormalized phonon spectrum and makes two-phonon processes a major scattering mechanism for adiabatic polarons [12].

There are several factors to the theory of a large polaron that attracted the close attention of theorists with a wide range of scientific interests. First, it presented a field theory model without divergences that enabled a consistent analysis at an arbitrary coupling constant and became a prototype for a number of self-localized states in nonlinear field theories. Second, the electron-phonon interaction emerged as a prospective mechanism of superconductivity because of the discovery of the isotopic effect and different arguments [13]. Third, electronic transport in polar conductors was of fundamental interest *per se*. Meanwhile, although the adiabatic limit is highly instructive in clarifying the essential differences between free electrons and polarons, there exist essential constraints on its applicability. Indeed, the Fröhlich electron-phonon coupling constant is $\alpha = (e^2/\hbar\kappa)\sqrt{m/2\hbar\omega}$, m being the electron effective mass, and the ground state energy of an adiabatic polaron E_0 and its effective mass m_p are $E_0 \approx -0.1\alpha^2\hbar\omega$ [1] and $m_p \approx 0.02\alpha^4 m$ [5]. The small numerical coefficients in both equations imply a strict criterion for the adiabatic limit, $\alpha \gtrsim 10$. With $m \approx m_0$ and $\kappa \sim 1$ this inequality can be fulfilled because α is about $(M/m_0)^{1/4}$ and therefore large (here M is the ion mass). However, under these conditions macroscopic description fails because the polaron radius becomes approximately a lattice constant. Fortunately, in many crystals $m \ll m_0$ and $\kappa \gg 1$, hence, the macroscopic description that is central to a large polaron theory is justified.

At this point, polaron theory splits into two branches. The first one deals with small (Holstein [14]) polarons where the detailed mechanism of a strong electron-phonon coupling is not of primary importance, while the second deals with large (Pekar–Fröhlich) polarons with polar electron-phonon interaction that is not necessarily strong. The difference in their properties may be significant, e.g., the effective mass of small polarons increases with the coupling constant exponentially, while for large adiabatic polarons by a power law.

Fröhlich et al. [15] developed a theory of weakly coupled large polarons, $\alpha \lesssim 1$. The next step in bridging the gap between weakly and strongly coupled polarons was made by Lee, Low, and Pines whose variational approach works for $\alpha \lesssim 5$ [16]. Among the different variational schemes, the Feynman technique that allowed the finding of E_0 and m_p with high accuracy for all α values had the largest impact [17]. Pekar and his collaborators generalized Feynman's approach for calculating thermodynamical functions [18] and proposed an independent, simple, and efficient variational procedure [19].

Comparing the results of polaron theory with experimental data is a challenging task. Historically, such a comparison began with multiphonon spectra of impurity centers that are conceptually closely related to the theory of strongly coupled polarons. The theory of such spectra was initiated independently by Huang and Rhys [20] and Pekar [21]. It is seen from the review paper by Markham [22] that some of Pekar's papers on this subject, published in Russian, were translated into English by different researchers and circulated in the West long before the regular translation of Soviet journals by the American Institute of Physics began.

Polaron effects manifest themselves even more spectacularly in the co-existence of free and self-trapped states of excitons that was observed by optical techniques. While Landau envisioned a barrier for self-trapping [3], Pekar has shown [9] that such a barrier is absent for polarons that are formed by a gradual lowering of their energy, which is tantamount to a sequence of single-phonon processes. However, for short range coupling to phonons the free states may persist as metastable states even in the presence of deep self-trapped states [7]. Such free states are protected by a barrier and decay through a collective (instanton) tunneling of a coupled electron-phonon system [23]. Conditions for the existence of the barrier for a Wannier–Mott exciton in a polar medium are nontrivial because the whole particle is neutral but each component of it is coupled to the lattice by polar interaction. They were clarified in [24], which was the very last of Pekar's papers related to polarons. Since 1957, he directed his attention mostly to additional light waves near exciton resonances that he predicted [25], and to some other problems of the solid state theory.

I am grateful to M. I. Dykman and V. A. Kochelap for their help and advice.

References

1. S. I. Pekar, Journal of Physics USSR **10**, 341 (1946).
2. S. Pekar, Zh. Eksp. Teor. Fiz. **16**, 341 (1946).
3. L. D. Landau, Sow. Phys. **3**,664 (1933).
4. S. I. Pekar, Zh. Eksp. Teor. Fiz. **18**, 105 (1948).
5. L. D. Landau and S. I. Pekar, Zh. Eksp. Teor. Fiz. **18**, 419 (1948).
6. M. F. Deigen and S. I. Pekar, Zh. Eksp. Teor. Fiz. **21**, 803 (1951).
7. E. I. Rashba, Opt. Spektrosk. **2**, 75 and 88 (1957).
8. I. M. Dykman and S. I. Pekar, Dokl. Acad. Nauk SSSR **83**, 825 (1952).
9. S. I. Pekar, Research in Electron Theory of Crystals, US AEC Transl. AEC-tr-555 (1963) [Russian edition 1951, German edition 1954].
10. N. N. Bogoliubov, Ukr. Mat. Zh. **2**, 3 (1950).
11. S. V. Tyablikov, Zh. Eksp. Teor. Fiz. **21**, 377 (1951).
12. G. E. Volovik, V. I. Mel'nikov, and V. M. Edel'stein, JETP Lett. **18**, 81 (1973).
13. R. A. Ogg, Jr., Phys. Rev. **69**, 243 (1946), where Bose condensation of trapped electron pairs in metal-ammonia solutions was proposed.
14. T. Holstein, Annals of Physics **8**, 325 and 243 (1959).
15. H. Frohlich, H. Pelzer, and S. Zienau, Phil. Mag. **41**, 221 (1950).
16. T. D. Lee, F. E. Low, and D. Pines, Phys. Rev. **90**, 297 (1953).
17. R. P. Feynman, Phys. Rev. **97**, 660 (1955).
18. M. A. Krivoglaz and S. I. Pekar, Izv. AN SSSR, ser. fiz., **21**, 3 and 16 (1957).
19. V. M. Buimistrov and S. I. Pekar, Zh. Eksp. Teor. Fiz. **33**, 1193 and 1271 (1957).
20. K. Huang and A. Rhys, Proc. Roy. Soc. (London) **A204**, 406 (1950).
21. S. I. Pekar, Zh. Eksp. Teor. Fiz. **20**, 510 (1950).
22. J. J. Markham, Rev. Mod. Phys. **31**, 956 (1959).
23. A. S. Ioselevich and E. I. Rashba, in: *Quantum Tunneling*, ed. by Yu. Kagan and A. J. Leggett (Elsevier) 1992, p. 347.
24. S. I. Pekar, E. I. Rashba, and V. I. Sheka, Sov. Phys. JETP, **49**, 129 (1979).
25. S. I. Pekar, Zh. Eksp. Teor. Fiz. **33**, 1022 (1957) [1958, Sov. Phys. JETP **6**, 785].

Contents

Part III Strongly Correlated Polarons

Part IV Polarons in Contemporary Materials

List of Contributors

Serge Aubry
Laboratoire Léon Brillouin, CEA
Saclay (CEA-CNRS),
91191 Gif-sur-Yvette (France)
saubry@dsm-mail.saclay.cea.fr

A. S. Alexandrov
Department of Physics, Loughborough University, Loughborough
LE11 3TU, United Kingdom
a.s.alexandrov@lboro.ac.uk

A. M. Bratkovsky
Hewlett-Packard Laboratories, 1501
Page Mill Road, Palo
Alto, California 94304
alex.bratkovski@hp.com

Harald Böttger
Institute for Theoretical Physics,
Otto-von-Guericke-University PF
4120, D-39016 Magdeburg, Germany
Harald.Boettger@Physik.
Uni-Magdeburg.DE

Valerij V. Bryksin
A. F. Ioffe Physico-Technical
Institute, Politekhnicheskaya 26,
19526 St. Petersburg Russia

Annette Bussmann-Holder
Max-Planck-Institut für
Festkörperforschung, Heisenbergstr.
1, D-70569 Stuttgart, Germany
A.Bussmann-Holder@fkf.mpg.de

V. Cataudella
CNR-INFM Coherentia and
University of Napoli,
V. Cintia 80126 Napoli, Italy
vit@na.infn.it

Thomas Damker
Institute for Theoretical Physics
Otto-von-Guericke-University PF
4120, D-39016 Magdeburg Germany

Jozef T. Devreese
University of Antwerp,
Groenenborgerlaan 171, B-2020
Antwerpen, Belgium
jozef.devreese@ua.ac.be

H. Fehske
Institut für Physik, Ernst-Moritz-
Arndt-Universität Greifswald,
D-17487 Greifswald, Germany
holger.fehske@physik.uni-greifswald.de

G. De Filippis
CNR-INFM Coherentia and
University of Napoli,
V. Cintia 80126 Napoli, Italy
vit@na.infn.it

Yu. A. Firsov
Solid State Physics Division, Ioffe
Institute, 26 Polytekhnicheskaya,
194021 St. Petersburg, Russia
yuaf@icomefrom.ru

Martin Hohenadler
Institute for Theoretical and
Computational Physics, TU Graz,
Austria
hohenadler@itp.tu-graz.ac.at

V. V. Kabanov
J. Stefan Institute, Jamova 39, 1001,
Ljubljana, Slovenia
viktor.kabanov@ijs.si

Hugo Keller
Physik-Institut der
Universität Zürich, Winterthurerstr.
190, CH-8057 Zürich,
Switzerland

Pavel Kornilovitch
Hewlett-Packard, Corvallis, Oregon,
97330, USA
pavel.kornilovich@hp.com

Dragan Mihailovic
Jozef Stefan Institute and
International Postgraduate School,
SI-1000 Ljubljana, Slovenia
dragan.mihailovic@ijs.si

A. S. Mishchenko
CREST, Japan Science and
Technology Agency (JST),
AIST, 1-1-1, Higashi, Tsukuba,
Ibaraki 305-8562, Japan
andry.mishenko@aist.co.jp

N. Nagaosa
Department of Applied Physics, The
University of Tokyo, 7-3-1 Hongo,
Bunkyo-ku, Tokyo 113, Japan
nagaosa@appi.t.u-tokyo.ac.jp

C. A. Perroni
Institut für Festkörperforschung
(IFF), Forschungszentrum Jülich,
52425 Jülich, Germany
a.perroni@fz-juelich.de

E. I. Rashba
Department of Physics, Harvard
University, Cambridge, Massachusetts 02138, U. S. A.
erashba@physics.harvard.edu

S. A. Trugman
Theoretical Division, Los Alamos
National Laboratory, Los Alamos,
New Mexico 87545, U. S. A.
sat@lanl.gov

Wolfgang von der Linden
Institute for Theoretical and
Computational Physics, TU Graz,
Austria
wvl@itp.tu-graz.ac.at

Guo-meng Zhao
Department of Physics and Astronomy, California State University,
Los Angeles, CA 90032
gzhao2@calstatela.edu

Marco Zoli
Istituto Nazionale Fisica della
Materia - Dipartimento di Fisica,
Universitá di Camerino, 62032
Camerino,
Italy   marco.zoli@unicam.it

Part I

Large and Small Polarons

Optical Properties of Few and Many Fröhlich Polarons from 3D to 0D

Jozef T. Devreese[1,2]

[1] Universiteit Antwerpen, T.F.V.S., Groenenborgerlaan 171, B-2020 Antwerpen, Belgium jozef.devreese@ua.ac.be

[2] Technische Universiteit Eindhoven, P. O. Box 513, NL-5600 MB Eindhoven, The Netherlands

Summary. In this chapter I treat basic concepts and recent developments in the field of optical properties of few and many Fröhlich polarons in systems of different dimensions and dimensionality. The key subjects are: comparison of the optical conductivity spectra for a Fröhlich polaron calculated within the all-coupling path-integral based theory with the results obtained using the numerical Diagrammatic Quantum Monte Carlo method and recently developed analytical approximations. The polaron excited state spectrum and the mechanism of the optical absorption by Fröhlich polarons are analysed in the light of early theoretical models (from 1964 on) and of recent results. Further subjects are the scaling relations for Fröhlich polarons in different dimensions; the all-coupling path-integral based theory of the magneto-optical absorption of polarons; Fröhlich bipolarons and their stability; the many-body problem (including the electron-electron interaction and Fermi statistics) in the few- and many-polaron theory; the theory of the optical absorption spectra of many-polaron systems; the ground-state properties and the optical response of interacting polarons in quantum dots; non-adiabaticity of polaronic excitons in semiconductor quantum dots. Numerous examples are shown of comparison between Fröhlich polaron theory and experiments in high-T_c materials, manganites, silver halides, semiconductors and semiconductor nanostructures, including GaAs/AlGaAs quantum wells, various quantum dots etc. Brief sections are devoted to the electronic polaron, to small polarons and to recent extensions of Landau's concept, including ripplopolarons.

1 Introduction

As is generally known, the polaron concept was introduced by Landau in 1933 [1]. Initial theoretical [2–8] and experimental [9] works laid the foundation of polaron physics. Among the comprehensive review papers and books covering the subject, I refer to [10–17].

Significant extensions and recent developments of the polaron concept have been realised (see, for example, [17–21] and references therein). Polarons have been invoked, e.g., to study the properties of conjugated polymers, colossal magnetoresistance perovskites, high-T_c superconductors, layered MgB_2 superconductors, fullerenes, quasi-1D conductors and semiconductor nanostructures.

A distinction was made between polarons in the continuum approximation where long-range electron-lattice interaction prevails (*"Fröhlich"* polarons) and polarons for which the short-range interaction is essential (*Holstein, Holstein-Hubbard, Su-Schrieffer-Heeger* models).

The chapter starts with a review of basic concepts and recent developments in the study of the optical absorption of Fröhlich polarons in three dimensions. Scaling relations are discussed for Fröhlich polarons in different dimensions. The scaling relation for the polaron free energy is checked for the path integral Monte Carlo results. The next section is devoted to the all-coupling path-integral based theory of the magneto-optical absorption of polarons, which allows for an interpretation – with high spectroscopic precision – of cyclotron resonance experiments in various solid structures of different dimensionality. In particular, the analysis of the cyclotron resonance spectra of silver halides provided one of the most convincing and clearest demonstrations of polaron features in solids. Fröhlich bipolarons, small bipolarons and their extensions are represented in the context of applications of bipolaron theory to high-T_c superconductivity.

Furthermore, recent results on the many-body problem ("the N-polaron problem") are discussed. The related theory of the optical absorption spectra of many-polaron systems has been applied to explain the experimental peaks in the mid-infrared optical absorption spectra of cuprates and manganites. The ground-state properties, the optical response of interacting polarons and the non-adiabaticity of the polaronic excitons in quantum dots are discussed in the concluding sections.

1.1 Fröhlich Polarons

A conduction electron (or hole) in an ionic crystal or a polar semiconductor is the prototype of a polaron. Fröhlich proposed a model Hamiltonian for this polaron through which its dynamics is treated quantum mechanically (the "Fröhlich Hamiltonian" [6]). The polarization, carried by the longitudinal optical (LO) phonons, is represented by a set of quantum oscillators with frequency ω_{LO}, the long-wavelength LO-phonon frequency, and the interaction between the charge and the polarization field is linear in the field. The strength of the electron–phonon interaction is expressed by a dimensionless coupling constant α. Polaron coupling constants for selected materials are given in Table 1.

This model has been the subject of extensive investigations. The first studies on polarons were devoted to the calculation of the self-energy and the effective mass of polarons in the limit of large α, or "strong coupling" [2–4].

Table 1. *Electron-phonon coupling constants* (Reprinted with permission after [22]. ©2003 by the American Institute of Physics.)

Material	α	Material	α
InSb	0.023	AgCl	1.84
InAs	0.052	KI	2.5
GaAs	0.068	TlBr	2.55
GaP	0.20	KBr	3.05
CdTe	0.29	$Bi_{12}SiO_{20}$	3.18
ZnSe	0.43	CdF_2	3.2
CdS	0.53	KCl	3.44
α-Al_2O_3	1.25	CsI	3.67
AgBr	1.53	$SrTiO_3$	3.77
α-SiO_2	1.59	RbCl	3.81

The "weak-coupling" limit, first explored by H. Fröhlich [6], is obtained from the leading terms of the perturbation theory for $\alpha \to 0$. Inspired by the work of Tomonaga, Lee et al. [7] derived the self-energy and the effective mass of polarons from a canonical-transformation formulation; the range of validity of their approximation is in principle not larger than that of the weak-coupling approximation. The main significance of the approximation of [7] is in the elegance of the used canonical transformation, together with the fact that it puts the Fröhlich result [6] on a variational basis.

An all-coupling polaron approximation was developed by Feynman using his path-integral formalism [8]. In a trial action he simulated the interaction between the electron and the polarization modes by a harmonic interaction between a hypothetical particle and the electron and introduced a variational principle for path integrals. Feynman derived first the self-energy E_0 and the effective mass m^* of the polaron [8]. The analysis of an exactly solvable ("symmetrical") 1D-polaron model [23, 24] demonstrated the accuracy of Feynman's path-integral approach to the polaron ground-state energy. Later Feynman et al. formulated a response theory for path integrals, derived a formal expression for the impedance and studied the mobility of all-coupling polarons [25, 26].

Subsequently the path-integral approach to the polaron problem was generalised and developed to become a tool to study optical absorption, magnetophonon resonance, cyclotron resonance etc.

1.2 Optical Absorption of Fröhlich Polarons at Arbitrary Coupling. Analytical Theory

The study of the internal excitations of Fröhlich polarons and their optical absorption started in 1964 [23, 27] with the analysis of the spectrum of an exactly solvable "symmetrical" 1D-polaron model. It was also shown that two types of excitations exist for this polaron model: a) scattering states (called "diffusion" states in [23], and b) "relaxed excited states" (RES). It was argued in [23, 27] that the RES is only stable for a sufficiently large electron-LO-phonon coupling constant.

In 1969, starting from [23, 27, 28], a mechanism for the optical absorption of strong-coupling Fröhlich polarons was proposed in [29]. This mechanism consists of transitions to a RES and to its LO-phonon sidebands that constitute a Franck-Condon band. In 1972, Devreese et al. (DSG; [30]) published all-coupling results for the optical absorption of the Fröhlich polaron. Reference [30] uses the Feynman ground-state polaron model and the path-integral response formalism [25] as its starting point. For $\alpha \lesssim 1$ the DSG-spectrum consists of a one-LO-phonon sideband (along with, for $T = 0$, a δ-peak at zero frequency). This result confirms the one-polaron limit of the perturbative treatment in [31]. For intermediate coupling ($3 \lesssim \alpha \lesssim 6$) DSG predict a transition to a RES and its LO-phonon sidebands (FC band). For $\alpha \gtrsim 6$, DSG identify a narrow RES-transition with a narrow sideband (as already stated in [30, 32] the resulting RES-peak is too narrow and – at sufficiently large α – inconsistent with the Heisenberg uncertainty principle).

Recent numerical [33] and analytical [34] studies have allowed for a more complete understanding of the optical absorption of the Fröhlich polaron at all α, as discussed in [35]. Later in this chapter I will analyse to what extent recent calculations confirm the mechanisms for the polaron optical absorption proposed in [29] and by DSG [30]. I refer also to the chapters by Mishchenko and Nagaosa [36] and Cataudella et al [37] in the present volume.

A path-integral Monte Carlo scheme was presented [38] to study the Fröhlich polaron model in three and two dimensions. The ground state features of the Fröhlich polaron model were revisited numerically using a Diagrammatic Quantum Monte-Carlo (DQMC) method [39] and analytically using an "all-coupling" variational Hamiltonian approach [34]. The three aforementioned schemes confirm the remarkable accuracy of the Feynman path-integral model [8] to calculate the polaron ground-state energy. The dependence of the calculated polaron ground state properties on the electron-phonon coupling strength supports the earlier conclusion [40–42] that the crossover between the two asymptotic regimes characterizing a polaron occurs smoothly and do not suggest any sharp "self-trapping" transition.

1.3 Small Polarons. Recent Extensions of the Polaron Concept

Holstein, using a 1D model, has pioneered the study of what are often called "small polarons", for which the lattice polarisation, induced by a charge carrier, is essentially confined to a unit cell [43, 44]. Hopping of electrons from one lattice site to another in the presence of the electron-phonon interaction is the key process determining the dynamical properties of small polarons (see e.g. [45–48]) and spin polarons (cf. [49]). A crossover regime of the Holstein polaron has been studied using a variational analysis based on a superposition of Bloch states that describe large polarons and small polarons by V. Cataudella et al. [50] and within a numerical variational approach [51]. Dynamical polaron solutions, which are characterised by very long lifetime at low temperatures, have been proposed for the Holstein model on a lattice with anharmonic local potential [52].

The first identification of small polarons in solids was made for non-stoichiometric uranium dioxide by the present author [53, 54]. The mechanisms of self-trapping, static and dynamic properties of small polarons in alkali halides and in several other ionic crystals were analysed e. g. in [55, 56]. Quantitative evidence for critical quantum fluctuations and superlocalisation of the small polarons in one, two and three dimensions was presented on the basis of the Quantum Monte Carlo approach in [57, 58]. It was demonstrated that for all lattice dimensionalities there exists a critical value of the electron-lattice coupling constant, below which self-trapping of Holstein polarons does not occur [59]. Several recent experimental and theoretical investigations have provided convincing evidence for the occurrence of small (bi)polarons in "contemporary" materials. In the case of short-range electron-phonon interaction, when a small (bi)polaron hops between lattice sites, the total lattice deformation vanishes at one site and then re-appears at a new one. The effective mass of a one-site bipolaron is then very large, and the predicted critical temperature T_c is very low (see [60]). A two-site small bipolaron model by A. Alexandrov and N. Mott [14] provides a parameter-free estimate of T_c for high-T_c superconducting cuprates [61]. A long-range Fröhlich-type, rather than short-range, electron-phonon interaction on a discrete ionic lattice [62] is assumed within the *"Fröhlich-Coulomb"* model of the high-T_c superconductivity proposed by A. Alexandrov [63]. For a long-range interaction, only a fraction of the total deformation changes as a (bi)polaron hops between the lattice sites. This leads to a dramatic mass reduction as compared to that of the Holstein small (bi)polaron. It was then proposed that in the superconducting phase the carriers are "superlight mobile bipolarons". As distinct from the conventional continuum Fröhlich polaron, a multipolaron lattice model is used with electrostatic forces taking into account the discreteness of the lattice, finite electron bandwidth and the quantum nature of phonons. This model is applied in an attempt to explain the physical properties of superconducting cuprates such as their T_c-values, the isotope effects, the normal-state diamagnetism, the pseudogap and spectral functions measured in tunnelling and photoemission (see [17, 64] for an extensive review). Experiments have been interpreted as due to small polarons in the paramagnetic (see e.g. [65]), ferromagnetic [66] and antiferromagnetic [67] states of manganites. The magnetization and resistivity of manganites near the ferromagnetic transition were interpreted in terms of pairing of oxygen holes into heavy bipolarons in the paramagnetic phase and their magnetic pair breaking in the ferromagnetic phase [68]. These studies do not preclude the occurrence of Fröhlich polarons in manganites, as evidenced in the work of Hartinger et al. [69]

1.4 Electronic Polarons

An extension of the polaron concept arises by considering the interaction between a carrier and the exciton field. One of the early formulations of this model was developed by Toyozawa [70]. The resulting quasi-particle is called the *electronic polaron.*

The self-energy of the electronic polaron (which is almost independent of wave number) must be taken into account when the bandgap of an insulator or semiconductor is calculated using pseudopotentials. For example if one calculates, with Hartree-Fock theory, the bandgap of an alkali halide, one is typically off by a factor of two. This was the original problem which was solved conceptually with the introduction of the electronic polaron [70]. Also in the soft X-ray spectra of alkali halides exciton sidebands have been observed which we attributed to the electronic polaron coupling [71] (see also [72]). For a review of the current experimental status of the "electronic polaron complexes", as predicted in [71, 73], I refer to [74].

Using the all-coupling theory of the polaron optical absorption [30, 32], we found [73] that the electronic polaron produces peaks in the optical absorption spectra beginning about an exciton energy above the absorption edge, allowing for the interpretation of the experiments on LiF, LiCl, and LiBr. This theory has been invoked recently e. g. for the interpretation of the experimental data on inelastic soft X-ray scattering in solid LiCl, resonantly enhanced at states with two Li $1s$ vacancies [75].

2 Optical Absorption of Fröhlich Polarons in 3D

2.1 Optical Absorption at Weak Coupling. The Role of Many-Polarons

At zero temperature and in the weak-coupling limit, the optical absorption of a Fröhlich polaron is due to the elementary polaron scattering process, schematically shown in Fig. 1.

In the weak-coupling limit ($\alpha \ll 1$) the polaron absorption coefficient for a many-polaron gas was first obtained by V. Gurevich, I. Lang and Yu. Firsov [31]. Their optical-absorption coefficient is equivalent to a particular case of the result of J. Tempere and J. T. Devreese [76], with the dynamic structure factor $S(\mathbf{q}, \Omega)$ corresponding to the Hartree-Fock approximation. In [76] the optical absorption coefficient of a many-polaron gas was shown to be given, to order α, by

$$\mathrm{Re}[\sigma(\Omega)] = n_0 e^2 \frac{2}{3}\alpha\frac{1}{2\pi\Omega^3} \int\limits_0^\infty dq q^2 S(\mathbf{q}, \Omega - \omega_{LO}), \tag{1}$$

where n_0 is the density of charge carriers.

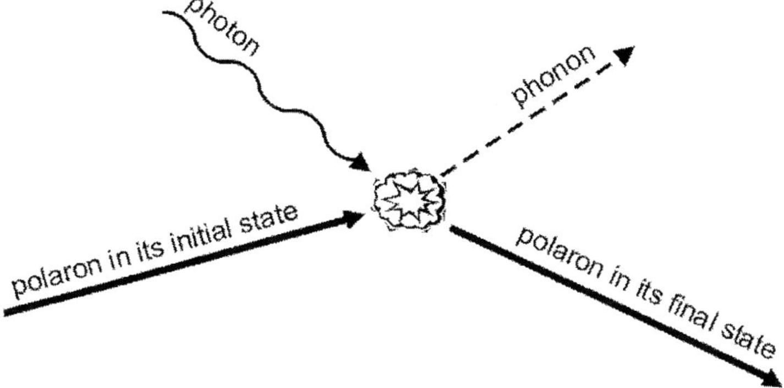

Fig. 1. Elementary polaron scattering process describing the absorption of an incoming photon and the generation of an outgoing phonon. (Reprinted with permission from [22]. ©2003, American Institute of Physics.)

In the zero-temperature limit, starting from the Kubo formula ([77], p. 165), the optical conductivity of a *single* Fröhlich polaron can be represented in the form

$$\sigma(\Omega) = i\frac{c^2}{m_b(\Omega+i\varepsilon)} + \frac{e^2}{m_b^2\hbar}\frac{1}{(\Omega+i\varepsilon)^3}\int_0^\infty e^{-\varepsilon t}\left(e^{i\Omega t}-1\right)\sum_{\mathbf{q},\mathbf{q}'} q_x q_x'$$

$$\times\left\langle \Psi_0 \left|\left[\begin{array}{c}\left(V_{\mathbf{q}}b_{\mathbf{q}}(t) + V_{-\mathbf{q}}^* b_{-\mathbf{q}}^+(t)\right)e^{i\mathbf{q}\cdot\mathbf{r}(t)},\\ \left(V_{-\mathbf{q}'}b_{-\mathbf{q}'} + V_{\mathbf{q}'}^* b_{\mathbf{q}'}^+\right)e^{-i\mathbf{q}'\cdot r}\end{array}\right]\right| \Psi_0 \right\rangle dt, \qquad (2)$$

where $\varepsilon = +0$ and $|\Psi_0\rangle$ is the ground-state wave function of the electron-phonon system. Within the weak coupling approximation, the following analytic expression for the real part of the polaron optical conductivity results from (2):

$$\mathrm{Re}\,\sigma(\Omega) = \frac{\pi e^2}{2m^*}\delta(\Omega) + \frac{2e^2}{3m_b}\frac{\omega_{\mathrm{LO}}\alpha}{\Omega^3}\sqrt{\Omega-\omega_{\mathrm{LO}}}\,\Theta(\Omega-\omega_{\mathrm{LO}}), \qquad (3)$$

where

$$\Theta(\Omega-\omega_{\mathrm{LO}}) = \begin{cases} 1 \text{ if } \Omega > \omega_{\mathrm{LO}}, \\ 0 \text{ if } \Omega < \omega_{\mathrm{LO}}. \end{cases}$$

The spectrum of the real part of the polaron optical conductivity (3) is represented in Fig. 2.

According to (3), the absorption coefficient for absorption of light with frequency Ω by free polarons for $\alpha \longrightarrow 0$ takes the form

$$\Gamma_p(\Omega) = \frac{1}{\epsilon_0 cn}\frac{2n_0 e^2\alpha\omega_{\mathrm{LO}}^2}{3m_b\Omega^3}\sqrt{\frac{\Omega}{\omega_{\mathrm{LO}}}-1}\;\;\Theta(\Omega-\omega_{\mathrm{LO}}), \qquad (4)$$

where ϵ_0 is the dielectric permittivity of the vacuum, n is the refractive index of the medium, n_0 is the concentration of polarons. A simple derivation in

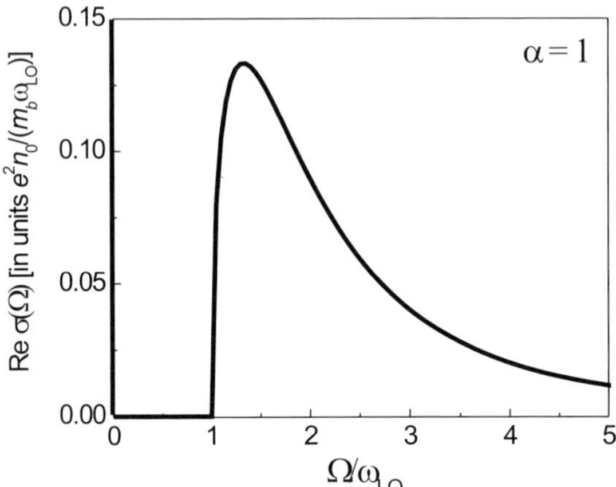

Fig. 2. Polaron optical conductivity for $\alpha = 1$ in the weak-coupling approximation, according to [32], p. 92. A δ-like central peak (at $\Omega = 0$) is schematically shown by a vertical line. (Reprinted with permission from [78]. ©2006, Società Italiana di Fisica.)

[79] using a canonical transformation method gives the absorption coefficient of free polarons, which coincides with the result (4). The step function in (4) reflects the fact that at zero temperature the absorption of light accompanied by the emission of a phonon can occur only if the energy of the incident photon is larger than that of a phonon ($\Omega > \omega_{LO}$). In the weak-coupling limit, according to (4), the absorption spectrum consists of a "one-phonon line". At nonzero temperature, the absorption of a photon can be accompanied not only by emission, but also by absorption of one or more phonons. Similarity between the temperature dependence of several features of the experimental infrared absorption spectra in high-T_c superconductors and the temperature dependence predicted for the optical absorption of a single Fröhlich polaron [30, 32] has been revealed in [80].

Experimentally, this one-phonon line has been observed for free polarons in the infrared absorption spectra of CdO-films, see Fig. 3. In CdO, which is a weakly polar material with $\alpha \approx 0.74$, the polaron absorption band is observed in the spectral region between 6 and 20 μm (above the LO phonon frequency). The difference between theory and experiment in the wavelength region where polaron absorption dominates the spectrum is due to many-polaron effects, see [76].

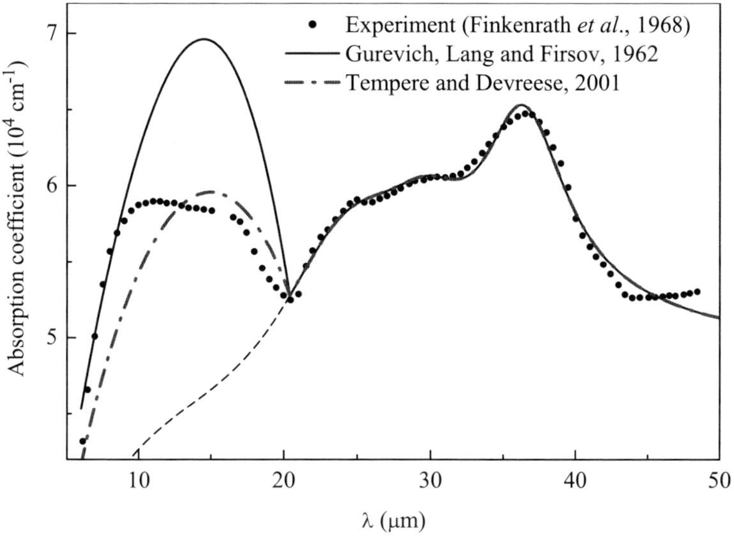

Fig. 3. Optical absorption spectrum of a CdO-film with carrier concentration $n_0 = 5.9 \times 10^{19}$ cm^{-3} at $T = 300$ K. The experimental data (solid dots) of [81] are compared to different theoretical results: with (solid curve) and without (dashed line) the single-polaron contribution of [31, 79] and for many polarons (dash-dotted curve) of [76]. The following values of material parameters of CdO were used for the calculations: $\alpha - 0.74$ [81], $\omega_{LO} = 490$ cm^{-1} (from the experimental optical absorption spectrum, Fig. 2 of [81]), $m_b = 0.11 m_e$ [82], $\varepsilon_0 = 21.9$, $\varepsilon_\infty = 5.3$ [82]. (Reprinted with permission from [22]. ©2003, American Institute of Physics.)

2.2 Optical Absorption at Strong Coupling

The problem of the structure of the Fröhlich polaron excitation spectrum constituted a central question in the early stages of the development of polaron theory. The exactly solvable polaron model of [27] was used to demonstrate the existence of the so-called "relaxed excited states" of Fröhlich polarons [23]. In [28], and after earlier intuitive analysis, this problem was studied using the classical equations of motion and Poisson-brackets. The insight gained as a result of those investigations concerning the structure of the excited polaron states, was subsequently used to develop a theory of the optical absorption spectra of polarons. The first work was limited to the strong coupling limit [29]. Reference [29] is the first work that reveals the impact of the internal degrees of freedom of polarons on their optical properties.

The optical absorption of light by free Fröhlich polarons was treated in [29] using the polaron states obtained within the adiabatic strong-coupling approximation. It was argued in [29], that for sufficiently large α ($\alpha \gtrsim 3$), the (first) relaxed excited state (RES) of a polaron is a relatively stable state, which gives rise to a "resonance" in the polaron optical absorption spectrum.

This idea was necessary to understand the polaron optical absorption spectrum in the strong-coupling regime. The following scenario of a transition, which leads to a *"zero-phonon" peak* in the absorption by a strong-coupling polaron, was then suggested. If the frequency of the incoming photon is equal to $\Omega_{\mathrm{RES}} = 0.065\alpha^2\omega_{\mathrm{LO}}$, the electron jumps from the ground state (which, at large coupling, is well-characterised by "s"-symmetry for the electron) to an excited state ("$2p$"), while the lattice polarization in the final state is adapted to the "$2p$" electronic state of the polaron. In [29], considering the decay of the RES with emission of one real phonon, it is argued that the "zero-phonon" peak can be described using the Wigner-Weisskopf formula valid when the linewidth of that peak is much smaller than $\hbar\omega_{\mathrm{LO}}$.

For photon energies larger than $\Omega_{\mathrm{RES}} + \omega_{\mathrm{LO}}$, a transition of the polaron towards the *first scattering state*, belonging to the RES, becomes possible. The final state of the optical absorption process then consists of a polaron in its lowest RES plus a free phonon. A "one-phonon sideband" then appears in the polaron absorption spectrum. This process is called *one-phonon sideband absorption*. The one-, two-, ... K-, ... phonon sidebands of the zero-phonon peak give rise to a broad structure in the absorption spectrum. It turns out that the *first moment* of the phonon sidebands corresponds to the Franck-Condon (FC) frequency $\Omega_{\mathrm{FC}} = 0.141\alpha^2\omega_{\mathrm{LO}}$.

To summarise, following [29], the polaron optical absorption spectrum at strong coupling is characterised by the following features (at $T = 0$):

a) An absorption peak ("zero-phonon line") appears, which corresponds to a transition from the ground state to the first RES at Ω_{RES}.
b) For $\Omega > \Omega_{\mathrm{RES}} + \omega_{\mathrm{LO}}$, a phonon sideband structure arises. This sideband structure peaks around Ω_{FC}. Even when the zero-phonon line becomes weak, and most oscillator strength is in the LO-phonon sidebands, the zero-phonon line continues to determine the onset of the phonon sideband structure.

The basic qualitative strong coupling behaviour predicted in [29], namely, zero-phonon (RES) line with a broader sideband at the high-frequency side, was confirmed by later studies, as will be discussed below.

2.3 Optical Absorption of Fröhlich Polarons at Arbitrary Coupling (DSG, [30])

In 1972 the optical absorption of the Fröhlich polaron was calculated by the present author et al. ([30, 32] ("DSG")) for the Feynman polaron model (and using path integrals). Until recently DSG (combined with [29]) constituted the basic picture for the optical absorption of the Fröhlich polaron. In 1983 [83] the DSG-result was rederived using the memory function formalism (MFF). The DSG-approach is successful at small electron-phonon coupling and is able to identify the excitations at intermediate electron-phonon coupling ($3 \lesssim \alpha \lesssim 6$).

In the strong coupling limit DSG still gives an accurate first moment for the polaron optical absorption but does not reproduce the broad phonon sideband structure (cf. [29] and [84]). A comparison of the DSG results with the OC spectra given by recently developed "approximation-free'" numerical [33] and approximate analytical [34, 35] approaches was carried out recently in [35], see also the chapters by V. Cataudella et al. and A. Mishchenko and N. Nagaosa in the present volume.

The polaron absorption coefficient $\Gamma(\Omega)$ of light with frequency Ω at arbitrary coupling was first derived in [30]. It was represented in the form

$$\Gamma(\Omega) = -\frac{1}{n\epsilon_0 c}\frac{e^2}{m_b}\frac{\mathrm{Im}\Sigma(\Omega)}{[\Omega - \mathrm{Re}\Sigma(\Omega)]^2 + [\mathrm{Im}\Sigma(\Omega)]^2} \ . \tag{5}$$

This general expression was the starting point for a derivation of the theoretical optical absorption spectrum of a single Fröhlich polaron at *all electron-phonon coupling strengths* by DSG in [30]. $\Sigma(\Omega)$ is the so-called "memory function", which contains the dynamics of the polaron and depends on Ω, α, temperature and applied external fields. The key contribution of the work in [30] was to introduce $\Gamma(\Omega)$ in the form (5) and to calculate $\mathrm{Re}\Sigma(\Omega)$, which is essentially a (technically not trivial) Kramers–Kronig transform of the more simple function $\mathrm{Im}\Sigma(\Omega)$. Only function $\mathrm{Im}\Sigma(\Omega)$ had been derived for the Feynman polaron [25] to study the polaron mobility μ from the impedance function, i. e. the static limit

$$\mu^{-1} = \lim_{\Omega\to 0}\left(\frac{\mathrm{Im}\Sigma(\Omega)}{\Omega}\right).$$

The basic nature of the Fröhlich polaron excitations was clearly revealed through this polaron optical absorption obtained in [30]. It was demonstrated in [30] that the Franck-Condon states for Fröhlich polarons are nothing else but a superposition of phonon sidebands. It was also established in [30] that a relatively large value of the electron-phonon coupling strength ($\alpha > 5.9$) is needed to stabilise the relaxed excited state of the polaron. It was, further, revealed that at weaker coupling only "scattering states"of the polaron play a significant role in the optical absorption [30, 85].

2.4 The Structure of the Polaron Excitation Spectrum

In the weak coupling limit, the optical absorption spectrum (5) of the polaron is determined by the absorption of radiation energy, which is re-emitted in the form of LO phonons. As α increases between approximately 3 and 6, a resonance with increasing stability appears in the optical absorption of the Fröhlich polaron of [30] (see Fig. 4). The RES peak in the optical absorption spectrum also has a phonon sideband-structure, whose average transition frequency can be related to an FC-type transition. Furthermore, at zero temperature, the optical absorption spectrum of one polaron also exhibits a zero-frequency "central peak" $[\propto \delta(\Omega)]$. For nonzero temperature, this "central peak" smears out and gives rise to an "anomalous" Drude-type low-frequency component of the optical absorption spectrum.

For $\alpha > 6.5$ the polaron optical absorption gradually develops the structure *qualitatively* proposed in [29]: a broad LO-phonon sideband structure appears with the zero-phonon ("RES") transition as onset. Reference [30] does not predict the broad LO-phonon sidebands at large coupling constant, although it still gives an accurate first Stieltjes moment of the optical absorption spectrum. Reference [35], discussed further in this chapter, sheds new light on the polaron optical absorption.

In Fig. 4 (from [30]), the main peak of the polaron optical absorption for $\alpha = 5.25$ at $\Omega = 3.71\omega_{\mathrm{LO}}$ is interpreted as due to transitions to a RES. The "shoulder" at the low-frequency side of the main peak is attributed as mainly due to one-phonon transitions to polaron "scattering states". The broad structure centred at about $\Omega = 6.6\omega_{\mathrm{LO}}$ is interpreted as an FC band (composed of LO-phonon sidebands). As seen from Fig. 4, when increasing the electron-phonon coupling constant to $\alpha = 6$, the RES peak at $\Omega = 4.14\omega_{\mathrm{LO}}$ stabilises. It is in [30] that an all-coupling optical absorption spectrum of a Fröhlich polaron, together with the role of RES-states, FC-states and scattering states, was first presented. Up to $\alpha = 6$, the DQMC results of [33] reproduce the main features of the optical absorption spectrum of a Fröhlich polaron as found in [30].

Based on [30], it was argued that it is Holstein polarons that determine the optical properties of the charge carriers in oxides like $SrTiO_3$, $BaTiO_3$ [86], while Fröhlich weak-coupling polarons could be identified e.g. in CdO [79].

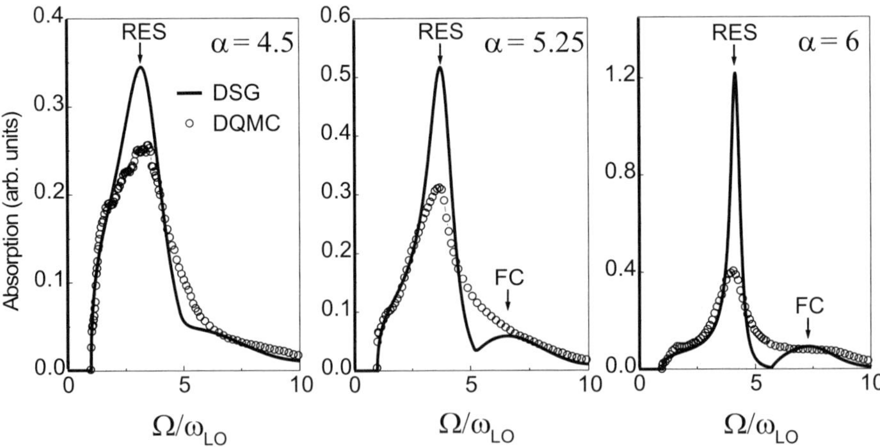

Fig. 4. Optical absorption spectrum of a Fröhlich polaron for $\alpha = 4.5$, $\alpha = 5.25$ and $\alpha = 6$ after [30] (DSG). The RES peak is very intense compared with the FC peak. The δ-like central peaks (at $\Omega = 0$) are schematically shown by vertical lines. The DQMC results of [33] are shown with open circles.

The Fröhlich coupling constants of polar semiconductors and ionic crystals are generally too small to allow for a static "RES". In [87] the RES-peaks of [30] were involved to explain the optical absorption spectrum of $Pr_2NiO_{4.22}$. Further study of the spectra of [87] is called for. The RES-like resonances in $\Gamma(\Omega)$, (5), due to the zero's of $\Omega - \mathrm{Re}\Sigma(\Omega)$, can effectively be displaced to smaller polaron coupling by applying an external magnetic field B, in which case the contribution for what is formally a "RES-type resonance" arises at $\Omega - \omega_c - \mathrm{Re}\Sigma(\Omega) = 0$ ($\omega_c = eB/m_b c$ is the cyclotron frequency). Resonances in the magnetoabsorption governed by this contribution have been clearly observed and analysed in many solids and structures, see Sect. 4.

2.5 Optical Absorption at Arbitrary Coupling. DQMC and DSG

Accurate numerical methods have been developed for the calculation of spectral characteristics and correlation functions of the Holstein polaron (see e.g. [51, 57–59, 88, 89]), of the Fröhlich polaron [39], and of the long-range discrete Fröhlich model [90]. The numerical calculations of the optical conductivity for the Fröhlich polaron performed within the DQMC method by Mishchenko et al. [33], see [36], confirm the analytical results derived in [30] for $\alpha \lesssim 3$. In the intermediate coupling regime $3 < \alpha < 6$, the low-energy behaviour and the position of the RES-peak in the optical conductivity spectrum of [33] follow closely the prediction of [30]. There are some minor quantitative differences between the two approaches in the intermediate coupling regime: in [33], the dominant ("RES") peak is less intense in the Monte-Carlo numerical simulations and the second ("FC") peak develops less prominently. The following qualitative differences exist between the two approaches: in [33], the dominant peak broadens for $\alpha \gtrsim 6$ and the second peak does not develop, but gives rise to a flat shoulder in the optical conductivity spectrum at $\alpha \approx 6$. As α increases beyond $\alpha \approx 6$, the DSG results for the OC do not produce the broad phonon sideband spectrum of the RES-transition that was *qualitatively* predicted in [29] and obtained with DQMC.

Figure 5 shows that already for $\alpha = 1$ noticeable differences arise between $\mathrm{Re}\sigma(\Omega)$ calculated with perturbation theory to $O(\alpha)$, resp. $O(\alpha^2)$, and DSG or DQMC. Remarkably, the DQMC results for $\alpha = 1$ seem to show a somewhat more pronounced two-phonon-scattering contribution than the perturbation theory result to $O(\alpha^2)$. This point deserves further analysis.

An instructive comparison between the positions of the main peak in the optical absorption spectra of Fröhlich polarons obtained within the DSG and DQMC approaches has been performed recently [92]. In Fig. 6 the frequency of the main peak in the OC spectra calculated within the DSG approach [30] is plotted together with that given by DQMC [33, 35]. As seen from the figure, the main-peak positions, obtained within DSG, are in good agreement with the results of DQMC for all considered values of α. At large α the positions of the main peak in the DSG spectra are remarkably close to those given by DQMC. The difference between the DSG and DQMC results is relatively

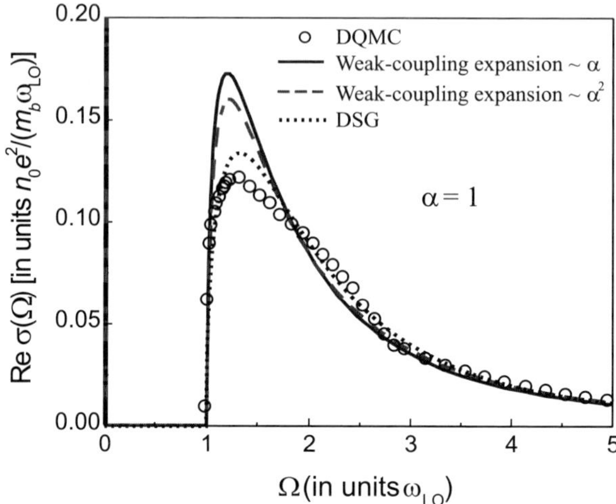

Fig. 5. One-polaron optical conductivity $\mathrm{Re}\sigma\,(\Omega)$ for $\alpha = 1$ calculated within the DQMC approach [33] (open circles), derived using the expansion in powers of α up to α [79] (solid line), up to α^2 [91] (dashed line) and within the DSG approach [30] (dotted line). A δ-like central peak (at $\Omega = 0$) is schematically shown by a vertical line.

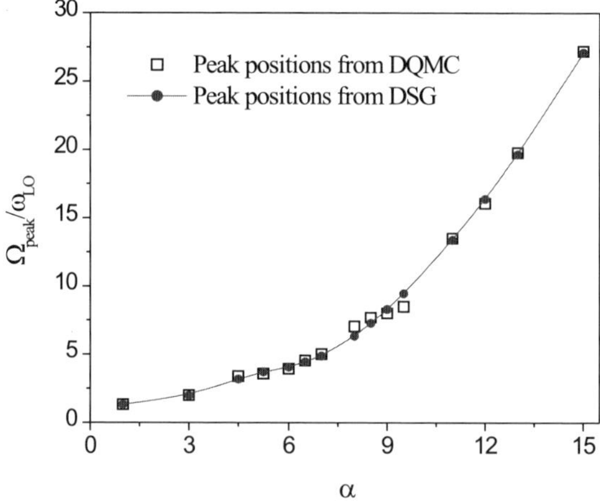

Fig. 6. Main peak positions from DQMC optical conductivity spectra of Fröhlich polarons [35] compared to those of the analytical DSG approach [30]. (From [92].)

larger at $\alpha = 8$ and for $\alpha = 9.5$, but even for those values of the coupling constant the agreement is quite good.

I suggest that the RES-peak at $\alpha \approx 6$ in the DSG-treatment, as α increases, gradually transforms into an FC-peak. As stated above and in [30], DSG predicts a much too narrow FC-peak in the strong coupling limit, but still at the "correct" frequency.

The DSG spectrum also satisfies the *zero* and *first* moment sum rules at all α as will be discussed further in the present chapter.

2.6 Extended Memory Function Formalism

In order to describe the OC main peak line width at intermediate electron-phonon coupling, the DSG approach was modified [35] to include additional dissipation processes, the strength of which is fixed by an exact sum rule, see the chapter by Cataudella et al [37].

To include dissipation [35], a finite lifetime for the states of the relative motion, which can be considered as the result of the residual e-ph interaction not included in the Feynman variational model was introduced. If broadening of the oscillator levels is neglected, the DSG results [30, 83] are recovered.

2.7 The Extended Strong-Coupling Expansion (SCE) of the Polaron Optical Conductivity [92]

Using the Kubo formula (at $T = 0$) the strong coupling OC of the polaron can be expressed in terms of the dipole-dipole two-point correlation function $f_{zz}(t)$:

$$\mathrm{Re}\,\sigma\left(\Omega\right) = \frac{\Omega}{2} \int_{-\infty}^{\infty} e^{i\omega t} f_{zz}\left(t\right)\, dt, \tag{6}$$

$$f_{zz}\left(t\right) = \langle z\left(t\right) z\left(0\right)\rangle. \tag{7}$$

The polaron optical conductivity within the strong-coupling approach can now be calculated beyond the Landau-Pekar approximation [2] in order to obtain rigorous results in the strong-coupling limit. The electron-phonon system is described by the Hamiltonian

$$H = \frac{\mathbf{p}^2}{2} + H_{ph} + \frac{1}{\sqrt{V}} \sum_{\mathbf{k}} \frac{\sqrt{2\sqrt{2}\pi\alpha}}{k}\left(b_{\mathbf{k}} + b^{+}_{-\mathbf{k}}\right) e^{i\mathbf{k}\cdot\mathbf{r}}, \tag{8}$$

$$H_{ph} = \sum_{\mathbf{k}}\left(b^{+}_{\mathbf{k}} b_{\mathbf{k}} + \frac{1}{2}\right), \tag{9}$$

where $m_b = 1$, $\hbar = 1$, $\omega_{\mathrm{LO}} = 1$. In the representation where the phonon coordinates and momenta

$$Q_{\mathbf{k}} = \frac{b_{\mathbf{k}} + b^{+}_{-\mathbf{k}}}{\sqrt{2}},\ P_{\mathbf{k}} = \frac{b_{-\mathbf{k}} - b^{+}_{\mathbf{k}}}{\sqrt{2i}},$$
$$Q^{+}_{\mathbf{k}} = Q_{-\mathbf{k}},\ P^{+}_{\mathbf{k}} = P_{-\mathbf{k}}$$

are used, this Hamiltonian is

$$H = \frac{\mathbf{p}^2}{2} + \frac{1}{2}\sum_{\mathbf{k}}(P_{\mathbf{k}}P_{-\mathbf{k}} + Q_{\mathbf{k}}Q_{-\mathbf{k}}) + \frac{1}{\sqrt{V}}\sum_{\mathbf{k}}\frac{2\sqrt{\sqrt{2}\pi\alpha}}{k}Q_{\mathbf{k}}e^{i\mathbf{k}\cdot\mathbf{r}}. \tag{10}$$

In order to develop a strong-coupling approach for the polaron OC, a scaling transformation of the coordinates and momenta of the electron-phonon system is made following Allcock [93] (p. 48):

$$\mathbf{r} = \alpha^{-1}\mathbf{x}, \ \mathbf{p} = -i\alpha\frac{\partial}{\partial\mathbf{x}},$$

$$\mathbf{k} = \alpha\boldsymbol{\kappa},$$

$$Q_{\mathbf{k}} = \alpha q_{\boldsymbol{\kappa}}, \ P_{\mathbf{k}} = \alpha^{-1}p_{\boldsymbol{\kappa}}, \tag{11}$$

$$\sum_{\boldsymbol{\kappa}}\ldots = \sum_{\mathbf{k}}\ldots = \frac{V}{(2\pi)^3}\int\ldots d\mathbf{k} = \frac{V\alpha^3}{(2\pi)^3}\int\ldots d\boldsymbol{\kappa} = \frac{\mathcal{V}}{(2\pi)^3}\int\ldots d\boldsymbol{\kappa}$$

$$\left(\mathcal{V} \equiv V\alpha^3\right)$$

This transformation is necessary in order to see explicitly the order of magnitude of the different terms in the Hamiltonian. Expressed in terms of the new variables, the Hamiltonian (10) is

$$H = -\frac{\alpha^2}{2}\frac{\partial^2}{\partial\mathbf{x}^2} + \frac{1}{2}\sum_{\boldsymbol{\kappa}}\left(\alpha^{-2}p_{\boldsymbol{\kappa}}p_{-\boldsymbol{\kappa}} + \alpha^2 q_{\boldsymbol{\kappa}}q_{-\boldsymbol{\kappa}}\right) + \alpha^2\frac{2\sqrt{\sqrt{2}\pi}}{\sqrt{\mathcal{V}}}\sum_{\boldsymbol{\kappa}}\frac{1}{\kappa}q_{\boldsymbol{\kappa}}e^{i\boldsymbol{\kappa}\cdot\mathbf{x}}. \tag{12}$$

This Hamiltonian can be written as a sum of two terms, which are of different order in powers of α:

$$H = H_1 + H_2,$$

where $H_1 \sim \alpha^2$ is the leading term,

$$H_1 = \alpha^2\left(-\frac{1}{2}\frac{\partial^2}{\partial\mathbf{x}^2} + \frac{1}{2}\sum_{\boldsymbol{\kappa}}q_{\boldsymbol{\kappa}}q_{-\boldsymbol{\kappa}} + \frac{2\sqrt{\sqrt{2}\pi}}{\sqrt{\mathcal{V}}}\sum_{\boldsymbol{\kappa}}\frac{1}{\kappa}q_{\boldsymbol{\kappa}}e^{i\boldsymbol{\kappa}\cdot\mathbf{x}}\right), \tag{13}$$

and $H_2 \sim \alpha^{-2}$ is the kinetic energy of the phonons,

$$H_2 = \alpha^{-2}\frac{1}{2}\sum_{\boldsymbol{\kappa}}p_{\boldsymbol{\kappa}}p_{-\boldsymbol{\kappa}}. \tag{14}$$

The total ground-state wave function of the electron-phonon system in the adiabatic approximation is given by the strong-coupling Ansatz

$$|\Psi_0\rangle = |\Phi_0\rangle|\psi_0\rangle, \tag{15}$$

where $|\Phi_0\rangle$ and $|\psi_0\rangle$ are, respectively, the phonon and electron wave functions. The phonon wave function is related to the phonon vacuum $|0_{ph}\rangle$ by

$$|\Phi_0\rangle = U\,|0_{ph}\rangle\,, \tag{16}$$

where U is the unitary transformation:

$$U = e^{\sum_{\mathbf{k}}\left(f_{\mathbf{k}}b_{\mathbf{k}} - f_{\mathbf{k}}^{*}b_{\mathbf{k}}^{+}\right)}. \tag{17}$$

The optimal values of the variational parameters $f_{\mathbf{k}}$ are

$$f_{\mathbf{k}} = \frac{\sqrt{2\sqrt{2}\pi\alpha}}{k\sqrt{V}}\rho_{\mathbf{k}}, \tag{18}$$

where $\rho_{\mathbf{k}}$ is the average

$$\rho_{\mathbf{k}} = \langle\psi_0\,|e^{i\mathbf{k}\cdot\mathbf{r}}|\,\psi_0\rangle\,. \tag{19}$$

Using the fact that $\rho_{-\mathbf{k}} = \rho_{\mathbf{k}}$ (due to the inversion symmetry of the ground state), we express the unitary operator (17) in the new variables:

$$U = \exp\left(i\sum_{\boldsymbol{\kappa}}g_{-\boldsymbol{\kappa}}p_{\boldsymbol{\kappa}}\right)$$

with

$$g_{\boldsymbol{\kappa}} = \frac{2\sqrt{\sqrt{2}\pi}}{\sqrt{V}\kappa}\rho_{\boldsymbol{\kappa}},\ \ \rho_{\boldsymbol{\kappa}} = \langle\psi_0\,|e^{i\boldsymbol{\kappa}\cdot\mathbf{x}}|\,\psi_0\rangle\,.$$

$\widetilde{H} = U^{-1}HU$ is the transformed Hamiltonian:

$$\widetilde{H} = -\frac{\alpha^2}{2}\frac{\partial^2}{\partial\mathbf{x}^2} + U_a\left(\mathbf{x}\right) + \Delta E + \frac{1}{2}\sum_{\boldsymbol{\kappa}}\left(\frac{1}{\alpha^2}p_{\boldsymbol{\kappa}}p_{-\boldsymbol{\kappa}} + \alpha^2 q_{\boldsymbol{\kappa}}q_{-\boldsymbol{\kappa}}\right) + \sum_{\boldsymbol{\kappa}}W_{\boldsymbol{\kappa}}\left(\mathbf{x}\right)q_{\boldsymbol{\kappa}} \tag{20}$$

with the notations

$$\Delta E \equiv \frac{\alpha^2}{2}\sum_{\boldsymbol{\kappa}}|g_{\boldsymbol{\kappa}}|^2\,, \tag{21}$$

$$U_a\left(\mathbf{x}\right) = -\alpha^2\frac{2\sqrt{\sqrt{2}\pi}}{\sqrt{V}}\sum_{\boldsymbol{\kappa}}\frac{1}{\kappa}g_{-\boldsymbol{\kappa}}e^{i\boldsymbol{\kappa}\cdot\mathbf{x}}, \tag{22}$$

$$W_{\boldsymbol{\kappa}}\left(\mathbf{x}\right) \equiv \alpha^2\frac{2\sqrt{\sqrt{2}\pi}}{\sqrt{V}}\frac{1}{\kappa}\left(e^{i\boldsymbol{\kappa}\cdot\mathbf{x}} - \rho_{\boldsymbol{\kappa}}\right)\,. \tag{23}$$

Here, $W_{\boldsymbol{\kappa}}\left(\mathbf{x}\right)$ are the amplitudes of the renormalised electron-phonon interaction and $U_a\left(\mathbf{x}\right)$ is the self-consistent adiabatic potential energy for the electron.

As a result, the correlation function $f_{zz}\left(t\right)$ takes the form

$$f_{zz}\left(t\right) = \left\langle0_{ph}\,\left|\left\langle\psi_0\,\left|e^{it\widetilde{H}}ze^{-it\widetilde{H}}z\right|\,\psi_0\right\rangle\right|\,0_{ph}\right\rangle\,. \tag{24}$$

The transformed Hamiltonian $\widetilde{H}$ is the sum of two terms:

$$\widetilde{H} = \widetilde{H}_0 + W \tag{25}$$

with

$$\widetilde{H}_0 = -\frac{\alpha^2}{2}\frac{\partial^2}{\partial \mathbf{x}^2} + U_a\left(\mathbf{x}\right) + \Delta E + \frac{1}{2}\sum_{\kappa}\left(\frac{1}{\alpha^2}p_\kappa p_{-\kappa} + \alpha^2 q_\kappa q_{-\kappa}\right), \tag{26}$$

$$W = \sum_{\kappa} W_\kappa\left(\mathbf{x}\right) q_\kappa. \tag{27}$$

The unperturbed Hamiltonian $\widetilde{H}_0$ and the renormalised electron-phonon interaction are, respectively,

$$\widetilde{H}_0 = \frac{\mathbf{p}^2}{2} + \sum_{\mathbf{k}} |f_{\mathbf{k}}|^2 + V_a\left(r\right) + \sum_{\mathbf{k}}\left(b_{\mathbf{k}}^+ b_{\mathbf{k}} + \frac{1}{2}\right), \tag{28}$$

$$W = \sum_{\mathbf{k}}\left(W_{\mathbf{k}} b_{\mathbf{k}} + W_{\mathbf{k}}^* b_{\mathbf{k}}^+\right). \tag{29}$$

Here, the $W_{\mathbf{k}}$ are the amplitudes of the renormalised electron-phonon interaction

$$W_{\mathbf{k}} = \frac{\sqrt{2\sqrt{2}\pi\alpha}}{k\sqrt{V}}\left(e^{i\mathbf{k}\cdot\mathbf{r}} - \rho_{\mathbf{k}}\right) \tag{30}$$

and $V_a\left(r\right)$ is the self-consistent adiabatic potential energy for the electron

$$V_a\left(r\right) = -\sum_{\mathbf{k}} \frac{4\sqrt{2}\pi\alpha}{k^2 V}\rho_{-\mathbf{k}}e^{i\mathbf{k}\cdot\mathbf{r}}. \tag{31}$$

Further on, a complete orthogonal basis consisting of the Franck-Condon (FC) states $|\psi_{n,l,m}\rangle$ is used, with l the quantum number of the angular momentum, m the z-projection of the angular momentum, n the radial quantum number. (In this classification, the ground-state wave function is $|\psi_{0,0,0}\rangle \equiv |\psi_0\rangle$.) The FC wave functions $|\psi_{n,l,m}\rangle$ are the exact eigenstates of the Hamiltonian $\widetilde{H}_0$.

Up to this point, the only approximation made in $f_{zz}\left(t\right)$ was the strong-coupling Ansatz for the polaron ground-state wave function. The next step is to apply the Born-Oppenheimer (BO) approximation [93], which neglects the non-adiabatic transitions between different polaron levels for the renormalised operator of the electron-phonon interaction W. The dipole-dipole correlation function $f_{zz}\left(t\right)$ in the BO approximation is [35, 92]

$$f_{zz}\left(t\right) = \sum_{n,l,m} |\langle\psi_0\left|z\right|\psi_{n,1,0}\rangle|^2 \, e^{it\left(E_0 - E_{n,1}\right)}$$

$$\times \left\langle 0_{ph}\left|\left\langle\psi_{n,1,0}\left|\mathrm{T}\exp\left(-i\int_0^t dsW\left(s\right)\right)\right|\psi_{n,1,0}\right\rangle\right|0_{ph}\right\rangle \tag{32}$$

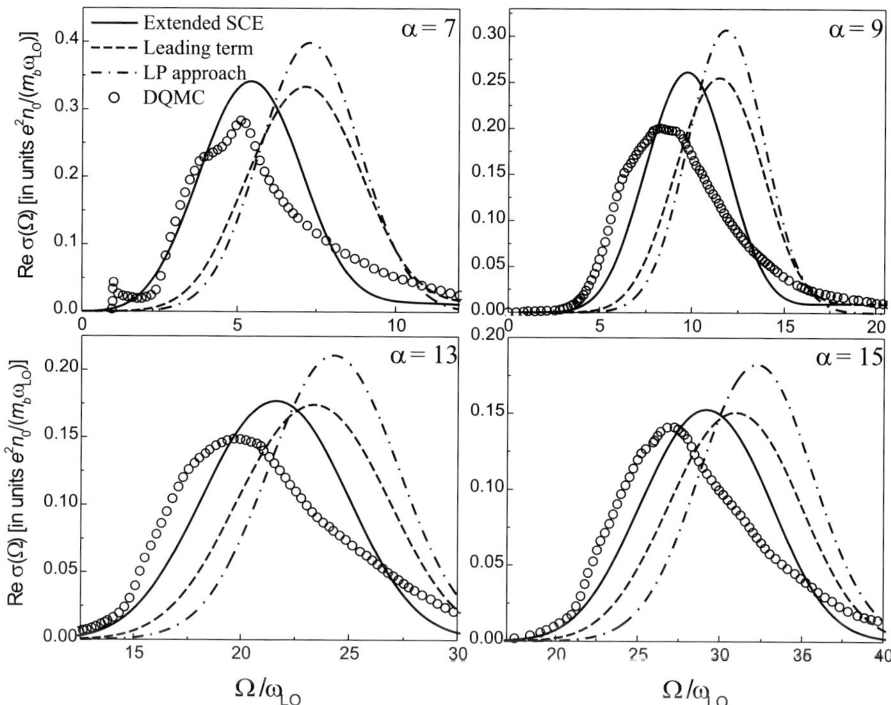

Fig. 7. The polaron OC calculated within the extended SCE taking into account corrections of order α^0 (solid curve), the OC calculated within the leading-term strong-coupling approximation (dashed curve), with the leading term of the Landau-Pekar (LP) adiabatic approximation (dash-dotted curve), and the numerical DQMC data (open circles) for $\alpha = 7, 9, 13$ and 15. (From [92].)

with the time-dependent interaction Hamiltonian

$$W\left(s\right) \equiv e^{is\widetilde{H}_0} W\left(s\right) e^{-is\widetilde{H}_0}. \tag{33}$$

The polaron energies E_0, $E_{n,1}$ and the wave functions ψ_0, $\psi_{n,1,0}$ are calculated taking into account the corrections of order of α^0.

Figure 7 shows the polaron OC spectra for different values of α calculated numerically using (32) within different approximations. The OC spectra calculated within the extended SCE approach taking into account both the Jahn-Teller effect – related to the degeneracy with respect to the quantum number m – and the corrections of order α^0 are shown by the solid curves. The OC obtained with the leading-term strong-coupling approximation taking into account the Jahn-Teller effect and with the leading term of the Landau-Pekar adiabatic approximation are plotted as dashed and dash-dotted curves, respectively. The open circles show the DQMC data [33, 35].

The polaron OC band of Fig. 7 obtained within the extended SCE generalises the Gaussian-like polaron OC band (as given e.g. by (3) of [35]) thanks to (i) the use of the numerically exact strong-coupling polaron wave functions [94] and (ii) the incorporation of both static and dynamic Jahn-Teller effects. The polaron OC broad structure calculated within the present extended SCE consists of a series of LO-phonon sidebands and provides a realisation – with all LO-phonons involved for a given α – of the scheme proposed by KED [29].

As seen from Fig. 7, the polaron OC spectra calculated within the asymptotically exact strong-coupling approach are shifted towards lower frequencies as compared with the OC spectra calculated within the LP approximation. This shift is due to the use of the numerically exact (in the strong-coupling limit) energy levels and wave functions of the internal excited polaron states, as well as the numerically exact self-consistent adiabatic polaron potential. Furthermore, the inclusion of the corrections of order α^0 leads to a shift of the OC spectra to lower frequencies with respect to the OC spectra calculated within the leading-term approximation. The value of this shift $\Delta\Omega_{n,0}/\omega_{LO} \approx -1.8$ obtained within the extended SCE, is close to the LP value $\Delta\Omega_{n,0}^{(LP)}/\omega_{LO} = -(4\ln 2 - 1) \approx -1.7726$ (cf. [95, 96]). The distinction between the OC spectra calculated with and without the Jahn-Teller effect is very small.

Starting from $\alpha \approx 9$ towards larger values of α, the agreement between the extended SCE polaron OC spectra and the numerical DQMC data becomes gradually better, consistent with the fact that the extended SCE for the polaron OC is asymptotically exact in the strong-coupling limit.

The results of the extended SCE as treated in the present section are qualitatively consistent with the interpretation advanced in [29]. In [29] only the 1-LO-phonon sideband was taken into account, while in [84] 2-LO-phonon emission was included.

The extended SCE carries on the program started in [29]. The spectra in Fig. 7, in the strong coupling approximation, consist of LO-phonon sidebands to the RES (which itself has negligible oscillator strength in this limit, similar to the optical absorption for some colour centres in alkali halides). These LO-phonon sidebands form a broad FC-structure.

2.8 Comparison Between the Optical Conductivity Spectra Obtained Within Different Approaches

A comparison between the optical conductivity spectra obtained with the DQMC method, extended MFF, SCE and DSG for different values of α is shown in Figs. 8 and 9, taken from [35]. The key results of the comparison are the following.

First, as expected, in the weak-coupling regime, both the extended MFF with phonon broadening and DSG [30] are in very good agreement with the DQMC data [33], showing significant improvement with respect to the weak-coupling perturbation approach [31, 79] which provides a good description of the OC spectra only for very small values of α. For $3 \leq \alpha \leq 6$, DSG predicts the essential structure of the OA, with a RES-transition gradually building up for increasing α, but underestimates the peak width. The damping, introduced in the extended MFF approach, becomes crucial in this coupling regime.

Second, comparing the peak and shoulder energies, obtained by DQMC, with the peak energies, given by MFF, and the FC transition energies from the SCE, it is concluded [35] that as α increases from 6 to 10 the spectral weights rapidly switch from the dynamic regime, where the lattice follows the electron motion, to the adiabatic regime dominated by FC transitions. In the intermediate electron-phonon coupling regime, $6 < \alpha < 10$, both adiabatic FC and non-adiabatic dynamical excitations coexist.

For still larger coupling ($\alpha \gtrsim 10$), the polaron OA spectrum consists of a broad FC-structure, built of LO-phonon sidebands.

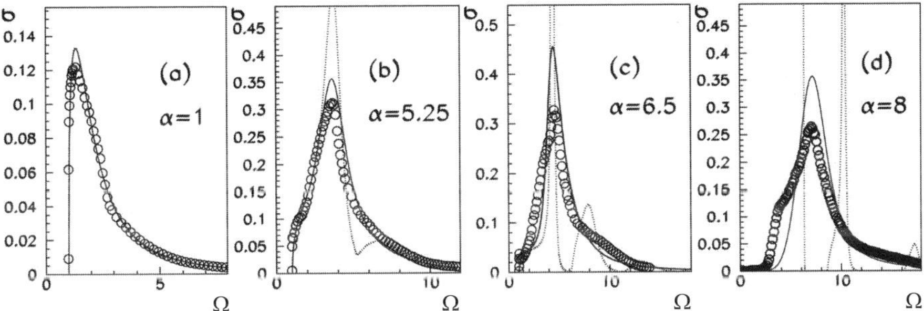

Fig. 8. Comparison of the optical conductivity calculated with the DQMC method (circles), extended MFF (solid line) and DSG [30, 83] (dotted line), for four different values of α. (Reprinted with permission from [35]. ©2006 by the American Physical Society.)

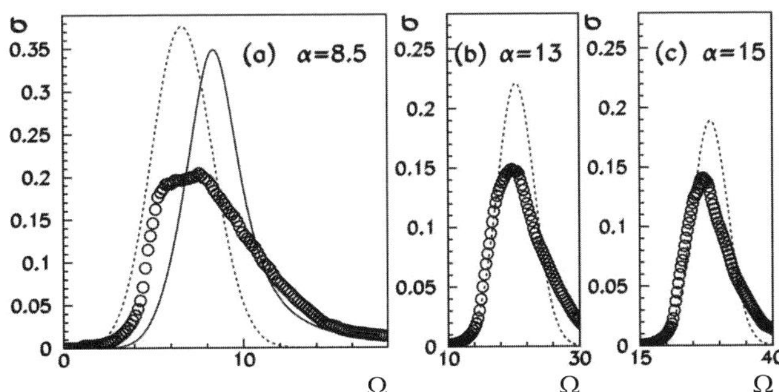

Fig. 9. Comparison of the optical conductivity calculated with the DQMC method (circles), the extended MFF (solid line) and SCE (dashed line) for three different values of α. (Reprinted with permission from [35]. ©2006 by the American Physical Society.)

24   Jozef T. Devreese

In summary, the accurate numerical results obtained from DQMC – modulo the linewidths for $\alpha > 6$ – and from the recent analytical approximations [34, 35] confirm the essence of the mechanism for the optical absorption of Fröhlich polarons, proposed in [30, 32] combined with [29] and do add important new extensions and new insights (see the chapters by V. Cataudella et al. and by A. Mishchenko and N. Nagaosa in the present volume).

2.9 Sum Rules for the Optical Conductivity Spectra of Fröhlich Polarons

In this section several sum rules for the optical conductivity spectra of Fröhlich polarons are applied to test the DSG approach [30] and the DQMC results [33]. The values of the polaron effective mass for the DQMC approach are taken from [39]. In Tables 2 and 3, we show the polaron ground-state E_0 and the following parameters calculated using the optical conductivity spectra:

$$M_0 \equiv \int_1^{\Omega_{\max}} \mathrm{Re}\sigma\left(\Omega\right) d\Omega, \tag{34}$$

$$M_1 \equiv \int_1^{\Omega_{\max}} \Omega \mathrm{Re}\sigma\left(\Omega\right) d\Omega, \tag{35}$$

where $\Omega_{\max}$ is the upper value of the frequency available from [33],

$$\widetilde{M_0} \equiv \frac{\pi}{2m^*} + \int_1^{\Omega_{\max}} \mathrm{Re}\sigma\left(\Omega\right) d\Omega. \tag{36}$$

Here m^* is the polaron mass, the optical conductivity is calculated in units $n_0 e^2/(m_b \omega_{\mathrm{LO}})$, m^* is measured in units of the band mass m_b, and the frequency is measured in units of ω_{LO}. The values of $\Omega_{\max}$ are: $\Omega_{\max} = 10$ for $\alpha = 0.01, 1$ and 3, $\Omega_{\max} = 12$ for $\alpha = 4.5, 5.25$ and 6, $\Omega_{\max} = 18$ for $\alpha = 6.5$, 7 and 8.

The parameters corresponding to the DQMC calculation are obtained using the numerical data kindly provided to the author by A. Mishchenko [97]. The optical conductivity derived by DSG [30] exactly satisfies the sum rule [98]

$$\frac{\pi}{2m^*} + \int_1^{\infty} \mathrm{Re}\sigma\left(\Omega\right) d\Omega = \frac{\pi}{2}. \tag{37}$$

Since the optical conductivity obtained from the DQMC results [33] is known only within a limited interval of frequencies $1 < \Omega < \Omega_{\max}$, the integral in (36) for the DSG-approach [30] is calculated over the same frequency interval as for the Monte Carlo results [33].

The comparison of the resulting zero frequency moments $\widetilde{M_0}^{(\mathrm{DQMC})}$ and $\widetilde{M_0}^{(\mathrm{DSG})}$ with each other and with the value $\pi/2 = 1.5707963...$ corresponding to the right-hand-side of the sum rule (37) shows that the difference

Table 2. Polaron parameters $M_0, M_1, \widetilde{M}_0$ obtained from the diagrammatic Monte Carlo results (Reprinted with permission from [78]. ©2006, Società Italiana di Fisica.)

α	$M_0^{\text{(DQMC)}}$	$m^{*\text{(DQMC)}}$	$\widetilde{M}_0^{\text{(DQMC)}}$	$M_1^{\text{(DQMC)}}/\alpha$	$E_0^{\text{(DQMC)}}$
0.01	0.00249	1.0017	1.5706	0.634	−0.010
1	0.24179	1.1865	1.5657	0.65789	−1.013
3	0.67743	1.8467	1.5280	0.73123	−3.18
4.5	0.97540	2.8742	1.5219	0.862	−4.97
5.25	1.0904	3.8148	1.5022	0.90181	−5.68
6	1.1994	5.3708	1.4919	0.98248	−6.79
6.5	1.30	6.4989	1.5417	1.1356	−7.44
7	1.3558	9.7158	1.5175	1.2163	−8.31
8	1.4195	19.991	1.4981	1.3774	−9.85

Table 3. Polaron parameters $M_0, M_1, \widetilde{M}_0$ obtained within the path-integral approach (Reprinted with permission from [78]. ©2006, Società Italiana di Fisica.)

α	$M_0^{\text{(DSG)}}$	$m^{*\text{(Feynman)}}$	$\widetilde{M}_0^{\text{(DSG)}}$	$M_1^{\text{(DSG)}}/\alpha$	$E_0^{\text{(Feynman)}}$
0.01	0.00248	1.0017	1.5706	0.633	−0.010
1	0.24318	1.1957	1.5569	0.65468	−1.0130
3	0.69696	1.8912	1.5275	0.71572	−3.1333
4.5	1.0162	3.1202	1.5196	0.83184	−4.8394
5.25	1.1504	4.3969	1.5077	0.88595	−5.7482
6	1.2608	6.8367	1.4906	0.95384	−6.7108
6.5	1.3657	9.7449	1.5269	1.1192	−7.3920
7	1.4278	14.395	1.5369	1.2170	−8.1127
8	1.4741	31.569	1.5239	1.4340	−9.6953

$\left| \widetilde{M}_0^{\text{(DQMC)}} - \widetilde{M}_0^{\text{(DSG)}} \right|$ on the interval $\alpha \leq 8$ is smaller than the absolute value of the contribution of the "tail" of the optical conductivity for $\Omega > \Omega_{\max}$ to the integral in the sum rule (37):

$$\int_{\Omega_{\max}}^{\infty} \text{Re}\sigma^{\text{(DSG)}}\left(\Omega\right) d\Omega \equiv \frac{\pi}{2} - \widetilde{M}_0^{\text{(DSG)}}. \tag{38}$$

Within the accuracy determined by the neglect of the "tail" of the part of the spectrum for $\Omega > \Omega_{\max}$, the contribution to the integral in the sum rule (37) for the optical conductivity obtained from the DQMC results [33] *agrees well* with that for the optical conductivity found within the path-integral approach in [30]. Hence, the conclusion follows that *the optical conductivity obtained from the DQMC results [33] satisfies the sum rule (37)* within the aforementioned accuracy.

We analyse the fulfilment of the "LSD" polaron ground-state theorem introduced in [99]:

$$E_0\left(\alpha\right) - E_0\left(0\right) = -\frac{3}{\pi} \int\limits_0^\alpha \frac{d\alpha'}{\alpha'} \int\limits_0^\infty \Omega \mathrm{Re}\sigma\left(\Omega, \alpha'\right) d\Omega \qquad (39)$$

using the first-frequency moments $M_1^{(\mathrm{DQMC})}$ and $M_1^{(\mathrm{DSG})}$. The results of this comparison are presented in Fig. 10. The solid dots indicate the polaron ground-state energy calculated by Feynman using his variational principle for path integrals. The dotted curve is the value of $E_0\left(\alpha\right)$ calculated numerically using the optical conductivity spectra and the ground-state theorem with the DSG optical conductivity [30] for the polaron,

$$E_0^{(\mathrm{DSG})}\left(\alpha\right) \equiv -\frac{3}{\pi} \int\limits_0^\alpha \frac{d\alpha'}{\alpha'} \int\limits_0^\infty \Omega \mathrm{Re}\sigma^{(\mathrm{DSG})}\left(\Omega, \alpha'\right) d\Omega. \qquad (40)$$

The solid curve and the open circles are the values obtained using $M_1^{(\mathrm{DSG})}\left(\alpha\right)$ and $M_1^{(\mathrm{DQMC})}\left(\alpha\right)$, respectively:

$$\widetilde{E}_0\left(\alpha\right) \equiv -\frac{3}{\pi} \int\limits_0^\alpha \frac{d\alpha'}{\alpha'} \int\limits_0^{\Omega_{\max}} \Omega \mathrm{Re}\sigma\left(\Omega, \alpha'\right) d\Omega = -\frac{3}{\pi} \int\limits_0^\alpha d\alpha' \frac{M_1\left(\alpha'\right)}{\alpha'}. \qquad (41)$$

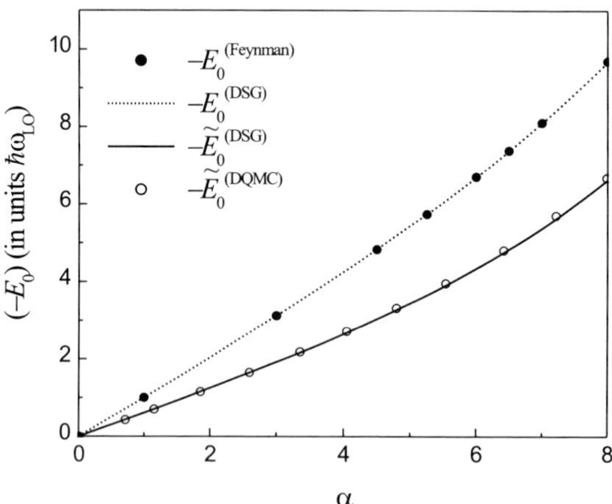

Fig. 10. Test of the ground-state theorem for a Fröhlich polaron from [99] using different optical conductivity spectra, DSG from [30] and DQMC from [33]. The notations are explained in the text. (Reprinted with permission after [78]. ©2006, Società Italiana di Fisica.)

As seen from the figure, $E_0^{(\mathrm{DSG})}(\alpha)$ coincides, to a high degree of accuracy, with the variational polaron ground-state energy.

From the comparison of $\widetilde{E}_0^{(\mathrm{DSG})}(\alpha)$ with $\widetilde{E}_0^{(\mathrm{DQMC})}(\alpha)$ it follows that the contribution to the integral in (41), with the given limited frequency region, which approximates the integral in the right-hand side of the "LSD" ground state theorem (39), for the optical conductivity obtained from the DQMC results [33] *agrees with a high accuracy* with the corresponding contribution to the integral in (41) for the optical conductivity derived from the path-integral approach of [30]. Because for the path-integral result, the integral $\widetilde{E}_0^{(\mathrm{DSG})}(\alpha)$ noticeably differs from the integral $E_0^{(\mathrm{DSG})}(\alpha)$, a comparison between the Feynman polaron ground state energy E_0 and the integral $\widetilde{E}_0^{(\mathrm{DSG})}(\alpha)$ is not justified. Similarly, a comparison between the polaron ground state energy obtained from the DQMC results and the integral $\widetilde{E}_0^{(\mathrm{DQMC})}(\alpha)$ would require us to overcome the limited frequency domain of the available optical conductivity spectrum [33].

The DQMC optical conductivity spectrum for higher frequencies than $\Omega_{\max}$ of [33] is needed in order to check the fulfilment of the sum rules (37) and (39) with a higher accuracy.

3 Polaron Scaling Relations

3.1 Derivation of the Scaling Relations

The form of the Fröhlich Hamiltonian in n dimensions is the same as in 3D,

$$H = \frac{\mathbf{p}^2}{2m_b} + \sum_{\mathbf{k}} \hbar\omega_{\mathbf{k}} a_{\mathbf{k}}^{\dagger} a_{\mathbf{k}} + \sum_{\mathbf{k}} \left(V_{\mathbf{k}} a_{\mathbf{k}} e^{i\mathbf{k}\cdot\mathbf{r}} + V_{\mathbf{k}}^{*} a_{\mathbf{k}}^{\dagger} e^{-i\mathbf{k}\cdot\mathbf{r}} \right), \qquad (42)$$

except that now all vectors are n-dimensional. In this subsection, dispersionless longitudinal phonons are considered, i.e., $\omega_{\mathbf{k}} = \omega_{\mathrm{LO}}$, and units are chosen such that $\hbar = m_b = \omega_{\mathrm{LO}} = 1$.

In 3D the interaction coefficient is well known, $|V_{\mathbf{k}}|^2 = 2\sqrt{2}\pi\alpha/V_3 k^2$. The interaction coefficient in n dimensions becomes [100]

$$|V_{\mathbf{k}}|^2 = \frac{2^{n-3/2}\pi^{(n-1)/2}\Gamma\left(\frac{n-1}{2}\right)\alpha}{V_n k^{n-1}} \qquad (43)$$

with V_n the volume of the n-dimensional crystal.

The only difference between the model system in n dimensions and the model system in 3D is that now one deals with an n-dimensional harmonic oscillator. Directly following [8], the variational polaron energy was calculated in [100]

$$E = \frac{n(v-w)}{2} - \frac{n\left(v^2-w^2\right)}{4v} - \frac{2^{-3/2}\Gamma\left(\frac{n-1}{2}\right)\alpha}{\Gamma\left(\frac{n}{2}\right)} \int_0^\infty \frac{e^{-t}}{\sqrt{D_0(t)}}\,dt$$

$$= \frac{n(v-w)^2}{4v} - \frac{\Gamma\left(\frac{n-1}{2}\right)\alpha}{2\sqrt{2}\,\Gamma\left(\frac{n}{2}\right)} \int_0^\infty \frac{e^{-t}}{\sqrt{D_0(t)}}\,dt, \tag{44}$$

where

$$D_0(t) = \frac{w^2}{2v^2}t + \frac{v^2-w^2}{2v^3}\left(1 - e^{-vt}\right). \tag{45}$$

In order to facilitate a comparison of E for n dimensions with the Feynman result [8] for 3D,

$$E_{3\mathrm{D}}(\alpha) = \frac{3\left(v-w\right)^2}{4v} - \frac{1}{\sqrt{2\pi}}\alpha \int_0^\infty \frac{e^{-t}}{\sqrt{D_0(t)}}\,dt, \tag{46}$$

it is convenient to rewrite (44) in the form

$$E_{n\mathrm{D}}(\alpha) = \frac{n}{3}\left[\frac{3\left(v-w\right)^2}{4v} - \frac{1}{\sqrt{2\pi}}\frac{3\sqrt{\pi}\,\Gamma\left(\frac{n-1}{2}\right)}{2n\Gamma\left(\frac{n}{2}\right)}\alpha \int_0^\infty \frac{e^{-t}}{\sqrt{D_0(t)}}\,dt\right]. \tag{47}$$

The parameters w and v must be determined by minimizing E. In the case of (47) one should minimise the expression in the brackets. The only difference of this expression from the rhs of (46) is that α is multiplied by the factor

$$a_n = \frac{3\sqrt{\pi}\,\Gamma\left(\frac{n-1}{2}\right)}{2n\Gamma\left(\frac{n}{2}\right)}. \tag{48}$$

This means that the minimizing parameters w and v in nD at a given α will be exactly the same as those calculated in 3D with the Fröhlich constant chosen as $a_n\alpha$:

$$v_{n\mathrm{D}}(\alpha) = v_{3\mathrm{D}}(a_n\alpha), \quad w_{n\mathrm{D}}(\alpha) = w_{3\mathrm{D}}(a_n\alpha). \tag{49}$$

Comparing (47) to (46), the following scaling relation [100–102] is obtained:

$$E_{n\mathrm{D}}(\alpha) = \frac{n}{3}E_{3\mathrm{D}}(a_n\alpha), \tag{50}$$

where a_n is given by (48). As discussed in [100], the above scaling relation is not an exact relation. It is valid for the Feynman polaron energy and also for the ground-state energy to order α. The next-order term (i.e., α^2) no longer satisfies (50). The reason is that in the exact calculation (to order α^2) the electron motions in different space directions are coupled by the electron-phonon interaction. No such coupling appears in the Feynman polaron model; this is the underlying reason for the validity of the scaling relation for the Feynman approximation.

In [83, 98, 100, 102], scaling relations were obtained also for the impedance function,

$$Z_{n\mathrm{D}}\left(\alpha;\Omega\right) = Z_{3\mathrm{D}}\left(a_{n}\alpha;\Omega\right),\tag{51}$$

the effective mass and the mobility of a polaron.

In the important particular case of 2D, the scaling relations take the form [100–102]:

$$E_{2\mathrm{D}}\left(\alpha\right) = \frac{2}{3}E_{3\mathrm{D}}\left(\frac{3\pi}{4}\alpha\right),\tag{52}$$

$$Z_{2\mathrm{D}}\left(\alpha;\Omega\right) = Z_{3\mathrm{D}}\left(\frac{3\pi}{4}\alpha;\Omega\right),\tag{53}$$

$$\frac{m^{*}_{2\mathrm{D}}\left(\alpha\right)}{\left(m_{b}\right)_{n\mathrm{D}}} = \frac{m^{*}_{3\mathrm{D}}\left(\frac{3\pi}{4}\alpha\right)}{\left(m_{b}\right)_{3\mathrm{D}}},\tag{54}$$

$$\mu_{2\mathrm{D}}\left(\alpha\right) = \mu_{3\mathrm{D}}\left(\frac{3\pi}{4}\alpha\right).\tag{55}$$

3.2 Check of the Polaron Scaling Relation for the Path Integral Monte Carlo Result for the Polaron Free Energy

The fulfilment of the PD-scaling relation [102] is checked here for the path integral Monte Carlo results [38] for the polaron free energy.

The path integral Monte Carlo results of [38] for the polaron free energy in 3D and in 2D are given for a few values of temperature and for some selected values of α. For a check of the scaling relation, the values of the polaron free energy at $\beta = 10$ ($\beta = \hbar\omega_{\mathrm{LO}}/k_{\mathrm{B}}T$, T is the temperature) are taken from [38] in 3D (Table I, for 4 values of α) and in 2D (Table II, for 2 values of α) and plotted in Fig. 11, upper panel, with open circles and squares, respectively.

The PD-scaling relation for the polaron ground-state energy as derived in [102] is given by (52).

In Fig. 11, lower panel, the available data for the free energy from [38] are plotted in the following form, *inspired by the lhs and the rhs parts of (52)*: $F_{2D}\left(\alpha\right)$ (squares) and $\frac{2}{3}F_{3D}\left(\frac{3\pi\alpha}{4}\right)$ (open triangles). As follows from the figure, *the path integral Monte Carlo results for the polaron free energy in 2D and 3D very closely follow the PD-scaling relation of the form given by (52):*

$$F_{2D}\left(\alpha\right) \equiv \frac{2}{3}F_{3D}\left(\frac{3\pi\alpha}{4}\right).\tag{56}$$

4 Magneto-Optical Absorption of Polarons

This section is based on the work of F. M. Peeters and J. T. Devreese et al.

The results on the polaron optical absorption [30, 32] paved the way for an all-coupling path-integral based theory of the magneto-optical absorption

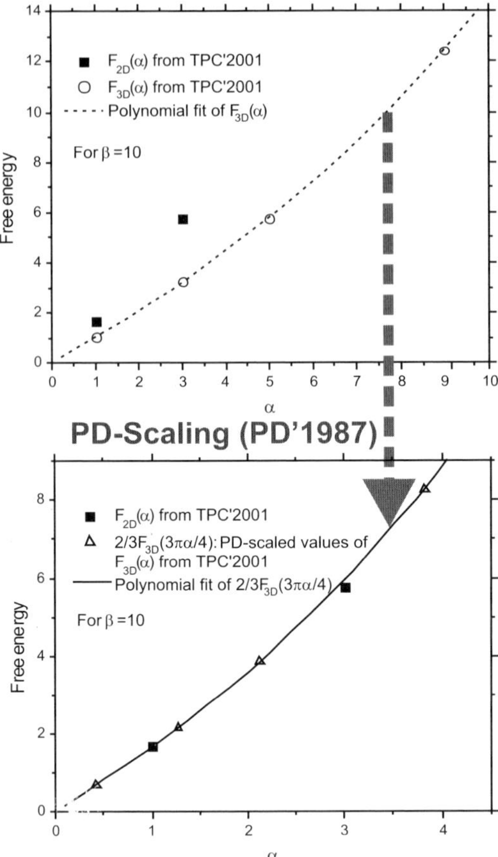

Fig. 11. *Upper panel:* The polaron free energy in 2D (squares) and 3D (open circles) obtained by TPC'2001 [38] for $\beta = 10$. The data for $F_{3D}(\alpha)$ are interpolated using a polynomial fit to the available four points (dotted line). *Lower panel:* Demonstration of the PD-scaling cf. PD'1987. The polaron free energy in 2D obtained by TPC'2001 [38] for $\beta = 10$ (squares). The **PD-scaled** according to PD'1987 [102] polaron free energy in 3D from TPC'2001 for $\beta = 10$ (open triangles). The data for $\frac{2}{3}F_{3D}\left(\frac{3\pi\alpha}{4}\right)$ are interpolated using a polynomial fit to the available four points (solid line). (Reprinted with permission from [78]. ©2006, Società Italiana di Fisica.)

of polarons (see [103]) with the aim to explain existing - and predict new - experimental magneto-optical polaron effects in solids, in systems of reduced dimensions and reduced dimensionalities. This work was also partly motivated by the insight that magnetic fields can stabilise the relaxed excited polaron states, so that information on the nature of relaxed excited states might be gained from the cyclotron resonance of polarons.

Some of the subsequent developments in the field of polaron cyclotron resonance are briefly reviewed below.

Evidence for the polaron character of charge carriers in AgBr and AgCl was obtained through high-precision cyclotron resonance experiments in external magnetic fields up to 16 T (see Fig. 12). Several polaron theories were compared in analysing the cyclotron resonance data. It turns out that the weak-coupling theories (Rayleigh-Schrödinger perturbation theory, Wigner-Brillouin perturbation theory and its improvements) fail (and are all off by at least 20% at 16 T) to describe the experimental data shown in Fig. 12 for the silver halides. The approach of [104] underestimates the polaron cyclotron mass by 2.5% at 15.3 T. The magneto-absorption calculated in [103], which is an all-coupling derivation, leads to the best quantitative agreement between theory and experiment for AgBr and AgCl as can be seen from Fig. 12. This quantitative interpretation of the cyclotron resonance experiment in AgBr and AgCl [105] by the theory of [103] provided one of the most convincing and clearest demonstrations of Fröhlich polaron features in solids.

The analysis in [105] on the basis of the theory for polaron magneto-absorption of [103] leads to the following polaron coupling constants (indicated in Table 1): $\alpha = 1.53$ for AgBr and $\alpha = 1.84$ for AgCl. The corresponding polaron masses are: $m^* = 0.2818$ for AgBr and $m^* = 0.3988$ for AgCl. For

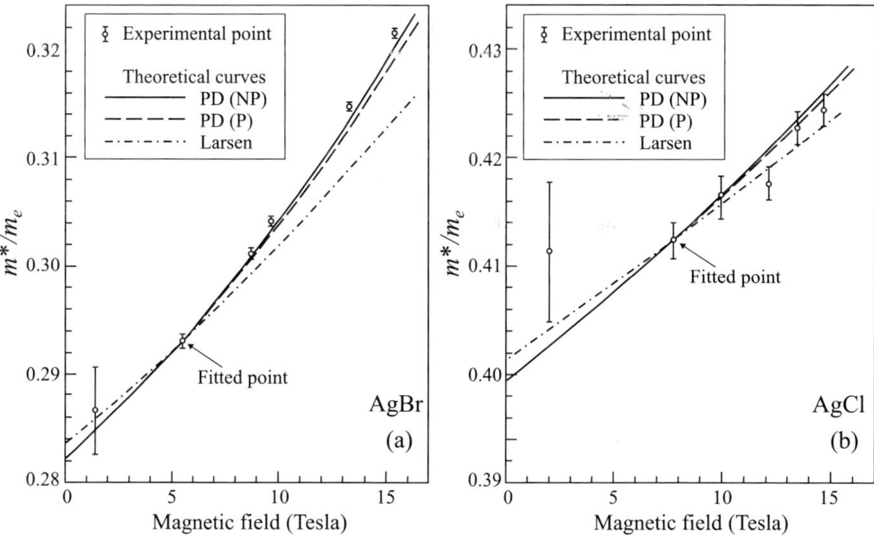

Fig. 12. The polaron cyclotron mass in AgBr (a) and in AgCl (b): comparison of experiment and theory (Larsen: [104]; PD: [103]); (P) – with parabolic band, (NP) – with corrections of a two-band Kane model. In each case the band mass was adjusted to fit the experimental point at 525 GHz. (Reprinted with permission from [105]. ©1987 by the American Physical Society.)

most materials with relatively large Fröhlich coupling constant, the band mass m_b is not known. The study in [105] is an example of the detailed analysis of the cyclotron resonance data that is necessary to obtain accurate polaron data like α and m_b for a given material. I refer to [105] for further details like, e.g., the role of the "fitted point" in Fig. 12.

Early infrared-transmission studies of hydrogen-like shallow-donor-impurity states in n-CdTe were reported in [106]. By studying the Zeeman splitting of the $(1s \rightarrow 2p, m = \pm 1)$ transition in the Faraday configuration at magnetic fields up to ~ 16 T, a quantitative determination of polaron shifts of the energy levels of a bound electron was made. The experimental data were shown to be in fair agreement with the weak-coupling theory of the polaron Zeeman effect. In this comparison, however, the value $\alpha = 0.4$ had to be used instead of $\alpha = 0.286$, which would follow from the definition of the Fröhlich coupling constant. Similarly, the value $\alpha \sim 0.4$ was suggested (see [107]) for the explanation of the measured variation of the cyclotron mass with magnetic field in CdTe. This discrepancy gave rise to some discussion in the literature (see, e.g., [108, 109] and references therein).

In [110], far-infrared photoconductivity techniques were applied to study the energy spectrum of shallow In donors in CdTe layers and experimental data were obtained on the magnetopolaron effect, as shown in Fig. 13.

A good overall agreement is found between experiment and a theoretical approach, in which the electron-phonon interaction is treated within second-order improved Wigner-Brillouin perturbation theory and a variational calculation is performed for the lowest-lying donor states ($1s$, $2p^{\pm}$, $2s$, $2p_z$, $3d^{-2}$, $4f^{-3}$). This agreement is obtained with the coupling constant $\alpha = 0.286$.

Nicholas et al. [111] demonstrated polaron coupling effects using cyclotron resonance measurements in a 2DEG, which naturally occurs in the polar semiconductor InSe. One clearly sees, over a wide range of magnetic fields ($B = 18$ to 34 T), two distinct magnetopolaron branches separated by as much as 11 meV ($\sim 0.4\omega_{LO}$) at resonance (Fig. 14). The theoretical curves show the results of calculations for coupling to the LO phonons in bulk (3D), sheet (2D) and after correction for the quasi-2D character of the system, using $\alpha = 0.29$ calculated in [111]. The agreement between theory and experiment is reasonable for the 3D case, but better for the quasi-2D system, if the finite spatial extent of the 2DEG in the symmetric planar layer is taken into account.

The energy spectra of polaronic systems such as shallow donors ("bound polarons"), e. g., the D_0 and D^- centres, constitute the most complete and detailed polaron spectroscopy realised in the literature, see for example Fig. 15.

In GaAs/AlAs quantum wells with sufficiently high electron density, anti-crossing of the cyclotron-resonance spectra has been observed near the GaAs transverse optical (TO) phonon frequency ω_{TO} rather than near the GaAs LO-phonon frequency ω_{LO} [114]. This anti-crossing near ω_{TO} was explained in the framework of many-polaron theory in [115].

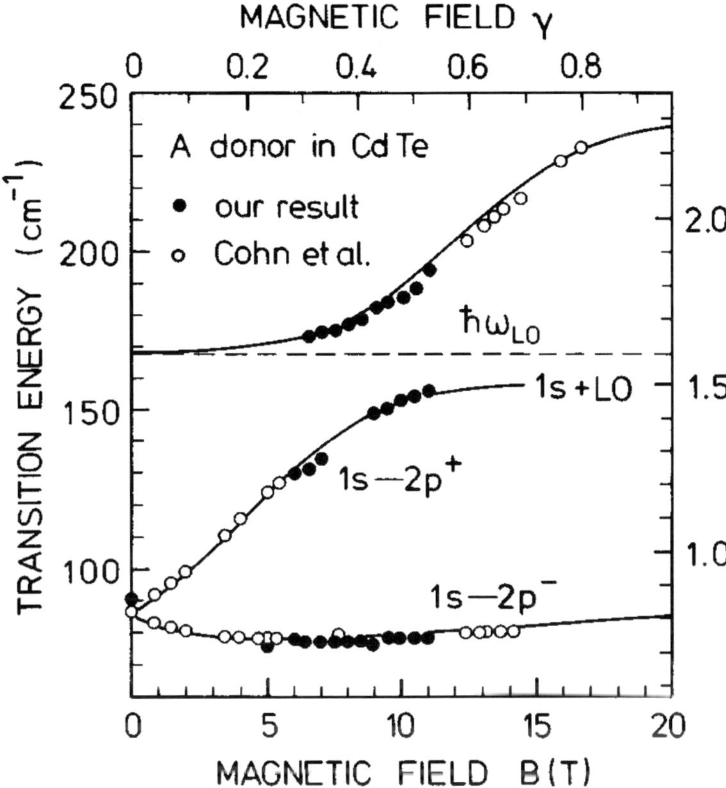

Fig. 13. Plot of the experimentally determined magnetic field dependence of the $1s \to 2p^{\pm 1}$ transition energies of shallow In-donors in CdTe layers grown by molecular-beam epitaxy. The solid lines represent the results of the calculation described in the text without any fitting parameters. The solid dots are the experimental data of [110] and the open circles represent the data of [106]. (Reprinted with permission from [110]. ©1996 by the American Physical Society.)

For further review on polaron cyclotron resonance the reader is referred to [13, 16, 22, 116] and references therein.

5 Fröhlich Bipolarons

When two electrons (or two holes) interact with each other simultaneously through the Coulomb force and via the electron-phonon-electron interaction, either two independent polarons can occur or a bound state of two polarons — the *bipolaron* — can arise (see [117–123] and the reviews [124, 125]). Whether bipolarons originate or not depends on the competition between the repul-

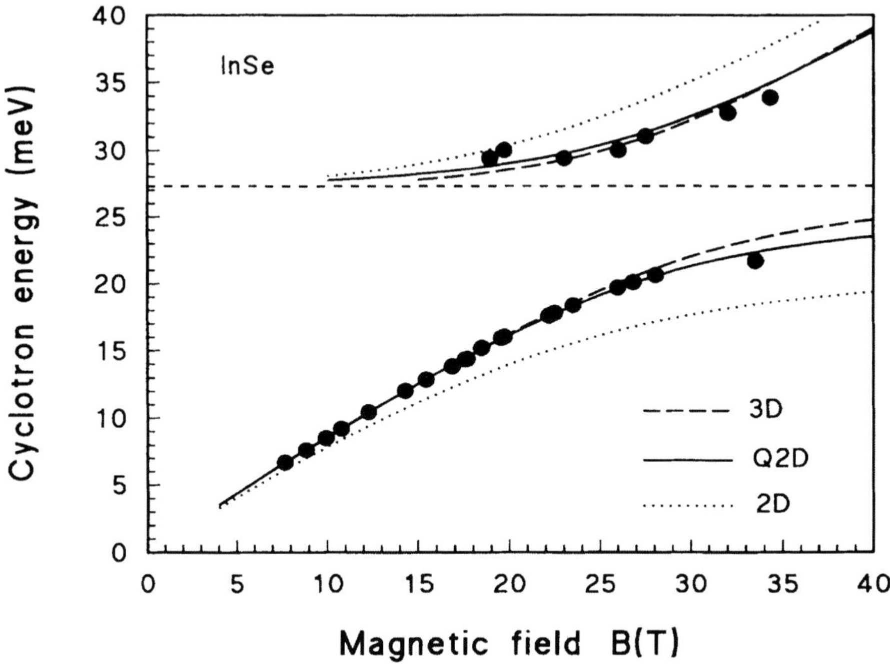

Fig. 14. The cyclotron resonance peak-position plotted as a function of magnetic field for InSe. (Reprinted with permission from [111]. ©1992 by the American Physical Society.)

sive forces (direct Coulomb interaction) and the attractive forces (mediated through the electron-phonon interaction).

The bipolaron can be *free* and characterised by translational invariance, or it can be *localised*. Bipolarons consisting of electrons or holes interacting with LO phonons in the continuum limit are referred to as *Fröhlich bipolarons*. Besides the electron-phonon coupling constant, the Fröhlich bipolaron energy depends also on the dimensionless parameter U, a measure for the strength of the Coulomb repulsion between the two electrons

$$U = \frac{1}{\hbar\omega_{LO}} \frac{e^2}{\varepsilon_\infty} \sqrt{\frac{m_b\omega_{LO}}{\hbar}}. \tag{57}$$

In the discussion of bipolarons, the ratio

$$\eta = \frac{\varepsilon_\infty}{\varepsilon_0} \tag{58}$$

of the electronic and static dielectric constant is often considered ($0 \leq \eta \leq 1$). The following relation exists between U and α:

$$U = \frac{\sqrt{2}\alpha}{1 - \eta}. \tag{59}$$

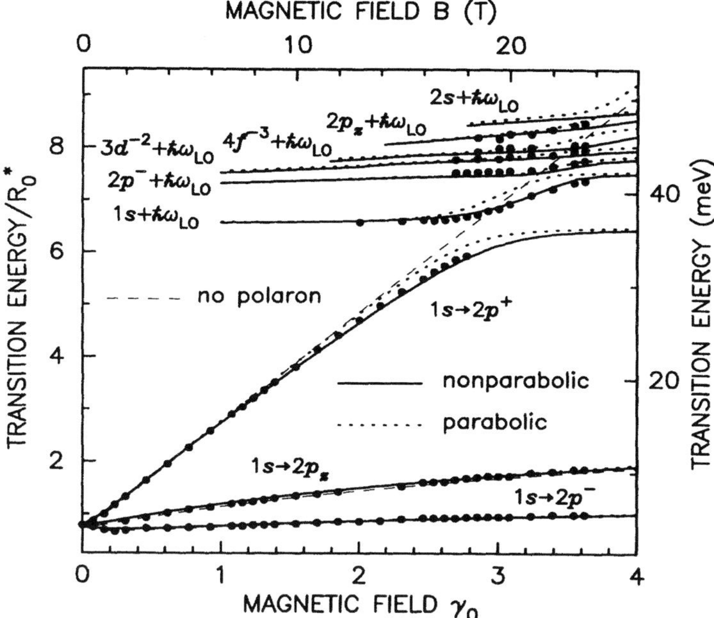

Fig. 15. The $1s \rightarrow 2p^{\pm}, 2p_z$ transition energies as a function of a magnetic field for a donor in GaAs. The authors of [113] compare their theoretical results for the following cases: (a) without the effect of polaron and band non parabolicity (thin dashed curves); (b) with polaron correction (dotted curves); (c) including the effects of polaron and band non-parabolicity (solid curves) to the experimental data of [112] (solid dots). (Reprinted with permission from [113]. ©1993 by the American Physical Society.)

Only values of U satisfying the inequality $U \geq \sqrt{2}\alpha$ have a physical meaning. It was shown that bipolaron formation is favoured by larger values of α and by smaller values of η.

Intuitive arguments suggesting that the Fröhlich bipolaron is stabilised in going from 3D to 2D had been given before, but the first quantitative analysis based on the Feynman path integral was presented in [122, 123]. The conventional condition for bipolaron stability is

$$E_{\mathrm{bip}} \leq 2E_{\mathrm{pol}}, \tag{60}$$

where E_{pol} and E_{bip} denote the ground-state energies of the polaron and bipolaron at rest, respectively. From this condition it follows that a Fröhlich bipolaron with zero spin is stable (given the effective Coulomb repulsion between electrons) if the electron-phonon coupling constant is larger than a certain *critical value*: $\alpha \geq \alpha_c$.

A "phase-diagram" for the two continuous polarons—bipolaron system was introduced in [122, 123]. It is based on the generalised trial action. This phase

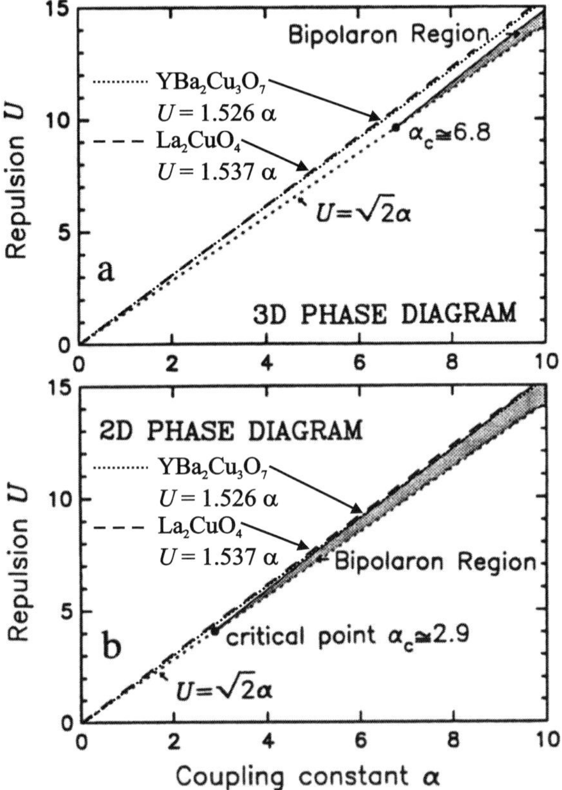

Fig. 16. The stability region for bipolaron formation in 3D (a) and in 2D (b). The dotted line $U = \sqrt{2}\alpha$ separates the physical region ($U \geq \sqrt{2}\alpha$) from the non-physical ($U < \sqrt{2}\alpha$). The shaded area is the stability region in physical space. The dashed (dotted) "characteristic line" $U = 1.537\alpha$ ($U = 1.526\alpha$) is determined by $U = \sqrt{2}\alpha/(1 - \varepsilon_\infty/\varepsilon_0)$ where we took the experimental values $\varepsilon_\infty = 4$ and $\varepsilon_0 = 50$ for La₂CuO₄ ($\varepsilon_\infty = 4.7$ [126] and $\varepsilon_0 = 64.7$ calculated using the experimental data of [126, 127] for YBa₂Cu₃O₇). The critical points $\alpha_c = 6.8$ for 3D and $\alpha_c = 2.9$ for 2D are represented as full dots. (Reprinted with permission after [122]. ©1990 by Elsevier.)

diagram is shown in Fig. 16 for 3D and for 2D. A Fröhlich coupling constant as high as 6.8 is needed to allow for bipolaron formation in 3D. No conclusive experimental evidence has been provided for the existence of materials with such a high Fröhlich coupling constant. The confinement of the bipolaron in two dimensions facilitates bipolaron formation at smaller α. This has been shown in [122, 123], where a *scaling relation* between the free energies F in two dimensions $F_{2D}(\alpha, U, \beta)$ and in three dimensions $F_{3D}(\alpha, U, \beta)$ is derived:

$$F_{2D}(\alpha, U, \beta) = \frac{2}{3} F_{3D}\left(\frac{3\pi}{4}\alpha, \frac{3\pi}{4}U, \beta\right). \tag{61}$$

According to (61), the critical value of the coupling constant for bipolaron formation α_c turns out to scale with a factor $3\pi/4 \approx 2.36$ or $\alpha_c^{(2D)} = \alpha_c^{(3D)}/2.36$. From Fig. 16b it is seen that bipolarons in 2D can be stable for $\alpha \geq 2.9$, a domain of coupling constants which is definitely realised in several solids. The "characteristic line" $U = 1.526\alpha$ for the material parameters of $YBa_2Cu_3O_7$ enters the region of bipolaron stability in 2D at a value of α which is appreciably smaller than in the case of La_2CuO_4. This fact suggests $YBa_2Cu_3O_7$ as a good candidate for the occurrence of stable Fröhlich bipolarons. Note, however, that the precision of the coefficients U/α (1.526, 1.537) used in Fig. 16 is interpreted here too optimistically: the estimated precision of these coefficients is ± 0.01 for $YBa_2Cu_3O_7$ and ± 0.03 for La_2CuO_4. A fair statement is that, on the basis of the available data for $\varepsilon_\infty, \varepsilon_0$, the "characteristic lines" for La_2CuO_4 and for $YBa_2Cu_3O_7$, for sufficiently large α, lie close to the bipolaron stability area. (This is in contrast to the "characteristic lines" of most conventional polar materials.) Further exploration of this point is needed.

An analytical strong-coupling asymptotic expansion in inverse powers of the electron-phonon coupling constant for the large bipolaron energy at $T = 0$ was derived in [128]

$$E_{3D}(\alpha, u) = -\frac{2\alpha^2}{3\pi} A(u) - B(u) + O(\alpha^{-2}), \tag{62}$$

where the coefficients are closed analytical functions of the ratio $u = U/\alpha$:

$$A(u) = 4 - 2\sqrt{2}u\left(1 + \frac{u^2}{128}\right)^{3/2} + \frac{5}{8}u^2 - \frac{u^4}{512} \tag{63}$$

and for $B(u)$ see the above-cited paper. The scaling relation (61) allows one to find the bipolaron energy in two dimensions as

$$E_{2D}(\alpha, u) = \frac{2}{3} E_{3D}\left(\frac{3\pi}{4}\alpha, u\right). \tag{64}$$

The stability of bipolarons has also been examined with the use of operator techniques [121]. The results of [121] and [122, 123] tend to confirm each other. In the framework of the renewed interest in bipolaron theory [13, 14] after the discovery of high-T_c superconductivity, an analysis of the optical absorption by large bipolarons was given in [129]. For a review of the recent work in the field of bipolarons see the present book.

6 Ground-State Properties of a Translational Invariant N-Polaron System

Thermodynamic and optical properties of interacting many-polaron systems are intensely investigated because they might play an important role in physical phenomena in high-T_c superconductors, see, e.g., [13–15] and references therein. The density functional theory and its time-dependent extension is exploited to construct an appropriate effective potential for studying the properties of the interacting polaron gas beyond the mean-field theory [130]. The main assumptions of this work are to consider the coupled system of electrons and ions as a continuum and to take the weak electron-phonon coupling limit.

At arbitrary electron-phonon coupling strength, the many-body problem (including the electron-electron interaction and Fermi statistics) in the N-polaron theory is not yet fully developed. Within the random-phase approximation, the optical absorption of an interacting polaron gas was studied in [131], taking over the variational parameters of Feynman's polaron model [8], which are derived for a single polaron without many-body effects. For a dilute arbitrary-coupling polaron gas, the equilibrium properties [132, 133] and the optical response [134] have been investigated using the path-integral approach and taking into account the electron-electron interaction but neglecting the Fermi statistics. The formation of many-polaron clusters was investigated in [135] using the Vlasov kinetic equations [136]. Also this approach does not take into account the Fermi statistics of the electrons, and therefore it is only valid for sufficiently high temperatures.

In [137], the ground-state properties of a translation invariant N-polaron system are theoretically studied in the framework of the variational path-integral method for identical particles [138–141], using a further development of the model introduced in [142–144]. An upper bound for the ground state energy is found as a function of the number of spin-up and spin-down polarons, taking the electron-electron interaction and the Fermi statistics into account.

6.1 Variational Principle

For distinguishable particles, it is well known that the Jensen-Feynman inequality [8, 96] provides a lower bound on the partition function Z (and consequently an upper bound on the free energy F)

$$Z = \oint e^S \mathcal{D}\bar{\mathbf{r}} = \left(\oint e^{S_0} \mathcal{D}\bar{\mathbf{r}} \right) \langle e^{S-S_0} \rangle_0 \geq \left(\oint e^{S_0} \mathcal{D}\bar{\mathbf{r}} \right) e^{\langle S-S_0 \rangle_0} \tag{65}$$

$$\text{with } \langle A \rangle_0 \equiv \frac{\oint A(\bar{\mathbf{r}})\, e^{S_0} \mathcal{D}\bar{\mathbf{r}}}{\oint e^{S_0} \mathcal{D}\bar{\mathbf{r}}},$$

$$e^{-\beta F} \geq e^{-\beta F_0} e^{\langle S-S_0 \rangle_0} \implies F \leq F_0 - \frac{\langle S - S_0 \rangle_0}{\beta} \tag{66}$$

for a system with real action S and a real trial action S_0. The many-body extension ([145], p. 4476) of the Jensen-Feynman inequality, analysed in more detail in [146], requires (of course) that the potentials are symmetric with respect to all particle permutations, and that the exact propagator as well as the model propagator are defined on the same state space. This means that both the exact and the model propagator should be antisymmetric for fermions (symmetric for bosons) at any time. The path integrals in (65) therefore have to be interpreted based on an antisymmetric state space. Within this interpretation, a generalization of Feynman's trial action is used in [137, 142–144]. This allows one to obtain a rigorous upper bound for the ground-state energy of an N-polaron system. In [137], a translation invariant generalization of Feynman's trial action is proposed.

6.2 Numerical Results

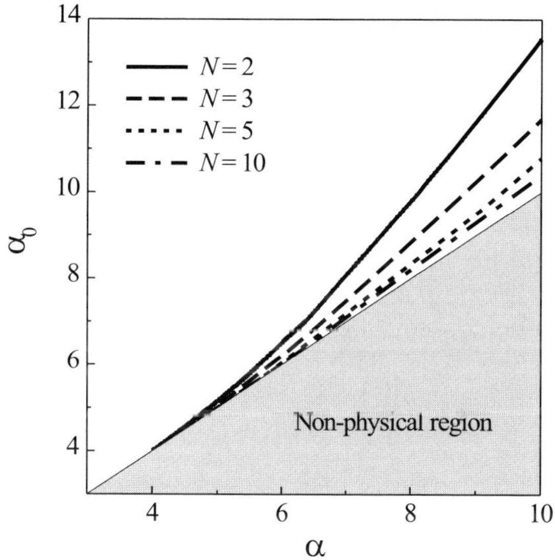

Fig. 17. The "phase diagrams" of a translation invariant N-polaron system. The grey area is the non-physical region, for which $\alpha > \alpha_0$. The stability region for each number of electrons is determined by the equation $\alpha_c < \alpha < \alpha_0$. (Reprinted with permission from [137]. ©2005 by the American Physical Society.)

In Fig. 17, "phase diagrams" analogous to that of [123] are plotted for an N-polaron system in bulk with $N = 2, 3, 5$, and 10. The area where $\alpha > \alpha_0$ (with $\alpha_0 = \frac{U}{\sqrt{2}}$) is the non-physical region. For $\alpha < \alpha_0$, each sector between a curve corresponding to a well defined N and the line $\alpha_0 = \alpha$ shows a stability region. When comparing the stability region for $N = 2$ from Fig. 17 with the bipolaron "phase diagram" of [123], the stability region in the present work starts from the value $\alpha_c \approx 4.1$ (instead of $\alpha_c \approx 6.9$ in [123]). The width of the stability region within the present model is also larger than the width of the stability region within the model of [123]. Also, the ground-state energy of a two-polaron system given by the present model is lower than that given by the approach of [123].

The "phase diagrams" for $N > 2$ demonstrate the existence of stable multipolaron states (see also [147]). As distinct from [147], here the ground state of an N-polaron system is investigated taking into account the fermionic nature of the electrons. As seen from these figures, for $N > 2$, the stability region for a multipolaron state is narrower than the stability region for $N = 2$, and its width decreases with increasing N. The critical value α_c for the electron-phonon coupling constant increases with increasing N. From this behaviour we can deduce the following general trend. For fixed values of α and $\eta = \varepsilon_\infty/\varepsilon_0$, the width of the stability region for a multipolaron state is a decreasing function of the number of electrons. Therefore, for any (α, η) there exists a critical number of electrons $N_0 (\alpha, \eta)$ such that a multipolaron state exists for $N \leq N_0 (\alpha, \eta)$ and does not exist for $N > N_0 (\alpha, \eta)$.

In Fig. 18, the ground-state energy per polaron, the confinement[3] frequency ω_{op} and the total spin S are plotted as a function of the coupling constant α for $\alpha_0/\alpha = 1.05$ and for different numbers of polarons. The ground-state energy turns out to be a continuous function of α, while ω_{op} and S reveal discontinuous transitions. For all considered $N > 2$, there exists a region of α in which S takes its maximal value, while $\omega_{op} \neq 0$. When lowering α, this spin-polarised state precedes the transition from the regime with $\omega_{op} \neq 0$ to the regime with $\omega_{op} = 0$ (the break-up of the multipolaron state). For $N = 2$ (bipolaron), we see from Fig. 18 that the ground state has a total spin $S = 0$ for all values of α, i. e., the ground state of a bipolaron is a singlet. This result is in agreement with earlier investigations on the Fröhlich bipolaron problem (see, e.g., [148]).

For sufficiently large values of the electron-phonon coupling constant and of the ratio $1/\eta = \varepsilon_0/\varepsilon_\infty$, the system of N interacting polarons can occur in a stable multipolaron ground state. When this state is formed, the total spin of the system takes its minimal possible value. The larger the number of electrons, the narrower the stability region of a multipolaron state becomes. So, when adding electrons one by one to a stable multipolaron state, it breaks up for a definite number of electrons N_0, which depends on the coupling constant and on the ratio of the dielectric constants. This break-up is preceded by the change from a spin-mixed ground state with a minimal possible spin to a spin-polarised ground state with parallel spins.

For a stable multipolaron state, the addition energy reveals peaks corresponding to closed shells. At $N = N_0$, the addition energy has a pronounced minimum. These features of the addition energy, as well as the total spin as a function of the number of electrons, might be resolved experimentally using, e.g., capacity and magnetization measurements.

---

[3] The confinement frequency characterises the degree of "localisation" of an N-polaron cluster. The cluster itself exhibits translational invariance.

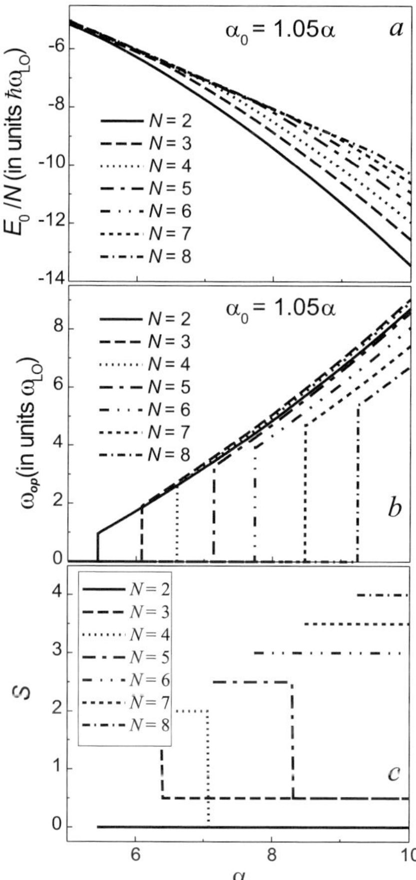

Fig. 18. The ground-state energy per particle (a), the optimal value ω_{op} of the confinement frequency (b), and the total spin (c) of a translation invariant N-polaron cluster as a function of the coupling strength α for $\alpha_0/\alpha = 1.05$. The vertical dashed lines in the panel c indicate the critical values α_c separating the regimes of $\alpha > \alpha_c$, where the multipolaron ground state with $\omega_{op} \neq 0$ exists, and $\alpha < \alpha_c$, where $\omega_{op} = 0$. (Reprinted with permission from [137]. ©2005 by the American Physical Society.)

7 Optical Absorption Spectra of Many-Polaron Systems

In [76], starting from the many-polaron canonical transformations and the variational many-polaron wave function introduced in [149], the optical absorption coefficient of a many-polaron gas has been derived. The real part of the optical conductivity of the many-polaron system is obtained in an intuitively appealing form, given by (1).

This approach to the many-polaron optical absorption allows one to include the many-body effects to order α in terms of the dynamical structure factor $S(\mathbf{k}, \Omega - \omega_{LO})$ of the electron (or hole) system. The experimental peaks in the mid-infrared optical absorption spectra of cuprates (Lupi et al., Fig. 19) and manganites (Hartinger et al., Fig. 20) have been adequately interpreted within this theory. As seen from Fig. 20, the many-polaron approach describes the experimental optical conductivity better than the single-polaron methods [31, 151]. Note that in [76], like in [87], coexistence of small and Fröhlich polarons in the same solid seems to be involved.

The optical conductivity of a many-polaron gas was further investigated in [131] in a different way by calculating the correction to the dielectric function of the electron gas, due to the electron-phonon interaction with variational parameters of a single-polaron Feynman model. A suppression of the optical absorption from the one-polaron optical absorption of [30, 32] with increasing

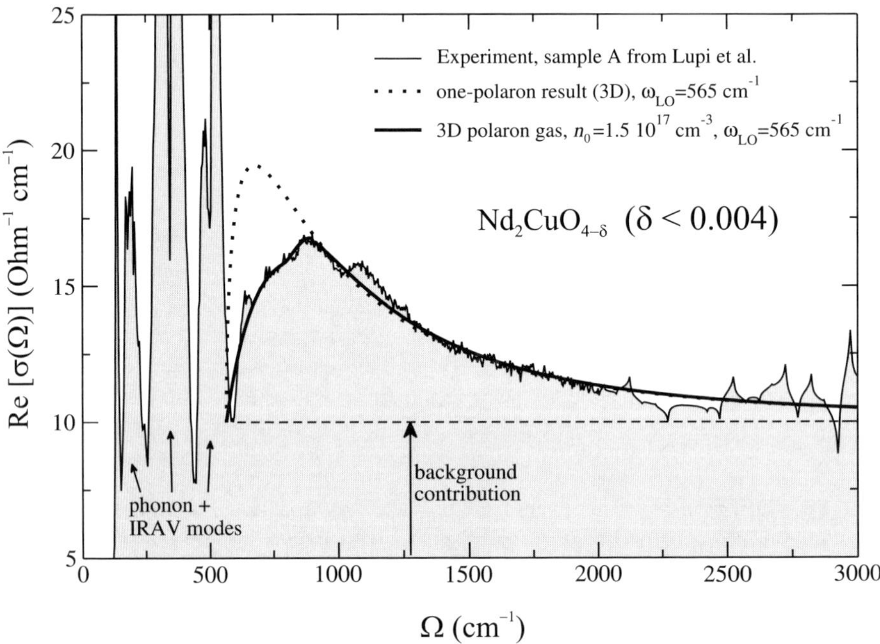

Fig. 19. The infrared optical absorption of $Nd_2CuO_{2-\delta}$ ($\delta < 0.004$) as a function of frequency. The experimental results of [150] are presented by the thin full curve. The experimental 'd-band' is clearly identified, rising in intensity at about 600 cm^{-1}, peaking around 1000 cm^{-1}, and then decreasing in intensity above that frequency. The dotted curve shows the single polaron result calculated according to [30]. The bold full curve presents the theoretical results of [76] for the interacting many-polaron gas with the following choice of parameters: $n_0 = 1.5 \times 10^{17}$ cm^{-3}, $\alpha = 2.1$ and $m_b = 0.5m_e$. (Reprinted with permission from [76]. ©2001 by the American Physical Society.)

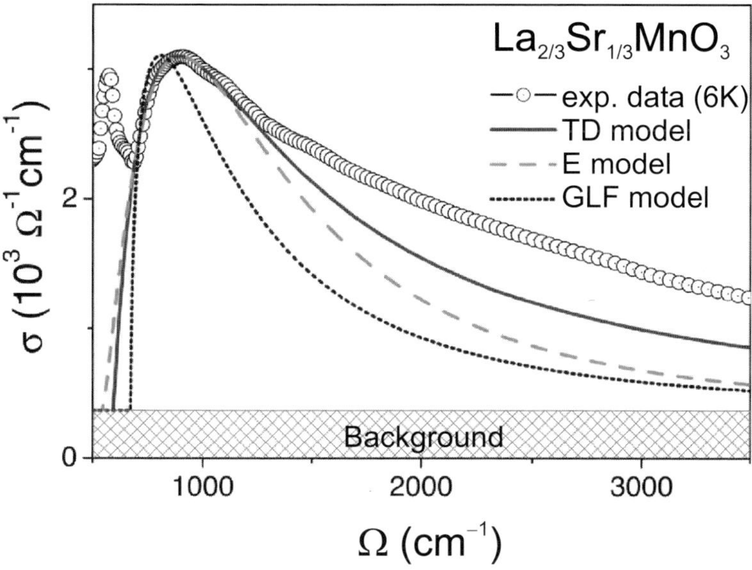

Fig. 20. Comparison of the measured mid-infrared optical conductivity in La$_{2/3}$Sr$_{1/3}$MnO$_3$ at $T = 6$ K to that given by several model calculations for $m_b = 3m_e$, α of the order of 1 and $n_0 = 6 \times 10^{21}$ cm^{-3}. The one-polaron approximations [the weak-coupling approach by V. L. Gurevich, I. G. Lang, and Yu. A. Firsov [31] (GLF model) and the phenomenological approach by D. Emin [151] (E model)] lead to narrower polaron peaks than a peak with maximum at $\Omega \sim 900$ cm^{-1} given by the many-polaron treatment by J. Tempere and J. T. Devreese (TD model) of [76]. (Reprinted with permission after [69]. ©2004 by the American Physical Society.)

density is found as shown in Fig. 21. Such a suppression is expected because of the screening of the Fröhlich interaction with increasing polaron density.

8 Many Polarons in Quantum Dots

8.1 Ground-State Properties of Interacting Polarons in a Quantum Dot

For a spherical quantum dot, a system of N electrons (or holes), with mutual Coulomb repulsion and interacting with the bulk phonons is analysed in [142, 144] using the variational inequality for identical particles (see [145, 146] and Subsect. 6.1). A parabolic confinement potential, characterised by the frequency parameter Ω_0, is assumed. Further on, the zero-temperature case is considered.

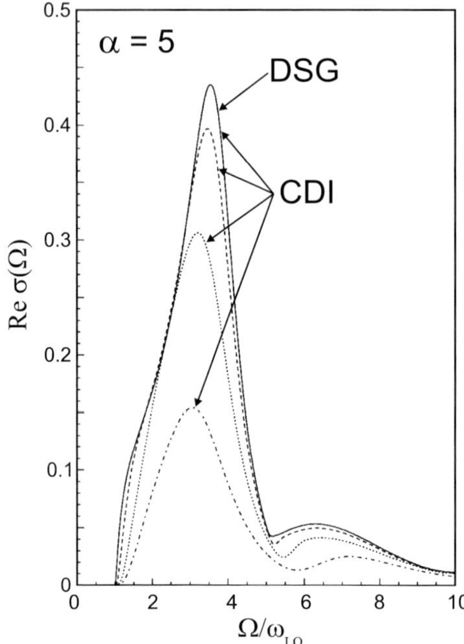

Fig. 21. Optical conductivity of a polaron gas at $T = 0$ as a function of the frequency as calculated in [131] (CDI) for different values of the electron density: $n_0 = 1.4 \times 10^{-5}$ (solid curve), $n_0 = 1.4 \times 10^{-4}$ (dashed curve), $n_0 = 1.4 \times 10^{-3}$ (dotted curve), and $n_0 = 1.4 \times 10^{-2}$ (dash-dotted curve). The electron density is measured per R_p^3, where R_p is the Fröhlich polaron radius. The value of $\varepsilon_0/\varepsilon_\infty$ is 3.4. The optical conductivity is expressed in units of $n_0 e^2/m_b \omega_{\mathrm{LO}}$. The solid curve practically coincides with the known optical conductivity of a single polaron [30] (DSG). (Reprinted with permission after [131]. ©1999, EDP Sciences, Società Italiana di Fisica, Springer.)

In Fig. 22, the total spin S of a system of interacting polarons in their ground state is plotted as a function of N for different values of the confinement frequency Ω_0, of the electron-phonon coupling constant α and of the parameter $\eta = \varepsilon_\infty/\varepsilon_0$. The parameter $\hbar\Omega_0$ is measured in effective Hartrees ($H^* = \left[m_b / \left(m_e \varepsilon_\infty^2 \right) \right] \times 1$ Hartree). Normally, for closed-shell systems $S = 0$, while for open-shell systems S takes its maximal value for a given shell filling (Hund's rule [152]). Hund's rule means that the electrons in the upper (partly filled) shell are distributed in such a way that the total spin takes its maximal possible value. As seen from Fig. 22 (a), for a quantum dot with $\hbar\Omega_0 = 0.5$ H^* at $\alpha = 0$ and at $\alpha = 0.5$, the shell filling does obey Hund's rule.

At sufficiently small Ω_0, a spin-polarised state for a system of interacting electrons in a quantum dot can become energetically more favourable than a state satisfying Hund's rule. For a quantum dot with $\hbar\Omega_0 = 0.1$ H^*, the

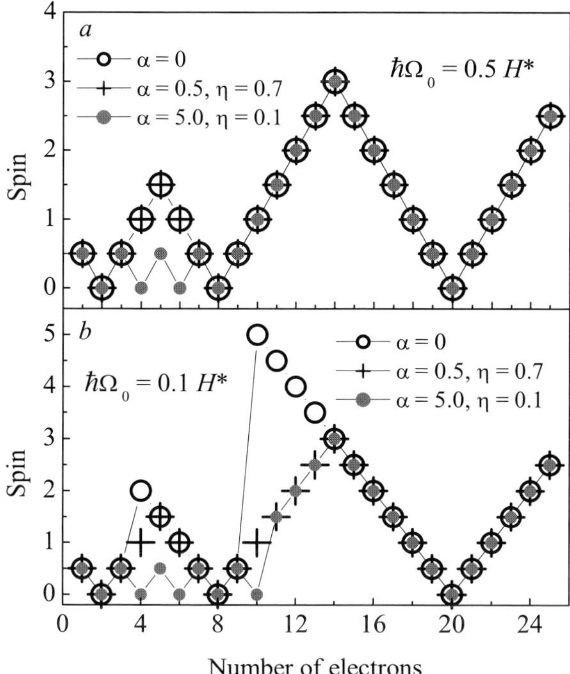

Fig. 22. Total spin of the system of interacting polarons in a parabolic quantum dot as a function of the number of electrons for $\hbar\Omega_0 = 0.5\ H^*$ (a) and for $\hbar\Omega_0 = 0.1$ H^* (b). (Reprinted with permission from [144]. ©2004 by the American Physical Society.)

spin-polarised state at $\alpha = 0$ appears to be energetically favourable for $N = 4$ and $N = 10$ (i.e. for a closed-shell spin-polarised system), as seen from Fig. 22 (b).

In the strong-coupling case ($\alpha \gg 1$ and $\eta \ll 1$), the total spin of an open-shell system for the ground state can take its minimal possible value, as seen from Fig. 22(a) for $\alpha = 5, \eta = 0.1$ at $N = 4$ to 6. This trend to minimise the total spin is a consequence of the electron-phonon interaction, presumably due to the fact that the phonon-mediated electron-electron attraction overcomes the Coulomb repulsion, so that a multipolaron state is formed.

Confined few-electron systems, without electron-phonon interaction, can exist in one of two phases: the spin-polarised state and a state obeying Hund's rule, depending on the confinement frequency (see, e.g., [153]). For interacting few-polaron systems, besides the above two phases, there may occur also a third phase — the state with minimal spin — in quantum dots of polar substances with sufficiently strong electron-phonon coupling (for instance, high-T_c superconductors).

8.2 Optical Properties of Interacting Polarons in Quantum Dots

To investigate the optical properties of the many-polaron system, in [144], the DSG-treatment of [30] is extended to the case of interacting polarons in a quantum dot.

The transitions between states with different values of the total spin, which occur when varying the confinement frequency, also manifest themselves in the spectra of the optical conductivity. In Fig. 23, optical conductivity spectra for $N = 14$ polarons in a quantum dot with $\alpha = 5$, $\eta = 0.1$ are represented for several values of the confinement energy $\hbar\Omega_0$.

For relatively weak confinement, the ground state is spin-polarised, like for $\hbar\Omega_0^{(1)} = 0.0085\ H^*$ (panel a). With increasing confinement, the transition from a spin-polarised state (with total spin $S = 7$) to a state obeying Hund's rule (with $S = 3$) occurs between $\hbar\Omega_0^{(1)} = 0.00883\ H^*$ (panel b) and $\hbar\Omega_0^{(2)} = 0.00884\ H^*$ (panel c). At still stronger confinement, like for $\hbar\Omega_0^{(1)} = 0.0092\ H^*$ (panel d), the ground state obeys Hund's rule.

In the inset to Fig. 23, the first frequency moment of the optical conductivity

$$\langle\Omega\rangle \equiv \frac{\int_0^\infty \Omega\,\mathrm{Re}\sigma\,(\Omega)\,\mathrm{d}\Omega}{\int_0^\infty \mathrm{Re}\sigma\,(\Omega)\,\mathrm{d}\Omega} \tag{67}$$

as a function of $\hbar\Omega_0$ shows a *discontinuity*, at the value of the confinement energy corresponding to the transition between the spin-polarised ground state and the ground state obeying Hund's rule. This discontinuity should be observable in optical measurements.

The shell structure for a system of interacting polarons in a quantum dot is clearly revealed when analysing both the addition energy and the first frequency moment of the optical conductivity. The addition energy $\Delta\,(N)$, needed to put a supplementary electron into a quantum dot containing N electrons, is defined as

$$\Delta\,(N) = E_0\,(N + 1) - 2E_0\,(N) + E_0\,(N - 1), \tag{68}$$

where $E_0\,(N)$ is the ground-state energy. In Figs. 24(a) and 24(b), we show both the function

$$\Theta\,(N) \equiv \langle\Omega\rangle|_{N+1} - 2\,\langle\Omega\rangle|_N + \langle\Omega\rangle|_{N-1}, \tag{69}$$

and the addition energy $\Delta\,(N)$. Distinct peaks appear in $\Theta\,(N)$ and $\Delta\,(N)$ at the "magic numbers" $N = 10$ and $N = 20$ for closed-shell configurations.

It follows that measurements of the addition energy and the first frequency moment of the optical absorption as a function of the number of polarons in

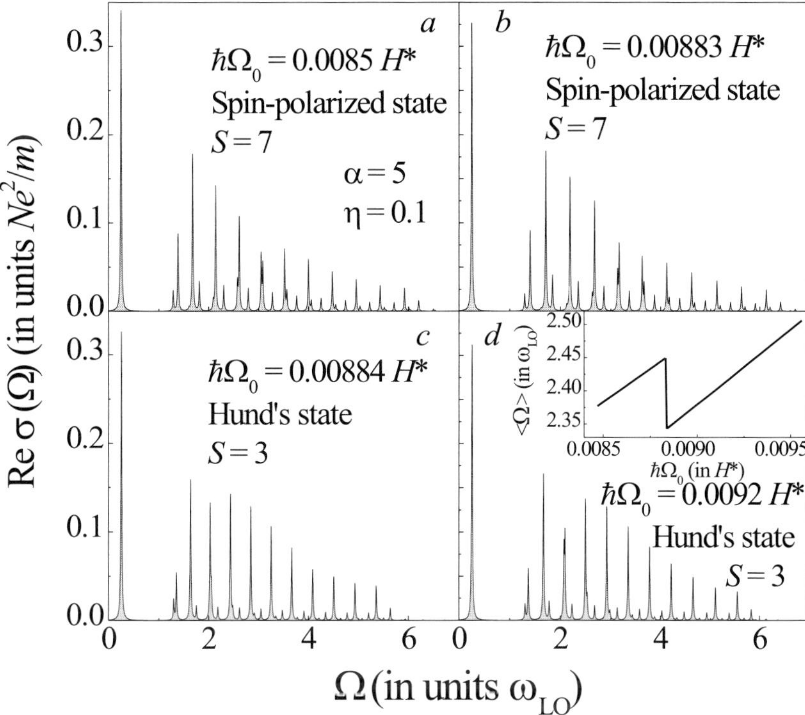

Fig. 23. Optical conductivity spectra of $N = 14$ interacting polarons in a quantum dot, for different confinement frequencies close to the transition from a spin-polarised ground state to a ground state obeying Hund's rule ($\alpha = 5$, $\eta = 0.1$). *Inset*: the first frequency moment $\langle\Omega\rangle$ of the optical conductivity as a function of the confinement energy. (Reprinted with permission from [22]. ©2003, American Institute of Physics.)

a quantum dot would allow for discrimination between open-shell and closed-shell configurations. In particular, the latter configurations may be revealed through peaks in the addition energy and the first frequency moment of the optical absorption in systems with sufficiently large α.

9 Non-Adiabaticity of Polaronic Excitons in Semiconductor Quantum Dots

A new aspect of the polaron concept has been investigated for semiconductor structures at the nanoscale: the exciton-phonon states are not factorisable into an adiabatic product Ansatz, so that a *non-adiabatic* treatment is needed [154].

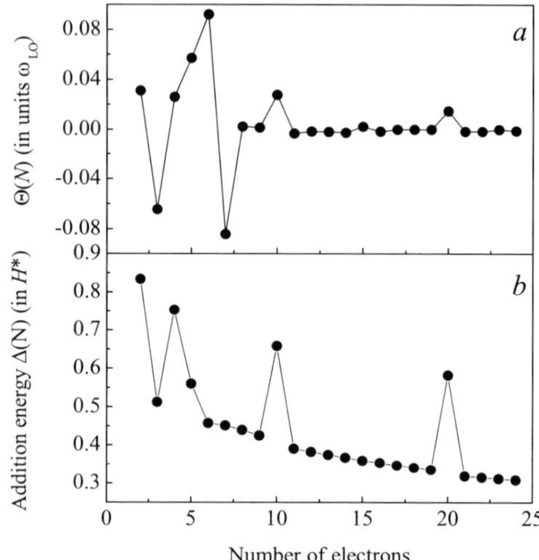

Fig. 24. The function $\Theta\left(N\right)$ (a) and the addition energy $\Delta\left(N\right)$ (b) for systems of interacting polarons in a quantum dot with $\alpha = 3$, $\eta = 0.3$ and $\Omega_0 = 0.5\omega_{\mathrm{LO}}$ ($\hbar\Omega_0 \approx 0.01361\ H^*$). (Reprinted with permission from [22]. ©2003, American Institute of Physics.)

It has been shown [154] that non-adiabaticity of exciton-phonon systems in some quantum dots drastically enhances the efficiency of the exciton-LO-phonon interaction, especially when the exciton levels are separated by energies close to the phonon energies. Also "intrinsic" excitonic degeneracy can lead to enhanced efficiency of the exciton-phonon interaction. The effects of non-adiabaticity are important to interpret the surprisingly high intensities of the phonon 'sidebands' observed in the optical absorption, the photoluminescence and the Raman spectra of many quantum dots. Considerable deviations of the oscillator strengths of the measured phonon-peak sidebands from the standard FC progression find a natural explanation within the non-adiabatic approach [154–156].

In [154], a method was proposed to calculate the optical absorption spectrum for a spherical quantum dot taking into account the non-adiabaticity of the exciton-phonon system. This approach has been further refined in [157]: for the matrix elements of the evolution operator a closed set of equations has been obtained using a diagrammatic technique. This set of equations describes the effect of non-adiabaticity both on the intensities and on the frequencies of the absorption peaks. The theory takes into account the Fröhlich interaction with all the phonon modes specific for a given quantum dot.

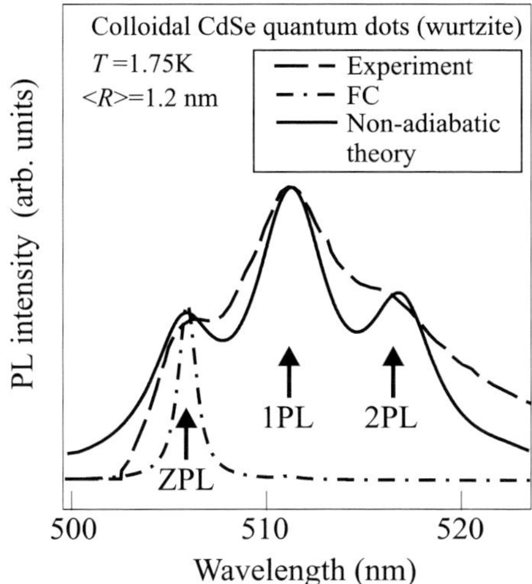

Fig. 25. Photoluminescence spectra of colloidal spherical CdSe quantum dots of wurtzite structure with average radius 1.2 nm. The dashed curve represents the experimental data of [158]. The dash-dotted curve displays the result of the adiabatic approximation – a FC progression with Huang-Rhys factor $S = 0.06$ as calculated in [159]. The solid curve results from the non-adiabatic theory. (Reprinted with permission after [154]. ©1998 by the American Physical Society.)

For some semiconductor quantum dots, where the adiabatic approximation predicts negligibly low intensities of the one- and two-phonon sidebands, the non-adiabatic theory allows for a quantitative interpretation of the observed high intensity of the phonon sidebands in the photoluminescence (Fig. 25) and the Raman (Fig. 26) spectra.

The conclusion about the large enhancement of the two-phonon sidebands in the luminescence spectra as compared to those predicted by the Huang-Rhys formula, which was explained in [154, 160] by non-adiabaticity of the exciton-phonon system in certain quantum dots, has been reformulated in [161] in terms of the Fröhlich coupling between product states with different electron and/or hole states.

Due to non-adiabaticity, multiple absorption peaks appear in spectral ranges characteristic for phonon satellites. From the states which correspond to these peaks, the system can rapidly relax to the lowest emitting state. Therefore, in the photoluminescence excitation (PLE) spectra of quantum dots, pronounced peaks can be expected in spectral ranges characteristic for phonon satellites. Experimental evidence of the enhanced phonon-assisted absorption due to effects of non-adiabaticity has been provided by PLE measurements on single self-assembled InAs/GaAs [162] and InGaAs/GaAs [163] quantum dots. The polaron concept was also invoked for the explanation of the

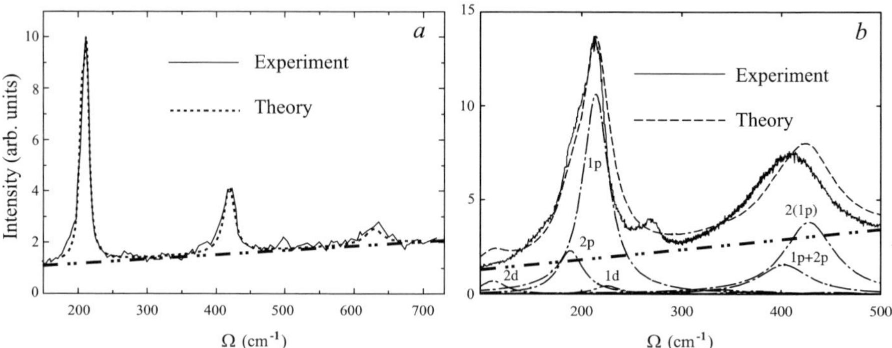

Fig. 26. Resonant Raman scattering spectra of an ensemble of CdSe quantum dots with average radius 2 nm at $T = 77$ K (panel a) and of PbS quantum dots with average radius 1.5 nm at $T = 4.2$ K (panel b). The dash-dot-dot curves show the luminescence background. The dash-dot curves in panel b indicate contributions into the Raman spectrum due to the phonon modes with different symmetry $1p$, $2p$, $1d$, $2d$ etc. (classified in analogy with electron states in a hydrogen atom), which are specific for the quantum dot. (Reprinted with permission after [155]. ©2002 by the American Physical Society.)

PLE measurements on self-organised $In_xGa_{1-x}As/GaAs$ [164] and CdSe/ZnSe [165] quantum dots.

10 Ripplopolarons in Multi-Electron Bubbles in Liquid Helium [166]

An interesting 2D system consists of electrons on films of liquid He [167, 168]. In this system the electrons couple to the ripplons of the liquid He, forming "ripplopolarons". The effective coupling can be relatively large and self-trapping can result. The acoustic nature of the ripplon dispersion at long wavelengths induces self-trapping.

Spherical shells of charged particles appear in a variety of physical systems, such as fullerenes, metallic nanoshells, charged droplets and neutron stars. A particularly interesting physical realization of the spherical electron gas is found in multielectron bubbles (MEBs) in liquid helium-4. These MEBs are $0.1\,\mu$m–$100\,\mu$m sized cavities inside liquid helium, that contain helium vapour at vapour pressure and a nanometre-thick electron layer anchored to the surface of the bubble [169, 170]. They exist as a result of equilibrium between the surface tension of liquid helium and the Coulomb repulsion of the electrons [171, 172]. Recently proposed experimental schemes to stabilize MEBs [173] have stimulated theoretical investigation of their properties (see e.g. [174]).

The dynamical modes of a MEB were described by considering the motion of the helium surface ("ripplons") and the vibrational modes of the electrons together. In particular, the case when the ripplopolarons form a Wigner lattice was analysed in [166]. The interaction energy between the ripplons and the electrons in the multielectron bubble is derived from the following considerations: (i) the distance between the layer electrons and the helium surface is fixed (the electrons find themselves confined to an effectively 2D surface anchored to the helium surface) and (ii) the electrons are subjected to a force field, arising from the electric field of the other electrons. To study the ripplopolaron Wigner lattice at finite temperature and for any value of the electron-ripplon coupling, we use the variational path-integral approach [8].

In their treatment of the electron Wigner lattice embedded in a polarisable medium such as a semiconductor or an ionic solid, Fratini and Quémerais [132] described the effect of the electrons on a particular electron through a mean-field lattice potential. The (classical) lattice potential V_{lat} is obtained by approximating all the electrons acting on one particular electron by a homogenous charge density, in which a hole is punched out; this hole is centred in the lattice point of the particular electron under investigation and has a radius given by the lattice distance d.

The Lindemann melting criterion [175, 176] states in general that a crystal lattice of objects (be it atoms, molecules, electrons, or ripplopolarons) will melt when the average motion of the objects around their lattice site is larger than a critical fraction δ_0 of the lattice parameter d. It would be a strenuous task to calculate, from first principles, the exact value of the critical fraction δ_0, but for the particular case of electrons on a helium surface, we can make use of an experimental determination. Grimes and Adams [177] found that the Wigner lattice melts when $\Gamma = 137 \pm 15$, where Γ is the ratio of the potential energy to the kinetic energy per electron. At temperature T the average kinetic energy in a lattice potential V_{lat}, characterized by the frequency parameter ω_{lat}, is

$$E_{kin} = \frac{\hbar \omega_{lat}}{2} \coth\left(\frac{\hbar \omega_{lat}}{2k_B T}\right), \tag{70}$$

and the average distance that an electron moves out of the lattice site is determined by

$$\langle \mathbf{r}^2 \rangle = \frac{\hbar}{m_e \omega_{lat}} \coth\left(\frac{\hbar \omega_{lat}}{2k_B T}\right) = \frac{2E_{kin}}{m_e \omega_{lat}^2}. \tag{71}$$

From this one finds that for the melting transition in Grimes and Adams' experiment [177], the critical fraction equals $\delta_0 \approx 0.13$. This estimate is in agreement with previous (empirical) estimates yielding $\delta_0 \approx 0.1$ [178].

Within the approach of Fratini and Quémerais [132], the Wigner lattice of (ripplo)polarons melts when at least one of the two following Lindemann criteria are met:

$$\delta_r = \frac{\sqrt{\langle \mathbf{R}_{cms}^2 \rangle}}{d} > \delta_0, \tag{72}$$

$$\delta_\rho = \frac{\sqrt{\langle \rho^2 \rangle}}{d} > \delta_0, \tag{73}$$

where ρ and $\mathbf{R}_{cms}$ are, respectively, the relative coordinate and the centre-of-mass coordinate of the model system: if $\mathbf{r}$ is the electron coordinate and $\mathbf{R}$ is the position coordinate of the fictitious ripplon mass M, then

$$\mathbf{R}_{cms} = \frac{m_e \mathbf{r} + M\mathbf{R}}{m_e + M}; \rho = \mathbf{r} - \mathbf{R}. \tag{74}$$

The appearance of two Lindemann criteria takes into account the composite nature of (ripplo)polarons. As follows from the physical sense of the coordinates ρ and $\mathbf{R}_{cms}$, the first criterion (72) is related to the melting of the ripplopolaron Wigner lattice towards a ripplopolaron liquid, where the ripplopolarons move as a whole, the electron together with its dimple. The second criterion (73) is related to the dissociation of ripplopolarons: the electrons shed their dimple.

The path-integral variational formalism allows us to calculate the expectation values $\langle \mathbf{R}_{cms}^2 \rangle$ and $\langle \rho^2 \rangle$ with respect to the ground state of the variationally optimal model system.

Numerical calculation shows that for ripplopolarons in a MEB the inequality

$$\langle \mathbf{R}_{cms}^2 \rangle \ll \langle \rho^2 \rangle \tag{75}$$

is realized. As a consequence, the destruction of the ripplopolaron Wigner lattice in a MEB occurs through the dissociation of ripplopolarons, since the second criterion (73) will be fulfilled before the first criterion (72). The results for the melting of the ripplopolaron Wigner lattice are summarized in the phase diagram found by J. Tempere et al. [166], which is shown in Fig. 27.

For every value of N, pressure p and temperature T in an experimentally accessible range, figure 27 shows whether the ripplopolaron Wigner lattice is present (points above the surface) or the electron liquid (points below the surface). Below a critical pressure (on the order of 10^4 Pa) the ripplopolaron solid will melt into an electron liquid. This critical pressure is nearly independent of the number of electrons (except for the smallest bubbles) and is weakly temperature dependent, up to the helium critical temperature 5.2 K. This can be understood since the typical lattice potential well in which the ripplopolaron resides has frequencies of the order of THz or larger, which correspond to ~ 10 K.

The new phase that was predicted [166], the ripplopolaron Wigner lattice, will not be present for electrons on a flat helium surface. At the values of the pressing field necessary to obtain a strong enough electron-ripplon coupling, the flat helium surface is no longer stable against long-wavelength deformations [179]. Multielectron bubbles, with their different ripplon dispersion and the presence of stabilizing factors such as the energy barrier against fissioning [180], allow for much larger electric fields pressing the electrons against the helium surface. The regime of N, p, T parameters, suitable for the creation of

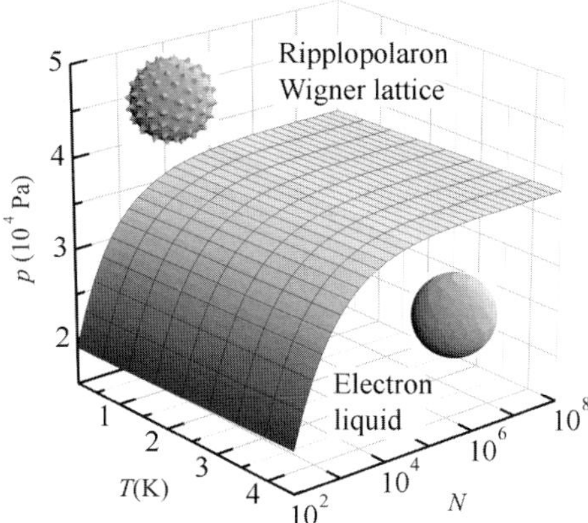

Fig. 27. The phase diagram for a spherical 2D layer of electrons in a MEB. Above a critical pressure, a ripplopolaron solid (a Wigner lattice of electrons with dimples in the helium surface underneath them) is formed. Below the critical pressure, the ripplopolaron solid melts into an electron liquid through dissociation of ripplopolarons. (Reprinted with permission from [166]. ©2003, EDP Sciences, Società Italiana di Fisica, Springer.)

a ripplopolaron Wigner lattice, lies within the regime that would be achievable in recently proposed experiments aimed at stabilizing multielectron bubbles [173]. The ripplopolaron Wigner lattice and its melting transition might be detected by spectroscopic techniques [177, 181] probing for example the transverse phonon modes of the lattice [182].

11 Conclusions

It is remarkable how the Fröhlich polaron, one of the simplest examples of a *Quantum Field Theoretical* problem, as it basically consists of a single fermion interacting with a scalar Bose-field, has resisted full analytical solution at all coupling since $\sim$ 1950, when its Hamiltonian was first written.

The understanding of its response properties, and in particular the optical absorption, is a case in point. Although a mechanism for the optical absorption of Fröhlich polarons was already proposed between 1969 (KED – [29]) and 1972 (DSG – [30]), some subtle characteristics were only clarified in 2006 [35] by combining numerical studies (DQMC – [33]) and improved (variational) approximate analytical methods (see [37]). The basic mechanism proposed in [29] (strong coupling) in combination with DSG (who start from an all-coupling path-integral formalism) was basically correct: the polaron optical absorption spectrum consists of a combination of transitions towards the RES and towards scattering states (i.e., scattering states of the ground state and of the RES; the latter transitions resulting in FC sidebands). However, refinements were necessary: in [29] KED proposed the strong-coupling mechanism (that – as shown by DSG in [30] – qualitatively survives at intermediate coupling) but only one- and two-LO-phonon sidebands could be calculated at the time [29, 84]. Only recently (see [35] and references therein) the KED-program was completed by calculating as many LO-phonon sidebands as necessary (~ 25 for $\alpha = 15$ e.g.). Thanks to the availability of the DQMC-results and the SCE ([34, 92] and Sect. 2.7) it could be suggested (as I did in this chapter) that the main peak in the DSG optical absorption gradually *changes its character from RES to FC-peak as α increases*. The transition from the coupling regime where "the lattice follows the electron" to the "Franck-Condon" regime, although implicitly present in DSG (without the correct stronger coupling linewidth) can only be analysed in quantitative detail from the work in [35]. The subtlety of the response properties of Fröhlich polarons suggests that equally the response properties of high-T_c materials, however subtle, might become subject of quantitative analysis, but only after disentangling highly intricate phenomena like this RES-to-FC transition.

From the comparison between theory and experiment (optical absorption, cyclotron resonance, photoluminescence, Raman scattering...) in this chapter, it is also striking to see how many phenomena and systems can be understood in detail, on the basis of the Fröhlich interaction.

I discussed the stability region of the Fröhlich bipolaron (cf. [123, 137]). Here the surprise is double: a) only in a very limited sector of the phase diagram (Coulomb repulsion versus α) the bipolaron is stable, b) most traditional Fröhlich polaron materials (alkali halides and the like) lie completely outside (and "far" from) this bipolaron stability sector, but... several high-T_c materials lie very close and even inside this very restricted area of the stability diagram. This should be a very hopeful sign, for physicists (see the chapter by A. S. Alexandrov [64]) who propose bipolarons as embodiment of the superconducting quasiparticles of the high-T_c materials.

Also the interpretation of the optical spectra of high-T_c materials (measured by Calvani et al., [15, 150]) in the normal phase, and of manganites – measured by Hartinger et al., [69] – as due to many polaron absorption using the theory developed in [76] strengthens the view that Fröhlich polarons play a substantial role in many solid structures.

Many-polaron effects can be treated to order α to the same degree of accuracy as the electron gas, using the structure factor [76]. For larger coupling the problem remains highly cumbersome. Progress has been made using path integral approaches to the many fermion system, that – inherently – is intricate to treat because of the "sign problem" that goes with it [138–141, 145, 146].

The richness and profundity of Landau's polaron concept is further illustrated by its extensions e.g. to the electronic polaron, to the Holstein polaron, to ripplopolarons.

Polaron effects play a role in many systems of reduced dimension and reduced dimensionality, that are significant in present day nanoscience, including the study of quantum dots. Of special importance I find the pronounced role of non-adiabaticity for polaronic excitons, which has been revealed through PL, PLE, Raman scattering spectra. Advances in the theoretical understanding of this non-adiabaticity were only made possible by the use of highly specialised techniques (such as the Feynman ordered operator calculus, a cousin of his path integrals, see [154] for this application), which are familiar from QED, but also from polaron theory. It seems highly significant that some of the most sophisticated theoretical tools of Quantum Field Theory are playing a key role in our understanding of state-of-the-art nanodevices, that are key elements for applications and for manipulation of materials at the nanoscale.

Acknowledgement. I would like to thank V. M. Fomin for discussions during the preparation of this manuscript. Thanks are due to S. N. Klimin for discussions and for numerical computations. I acknowledge discussions with V. N. Gladilin, Sasha Alexandrov, A. S. Mishchenko, V. Cataudella, G. De Filippis, R. Evrard, F. Brosens, L. Lemmens and J. Tempere. This work has been supported by the GOA BOF UA 2000, IUAP, FWO-V projects G.0306.00, G.0274.01N, G.0435.03, the WOG WO.035.04N (Belgium) and the European Commission SANDiE Network of Excellence, contract No. NMP4-CT-2004-500101.

References

1. L. D. Landau: Phys. Z. Sowjetunion **3**, 664 (1933)
2. L. D. Landau, S. I. Pekar: Zh. Eksp. i Teor. Fiz. **18**, 419 (1948)
3. S. I. Pekar: *Issledovanija po Ekektronnoj Teorii Kristallov* (Gostekhizdat, Moskva 1951) [German translation: *Untersuchungen über die Elektronentheorie der Kristalle* (Akademie Verlag, Berlin 1951)]
4. N. N. Bogolubov: Ukr. Matem. Zh. **2**, 3 (1950)
5. N. N. Bogolubov, S. V. Tyablikov: Zh. Eksp. i Teor. Fiz. **19**, 256 (1949)
6. H. Fröhlich: Adv. Phys. **3**, 325 (1954)
7. T. D. Lee, F. E. Low, D. Pines: Phys. Rev. **90**, 297 (1953)
8. R. P. Feynman: Phys. Rev. **97**, 660 (1955)
9. F. C. Brown: Experiments on the polaron. In *Polarons and Excitons*, ed by C. G. Kuper, G. D. Whitfield (Oliver and Boyd, Edinburgh 1963) pp 323-355
10. *Polarons and Excitons*, ed by G. C. Kuper, G. D. Whitfield (Oliver and Boyd, Edinburgh 1963)
11. *Polarons in Ionic Crystals and Polar Semiconductors*, ed by J. T. Devreese (North-Holland, Amsterdam 1972)
12. J. Appel: Polarons. In *Solid State Physics*, vol. 21, ed by F. Seitz, D. Turnbull, H. Ehrenreich (Academic Press, New York 1968) pp 193-391
13. J. T. Devreese. In *Encyclopedia of Applied Physics*, vol 14, ed by G. L. Trigg (VCH, Weinheim 1996) pp 383-413
14. A. S. Alexandrov, N. Mott: *Polarons and Bipolarons* (World Scientific, Singapore 1996)
15. P. Calvani: *Optical Properties of Polarons* (Editrice Compositori, Bologna 2001)

16. F. M. Peeters: Theory of electron-phonon interactions in semiconductors. In *High Magnetic Fields: Science and Technology*, vol 2, ed by F. Herlach and N. Miura (World Scientific, Singapore 2003), vol. 2, pp 23-45

17. A. S. Alexandrov: *Theory of Superconductivity. From Weak to Strong Coupling* (IOP Publishing, Bristol 2003)

18. A. S. Alexandrov, N. F. Mott: *High Temperature Superconductors and Other Superfluids* (Taylor and Francis, London 1994)

19. *Proceedings of the International School of Physics "Enrico Fermi", Course CXXXVI, Varenna, 1997, Models and Phenomenology for Conventional and High-temperature Superconductivity* ed by G. Iadonisi, J. R. Schrieffer, M. L. Chiofalo (IOS Press, Amsterdam 1998)

20. J. T. Devreese: Polaron. In *Encyclopedia of Physics*, vol. 2, ed by R.G. Lerner, G. L. Trigg (Wiley-VCH, Weinheim 2005) pp 2004–2027

21. *Proceedings of the "Enrico Fermi' Summer School, Course CLXI – Polarons in Bulk Materials and Systems with Reduced Dimensionality, Varenna, June 21 - July 1, 2005* ed by G. Iadonisi, J. Ranninger and G. De Filippis (IOS Press, Amsterdam 2006).

22. J. T. Devreese: Polarons. In *Lectures on the Physics of Highly Correlated Electron Systems VII* ed by A. Avella, F. Mancini (AIP, Melville 2003) pp 3-56

23. J. T. Devreese, R. Evrard: Phys. Letters **11**, 278 (1964)

24. J. T. Devreese, R. Evrard: Proceedings of the British Ceramic Society **10**, 151 (1968)

25. R. P. Feynman, R. W. Hellwarth, C. K. Iddings et al: Phys. Rev. **127**, 1004 (1962)

26. K. K. Thornber, R. P. Feynman: Phys. Rev. B **1**, 4099 (1970)

27. J. T. Devreese: Bijdrage tot de polarontheorie. Proefschrift tot het verkrijgen van de graad van Doctor in de Wetenschappen, KU Leuven (1964)

28. R. Evrard: Phys. Letters **14**, 295 (1965)

29. E. Kartheuser, R. Evrard, J. Devreese: Phys. Rev. Lett. **22**, 94 (1969)

30. J. T. Devreese, J. De Sitter, M. Goovaerts: Phys. Rev. B **5**, 2367 (1972)

31. V. L. Gurevich, I. G. Lang, Yu. A. Firsov: Fiz. Tverd. Tela **4**, 1252 (1962) [English translation: Sov. Phys. — Solid St. **4**, 918 (1962)]

32. J. T. Devreese: Internal structure of free Fröhlich polarons, optical absorption and cyclotron resonance. In *Polarons in Ionic Crystals and Polar Semiconductors* (North-Holland, Amsterdam 1972) pp 83-159

33. A. S. Mishchenko, N. Nagaosa, N. V. Prokof'ev et al: Phys. Rev. Lett. **91**, 236401 (2003)

34. G. De Filippis, V. Cataudella, V. Marigliano Ramaglia et al: Eur. Phys. J. B **36**, 65 (2003)

35. G. De Filippis, V. Cataudella, A. S. Mishchenko, C. A. Perroni, J. T. Devreese: Phys. Rev. Lett. **96**, 136405 (2006)

36. A. S. Mishchenko, N. Nagaosa, this volume

37. V. Cataudella, G. De Filippis, C. A. Perroni, this volume

38. J. T. Titantah, C. Pierleoni, S. Ciuchi: Phys. Rev. Lett. **87**, 206406 (2001)

39. A. S. Mishchenko, N. V. Prokof'ev, A. Sakamoto et al: Phys. Rev. B **62**, 6317 (2000)

40. F. M. Peeters, J. T. Devreese: Phys. Stat. Sol. (b) **112**, 219 (1982)

41. B. Gerlach, H. Löwen: Phys. Rev. B **35**, 4291 (1987)

42. B. Gerlach, H. Löwen: Phys. Rev. B **35**, 4297 (1987)

43. T. Holstein: Ann. Phys. (USA) **8**, 325 (1959)
44. T. Holstein: Ann. Phys. (USA) **8**, 343 (1959)
45. I. G. Lang, Yu. A. Firsov: Zh. Eksp. Teor. Fiz. **43**, 1843 (1962) [Sov. Phys. JETP **16**, 1301 (1963)]
46. Yu. A. Firsov: *Polarons* (Nauka, Moscow 1975)
47. Yu. A. Firsov, V. V. Kabanov, E. K. Kudinov, A. S. Alexandrov: Phys. Rev. B **59**, 12132 (1999)
48. M. Zoli: Phys. Rev. B **61**, 14523 (2000)
49. T. Damker, H. Böttger, V. V. Bryksin: Phys. Rev. B **69**, 205327 (2004)
50. V. Cataudella, G. De Filippis, G. Iadonisi: Phys. Rev. B **62**, 1496 (2000)
51. L.-C. Ku, S. A. Trugman, J. Bonča: Phys. Rev. B **65**, 174306 (2002)
52. S. Aubry: Physica D **216** 1 (2006)
53. J. Devreese: Bull. Soc. Belge de Phys., Ser. III, no. 4, 259 (1963)
54. P. Nagels, M. Denayer, J. Devreese: Solid State Commun. **1**, 35 (1963)
55. A. M. Stoneham: Advances in Physics **28**, 457 (1979)
56. A. L. Shluger, A. M. Stoneham: J. Phys. Cond. Mat. **5**, 3049 (1993)
57. H. de Raedt, A. Lagendijk, Phys. Rev. Lett. **49**, 1522 (1982)
58. H. de Raedt, A. Lagendijk, Phys. Rev. B **27**, 6097 (1983)
59. H. de Raedt, A. Lagendijk, Phys. Rev. B **30**, 1671 (1984)
60. A. S. Alexandrov, P. E. Kornilovitch: Journal of Superconductivity **15**, 403 (2002)
61. A. S. Alexandrov: Phys. Rev. Lett. **82**, 2620 (1999)
62. A. S. Alexandrov, P. E. Kornilovitch: Phys. Rev. Lett. **82**, 807 (1999)
63. A. S. Alexandrov: Phys. Rev. B **61**, 12315 (2000).
64. A. S. Alexandrov, this volume
65. G.-M. Zhao, M. B. Hunt, H. Keller: Phys. Rev. Lett. **78**, 955 (1997)
66. G. M. Zhao, V. Smolyaninova, W. Prellier, H. Keller: Phys. Rev. Lett. **84**, 6086 (2000)
67. T. Mertelj, D. Kuščer, M. Kosec, D. Mihailovic: Phys. Rev. B **61**, 15107 (2000)
68. A. S. Alexandrov, A. M. Bratkovsky, V. V. Kabanov: Phys. Rev. Lett. **96**, 117003 (2006)
69. Ch. Hartinger, F. Mayr, J. Deisenhofer et al: Phys. Rev. B **69**, 100403(R) (2004)
70. Y. Toyozawa: Prog. Theor. Phys. **12**, 421 (1954)
71. J. T. Devreese, A. B. Kunz, T. C. Collins: Solid State Commun. **11**, 673 (1972)
72. F. C. Brown: Ultraviolet spectroscopy of solids with the use of synchrotron radiation. In *Solid State Physics*, vol. 29, ed by H. Ehrenreich, F. Seitz, D. Turnbull, (Academic Press, New York 1974) pp 1-73
73. A. B. Kunz, J. T. Devreese, T. C. Collins: J. Phys. C **5**, 3259 (1972)
74. G. Zimmerer: Journal of Luminescence **119-120**, 1 (2006)
75. M. Agåker, T. Käämbre, C. Glover et al.: Phys. Rev. B **73**, 245111 (2006)
76. J. Tempere, J. T. Devreese: Phys. Rev. B **64**, 104504 (2001)
77. G. D. Mahan: Many-Particle Physics (Kluwer/Plenum, New York 2000)
78. J. T. Devreese, Fröhlich polarons from 3D to 0D: Concepts and recent developments. In *Proceedings of the "Enrico Fermi' Summer School, Course CLXI - "Polarons in Bulk Materials and Systems with Reduced Dimensionality"*, *Varenna, June 21 - July 1, 2005*, ed by G. Iadonisi, J. Ranninger and G. De Filippis (IOS Press, Amsterdam et al 2006), pp 27-52
79. J. Devreese, W. Huybrechts, L. Lemmens: Phys. Stat. Sol. (b) **48**, 77 (1971)

80. J. T. Devreese, J. Tempere: Solid State Commun. **106**, 309 (1998)
81. H. Finkenrath, N. Uhle, W. Waidelich: Solid State Commun. **7**, 11 (1969)
82. *Landolt-Börnstein Numerical Data and Functional Relationships in Science and Technology, New Series*, Group III, vol. 17, subvolume b, ed by K.-H. Hellwege (Springer, New York 1982) p 161
83. F. M. Peeters, J. T. Devreese: Phys. Rev. B **28**, 6051 (1983)
84. M. J. Goovaerts, J. De Sitter, J. T. Devreese: *Phys. Rev. B* **7**, 2639 (1973)
85. J. Devreese, J. De Sitter, M. Goovaerts: Solid State Commun. **9**, 1383 (1971)
86. W. Huybrechts, J. Devreese: Solid State Commun. **17**, 401 (1975)
87. D. M. Eagles, R. P. S. M. Lobo, F. Gervais: Phys. Rev. B **52**, 6440 (1995)
88. M. Hohenadler, H.G. Evertz, W. von der Linden: Phys. Rev. B **69**, 024301 (2004)
89. A. Weiße, G. Wellein, A. Alvermann, H. Fehske: Rev. Mod. Phys. **78**, 275 (2006)
90. P. E. Spencer, J. H. Samson, P. E. Kornilovitch, A. S. Alexandrov: Phys. Rev. B **71**, 184310 (2005)
91. W. Huybrechts, J.T. Devreese: Phys. Rev. B **8**, 5754 (1973)
92. J. T. Devreese, S. N. Klimin: in preparation
93. G. R. Allcock: Strong-coupling theory of the polaron. In *Polarons and Excitons*, ed by C. G. Kuper, G. D. Whitfield (Oliver and Boyd, Edinburgh 1963), pp 45-70
94. S. J. Miyake: J. Phys. Soc. Jap. **38**, 181 (1975)
95. G. R. Allcock: Adv. Phys. **5**, 412 (1956)
96. R. P. Feynman: *Statistical Mechanics* (Benjamin, Reading 1972)
97. A. S. Mishchenko: private communication
98. J. T. Devreese, L. Lemmens, J. Van Royen: Phys. Rev. B **15**, 1212 (1977)
99. L. F. Lemmens, J. De Sitter, J. T.Devreese: Phys. Rev. B **8**, 2717 (1973)
100. F. M. Peeters, Wu Xiaoguang, J. T. Devreese: Phys. Rev. B **33**, 3926 (1986)
101. Wu Xiaoguang, F. M. Peeters, J. T. Devreese: Phys. Rev. B **31**, 3420 (1985)
102. F. M. Peeters, J. T. Devreese: Phys. Rev. B **36**, 4442 (1987).
103. F. M. Peeters, J. T. Devreese: Phys. Rev. B **34**, 7246 (1986)
104. D. Larsen: Journal of Physics C **7**, 2877 (1974)
105. J. W. Hodby, G. P. Russell, F. M. Peeters, J. T. Devreese, D. M. Larsen: Phys. Rev. Lett. **58**, 1471 (1987)
106. D. R. Cohn, D. M. Larsen, B. Lax: Phys. Rev. B **6**, 1367 (1972)
107. D. M. Larsen: Polaron energy levels in magnetic and Coulomb fields. In *Polarons in Ionic Crystals and Polar Semiconductors*, ed by J. T. Devreese (North-Holland, Amsterdam 1972) pp 237-287
108. P. Pfeffer, W. Zawadzki: Phys. Rev. B **37**, 2695 (1988)
109. N. Miura, H. Nojiri, P. Pfeffer et al: Phys. Rev. B **55**, 13598 (1997)
110. M. Grynberg, S. Huant, G. Martinez, J. Kossut, T. Wojtowicz, G. Karczewski, J. M. Shi, F. M. Peeters, J. T. Devreese Phys. Rev. B **54**, 1467 (1996)
111. R. J. Nicholas, M. Watts, D. F. Howell, F. M. Peeters,X.-G. Wu, J. T. Devreese, L. van Bockstal, F. Herlach, C. J. G. M. Langerak, J. Singleton, A. Chery: Phys. Rev. B **45**, 12144 (1992)
112. J.-P. Cheng J.-P., B. D. McCombe, J. M. Shi, F. M. Peeters, J. T. Devreese: Phys. Rev. B **48**, 7910 (1993)
113. J. M. Shi, F. M. Peeters, J. T. Devreese: Phys. Rev. B **48**, 5202 (1993)
114. A. J. L. Poulter, J. Zeman, D. K. Maude et al: Phys. Rev. Lett. **86**, 336 (2001)

115. S. N. Klimin, J. T. Devreese, Phys. Rev. B **68**, 245303 (2003)

116. Y.J. Wang, H.A. Nickel, B.D. McCombe et al: Phys. Rev. Lett. **79**, 3226 (1997)

117. V. L. Vinetskii, M. S. Giterman: Zh. Eksp. i. Teor. Fiz. **33**, 730 (1957)

118. V. L. Vinetskii: Zh. Eksp. i. Teor. Fiz. **40**, 1459 (1961) [English translation: Sov. Phys.—JETP **13**, 1023 (1961)].

119. H. Hiramoto, Y. Toyozawa: J. Phys. Soc. Japan **54**, 245 (1985)

120. J. Adamowski: Phys. Rev. B **39**, 3649 (1989)

121. F. Bassani, M. Geddo, G. Iadonisi et al: Phys. Rev. B **43**, 5296 (1991)

122. G. Verbist, F. M. Peeters, J. T. Devreese: Solid State Commun. **76**, 1005 (1990)

123. G. Verbist, F. M. Peeters, J. T. Devreese: Phys. Rev. B **43**, 2712 (1991)

124. M. A. Smondyrev, V. M. Fomin: Pekar-Fröhlich bipolarons. In: *Polarons & Applications (Proceedings in nonlinear science)*, ed by V. D. Lakhno (John Wiley & Sons, Chichester 1994), pp 13 - 70.

125. J. T. Devreese: Bipolarons and high-T_c superconductivity. In: *Proceedings of the International School of Physics "Enrico Fermi", Course CXXXVI, Varenna, 1997, Models and Phenomenology for Conventional and High-Temperature Superconductivity* ed by G. Iadonisi, J. R. Schrieffer and M. L. Chiofalo, (IOS Press, Amsterdam 1998), pp 287-308

126. L. Genzel, A. Wittlin, M. Bauer, M. Cardona, E. Schönherr, A. Simon: Phys. Rev. B **40**, 2170 (1989).

127. W. Kress, U. Schröder, J. Prade et al: Phys. Rev. B **38**, 2906 (1988).

128. M. A. Smondyrev, J. T. Devreese, F. M. Peeters: Phys. Rev. B **51**, 15008 (1995)

129. J. T. Devreese, G. Verbist, F. Peeters: Large Bipolarons and High-T_c Materials. In: *Polarons and Bipolarons in High-T_c Superconductors and Related Materials* ed by E. Salje, A. Alexandrov and W. Liang (Cambridge University Press 1995), pp. 385-391.

130. F. G. Bassani, V. Cataudella, M. L. Chiofalo, G. De Filippis, G. Iadonisi, C. A. Perroni: Phys. Stat. Sol. (b), **237**, 173 (2003)

131. V. Cataudella, G. De Filippis, G. Iadonisi: Eur. Phys. J. B **12**, 17 (1999)

132. S. Fratini, P. Quémerais: Eur. Phys. J. B **14**, 99 (2000)

133. S. Fratini, P. Quémerais: Eur. Phys. J. B **29**, 41 (2002)

134. S. Fratini, F. de Pasquale, S. Ciuchi: Phys. Rev. B **63**, 153101 (2001)

135. C. A. Perroni, G. Iadonisi, V. K. Mukhomorov: Eur. Phys. J. B **41**, 163 (2004)

136. A.A. Vlasov, *Many-particle theory and its application to plasma* (Gordon and Breach, New York 1961); A.A. Vlasov, *Nelokal'naia statiaticheskaia mehanica* (*Nonlocal statistical mechanics*) [in Russian] (Nauka, Moscow 1978).

137. F. Brosens, S. N. Klimin, J. T. Devreese: Phys. Rev. B **71**, 214301 (2005)

138. F. Brosens, J. T. Devreese, L. F. Lemmens: Phys. Rev. E **55**, 227 (1997)

139. F. Brosens, J. T. Devreese, L. F. Lemmens: Phys. Rev. E **55**, 6795 (1997)

140. F. Brosens, J. T. Devreese, L. F. Lemmens: Phys. Rev. E **58**, 1634 (1998)

141. L. F. Lemmens, F. Brosens, J. T. Devreese: Solid State Commun. **109**, 615 (1999)

142. J. T. Devreese, S. N. Klimin, V. M. Fomin, F. Brosens: Solid State Commun. **114**, 305 (2000)

143. S. N. Klimin, V. M. Fomin, F. Brosens, J. T. Devreese: Physica E **22**, 494 (2004)

144. S. N. Klimin, V. M. Fomin, F. Brosens, J. T. Devreese: Phys. Rev. B **69**, 235324 (2004)

145. L. F. Lemmens, F. Brosens, J. T. Devreese:, Phys. Rev. E **53**, 4467 (1996)
146. J. T. Devreese: Note on the path-integral variational approach in the many-body theory. In: *Fluctuating Paths and Fields*, ed by W. Janke, A. Pelster, H.-J. Schmidt, M. Bachmann (World Scientific, Singapore 2001), pp 289-304
147. M. A. Smondyrev, G. Verbist, F. M. Peeters, J. T. Devreese: Phys. Rev. B **47**, 2596 (1993)
148. N. I. Kashirina, V. D. Lakhno, V. V. Sychyov: Phys. Stat. Sol. (b) **239**, 174 (2003)
149. L. F. Lemmens, J. T. Devreese, F. Brosens: Phys. Status Solidi (b) **82**, 439 (1977)
150. S. Lupi, P. Maselli, M. Capizzi P. Calvani, P. Giura, P. Roy: Phys. Rev. Lett. **83**, 4852 (1999)
151. D. Emin: Phys. Rev. B **48**, 13691 (1993)
152. A. Messiah: *Quantum Mechanics*, vol. 2, (North-Holland, Amsterdam 1966), p 702
153. S. M. Reimann, M. Koskinen, M. Manninen: Phys. Rev. B **62**, 8108 (2000)
154. V. M. Fomin, V. N. Gladilin, J. T. Devreese, E. P. Pokatilov, S. N. Balaban, S. N. Klimin: Phys. Rev. B **57**, 2415 (1998)
155. E. P. Pokatilov, S. N. Klimin, V. M. Fomin, J. T. Devreese, F. W. Wise: Phys. Rev. B **65**, 075316 (2002)
156. V. A. Fonoberov, E. P. Pokatilov, V. M. Fomin, J. T. Devreese: Phys. Rev. Lett. **92**, 127402 (2004)
157. V. N. Gladilin, S. N. Klimin, V. M. Fomin, J. T. Devreese: Phys. Rev. B **69**, 155325 (2004)
158. M. Nirmal, C. B. Murray, D. J. Norris et al: Z. Phys. D **26**, 361 (1993)
159. S. Nomura, S. Kobayashi: Phys. Rev. B **45**, 1305 (1992)
160. V. N. Gladilin, S. N. Balaban, V. M. Fomin, J. T. Devreese: Photoluminescence in semiconductor quantum dots: enhanced probabilities of phonon-assisted transitions. In *Proc. 25th Int. Conf. on the Physics of Semiconductors, Osaka, Japan, 2000* ed by N. Miura and T. Ando, Part II, (Springer, Berlin 2001) pp 1243-1244
161. O. Verzelen, R. Ferreira, G. Bastard: Phys. Rev. Lett. **88**, 146803 (2002)
162. A. Lemaître, A. D. Ashmore, J. J. Finley, D. J. Mowbray, M. S. Skolnick, M. Hopkinson, T. F. Krauss: Phys. Rev. B **63**, 161309(R) (2001)
163. A. Zrenner, F. Findeis, M. Baier, M. Bichler, G. Abstreiter: Physica B **298**, 239 (2001)
164. R. Heitz, H. Born, F. Guffarth, O. Stier, A. Schliwa, A. Hoffmann, D. Bimberg: Phys. Rev. B **64** 241305 (2001)
165. U. Woggon, D. Miller, F. Kalina et al: Phys. Rev. B **67**, 045204 (2003)
166. J. Tempere, S. N. Klimin, I. F. Silvera, J. T. Devreese: Eur. Phys. J. **32**, 329 (2003)
167. V. B. Shikin, Yu. P. Monarkha: Zh. Exp. Theor. Phys. **65**, 751 (1973) [English translation: Sov. Phys. – JETP **38**, 373 (1973)]
168. S. A. Jackson, P. M. Platzman: Phys. Rev. B **24**, 499 (1981)
169. A. P. Volodin, M. S. Khaikin, V. S. Edelman: JETP Lett. **26**, 543 (1977).
170. U. Albrecht, P. Leiderer: Europhys. Lett. **3**, 705 (1987)
171. V. B. Shikin: JETP Lett. **27**, 39 (1978)
172. M. M. Salomaa, G. A. Williams: Phys. Rev. Lett. **47**, 1730 (1981)
173. I. F. Silvera: Bull. Am. Phys. Soc. **46**, 1016 (2001)

174. J. Tempere, I. F. Silvera, J. T. Devreese: Phys. Rev. Lett. **87**, 275301 (2001)
175. F. Lindemann: Z. Phys. **11**, 609 (1910)
176. C. M. Care, N. H. March: Adv. Phys. **24**, 101 (1975)
177. C. C. Grimes, G. Adams: Phys. Rev. Lett. **42**, 795 (1979)
178. V. M. Bedanov, F. M. Peeters: Phys. Rev. B **49**, 2667 (1994)
179. L. P. Gorkov, D. M. Chernikova: Pis'ma Zh. Eksp. Teor. Fiz. **18**, 119 (1973) [English Translation: JETP Lett. **18**, 68 (1973)]
180. J. Tempere, I. F. Silvera, J. T. Devreese: Phys. Rev. B **67**, 035402 (2003)
181. D. S. Fisher, B. I. Halperin, P. M. Platzman: Phys. Rev. Lett. **42**, 798 (1979)
182. G. Deville, A. Valdes, E. Y. Andrei et al: Phys. Rev. Lett. **53**, 588 (1984)

Small Polarons: Transport Phenomena

Yurii A. Firsov

Ioffe Physico-Technical Institute, 194021, St.Petersburg, Russia,
yuaf@icomefrom.ru

1 Introduction

The theory of small polarons is a huge subfield of solid state physics. It is
currently enjoying a new burst of activity, particularly in Europe. Not the
least of the reasons is research on T_c superconductors, fullerenes, correlated
nanowires and nanotubes, quantum dots and other advanced materials.

No claim is made that this chapter completely covers all aspects and applications of small polaron theory, nor has an effort been made to compile
an exhaustive bibliography. The topics of this review are the principles for
constructing a theory of transport phenomena in a situation in which the
mobility of charge carriers in a crystal is so low that transport cannot be described by the Boltzmann equation. The effectiveness of this approach will be
demonstrated for small polarons. Abram Fedorovich Ioffe [1] was the first to
draw the attention of physicists to the low mobility problem. To illustrate his
idea we consider the standard expression for the mobility which follows from
the kinetic equation

$$\mu = \frac{e\tau}{m^*} = \xi \frac{e}{\hbar} \lambda l. \tag{1}$$

Here τ is a relaxation time, $m^\star$ and λ are the effective mass and the de
Broglie wavelength of the charge carriers, l is the mean free path, and the
dimensionless factor ξ is of the order of unity.

The condition for the applicability of the kinetic equation (in the case of
Boltzmann statistics) is

$$\hbar/\tau kT < 1, \tag{2}$$

where k is the Boltzmann constant, or $\lambda/l < 1$. This condition means that the
uncertainty in the carrier energy due to scattering must be smaller than the
average energy of the carrier. Equivalently, the mean free path of the carrier,
l, must be larger than λ.

Using (1), we can rewrite condition (2) as

$$\frac{\hbar}{\tau k T} \approx 20 \frac{m}{m^*} \frac{500\text{K}}{T} \frac{\text{cm}^2/\text{Vs}}{\mu} < 1. \tag{3}$$

Here m is the mass of a free electron. We see that condition (3) does not hold with $m^* \approx m$ and $\mu = 1\text{cm}^2/(\text{V} \cdot \text{s})$. However, there are many materials with $\mu \leq 1\text{cm}^2/(\text{V} \cdot \text{s})$. How can we derive a theory of transport phenomena in the case $l \sim \lambda$? This condition means that the scattering is very strong, and that the interaction with the scatterers (phonons) is therefore very strong. For long-wave longitudinal (optical) phonons this interaction is characterized by the dimensionless coupling constant α:

$$\alpha = \left(\frac{1}{\epsilon_\infty} - \frac{1}{\epsilon_0} \right) \frac{e^2}{2l_0 \hbar \omega_0}. \tag{4}$$

Here ϵ_0 and ϵ_∞ are the static and high frequency dielectric constants, and $l_0 = (\hbar/2m^*\omega_0)^{1/2}$ is the length scale found from the condition $\hbar^2/2m^* l_0^2 = \hbar \omega_0$, where ω_0 is the limiting frequency of longitudinal phonons. Here we have $l_0 \gg a$, where a is the lattice constant.

For many ionic crystals the relation $\alpha \gg 1$ holds. In this case the charge carriers are dressed in a "phonon cloud." These carriers are called "polarons". They may have a large radius ($r_p \gg a$), in which case they are "large polarons" or a small one ($r_\rho \leq a$), in which case they are "small polarons." Research on large polarons began long before research on small polarons, in the pioneering study by Landau [2]. The theory of large polarons was pursued actively by Pekar ([3], for example), Bogolyubov [4], Tyablikov [5], Fröhlich [6], and Feynman [7,8] (for further references see the preceding chapter by Prof. Jozef T. Devreese [9]). Looking at (1), one might hope that by replacing m^* by a larger polaron mass M_p it would become easier to satisfy condition (3) [or (2)], even at values $\mu < 1\text{cm}^2/(\text{V} \cdot \text{s})$. However, [6–8] have shown that at $\alpha < 10$ the mass of a large polaron does not change appreciably: $M_p \simeq m^*(1 + \alpha/6)$. In the adiabatic limit, $\alpha > 10$, renormalization effects are pronounced:

$$\begin{aligned} E_p &\simeq -0.1\alpha^2 \hbar \omega_0, \\ M_\rho &= 0.025\alpha^4 m^*, \\ r_\rho &= l_0 \frac{10}{\alpha} = 10\left(\frac{1}{\epsilon_\infty} - \frac{1}{\epsilon_0} \right)^{-1} \frac{m}{m^*} a_B. \end{aligned} \tag{5}$$

Here E_p is the polaron's binding energy, and $a_B = \hbar^2/me^2$ is the first Bohr radius. It can be seen from (5) that the polaron radius decreases with increasing α, but in the limit $r_p \to a$ the continuum approximation used in the theory of large polarons is no longer valid. It becomes necessary to construct a theory for small polarons in which the discrete nature of the lattice is taken into account. What does the large-polaron theory yield in terms of a description of low mobilities? For high temperatures $kT > \hbar \omega_0$ Pekar [3] derived expressions for the mobility for single–phonon scattering (involving the emission or absorption of one phonon) and two–phonon scattering:

$$\mu^{(1)} = \mu_0 \left(\frac{m^*}{M_\rho}\right)^{1/2} \left(\frac{kT}{\hbar\omega_0}\right)^{1/2} 0.77\alpha \simeq \frac{5}{\alpha}\left(\frac{kT}{\hbar\omega_0}\right)^{1/2}\mu_0,$$

$$\mu^{(2)} = \mu^{(1)}\frac{\hbar\omega_0}{kT}\frac{200}{\alpha^2},$$

$$(6)$$

respectively, where μ_0 is the characteristic scale of the mobility, given by

$$\mu_0 = \frac{el_0^2}{\hbar} = \frac{e}{2m^*\omega_0} = 20\frac{m}{m^*}\frac{4\times 10^{13}\mathrm{s}^{-1}}{\omega_0}(\mathrm{cm}^2/\mathrm{Vs}). \qquad (7)$$

In practice, μ_0 is always greater than $1\ \mathrm{cm}^2/\mathrm{Vs}$, and the quantities $\mu^{(1)}$ and $\mu^{(2)}$ cannot be greatly different from μ_0. Furthermore, Pekar's theory does not apply when the two-phonon scattering becomes more effective than the one-phonon scattering, i.e., at $\alpha^2 > 200\hbar\omega/kT$, since in this case scattering processes involving progressively more and more phonons must be taken into account. Figuratively speaking, the cloud shrinks with increasing α, the polaron radius (the size of the cloud) decreases according to [5], and we have the case of small polarons, i.e., $r_\rho \leq a$, in which the transport mechanism might be fundamentally non-band-like in nature. Ioffe suggested [1] that when condition (2) is violated, the transport may involve hops from one site to another. Heiks and Johnston [10] analyzed experimental data on NiO and came to the conclusion that the carrier mobility in NiO is of an activation-law nature at high temperatures. On this basis they concluded that there is a phonon activated hopping mechanism in this case, as for the diffusion of ions along interstitial positions.

But how can all this be described with mathematical rigor? At first glance, these arguments would seem to contradict Bloch's theorem, according to which a wave packet describing a charge carrier localized at a lattice site necessarily spreads out over the entire crystal, and this situation is a stationary state. However, some pioneering studies [5, 10, 11] of the energy *spectrum* of small polarons (rather than of the transport mechanism) showed that the mass of a small polaron can be very large, while the width of the allowed band for a small polaron, ΔE_p, can be very small:

$$\Delta E_p \simeq \Delta E \exp\left(-\gamma \coth\frac{\hbar\omega_0}{2kT}\right). \qquad (8)$$

Here ΔE is the width of the original (unrenormalized) band, and the dimensionless coupling constant γ is larger than α (see below). As the temperature is raised ($kT > \hbar\omega_0/2$), the band becomes narrower, and one might suggest that the uncertainty in the energy due to the multiphonon interaction becomes larger than ΔE_ρ. Then the concept of the polaron band becomes meaningless, and the situation should be described in terms of states of small polarons localized at lattice sites (this is the lattice–site representation), while the "residual" polaron–phonon coupling (after the formation of the cloud) would lead to transitions from site to site by an over-barrier transport mechanism or a tunnelling. Holstein [12] was the first to attempt to put these ideas in mathematical form. He postulated that the motion of a small polaron at high temperatures is a random walk consisting of steps from site to site. Making use of the form of the wave functions of small polarons localized at lattice sites, and singling out the terms of the Hamiltonian which were not diagonal in terms of sites (which were proportional to J, where J is the overlap integral for nearest neighbours), Holstein calculated the probability $W(T)$ for the hop of a small polaron to a neighbouring site. He then suggested that the random walk was a Markovian process. Taking this approach, he was able to write the diffusion coefficient as

$$D = \frac{1}{2}a^2 W(T). \tag{9}$$

At high temperatures, $kT > \hbar\omega_0/2$, the probability for a hop is, according to [12],

$$W(T) = \frac{\omega_0}{2\pi} f(\eta_2) e^{-E_a/kT}. \tag{10}$$

Here $E_a \simeq (\gamma/2)\hbar\omega_0$ is the activation energy (see Sect.2 and 3 for more details), $f(\eta_2)$ is a dimensionless function of the dimensionless parameter $\eta_2 = J^2/\hbar\omega_0(E_a kT)^{1/2}$, and J is an overlap integral characterizing the width ΔE of the original electron band. For cubic crystals we would have $\Delta E = 2zJ$, where z is the number of nearest neighbours. The function $f(\eta_2)$ does not exceed unity,

$$f(\eta_2) \simeq \begin{cases} \pi^{3/2}\eta_2, \eta_2 \ll 1 \\ 1, \eta_2 > 1. \end{cases} \tag{11}$$

For hops of ions between interstitial positions, the result for $W(T)$ is similar to (10), but (10) contains the electron characteristic J. From the Einstein relation $\mu = eD/kT$ we find the following expression for the hopping mobility μ_h :

$$\mu_h = u_0 \frac{1}{2\pi} \frac{\hbar\omega_0}{kT} f(\eta_2) e^{-E_a/kT}, \tag{12}$$

where u_0 is the characteristic scale of the mobility, which satisfies $u_0 \ll \mu_0$ [cf. (7)] given by

$$u_0 = \frac{ea^2}{\hbar} \approx 1.6(a/3A^0)^2 (cm^2/Vs) \tag{13}$$

From the activation-law factor in (12) we have $\mu \ll 1\mathrm{cm}^2/(\mathrm{V}\cdot\mathrm{s})$. The mobility of small polarons in the hopping regime is thus indeed small. If we assume that the uncertainty in the energy is $\hbar W(T)$, then it is greater then ΔE_ρ, and the switch to the lattice-site representation (as discussed above) is justified. Holstein suggested that at low temperatures $kT < \hbar\omega_0/2\ln\gamma$, there is the ordinary Boltzmann transport in momentum (**k**) space, but in a narrow polaron band. He found the result

$$\mu_B = \frac{e}{kT}\langle v_p^2(\mathbf{k})\tau(\mathbf{k})\rangle = u_0 \frac{1}{2} \frac{\Delta E_p}{zkT} \frac{\Delta E_p}{z\hbar}. \tag{14}$$

Here $v_p(\mathbf{k})$ is the velocity of the polaron, and $\tau(\mathbf{k})$ is the relaxation time. For τ he suggested using $W_2^{-1}(T)$ in the lowest order in J, but calculated for $kT < \hbar\omega_0/2$. For $kT_0 < kT < \hbar\omega_0/2$, for example, where $kT_0 = \hbar\omega_0/2\ln\gamma$, we would have

$$\tau_2^{-1} = W_2 = \pi^{3/2}\frac{J^2}{\hbar(E_a\hbar\omega_0)^{1/2}}(2\sinh\frac{\hbar\omega_0}{2kT})^{1/2}\exp(-\frac{4E_a}{\hbar\omega_0}\tanh\frac{\hbar\omega_0}{4kT}) \quad (15)$$

At $kT > \hbar\omega_0/2$, expression (15) becomes (10) (with $\eta_2 < 1$). Substituting (15) into (9), and using the Einstein relation, we find an expression for μ_h which is useful over a broader temperature range than (12) (T > T$_0$). In the interval $T_1 < T < T_0$, where T_1 is found from the condition $kT_1 \approx (1/z)\Delta E_p$, we have

$$\tau_2^{-1} \simeq W = \frac{J^2}{\hbar^2\Delta\overline{\omega}}\gamma^2 e^{-2\gamma}\sinh^2(\frac{\hbar\omega_0}{2kT}), \quad (16)$$

where $\Delta\omega$ is the width of the dispersion band for longitudinal phonons. Under the condition $T < T_0$ we find

$$\mu_B = u_0\frac{\hbar\Delta\omega}{kT}\gamma^{-2}\sinh^2(\frac{\hbar\omega_0}{2kT}) > u_0, \quad (17)$$

i.e. the mobility is no longer small. With increasing T, the mobility thus first decreases and then begins to increase. Holstein suggested that the crossover temperature from the Bloch–type band motion to a hopping mechanism, T_3, can be found from the condition

$$\mu_h(T_3) = \mu_B(T_3). \quad (18)$$

Using (12), (14), and (15), we find the following expression from (18):

$$\Delta E_p/z = \frac{\hbar}{\tau_2(T_3)} = \hbar W_2(T_3). \quad (19)$$

According to Holstein, the transition from the band regime to the hopping regime occurs when the uncertainty in the polaron energy becomes comparable to the width of the polaron band. We will see below that we need to replace (15) by another expression for τ. Then we find a broad intermediate temperature range in which a fundamentally different transport mechanism operates, which we call the "tunneling" mechanism (Sect. 3). Holstein's paper [12], whose basic results are presented below, had a huge influence on all subsequent research. However, there was still a very extensive program of research to be carried out.

1. It was necessary to construct, from general principles, a unified mathematical formalism for describing transport processes in configuration space (lattice–site space).

2. It was necessary to prove that the individual hops are uncorrelated, i.e., that the process is indeed Markovian; only in this case would (9) be valid. Under what conditions would the situation cease to be Markovian?

3. It was necessary to identify the transport mechanism at intermediate temperatures, where neither a band–like mechanism nor a hopping mechanism is operating.

4. It was necessary to find the basic dimensionless parameters of the theory and to attempt to extend the theory to the region of parameter values in which the latter are not small.

5. It was necessary to find a rigorous method for calculating different kinetic coefficients.

Clearly, it was necessary to work from the "Kubo formula" for the electrical conductivity:

$$\sigma_{xx} = \beta \frac{1}{V} \lim_{s \to 0} \mathrm{Re} \int_0^\infty e^{-st} \frac{1}{Z} \mathrm{Tr}\{e^{-\beta H} j_x(t) j_x(0)\} dt. \tag{20}$$

Here $\beta = 1/kT$, V is the volume of the system, Z is the partition function, H is the total Hamiltonian of the system, and $\mathbf{j}(t)$ is the current operator in the Heisenberg representation. Konstantinov and Perel [14] suggested a graphical technique (the KP method) for deciphering the Kubo formula. That technique is based on an expansion in powers of $1/s$ in (20). They found a transport equation in the $\mathbf{k}$ representation. Gurevich and Firsov [15] used that technique and obtained from (20) the Titeica formula, which describes the hopping mechanism (in a continuous medium) in a strong magnetic field, but for a weak interaction with phonons. That study, incidentally, resulted in the prediction of a new effect, the magnetophonon resonance, which has been observed and studied by Parfen'ev and Shalyt [16, 17].

By the end of 1960 the plan of action was thus clear. A. F. Ioffe died at that time, and A. R. Regel became the director of the Institute of Semiconductors, Academy of Sciences of the USSR. Regel too called on theoreticians to derive a systematic theory of low mobility. Many researchers working independently looked for a solution of the problem. At the Institute of Semiconductors, Lang and Firsov [18–20] proposed a special canonical transformation (see below) modifying the KP method.

2 Canonical Transformation in the Small Polaron Theory

The second quantization Hamiltonian of the electron–phonon system takes the following form in the $\mathbf{k}$-representation:

$$H = \sum_{\mathbf{k}} \epsilon(\mathbf{k}) a_{\mathbf{k}}^+ a_{\mathbf{k}} + \sum_{\mathbf{q}} \hbar \omega_{\mathbf{q}} \left(b_{\mathbf{q}}^+ b_{\mathbf{q}} + \frac{1}{2}\right)$$
$$+ \frac{1}{\sqrt{2N}} \sum_{\mathbf{k},\mathbf{q}} \hbar \omega_{\mathbf{q}} (\gamma_{\mathbf{q}} b_{\mathbf{q}}^+ a_{\mathbf{k}}^+ a_{\mathbf{k}+\mathbf{q}} + \gamma_{\mathbf{q}}^* b_{\mathbf{q}} a_{\mathbf{k}}^+ a_{\mathbf{k}-\mathbf{q}}). \tag{21}$$

The operators a^+ (a) and b^+ (b) create (annihilate) electrons and phonons. The first two terms in (21) describe free electrons and phonons, $\epsilon(\mathbf{k})$ is the band energy of the electron, and $\omega_{\mathbf{q}}$ is the frequency of an optical (polarization) phonon. The third term in (21) is the interaction Hamiltonian which describes the scattering of an electron accompanied by the creation (annihilation) of one phonon. Here N is the number of unit cells in the volume V. The dimensionless function $\gamma_{\mathbf{q}}$ characterizes the interaction

$$|\gamma_{\mathbf{q}}|^2 = 8\pi\alpha\frac{l_0}{\mathbf{q}^2\Omega}f(\mathbf{q}\cdot\mathbf{g}),\tag{22}$$

where Ω is the volume of the unit cell, $\boldsymbol{q}$ is the phonon wave vector, $\boldsymbol{g}$ is the vector to the nearest neighbor in the lattice, and the Fröhlich coupling constant α and the length l_0 were defined above [see (4)]. Under the condition $\boldsymbol{q}\cdot\boldsymbol{g} < 1$ we have a dimensionless function $f \simeq 1$. For large polarons we have $q \sim r_p^{-1} \ll a^{-1}$ and the function f in (22) can be replaced by 1. This cannot be done in the case of small polarons. Since the coupling constant $\gamma_{\boldsymbol{q}}$ is large, it is useless to expand Kubo formula (20) in powers of $\gamma_{\boldsymbol{q}}$. Lang and Firsov [18–20] proposed a canonical transformation which "dressed electrons up in a cloud of virtual phonons." The interaction Hamiltonian becomes a multiphonon Hamiltonian as a result, and some small parameters arise in the theory. These parameters can be used for an expansion of the Kubo formula. It is more convenient to go through this procedure when the initial Hamiltonian, (21), is written in the lattice–site representation (for electrons):

$$H = \sum_{\mathbf{m},g} J(\mathbf{g})a_{\mathbf{m}+g}^+ a_{\mathbf{m}} + \sum_{\mathbf{q}}{}^{\!\!\prime} \hbar\omega_{\mathbf{q}}(b_{\mathbf{q}}^+ b_{\mathbf{q}} + 1/2)\cdots$$
$$- \sum_{\mathbf{m}} a_{\mathbf{m}}^+ a_{\mathbf{m}} \sum_{\mathbf{q}} \hbar\omega_{\mathbf{q}}[U_{\mathbf{m}}^*(\mathbf{q})b_{\mathbf{q}} + U_{\mathbf{m}}(\mathbf{q})b_{\mathbf{q}}^+].\tag{23}$$

Here $\mathbf{m}$ is the vector index of the lattice site, and the operators $a_{\mathbf{m}}^+$ and $a_{\mathbf{m}}$ create (annihilate) an electron at site $\mathbf{m}$. The dimensionless quantity $U_{\mathbf{m}}(\mathbf{q})$ characterizes a displacement which results from the polarization of the lattice by an electron on site $\mathbf{m}$:

$$U_{\mathbf{m}}(\mathbf{q}) = -\gamma_{\mathbf{q}}(1/2N)^{1/2}e^{-\iota \mathbf{q}\mathbf{R_m}}.\tag{24}$$

where $\mathbf{R_m}$ is the position vector of site $\mathbf{m}$ in the lattice. The quantity $J(\mathbf{g})$ is the overlap integral between sites $\mathbf{m}+\mathbf{g}$ and $\mathbf{m}$. We restrict the discussion below to the case in which $\mathbf{g}$ corresponds to nearest neighbours. Going back to the $\mathbf{k}-$ representation by means of the substitution

$$a_{\mathbf{m}} = (\frac{1}{N})^{1/2}\sum_{\mathbf{k}} a_{\mathbf{k}}e^{-i\mathbf{k}\mathbf{R_m}}.\tag{25}$$

and summing over $\mathbf{m}$, we obtain (21). In other words, expressions (21) and (23) are equivalent. Here we have $\epsilon(\mathbf{k}) = \sum_{g} J(\boldsymbol{g})e^{i\mathbf{k}\boldsymbol{g}}$. An exact canonical transformation can be found more easily for (23), since the largest term in it (the last term) is diagonal in the electron operators. We carry out this transformation by means of the operator $exp(-S)$, where

$$S = \sum_{\mathbf{m}} S_{\mathbf{m}} a_{\mathbf{m}}^+ a_{\mathbf{m}}$$

$$S_{\mathbf{m}} = \sum_{\mathbf{m}} [b_{\mathbf{q}}^+ U_{\mathbf{m}}(\boldsymbol{q}) - b_{\mathbf{q}} U_{\mathbf{m}}^*(\boldsymbol{q})]. \tag{26}$$

The renormalized Hamiltonian $\widetilde{H} = e^{-S} H e^{S}$ takes the form

$$\widetilde{H} = -E_b \sum_{\mathbf{m}} a_{\mathbf{m}}^+ a_{\mathbf{m}} + \sum_{\mathbf{q}} \hbar\omega_{\mathbf{q}}(b_{\mathbf{q}}^+ b_{\mathbf{q}} + \frac{1}{2}) + \sum_{\mathbf{m},\boldsymbol{g}} J(\mathbf{g}) a_{\mathbf{m}+\boldsymbol{g}}^+ a_{\mathbf{m}} \Phi_{\mathbf{m},\boldsymbol{g}} + \Delta H. \tag{27}$$

Here E_b is the polaron binding energy, given by

$$E_b = \frac{1}{2N} \sum_{\mathbf{q}} \hbar\omega_{\mathbf{q}} |\gamma_{\mathbf{q}}|^2. \tag{28}$$

The operator ΔH describes an effective attraction between charge carriers due to a virtual exchange of phonons and is given by

$$\Delta H = -\frac{1}{2} \sum_{\mathbf{m}_1 \neq \mathbf{m}_2} a_{\mathbf{m}_1}^+ a_{\mathbf{m}_1} a_{\mathbf{m}_2}^+ a_{\mathbf{m}_2} \sum_{\mathbf{q}} \hbar\omega_{\mathbf{q}} |\gamma_{\mathbf{q}}|^2 \times \cos[\mathbf{q} \cdot (\mathbf{R}_{\mathbf{m}_1} - \mathbf{R}_{\mathbf{m}_2})]. \tag{29}$$

If conditions do not favor the formation of bipolarons (Part II), this term can be omitted at low carrier concentrations. The third term in (27) describes a multiphonon interaction. Due to this interaction a charge carrier (a polaron) is displaced from site $\mathbf{m}$ to site $\mathbf{m} + \boldsymbol{g}$ by means of the operator

$$\Phi_{\mathbf{m},\boldsymbol{g}} = \exp(S_{\mathbf{m}+\boldsymbol{g}} - S_{\mathbf{m}}). \tag{30}$$

Expression (30) has parts which are diagonal in the phonon operators, i.e. parts which depend only on $N_{\mathbf{q}} = b_{\mathbf{q}}^+ b_{\mathbf{q}}$, and also some nondiagonal parts. Adding a diagonal contribution $< \Phi_{\boldsymbol{g}} >$, averaged over the statistical base, to (30), and also subtracting this contribution $\{i.e. N_{\mathbf{q}} \to \langle N_{\mathbf{q}} \rangle = [exp(\hbar\omega_{\mathbf{q}}/kT) - 1]\}$ ([17, 18]) we find

$$\langle \Phi_{\mathbf{m},\boldsymbol{g}} \rangle = e^{-S_T(\boldsymbol{g})}$$

$$S_T \langle \mathbf{g} \rangle = \frac{1}{2N} \sum_{\mathbf{q}} |\gamma_{\mathbf{q}}|^2 [1 - \cos(\mathbf{q} \cdot \mathbf{g})] \coth \frac{\hbar\omega_{\mathbf{q}}}{2kT}. \tag{31}$$

Again we use the $\mathbf{k}-$ representation. As a result, we find

$$\widetilde{H} = \widetilde{H}_0 + H_{int}, \tag{32}$$

where

$$\widetilde{H}_0 = \sum_{\mathbf{k}} a_{\mathbf{k}}^+ a_{\mathbf{k}} \epsilon_p(\mathbf{k}) + \sum_{\mathbf{q}} \hbar\omega_{\mathbf{q}}(b_{\mathbf{q}}^+ b_{\mathbf{q}} + \frac{1}{2}), \tag{33}$$

and $\epsilon_p(\mathbf{k})$ is the renormalized energy of the polaron band, given by

$$\epsilon_p(\mathbf{k}) = -E_b + \sum_{g} J(\mathbf{g})\mathrm{e}^{-i\mathbf{k}\cdot\mathbf{g}} \cdot \mathrm{e}^{-S_T(\mathbf{g})}. \tag{34}$$

For a cubic crystal, for example, we have $\epsilon_p(\mathbf{k}) = -E_p + \epsilon(\mathbf{k})e^{-S_T}$,where $\epsilon(\mathbf{k}) = 2J[\cos ak_x + \cos ak_y + \cos ak_z]$ is the energy of the electron. The width of the initial (electron) band is $\Delta J = 12J$, and the width of the polaron band is $\Delta E_p = \Delta E e^{-S_T}$. In other words, the latter band is narrower than the initial one by a factor of $e^{-S_T} \ll 1$. With increasing T, there is an increase in S_T [see (31)], and the polaron band becomes narrower. The polaron state is more favorable if the polaron band lies below the lower edge of the electron band, i.e., if

$$E_p > \frac{\Delta E}{2} \to 6J. \tag{35}$$

This is the basic condition for the formation of small polarons. A necessary condition for the formation of small polarons is thus that the initial bands are narrow, i.e. that the quantities J have intermediate values and that the coupling constant is large (see (28)). The interaction Hamiltonian H_{int} in (32) is

$$H_{int} = \sum_{\mathbf{k},\mathbf{k}'} a_{\mathbf{k}'}^+ a_{\mathbf{k}} \frac{1}{N} \sum_{\mathbf{m},\mathbf{g}} J(\mathbf{g})(\varPhi_{\mathbf{m},\mathbf{g}} - \langle\varPhi_{\mathbf{m},\mathbf{g}}\rangle)\mathrm{e}^{i(\mathbf{k}-\mathbf{k}')\cdot\mathbf{R_m}-i\mathbf{k}'\cdot\mathbf{g}} \tag{36}$$

The main goal has thus been reached: the polaron has been separated, and the operator H_{int}, which describes the interaction of the polaron with lattice vibrations, has been found. Bryksin[22] showed that exactly the same canonical transformation can be carried out for an initial Hamiltonian more complex than (23). For example, in place of the constant $J(\mathbf{g})$ one could introduce in (23) an overlap integral $J_{\mathbf{mm}'}$, which depends exponentially on the nonequilibrium atomic displacements $U_{\mathbf{m}}$

$$J_{\mathbf{mm}'} = J_{\mathbf{mm}'}^{(0)}\exp(-\frac{1}{\rho}|\mathbf{R_m} + \mathbf{U_m} - \mathbf{R_{m'}} - \mathbf{U_{m'}}|) \tag{37}$$

Here ρ is the radius of a state localized at a site (i.e. of the Wannier function) satisfying $\rho < a$, where a is the lattice constant. Assuming $|\mathbf{U_m} - \mathbf{U_{m'}}| \ll |\mathbf{R_m} - \mathbf{R_{m'}}|$, restricting the analysis to nearest neighbours ($\mathbf{m}' = \mathbf{m} + \mathbf{g}$), and expanding $\mathbf{U_m}$ in series in the phonon displacements, we find

$$\begin{aligned} J_{\mathbf{mm}'} &= J(\mathbf{g})\exp(-\frac{\mathbf{g}(\mathbf{U_m} - \mathbf{U_{m'}})}{\rho a}) = \\ &\quad J(\mathbf{g})\exp(-\sum_{\mathbf{q}}(V_{\mathbf{mm}'}^\star b_{\mathbf{q}}^+ + V_{\mathbf{mm}'} b_{\mathbf{q}})), \end{aligned} \tag{38}$$

where

$$V_{\mathbf{mm'}}(\mathbf{q}) = (\frac{1}{2N})^{1/2}\delta(\mathbf{g},\mathbf{q})[e^{-i\mathbf{qR_m}} - e^{-i\mathbf{qR_{m'}}}],$$

$$\delta(\mathbf{g},\mathbf{q}) = -(\frac{\hbar}{M\omega_{\mathbf{q}}})^{1/2}(\mathbf{e_q}\cdot\mathbf{g})\frac{1}{a\rho}, \tag{39}$$

Here M is the mass of a lattice atom, and $\mathbf{e_q}$ is the eigenvector of a longitudinal (optical) phonon. Replacing $J(\mathbf{g})$ in (23) by (38), we thus obtain a highly nonlinear dependence on the operators b and b^+ in the initial Hamiltonian. Carrying out the canonical transformation in (26), we can put $\widetilde{H}$ in the form of (27), but with the substitution

$$\Phi_{\mathbf{m},g} \to \Psi_{\mathbf{m},g} = \exp(\sum_{\mathbf{q}}[\Gamma^*_{\mathbf{mm'}}(\mathbf{q})b^+_{\mathbf{q}} + \Gamma_{\mathbf{m'm}}(\mathbf{q})b_{\mathbf{q}}]), \tag{40}$$

where $\Gamma_{\mathbf{mm'}} = -U_{\mathbf{m'}}(\boldsymbol{q}) + U_{\mathbf{m}}(\boldsymbol{q}) - V_{\mathbf{m'm}}(\boldsymbol{q})$. Since the operator structure of the multiphonon operator $\Psi_{\mathbf{m},g}$ is the same as for $\Phi_{\mathbf{m},g}$, all the calculations are carried out in precisely the same way. In expression (31) for $S_T(\mathbf{g})$, for example, we need to make the substitution $[\gamma_{\mathbf{q}}]^2 \mapsto [\gamma_{\mathbf{q}}]^2 - \delta^2(\boldsymbol{g},\mathbf{q})$. In other words, the temperature–dependent part of the renormalization of the polaron band consists of two contributions, which differ in sign. The positive contribution to $S_T(\boldsymbol{g})$ stems from the mass enhancement of the particle due to polaron effects, while the negative contribution is due to an increase in the probability for tunneling from site to site with increasing amplitude of the phonon vibrations. At high temperatures $(kT > \hbar\omega_0)$, it is given by $(\frac{1}{2\rho^2})\langle U^2_{\mathbf{mm'}}\rangle$ where $\langle U^2_{\mathbf{mm'}}\rangle$ is the mean square displacement of the atoms, which is of the order of $kT/M\omega_0^2\rho^2$ and can reach values of the order of unity. When the coupling constants are large, however, the overall sign of S_T does not change. The consequences of substitution (40) in the expression for the mobility will be discussed below.

3 Conditional-Probability Function Description of Polaron Motion

We can now use the Kubo formula (20), which is invariant under canonical transformation (26) carrying out an expansion in powers of H_{int}, in (36). In other words, we can carry out an expansion in powers of the small parameter $J(\mathbf{g})$, rather than in powers of the large coupling constant γ [21, 23, 24]. Since it has the dimensionality of energy, the actual dimensionless parameters of the theory are expressed as ratios of J to $E_a, kT, \hbar\omega_0$, etc. All these questions were discussed in detail in [18–20, 25, 26] by Lang and Firsov. They developed a fundamentally new graphical technique for multiphonon operators H_{int} [18, 23] and presented rigorous derivations of the expressions for μ_h and μ_t, the hopping and tunneling components of the mobility, respectively. These calculations were carried out in the $\mathbf{k}-$ representation, i.e. using expressions (32), (33), and (36) for the Hamiltonian $\widetilde{H}$.

Below we describe a different approach, which we regard as more general and more graphic. It was developed in some later papers by Firsov and Kudinov [23, 24]. This approach makes it possible to reproduce the results of [18] and [20] and furthermore to derive many new results. The lattice–site representation was used in [23] and [24], i.e. the Hamiltonian was used in the form (27) (the term ΔH was omitted). The lattice–site representation is effective at high ($kT > \hbar\omega_0$) and intermediate temperatures, where the width of the polaron band is small,

$$\Delta E_p < kT, \Delta E_p < \hbar/\tau. \tag{41}$$

The second condition in (41) rules out the possibility of a band–like transport describable by the Boltzmann equation (see the Introduction). In this case, the expression for the mobility found from the Kubo formula after some appropriate simplifications becomes [23]

$$\mu = \frac{e}{2kT} \lim_{t\to\infty} \frac{1}{t} \sum_{\mathbf{m}} X_{\mathbf{m}}^2 P(\mathbf{m}, t) = \frac{e}{2kT} \lim_{s\to o} s^2 \sum_{\mathbf{m}} X_{\mathbf{m}}^2 P(\mathbf{m}, s). \tag{42}$$

Here $P(\mathbf{m}, s)$ and $P(\mathbf{m}, t))$ are shorthand for the diagonal components $P_{\mathbf{m}0}^{\mathbf{m}0}(s)$ and $P_{\mathbf{m}0}^{\mathbf{m}0}(t)$ of the conditional–probability functions:

$$P_{\mathbf{m}_2\mathbf{m}'}^{\mathbf{m}_1\mathbf{m}'}(t) = \frac{1}{\mathrm{Tr}(e^{-\beta H_0})} \mathrm{Tr}\{e^{-\beta H_0} \langle 0|a_{\mathbf{m}_2}(a_{\mathbf{m}'}^{+} a_{\mathbf{m}'})_t a_{\mathbf{m}_1}^{+}|0\rangle\} \tag{43}$$

where H_0 is the part of Hamiltonian (27) which is diagonal in the operators a and b, and the operator $(a_{\mathbf{m}'}^{+} a_{\mathbf{m}'})_t$ is the position operator of a carrier at site $\mathbf{m}'$ at time t in the Heisenberg picture [with operator (27)],

$$P_{\mathbf{m}_2\mathbf{m}'}^{\mathbf{m}_1\mathbf{m}'}(s) = \int_0^{\infty} e^{-st} P_{\mathbf{m}_2\mathbf{m}'}^{\mathbf{m}_1\mathbf{m}'}(t)dt. \tag{44}$$

In other words, $P(s)$ is the Laplace transform of $P(t)$. For the quantities $P(s)$ we have the normalization condition

$$\sum_{\mathbf{m}'} P_{\mathbf{m}_2\mathbf{m}'}^{\mathbf{m}_1\mathbf{m}'}(s) = \frac{1}{s}\delta_{\mathbf{m}_1\mathbf{m}_2}. \tag{45}$$

The diagonal components $P_{\mathbf{mm}'}^{\mathbf{mm}'}(t)$ are probability functions which give the probability for finding a polaron at site $\mathbf{m}'$ at time t, if we reliably know that this polaron was at site $\mathbf{m}$ at the time $t = 0$. These functions have the standard meaning in the theory of stochastic processes. The nondiagonal components $P_{\mathbf{m}_2\mathbf{m}'}^{\mathbf{m}_1\mathbf{m}'}(t)$ describe not only a displacement of the center of gravity of the wave packet but also the extending in size ("melting away") and dephasing of the wave packet due to tunneling effects which couple sites $\mathbf{m_1}$ and $\mathbf{m_2}$

($\mathbf{m_1} \neq \mathbf{m_2}$) in the final state (entanglement) [23]. The diagonal and nondiagonal components of the conditional–probability functions obey the system of equations

$$
\begin{aligned}
sP_{\mathbf{mm'}}^{\mathbf{mm'}}(s) =&\, \delta_{\mathbf{mm'}} + \sum_{\mathbf{m_1} \neq \mathbf{m}} W_{\mathbf{mm_1}}^{\mathbf{mm_1}} [P_{\mathbf{m_1 m'}}^{\mathbf{m_1 m'}}(s) - P_{\mathbf{mm'}}^{\mathbf{mm'}}(s)] + \cdots \\
&+ \sum_{\mathbf{m_1} \neq \mathbf{m_2}} W_{\mathbf{mm_2}}^{\mathbf{mm_1}} P_{\mathbf{m_2 m'}}^{\mathbf{m_1 m'}}(s), \\
sP_{\mathbf{m_2 m'}}^{\mathbf{m_1 m'}}(s) =&\, W_{\mathbf{m_2 m_2}}^{\mathbf{m_1 m_1}} P_{\mathbf{m_2 m'}}^{\mathbf{m_1 m'}}(s) + \sum_{\mathbf{m_3}} W_{\mathbf{m_2 m_3}}^{\mathbf{m_1 m_3}} P_{\mathbf{m_3 m'}}^{\mathbf{m_3 m'}} + \cdots \\
&+ \sideset{}{'}\sum_{\mathbf{m_3}} W_{\mathbf{m_2 m_4}}^{\mathbf{m_1 m_3}} P_{\mathbf{m_4 m'}}^{\mathbf{m_3 m'}}(s).
\end{aligned}
\tag{46}
$$

The prime on the summation symbol in (46) means $\mathbf{m_3} \neq \mathbf{m_4}$ and, simultaneously, $\mathbf{m_1} \neq \mathbf{m_3}$ or $\mathbf{m_2} \neq \mathbf{m_4}$. The probabilities W depend on the form of H_{int} and are found by a graphical technique. They describe various processes in configuration space; $W_{\mathbf{mm_2}}^{\mathbf{mm_1}}$ describes hops from site $\mathbf{m_1}$ to site $\mathbf{m}$ and leads to a hopping component of the mobility while $W_{\mathbf{m_2 m_2}}^{\mathbf{m_1 m_1}}$ and the corresponding "relaxation times"

$$
\tau(\mathbf{m_1} - \mathbf{m_2}) = (W_{\mathbf{m_2 m_2}}^{\mathbf{m_1 m_1}})^{-1}
\tag{47}
$$

describe a certain process by which equilibrium in lattice–site space is approached [21, 23]. Accordingly, these two diagonal components of W are generally not equal. Working from hypothesis (15), Holstein suggested that they are equal, and it was that step which caused him to "lose" the tunneling component of the mobility.

The various nondiagonal components of W describe the probability for a displacement of a wave packet and its "melting away". The motion of a quantum object (wave packet) along lattice sites is thus more complex than a classical random walk. None of the arguments above are restricted to the case of the small-polaron theory. Using Hamiltonian H in the form (27), including the first two terms in the definition of H_0, omitting the fourth term, and taking the third term in (27) as H_{int}, we can carry out specific calculations of W for small polarons using the graphical technique developed by Lang and Firsov in [18–20] to determine the products of the multiphonon operators (30). It turns out that for intermediate and high temperatures the nondiagonal components of W are always smaller than the diagonal ones. In a zero approximation we can thus omit the last term from the first equation in (46). As a result we find a closed system of equations for the diagonal $P(s)$ exclusively. This system is of classical form and can be solved easily,

$$
sP(\mathbf{m}, s) = \delta_{\mathbf{m}0} + \sum_{\mathbf{G} \neq \mathbf{0}} W(\mathbf{G}, s)[P(\mathbf{m} - \mathbf{G}, s) - P(\mathbf{m}, s)],
\tag{48}
$$

where $\mathbf{G}$ is a vector in the direct lattice (which does not necessarily link nearest neighbours). Although all W remain nonzero in the limit $s \to 0$, they do depend on s, so the time evolution of P is described by

$$\frac{\partial P\langle \mathbf{m}, t\rangle}{\partial t} = \sum_{\mathbf{G} \neq 0} \int_0^\infty \mathbf{W}(\mathbf{G}, \tau)[P(\mathbf{m}, t) - P(\mathbf{m} - \mathbf{G}, t - \tau)]d\tau \qquad (49)$$

with the initial condition $P(\mathbf{m}, t)|_{t=0} = \delta_{\mathbf{m0}}$. The quantity $P(\mathbf{m}, t)$ thus generally satisfies an equation which is an integrodifferential equation in time. The motion of the small polarons is therefore generally not a Markovian process. In other words, the walk of the small polaron has memory, since $\partial P/\partial t$ at the time t is determined by the values of $P(t')$ at all preceding times, starting at $t' = 0$. It turns out [23] that if it is sufficiently small, and the phonon dispersion $\Delta\omega$ is moderately weak, the quantity $W(\boldsymbol{G}, \tau)$ is nonzero within only a narrow interval

$$0 < \tau < t_0 \simeq \hbar/(E_a kT)^{1/2} \ll \omega_0^{-1} \qquad (50)$$

In this case the process is Markovian (the non Markovian process in the case $\Delta\omega \ll \omega_0$ was analyzed in [27] and [28], but this problem was not completely resolved). In the Markovian case, (44) becomes a differential equation, i.e. (48) is solved in the limit $s \to 0$. From (42) we find

$$\mu = \mu_h = \frac{e}{2kT} \sum_{\mathbf{G}} G_x^2 \mathbf{W}(\mathbf{G}). \qquad (51)$$

This is a typical expression for the hopping component [cf. (12)], but it has contributions from hops not exclusively to the nearest sites. This expression will be analyzed below, while at this point we wish to derive another component of the mobility, the so-called "tunneling" component. Equations (46), which are the linear system of equations, can be used to express the nondiagonal components of P in terms of the diagonal ones:

$$P_{\mathbf{m}_2 \mathbf{m}'}^{\mathbf{m}_1 \mathbf{m}'} = \sum_{\mathbf{m}_3} L_{\mathbf{m}_2 \mathbf{m}_3}^{\mathbf{m}_1 \mathbf{m}_3} P_{\mathbf{m}_3 \mathbf{m}'}^{\mathbf{m}_3 \mathbf{m}'}. \qquad (52)$$

Here $L_{\mathbf{m}_2 \mathbf{m}_3}^{\mathbf{m}_1 \mathbf{m}_3}$ are certain matrices. Substituting (52) into the last term of the first equation in (46) we find a system of linear equations for the diagonal quantities P. This system of equations has the same form as (48), but the probabilities are renormalized:

$$\overline{W}(\mathbf{m} - \mathbf{m}_1) = W_{\mathbf{mm}_1}^{\mathbf{mm}_1} + \sum_{\mathbf{m}_2 \neq \mathbf{m}_3} W_{\mathbf{mm}_3}^{\mathbf{mm}_2} L_{\mathbf{m}_3 \mathbf{m}_1}^{\mathbf{m}_2 \mathbf{m}_1}. \qquad (53)$$

We thus find an expression for μ like (51) with $W \to \overline{W}$. Although it is not possible to derive a general expression for the matrix L, we can write this matrix as a power series in the ratios of the nondiagonal components of W to the diagonal ones:

$$L_{\mathbf{m_3 m_1}}^{\mathbf{m_2 m_1}} = \tau(\mathbf{m_2 - m_3}) W_{\mathbf{m_3 m_1}}^{\mathbf{m_2 m_1}} + \tau(\mathbf{m_2 - m_3}) \sum_{\mathbf{m_4, m_5}}' W_{\mathbf{m_3 m_5}}^{\mathbf{m_2 m_4}} \tau(\mathbf{m_4 - m_5}) W_{\mathbf{m_5 m_1}}^{\mathbf{m_4 m_1}}.$$

$$(54)$$

The prime on the summation symbol here means $\mathbf{m_4} \neq \mathbf{m_5}$ and, simultaneously, $\mathbf{m_4} \neq \mathbf{m_2}$ or $\mathbf{m_5} \neq \mathbf{m_3}$. Here $\tau(\mathbf{m_2} - \mathbf{m_3}) = (\mathbf{W}_{\mathbf{m_3 m_3}}^{\mathbf{m_2 m_2}})^{-1}$ is the reciprocal probability which we call the "relaxation time." It characterizes the time evolution of the nondiagonal components of the conditional–probability function P with allowance for the friction due to multiphonon interaction processes ([21], Chap. 1, Part 2, §7). All probabilities are calculated by a modified graphical technique (see the mathematical appendices in [21]) and are written as power series in the dimensionless parameters $\eta_1 = J/E_a, \eta_3 = J/kT, \eta_2 = J^2/\hbar\omega_0(E_a kT)^{1/2}$ (see Sect. 5). In the lowest approximation in J the nondiagonal probabilities are proportional to $\Delta E_p/\hbar z$, i.e. expression (54) is a power series in the parameter $(\Delta E_p/\hbar z)\tau$. If this parameter is small, i.e., if the uncertainty in the energy, h$/\tau$, is larger than the width of the polaron band [this case corresponds to condition (41)], then we can truncate series (54), retaining only its first term. As a result, we find

$$\mu = \mu_h + \mu_t = \frac{e}{2kT} \sum_{\mathbf{m}} X_{\mathbf{m}}^2 W(\mathbf{m}) +$$
$$+ \frac{e}{2kT} \sum_{\mathbf{m, m_1} \neq \mathbf{m_2}} X_{\mathbf{m}}^2 W_{\mathbf{mm_2}}^{\mathbf{mm_1}} W_{\mathbf{m_2 0}}^{\mathbf{m_1 0}} \tau(\mathbf{m_1 - m_2}).$$

$$(55)$$

Here the first term corresponds to the hopping component, and the second one is the tunneling component. We are using the word "tunneling" in order to distinguish this mechanism from the band-like transport mechanism which operates under the opposite condition $\Delta E_p/z > kT$. This term reflects the physical essence of the process better ([21], Chap. 1, Part 2, §7). If J is not too large, the following chain of inequalities holds for small polarons at intermediate temperatures:

$$\tau^{-1}(\mathbf{m_1 - m_2}) = |W_{\mathbf{m_2 m_2}}^{\mathbf{m_1 m_1}}| > W_{\mathbf{m_1 m_2}}^{\mathbf{m_1 m_2}} = W_h(\mathbf{m_1 - m_2}) >> W_{\mathbf{m_2 m_4}}^{\mathbf{m_1 m_3}}. \quad (56)$$

The subscript h is used here in order to distinguish the diagonal probability $W_{\mathbf{m_1 m_2}}^{\mathbf{m_1 m_2}}$, which describes hops from site to site, from the other diagonal probability $W^{\mathbf{m_1 m_1}}$, which corresponds to the reciprocal relaxation time for nondiagonal components of the conditional probability function. We recall that Holstein did not distinguish between τ^{-1} and W_h. The temperature T_3, at which $\mu_h = \mu_t$, is found from the condition

$$W(\mathbf{m}) = \sum_{\mathbf{m_1} \neq \mathbf{m_2}} W_{\mathbf{mm_2}}^{\mathbf{mm_1}} W_{\mathbf{m_2 0}}^{\mathbf{m_1 0}} \tau(\mathbf{m_1 - m_2}), \quad (57)$$

which takes the following form in the limit of small values of J:

$$\left|\frac{2J\mathrm{e}^{-S_T}}{\hbar}\right|^2 = W_h \tau^{-1}. \tag{58}$$

By virtue of the condition $\tau^{-1} > W_h$, this temperature is higher than that found by Holstein using the condition (19). Thus there is a temperature interval $T_2 < T < T_3$ (here T_2 is the temperature below which the condition $\Delta E_p/z < \hbar/\tau$ holds) in which the mobility is described by the second term in (55) and it is a consequence of the tunneling transport mechanism. At $T < T_2$, there is a band-like transport in the narrow band. At $T > T_3$ hopping transport is predominant. The program formulated in step 1 in the Introduction has thus been executed. Other results pertinent to step 1 of the Introduction are described in Sect. 4.

4 Static Small Polaron Conductivity in the Small J Limit

At small values of J the summation over $\mathbf{m}$ in (55) can be restricted to nearest neighbours ($\mathbf{m} = \mathbf{g}$). At $T > T_0$, where

$$kT_0 = \frac{\hbar\omega_0}{2\,\mathrm{arc\,sinh}(2\gamma)}, \tag{59}$$

the expression for $W_h(\mathbf{g})$ takes the form of (15) with the following argument of the exponential function

$$\frac{\widetilde{E_a}(T)}{kT} = \Omega \int \frac{d^3q}{(2\pi)^3} |\gamma_\mathbf{q}|^2 [1 - \cos(\mathbf{q}\cdot\mathbf{g})] \tanh\frac{\hbar\omega_\mathbf{q}}{4kT} = \frac{4E_a'}{\hbar\omega_0} \tanh\frac{\hbar\omega_0}{4kT}. \tag{60}$$

Here and below, the integration over $\mathbf{q}$ is carried out within the first Brillouin zone. The expression on the right in (60) is derived by ignoring the dispersion of optical phonons. This simplification is legitimate under the condition [12]

$$2\pi \frac{\Delta\omega}{\omega_0}\left(\frac{E_a}{\hbar\omega_0}\cosh\frac{\hbar\omega_0}{2kT}\right)^{1/2} > 1 \tag{61}$$

The activation energy E_a is given by

$$E_a = \Omega \int \frac{d^3q}{(2\pi)^3} |\gamma_\mathbf{q}|^2 [1 - \cos(\mathbf{q}\cdot\mathbf{g})]\frac{\hbar\omega_\mathrm{B}}{4} \approx \frac{\gamma}{2}\hbar\omega_0, \tag{62}$$

where the coupling constant is

$$\gamma \approx \frac{1}{2}\Omega \int \frac{d^3q}{(2\pi)^3} |\gamma_\mathbf{q}|^2 [1 - \cos(\mathbf{q}\cdot\mathbf{g})]. \tag{63}$$

In the same approximation the expression for S_T is

$$S_T \simeq \gamma \coth\left(\frac{\hbar\omega_0}{2kT}\right). \tag{64}$$

The factor η_2 in (11) should be replaced as follows at $T > T_0$:

$$\eta_2 \to \frac{J^2}{\hbar\omega_0}[1/2\Omega \int \frac{d^3q}{(2\pi)^3} |\gamma_q|^2 \omega_q^2 [1 - \cos(\mathbf{q} \cdot \mathbf{g})] \cosh\frac{\hbar\omega_q}{2kT}]^{-1/2}$$

$$\simeq 2\frac{J^2}{E_a^{1/2}(\hbar\omega_0)^{3/2}} \sinh^{1/2}(\frac{\hbar\omega_0}{2kT}) \tag{65}$$

At small values of J the expression for the hopping component of the mobility is thus indeed the same as the Holstein expression. However, the expressions for the second component, which Holstein associated with the band component, and which we call the "tunneling" component, are markedly different. The expression for τ_4^{-1} in our case does not contain the exponentially small factor $\exp(-E_a/kT)$ or $\exp(-2S_T)$, although it is proportional to J^4 [19, 20]:

$$\tau_4^{-1} = \eta_1^4 \frac{\omega_0^2}{\Delta\omega}[sinh(\frac{\hbar\omega_0}{2kT})]^{-2}. \tag{66}$$

The temperature dependence in (66) corresponds to two-phonon scattering of small polarons (accompanied by the emission and absorption of a phonon, and vice versa). In the lowest approximation in J, and also in powers of $\exp(-2S_T)$ and $\exp(-E_a/kT)$ we have [19, 20]

$$\mu_t = u_0 \frac{J^2}{\hbar\omega_0 kT} \frac{\Delta\omega}{\omega_0} \eta_1^{-4} \sinh^2(\frac{\hbar\omega_0}{2kT})\exp(-2\gamma \coth\frac{\hbar\omega_0}{2kT}). \tag{67}$$

This expression is valid under the condition $\eta_1 = J/E_a > \gamma^2 e^{-\gamma}$. Under the opposite inequality Holstein's result, derived through the substitution $\tau^{-1} \to \tau_2^{-1} = W_h^{(2)}$, is valid. Condition (57) for determining T_3, equivalent to (18), becomes

$$(\sinh\alpha_3)^{3/2} \exp[-\frac{\sinh\alpha_0}{\sinh\alpha_3}] = \frac{\omega_0}{\Delta\omega}\gamma^{-1/2}\eta_1^4, \tag{68}$$

where $\alpha_0 = \hbar\omega_0/2kT_0$, i.e. $\sinh\alpha_0 = 2\gamma$, and $\alpha_3 = \hbar\omega_0/2kT_3$. Since the right side is less than one, we find $T_3 > T_0$ from (68).

5 Expanding the Range of Applicability of the Theory

Lifting the restrictions on the values of the various dimensionless parameters of the theory implies more than simply changes in the form of the equations, it may also reflect a change in the physical nature of the phenomenon itself.

We begin with the region of high temperatures, $kT > \hbar\omega_0/2$. Here the basic small parameters containing J are

$$\eta_1 = \frac{J}{E_a}, \eta_2 = \frac{J^2}{\hbar\omega_0(E_akT)^{1/2}}, \eta_3 = \frac{J^2}{kTE_a}, \eta_4 = \frac{J}{kT}. \tag{69}$$

Violation of the condition $\eta_1 < 1$ means that we are now dealing with large polarons [12].

Lifting the restriction $\eta_2 < 1$ (at $T > T_0$) means a transition to "adiabatic" hops [12, 26]. A hop of the small polaron from site to site occurs in two steps. At the neighboring site, where there is originally no polaron, an "empty" polaron polarization well (due to a virtual phonon cloud) arises in a fluctuation manner. The probability for this process is proportional to $\exp(-E_a/kT)$. An electron in a neighboring occupied polaron well tunnels into this empty well. The "collapse" of the emptied well is usually not considered. The tunneling transition occurs only under conditions of a symmetric resonance, such that the electron energy levels in the two wells are identical. The tunneling time t_1 is of the order of $\hbar/2J$. The time $\bar{t}$, over which conditions favorable for a symmetric resonance prevail, is equal to $\hbar(\hbar\omega_0)^{-1/2}(E_akT)^{-1/4}$. The time $\bar{t}$ thus characterizes the response time of the phonon system, and t_1 that of the electron system. Under the condition $\bar{t} < t_1$, an electron is clearly unable to tunnel, i.e. the transition probability is small. At $\bar{t} > t_1$, on the contrary, the process has its maximum probability [21].

The parameter η_2 is the square of the ratio of these two times, $\eta_2 = (\bar{t}/t_1)^2$. Lang and Firsov [25, 26] calculated the probabilities $W(g)$ for $kT > \hbar\omega_0/2$ and arbitrary η_2:

$$W(g) = \frac{\omega_0}{2\pi}\exp(-E_a/kT)\int_{x_0}^{\infty}\exp(-x) \tag{70}$$

$$\times 2\{1 - \exp[-(\pi/2)\eta_2(1/x)^{1/2}]\} \times \{2 - \exp[-(\pi/2)\eta_2(1/x)^{1/2}]\}^{-1}dx$$

Here $x = (E - E_a)/kT$ is the dimensionless energy of some "effective" particle, reckoned from the height of the barrier,

$$x_0 = [\frac{E_a}{kT}(\frac{\hbar\omega_0}{kT})^2]^{1/3} \approx (\frac{\hbar}{\bar{t}kT})^{4/3}. \tag{71}$$

If the uncertainty in the energy $(\hbar/\bar{t})$ due to the finite length of the interval $\bar{t}$ (see the discussion above) is smaller than kT, then we have $x_0 < 1$, and the quantity x_0 does not appear in the result. The limiting cases $\eta_2 \ll 1$ and $\eta_2 > 1$ in (70) correspond to (11). The case $x_0 > 1$ was analyzed by Arnold and Holstein [29, 30]. They were unable to construct a series in J, but they asserted that under the condition $x_0 > 1$ the ratio of the contribution $\sim J^4$ to the contribution $\sim J^2$ is equal not to η_2 but to

$$\eta_2 \frac{1}{x^{3/2}} \simeq (\frac{J}{\hbar\omega_0})^2 \frac{kT}{E_a} < 1. \tag{72}$$

It was shown in [22] that incorporating the dependence of the overlap integral on nonequilibrium atomic displacements (see (37)) leads to a change in the argument of the exponential function in expression (10) for $W_h^{(2)}$ in the high–temperature limit $kT > \hbar\omega_0/2$ under the condition $\eta_2 < 1$.

$$-\frac{E_a}{kT} \rightarrow -\frac{E_a}{kT} + \frac{kT}{\epsilon}, \tag{73}$$

where

$$\begin{aligned}
\epsilon^{-1} =& 4\Omega \int \frac{d^3q}{(2\pi)^3} \frac{1 - \cos(\mathbf{q}\cdot\mathbf{g})}{\hbar\omega_\mathbf{q}} \delta_\mathbf{q}^2 \\
& - \frac{1}{E_a'} \{\Omega \int \frac{d^3q}{(2\pi)^3} [1 - \cos(\mathbf{q}\cdot\mathbf{g})]\delta_\mathbf{q}\gamma_\mathbf{q}\}^2,
\end{aligned} \tag{74}$$

δ_q and γ_q are defined in Sect. 2, and E_a' differs from E_a by the substitution $|\gamma_\mathrm{q}|^2 \rightarrow |\gamma_\mathrm{q}|^2 + |\delta_\mathrm{q}|^2$ in (62).

Expression (74) can be rewritten approximately as

$$\epsilon^{-1} \simeq \frac{1}{M\omega_0^2\rho^2} (\xi_1 - \xi_2 \frac{E_a}{E_a'}) > 0, \tag{75}$$

where ξ_1 and ξ_2 are numbers of the order of unity.

The second term in (73) is thus equal in order of magnitude to the ratio of the mean square displacement of the atoms to the square, ρ^2, of the length over which the overlap integral J decreases (see the end of Sect. 2).

We should also make the substitution $E_a \rightarrow E_a'$ in the pre-exponential factor, i.e. in the expression for η_2. If $\eta_1 < 1$ and $\eta_3 = J^2/E_a kT < 1$ but $\eta_2 > 1$ and $J/kT > 1$, we should make the substitution $E_a \rightarrow E_a - J$ in expression (10) (see (73)). In other words, the hopping mobility increases.

Let us look at the contribution to μ_h from hops to more distant sites, with a vector $\boldsymbol{G}$ $(G > g)$. If the quantity $|\gamma_\mathbf{q}^2| \hbar\omega_\mathbf{q}$ falls smoothly with increasing $\mathbf{q}$, then $E_a(\boldsymbol{G})$ increases with increasing $\boldsymbol{G}$:

$$E_a(\mathbf{G}) = \frac{1}{4}\Omega \int \frac{d^3q}{(2\pi)^3} |\gamma_\mathrm{q}|^2 \hbar\omega_\mathrm{q} [1 - \cos(\mathbf{q}\cdot\mathbf{G})]. \tag{76}$$

From (76) we see that $\lim_{G\rightarrow 0} E_a(\boldsymbol{G}) \rightarrow E_b/2 > E_a(\boldsymbol{g})$. For a hop to a site at a distance $\mathbf{G} = \mathbf{i}n_1 g_1 + \mathbf{j}n_2 g_2 + \mathbf{k}n_3 g_3$, the expansion of the function $F(\mathbf{G})$ in front of the activation exponent (for hops between nearest neighbors it is denoted as $f(\eta_2)$) begins with the term $\eta_3^{n_1+n_2+n_3-1}$. Correspondingly, the contribution $\sim J^{2n}$ to the probability for a hop over a large distance contains $\eta_3^m \eta_1^{2p} \ll \eta_2^n (m = n_1 + n_2 + n_3 - 1)$ instead of η_2^{n-m-p}. At $\eta_3 \ll 1$ the probability for a hop to more distant sites falls off because of the decrease in $F(\boldsymbol{G})$ and the increase in $E_a(\boldsymbol{G})$. However, it can be seen from (51) that G_x^2 increases with increasing G for the corresponding partial contributions to the mobility, and the number of possible final states for the hop increases.

Accordingly, even at $\eta_2 \lesssim 1$ the temperature dependence of the hopping component of the mobility can be quite complex, and the component μ_h itself can be far larger than in the case in which hops between nearest neighbors exclusively are taken into account. The case $\eta_3 > 1$ has not been studied at all for the hopping mechanism.

To estimate the tunneling component of the mobility, μ_t, at intermediate values $\eta_3 \lesssim 1$, we need to examine the behavior of the "nondiagonal" probabilities $W_{\mathbf{mm}_2}^{\mathbf{mm}_1}$ and of the relaxation times $\tau(\boldsymbol{m}_1 - \boldsymbol{m}_2)$ as a function of the parameter η_3.

It was shown in [21, 23] that at $T > T_0$ the contribution $\sim J^{2n+1}$ is

$$W_{\mathbf{mm}_2}^{\mathbf{mm}_1} = J/\hbar\eta_3^n e^{-a_n S_T} e^{-\gamma_n \cdot (E_a/kT)}. \tag{77}$$

In the lowest order, $n = 0$ we have $a_0 = 1$ and $\gamma_0 = 0$. With increasing n, the numbers a_n fall off, while γ_ns increase, but they do not exceed values of the order of unity. In other words, the exponential small factor weakens, and as η_3 increases the contributions with a larger index n become progressively larger. The tunneling component μ_t may become predominant even at fairly high temperatures. Bryksin and Firsov [31] analyzed the behavior of $[\tau(\boldsymbol{G})]^{-1}$ for arbitrary η_3. The terms of the series (in powers of J) are separated into two groups: the terms of the first group (of type τ_4^{-1}) do not contain the activation factor, while those of the second group (type τ_2^{-1}) are exponentially small ($\sim e^{-2\gamma}$). The basic parameters of the expansion sum out are η_1, γ^{-1} and $\xi = J^2/E_a\hbar\Delta\omega$.

It was found to be possible to sum the first set of terms for arbitrary η_3 and ξ. As before, it is proportional to η^4 (like τ_4^{-1}), but it has an additional small factor on the order of $e^{-a\xi}$, where $a \geq 1$. In other words, the values of T_3 move to higher temperatures. For the set of terms of the type τ_4^{-1}, to make a larger contribution than the set of the type τ_2^{-1}, the condition $a\xi < 4E_a/\hbar\omega_0$ must hold. This condition is equivalent to the condition that there are no local vibrations in the phonon spectrum. Such vibrations may arise because a charge carrier spends a long time (in comparison with ω_0^{-1}) at a lattice site [21].

We recall that the parameter η_3 is equal to the square of the ratio of two times, $\eta_3 = t_0^2/t_1^2$, the hopping time $t_0 = \hbar(E_a kT)^{-1/2}$ and the time $t_1 \simeq \hbar/J$ characterising the rate at which the wave packet spreads out for a "bare" electron at a lattice site. In the case $\eta_3 > 1$, an electron wave packet is able to spread out over a large number of sites before it self–localizes again. An electron escapes from a polaron well and "runs over" many unit cells before it localizes in another polarization well. The term "relay–race" has been proposed [21] for this mechanism. Unfortunately, specific expressions for μ_h and μ_t have not been derived for the case $\eta_3 > 1$ (it is difficult to sum power series in the parameter η_3).

At low temperatures, where the mobility is dominated by tunneling, the results are even scantier. There has been a more comprehensive study of the behavior of $\tau^{-1}(G)$ as a function of η_1, γ^{-1} and $\xi = J^2/E_a\hbar\Delta\omega$, see [19–21, 31]. There has been essentially no study of the nondiagonal probabilities $W_{\mathbf{mm}_2}^{\mathbf{mm}_1}$ at $T < T_0$. Lang and Firsov [19, 20] studied the structure of the series for "vertices", which is essentially the same thing under the condition $\frac{\Delta E_p}{z} < kT$. The basic dimensionless parameters at $T < T_0$ are [19, 20]

$$\eta_1 = \frac{J}{E_a}, \eta_2' = \frac{J^2}{2E_a\hbar\omega_0}, \eta_2'' = \eta_2' \ln\gamma, \xi = \frac{J^2}{E_a\Delta\omega\hbar}. \tag{78}$$

The structure of the series (in powers of these parameters) for vertices, non-diagonal probabilities, and the width of the polaron band, ΔE_p, should be similar. Holstein[12] calculated the half-splitting of two polaron levels (δE) in the two–site–cluster model under the condition $\eta_1 \ll 1$. To some extent, this quantity characterizes ΔE_p. Holstein's results can be written as

$$\delta E = Je^{-\gamma}(\pi\eta_2')^{1/2} \exp[\eta_2' \ln(\frac{1}{\eta_1} 4e^{1/2})]. \tag{79}$$

The first two factors in (79) correspond to the results for ΔE_ρ in the limit of small J. The last two factors describe a renormalization (a broadening of the band) in the case $\eta_2' > 1$. The argument of the exponential function in (79) can be written as $\eta_2'(1/2+\ln 2)+\eta_2''+\eta_2' \ln(\hbar\omega_0/J)$. In other words, the independent parameter $J/\hbar\omega_0$ may arise in the theory along with the parameters γ, η_1, and η_2'. In the case discussed by Holstein, however, that new parameter does not arise explicitly, and it can be removed in the determination of η_1, η_2' and η_2'' (see (79)).

An even more accurate expression for δE was recently reported [32] but that expression does not include the ratio $J/\hbar\omega_0$ as an independent parameter. Unfortunately, a result analogous to (79) for the quantity W, which appears in the expression for μ_t, has yet to be derived for arbitrary parameter values [see (78)]. If the result of the renormalization is similar to (79), then the quantity μ_t increases, and the transition from a tunneling transport to a band–like transport may set in earlier.

According to [32], the following refinements must be made in expression (79):

1) an additional factor $\beta^{5/2}\lambda^{1-\beta}(1+\beta)^{-\beta}$, where $\beta = (1-\lambda^{-2})^{1/2}$ should be introduced in the pre–exponential expression in (79).

2) An additional factor of $(1/2)(1+\beta)$ must be introduced inside the logarithm in the argument of the exponential function.

Since we have $\lambda^{-1} = \eta_1$ in our notation these corrections reduce to merely a refinement of the dependence of δE on the small parameter η_1 and they do not introduce an explicit dependence on the ratio $J/\hbar\omega_0$. We accordingly restrict the analysis to expression (79).

One sometimes encounters an assertion that the perturbation theory cannot be used even in the limit $J \to 0$. That assertion is backed up by citing expression (79), which gives us the following result in the limit $J \to 0$:

$$\delta E = \hbar \omega_0 (\frac{\gamma}{\pi})^{1/2} e^{-\gamma}.$$

We recall that the adiabatic approximation itself is not valid in the small J limit. It is interesting to determine at which values of J the results of the two approximations join. According to [19, 20] it follows from (78) that the nonadiabatic perturbation theory in powers of J is valid up to the values $\eta_2'' \simeq 1$, i.e., under the conditions $J/\hbar \omega_0 \lesssim \gamma / \ln \gamma > 1$. In the region $\eta_2'' \simeq 1$, the two solutions for δE join. Let us demonstrate this assertion. A joining occurs when the additional factor in (79) becomes comparable to one. We denote by x the value of η_2' at which this occurs:

$$(\frac{1}{\pi x})^{1/2} \exp[\frac{x}{2}(1 + \ln 4\gamma) + \frac{1}{2} x \ln \frac{1}{x}] \approx 1. \tag{80}$$

We thus need to solve the transcendental equation

$$x(1 + \ln 4\gamma) + (1 + x)\ln \frac{1}{x} - \ln \pi = 0. \tag{81}$$

Holstein [12] suggested that x $\simeq$ 1 corresponds to $(J/\hbar \omega_0)^2 \simeq \gamma \gg 1$. In this case the left side of (80) is equal to $(4\gamma/\pi)^{1/2}$, but not unity. The accuracy can be improved by setting $x = \ln \pi (1 + \ln 4\gamma)^{-1}$. The left side of (80) then becomes equal to $(\ln 4\gamma / \ln \pi)^{1/2}$, which is not greatly different from one even at fairly large values of γ. Such a value of x actually corresponds to a value $\eta_2'' = \eta_2' \ln 4\gamma \simeq 1$, in agreement with an assertion by Lang and Firsov [19, 20]. The adiabatic approximation thus becomes valid not under the condition $(J/\hbar \omega_0)^2 \geq 1$ but at larger values of the parameter J, found from the condition

$$(\frac{J}{\hbar \omega_0})^2 > \frac{\gamma}{\ln 4\gamma} > 1. \tag{82}$$

At smaller values of J the first few terms in the perturbation–theory series in J are sufficient.

Solving (81) numerically, we find $x = f(\gamma)$, where $f(\gamma) < 1$ for $\gamma > 1$, but the main assertion, $J/\hbar \omega_0 > 1$, remains in force. We recall that (79) was derived in the two-site model. Alexandrov et al. [32] carried out numerical calculations not only for the two-site cluster model, but also for the four and six-site models. They showed a plot of δE versus γ ($\gamma = g^2$ in the notation of [32]) for various values of $J/\hbar \omega_0$ ($J - t$ in the notation of [32]) for the two-site-cluster model. For the value $J/\hbar \omega_0 \approx 1.1$, the exact results are closer to the perturbation–theory results, while at $J/\hbar \omega_0 \simeq 2$ and with $4 < \gamma < 8$ the results agree better with the adiabatic approximation. These results do not contradict condition (82).

6 General Expressions for the Mobility in "k-R" Representation

As was mentioned in Sect. 2, a description of the motion in lattice–site space is instrumental as long as the condition $(1/z)\Delta E_p < kT$, holds, i.e. as long as the polaron distribution function $n(\boldsymbol{k}) \simeq constant$ over the polaron band. As the temperature is lowered, or J is increased and γ is reduced (but still $\gamma > 1$), the low-temperature region in which that approach is inadequate becomes wider. The equations presented below [23, 24] are valid in this temperature region allowing for a unified description of the band–like tunneling and hopping mechanisms of transport.

We denote as $n(\boldsymbol{k})$ the exact equilibrium one–particle distribution function for small polarons. We choose the normalization (per small polaron)

$$\frac{1}{N}\sum_{\boldsymbol{k}} n(\mathbf{k}) = \Omega \int \frac{d^3\boldsymbol{k}}{(2\pi)^3} n(\mathbf{k}) = 1. \tag{83}$$

According to [21, 24] the exact expression for the mobility is

$$\begin{aligned}
\mu_{xx} &= \frac{e}{2kT} \lim_{t\to\infty} \frac{1}{t} \sum_{\boldsymbol{R}} R_x^2 \frac{1}{N} \sum_{\mathbf{k}} n((\mathbf{k})F(\mathbf{k}, \boldsymbol{R}; t) \\
&= -\frac{e}{kT} \lim_{s\to 0} s^2 \sum_{\mathbf{k}} n(\mathbf{k}) \frac{\partial^2 F(\mathbf{k}, \boldsymbol{\kappa}, s)}{\partial \kappa_x^2}\Big|_{\kappa=0}
\end{aligned} \tag{84}$$

The sums over $\mathbf{k}$ in (84) are defined as in (83), and the integration over $\mathbf{k}$ is carried out within the first Brillouin zone. The function $F(\mathbf{k}, \boldsymbol{R}; t)$ is defined by

$$F(\mathbf{k}, \boldsymbol{R}; t) = \sum_{\Delta} \mathrm{e}^{-2\iota \mathbf{k}\Delta} P_{\mathbf{R}+\Delta,0}^{\mathbf{R}-\Delta,0}(t). \tag{85}$$

The conditional–probability function in lattice–site space, P, is defined by (43). In the case of narrow (renormalized) bands we have

$$\frac{1}{N}\sum_{\mathbf{k}} n(\mathbf{k})e^{-2ik\Delta} = \delta_{\Delta,0} \tag{86}$$

and we return to (42). The quantity $F(\mathbf{k}, \boldsymbol{\kappa}; s)$ is the Fourier component along the coordinate $\mathbf{R}$, and it is also the Laplace (t) transform of the function $F(\mathbf{k}, \boldsymbol{R}; t)$. This Fourier component has much in common with the Wigner one–particle density matrix, [24] and [21] (Chap. 1, Part 2, §10).

The equation for $F(\mathbf{k}, \boldsymbol{\kappa}; s)$ is typical of the KP method [14],

$$\{s + \frac{i}{\hbar}[\epsilon(\mathbf{k} + \frac{\boldsymbol{\kappa}}{2}) - \epsilon(\mathbf{k} - \frac{\boldsymbol{\kappa}}{2})]\}F(\mathbf{k}, \boldsymbol{\kappa}; s) = 1 + \sum_{\mathbf{k'}} W(\mathbf{k}, \mathbf{k'}; \boldsymbol{\kappa})F(\mathbf{k'}, \boldsymbol{\kappa}; s). \tag{87}$$

Here $W(\mathbf{k}, \mathbf{k'}; \boldsymbol{\kappa})$ is the probability for the scattering of a quasiparticle out of state $\mathbf{k}$ into $\mathbf{k'}$, when there is a slight spatial inhomogeneity ($\boldsymbol{\kappa} \neq 0$). These quantities can be determined by a graphical technique. One can avoid the task of solving (87) when there is a slight spatial inhomogeneity ($\kappa \neq 0, \kappa \to 0$).

Kudinov and Firsov [24] showed that one could use (84) and (87) to rewrite σ_{xx} as

$$\sigma_{xx} = \frac{e^2}{2kT}\frac{1}{V}\sum_{\mathbf{k_1},\mathbf{k_2}} n(\mathbf{k_1})W_2(\mathbf{k_1},\mathbf{k_2}) + \frac{e^2}{2kT}lim_{s\to 0}\frac{1}{V}\sum_{\mathbf{k_1},\mathbf{k_2},\mathbf{k_3},\mathbf{k_4}} n(\mathbf{k_1})\times$$

$$\times[\mathrm{v}_x(\mathbf{k_1})\delta_{\mathbf{k_1}\mathbf{k_2}} - iW_1(\mathbf{k_1},\mathbf{k_2})]P(\mathbf{k_2},\mathbf{k_3};s)[\mathrm{v}_x(\mathbf{k_3})\delta_{\mathbf{k_3}\mathbf{k_4}} - iW_1(\mathbf{k_3},\mathbf{k_4})].$$

$$(88)$$

Here

$$W_1(\mathbf{k},\mathbf{k'}) = \{\frac{\partial W(\mathbf{k},\mathbf{k'};\boldsymbol{\kappa})}{\partial\kappa_x}\}_{\boldsymbol{\kappa}=0};$$

$$W_2(\mathrm{k},\mathrm{k'}) = \{\frac{\partial^2 W(\mathbf{k},\mathbf{k'};\boldsymbol{\kappa})}{\partial\kappa_x^2}\}_{\boldsymbol{\kappa}=0}.$$

$$(89)$$

and $\mathrm{v}_x(\mathbf{k}) = 1/\hbar[d\epsilon(\mathbf{k})/dk_x]$, where $\epsilon(\mathbf{k})$ is the exact (renormalized) energy of the quasiparticle. The quantity $P(\mathbf{k},\mathbf{k'};s)$ is a function obtained by taking the Laplace transform of the function $P(\mathbf{k},\mathbf{k'};t)$ which is the conditional probability for finding a charge carrier in state $\mathbf{k'}$ at the time $t > 0$, if this carrier was in state $\mathbf{k}$ at the time $t = 0$, and the lattice was at thermodynamic equilibrium. Its formal definition is similar to (43) with $\boldsymbol{m} \to \mathbf{k}$ and $\boldsymbol{m_1} = \boldsymbol{m_2}$ $\to \mathbf{k'}$. The equation for $P(\mathbf{k},\mathbf{k'};s)$ is (see [24])

$$sP(\mathbf{k},\mathbf{k'};\mathrm{s}) = \delta_{\mathbf{k},\mathbf{k'}}, + \sum_{k_1} W(\mathbf{k},\mathbf{k_1})[P(\mathbf{k_1},\mathbf{k'};\mathrm{s}) - P(\mathbf{k},\mathbf{k'};\mathrm{s})]. \qquad (90)$$

The quantity W in (87) with $\boldsymbol{\kappa} = 0$ (i.e. in the absence of the spatial inhomogeneity) corresponds to the probability for the scattering of quasiparticles from state $\mathbf{k}$ to state $\mathbf{k'}$. It was shown in [24] how existing results can be obtained from (84) and (88) in the case of a weak coupling with phonons and in the limit of a very strong coupling for small polarons. The procedure for going from (84) to (42) for small polarons was explained above. In the weak–coupling case, $\lambda \ll 1$, we can ignore the quantities W_1 and W_2 in (88), since they are proportional to λ^2. It follows from (90) that we have $P(\mathbf{k},\mathbf{k'}) \sim \tau \sim \lambda^{-2}$, so that we obtain the usual expression for σ_{xx}.

In the strong–coupling case the first term in (88) describes the hopping component. For small polarons at $T > T_2$ we have $n(\mathbf{k_1}) \to 1$ i.e. the first term in (88) is proportional to

$$\frac{1}{N}\sum_{\mathbf{k_1},\mathbf{k_2}} W_2(\mathbf{k_1},\mathbf{k_2}) = \sum_{\mathbf{m}} X_{\mathbf{m}}^2 W_h(\mathbf{m}). \qquad (91)$$

It was shown in [24] and [21] (Chap. 1, Part 2. §10) that the second term in (88) describes the tunneling component in the case $\Delta E_p < \hbar/\tau$. The stochastic interpretation of the functions F and P can be found in the same papers. That approach is particularly effective in the description of transport in a strong electric field E and of the Hall effect, since in the case of small polarons the effect of E and of a magnetic field H is exerted primarily through their effect on transition probabilities.

7 High-Frequency Conductivity

To calculate the dimensionless absorption coefficient $\alpha(\omega)$ or the coefficient $K(\omega)$, which has the dimensionality of a reciprocal length, it is sufficient to find the complex conductivity $\sigma(\omega)$ [20, 33]. The total dielectric constant of the medium is

$$\epsilon(\omega) = \bar{\epsilon}(\omega) + \Delta\epsilon(\omega), \tag{92}$$

where $\Delta\epsilon(\omega)$ corresponds to the primary absorption mechanism in the frequency range of interest, and $\bar{\epsilon}(\omega)$ includes all other mechanisms. We break up all functions in (92) into real parts ($'$) and imaginary parts ($''$). In the case of a weak absorption, with $\epsilon''/\epsilon' \ll 1$ we find (under the assumption $\bar{\epsilon}'' = 0$)

$$K(\omega) = \frac{\omega}{c}\alpha(\omega), \alpha(\omega) = \frac{\Delta\epsilon''(\omega)}{2(\epsilon')^{1/2}}, \tag{93}$$

where

$$\Delta\epsilon(\omega) = i\frac{4\pi\sigma(\omega)}{\omega}. \tag{94}$$

If $\Delta\epsilon'/\epsilon' \ll 1$, the refractive index $n(\omega) = (\epsilon'(\omega))^{1/2}$ is

$$\begin{aligned} n &= \bar{n} + \Delta n, \bar{n} = (\bar{\epsilon})^{1/2}, \\ \Delta n &= \Delta\epsilon'/2n'. \end{aligned} \tag{95}$$

The Kubo formula for $\sigma_{xx}(\omega)$ is conveniently rewritten as [33]

$$\begin{aligned} \sigma_{xx}(\omega) = &-\frac{1}{V\hbar\omega}\int_0^\infty e^{-i\omega t - st}\langle j_x(t)j_x(0)\rangle - \langle j_x(0)j_x(t)\rangle dt \\ &+\frac{i}{V\omega}\int_0^\beta \langle j_x(-i\hbar\lambda)j_x(0)\rangle d\lambda. \end{aligned} \tag{96}$$

Only the first term in (96) contributes to $\sigma'(\omega) = \mathrm{Re}\sigma_{xx}(\omega)$. A graphical technique for calculating the current correlation functions in (96) has been developed for the static electrical conductivity. The only distinctive feature is the replacement $s \to s + i\omega$ in the evaluation of the integrals over time. Below we summarize the results for the intraband absorption and for the interband optical transitions.

7.1 Intraband Absorption

By "intraband absorption" here we mean a nonactivation transition of the polaron from site to site induced by light. This topic was first taken up by Eagles [36]. The Kubo formula for $\sigma_{xx}(\omega)$ was used by Reik [37] and Klinger [38]. In the hopping region $(T > T_0)$ at small values of $\eta_3 = J^2/kTE_a$, we find from (96) in accordance with [35–40],

$$\sigma'(\omega) = neu_0 \frac{\pi^{1/2}}{2} \frac{J^2}{\sqrt{(E_a kT)}}$$
$$\times \exp[-\frac{(\hbar\omega - 4E_a)^2}{16kTE_a}]. \tag{97}$$

In the limit $\omega \to 0$ it is (12). Equation (97) is not exact. Taking the approach developed for calculating optical absorption spectra at F centers, Bogomolov et al. [35] derived a more accurate formula for $\sigma'(\omega)$, incorporating a slight frequency asymmetry:

$$\sigma'(\omega) = \sigma(0)f(\omega), \tag{98}$$

where

$$f(\omega) = \frac{\sinh \alpha}{\alpha}(1 + x^2)^{-1/4} \exp\{\Gamma[-x \cdot \text{arc}\sinh x + (1 + x^2)^{1/2} - 1]\},$$

and $\alpha = \dfrac{\hbar\omega}{2kT}$, $x = \dfrac{\omega}{\Gamma\omega_0}$, $\Gamma = \dfrac{8E_u kT}{(\hbar\omega_0)^2}$.

Expanding in series in x < 1, we go from (98) to (97). The case T < T_0 with a slight dispersion, $\Delta\omega/\omega_0 \ll 1$ was discussed in [39, 40], while the case $\Delta\omega/\omega_0 \simeq 1$ was discussed in [35]. When the possibility of induced hops to more remote sites is taken into account ((97) and (98) incorporate transitions between nearest neighbors exclusively), there is an additional asymmetry of the "bell". The resultant $\sigma(\omega)$ curve is a set of gaussians, lying on the interval from $E_a(\mathbf{g})$ to $E_p/2$ and decreasing in amplitudes [21] if $\eta_3 < 1$. With increasing η_3 an additional broadening arises, since there is a continuous spectrum in the energy interval $2zJ$ at a distance $4E_a$ above the lowest polaron level (z is the number of nearest neighbours). Under the condition $\eta_3 > 1$ an electron ejected from the polaron well by light can move a large distance by tunneling before it self–localizes again (the relay–race mechanism). This topic is discussed in more detail in [40] and [21](Chap. 2, Part 2, §2).

A bell-shaped curve for $K(\omega)$ in the case of absorption by free carriers, rather than by F centers, has been observed in many materials. It has been analyzed in detail for the particular case of rutile [35] TiO_2. Taken along with other experimental facts, this result made it possible to demonstrate reliably, for the first time, that the charge carriers in rutile are small polarons [41].

Let us briefly discuss the intraband absorption at low frequencies $\hbar\omega \ll 4E_a$, i.e. to the left of the bell. The hopping component here increases slowly with increasing ω, while the tunneling component is described by an expression like the Drude–Lorentz formula,

$$\sigma'_t(\omega) = \sigma_t(0)[1 + \omega^2\tau^2(\omega)]^{-1}. \tag{99}$$

At $\omega > \omega_0$, the quantity $\tau(\omega)$ begins to vary in a nonmonotonic way as a function of ω. There has been no detailed study of the region $\omega < \omega_0 < 4E_a/\hbar$.

7.2 Interband Optical Transitions

Eagles [36] studied optical transitions between two different polaron bands (in the case $\eta_3 \ll 1$) and a transition from a wide valence band to a narrow polaron band. Kudinov and Firsov [33, 40] studied transitions from a deep atomic level (or from a narrow electron band) whose electrons are coupled weakly with phonons to a narrow polaron band in the cases $\eta_3 < 1$ and $\eta_3 > 1$. They also studied a transition from a narrow polaron valence band to a wide conduction band in which the electron–phonon coupling is weak [40]. As an example we consider a transition from a deep atomic level to a narrow polaron band [33]. If the transition is allowed, and if the coupling of electrons in the lowest level with phonons is weak, then we find from (96),

$$\sigma(\omega) = \Delta\sigma(\omega) + \frac{ie^2 n_0}{2m} f_{01} \frac{2\Omega + \omega}{\Omega(\Omega + \omega)} \tag{100}$$

where f_{01} is a dimensionless oscillator strength, m is the mass of the free electron, $\hbar\Omega_0$ is the distance between the deep level (subscript 0) and the center of the unrenormalized upper band (subscript 1), $\hbar\Omega = \hbar\Omega_0 - E_p$ is the distance between the deep level and the center of the polaron band formed from band 1, and n_0 is the electron concentration in the lower level 0.

The second term in (100) is a smooth function of the frequency. All structural features are in the first term:

$$\Delta\sigma(\omega) = \frac{e^2 n_0}{2m} f_{01} \frac{\Omega_0}{\omega} \int_0^\infty \exp[-S_T^{(0)} + F_0^*(t + \frac{i\hbar\beta}{2})] \times \exp[-i(\omega - \Omega)t]\exp(-\frac{t}{\tau})dt, \tag{101}$$

where $S_T^{(0)} = \frac{1}{2N}\sum_{\mathbf{q}} |\gamma_{\mathbf{q}}|^2 \coth(\frac{\hbar\omega_{\mathbf{q}}\beta}{2})$,

$$F_0^*(t + \frac{i\hbar\beta}{2}) = \frac{1}{2N}\sum_{\mathbf{q}} \frac{|\gamma_{\mathbf{q}}|^2}{\sinh(\hbar\omega_{\mathbf{q}}\beta/2)} \cos[\omega_{\mathbf{q}}(t - \frac{i\hbar\beta}{2})],$$

and τ^{-1} is the characteristic damping. In contrast with the case of intraband transitions between neighboring sites, expression (101) does not have the factor of $(1 - \cos(\mathbf{q} \cdot \mathbf{g}))$ in the summation over $\mathbf{q}$.

The function $F_0^*(t + i\hbar\beta/2)$ falls off less slowly than $O(t^{-3/2})$ with increasing t. Hence we find $\exp(F_0^\star) \to 1$ in the limit $t \to \infty$, so that at $\omega = \Omega$ there is a singularity of the type $(\omega - \Omega)^{-1}$ in (101). This singularity corresponds to a "zero–phonon peak". The total intensity of this peak is of the order of

$\exp(-S_T^{(0)})$. This factor reflects the finite probability that, by the time of the electron transition, the phonon system will undergo a displacement to a new position corresponding to the presence of the electron at the upper level ([21], Chap. 3, §3). This factor is analogous to the Debye–Waller factor.

In addition to the "zero–phonon peak" two spikes differing in intensity arise at the frequencies $\Omega \pm \omega_0$. A "bell–shaped" hill of the same shape as in the case of intraband absorption and of the same nature arises to the right of the zero–phonon peak. Curves of $K(\omega)$ and $\Delta n(\omega)$ are given in [33] and [21].

Studies of interband optical transitions from a narrow polaron valence band to a wide conduction band [40] and from a wide valence band to a narrow polaron conduction band [36] gave qualitatively similar results: the frequency dependence at the absorption edge is not a power law but nearly a Gaussian, and the absorption edge undergoes a large shift as a function of the temperature at $kT > \hbar\omega_0/2$.

Theoretical studies of optical absorption by small polarons in the magnetic field [42] and in the strong electric field [43] have opened up some interesting experimental possibilities. Even in the static case, however, the effect of fields H and E requires a special analysis as discussed below.

8 Transport Phenomena in a Strong Electric Field

The effect of a strong electric field $\mathbf{E}$ on the motion of small polarons can be taken into account most easily in the hopping regime [44–48]. If we assume that the difference between the energies of a small polaron at two neighboring sites is $e\mathbf{g}\cdot\mathbf{E}$, then this quantity (divided by $\hbar$) serves as a frequency in a calculation of the hopping probability. The result for $\sigma_h(E)$ is therefore similar to (97):

$$j = E\sigma_h(E),$$

$$\sigma_h(E) = \sigma_h(0)\frac{\sinh(eaE/2kT)}{(eaE/2kT)}\exp[-\frac{(eaE)^2}{16kTE_a}]. \tag{102}$$

Here $\sigma_h(\omega)$ is the static electrical conductivity in the hopping regime in a weak electric field. This question is discussed in more detail, at a qualitative level, in [50] and [51]. Below we derive result (102) as a particular case of the general theory of transport phenomena in the strong electric field E. It is possible to describe the effect of E on the tunneling transport mechanism and to predict a new effect, an *electrophonon resonance*.

It was shown in [45] that the most common definition for the current in an electric field of arbitrary strength is (if we ignore the electron–phonon interaction)

$$j = -n \lim_{s\to 0} \lim_{V\to\infty} s^2 \int_0^\infty dt\, e^{-st}\frac{1}{Z_0}\mathrm{Tr}\{e^{-\beta\mathcal{H}}[D(t) - D(0)]\}, \tag{103}$$

where $\beta = (kT)^{-1}$, $Z_0 = \mathrm{Tr}\{\exp(-\beta\mathcal{H}_{\mathrm{ph}})\}$, and the Hamiltonian of the system is

$$\mathcal{H} = \mathcal{H}_{ph} + \mathcal{H}_e^{(0)} + \mathcal{H}_{int}. \tag{104}$$

The first term here is the Hamiltonian of the phonon field and $\mathcal{H}_e^{(0)}$ is the Hamiltonian of the renormalized charged carriers (small polarons in the case at hand). The dipole moment operator $\mathbf{D} = \mathbf{d} + \mathbf{d}'$ contains parts which are diagonal ($\mathbf{d}$) and nondiagonal ($\mathbf{d}'$) with respect to sites,

$$\mathbf{d} = e\sum_{\mathbf{m}} R_{\mathbf{m}}a_{\mathbf{m}}^{+}a_{\mathbf{m}}, \mathbf{d}' = e\sum_{\mathbf{m},g} \mathbf{L}(\mathbf{g})a_{\mathbf{m}+g}^{+}a_{\mathbf{m}}. \tag{105}$$

In the presence of an external electric field we have the Hamiltonian

$$H = \mathcal{H} - \mathbf{D}{\cdot}\boldsymbol{E} \tag{106}$$

The quantity $D(t)$ in (103) is the dipole moment Heisenberg operator defined with the Hamiltonian H. It was proved rigorously in [45] that the operator $\mathbf{D}$ in square brackets in (103) can be replaced by its diagonal part, $\mathbf{d}$. Operator $\mathbf{D}$ in expression (106) for H cannot be replaced by $\mathbf{d}$ but we can include $\mathbf{d}' = \Sigma_{\mathbf{k}}\mathbf{L}(\mathbf{k})a_{\mathbf{k}}^{+}a_{\mathbf{k}}$ (in the $\mathbf{k}$ representation) in the expression for $\mathcal{H}_e$:

$$\mathcal{H}_e = \mathcal{H}_e^{(0)} - \boldsymbol{d}' \cdot \boldsymbol{E} = \sum_{\mathbf{k}} \epsilon(\mathbf{k})a_{\mathbf{k}}^{+}a_{\mathbf{k}}, \tag{107}$$

where $\epsilon(\mathbf{k}) = \epsilon^{(0)}(\mathbf{k}) - e\mathbf{K}\cdot\mathbf{L}(\mathbf{k})$. The limits s $\to$ 0 and $V \to \infty$ must be taken in the order specified, the opposite order corresponds to a field effect.

The calculations can be carried out most conveniently in the representation of a "Stark ladder" (the α representation). This is a particular case of the Houston representation [52]. In the α representation the Hamiltonian $\mathcal{H}_e - \boldsymbol{d}\cdot\boldsymbol{E}$ diagonalises as $\mathcal{H}_e - \boldsymbol{d} \cdot \boldsymbol{E} = \sum_{\alpha} \epsilon_{\alpha}a_{\alpha}^{+}a_{\alpha}$

$$\epsilon_{\alpha} = -eEX_{\mathbf{m}} + \epsilon(\mathbf{k}_{\perp}), \epsilon(\mathbf{k}_{\perp}) = \frac{1}{N_x}\sum_{k_x} \epsilon(\mathbf{k}). \tag{108}$$

Here N_x is a normalization number for the summation over k_x. For example, this number is equal to the number of lattice atoms along the $\mathbf{x}$ axis, if $\mathbf{E}||\mathbf{x}$ and if $\mathbf{x}$ is one of the symmetry axes of the crystal. It can be seen from (108) that the set of quantum numbers contains $\mathbf{k}_{\perp} = \{k_y, k_z\}$ and $X_{\mathbf{m}} = (R_{\mathbf{m}})_x$. The relationship between α and $\mathbf{k}$ representations is described by the linear relations

$$a_{\alpha} = \frac{1}{N_x^{1/2}}\sum_{k_x} a_{\mathbf{k}}exp[ik_x X_{\mathbf{m}} + i\chi(\mathbf{k})], \tag{109}$$

where $\chi(\mathbf{k})$ satisfies the differential equation

$$eE\frac{\partial \chi}{\partial \mathrm{k}_{x}} = \epsilon(\mathbf{k}) - \epsilon(\mathbf{k}_{\perp}). \tag{110}$$

The expression in square brackets in (109) can then be rewritten

$$ik_{x}X_{\mathbf{m}} + i\chi(\mathbf{k}) = -\frac{i}{eE}\int_{0}^{k_{X}}[\epsilon_{\alpha} - \epsilon(\mathbf{k}')]dk'_{x}. \tag{111}$$

We can rewrite expression (103) for the current in this α representation as

$$\lim_{s\to 0}\lim_{V\to\infty} \mathrm{s}^2 \sum_{a,a'}(X_{\mathbf{m}'} - X_{\mathbf{m}})M_{\mathrm{aa}'}^{\mathrm{aa}'}. \tag{112}$$

The problem of calculating the current has thus been reduced to one of disentanglement of the quantity $M_{\mathrm{aa}'}^{\mathrm{aa}'}$, which can now be written as an infinite series in powers of $[\mathrm{s} - \frac{1}{h}eE(X_{\mathbf{m}_1} - X_{\mathbf{m}_2})]^{-1}$ (which corresponds to a contribution from "quasi-free" cross sections) and in powers of s^{-1} (which, as in the case of the weak field, corresponds to a contribution from "free" cross sections). The summation over powers of s^{-1} leads to a transport equation in the α representation ([45] and [21], Chap. 4), which is valid for an arbitrary strength of the electric field. We then write an exact expression for the current as a power series in E^{-1}, which is particularly convenient for a further analysis in the case of small polarons:

$$j_{x} = en \sum_{\mathbf{k}'_{\perp}} n(\mathbf{k}'_{\perp}) \sum_{X_{m},\mathbf{k}_{\perp}} X_{\mathbf{m}}\widetilde{W}_{0X_{\mathbf{m}}}^{0X_{\mathbf{m}}}(\mathbf{k}'_{\perp},\mathbf{k}_{\perp}). \tag{113}$$

Here $\widetilde{W}$ is an "effective" probability for a transition from state $\alpha' = (0,\mathbf{k}'_{\perp})$ to state $\alpha = (X_{\mathbf{m}},\mathbf{k}_{\perp})$. This probability is written as a series "in powers of $1/E$" :

$$\widetilde{W}_{X_{m}X_{m'}}^{X_{m}X_{m'}}(\mathbf{k}_{\perp},\mathbf{k}'_{\perp}) = W_{X_{m}X_{m'}}^{X_{m}X_{m'}}(\mathbf{k}_{\perp}\mathbf{k}'_{\perp})+$$

$$+\sum_{\mathbf{k}''_{\perp}} \sum_{X_{m_1},X_{m_2},(X_{m_1}\neq X_{m_2})} W_{X_{m}X_{m_2}}^{X_{m}X_{m_1}}(\mathbf{k}_{\perp},\mathbf{k}''_{\perp})\times$$

$$\times \frac{\hbar}{ieE(X_{m_2} - X_{m_1})}W_{X_{m_2}X_{m'}}^{X_{m_1}X_{m'}}(\mathbf{k}''_{\perp},\mathbf{k}'_{\perp}) + \dots \tag{114}$$

Here W are the ordinary probabilities for a transition from state α to α' (these probabilities do not contain quasi-free cross sections). They are determined by a graphical technique and depend on the electric field. The distribution $n(\mathbf{k}_\perp)$ with respect to the component $\mathbf{k}_\perp$ of the wave vector $\mathbf{k}$, which is perpendicular to the field, depends on the strength of the electric field. It is found from the equation

$$\sum_{\mathbf{k}'_\perp} n(\mathbf{k}'_\perp)\widetilde{W}(\mathbf{k}'_\perp,\mathbf{k}_\perp) = 0, \tag{115}$$

where $\widetilde{W}$ is the probability for a transition from state $\mathbf{k}_\perp$ to $\mathbf{k}'_\perp$, given by

$$\widetilde{W}(\mathbf{k}'_\perp,\mathbf{k}_\perp) = \sum_{X_m} \widetilde{W}^{0X_{\mathbf{m}}}_{0X_{\mathbf{m}}}(\mathbf{k}'_\perp,\mathbf{k}_\perp). \tag{116}$$

The normalization condition on $n(\mathbf{k}_\perp)$ is

$$\sum_{\mathbf{k}_\perp} n(\mathbf{k}_\perp) = 1. \tag{117}$$

Equations (113) - (117) thus completely determine the current in a strong electric field. The expansion in powers of $1/E$ in (114) looks a bit unusual. Bryksin [49] showed that a summation of this series generally leads to a transport equation of the kinetic type, in which the electric field appears not only on the left side (as it usually does) but also on the right side, through a dependence of the kernel of this integral equation on E. In the small *polaron* case of interest here, at $T > T_2$, we can set $n(\mathbf{k}_\perp) = constant$, and the quantities W and $\widetilde{W}$ in (114) are related to the corresponding quantities in the lattice-site representation through a simple Fourier transformation (since the bands for small polarons are narrow at $T > T_2$). As a result we find from (113)[44, 48]:

$$j_x = en\sum_{\mathbf{m}} X_{\mathbf{m}}\widetilde{W}^{0\mathbf{m}}_{0\mathbf{m}} = en\sum_{\mathbf{m}} X_{\mathbf{m}}\sinh(\frac{eEX_{\mathbf{m}}}{2kT})\widehat{W}^{0\mathbf{m}}_{0\mathbf{m}}. \tag{118}$$

where $\widehat{W}^{0\mathbf{m}}_{0\mathbf{m}}$ and $\widetilde{W}^{0\mathbf{m}}_{0\mathbf{m}}$ are effective probabilities for a transition from site 0 to site $\mathbf{m}$ [cf. (53) in Sect. 3]. These probabilities are related by

$$\widetilde{W}^{0\mathbf{m}}_{0\mathbf{m}} = \widehat{W}^{0\mathbf{m}}_{0\mathbf{m}}\exp(\frac{eEX_{\mathbf{m}}}{2kT}). \tag{119}$$

The quantity $\widehat{W}$ is symmetric under the substitution $\mathbf{m} \to -\mathbf{m}$ or $E \to -E$. From (114) we find

$$\widehat{W}^{0\mathbf{m}}_{0\mathbf{m}} = {}^{(s)}W^{0\mathbf{m}}_{0\mathbf{m}} +$$

$$+ \sum_{\mathbf{m}_1,\mathbf{m}_2,(X_{m_1}\neq X_{m_2})} {}^{(s)}W^{0\mathbf{m}_1}_{0\mathbf{m}_2}\frac{\hbar}{ieE(X_{m_1} - X_{m_2})}{}^{(s)}W^{\mathbf{m}_1\mathbf{m}}_{\mathbf{m}_2\mathbf{m}} + \cdots. \tag{120}$$

The quantities $^{(s)}W$ in (120) are symmetric with respect to E. In other words, we can also single out a factor of $\exp(eEX_\mathbf{m}/2kT)$ from them. Under approximations similar to those used in the weak–field case in Sect. 3 (see inequalities (56)), we find [44, 48],

$$\widehat{W}^{0\mathbf{m}}_{0\mathbf{m}} = {}^{(s)}W^{0\mathbf{m}}_{0\mathbf{m}} +$$

$$+ \sum_{\mathbf{m}_1,\mathbf{m}_2,(\mathbf{m}_1 \neq \mathbf{m}_2)} \frac{{}^{(s)}W^{0\mathbf{m}_1}_{0\mathbf{m}_2}\,{}^{(s)}W^{\mathbf{m}_1\mathbf{m}}_{\mathbf{m}_2\mathbf{m}}\tau_E(\mathbf{m}_1 - \mathbf{m}_2)}{1 + (eE/\hbar)(X_{\mathbf{m}_1} - X_{\mathbf{m}_2})^2\tau_E^2(\mathbf{m}_1 - \mathbf{m}_2)}. \tag{121}$$

In the limit $E \to 0$, expression (121) yields the result given in Sect. 3 (see (53) and (54)). The quantity $\sigma_h(E) = j(E)/E$ can thus also be divided into two parts, a hopping part $\sigma_h(E)$ and a tunneling part $\sigma_t(E)$:

$$\sigma_h(E) = \frac{e^2 n}{2kT} \sum_\mathbf{m} X_\mathbf{m}^2 \frac{\sinh(eEX_\mathbf{m}/2kT)}{(eEX_\mathbf{m}/2kT)}\,{}^{(s)}W^{0\mathbf{m}}_{0\mathbf{m}}, \tag{122}$$

$$\sigma_t(E) = \frac{e^2 n}{2kT} \sum_\mathbf{m} X_\mathbf{m}^2 \frac{\sinh(eEX_\mathbf{m}/2kT)}{(eEX_\mathbf{m}/2kT)} +$$

$$+ \sum_{\mathbf{m}_1,\mathbf{m}_2,(\mathbf{m}_1 \neq \mathbf{m}_2)} \frac{{}^{(s)}W^{0\mathbf{m}_1}_{0\mathbf{m}_2}\,{}^{(s)}W^{\mathbf{m}_1\mathbf{m}}_{\mathbf{m}_2\mathbf{m}}\tau_E(\mathbf{m}_1 - \mathbf{m}_2)}{1 + [(eE/\hbar)(X_{\mathbf{m}_1} - X_{\mathbf{m}_2})]^2\tau_E^2(\mathbf{m}_1 - \mathbf{m}_2)}. \tag{123}$$

When hops between only nearest neighbours are taken into account the part of W in (122), which is symmetric with respect to E, is

$$^{(s)}W_h(E) = W_h(0)\exp\left[-\frac{(eaE)^2}{16kTE_a}\right], \tag{124}$$

and the corresponding contribution is the same as in (102). For electric fields which are not too strong ($\tau(E)/\tau(0) = 1, eaE/2kT \ll 1$), E-dependence of σ_t is reminiscent of the magnetic-field dependence of the transverse electrical conductivity in strong but nonquantizing magnetic fields. In stronger electric fields, the E-dependence of τ arises. While the main contribution to τ at $E = 0$ comes from the two–phonon processes ($\tau^{-1} \sim \sinh^{-2}(\hbar\omega_0/2kT)$, this situation corresponds to the presence of a δ–function of the type $\delta(\omega_{\mathbf{q}2} - \omega_{\mathbf{q}1})$ in the expression for τ^{-1}, one-phonon ($\delta(\omega_\mathbf{q} - \Omega_E)$), three-phonon and other multiphonon processes come into play with increasing E. Here $\Omega_E = eaE/\hbar$ is the distance between the closest Stark levels. As a result, $\tau(E)$ and therefore $\sigma_t(E)$ vary in a nonmonotonic way as a function of the electric field. If there is a slight dispersion of optical phonons ($\Delta\omega \ll \omega_0$), this nonmonotonic behavior may include oscillations with a period determined by the resonance condition $MeaE = N\hbar\omega_0(M, N = 1, 2, 3, \ldots)$. Here M specifies the number of Stark levels which fit between the initial and final states of the electron, and N is the number of phonons emitted in the course of the transition.

The oscillation amplitude falls off with increasing N. This effect, now known as the "electrophonon resonance", was predicted [44] theoretically in 1970 within the framework of small–polaron theory. The effect was first observed experimentally, in the same year, by Maekawa [53] in ZnS for an intermediate coupling of electrons with phonons ($\alpha \simeq 1-2$). In general, the electrophonon resonance can occur for an arbitrary coupling strength α.

Bryksin and Firsov [47,54] derived a systematic theory for the electrophonon resonance in the case $\alpha < 1$. They started from general expression (113) and truncated series (114), i.e. they assumed $\tau \Omega_E >> 1$. Bogomolov et al. [55] observed the strong electrophonon resonance in comparatively weak fields, $E < 10^3 \mathrm{V/cm}$, in Na-X zeolite matrices filled with tellurium (Te_{16}). The period of the superlattice there was ten times the usual values. A detailed theory of the electrophonon resonance for this case was derived in [56–58].

9 Hall Effect

The theory of the Hall effect is one of the most complex questions in the small–polaron theory. In general, the relationships among the Hall coefficient R, the Hall mobility μ_H, the drift mobility μ_D, and the nondiagonal components of the antisymmetric tensor $\mu_{ij}^{(a)}$(i, j = x, y, z) for the cubic crystal in a weak magnetic field H||z are described by

$$\mu_H = cR\sigma = \frac{c}{H}\frac{\mu_{xy}}{\mu}, R = \frac{1}{H}\frac{\mu_{xy}}{\mu\sigma} = \frac{1}{enc}\frac{\mu_H}{\mu_D} \qquad (125)$$

For tetragonal and hexagonal crystals there are two Hall coefficients and two Hall mobilities. The conventional relations

$$R = \frac{1}{enc}, \mu_H \simeq \mu_D \qquad (126)$$

are the consequence of the Lorentz correlation between the x and y components of the velocity of the electron moving in a wide energy band. In this case the kinetic equation yields

$$\mu_{xy} \simeq \mu\Omega\tau, \qquad (127)$$

where τ is a relaxation time, and $\Omega = eH/m^*c$ is the Larmor frequency. Substituting (127) into (125) we obtain (126).

In general, however, we would not expect (126) to hold. In other words we cannot draw conclusions about the temperature dependence of μ_D from the temperature dependence of the Hall mobility. For example, even in the weak-coupling case $\alpha \ll 1$, we find [59,60] $\mu_H/\mu_D \approx kT/\Delta\epsilon$ for the case of narrow bands ($\Delta\epsilon \ll kT$). The first nontrivial results were derived for small polarons in the hopping regime. For hexagonal crystals in the field H||c_3 Friedman and Holstein [61] found in the small-J limit,

$$\mu_{xy} \sim \frac{J^3}{T^2} \exp(-\frac{4E_a}{3kT}),$$

$$\mu_H \sim \frac{J}{T^{1/2}} \exp(-\frac{E_a}{3kT}), \qquad (128)$$

$$\frac{\mu_H}{\mu_D} \simeq \frac{kT}{J} \exp(\frac{2E_a}{3kT})$$

The same result was obtained from the Kubo formula by Firsov [62]. For the cubic crystals, Bryksin and Firsov found [63]

$$\mu_{xy} \sim \frac{J^4}{T^2} \exp(-\delta\frac{E_a}{kT}),$$

$$\mu_H \sim J^2/T^{1/2} \exp(-(\delta - 1)\frac{E_a}{kT}), \qquad (129)$$

$$\frac{\mu_H}{\mu_D} \sim \frac{kT}{E_a} \exp[(2 - \delta)\frac{E_a}{kT}].$$

Here $\delta - (1 \quad E_a'/4E_a)$, and E_a and E_a' are the activation energies for hops along the edge and the face diagonal, respectively, of the cubic cell. It can be shown [63] that we have $4/3 < \delta < 2$. Expressions (129) are valid under the condition $kT < E_a(2 - \delta)(1 - 1/\delta)$. Results which are valid to $kT \approx E_a$ are given in [64].

For small polarons at high temperatures, where the drift mobility increases as $\exp(-E_a/kT)$ with increasing T, the Hall mobility thus increases far more slowly, and numerically it is far larger than μ_D. However, according to [69, 70] for adiabatic correlated (non-Markovian) hops the ratio μ_H to μ_D (cf. (128) and (129)), in general, becomes far closer to one and may be even less than one. Methods developed by Firsov and Bryksin in [62, ?, ?, ?, ?, ?, ?] for calculation of hopping magnetotransport were successfully generalized by Böttger, Bryksin and Damker (for instance, see their chapter in Part 1 of this book [71]), who took into account spin-orbit interaction and described spin accumulation and spin-Hall effect in the small–polaron hopping regime.

The tunneling component $\mu_{xy}^{(t)}$ also has an extremely nontrivial behavior. It became possible to calculate this component only after the derivation of a general theory of galvanomagnetic phenomena, derived in [65] and [66] by Bryksin and Firsov. The basic results on $\mu_{xy}^{(t)}$ are given in [67, 68]. Below we will try to give only a general picture of the approach presenting the final results on μ_{xy} and μ_H within the framework of the small–polaron theory.

In the presence of the magnetic field H, the resonance integral $J(\boldsymbol{g})$ in the Hamiltonian (27) acquires an additional phase factor:

$$J(\mathbf{g}) \to J(\mathbf{m}+\mathbf{g}, \mathbf{m}) = J(\mathbf{g})exp[i\alpha(\mathbf{m}+\mathbf{g}, \boldsymbol{m})], \tag{130}$$

where

$$\alpha(\mathbf{m}+\mathbf{g}, \mathbf{m}) \doteq \mathbf{A}(\mathbf{m}) \cdot \mathbf{g} = \frac{e}{2\hbar c}\mathbf{H} \cdot [\mathbf{m} \times \mathbf{g}]. \tag{131}$$

The effective interaction Hamiltonian thus becomes (cf. the third term in (27))

$$H_{int} = \sum_{\mathbf{m},\mathbf{g}} J(\mathbf{g})\mathbf{a}^{+}_{\mathbf{m}+\mathbf{g}}\mathbf{a}_{\mathbf{m}}exp[S(\mathbf{m}+\mathbf{g}) - S(\mathbf{m})]exp[i\alpha(\mathbf{m}+\mathbf{g}, \mathbf{m})] \tag{132}$$

Calculating the probability for the hop to a neighbouring site in the second order in J the product $J(\mathbf{m}+\mathbf{g}, \mathbf{m})J(\mathbf{m}, \mathbf{m}+\mathbf{g}) = J(\mathbf{g})J(-\mathbf{g})$ arises. The phase factors α, which contain the magnetic field, cancel out in this product, and the same happens in any order in J for a two–site cluster. Friedman and Holstein [61] thus suggested incorporating a "three–site" probability for $H \neq 0$. For hexagonal crystals in the field $\mathbf{H}||\mathbf{c_3}$, for example, three neighbouring sites lying in the plane perpendicular to H form an isosceles triangle: $\mathbf{m}_1, \mathbf{m}_2 = \mathbf{m}_1 + \mathbf{g}_1, \boldsymbol{m}_3 = \boldsymbol{m}_2 + \boldsymbol{g}_2 = \boldsymbol{m}_1 + \boldsymbol{g}_1 + \boldsymbol{g}_2$. Here we have $\mathbf{m}_3 + \mathbf{g}_3 = \boldsymbol{m}_1$, i.e $\boldsymbol{g}_1 + \boldsymbol{g}_2 + \boldsymbol{g}_3 = \mathbf{0}$. It is easy to see that in this case the quantity

$$-i\frac{e}{\hbar c}\mathbf{H} \cdot \mathbf{S_{321}}, \tag{133}$$

will appear in the argument of the exponential function for the product of the three phase factors. Here $\mathbf{S_{321}}$ is a vector (parallel to $\mathbf{c_3}$), which is numerically equal to the area of the triangle formed by these three sites $(\mathbf{m}_1, \mathbf{m}_2, \mathbf{m}_3)$ given by

$$\mathbf{S_{321}} = 1/2\{[\mathbf{m}_1 \cdot \mathbf{m}_2]+[\mathbf{m}_2 \cdot \mathbf{m}_3]+[\mathbf{m}_3 \cdot \mathbf{m}_1]\} = 1/2[\mathbf{g_1} \cdot \mathbf{g_2}] = \frac{3^{1/2}}{4}a^2\mathbf{k_0}, \tag{134}$$

where $\mathbf{k_0}$ is the unit vector along the axis $\mathbf{c_3}||\mathbf{z}$.

The argument in (133) is thus equal to the ratio $-i2\pi\Phi/\Phi_0$ where Φ is the magnetic flux through the triangle, and $\Phi_0 = 2\pi\hbar(c/e)$ is the flux quantum. In the hopping regime the functional dependence $\sigma_h(H)$ is determined not by the parameter $\Omega\tau$ (as in the case of a band–like transport) but by the ratio $2\pi\Phi/\Phi_0 \sim u_0 H/c$. This assertion also applies to the tunneling component.

It was shown in [65, 66] that the current can in general be written as

$$\mathbf{j} = en \sum \mathbf{v}^{eff} f(\mathbf{k}). \tag{135}$$

Here

$$\mathbf{v}^{\text{eff}}(\mathbf{k}) = \mathbf{v}(\mathbf{k}) - i \sum_{\mathbf{k}',} W_1(\mathbf{k}, \mathbf{k}'), \tag{136}$$

where W is defined in the same way as in Sect. 6 (see (89)), but it depends on H. The function $f(\mathbf{k})$, normalized to unity, is found from the transport equation,

$$eE\nabla_{\mathbf{k}}f(\mathbf{k})-i\{\epsilon[\mathbf{k}+\mathbf{A}(i\nabla_{\mathbf{k}})]-\epsilon[(\mathbf{k}-\mathbf{A}(i\nabla_{\mathbf{k}})]\}f(\mathbf{k}) = \hbar\sum_{\mathbf{k}'} f(\mathbf{k}')W(\mathbf{k}',\mathbf{k};\mathrm{E},\mathrm{H}).$$

$$(137)$$

Here $\epsilon(\mathbf{k})$ is the renormalized energy, $\nabla_{\mathbf{k}} = \partial/\partial\mathbf{k}$, and

$$\mathbf{A}(i\nabla_{\mathbf{k}}) = i\frac{e}{2\hbar c}\mathbf{H}\times\nabla \tag{138}$$

The probabilities W in (137) depend on H and are determined by a graphical technique. If we linearize the operator on the left side of (137) with respect to H, we obtain the usual expression on the left $\hbar^{-1}(e\mathbf{E}+\frac{e}{c}\mathbf{v}\times\mathbf{H})\nabla_{\mathbf{k}}f(\mathbf{k})$. In this expression the effect of the magnetic field is described by the Lorentz force, but we should bear in mind that W itself depends on H (through the ratio Φ/Φ_0). Linearizing f and W in (137) with respect to E and H we find the hopping component $j_y^{(h)}$ and the tunneling component $j_y^{(t)}$ of the current (see [66, 67], and [21], Chap. 5, Part 2, §5). For small polarons we find

$$j_y^{(h)} = en\sum_{\mathbf{m}} Y_{\mathbf{m}}\widehat{W}_{0\mathbf{m}}^{0\mathbf{m}}(\mathrm{H},\mathrm{E}). \tag{139}$$

Here $\widehat{W}$ depends on E (as was explained in Sect. 8) and must be linearized with respect to E. Friedman and Holstein used this expression in their first paper [61]. This expression has a simple physical meaning, and they wrote it on the basis of purely intuitive considerations. Formally, one can derive component (139) by replacing $f(\boldsymbol{k})$ in (135) by $f(\mathbf{k}) = n(\mathbf{k})$. The contribution associated with the first term, $\mathbf{v}(\mathbf{k}) = (1/\hbar)\nabla_{\mathbf{k}}\epsilon(\mathbf{k})$, in the expression for $\mathbf{v}^{eff}(\mathbf{k})$ drops out, since we have $n(-\mathbf{k}) = n(\mathbf{k})$ and $\mathbf{v}(-\mathbf{k}) = -\mathbf{v}(\mathbf{k})$. We are left with only the contribution containing W_1. Setting $n(\mathbf{k}) = constant$ (for small polarons at $T > T_2$), and switching from the $\mathbf{k}$ representation to the lattice–site representation, we find (139). The result is nonzero only because W depends on E and H.

In contrast with $j_y^{(h)}$, the tunneling component $j_y^{(t)}$ is due entirely to the difference between $f(\mathbf{k})$ and $n(\mathbf{k})$. Since we are seeking the contribution linear in H, we expand all the functions in (137) in series in H. To avoid any mathematical difficulties, which are not of a fundamental nature, and to improve the clarity of the analysis, we solved (137) in the relaxation-time approximation. The results of the derivation are no less general because of this approximation. Furthermore, for small polarons at $T > T_0$ the function τ can be determined accurately (Sect. 3). After simple but tedious calculation (see [66, 67]), Bryksin and Firsov have shown that the tunneling contribution, $\mu_{xy}^{(t)}$, to μ_{xy} is the sum of three terms: the Lorentz-type($\mu_{xy,l}^{(t)}$), the "quasi-Lorentz" ($\mu_{xy,ql}^{(t)}$), and "non-Lorentz" ($\mu_{xy,nl}^{(t)}$):

$$\mu_{xy,ql}^{(t)} = e\beta\langle\mathrm{v}_x^{eff}(\mathbf{k})\tau(\mathbf{k})\frac{eH_z}{\hbar c}\langle[(\mathbf{v}(\mathbf{k})+\Delta\mathbf{v}(\mathbf{k}))\times\nabla_{\mathbf{k}}]_z n(\mathbf{k})\widetilde{\mathbf{v}}_y(\mathbf{k})\tau(\mathbf{k})\}\rangle. \tag{140}$$

The angle brackets here mean an average over $\mathbf{k}$, i.e. the integration within the first Brillouin zone. Expression (140) differs from the ordinary Lorentz contribution in that all three velocities in (140) are different. For Lorentz contribution all three velocities should be put equal to $\mathbf{v}$. These differences stem from our incorporation of corrections to the transition probabilities which are linear in H and E.

The velocity is found from the following relation (at $H = 0$)

$$n(\mathbf{k})\widetilde{\mathbf{v}}(\mathbf{k}, H) = n(\mathbf{k})\mathbf{v}(\mathbf{k}) - i \sum_{\mathbf{k}'} n(\mathbf{k}')W_1(\mathbf{k}', \mathbf{k}; H) \qquad (141)$$

The correction to the velocity, $\Delta\mathbf{v}(\mathbf{k})$ is found from the relation

$$\frac{e}{hc}\Delta\mathbf{v}(\mathbf{k})\mathbf{H} \times \nabla_{\mathbf{k}} f(\mathbf{k}) = -\sum_{\mathbf{k}'} f(\mathbf{k}')\delta W(\mathbf{k}', \mathbf{k}; H). \qquad (142)$$

Here $f(\mathbf{k}) = -e\beta n(\mathbf{k})\tau(\mathbf{k})\mathbf{E}\widetilde{\mathbf{v}}(\mathbf{k})$.

The "non-Lorentz" contribution is given by

$$\mu^{(t)}_{xy,nl} = \frac{e\beta}{2}\langle n(\mathbf{k})\tau(\mathbf{k})[\mathbf{v}^{eff}(\mathbf{k}, H) \times \widetilde{\mathbf{v}}(\mathbf{k}, H)]^{(1)}_z\rangle. \qquad (143)$$

Here $\mathbf{v}^{eff}(\mathbf{k})$ is defined by (136). The superscript (1) on the square bracket means that we need to use the part of the quantity in square brackets which is linear in H. It can be seen from (136) and (141) that the quantity in square brackets in (143) vanishes if we ignore the corrections due to W. In the weak-coupling case, $\alpha \ll 1$, the contribution in (140) is proportional to α^{-4}, while the contribution in (143) is proportional to α^{-2} and can be ignored. In the case of small polarons, contributions (140) and (143) may be comparable. We rewrite the non-Lorentz contribution in (143) in the lattice-site representation [67] as

$$\mu^{(t)}_{xy,nl} = \frac{e}{2kT} \sum_{\mathbf{m},\mathbf{m}'} [\mathbf{m} \times \mathbf{m}']_z \mathrm{Re} \sum_{\mathbf{m}_1,\mathbf{m}_2} in(\mathbf{m}_1)$$
$$\times \exp[i\mathbf{A}(\mathbf{m}_1) \times \mathbf{m}]W^{\mathbf{m}0}_{\mathbf{m}_1+\mathbf{m},\mathbf{m}_2}\tau(\mathbf{m}_2)W^{0\mathbf{m}'}_{\mathbf{m}_2\mathbf{m}'}\cdots \qquad (144)$$

We should linearize this expression with respect to H, allowing for the H dependence of the nondiagonal probabilities W_{nd} in the case $J = 0$. The quantities $n(\mathbf{m})$ and $n(\mathbf{k})$ are related by the Fourier transformation. For hexagonal crystals in a field $H\|c_3$, in the lowest approximation in $\exp(-S_T)$ we assume $\mathbf{m}_1 = 0$, i.e. $n(\mathbf{m}_1) \to 1$ in (144)], and in lowest order in J we find, taking a combination of blocks of type $W^{(1)}_{nd}W^{(2)}_{nd}$, a contribution about τJ^3. At $T > T_0$ this contribution is given by

$$\mu^{(t)}_{xy,nl} = u_0(u_0 H/c)(9/2)\pi^{1/2} J^3[(E_a\hbar\omega_0)^{1/2}kT\hbar/\tau]^{-1}$$
$$\times \sinh^{1/2}\{\frac{\hbar\omega_0}{2kT}\}\exp\{-2S_T - \frac{E_a}{2kT}\}. \qquad (145)$$

Expressions for $\mu_{xy,l}^{(t)}$ (the Lorentz contribution) and for $\mu_{xy,ql}^{(t)}$ in the lattice − site representation can be found in [67] and [21](Chap. 5, Part 2, §6). Here we write only the final expressions for these contributions for hexagonal crystals in the lowest order in $J(J^3)$,

$$\mu_{xy,l}^{(t)} = u_0 \frac{u_0 H}{c} \frac{9}{2} \frac{J^3}{kT(\hbar/\tau)^2} e^{-3S_T}, \mu_{xy,ql}^{t)} = 0. \tag{146}$$

At sufficiently high temperatures both are smaller than the hopping contribution. However, in addition to the high-temperature region, where we can assume $\mu_H = (c/H)(\mu_{xy}^{(h)}/\mu_h)$, there is a broad temperature interval in which both of the contributions (hopping and tunneling) to μ_{xy} and μ_{xx} are important, and it becomes necessary to use the following general expression for μ_H:

$$\mu_H = \frac{c}{H} \frac{\mu_{xy,l}^{(t)} + \mu_{xy,ql}^{(t)} + \mu_{xy,nl}^{(t)} + \mu_{xy}^{(h)}}{\mu_h + \mu_t} \tag{147}$$

It is assumed in (147) that the purely Lorentz contribution has been singled out of the quasi–Lorentz contribution [21]. Since each of the functions in (147) has a distinctive temperature dependence, and since the absolute values of these functions depend on the numerical values of the basic dimensionless parameters of the theory, the temperature dependence of μ_H in the intermediate temperature region may be extremely complex. Various versions of this behavior are discussed in [21,68]. In principle, μ_H may decrease where μ_D increases with increasing T, and vice versa. Accordingly, measurements of μ_H alone are insufficient for determining the transport mechanism. Furthermore, we cannot determine the sign of the charge carriers (an electron or hole small polaron) from the sign of the Hall effect. We would thus like to know what sort of information we could extract from measurements of other kinetic coefficients, e.g. from data on the thermal emf.

10 Thermal EMF

Morin [72] suggested using the formula

$$S = \frac{k}{e} \ln \frac{n}{N-n} \tag{148}$$

for the Seebeck coefficient S in the hopping regime. Here n is the carrier concentration, and N is the concentration of lattice sites. In principle, measurements of $S(T)$ may reveal the dependence $n(T)$. Knowing $\sigma(T)$, one can determine the temperature dependence of the drift mobility μ_D. However, (148) requires a rigorous foundation. Cutler and Mott's arguments [72] in favor of (148) look convincing only at low temperatures, where there might be a standard transport in the narrow polaron band ($E_p > kT$). In an attempt to justify (148) in the hopping region, Schotte [74], Efros [75] and Emin [76] explicitly or implicitly assigned a certain temperature to each unit cell. This approach is unconvincing from the thermodynamic standpoint. Using a general formula for S, which looks like the Kubo formula, runs into fundamental difficulties because of the ambiguity in the determination of the energy flux operator in the region of the strong electron–phonon interaction. A completely different approach to the linear theory of the thermal response was proposed by Kudinov [77]. He offered some arguments in favor of the use of Morin's formula, (148), in the hopping region. Nevertheless, the validity of (148) is still somewhat questionable. There has been no discussion at all of the validity of (148) in the case of the tunneling transport mechanism.

11 Conclusion

We can put the results described above in two categories:

1. The first category consists of conclusions which do not depend on specific limitations on the values of the dimensionless parameters of the theory, with the possible exception of the parameter $\eta_1 = J/E_a$ (the restriction $\eta < 1$ is the condition for the existence of small polarons). These conclusions include the method for describing the motion in the configuration space by means of the conditional probability functions (Sect. 3) and basic formulas (42),(43), and (55) (the latter is restricted only by condition (56)), the important conclusion that hopping and tunnelling transport mechanisms are operating, all results in Sect. 6, many of the general formulas in Sects. 8 and 9, and, finally, the conclusion that there are quasi–Lorentz and non–Lorentz components of μ_{xy} and therefore of μ_H. Many of these results are limited only by the condition that the renormalized band for the quasiparticle should be narrow.

2. The second category includes specific results for μ_h, μ_t, μ_{xy}, μ_H etc. found under certain restrictions on the values of the dimensionless parameters of the theory (see (69) and (78)). The physical meaning of these restrictions was discussed in detail in Sect. 3. Some of the results in Sects. 4, 5 and 9 for the case $kT > \hbar\omega_0/2$ are limited only by the condition $\eta_1 < 1$. At low temperatures, $kT < \hbar\omega_0/2\ln2\gamma$, the results are limited by the condition $\eta_2^{''} = (J/\hbar\omega_0)^2 \ln(\gamma)/\gamma < 1$, which corresponds to the nonadiabatic limit (see Sect. 5 for more details). How do we correctly determine the ground state of the adiabatic small polaron under the condition $\eta_2^{''} > 1$? With this knowledge we would be in a better position to distinguish the term H_{int}, responsible for transitions from site to site, and to construct a corresponding graphical technique for finding the transition probability.

11.1 Canonical Transformation

A few remarks concerning the Lang-Firsov canonical transformation (see Sect. 2). It is destined for calculation of different correlators in the definition of kinetic and optical coefficients. They are represented as expansions in powers of J, which is the overlap integral for unrenormalized electrons. In reality, it is an expansion in powers of dimensionless parameters $\eta_1, \eta_2, \eta_2', \eta_2'', \eta_3, \ \gamma^{-1}$ (see (69), Sect. 5). Sometimes it is possible to sum the expansion with respect to some of these parameters and get results, which are true for arbitrary values of η_2 (see (70)), η_2' (see (79)).

Usually different parts of the parameter space correspond to different mechanisms of transport. For instance, $\eta_2 < 1$ corresponds to the nonadiabatic hopping and $\eta_2 > 1$ to the adiabatic hopping. There are many papers now in which the authors numerically calculate the polaron shift, E_p, and the polaron mass M_p(or the width of the polaron band ΔE_p, which is about the same) and compare them with the results obtained using the standard canonical transformation in the zero order in J. In some regions of parameters they observe appreciable difference between numerical and analytical results. However, we should keep in mind that at high and intermediate temperatures small polarons cannot be considered as good "quasiparticles" moving with a "weak friction" $(l \gg a)$. The hopping and the tunneling regimes of charge transfer are quite different (see Sects. 4 and 5) as clearly seen from the expression for the hopping mobility (12) and (70). Such quantities as E_p and $\Delta E_p \sim e^{-s_T} \ll 1$ do not appear in (12) and (70), but J characterising the size of the electron packet and E_a (*rather than* E_p) characterising the activation energy of creation another empty well on the nearest site do appear. Thus, we can figuratively say that the polarons disappear on one site and arise on the neighbouring site ("teleportation" of some sort). So, it is more reasonable to compare results for W_h (or for μ_h) obtained by both methods in order to come to the right conclusions about the range of applicability of the canonical transformation for calculations of the kinetic and optical coefficients.

Let us discuss some applications of the Lang-Firsov canonical transformation in other physical fields. The need of a theoretical description of bipolarons arose a long time ago. The active experimental research of bipolaron properties of certain oxides of transition metals began in 1976. However, serious theoretical work on small bipolarons began after the observation that the size of the Cooper pairs in the direction perpendicular to the superconducting planes in the high-T_c superconductors can be of the order of the lattice constant. It was suggested that it might be worthwhile reviving the "Schafroth model" of a superconducting charged Bose gas, in which the size of the Bose pair (with a charge 2e) is assumed to be smaller than the average distance between pairs. A mathematical description of the bipolaronic model of high-T_c superconductors based on the canonical transformation (Sect. 2) was proposed by Alexandrov. Discussion of that topic goes beyond the scope of the present review. There is comprehensive discussion of the bipolaron model of superconductivity and its applicability to real high-T_c superconductors in the review by Alexandrov and Mott [79] and in the present volume [80] with a fairly comprehensive bibliography.

It is worthwhile mentioning another possible application of all machinery developed in the small polaron theory based on the canonical transformation, which dresses electrons up in a boson (phonon) cloud. The problem, which at first sight has nothing to do with polarons, is the charge/spin current in strongly correlated quantum wires, where the 1D Luttinger-liquid (LL) model is capable of describing the charge-spin separation phenomenon. Electronic degrees of freedom in the LL model may be described using bosonization representation of the fermionic fields as

$$\Psi_{r\sigma}(x) = lim_{a\to 0} \exp(irk_F x)/(2\pi a)^{1/2} F_{r\sigma} e^{ir\Phi_{r\sigma}(x)}. \tag{149}$$

Here $\Psi_{r\sigma}$ describes fermions with spin $\sigma = \uparrow, \downarrow$ on two branches ($r = \pm$) with the linear dispersion, $\epsilon(\mathbf{k}) = v_F(rk - k_F)]$, near two Fermi points $-k_F, +k_F$. The boson field $\Phi(x)$ is defined as

$$\Phi_{r\sigma}(x) = \frac{2\pi x}{L}N_{r\sigma} + i\sum_{q>0}(\frac{2\pi}{qL})^{1/2}(b_{qr\sigma}e^{-iqx} - b_{qr\sigma}^{+}e^{iqx})e^{-qa/2}, \tag{150}$$

where the Tomonaga bosons, $b_{qr\sigma}$, are related to the original electron operators by $b_{qr\sigma} = (2\pi/qL)^{1/2}\sum_{\mathbf{k}}\Psi_{\mathbf{k}r\sigma}^{+}\Psi_{\mathbf{k}+qr\sigma}$ and $x \in [-L/2, L/2]$. $N_{r\sigma}$ and $F_{r\sigma}$ represent the zero modes. $N_{r\sigma}$ is the deviation of the electron occupation number from the chosen reference state value. It represents $q = 0$ counterpart of the finite q Tomonaga boson occupation number $n_{qr\sigma}$, while the Klein factor $F_{r\sigma}^{+}$ ($F_{r\sigma}$) raises (lowers) $N_{r\sigma}$ by one within the Hilbert space (see, for instance, [81]).

It can be easily seen that the second term in $\Phi_{r\sigma}$ substituted in the exponent in $\Psi_{r\sigma}$ makes it look like the canonically transformed electron operator $\tilde{a}_{\mathbf{m}}$:

$$\tilde{a}_{\mathbf{m}} = e^{-S}a_{\mathbf{m}}e^{S} = a_{\mathbf{m}}e^{-S_{\mathbf{m}}}. \tag{151}$$

Linear in $b_{\mathbf{q}}$ operator $S_{\mathbf{m}}$ is defined in (26). It opens the possibility of using the methods of the small polaron theory for calculations of the current in correlated 1D quantum wires.

Now, a few words about the two-site small polaron model. It proved to be very useful for a qualitative understanding of nontrivial features of the polaron problem [12, 78], and even for obtaining some quantitative results (see, for example, [25, 26, 78]). It can be applied, for instance, to investigation of (undesirable) effects of decoherence in open quantum system. This is an important issue when it comes to implementation of quantum algorithms into real systems such as 2–level qubits. In this field the model is often called the $spin-boson$ model, and different unitary transformations have been proposed to treat it. The experience gathered in the small–polaron theory could be useful in this field as well.

Acknowledgement. This work was supported by the Russian Foundation for Basic Research (project 05-02-17804a).

References

1. A. F. Ioffe, Journ. of Can. Phys. **26**, 582 (1957).
2. L. D. Landau, Sow. Phys. **3**, 664 (1933).
3. S. I. Pekar, *Research on the Electronic Theory of Crystals* [in Russian] (Gostekhnizdat, 1951).
4. N. N. Bogolyubov, Ukr. Mat Zh. **2**, 3 (1950).
5. S. V. Tyablikov, Zh. Eksp. Teor. Fiz. **21**, 377 (1951); **23**, 381 (1952).
6. H. Fröhlich, Adv. Phys. **3**, 325 (1954).
7. R. P. Feynman, Phys. Rev. **97**, 660 (1955).
8. R. P. Feynman, R. W. Hellwarth, C K. Iddings, and R. M. Platzman, Phys. Rev. **127**, 1004(1962).
9. J. T. Devreese, in the present volume.
10. R. Heiks and W. Johnston, J. Chem. Phys. **26**, 582 (1957).
11. J. Yamashita and T. Kurosawa, J. Phys. Chem. Solids **5**, 34 (1958).
12. T. Holstein, Ann. Phys. **8**, 325, 343 (1959).
13. R. Kubo, J. Phys. Soc. Jpn. **12**, 570 (1957).
14. O. V. Konstantinov and V. I. Perel, Zh. Eksp. Teor. Fiz. **39**, 197 (1960).
15. V. L. Gurevich and Yu. A. Firsov, Zh. Eksp. Teor. Fiz. **40**, 199 (1961).
16. S. S. Shalyt, R. V. Parfen'ev, and V. M. Muzhdaba, Sov. Phys.- Solid State **6**, 508 (1964).
17. Yu. A. Firsov, V. L. Gurevich, R. V. Parfen'ev, and I. M. Tsidil'kovskii, in *Landau Levels Spectroscopy,* ed by E. I. Rashba and G. Landhwor (Elscvicr Science Publishers, 1991), p. 1183.
18. I. G. Lang and Yu. A. Firsov, Zh. Eksp. Teor. Fiz. **43**, 1843 (1962) [Sov. Phys. JETP 16, 1301 (1962)].
19. I. G. Lang and Yu. A. Firsov, Zh. Eksp. Teor. Fiz. **45**, 378 (1963) [Sov.Phys. JETP 18, 262 (1963)].
20. I. G. Lang and Yu. A. Firsov, Fiz. Tverd. Tela (Leningrad) **5**, 2799 (1963)[Sov. Phys. Solid State **5**, 2049 (1963)].
21. *Polarons* [in Russian], ed. by Yu. A. Firsov (Nauka, Moscow, 1975).
22. V. V. Bryksin, Zh. Eksp. Teor. Fiz. **100**, 1556 (1991) [Sov. Phys. JETP **73**, 861 (1991)].
23. E. K. Kudinov and Yu. A. Firsov, Zh. Eksp. Teor. Fiz. **49**, 867 (1965) [Sov. Phys. JETP **22**, 603(1965)].
24. E. K. Kudinov and Yu. A. Firsov, Fiz. Tverd. Tela (Leningrad) **8**, 666 (1966) [Sov. Phys. Solid State **8**, 536 (1966)].
25. I. G. Lang and Yu. A. Firsov, Fiz. Tverd. Tela (Leningrad) **9**, 3422 (1967) [Sov. Phys. Solid State **9**, 2701 (1967)].
26. I. G. Lang and Yu. A. Firsov, Zh. Eksp. Teor. Fiz. 54, 826 (1968) [Sov.Phys. JETP **27**, 443 (1968)].
27. D. Emin, Phys. Rev. Lett. **25**, 1751 (1970).
28. D. Emin, Phys. Rev. **B 3**, 1321 (1971); **4**, 3639 (1971).
29. G. B. Arnold and T. Holstein, Ann. Phys. **132**, 163 (1981).
30. T. Holstein, Ann. Phys. **132**, 212 (1981).
31. V. V. Bryksin and Yu. A. Firsov, Fiz. Tverd. Tela (Leningrad) **10**, 2600 (1968) [Sov. Phys. Solid State **10**, 2049 (1968)].
32. A. S. Alexandrov, V. V. Kabanov, and D. K. Ray, Phys. Rev. B **49**, 9915 (1994).

33. E. K. Kudinov and Yu. A. Firsov, Zh. Eksp. Teor. Fiz. **47**, 601 (1964) [Sov. Phys. JETP **20**, 400(1964)].

34. E. K. Kudinov, D. N. Mirlin and Yu. A. Firsov, Fiz. Tverd. Tela (Leningrad) **11**, 2789 (1969) [Sov. Phys. Solid State, **11**, 2257 (1970)].

35. V. N. Bogomolov, E. K. Kudinov, D. N. Mirlin, and Yu. A. Firsov, Fiz. Tverd. Tela (Leningrad) **9**, 2077 (1967) [Sov. Phys. Solid State **9** 1630 (1967)].

36. D. M. Eagles, Phys. Rev. **130**, 1381 (1963); in *Polarons and Excitons* ed. by C. G. Kuper and G. S. Whitefield (Edinburgh and London, 1963) p.255.

37. H. G. Reik, Solid State Commun. **1**, 67 (1963).

38. M. I. Klinger, Phys. Lett. **7**, 102 (1963).

39. H. G. Reik and D. Hecse, J. Phys. Chem. Solids **28**, 581 (1967).

40. Yu. A. Frisov, Fiz. Tverd. Tela (Leningrad) **10**, 1950 (1968) [Sov. Phys. Solid State **10**, 1537(1969)].

41. V. N. Bogomolov, E. K. Kudinov, and Yu. A. Firsov, Fiz. Tverd. Tela (Leningrad) **9**, 3175 (1967) [Sov. Phys. Solid State **9**, 2502 (1967).

42. G. B. Arnold and T. Holstein, Phys. Rev. Lett. **33**, 1547 (1974).

43. I. G. Austin, J. Phys. **C 5**, 1687 (1972).

44. V. V. Bryksin and Yu. A. Firsov, *Proc. Xth Int. Conf. on the Physics of Semiconductors* (1970), p. 767.

45. V. V. Bryskin and Yu. A. Firsov, Fiz. Tverd. Tela (Leningrad)**, 13**, 3246(1971) [Sov. Phys. Solid State **13**, 2729 (1971)].

46. V. V. Bryksin and Yu. A. Firsov, Fiz. Tverd. Tela (Leningrad) **14**, 1857(1972) [Sov. Phys. Solid State 14, 1615 (1972)].

47. V. V. Bryksin and Yu. A. Firsov, Zh. Eksp. Teor. Fiz **61**, 2373 (1971) [Sov. Phys. JETP **34**, 1272 (1971)].

48. V. V. Bryksin and Yu. A. Firsov, Fiz. Tverd. Tela (Leningrad) **14**, 3599 (1972) [Sov. Phys. Solid State **14**, 3019 (1972)].

49. V. V. Bryksin, Fiz. Tverd. Tela (Leningrad) **14**, 2902 (1972) [Sov. Phys. Solid State 14, 2505 (1972)].

50. A. L. Efros, Fiz. Tverd. Tela (Leningrad) **9**, 1152 (1967) [Sov. Phys. Solid State **9**, 901 (1967)].

51. H. Boettger and V.V. Bryksin, *Hopping Conduction in Solids* (Academie-Verlag, Berlin, 1985).

52. W. V. Houston, Phys. Rev. **57**, 184 (1940).

53. S. Maekawa, Phys. Rev. Lett. **24**, 1175 (1970).

54. V. V. Bryksin and Yu. A. Firsov, Solid State Commun. **10**, 471 (1972).

55. V. N. Bogomolov, A. I. Zadorozhnyi, and T.M. Pavlova, Fiz. Tekh. Poluprovodn. **15**, 2029(1981) [Sov. Phys. Semicond. **15**, 1176 (1981)].

56. V. V. Bryksin, Yu. A. Firsov, and S. A. Ktitorov, Solid State Commun. **39**, 385 (1981).

57. V. V. Bryksin, R. de Dios, and Yu. A. Firsov, Fiz. Tverd. Tela (Leningrad) **29**, 1141 (1987) [Sov. Phys. Solid State **29**, 651 (1987)].

58. V. V. Bryksin, Fiz. Tverd. Tela (Leningrad) **29**, 2027 (1987) [Sov phvs. Solid State **29**, 1166(1987)].

59. Yu. A. Firsov, Fiz. Tverd. Tela (Leningrad) **5**, 2149 (1963) [Sov phvs. Solid State **5**, 1566(1963)].

60. L. Friedmann, Phys. Rev. **131**, 2455 (1963).

61. L. Friedmann and T. Holstein, Ann. Phys. **21**, 494 (1963).

62. Yu. A. Firsov, Fiz. Tverd. Tela (Leningrad) **10**, 1950 (1968) [Sov Ph>s. Solid State **10**, 1537 (1969)].

63. V. V. Bryksin and Yu. A. Firsov, Fiz. Tverd. Tela (Leningrad] **12**, 627 (1970) [Sov. Phys. Solid State **12**, 480 (1970)].
64. V. V. Bryksin and Yu. A. Firsov, Fiz. Tverd. Tela (Leningrad) **14**, 463(1972) [Sov. Phys. Solid State **14**, 384 (1972)].
65. V. V. Bryksin and Yu. A. Firsov, Fiz. Tverd. Tela (Leningrad) **15**. 3235 (1973) [Sov. Phys. Solid State **15**, 2158 (1973)].
66. V. V. Bryksin and Yu. A. Firsov, Fiz. Tverd. Tela (Leningrad) **15**. 3344 (1973) [Sov. Phys. Solid State **15**, 2224 (1973)].
67. V. V. Bryksin and Yu. A. Firsov, Fiz. Tverd. Tela (Leningrad) **16**, 811(1974) [Sov. Phys. Solid State **16**, 524 (1975)].
68. V. V. Bryksin and Yu. A. Firsov, Fiz. Tverd. Tela (Leningrad) **16**, 1941 (1974) [Sov. Phys. Solid State **16**, 1266 (1975)].
69. E. Emin and T. Holstein, Ann. Phys. **53**, 439 (1969).
70. I. G. Austin and N. F. Mott, Adv. Phys. **18**, 41 (1969).
71. H. Böttger, V. V. Bryksin and T. Damker, in the present volume.
72. F. J. Morin, Bell Syst. Techn. J. **37**, 1047 (1958).
73. M. Cutler and N. F. Mott, Phys. Rev. **181**, 1336 (1969).
74. K. Schotte, Z. Phys. **196**, 393 (1966)
75. A. L. Efros, Fiz. Tverd. Tela (Leningrad) **9**, 1152 (1967) [Sov. Phys. Solid State **9**,901 (1967)].
76. D. Emin, Phys. Rev. Lett. **35**, 882 (1975).
77. E. K. Kudinov, Fiz. Tverd. Teia (Leningrad) **18**, 1775 (1976) [Sov. Phys. Solid State **18**, 1035 (1976)].
78. Yu. A. Firsov and E. K. Kudinov, Fiz. Tver. Tela **39**, 2159 (1997) [Phys. Solid State **39**, 1930 (1997)].
79. A. S. Alexandrov and N. F. Mott, Rep. Prog. Phys. **57**, 1197 (1994); *'Polarons and Bipolarons'* (World Scientific, Singapore, 1995).
80. A. S. Alexandrov, in the present volume.
81. S. Rabello and Qimiao Si, Europhys. Lett. **60**, 882 (2002)

Magnetic and Spin Effects in Small Polaron Hopping

Harald Böttger[1], Valerij V. Bryksin[2], and Thomas Damker[1]

[1] Institute for Theoretical Physics, Otto-von-Guericke-University, PF 4120,
D-39016 Magdeburg, Germany `Harald.Boettger@Physik.Uni-Magdeburg.DE`
[2] A. F. Ioffe Physico-Technical Institute, Politekhnicheskaya 26, 19526 St.
Petersburg, Russia

1 Introduction

Magneto-transport in the hopping regime is an attractive field of research.
In particular, quantum-mechanical interference effects in hopping magneto-
transport have attracted a great deal of attention.

Holstein [1] was the first to prove that the impact of a magnetic field
on phonon-induced hopping transport of localized charge carriers is not due
to the Lorentz force but originates from quantum interferences. He pointed
out that a magnetic field-dependent contribution to the hopping probability
between two sites arises from the interference between the amplitude for a
direct transition between the initial and final states and the amplitude for
an indirect transition involving intermediate occupancy of a third site. Accordingly, in the presence of a magnetic field $\mathbf{H}$ the familiar two-site model
for studying hopping transport needs to be extended by including at least a
third site. A three-site model allows hopping transitions along paths directed
either clockwise or anticlockwise with respect to a magnetic field, with different transition probabilities, which deflects the movement of the charge carrier
from the direction of an applied electric field $\mathbf{E}$, therefore giving rise to a Hall
effect.

A three-site model was adopted in [2] for studying the hopping Hall effect
in polaronic materials with hexagonal structure. It is worth noticing that such
a model is only the simplest model for taking into account the interference effects in question. In the case of crystalline materials a three-site model may be
used only for crystals with hexagonal structure. In the more complicated case
of a cubic crystal, for instance, one needs to consider the more complicated
four-site model [3, 4].

For a long time, all calculations of hopping magneto-transport were carried out in the linear approximation with respect to $\mathbf{H}$, i.e. they were focused
on the transverse effect (Hall effect). Simultaneously, nonlinearity in hopping
magneto-transport was well known to be governed by the parameter SH/Φ_0,

where S is the area of the triangle shaped by the triad of sites (S is proportional to a^2, where a is the lattice constant) and $\Phi_0 = hc/e$ is the magnetic flux quantum. In contrast, nonlinearity in band transport is governed by the parameter $\omega_c \tau$, where $\omega_c = eH/m^*c$ is the cyclotron frequency and τ is the relaxation time.

Nonlinear hopping transport with respect to $\mathbf{H}$ was first theoretically studied in [5], where longitudinal (magneto-conductance) and transverse (Hall effect) currents were calculated in the small polaron model, on account of electric fields of arbitrary strength. Theoretical inspection of hopping magneto-transport, including hopping transport in systems with strong electron-phonon interaction, has been much stimulated by the recent great interest in disordered systems, especially in transport in disordered systems.

In disordered systems, theory of hopping transport provides rate equations, which in certain conditions can be reduced to random resistor networks, which in turn require suitable methods, such as percolation theory or effective-medium theory, for studying their properties. Hopping magneto-transport in disordered systems can be reduced to a percolation problem including three-site hops, which was first tackled in [6, 7] (cf. also [8]), where the hopping Hall mobility was calculated for strong (small polaron model) as well as weak interaction with phonons. The problem of calculating hopping magneto-resistance in disordered systems is more complicated than that of the hopping Hall effect. The former problem was first tackled in [9–11], not by means of ideas from percolation theory, but with the aid of the effective-medium theory suggested in [12].

Much attention has recently been devoted to the problem of utilizing electron spin in semiconductor electronics. An overview of this evolving subject of spintronics is given in [13]. One central aim of these efforts is the generation, manipulation, and detection of spin currents. A great deal of interest has been focused on the possibility of affecting spin behavior through electrical means by utilizing two-dimensional (2D) structures (semiconductor interfaces or heterostructures) showing Dresselhaus [14] and/or Rashba [15] spin-orbit interaction.

The bulk of spintronics-related literature is concerned with magneto-electrical effects of itinerant electron (band transport) systems. But as suggested by first theoretical studies, similar effects can also be expected for localized charge carriers in the hopping regime. It turns out, that in the presence of a linear Rashba spin-orbit interaction, each hopping path acquires a spin-dependent phase factor of the same form as that in a perpendicular (to the 2D system) magnetic field. Accordingly, in the case of such spin-orbit interaction, three-site hops give rise to fundamental effects, such as spin accumulation and spin-Hall effect [16, 17]. Moreover, ac spin-Hall effect [18] and effects of disorder [19] were recently studied in hopping systems with Rashba spin-orbit interaction.

The present review of magnetic and spin effects in small polaron hopping is organized in the following way: Section 2 is concerned with nonlinear effects with respect to magnetic and electric fields on transport of small polarons in crystals. Section 3 is devoted to transport in the presence of a magnetic field in disordered systems with localized charge carriers (including small polarons). Section 4 deals with effects of spin-orbit interaction on electronic and spin transport in the hopping regime.

2 Hopping Magneto-Transport in the Small Polaron Model

2.1 Rate Equations

On studying transport in the hopping regime we use a Fröhlich-like Hamiltonian, which after the Lang-Firsov small-polaron transformation [20] reads

$$H = \sum_m V_m a_m^\dagger a_m + \sum_\mathbf{q} \hbar\omega_\mathbf{q} \left(b_\mathbf{q}^\dagger b_\mathbf{q} + \frac{1}{2} \right) + \sum_{m,m'} J_{m'm}(\mathbf{H})\Phi_{m'm} a_{m'}^\dagger a_m. \quad (1)$$

Here $a_m^\dagger$ (a_m) is the creation (annihilation) operator of an electron at site m with radius vector $\mathbf{R}_m$ and site energy ϵ_m (for a crystal $\epsilon_m = 0$), $b_\mathbf{q}^\dagger$ ($b_\mathbf{q}$) is the creation (annihilation) operator of a phonon with wave vector $\mathbf{q}$ and frequency $\omega_\mathbf{q}$ (in the small polaron model only a branch of longitudinal optical phonons is usually taken into account),

$$J_{m'm}(\mathbf{H}) = J_{m'm} \exp \left\{ i\frac{e}{2\hbar c} \mathbf{H} \cdot [\mathbf{R}_m \times \mathbf{R}_{m'}] \right\} \quad (2)$$

is the resonance integral between sites m and m' renormalized by the magnetic field,

$$V_m = \epsilon_m - e\mathbf{E} \cdot \mathbf{R}_m, \quad (3)$$

and

$$\Phi_{m'm} = \exp \left\{ \frac{1}{\sqrt{2N}} \sum_\mathbf{q} [\exp(i\mathbf{q} \cdot \mathbf{R}_m) - \exp(i\mathbf{q} \cdot \mathbf{R}_{m'})] \gamma_\mathbf{q}^* b_\mathbf{q}^\dagger + \text{h.c.} \right\} \quad (4)$$

is the multi-phonon operator, where $\gamma_\mathbf{q}$ is the dimensionless electron-phonon coupling constant, and N the total number of sites in the system. Note that the magnetic field enters into the Hamiltonian (1) only through the phase factor in the resonance integral (2). This implies neglection of magnetic field induced deformation of the localized electronic wave functions, which is justified for not too high magnetic fields. Moreover, in the Hamiltonian (1) a term has been neglected which arises from attractive electron-electron interaction due to virtual exchange of phonons [20] and which governs bipolaron pairing [21].

A characteristic feature of the theory of hopping transport is its formulation in terms of the time dependent diagonal elements $\rho_m(t)$ of the density matrix with respect to the site indices. For $\rho_m(t)$ one gets the following rate equation

$$\frac{d\rho_m}{dt} = \sum_{m'} \left[\rho_{m'}(1-\rho_m)W_{m'm} - \rho_m(1-\rho_{m'})W_{mm'} \right]$$

$$+ \sum_{m_1 m_2} \left[\rho_m(1-\rho_{m_1})(1-\rho_{m_2})W'^{(e)}_{m_1 m_2 m} + \rho_{m_1}(1-\rho_m)(1-\rho_{m_2})W^{(e)}_{m_1 m_2 m} \right.$$

$$\left. - (1-\rho_m)\rho_{m_1}\rho_{m_2}W'^{(h)}_{m_1 m_2 m} - (1-\rho_{m_1})\rho_m\rho_{m_2}W'^{(h)}_{m_1 m_2 m} \right], \quad (5)$$

which was derived by using the familiar decoupling procedure [22] for many-particle correlators, neglecting the off-diagonal elements of the density matrix (which govern band transport), taking the Markovian limit, and taking into account two- and three-site hops. Here $W_{m'm}$ is the two-site hopping probability between the sites m' and m:

$$W_{m'm} = \frac{1}{2}\tilde{J}_{mm'}(\mathbf{H})\tilde{J}_{m'm}(\mathbf{H}) \int_{-\infty}^{\infty} dt \exp\left(\frac{i}{\hbar}V_{m'm}t\right) P_{m'm}\left(t + \frac{1}{2}i\hbar\beta\right), \quad (6)$$

where

$$V_{m'm} = V_m - V_{m'}, \quad \beta = 1/kT, \quad \mathbf{R}_{m'm} = \mathbf{R}_{m'} - \mathbf{R}_m,$$

$$P_{m'm}(t) = \exp\left\{ \sum_{\mathbf{q}} \frac{|\gamma_{\mathbf{q}}|^2[1 - \cos(\mathbf{q}\cdot\mathbf{R}_{m'm})]}{N\sinh(\hbar\omega_{\mathbf{q}}\beta/2)} \cos(\omega_{\mathbf{q}}t) \right\} - 1, \quad (7)$$

$$\tilde{J}_{m'm}(\mathbf{H}) = J_{m'm}(\mathbf{H})\exp[-S_T(m'm)], \quad (8)$$

$$S_T(m'm) = \sum_{\mathbf{q}} \frac{|\gamma_{\mathbf{q}}|^2[1 - \cos(\mathbf{q}\cdot\mathbf{R}_{m'm})]}{2N}\cosh(\hbar\omega_{\mathbf{q}}\beta/2). \quad (9)$$

The two-site probabilities (6) do not depend on the magnetic field and are subject to the condition of detailed equilibrium

$$W_{m'm} = W_{mm'}\exp(\beta V_{m'm}). \quad (10)$$

The factor $\exp(-S_T)$ describes the polaronic narrowing of the electronic band. Note that the subtraction of the unity in (7) is a matter of principle, because it ensures the convergence of the time-integrals in the limit as $t \to \infty$. On account of the off-diagonal elements of the density matrix this subtraction leads, in addition to the contribution to the current from hopping, to a contribution to the current from band transport. The latter dominates in polaronic materials at low temperatures.

In (5) the three-site hopping probabilities for electronic and hole transitions are marked with the indices (e) and (h), respectively. One type of electronic three-site probabilities is given by

$$W'^{(e)}_{m_1 m_2 m} = -\frac{2}{\hbar^3} \mathrm{Im}\, \widetilde{J}_{mm_2}(\mathbf{H}) \widetilde{J}_{m_2 m_1}(\mathbf{H}) \widetilde{J}_{m_1 m}(\mathbf{H})$$

$$\times \int_0^\infty dt \int_0^\infty dt' \exp\left[\frac{i}{\hbar}(V_{mm_2} t + V_{mm_1} t')\right]$$

$$\times P\left(t + t' + i\frac{\hbar\beta}{2}\right) P\left(t + i\frac{\hbar\beta}{2}\right) P\left(t' + i\frac{\hbar\beta}{2}\right). \qquad (11)$$

Here and hereafter the indices $m'm$ at the quantity P are omitted, because for nearest-neighbor hops in crystals these quantities (and likewise the quantities S_T (9)) do not depend on the site numbers. In contrast, in disordered systems there is in principle such a dependence. However, it is in general ignored, since it is irrelevant and, as usual, on assuming short-range interaction the term $\cos(\mathbf{q} \cdot \mathbf{R}_{m'm})$ in (7) and (9) may be neglected.

The other type of electronic three-site hopping probabilities is given by

$$W^{(e)}_{m_1 m_2 m} = \frac{2}{\hbar^3} \mathrm{Im}\, \widetilde{J}_{mm_2}(\mathbf{H}) \widetilde{J}_{m_2 m_1}(\mathbf{H}) \widetilde{J}_{m_1 m}(\mathbf{H})$$

$$\times \int_{-\infty}^\infty dt \int_0^\infty dt' \exp\left[\frac{i}{\hbar}(V_{m_1 m} t + V_{m_1 m_2} t')\right]$$

$$\times P\left(t + t' + i\frac{\hbar\beta}{2}\right) P\left(t + i\frac{\hbar\beta}{2}\right) P\left(t' + i\frac{\hbar\beta}{2}\right). \qquad (12)$$

The hole probabilities $W'^{(h)}_{m_1 m_2 m}$ and $W^{(h)}_{m_1 m_2 m}$ can be obtained from the corresponding electronic probabilities $W'^{(e)}_{m_1 m_2 m}$ and $W^{(e)}_{m_1 m_2 m}$, respectively, by replacing $\widetilde{J}_{mm'}(\mathbf{H}) \to -\widetilde{J}^*_{mm'}(\mathbf{H})$ and $V_m \to -V_m$.

The diagram technique for evaluating the transition probabilities is described in [5]. The triad of sites m, m_1, m_2 forms an elementary triangle in a hexagonal structure, so that $\mathbf{R}_{m_1 m_2}$, $\mathbf{R}_{m_2 m}$, $\mathbf{R}_{mm_1}$ are vectors between nearest neighbor sites and, therefore, $\mathbf{R}_{m_1 m_2} + \mathbf{R}_{m_2 m} + \mathbf{R}_{mm_1} = 0$. The magnetic field is perpendicular to the plane of the triangle and is directed along a sixfold axis.

The three-site probabilities possess symmetry properties, which are important in the following. Here we only specify symmetry relations for the electronic probabilities. The corresponding relations for the hole probabilities can be obtained by making use of the rules for the transition from electronic to hole probabilities. The three-site probabilities obey the detailed balance relation

$$W^{(e)}_{m_1 m_2 m}(\mathbf{H}) = W^{(e)}_{mm_2 m_1}(-\mathbf{H}) \exp(\beta V_{m_1 m}), \qquad (13)$$

and also the following relation

$$W'^{(e)}_{m_1 m_2 m}(\mathbf{H}) = W'^{(e)}_{m_2 m_1 m}(-\mathbf{H}). \qquad (14)$$

By comparing (11) and (12) we find that

$$W'^{(e)s}_{m_1 m_2 m} = -W^{(e)s}_{m m_2 m_1}, \tag{15}$$

where the index s designates the symmetrical part with respect to $\mathbf{H}$ of the transition probability. Only the combination $W'^{(e)}_{m_1 m_2 m} + W'^{(e)}_{m_2 m_1 m}$ enters into the rate equations (5). Because of this, with the aid of (14) and (15), the probabilities $W'^{(e)}_{m_1 m_2 m}$ may be replaced by $-W^{(e)s}_{m m_2 m_1}$ in the rate equations (5), which allows the probabilities $W'^{(e)}_{m_1 m_2 m}$ (and likewise $W'^{(h)}_{m_1 m_2 m}$) to be eliminated from (5).

It can be readily seen that the right-hand side of (5), summed over all sites m, yields zero, which implies the conservation of the number of particles in time, $\sum_m d\rho_m / dt = 0$.

The antisymmetric part (a) of the three-site probabilities obeys the following symmetry relationship

$$W^{(e)a}_{m_1 m_2 m} = -W^{(e)a}_{m_1 m m_2}. \tag{16}$$

Note that the relations of detailed balance (10) and (13) guarantee that for

$$\rho_m = f_m = \{\exp[\beta(V_m - \epsilon_F)] + 1\}^{-1}, \tag{17}$$

the right hand side of (5) vanishes. Here f_m is the Fermi distribution of electrons at the sites in an external electric field (ϵ_F is the Fermi energy).

2.2 Hopping Contribution to the Current in (Hexagonal) Crystalline Systems in the Presence of a Magnetic Field

The current density $\mathbf{j}$ in electric and magnetic fields of arbitrary strength is related to the density matrix by the relation

$$\mathbf{j} = \lim_{t \to \infty} \frac{e}{\Omega} \sum_m \mathbf{R}_m \frac{d\rho_m(t)}{dt}, \tag{18}$$

where Ω denotes the volume of the system. On replacing in (18) the term $d\rho_m / dt$ by the right hand side of (5) and, corresponding to the limit $t \to \infty$, ρ_m by f,

$$f = [\exp(-\beta \epsilon_F) + 1]^{-1} \tag{19}$$

(in a crystal, $\epsilon_m = 0$), we obtain

$$\mathbf{j} = \mathbf{j}_2 + \mathbf{j}_3, \tag{20}$$

where $\mathbf{j}_2$ and $\mathbf{j}_3$ are the two-site and three-site contributions, respectively,

$$\mathbf{j}_2 = enf(1 - f) \sum_m \mathbf{R}_m W_{0m}, \tag{21}$$

$$\mathbf{j}_3 = enf(1 - f) \sum_{m,m'} \mathbf{R}_m [(1 - f) W^{(e)}_{0m'm} - f W^{(h)}_{0m'm}], \tag{22}$$

where $n = N/\Omega$ is the concentration of sites.

For strong electron-phonon interaction the two-site probabilities, calculated by utilizing the steepest descent method, are given by

$$W_{0\mathbf{g}} = \frac{\sqrt{\pi}}{2}\frac{J^2}{\sqrt{E_a kT}}\exp\left(-\frac{E_a}{kT} + \frac{e\mathbf{E}\cdot\mathbf{g}}{2kT}\right), \tag{23}$$

where $\mathbf{g}$ is the vector to the nearest neighbor, $J \equiv J_{0\mathbf{g}}$ is the resonance integral between the nearest neighbors (being independent of $\mathbf{g}$ for the structure under consideration),

$$E_a = \sum_{\mathbf{q}} \frac{|\gamma_{\mathbf{q}}|^2 \hbar\omega_{\mathbf{q}}}{4N}(1 - \cos\mathbf{q}\cdot\mathbf{g}) \tag{24}$$

is the activation energy for a polaron hop between nearest neighbors. The relation (23) is valid in the limiting case of temperatures above the Debye temperature, $2kT > \hbar\omega_{\mathbf{q}}$, and in not too high electric fields, if $eEa < 4E_a$ ($a = |\mathbf{g}|$, lattice constant).

As a result we obtain the following expression for the two-site current

$$\mathbf{j}_2 = \frac{\sqrt{\pi}}{2}enf(1-f)\frac{J^2}{\sqrt{E_a kT}}\exp\left(-\frac{E_a}{kT}\right)\sum_{\mathbf{g}}\mathbf{g}\sinh\left(\frac{e\mathbf{E}\cdot\mathbf{g}}{2kT}\right). \tag{25}$$

In the nonlinear regime with respect to $\mathbf{E}$ the current depends on the angle between the electric field and the axes of symmetry of the crystal. The directions of $\mathbf{E}$ and $\mathbf{j}$ coincide only in the case of motion along axes of high symmetry.

For $\mathbf{E}$ directed along a sixfold axis in a hexagonal crystal we find

$$\sum_{\mathbf{g}}\mathbf{g}\sinh\left(\frac{e\mathbf{E}\cdot\mathbf{g}}{2kT}\right) = 2a\frac{\mathbf{E}}{E}\left\{\sinh\left(\frac{eEa}{2kT}\right) - \sinh\left(\frac{eEa}{4kT}\right)\right\} \tag{26}$$

and for $\mathbf{E}$ perpendicular to it

$$\sum_{\mathbf{g}}\mathbf{g}\sinh\left(\frac{e\mathbf{E}\cdot\mathbf{g}}{2kT}\right) = 2\sqrt{3}a\frac{\mathbf{E}}{E}\sinh\left(\frac{eEa\sqrt{3}}{4kT}\right). \tag{27}$$

In the Ohmic regime only, the directions of $\mathbf{E}$ and $\mathbf{j}$ always coincide and in the case of a hexagonal crystal we have

$$\sum_{\mathbf{g}}\mathbf{g}\sinh\left(\frac{e\mathbf{E}\cdot\mathbf{g}}{2kT}\right) \approx \frac{3eEa^2}{2kT}, \tag{28}$$

and, therefore, the two-site current takes the familiar form [22, 23]

$$\mathbf{j}_2 = \frac{3\sqrt{\pi}}{4}enu_0 f(1-f)\frac{J^2}{kT\sqrt{E_a kT}}\exp\left(-\frac{E_a}{kT}\right)\mathbf{E}, \tag{29}$$

where $u_0 = ea^2/\hbar$ is a quantity with the dimension of a mobility.

Let us now turn to the three-site contribution to the current. Its antisymmetric part (a) with respect to $\mathbf{H}$ describes the Hall effect and its symmetric part (s) the magneto-resistance. Note that in crystalline materials, the connection between $W^{(e)}$ and $W^{(h)}$ simply reads

$$W^{(h)}_{m_1 m_2 m}(\mathbf{E}, \mathbf{H}) = -W^{(e)}_{m_1 m_2 m}(-\mathbf{E}, -\mathbf{H}). \tag{30}$$

Thus, on decomposing the three-site current $\mathbf{j}_3$ into a symmetric and an antisymmetric part,

$$\mathbf{j}_3 = \mathbf{j}_s + \mathbf{j}_a, \tag{31}$$

from (22) on account of (30), we obtain

$$\mathbf{j}_s = enf(1-f)(1-2f) \sum_{m,m'} \mathbf{R}_m W^s_{0m'm}, \tag{32}$$

$$\mathbf{j}_a = enf(1-f) \sum_{m,m'} \mathbf{R}_m W^a_{0m'm}, \tag{33}$$

where

$$W^s_{0m'm} \equiv W^{(e)s}_{0m'm}, \quad W^a_{0m'm} \equiv W^{(e)a}_{0m'm}.$$

In this way the problem of evaluating the three-site current is reduced to that of the probabilities $W^{(e)}_{m_1 m_2 m}$ defined in (12). For strong interaction with phonons, the time integrals in (12) may be evaluated by means of the steepest descent method, as described in detail in [5]. The resulting three-site probabilities read

$$W^s_{0m'm} = -\frac{\sqrt{\pi} J^3}{2\hbar E_a \sqrt{E_a kT}} \exp\left(-\frac{E_a}{kT} + \frac{e\mathbf{E} \cdot \mathbf{g}}{2kT}\right) \cos\left(\frac{e}{2\hbar c}\mathbf{H} \cdot [\mathbf{g} \times \mathbf{g}']\right), \tag{34}$$

$$W^a_{0m'm} = \frac{\pi J^3}{\sqrt{3}\hbar E_a kT} \exp\left(-\frac{4E_a}{3kT} - \frac{e\mathbf{E} \cdot (\mathbf{g} + \mathbf{g}')}{3kT}\right) \sin\left(\frac{e}{2\hbar c}\mathbf{H} \cdot [\mathbf{g} \times \mathbf{g}']\right), \tag{35}$$

where $\mathbf{g} = \mathbf{R}_{m0}$, $\mathbf{g}' = \mathbf{R}_{m'0}$ are vectors connecting nearest neighbors in a triangle.

Substituting (34) into (32), we get the symmetric with respect to the magnetic field three-site current

$$\mathbf{j}_s = enf(1-f)(2f-1)\frac{\sqrt{\pi} J^3}{\hbar E_a \sqrt{E_a kT}} \exp\left(-\frac{E_a}{kT}\right) \cos\left(2\pi \frac{SH}{\Phi_0}\right)$$
$$\times \sum_{\mathbf{g}} \mathbf{g} \sinh\left(\frac{e\mathbf{E} \cdot \mathbf{g}}{2kT}\right), \tag{36}$$

where $S = \sqrt{3}a^2/4$ is the area of the triangle, and $\Phi_0 = hc/e$ is the magnetic flux quantum. Furthermore, substituting (35) into (33) we find the antisymmetric current (Hall current)

$$\mathbf{j}_a = enf(1-f)\frac{\pi J^3}{\sqrt{3}\hbar E_a kT}\exp\left(-\frac{4E_a}{3kT}\right)\sin\left(2\pi\frac{SH}{\Phi_0}\right)$$
$$\times\sum_{\mathbf{g},\mathbf{g}'}\mathbf{g}\cosh\left(\frac{e\mathbf{E}\cdot\mathbf{g}}{2kT}\right)\sinh\left(\frac{e\mathbf{E}\cdot(2\mathbf{g}'-\mathbf{g})}{6kT}\right). \quad (37)$$

It is interesting to remark that the temperature and $\mathbf{E}$-field dependencies of the two-site current (25) and the symmetric contribution to the three-site current (36) coincide for small polarons. Actually, the currents $\mathbf{j}_2$ and $\mathbf{j}_s$ have the same directions for any orientation of the field with respect to the crystallographic axes and their ratio is given by

$$\frac{j_s}{j_2} = \frac{2J}{E_a}(2f-1)\cos\left(2\pi\frac{SH}{\Phi_0}\right). \quad (38)$$

Let us now consider the antisymmetric contribution to the current (37). In the presence of strong electric fields it is directed along the vector $\mathbf{E}\times\mathbf{H}$ (i.e., it has purely transverse character), if the electric field is directed along axes of high symmetry. In particular if the field $\mathbf{E}$ is directed along a sixfold axis, which coincides with the direction of the x-axis, then $j_{ax}=0$ and

$$j_{ay} = -enf(1-f)\frac{\pi J^3 a}{\hbar E_a kT}\exp\left(\frac{4E_a}{3kT}\right)\sin\left(2\pi\frac{SH}{\Phi_0}\right)\sinh\left(\frac{eEa}{2kT}\right). \quad (39)$$

In the linear approximation with respect to $\mathbf{E}$ and $\mathbf{H}$ this equation turns into the familiar one for crystals [2, 22, 24] which has a purely transverse character and does not depend on the orientation of the vector $\mathbf{E}$ with respect to the crystallographic axes

$$j_{ay} = -enu_0 Ef(1-f)\frac{\sqrt{3}\pi J^3}{8E_a(kT)^2}\exp\left(-\frac{4E_a}{3kT}\right)\frac{u_0 H}{c}. \quad (40)$$

From the ratio between the Hall current (40) and the longitudinal current (29) we obtain for the Hall mobility, $u_H = -\frac{cj_{ay}}{Hj_{2x}}$, the following expression

$$u_H = u_0\frac{\sqrt{\pi}}{2\sqrt{3}}\frac{J}{\sqrt{E_a kT}}\exp\left(-\frac{E_a}{3kT}\right), \quad (41)$$

which differs markedly from the expression for the drift mobility which according to (29) is given by

$$u = u_0\frac{3\sqrt{\pi}}{4}\frac{J^2}{kT\sqrt{E_a kT}}\exp\left(-\frac{E_a}{kT}\right). \quad (42)$$

Thus, for small polarons in the three-site model the off-diagonal components of the conductivity tensor depend on the temperature as $\sigma_{xy}\propto\exp(-4E_a/3kT)$. Accordingly, the activation energy for three-site hops is equal to $\frac{4}{3}E_a$, which proves to be less than the activation energy for two uncorrelated two-site hops, $2E_a$. The activation energy for the Hall transport $\frac{4}{3}E_a = \frac{2}{3}E_p$ (E_p - polaron shift, $2E_a = E_p$) can be viewed as arising from a configuration (caused by a thermal fluctuation) in which the energy of each of the two other sites is reduced by $\frac{1}{3}E_p$, so that the levels are equalized and a resonance transition may occur. In this configuration, the three sites are surrounded by polarization clouds of the same kind.

The fact that for small polarons the Hall mobility (41) and the drift mobility (42) exhibit different activation energies needs to be taken into account in interpreting experimental findings. However, recall that the above results were obtained by means of a three-site model, which applies to hexagonal crystals. In the case of cubic crystals one needs to consider a four-site model, which yields, generally speaking, another ratio between the Hall and drift mobilities.

Let us draw attention to the so-called p-n anomaly. It is well known that in the case of band transport the sign of the Hall effect is altered on changing from electron to hole transport. However, in the case of hopping transport in the framework of the three-site model such change does not alter the sign of the Hall current (37), because a replacement of f by $(1-f)$ (i.e. $\epsilon_F \to -\epsilon_F$) does not affect (37). This important fact was first obtained in [25] (cf. also [26]). A similar p-n anomaly also exists for the hopping magneto-resistance in the three-site model [5]. On replacing f by $(1-f)$ the sign of the symmetrical part with respect to the magnetic field of the current (36) changes.

On studying the drift mobility, it turns out that below some temperature, the hopping mechanism ceases to dominate, the tunnelling mechanism becomes operative, and at very low temperatures the band mechanism is effective.

The situation is similar in the case of the Hall effect. In the band regime, where the width of the polaron band is given by $J\exp(-S_T)$, the Hall effect is described by means of the Boltzmann equation in the momentum representation [2]. An intricate problem is its theoretical description in the tunnelling regime [24, 25]. Here the Hall effect is due to quantum interferences and not due to the Lorentz force. In [27], the Hall mobility was theoretically studied in a large temperature range taking into account the various mechanisms contributing to it.

According to [27], the Hall mobility is given by

$$u_H = u_0 \frac{3\sqrt{\pi}J}{2\hbar\omega_{op}} \frac{\nu}{\Theta} e^{-\Theta\cosh\alpha} \frac{1 + \frac{\sqrt{2\Theta^3}}{\nu}e^{\Theta} + \frac{\Theta^3}{\nu^2\sqrt{27}}\frac{\exp(3\Theta\cosh(\alpha/3))}{\cosh(\alpha/3)}}{1 + \frac{\Theta^{3/2}}{2\nu}e^{2\Theta}}, \tag{43}$$

where ω_{op} is the frequency of an optical phonon, $\alpha = \hbar\omega_{op}/2kT$, $\Theta = \gamma/\sinh\alpha$, γ the electron-phonon coupling constant, $\nu = \gamma^2\frac{\Delta\omega}{\omega_{op}}\frac{J^4}{E_a^4}$, and $\Delta\omega$ the width of the optical phonon branch. In the temperature range, which is covered by (43), the drift mobility is described by

$$u = u_0 \frac{6\sqrt{\pi}J^2}{(\hbar\omega_{op})^2} \frac{\nu\alpha}{\Theta^2} e^{-2\Theta\cosh\alpha}\left[1 + \frac{\Theta^{3/2}}{2\nu}e^{2\Theta}\right]. \tag{44}$$

From a numerical inspection of (43) and (44) [22, 27] one can learn that there is a temperature range, where with increasing temperature, u_H rapidly decreases, while u increases, which is considered to be an important feature of the tunnelling regime.

Experimental examination of the hopping Hall effect in small polaron materials meets with the difficulty of measuring a small signal. In [28, 29] such an examination in TiO_2 in a large temperature range is reported, which exhibits, with increasing temperature, a decrease of the Hall mobility, while the drift mobility increases. Note that in the hopping regime of small polaron materials, as generally in low mobility materials, drift and Hall mobilities differ drastically, whereas they coincide in the case of band transport. Therefore, in such materials, an inspection of electronic conductivity and Hall effect does not allow drift mobility and charge carrier concentration to be determined. Additional methods are required in order to determine these quantities.

In [28] the concentration of polarons was obtained by measuring the Seebeck coefficient S, which, in the hopping regime, is related to the site occupation probability as [30]

$$S = \frac{k}{e} \ln \frac{1-f}{f}. \tag{45}$$

Time-of-flight experiments allow the drift mobility to be measured directly [22, 31]. In such an experiment, in a sample sandwiched between plane metal electrodes, a sheet of charge carriers is injected at one electrode, for instance, by a light flash, and the external current induced by the propagation of the excess carrier pulse is monitored.

Discarding the diffusive spreading of the carrier pulse, the transient time t_c of the pulse is given by $t_c = L/uE$ (L - distance between the electrodes, E - electric field), which provides the drift mobility u.

We close this section by presenting theoretical results on ultrasmall polarons in a magnetic field. The term ultrasmall is used for polarons with a radius as small as the root of the mean-square thermal ionic displacement [32, 33]. If the resonance integral between two sites is assumed to depend on the distance R between the sites as

$$J(R) = J_0 e^{-\alpha R}, \tag{46}$$

where α^{-1} is the ionic radius, then the condition of existence of ultrasmall polarons reads $\alpha^2 \langle x^2 \rangle \geq 1$, where $\langle x^2 \rangle$ is the mean-square thermal displacement. This condition requires sufficiently high temperatures and a small ionic radius (met, e.g., in transition metals).

Theoretical studies [32, 33] predict a transition in u from the familiar activated (Arrhenius type) behavior of small polarons to a (Berthelot type) dependence $\ln u \propto AT$, as a characteristic feature of ultrasmall polarons in the temperature dependence of the drift mobility. Such a transition has been experimentally observed in various materials, such as $Ti_n O_{2n-1}$, $V_n O_{2n-1}$ [32], $Ti_{1-x} Nb_x O_2$ [34], $Ta_2 O_5$ [35], $CuYr_2 S_4$ [36], $V_2 O_3$ [37]. Note that according to [38], the numerical estimation [32] of the compliance of the condition of existence of ultrasmall polarons is too optimistic.

In [33], a theory of charge carrier transport in the absence of a magnetic field is formulated in the model of ultrasmall polarons. This theory predicts for hexagonal crystals the following temperature dependence of the drift mobility

$$u = u_0 \frac{3\sqrt{\pi}}{4} \frac{J^2}{kT\sqrt{E_a kT}} \exp\left(-\frac{E_a}{kT} + \zeta kT\right), \tag{47}$$

where

$$\zeta = \sum_{\mathbf{q}} \frac{4\delta_{\mathbf{q}}}{N\hbar\omega_{\mathbf{q}}}(1 - \cos(\mathbf{q}\cdot\mathbf{g})), \tag{48}$$

$\delta_{\mathbf{q}} = \alpha^2\hbar/(M\omega_{\mathbf{q}})$, M is the ionic mass, and α^{-1} the radius of the ionic wave function defined in (46). Note that, at temperatures above the Debye temperature, $\zeta kT = 2\alpha^2\langle x^2\rangle$, which is the fundamental dimensionless parameter of the theory of ultrasmall polarons.

In [39] the Hall effect of ultrasmall polarons is theoretically studied in hexagonal crystals with the aid of the three-site model. This study predicts the following temperature dependence of the Hall mobility

$$u_H = u_0 \frac{\sqrt{\pi}}{2\sqrt{3}} \frac{J}{\sqrt{E_a kT}} \exp\left(-\frac{E_a}{3kT} + \frac{1}{8}\zeta kT\right). \tag{49}$$

It is interesting to compare the expressions for the drift mobilities of small (42) and ultrasmall (47) polarons and to note that the transition from Arrhenius to Berthelot type temperature dependence, characteristic of the model of ultrasmall polarons, occurs only at the very high temperature $T_c \approx \sqrt{E_a/\zeta}/k$. Accordingly, for experimental verification of this transition one should choose materials with not too large activation energy E_a.

For the Hall mobility, the analogous transition occurs at yet higher temperature $T_{cH} \approx \sqrt{8E_a/3\zeta}/k$.

Let us yet draw attention to the paper [40] devoted to a theoretical examination of light absorption and nonlinear current-voltage characteristic in the model of ultrasmall polarons.

3 Hopping Magneto-Transport in Disordered Systems

3.1 Hall Effect

Hopping transport in disordered systems is described by the rate equations (5), with the hopping probabilities (6), (11), and (12). However, it is rather difficult to find a solution to the rate equations (5), which is due to the fact that the intersite distances $R_{m'm}$ and, therefore, the resonance integrals $J_{m'm}$ (46), and, in addition, the site energies ϵ_m are random quantities in a disordered system.

In the absence of a magnetic field, and on assuming only two-site hops, the problem of evaluation of the hopping current in the linear approximation (with respect to the electric field $\mathbf{E}$) reduces to the problem of averaging a random resistor network (Miller-Abrahams network) [41] with resistors

$$Z_{m'm} = \left[e^2 \beta f_{m'}(1 - f_m) W^{(0)}_{m'm} \right]^{-1}, \tag{50}$$

where f_m is the thermal equilibrium occupation (17) of site m and $W^{(0)}_{m'm}$ is the two-site hopping probability (6) for $\mathbf{E} = 0$.

In the case of frequency dependent hopping transport (ac conductivity) the corresponding equivalent scheme includes, in addition to the resistors (50), capacitors

$$C_m = e^2 \beta f_m (1 - f_m), \tag{51}$$

connecting each site m with a voltage source $-\mathbf{E} \cdot \mathbf{R}_m$ [42].

The dc case may be related to the classical percolation problem as follows [43, 44]: The resistors $Z_{m'm}$ are all removed from the network and after that replaced one by one, the smallest first. The value of $Z_{m'm}$ at which the first infinite cluster occurs is Z_c, the threshold value (critical resistor) in the language of percolation theory.

A study of hopping Hall effect requires inclusion of three-site hops in (5), in addition to two-site ones, and linearization of (5) with respect to $\mathbf{E}$ and $\mathbf{H}$, in order to get ρ_m^{EH}, the linear-in-$\mathbf{E}$-and-$\mathbf{H}$ contribution to the density matrix ρ_m.

Introducing $\delta\mu_m^{EH}$, the change of the chemical potential in the linear approximation with respect to $\mathbf{E}$ and $\mathbf{H}$, as [6]

$$\rho_m^{EH} = \beta \delta\mu_m^{EH} f_m (1 - f_m), \tag{52}$$

then this quantity obeys the relation

$$-i\omega C_m \delta\mu_m^{EH} = \frac{1}{e} \sum_{m'} \frac{\delta\mu_{m'}^{EH} - \delta\mu_m^{EH}}{Z_{m'm}} + i_m, \tag{53}$$

where $Z_{m'm}$ and C_m are given by (50) and (51), respectively, and

$$i_m = e^2 \beta \sum_{m_1 m_2} (U_{m_1} - U_m) W^H_{m_1 m_2 m}, \tag{54}$$

$$W^H_{m_1 m_2 m} = f_{m_1}(1 - f_{m_2})(1 - f_m) W^{(e)H}_{m_1 m_2 m} + (1 - f_{m_1}) f_{m_2} f_m W^{(h)H}_{m_1 m_2 m}.$$

Here i_m has the dimension of a current, $U_{m_1} - U_m$ is the difference of the potentials in the Miller-Abrahams network, and $W^H_{m_1 m_2 m}$ is the linear (with respect to the magnetic field) three-site probability (cf. (12)). In the case of strong electron-phonon interaction, the latter quantity is given by

$$W^{H}_{123} = \frac{\pi}{16\sqrt{3}} \frac{\cosh\left\{\left[\frac{1}{3}(\epsilon_1 + \epsilon_2 + \epsilon_3) - \epsilon_F\right]/2kT\right\}}{\cosh[(\epsilon_1 - \epsilon_F)/2kT]\cosh[(\epsilon_2 - \epsilon_F)/2kT]\cosh[(\epsilon_3 - \epsilon_F)/2kT]}$$

$$\times \frac{e}{\hbar c}\mathbf{H}\cdot[(\mathbf{R}_3 - \mathbf{R}_1)\times(\mathbf{R}_2 - \mathbf{R}_1)]$$

$$\times \frac{J_0^3}{\hbar E_a kT}\exp\left\{-\alpha\left[|\mathbf{R}_1 - \mathbf{R}_2| + |\mathbf{R}_2 - \mathbf{R}_3| + |\mathbf{R}_3 - \mathbf{R}_1|\right]\right\}$$

$$\times \exp\left\{-\left[\frac{4}{3}E_a + \frac{1}{24E_a}\left((\epsilon_1 - \epsilon_2)^2 + (\epsilon_2 - \epsilon_3)^2 + (\epsilon_3 - \epsilon_1)^2\right)\right]/kT\right\}, \quad (55)$$

where the energy ϵ_i and the position $\mathbf{R}_i$ of site i are random quantities. On deriving (55), the analytical expression (46) was used for the spacing dependence of the resonance integral. Furthermore, it was ignored that in disordered systems the activation energy (cf. (24)) is in principle a fluctuation quantity, due to its dependence on the hopping distance and the site-number dependence of the electron-phonon coupling constant. Taking the fluctuation of E_a into account may be important on comparing theoretical results with experimental findings. However, here and henceforth we ignore the randomness of E_a.

For zero frequency (dc Hall effect), ω must be set to zero in (53), so that the capacitors C_m drop out of the problem.

Equation (53) reveals that in disordered systems the problem of the hopping Hall effect is reducible to the problem of calculating the Hall current of a random resistor network (including also capacitors, for $\omega \neq 0$) with "extrinsic" currents i_m (induced by the magnetic field due to quantum interferences) running into/away from site m. The latter problem needs to be solved in two steps.

At first, the Miller-Abrahams network (in the absence of a magnetic field) must be studied, in order to obtain the potentials U_m, which are needed for calculating the extrinsic currents with the aid of (54).

Thereafter, substituting these currents into (53), the quantities $\delta\mu_m^{EH}$ can be determined, which are required for calculating the Hall current. The latter step implies appropriate configuration averaging of the resistor network quoted.

Such a program was realized in [6, 7] (see also [22]). On using ideas of percolation theory, the main difference between the theory of the Hall effect and electrical conductivity is that the conductivity is governed by the critical resistor Z_c on the percolation path, while the Hall effect is governed by triads of resistors at branchings of percolation paths. The concentration of such branchings (triads) is small and the distance between them is of the order of the correlation length of the critical percolation cluster. According to [6], all three resistors in the triads governing the hopping Hall effect are of equal magnitude Z_c. This result allows the Hall mobility of small polarons to be written down readily.

Doing so, two limiting cases need to be discriminated. In the limit of high temperatures, where the relevant hops take place far away from the Fermi level, the hopping transport is governed by the random distance between the sites. The corresponding model is termed R-percolation. According to percolation theory, the site spacing (critical hopping length) characterizing a critical resistor in R-percolation is given by $\eta\nu^{-1/3}$, where ν is the concentration of sites, and $\eta = 0.85$ is a numerical coefficient. Thus, the Hall mobility of small polarons in disordered systems reads (cf. (41))

$$u_H = b_H \frac{e\nu^{-2/3}}{\hbar} \frac{J_0}{\sqrt{E_a kT}} (\alpha\nu^{-1/3})^{-3} \exp\left(-\eta\alpha\nu^{-1/3} - \frac{E_a}{3kT}\right), \qquad (56)$$

where b_H is a numerical factor whose magnitude remains open in percolation theory. According to [4] numerical calculations yield $b_H = 28$.

On comparing (56) with the Hall mobility of small polarons in crystals (41), we note that a switch from (41) to (56) implies replacement of the lattice constant a by the critical spacing $\eta\nu^{-1/3}$, introduction of an additional factor $(\alpha\nu^{-1/3})^{-3}$, and appearance of an exponential dependence on the concentration of sites ν, as $\propto \exp(-\eta\alpha\nu^{-1/3})$. Note that applicability of percolation theory requires $\alpha\nu^{-1/3} \gg 1$, i.e. a low concentration of sites. Let us yet draw attention to the papers [8, 46, 47], where hopping Hall mobility is theoretically investigated by other methods (Green function method, effective-medium theory). However, in these papers only models with weak electron-phonon interaction are studied. The obtained results are similar to those derived by means of percolation theory [6, 7].

Let us now turn to the limiting case of low temperatures, where hopping transport is governed by hops near the Fermi level. The corresponding percolation model is termed R-ϵ percolation, because random distributions of both site spacings and site energies are important for forming the percolation cluster. In this model, for weak interaction with phonons, the drift mobility depends on temperature according to Mott's law $\sigma \propto \exp[-(T_0/T)^{1/4}]$ [48], where $T_0 = c\alpha^3/kG(\epsilon_F)$ is a characteristic temperature, $c \approx 20$ is a numerical coefficient, and $G(\epsilon_F)$ is the density of states at the Fermi level. Furthermore, in this model the major contribution to the hopping Hall mobility comes from triads of sites forming an equilateral triangle with side length equal to $(3/8)\alpha^{-1}(T_0/T)^{1/4}$ and with the energy at each site equal to $(1/4)kT(T_0/T)^{1/4}$ [49] (see also [50]). For the related temperature dependence of the Hall mobility of small polarons one finds $\ln u_H \propto -\lambda(T_0/T)^{1/4} - E_a/3kT$, where $\lambda = 3/8$ is a numerical coefficient. For λ, other values are given in [51] ($\lambda = 0.354$) and in [52] ($\lambda = 0.36$, calculated numerically). However, in [6] (see also [53]) it was obtained that $\lambda = 0$. The reason of this discrepancy was elucidated in [11], where the Hall mobility was studied by means of an effective-medium theory. In this paper, it was shown that both appropriate effective-medium theory and proper application of percolation theory provide $\lambda = 0$.

In the papers [11, 53, 54], for weak interaction with phonons, the Hall mobility is studied in the R-ϵ percolation regime. The results obtained differ from the corresponding results for strong coupling with phonons by the absence of the factor $\exp(-E_a/3kT)$, which is characteristic of small polarons. In particular, the effective-medium theory [11] (if $\lambda = 0$) provides for weak coupling with phonons a power-like temperature dependence of the Hall mobility in the R-ϵ regime.

Let us now turn to the frequency dependence of hopping transport. As is well known, strong frequency dependence of the conductivity at low frequencies is a characteristic feature [22] (related to the phenomenon of dispersive transport) of disordered systems. To realize this, note that with increasing frequency during half a period of the external electric field, charge carriers can move through site clusters of decreasing size, so that with increasing frequency highly conducting regions of finite size become more and more effective, that is, the critical hopping distance decreases with increasing frequency, which gives rise to an increase of conductivity with rising frequency.

Effective-medium theory [12] (cf. [22]) provides for the ac hopping conductivity $\sigma_{xx}(\omega)$ the following transcendental equation

$$\frac{\sigma_{xx}(\omega)}{\sigma_{xx}(0)} \ln \frac{\sigma_{xx}(\omega)}{\sigma_{xx}(0)} = i \frac{\omega}{\omega_0}, \tag{57}$$

where ω_0 is a characteristic frequency of the order of the probability of a critical hop on a percolation path (cf. [12, 22, 55]),

$$\hbar \omega_0 = \frac{J_0^2}{\sqrt{E_a kT}} \exp\left(-\frac{E_a}{kT} - 2\alpha R_c\right),$$

where R_c is the critical hopping length. This relation describes very well the bulk of experimental findings.

For the frequency dependence of the Hall effect the effective-medium theory [55] provides the following relation [11]

$$\frac{\sigma_{xy}(\omega)}{\sigma_{xy}(0)} = \left[\frac{\rho_c(\omega)}{\rho_c(0)}\right]^\tau \exp\left\{c[\rho_c(0) - \rho_c(\omega)]\right\}. \tag{58}$$

Here, for R-percolation, the parameters τ, c, and $\rho_c(0)$ are given by $\tau = 7$, $c = 3/2$, and $\rho_c(0) = 2\alpha R_c = 2\alpha \eta \nu^{-1/3}$. For R-ϵ percolation, the parameters c and $\rho_c(0)$ are 1 and $(T_0/T)^{1/4}$, respectively. In this case, the parameter τ equals 0 for strong coupling with phonons and 1 for weak coupling. The quantity $\rho_c(0)$ is twice the dimensionless critical hopping distance in the infinite percolation cluster (i.e. for $\omega = 0$) and $\rho_c(\omega)$ describes its frequency dependence.

From (58) it is clearly evident that $\sigma_{xy}(\omega)$ increases with increasing frequency, similarly as $\sigma_{xx}(\omega) = \sigma_{xx}(0) \exp[\rho_c(0) - \rho_c(\omega)]$ (cf. [11, 12]).

Further inspection of (58) requires specification of the parameter ω/ω_0. Most interesting is the range $\omega/\omega_0 \gg 1$, where one gets [11]

$$\rho_c(0) - \rho_c(\omega) \approx \ln\left[\frac{i\omega/\omega_0}{\ln(i\omega/\omega_0)}\right]. \tag{59}$$

Substituting (59) into (58), one finds that for R-ϵ percolation the Hall mobility increases only weakly with frequency, as $u_H \propto \ln[\omega/\omega_0]$. However, the situation changes drastically in the case of R-percolation, where with increasing frequency the real part of the ratio $\sigma_{xy}(\omega)/\sigma_{xy}(0)$ changes sign, so that

$$\mathrm{Re}\frac{u_H(\omega)}{u_H(0)} = -\frac{\sqrt{2}}{\pi}\sqrt{\frac{\omega}{\omega_0}\ln\left(\frac{\omega}{\omega_0}\right)}. \tag{60}$$

A detailed examination of the frequency dependence of the hopping Hall effect is given in [11], including a numerical inspection of (58).

Owing to the lack of suitable experimental data it is hard to compare theoretical ac Hall results to experiments. The first theoretical study of the ac Hall effect was carried out in [1] (for weak coupling with phonons) in the framework of the three-site model. However, the experimental study [57] of frequency-dependent Hall effect in the R-percolation regime could not confirm the results of [1], not least as it was performed in a frequency range where the study [1] loses its applicability. In [58] ac Hall measurements are reported for the small-polaron material $80V_2O_5$-$20P_2O_5$. In this study, the activation energy of the Hall mobility was not found to be equal to a third of the activation energy of the drift mobility. However, the materials studied in [58] do not exhibit hexagonal structure. Therefore, the experimental findings [58] do not clearly disprove theoretical prediction.

3.2 Magneto-Resistance

Hopping magneto-resistance has been the subject of numerous theoretical studies. In disordered materials such a study is complicated by the necessity of performing a configuration average. But the first move is to linearize the rate equation (5) with respect to the electrical field. Above, we considered linearization of (5) also with respect to $\mathbf{H}$, which is sufficient for studying the Hall effect, but not for inspection of magneto-resistance.

In [5, 59] the rate equation (5) was obtained in the linear approximation with respect to $\mathbf{E}$, but for magnetic fields of arbitrary strength. The resulting equation is reminiscent of the familiar equation for the chemical potential in the case of ac hopping conductivity (in the absence of a magnetic field), which is equivalent to a random resistor network with capacitors

$$-i\omega C_m(\delta\mu_m^E + e\mathbf{E}\cdot\mathbf{R}_m) = \sum_{m'}\frac{\delta\mu_{m'}^E - \delta\mu_m^E}{\Theta_{m'm}}. \tag{61}$$

Here $\delta\mu_m^E$ is the linear contribution (with respect to $\mathbf{E}$) to the chemical potential at site m, $\Theta_{m'm}$ the resistance in the presence of $\mathbf{H}$ defined by $\Theta_{m'm}^{-1} = Z_{m'm}^{-1} + \widetilde{Z}_{m'm}^{-1}$, where $Z_{m'm}$ is the familiar resistance in the two-site model (50), and $\widetilde{Z}_{m'm}$ is the resistance of a resistor connected in parallel, which originates from three-site hops and depends on the magnetic field,

124   Harald Böttger, Valerij V. Bryksin, and Thomas Damker

$$\widetilde{Z}_{m'm}^{-1} = e^2 \beta \sum_{m_1} \left\{ f_{m'}(1 - f_m)(1 - f_{m_1}) W_{m'm_1m}^{0(e)} + (1 - f_{m_1}) f_m f_{m_1} W_{m'm_1m}^{0(h)} \right\},$$

$$(62)$$

where the three-site probabilities $W_{m'm_1m}^{0(e,h)}$ are defined in (12) and the superscript 0 indicates $\mathbf{E} = 0$.

Despite the familiar appearance, equation (61) is much more intricate than the corresponding one for hopping at $\mathbf{H} = 0$, because $\Theta_{m'm}$ does not only depend on the two sites m' and m, but also on the third sites m_1, which impedes application of standard methods of averaging to the network at hand.

For studying magneto-resistance it is sufficient to take into account only the symmetric part (34) with respect to $\mathbf{H}$ of the three-site probabilities. Thus, for small polarons in disordered systems, we get [59]

$$Z_{m'm}^{-1} = \frac{e^2}{16kT} \frac{\sqrt{\pi}}{\hbar\sqrt{E_a kT}} \frac{J_0^2 \exp\left[-2\alpha|\mathbf{R}_{m'm}| - \frac{E_a}{kT}\right]}{\cosh[(\epsilon_F - \epsilon_m)/2kT]\cosh[(\epsilon_F - \epsilon_{m'})/2kT]}. \tag{63}$$

$$\widetilde{Z}_{m'm}^{-1} = \frac{e^2}{8kT} \frac{J_0^3 \exp(-E_a/kT)}{\hbar E_a \sqrt{E_a kT}} \sum_{m_1} \exp\left\{-\alpha(|\mathbf{R}_{m'm}| + |\mathbf{R}_{m'm_1}| + |\mathbf{R}_{m_1m}|)\right\}$$

$$\times \frac{\tanh[(\epsilon_F - \epsilon_{m_1})/2kT]}{\cosh[(\epsilon_F - \epsilon_m)/2kT]\cosh[(\epsilon_F - \epsilon_{m'})/2kT]}$$

$$\times \cos\left\{\frac{e}{2\hbar c}\mathbf{H} \cdot [\mathbf{R}_{mm_1} \times \mathbf{R}_{mm'}]\right\}. \tag{64}$$

Here, the triad of vectors $\mathbf{R}_{mm'} + \mathbf{R}_{m'm_1} + \mathbf{R}_{m_1m} = 0$ forms a triangle.

As already noted, examination of (61), with the resistors (63) and (64) is a rather intricate problem. In particular, application of percolation theory to this problem is very complicated.

Corresponding results (see, e.g., [60–62], and also [9], which includes a detailed bibliography) are contradictory, even with respect to the sign of the magneto-resistance and its magnetic-field dependence at $\mathbf{H} \to 0$.

In [11], an effective-medium theory was proposed for examining equation (61). The basic idea of any effective-medium theory consists in replacing a disordered system by some effective ordered medium, whose parameters are subsequently chosen such that it describes the properties of the actual medium as well as possible. Such a theory was at first proposed in [63] for studying the conductivity in a disconnected network. An effective-medium theory requires in general a definition of an appropriate ordered reference lattice, which implies the definition of nearest neighbors.

However, this is an intricate problem in positionally disordered materials. Furthermore, such a lattice is difficult to find in the case of three-site hops in a magnetic field. The effective-medium theory [11] of hopping magneto-transport does not rely on the introduction of an ordered reference lattice. It avoids a reference lattice by utilizing continuous coordinates instead of discrete ones. To this end, in [11] the structural factor

$$\eta(\rho) = \sum_m \delta(\mathbf{R} - \mathbf{R}_m)\delta(\epsilon - \epsilon_m)$$

is introduced, a random quantity in the four-dimensional space of coordinates and energy, $\rho = \{\mathbf{R}, \epsilon\}$. This approach has proved successful in investigating various transport problems in the case of R-percolation as well as R-ϵ percolation, including hopping Hall effect.

In [11], on account of three-site hops, the self-consistency equation for determining the effective medium is formulated by utilizing a diagrammatic technique. This procedure yields the following expression for the hopping magneto-conductivity [65, 66]

$$\delta\sigma_{\Uparrow,\perp}(\mathbf{H}) = \frac{4\pi^2 e^2}{3kT} \int\limits_0^\infty dR_1 dR_2 \int\limits_{|R_1-R_2|}^{R_1+R_2} dR_3 \int\limits_{-\infty}^\infty d\epsilon_1 d\epsilon_2 d\epsilon_3 R_1 R_2 R_3$$
$$\times\, G(\epsilon_1)G(\epsilon_2)G(\epsilon_3)\gamma(\rho_1,\rho_2,\rho_3)D^{-2}(\rho_1,\rho_2,\rho_3)g_{\Uparrow,\perp}(h)$$
$$\times\, \{(b_{123} - b_{213})(R_1^2 b_{123} - R_2^2 b_{213}) + R_3^2 b_{123} b_{213}\}. \tag{65}$$

Here $\rho_i = \{R_i, \epsilon_i\}$, R_i is the length of side i of the triangle (from the triad of sites contributing to (65)), ϵ_i the energy of site i (vertex i of the triangle), $G(\epsilon)$ the density of states,

$$b_{ikl} - 1 + 2f\Gamma_{kl} + f\Gamma_{il},$$
$$D = 1 + 2f(\Gamma_{12} + \Gamma_{13} + \Gamma_{23}) + 3f^2(\Gamma_{12}\Gamma_{13} + \Gamma_{12}\Gamma_{23} + \Gamma_{13}\Gamma_{23}),$$
$$\Gamma_{ik} = \nu_p \exp\left(-2\alpha|\mathbf{R}_{lk}| - \frac{|\epsilon_F - \epsilon_i| + |\epsilon_F - \epsilon_k|}{2kT}\right),$$
$$\nu_p = \frac{\sqrt{\pi}}{2\hbar}\frac{J_0^2}{\sqrt{E_a kT}}\exp\left(-\frac{E_a}{kT}\right). \tag{66}$$

Here, the quantity ν_p is given for strong coupling with phonons (small polarons). The self-consistency parameter f, defined by $f\nu_p = \exp[2\alpha R_c(\omega)]$ depends on the frequency ω of the electric field through the critical hopping length $R_c(\omega)$. As already noted, for zero frequency, the critical hopping length is equal to $0.85\nu^{-1/3}$ for R-percolation and to $(2\alpha)^{-1}(T_0/T)^{1/4}$ for R-ϵ-percolation.

The decrease of the critical hopping length with increasing frequency, due to shrinkage of the critical cluster size, is governed by the self-consistency equation for determining the effective medium

$$2\alpha[R_c(0) - R_c(\omega)]\exp\{2\alpha[R_c(0) - R_c(\omega)]\} = i\frac{\omega}{\omega_0}, \tag{67}$$

which immediately results from (57), if one takes into consideration that $\sigma_{xx}(\omega)/\sigma_{xx}(0) = \exp[2\alpha(R_c(0) - R_c(\omega))]$.

The quantity γ, in (65), is related to the three-site hopping probability. For the small-polaron model, it is given by

$$\gamma(\rho_1, \rho_2, \rho_3) = \nu_p \frac{J_0}{E_a} \tanh\left(\frac{\epsilon_F - \epsilon_3}{2kT}\right)$$

$$\times \exp\left[-\alpha(|\mathbf{R}_{12}| + |\mathbf{R}_{13}| + |\mathbf{R}_{23}|) - \frac{|\epsilon_F - \epsilon_1| + |\epsilon_F - \epsilon_2| + |\epsilon_F - \epsilon_3|}{2kT}\right]. \tag{68}$$

Furthermore,

$$g_{\Uparrow}(h) = \frac{3}{2}\left(1 + \frac{d^2}{dh^2}\right)\frac{\sin(h)}{h} - 1, \tag{69}$$

$$g_{\perp}(h) = \frac{1}{2h}\frac{d}{dh}\left(h^2 g_{\Uparrow}(h)\right). \tag{70}$$

Here h is the dimensionless magnetic field

$$h = \frac{eH\sqrt{4R_1^2 R_2^2 - (R_1^2 + R_2^2 - R_3^2)^2}}{4\hbar c} = 2\pi\frac{HS}{\Phi_0},$$

which is equal to the number of flux quanta Φ_0 penetrating the area S of the triangle formed by the sides R_1, R_2, and R_3. The subscripts $\Uparrow$ and $\perp$, respectively, indicate parallel (longitudinal) and perpendicular (transverse) orientation of the magnetic field with respect to the electrical field.

According to (70), the longitudinal part of the magneto-conductivity $\delta\sigma_{\Uparrow}(H)$ differs from the transverse part $\delta\sigma_{\perp}(H)$. Equation (70) entails that

$$\delta\sigma_{\perp}(H) = \frac{1}{2H}\frac{d}{dH}\left(H^2 \delta\sigma_{\Uparrow}(H)\right). \tag{71}$$

This means, from a phenomenological point of view, that the magnetic-field-induced change of the current $\delta\mathbf{j}_H$ may be written as

$$\delta\mathbf{j}_H = \delta\sigma_{\perp}(H)\mathbf{E} + [\delta\sigma_{\Uparrow}(H) - \delta\sigma_{\perp}(H)]\frac{(\mathbf{E}\cdot\mathbf{H})\mathbf{H}}{H^2}, \tag{72}$$

where $\delta\sigma_{\perp}(H) \neq \delta\sigma_{\Uparrow}(H)$, according to (71).

To go ahead with the examination of (65), we need to consider separately R-percolation and R-ϵ percolation. Let us start with the static limiting case, where $\omega = 0$, and in (65) we have $f = \nu_p^{-1}\exp(2\alpha R_c(0))$. In the case of R-percolation, where we may choose the density of states as $G(\epsilon) = \nu\delta(\epsilon)$, from (65) we get

$$\frac{\delta\sigma_{\Uparrow}(H,0)}{\sigma} = -a\frac{\pi}{24}\frac{J_0}{E_a}\frac{\nu^{2/3}}{\alpha^2}\tanh\left(\frac{\epsilon_F}{2kT}\right)\left(\frac{\pi H}{\alpha^2\Phi_0}\right)^2, \tag{73}$$

where $\alpha \approx 13$ is a numerical coefficient and $\delta\sigma(H,0) \equiv \delta\sigma(H,\omega)|_{\omega=0}$. The quantity σ denotes the static conductivity of small polarons in disordered systems, in the case of R-percolation. In the framework of the effective-medium theory [12], it is given by

$$\sigma = \frac{\pi^{3/2}}{15}(0.85)^5 \frac{e^2 \nu^{1/3}}{\hbar} \frac{J_0^2}{kT\sqrt{E_a kT}} \exp\left(-\frac{E_a}{kT} - 1.7\alpha\nu^{-1/3}\right). \qquad (74)$$

On deriving (73), the magnetic field was assumed to be weak, $H/(\alpha^2\Phi_0) \ll 1$. It is worth noticing that, unlike the Hall mobility, the ratio $\delta\sigma(H)/\sigma$ (73) includes neither an activated temperature dependence nor an exponential dependence on the concentration of sites ν. With regard to the absence of activated dependence, recall that the same thing happens to small polarons in crystals (cf. (38)). Furthermore, the absence of an exponential concentration dependence originates from the fact that the leading contribution to the integration with respect to coordinates in (55) arises from isosceles triangles, with two sides of the order of the critical hopping length $R_c = 0.85\nu^{-1/3}$ and one side of the order of α^{-1} [66], in contrast to the Hall effect, which is governed by triads of sites forming equilateral triangles with side lengths of the order of the critical hopping length R_c.

For R-ϵ percolation, the analogue to (73) and (74) is given by

$$\frac{\delta\sigma_{\Uparrow}(H,0)}{\sigma} = -\frac{7\pi^3}{22}\alpha^{-3}\frac{J_0 T_0}{E_a T}\left(\frac{H}{80\alpha^2\Phi_0}\right)^2 \int\limits_{-\infty}^{\infty} d\epsilon\, G(\epsilon)\tanh\left(\frac{\epsilon_F - \epsilon}{2kT}\right), \quad (75)$$

$$\sigma = \frac{\pi^{3/2}}{630}\frac{e^2}{\hbar}\frac{J_0^2\sqrt{kT}}{\sqrt{E_a}}\left(\frac{T_0}{T}\right)^{7/4} G^2(\epsilon_F)\exp\left[-\frac{E_a}{kT} - \left(\frac{T_0}{T}\right)^{1/4}\right]. \qquad (76)$$

Note that, just as in (73), there is no activated temperature dependence in (75). It is worth noticing that both expressions (73) and (75) exhibit a p-n anomaly, i.e. the sign of the magneto-conductance changes if $\epsilon_F \to -\epsilon_F$ (transition from electron-like to hole-like transport).

Thus, in the region of weak magnetic fields, the magneto-conductivity depends quadratically on H, $\delta\sigma_{\Uparrow}(H) \propto H^2$. In this region, according to (71), it holds that

$$\delta\sigma_{\perp}(H) = 2\delta\sigma_{\Uparrow}(H). \qquad (77)$$

Consequently, in disordered systems the magneto-conductivity exhibits a dependence on the angle between electric and magnetic fields, i.e. an anisotropy, even in the region of weak magnetic fields. In [67] such an anisotropy was experimentally observed in n-type GaAs samples in a strong electric field. Here it was measured that $\delta\sigma_{\perp}/\delta\sigma_{\Uparrow} = 1.94$, which is well covered by (77). However, in other experimental studies (cf., e.g., [68–70]) such an anisotropy was not observed. For a detailed discussion of this problem, see [66].

Thus, for weak magnetic fields, magneto-conductivity depends quadratically on the magnetic field, as described by (73) and (75). For moderate fields, (65) predicts a linear field dependence, and for high fields a saturation of magneto-conductivity. On passing to the saturation regime, the magnetic flux through a critical triangle of sites approaches a flux quantum Φ_0. In this regime of "quantizing" magnetic fields the anisotropy of magneto-conductivity diminishes and, unlike the situation in crystals, the magneto-conductivity does not exhibit quantum oscillations in dependence on the strength of the magnetic field. This absence of oscillations is due to the averaging over the areas of the triangles, which in disordered systems are randomly oriented with respect to the magnetic field.

Let us now turn to the frequency dependence of the magneto-conductivity, which in the theoretical approach under consideration is solely governed by the frequency dependence of the critical hopping length $R_c(\omega)$. For not too high frequencies, if $|R_c(0) - R_c(\omega)| \ll R_c(0)$, the frequency dependence of the magneto-conductivity is described by the relation

$$\frac{\delta\sigma(H,\omega)}{\delta\sigma(H,0)} \approx \frac{\sigma(\omega)}{\sigma} - i\frac{n}{2\alpha R_c}\frac{\omega}{\omega_0} \tag{78}$$

and only weakly differs from the frequency dependence of the conductivity $\sigma(\omega)$ [66], as $\alpha R_c \gg 1$. Here $n = 1$, $R_c = 0.85\nu^{-1/3}$ for R-percolation, and $n = 4$, $R_c = (2\alpha)^{-1}(T_0/T)^{1/4}$ for R-ϵ percolation, and the frequency dependence of the conductivity $\sigma(\omega)$ is given by the relation (57).

In contrast, for high frequencies, if $\omega \gg \omega_0$, the frequency dependences $\delta\sigma(H,\omega)$ and $\sigma(\omega)$ differ markedly. In this frequency region displacement currents become decisive. Therefore, one may adopt the familiar two-site model [71] (cf. also [22]) for getting $\sigma(\omega)$ and the three-site model for magneto-transport. In this frequency region, and if $\omega < \nu_p$, the ratio $\mathrm{Re}\,\delta\sigma(H,\omega)/\mathrm{Re}\,\sigma(\omega)$ decreases with increasing frequency rather quickly like [65]

$$\frac{\mathrm{Re}\,\delta\sigma(H,\omega)}{\mathrm{Re}\,\sigma(\omega)} \propto \ln^4\left(\frac{\nu_p}{\omega}\right).$$

In closing this section let us stress that the prediction of anisotropy of magneto-conductivity in disordered systems is a success of the qualification of effective-medium theory for examining magneto-transport as suggested in [66]. In all the preceding theoretical studies, based on conventional effective-medium theory or percolation theory [59–61, 71–74], such anisotropy was not predicted. Anisotropy of magneto-conductivity was only observed in a numerical study [75], but it was ascribed to the finite size of the samples studied.

However, according to the rigorous high frequency results obtained by means of the three-site model [65], anisotropy of magneto-conductivity is expected even in isotropic disordered systems, which implies $\delta\sigma_\Uparrow \neq \delta\sigma_\perp$ in the phenomenological relation (72).

4 Small-Polaron Transport in the Presence of Spin-Orbit Interaction

4.1 Spin Transport in Ordered Hopping Systems

In recent years, control of spin dependent transport in systems with spin-orbit interaction by electrical (cf., e.g., [76, 77]) and optical [78, 79] means and of spin injection at ferromagnet-semiconductor (or nonmagnetic metal) interfaces [80, 81] have attracted a great deal of attention. Thereby, the focus has been on the elucidation of the microscopic mechanisms of the envisaged, phenomenologically well-known, magneto-optical effects and on their possible application in spin electronics [82]. This has stimulated the development of the theory of transport properties in systems with spin-orbit interaction, particularly in two-dimensional systems.

A characteristic feature of transport in such systems is the coupling between the evolution equations of particles (electrons) and spins (magnetic moments). The formulation of spin evolution equations encounters the difficulty of the absence of a conservation law of the total spin, which markedly discerns this problem from that of formulating particle evolution equations. From this difficulty emerge at present plenty of contradictory views and results in the relevant literature, even with regard to the definition of such a fundamental quantity as the spin current [83], which, in our mind, is due to the absence of this quantity in Maxwell's equations. The idea suggests itself to define the spin current through a spin continuity equation ("spin diffusion equation"). However, in spite of all corresponding efforts [17, 83–86], so far no consensus has been reached about the expression for the spin current.

The bulk of theoretical papers in this field of research is devoted to materials with a wide electron gap and weak coupling with phonons and impurities. The effect of spin-orbit interaction on hopping transport is the subject of the papers [16, 17, 89, 109] (small polarons in crystals), and also [18, 19] (disordered systems). As to spin-orbit interaction, there are three models under consideration in two-dimensional systems: the Dresselhaus model [14], the (linear) Rashba model [87, 88], and some combinations of these models.

Though all these models exhibit inversion symmetry breaking as a vital feature, each of them is also characterized by some specific features. Characteristic of the Dresselhaus model is bulk inversion asymmetry, which makes the response to an electric field dependent on the angle between field direction and the direction of the main axes of the corresponding tensor ellipsoid. On the other hand, the Rashba model is characterized by structure inversion asymmetry in two dimensions. In the linear Rashba model the effect of spin-orbit interaction is reminiscent of magnetic field action.

In what follows we adopt the Rashba spin-orbit Hamiltonian, which, for itinerant electrons with effective mass m^* and wave-vector $\mathbf{k}$, reads (except for some irrelevant constant $\hbar^2 K^2/(4m^*)$) as

$$H_{so} = \frac{\hbar^2}{2m^*} \left(\mathbf{k} - [\boldsymbol{\sigma} \times \mathbf{K}]\right)^2, \tag{79}$$

where $\boldsymbol{\sigma}$ denotes the vector of Pauli's spin matrices, and $\mathbf{K}$ is a vector, which has the dimension of an inverse length. It is perpendicular to the two-dimensional plane, and its length signifies the Rashba spin-orbit interaction strength.

For studying hopping transport, the Hamiltonian needs to be written in site (Wannier) representation. Taking advantage of the analogy between the Hamiltonian (79) and the Hamiltonian in the presence of a magnetic field (1), after small-polaron canonical transformation, the electron-phonon Hamiltonian in question is given by [17, 89] (cf. (1))

$$
H = \sum_{m,\lambda} V_m a_{m\lambda}^\dagger a_{m\lambda} + \sum_{\mathbf{q}} \omega_{\mathbf{q}} \left(b_{\mathbf{q}}^\dagger b_{\mathbf{q}} + \frac{1}{2} \right) + \sum_{m,m',\lambda,\lambda'} J_{m'm}^{\lambda'\lambda} \Phi_{m'm} a_{m'\lambda'}^\dagger a_{m\lambda},
$$

$$(80)$$

where $a_{m\lambda}^\dagger$ $(a_{m\lambda})$ is the creation (annihilation) operator of an electron at site m in spin state $\lambda = 1,\ 2$, and for the resonance integral between the states $m\lambda$ and $m'\lambda'$ one finds

$$
\begin{aligned}
J_{m'm}^{\lambda'\lambda} &= J_{m'm} \exp\left(i\boldsymbol{\sigma} \cdot [\mathbf{K} \times \mathbf{R}_{m'm}]\right)_{\lambda'\lambda} \\
&= J_{m'm} \left\{ \cos(K^2 R_{m'm}^2)\delta_{\lambda'\lambda} + i\frac{\boldsymbol{\sigma} \cdot [\mathbf{K} \times \mathbf{R}_{m'm}]}{K^2 R_{m'm}^2} \sin(K^2 R_{m'm}^2) \right\}
\end{aligned}
$$

$$(81)$$

(cf. (2)). Formally, the Hamiltonian in the presence of spin-orbit interaction emerges from the standard small-polaron Hamiltonian by replacing in the latter one $J_{m'm} \to J_{m'm}^{\lambda'\lambda}$ and ascribing two spin states λ to each site.

Accordingly, in the theory of hopping transport the site occupation probability now becomes a 2×2 matrix in spin space

$$
\rho_{\lambda'}^\lambda(m, t) = \langle a_{m\lambda'}^\dagger a_{m\lambda} \rangle_t,
$$

$$(82)$$

which is governed by the rate equation (on account of two-site hopping probabilities)

$$
\frac{d\rho_{\lambda'}^\lambda(m)}{dt} = \sum_{m',\lambda_1,\lambda_2} \rho_{\lambda_2}^{\lambda_1}(m) W_{\lambda_2\lambda'}^{\lambda_1\lambda}(m', m),
$$

$$(83)$$

where, unlike in (5), the one-particle approximation $\rho_{\lambda'}^\lambda(m) \ll 1$ was adopted.

The hopping probabilities $W_{\lambda_2\lambda'}^{\lambda_1\lambda}(m', m)$ may be obtained by means of the standard small-polaron diagram technique [22] by associating therein each interaction point with an additional factor

$$
\left\{ \cos(K^2 R_{m'm}^2)\delta_{\lambda'\lambda} + i\frac{\boldsymbol{\sigma}_{\lambda'\lambda} \cdot [\mathbf{K} \times \mathbf{R}_{m'm}]}{K^2 R_{m'm}^2} \sin(K^2 R_{m'm}^2) \right\}
$$

$$
\approx \delta_{\lambda'\lambda} + i\left(\boldsymbol{\sigma}_{\lambda'\lambda} \cdot [\mathbf{K} \times \mathbf{R}_{m'm}]\right). \quad (84)
$$

The latter approximation requires sufficiently small spin-orbit coupling, $Ka \ll 1$ (a is the lattice constant).

For disordered systems, it is a rather difficult problem to solve the set of rate equations (83) analytically, whereas for ordered systems, where $W(m', m) = W(m' - m)$, the problem is easier and can be tackled by transforming the quantities into wave vector space

$$\rho_{\lambda'}^{\lambda}(\boldsymbol{\kappa}) = \sum_m \rho_{\lambda'}^{\lambda}(m)\exp(i\boldsymbol{\kappa}\cdot\mathbf{R}_m), \tag{85}$$

so that (83) becomes

$$\frac{d\rho_{\lambda'}^{\lambda}(\boldsymbol{\kappa})}{dt} = \sum_{\lambda_1,\lambda_2} \rho_{\lambda_2}^{\lambda_1}(\boldsymbol{\kappa}) W_{\lambda_2\lambda'}^{\lambda_1\lambda}(\boldsymbol{\kappa}), \tag{86}$$

It is advantageous to introduce, instead of $\rho_{\lambda'}^{\lambda}$, the density matrices ρ and $\boldsymbol{\rho}$ for particles and spins, respectively,

$$\rho = \sum_{\lambda} \rho_{\lambda}^{\lambda},$$

$$\boldsymbol{\rho} = \sum_{\lambda,\lambda'} \rho_{\lambda'}^{\lambda}\boldsymbol{\sigma}_{\lambda\lambda'}. \tag{87}$$

Analogous to hopping in a magnetic field, hopping in the presence of spin-orbit interaction requires account of three-site hopping probabilities (in hexagonal crystals), which govern such important effects as spin accumulation and spin current.

Consider at first the right-hand side of (86) on account of two-site hopping probabilities. In the equation of the particle density matrix ρ spin-orbit interaction drops out (just this fact requires account of three-site hopping probabilities) and, ignoring quadratic and higher order corrections with respect to $\boldsymbol{\kappa}$, in the linear approximation with respect to the electric field $\mathbf{E}$ the expression in question becomes

$$-\left[D\kappa^2 + iu\boldsymbol{\kappa}\cdot\mathbf{E}\right]\rho, \tag{88}$$

where u is the drift mobility of small polarons, and $D = kTu/e$ is the diffusion constant. For the spin density matrix $\boldsymbol{\sigma}$, on account of two-site hopping probabilities, the right-hand side of (86) reads

$$-\left[D\kappa^2 + iu\boldsymbol{\kappa}\cdot\mathbf{E} + \frac{A}{\tau}\right]\boldsymbol{\rho} - 4D\left\{\left[\mathbf{K}\times\left(i\boldsymbol{\kappa} - \frac{e\mathbf{E}}{2kT}\right)\right]\times\boldsymbol{\rho}\right\}, \tag{89}$$

where $\tau^{-1} = 4DK^2$ is the spin relaxation time, and $A = 1$ for $\rho_{x,y}$ and $A = 2$ for ρ_z.

Consider now the three-site hopping probabilities, which in the absence of spin-orbit interaction take the form (11), (12). Generalizing these probabilities with the aid of (81), the right-hand side of the rate equations for ρ becomes (cf. [17])

$$-i\frac{u_H}{\tau}\left\{\frac{1}{e}\left(\boldsymbol{\kappa}\cdot[\mathbf{K}\times\boldsymbol{\rho}]\right) + \frac{(\boldsymbol{\kappa}\,[\mathbf{K}\times\mathbf{E}])\,(\mathbf{K}\cdot\boldsymbol{\rho})}{2kTK^2}\right\}. \tag{90}$$

Here u_H is the Hall mobility of small polarons (41). And, similarly, the right hand side for $\boldsymbol{\rho}$ takes the form

$$- \frac{u_H}{\tau} \left\{ \frac{1}{e} \left[\mathbf{K} \times \left(i\boldsymbol{\kappa} - \frac{e\mathbf{E}}{kT} \right) \right] + i \frac{(\boldsymbol{\kappa} \, [\mathbf{K} \times \mathbf{E}]) \, \mathbf{K}}{2kTK^2} \right\} \rho. \tag{91}$$

Here, contributions proportional to ρ, which are small for weak spin-orbit coupling K, are neglected. Note that in [17] only the first terms in the braces (90), (91) are taken into account, but the second terms of higher order in K, being proportional to K^3, are ignored. O. Bleibaum turned our attention to the important role of these ignored terms in [17]. In particular, these terms govern spin accumulation.

Now, on account of (88) to (91), the rate equations (86) for the particle and spin density matrices become

$$\frac{d\rho}{dt} = - \left[D\kappa^2 + iu\boldsymbol{\kappa} \cdot \mathbf{E} \right] \rho - i \frac{u_H}{\tau} \left\{ \frac{1}{e} (\boldsymbol{\kappa} \cdot [\mathbf{K} \times \boldsymbol{\rho}]) + \frac{(\boldsymbol{\kappa} \, [\mathbf{K} \times \mathbf{E}]) \, (\mathbf{K} \cdot \boldsymbol{\rho})}{2kTK^2} \right\}, \tag{92}$$

$$\frac{d\boldsymbol{\rho}}{dt} = - \left[D\kappa^2 + iu\boldsymbol{\kappa} \cdot \mathbf{E} + \frac{A}{\tau} \right] \boldsymbol{\rho} - 4D \left\{ \left[\mathbf{K} \times \left(i\boldsymbol{\kappa} - \frac{e\mathbf{E}}{2kT} \right) \right] \times \boldsymbol{\rho} \right\}$$
$$- \frac{u_H}{\tau} \left\{ \frac{1}{e} \left[\mathbf{K} \times \left(i\boldsymbol{\kappa} - \frac{e\mathbf{E}}{kT} \right) \right] + i \frac{(\boldsymbol{\kappa} \, [\mathbf{K} \times \mathbf{E}]) \, \mathbf{K}}{2kTK^2} \right\} \rho. \tag{93}$$

In the literature, there are a number of papers devoted to the derivation of evolution equations for electron and spin densities in the presence of an external electric field [83–86, 90, 91], considering itinerant electrons, which are only weakly scattered by impurities. In all of these papers, instead of switching on a true electric field, a particle concentration is introduced, i.e. the idea of an electro-chemical potential is adopted. However, such an approach is justified only in the case of a conserved quantity, i.e. for a particle density, but not for a spin density. In our study of hopping transport, we do not adopt the idea of an electro-chemical potential. Therefore, the form of equations (92), (93) differs from corresponding ones in the literature, and the electro-chemical potential occurs only in the last term on the right-hand side of (93).

Let us start our inspection of equations (92), (93) by considering homogeneous systems, i.e. $\boldsymbol{\kappa} = 0$. It is appropriate to study these equations by using the Laplace transform with respect to time. Thus, (92), (93) become

$$s\rho^{(0)} = n, \tag{94}$$

$$(s\tau + A)\boldsymbol{\rho}^{(0)} - 2u\tau \left[[\mathbf{K} \times \mathbf{E}] \times \boldsymbol{\rho}^{(0)} \right] - \frac{u_H}{kT} [\mathbf{K} \times \mathbf{E}] \rho^{(0)} = \boldsymbol{\rho}_0 \tau, \tag{95}$$

where s is the Laplace variable. Here $\boldsymbol{\rho}_0$ is the initial condition of the homogeneous spin distribution, n is the particle concentration, $\rho^{(0)} = \rho(\boldsymbol{\kappa} = 0)$. Solving these equations, we find the following expression for the spin density matrix

$$\boldsymbol{\rho}^{(0)}(s) = \boldsymbol{\rho}_{ac}(s) + \delta\boldsymbol{\rho}(s), \tag{96}$$

$$\boldsymbol{\rho}_{ac} = \frac{u_H}{kT}[\mathbf{K} \times \mathbf{E}]\frac{n}{s(1 + s\tau)}, \tag{97}$$

$$\delta\rho_x = \tau\frac{(s\tau + 2)\rho_{0x} + (\zeta/2)\rho_{0z}}{(s\tau + 1)(s\tau + 2) + (\zeta/2)^2}, \tag{98}$$

$$\delta\rho_y = \frac{\tau\rho_{0y}}{s\tau + 1}, \tag{99}$$

$$\delta\rho_z = \tau\frac{(s\tau + 1)\rho_{0z} - (\zeta/2)\rho_{0x}}{(s\tau + 1)(s\tau + 2) + (\zeta/2)^2}. \tag{100}$$

Here

$$\zeta = E/E_c, \qquad E_c = (4u\tau K)^{-1} = DK/u = KkT/e, \tag{101}$$

where E_c is a characteristic electric field, the physical meaning of which will be discussed below.

The coordinate system is fixed such that $\mathbf{e}_z \propto \mathbf{K}$, $\mathbf{e}_x \propto \mathbf{E}$ and $\mathbf{e}_y \propto \mathbf{K} \times \mathbf{E}$. Then the spin density matrix $\boldsymbol{\rho}_{ac}$ is directed along the y-axis and for large times, $t \gg \tau$, it evolves to the expression

$$\boldsymbol{\rho}_\infty = \frac{u_H}{kT}[\mathbf{K} \times \mathbf{E}]n = u_H[\mathbf{K} \times \mathbf{E}]\frac{dn}{d\epsilon_F}. \tag{102}$$

Here, $n/kT = dn/d\epsilon_F$ was adopted, which is true for Boltzmann statistics. On applying an alternating electric field $\mathbf{E}(t) = \mathbf{E}\exp(-i\omega t)$ to the system in question, then in the stationary regime, a homogeneous magnetic moment $\mathbf{M} = \mu_B\boldsymbol{\rho}$ (μ_B - Bohr magneton) arises, which exhibits a frequency dispersion and which is called spin accumulation in the relevant literature

$$\mathbf{M}_{ac}(\omega) = \frac{\mu_B\boldsymbol{\rho}_\infty}{1 - i\omega\tau}. \tag{103}$$

In electrodynamics this phenomenon is called the magneto-electric effect. The occurrence of a magnetic moment due to an applied electric field was pointed out at first in [92] by inspection of a model with itinerant electrons, Fermi statistics and elastic scattering off impurities. It is worthwhile noting that the expression for spin accumulation (102) utterly agrees with the result of the paper [92], despite the fact that we studied, unlike [92], hopping transport and Boltzmann statistics. To compare both results, the expression for the Hall mobility of itinerant electrons is needed (which coincides with the drift mobility), $u_H = e\tau/n$. The agreement of both results indicates universality of the expression for spin accumulation (102). As to the frequency dispersion $\mathbf{M}_{ac}(\omega)$, for hopping the dispersion is Drude-like, and for itinerary it is more complicated and exhibits resonant behavior [93].

Consider now relaxation of a homogeneous spin state governed by the relations (99) to (100). The component $\delta\rho_y$ exhibits the trivial behavior

$$\delta\rho_y(t) = \rho_{0y}\exp\left(-\frac{t}{\tau}\right). \tag{104}$$

The two other components of the magnetic moment exhibit a more complicated behavior. After inverse Laplace transformation, from (98), (100) one gets

$$\delta\rho_x = \frac{\rho_{0z}}{\sqrt{1-\zeta^{-2}}} \exp\left(-\frac{3t}{2\tau}\right) \sin\left(\frac{t}{2\tau}\sqrt{\zeta^2-1}\right),\tag{105}$$

$$\delta\rho_z = \frac{\rho_{0z}}{\sqrt{1-\zeta^{-2}}} \exp\left(-\frac{3t}{2\tau}\right) \sin\left(\frac{t}{2\tau}\sqrt{\zeta^2-1}+\phi\right),\tag{106}$$

where the difference between the phases of the two components is given by $\phi = \pi/2 + \zeta^{-1}$. The above results were obtained with the initial conditions $\rho_{0x} = 0$, $\rho_{0z} \neq 0$. However, with the initial conditions $\rho_{0x} \neq 0$, $\rho_{0z} = 0$, the result does not change fundamentally.

Thus, according to (105), (106), in an electric field that exceeds the critical one, $E > E_c$, the relaxation of magnetic moment, which at time $t = 0$ was in the x-z plane, is, apart from a decay factor, characterized by its rotation in the x-z plane with frequency $\sqrt{\zeta^2-1}/2\tau$. Thereby, the magnetic moment $\boldsymbol{\rho}(t)$ describes a slanted ellipse, given by $x^2 + 2xz\cos\phi + z^2 = \sin^2\phi$. This rotation is a distinct characteristic feature, if $E \gg E_c$. It is worth noticing that for itinerant electrons such a rotation of a homogeneous magnetic moment in an electric field was also theoretically obtained.

Let us now turn to the particle current $\mathbf{j}$, defined as follows

$$\mathbf{j}(t) = ie\boldsymbol{\nabla}_{\mathbf{k}}\frac{d}{dt}\rho(\boldsymbol{\kappa},t)|_{\boldsymbol{\kappa}=0}.\tag{107}$$

By differentiating (92) with respect to $\mathbf{k}$ and putting $\mathbf{k} = 0$, we obtain

$$\mathbf{j}(t) = eun\mathbf{E} + \frac{u_H}{\tau}\left\{\frac{e[\mathbf{K}\times\mathbf{E}]\left(\mathbf{K}\cdot\boldsymbol{\rho}^{(0)}(t)\right)}{2kTK^2} + [\mathbf{K}\times\boldsymbol{\rho}^{(0)}(t)]\right\},\tag{108}$$

where $\boldsymbol{\rho}^{(0)}(t)$ is given by the equations (97) to (100) (in Laplace representation). For the initial condition $\boldsymbol{\rho}_0 = 0$, when $\boldsymbol{\rho}^{(0)} = \boldsymbol{\rho}_{ac}(s)$, and assuming an alternating electric field $\mathbf{E}(t) = \mathbf{E}\exp(-i\omega t)$, the particle current becomes

$$\mathbf{j}(\omega) = \left\{1 - \frac{(2u_H/u_K)^2}{1-i\omega\tau}\right\}enu\mathbf{E},\tag{109}$$

where $u_K = e/(\hbar K^2)$ has the dimension of a mobility.

Accordingly, in the stationary regime the current is directed along the electric field and the conductivity σ_{xx} is only weakly affected by spin-orbit interaction, giving rise to a contribution $\propto K^4$. An important property of this contribution is its frequency dependence, which opens the possibility of its experimental detection. An analogous weak renormalization of the longitudinal stationary current due to spin-orbit interaction occurs also in the case of itinerant electrons [93, 94].

The stationary Hall current is determined by the first term in the braces in the right hand side of (108), which is only present for $\rho_z^{(0)} \neq 0$. The possible existence of such an anomalous Hall effect (without a magnetic field) in the presence of spin-orbit interaction was predicted in [95, 96]. In an infinite crystal and for zero magnetic field, the condition for anomalous Hall effect is compliant with magnetic moment initially (at $t = 0$) placed in the x-z plane. In particular, if at $t = 0$ the magnetic moment is directed along the z-axis, then, from (108) and (106), we find the following time-dependence of the Hall current

$$j_{Hy}(t) = u_H \rho_{0x} E \frac{2K^2 u}{\sqrt{1 - \zeta^{-2}}} \exp\left(-\frac{3t}{2\tau}\right) \sin\left(\frac{t}{2\tau}\sqrt{\zeta^2 - 1} + \tau\right). \qquad (110)$$

Accordingly, after switching on an electric field, the Hall current decreases with increasing time due to relaxation of the magnetic moment. Additionally, if $E > E_c$, this decrease is superimposed by oscillations caused by rotation of the magnetic moment in the electric field. In this respect, the phenomenon under consideration markedly differs from the anomalous Hall effect in ferromagnetic materials, where a time-independent Hall current is caused by time-independent spontaneous magnetic moments. Obviously, the time-independent anomalous Hall effect due to spin-orbit interaction requires the presence of spontaneous magnetic moments in the system due to an external magnetic field or due to an interface with a ferromagnet.

In the literature, there are a large number of papers devoted to theoretical studies of spin Hall current, which was predicted at first in [97]. Almost all of these papers are concerned with spin current in models with itinerant electrons. Characteristic of these papers is that the results obtained are to a high degree contradictory to each other [93]. In particular, there is no generally accepted definition of the spin current [88, 93, 98]. As already mentioned above, this is obviously due to the absence of the spin current concept in Maxwell's equations. Furthermore, spin current can be detected only indirectly through accumulation of magnetic moments close to a perpendicular contact [99, 100]. Here, we calculate spin current $j_{(s)i}^l$ by adopting a definition analogous to that of the particle current (107)

$$j_{(s)i}^l(t) = \frac{i}{2}\frac{d}{dt}\rho_i^l(t), \qquad (111)$$

$$\rho_i^l(t) = \frac{\partial}{\partial \kappa_i}\rho^{(l)}(\boldsymbol{\kappa}, t)|_{\boldsymbol{\kappa}=0}, \qquad (112)$$

where $l = x$, y, z denotes the projections of the spin density matrix and $i = x$, y indicates the projections on the coordinate axes of the two-dimensional system in question. The relation (112) defines the spin current through the velocity of motion of the spin packet center of gravity. Carrying out differentiation with respect to $\boldsymbol{\kappa}$ in (93) and putting $\boldsymbol{\kappa} = 0$, we find the following equations for the quantities $\rho_y^l(s)$ in Laplace representation

$$(s\tau + 1)\rho_y^x + \frac{\zeta}{2}\rho_y^z = iK\frac{u_H}{e}\frac{n}{s}, \tag{113}$$

$$(s\tau + 2)\rho_y^z - \frac{\zeta}{2}\rho_y^x = iK\frac{u_H}{e}\frac{\zeta}{2}\frac{1 - s\tau}{1 + s\tau}\frac{n}{s}. \tag{114}$$

This solution was obtained by assuming the initial condition $\boldsymbol{\rho}_0 = 0$ and by using (97). In this case we have $\rho_y^y = 0$. Solving the above set of equations, we find the spin current running perpendicular to the direction of the electric field

$$j_{(s)y}^x(s) = -K\frac{u_H}{2e}\frac{n}{1 + s\tau}\left\{1 - \frac{\zeta^2}{4}\frac{2 - s\tau}{(1 + s\tau)(2 + s\tau) + \zeta^2/4}\right\}, \tag{115}$$

$$j_{(s)y}^z(s) = K\frac{u_H n}{2e}\frac{\zeta}{2}\frac{2 - s\tau}{(1 + s\tau)(2 + s\tau) + \zeta^2/4}. \tag{116}$$

According to (115), spin Hall current also exists for $\mathbf{E} = 0$. Formally, this current arises from switching on spin-orbit interaction at $t = 0$ and it decreases with increasing time as

$$j_{(s0)y}^x(t) = -K\frac{u_H}{2e\tau}n\exp\left(-\frac{t}{\tau}\right). \tag{117}$$

It is interesting to note that this current is connected with spin accumulation (97) through the relation

$$\frac{d\rho_{ac}^y(t)}{dt}n = -2eEj_{(s0)y}^x(t)\frac{dn}{d\epsilon_F}. \tag{118}$$

This connection has universal character and it results from the concept of the electro-chemical potential. For a model with itinerant electrons it was obtained in [93].

In the presence of an electric field, an additional contribution to spin-Hall current $j_{(sE)y}^l$ occurs, which in time-representation reads

$$j_{(sE)y}^x(t) = K\frac{u_H n}{2e\tau}\left\{\exp\left(-\frac{t}{\tau}\right) - \exp\left(-\frac{3t}{2\tau}\right)\right.$$
$$\left. \times \left[\frac{1 + \zeta^2/6}{\sqrt{\zeta^2 - 1}}\sin\left(\frac{t\sqrt{\zeta^2 - 1}}{2\tau}\right) + \cos\left(\frac{t\sqrt{\zeta^2 - 1}}{2\tau}\right)\right]\right\}, \tag{119}$$

$$j_{(sE)y}^z(t) = -K\frac{u_H n}{2e\tau}\frac{\zeta}{2}\exp\left(-\frac{3t}{2\tau}\right)$$
$$\times \left\{\frac{7}{\sqrt{\zeta^2 - 1}}\sin\left(\frac{t\sqrt{\zeta^2 - 1}}{2\tau}\right) - \cos\left(\frac{t\sqrt{\zeta^2 - 1}}{2\tau}\right)\right\}. \tag{120}$$

Thus, in the course of relaxation of spin-Hall current, the vector of the magnetic moment rotates in the x-z plane with frequency $\sqrt{\zeta^2 - 1}/(2\tau)$ (if

$E > E_c$), which is analogous to the behavior of spin accumulation discussed above. Note that hitherto in all preceding papers, the theory of spin-Hall effect was focused on the linear response with respect to the electric field. Accordingly, in these papers $j^x_{(sE)} = 0$ and no rotation of the magnetic moment occurs. In this approximation, the frequency dependence of the spin current in the presence of an alternating electric field, $\mathbf{E} \exp(-i\omega t)$, is of interest. The frequency dispersion of the spin current in the linear approximation (with respect to $\mathbf{E}$) may be readily obtained from (116) by replacing $s \to i\omega$

$$j^z_{(s)y}(\omega) = \frac{u_H E n}{4kT} \frac{i\omega(2 + i\omega\tau)}{(1 - i\omega\tau)(2 - i\omega\tau)}. \tag{121}$$

Let us consider a characteristic feature of frequency dependence of spin current: at $\omega \to 0$, the current goes to zero and is proportional to ω^2, and with increasing frequency it crosses over to a plateau $u_H E n/(4kT\tau)$. The transition between these two regions occurs at $\omega \approx \tau^{-1} = 4DK^2$. Such frequency dispersion is also characteristic of itinerant electrons. For the latter case, theory predicts non-zero spin current at $\omega \to 0$ (cf., e.g., [93, 101–103]).

Let us now turn to particle and spin diffusion. Equations (92) and (93) allow us to study transport phenomena in spatially inhomogeneous systems. On assuming zero electrical field, $\mathbf{E} = 0$, after some transformations, these equations may be written as

$$\frac{d\rho}{dt} + D\kappa^2\rho + i\frac{u_H K}{e\tau}\nu_r = 0, \tag{122}$$

$$\frac{d\nu_r}{dt} + \left(D\kappa^2 + \frac{1}{\tau}\right)\nu_r + i\frac{u_H}{e\tau}K\kappa^2\rho = 0, \tag{123}$$

$$\frac{d\nu_d}{dt} + \left(D\kappa^2 + \frac{1}{\tau}\right)\nu_d + 4iD\kappa^2\rho_z = 0, \tag{124}$$

$$\frac{d\rho_z}{dt} + \left(D\kappa^2 + \frac{2}{\tau}\right)\rho_z - 4iDK\nu_d = 0, \tag{125}$$

where we introduced the notations

$$\nu_r = \frac{\mathbf{K} \cdot [\boldsymbol{\kappa} \times \boldsymbol{\rho}]}{K}, \qquad \nu_d = (\boldsymbol{\kappa} \cdot \boldsymbol{\rho}). \tag{126}$$

Hence, at $\mathbf{E} = 0$, the four connected equations governing the density matrices $\boldsymbol{\rho}$ and ρ fall into two pairs of equations for ρ, ν_r and ρ_z, ν_d, respectively. Accordingly, a spatially inhomogeneous particle distribution gives rise to a spin magnetic moment $\boldsymbol{\rho}(\mathbf{r}, t)$ lying in the x-y plane and satisfying the condition $\mathrm{div}\boldsymbol{\rho}(\mathbf{r}, t) = 0$. If an electric field is switched on, the spin magnetic moment starts to rotate around the y-axis, which gives rise to a component ρ_z of the moment perpendicular to the plane of the sample (x-y plane).

For the initial conditions $\rho(\boldsymbol{\kappa}, t = 0) = 1$, $\nu_r(\boldsymbol{\kappa}, t = 0) = 0$ the set of equations (122) and (123) in Laplace space, becomes

$$\rho(\boldsymbol{\kappa}, s) = \frac{s + D\kappa^2 + 1/\tau}{R_+(\boldsymbol{\kappa}, s)\, R_-(\boldsymbol{\kappa}, s)}, \tag{127}$$

$$\nu_r(\boldsymbol{\kappa}, s) = -i\frac{u_H}{e\tau} \frac{K\kappa^2}{R_+(\boldsymbol{\kappa}, s)\, R_-(\boldsymbol{\kappa}, s)}, \tag{128}$$

where

$$R_\pm(\boldsymbol{\kappa}, s) = s + D\kappa^2 + \frac{1}{2\tau}\left[1 \pm \sqrt{1 - \left(\frac{2\hbar u_H}{e}\right)^2 K^2\kappa^2}\right] \tag{129}$$

In the following we restrict ourselves to the lowest order with respect to the spin-orbit coupling constant K. In this approximation the particle diffusion function takes the form customary for a two-dimensional system

$$\rho(\mathbf{r}, t) = \frac{1}{4\pi Dt}\exp\left(-\frac{r^2}{4Dt}\right). \tag{130}$$

Furthermore, on account of the condition $\mathrm{div}\boldsymbol{\rho} = 0$, for the spin diffusion function we obtain

$$\boldsymbol{\rho}(\mathbf{r}, t) = \frac{u_H}{2eDt}[\mathbf{K} \times \mathbf{r}]\left[1 - \exp\left(-\frac{t}{\tau}\right)\right]\rho(\mathbf{r}, t). \tag{131}$$

According to (131), a spatially inhomogeneous particle distribution gives rise to a magnetic moment lying in the plane of the sample. For a punctiform source the equipotential lines of this moment are circles centered at the source. Note that the induced spin magnetic moment under consideration is closely associated with the spin-Hall current (117) occurring in the absence of an electric field.

Let us now study the effect of sample boundaries on particle and spin distributions. For doing so, we use (92), (93), consider the stationary state, when $d\boldsymbol{\rho}/dt = 0$ and $d\rho/dt = 0$, and restrict ourselves to the geometry of a semi-infinite sample in half-space $y > 0$. For such conditions, after going back to the coordinate space by replacing $-i\boldsymbol{\kappa} \to \boldsymbol{\nabla}_{\mathbf{r}}$, equations (92) and (93) become

$$\rho'' + 2\Lambda\rho'_x + 2\Lambda\zeta\rho'_z = 0, \tag{132}$$

$$\rho''_x - \rho_x + \rho_z\frac{\zeta}{2} - 2\Lambda\rho' = 0, \tag{133}$$

$$\rho''_y - \rho_y + 2\rho'_z + \Lambda\zeta\rho = 0, \tag{134}$$

$$\rho''_z - 2\rho_z - 2\rho'_y + 2\Lambda\zeta\rho' - \frac{\zeta}{2}\rho_x = 0. \tag{135}$$

Here, $\Lambda = u_H/u_K$, $u_K = e/(\hbar K^2)$ is a quantity with the dimension of a mobility, and the dashes indicate differentiation with respect to the dimensionless coordinate $\xi = y/\sqrt{D\tau} = 2Ky$.

In general, the solution of this set of equations requires the utilization of numerical methods. In order to obtain analytical solutions, we assume weak spin-orbit coupling, where $\Lambda \ll 1$. For $\Lambda = 0$, the particle density matrix trivially becomes $\rho = n$, where n is the spatially homogeneous equilibrium density of electrons. Hence, the spin components $\boldsymbol{\rho}$ may be written as

$$\rho_x = c \exp(-\xi) + \zeta \mathrm{Re} c_1 \exp(-\lambda\xi), \tag{136}$$

$$\rho_y = -\frac{\zeta}{4} c \exp(-\xi) + 4\mathrm{Re}[\lambda c_1 \exp(-\lambda\xi)] + \Lambda\zeta n, \tag{137}$$

$$\rho_z = -2\mathrm{Re}[(\lambda^2 - 1)c_1 \exp(-\lambda\xi)], \tag{138}$$

where c and c_1 are real and complex constants, respectively, which are determined through the boundary conditions, and

$$\lambda = \lambda_+ + i\lambda_-, \qquad \lambda_\pm = \frac{1}{2}\sqrt{\sqrt{8 + \zeta^2} \pm 1}. \tag{139}$$

The last term on the right hand side of (137), $\Lambda\zeta n = u_H E n/(kT)$, describes spin accumulation in an external electric field (102).

In view of the clumsy form of the final expressions, we do not specify solutions (136) to (138) for definite boundary conditions. Here we discuss the spatial dependence of spin moment in a sample adjoining to a ferromagnet at the interface (cf., e.g., [80]). In the case of such an interface, the magnetic moment is given at the boundary. According to (136) to (138), for $\rho_z \neq 0$ at the boundary, and $\mathbf{E} \neq 0$, the three components of the vector of the magnetic moment decrease in the sample exponentially, as $\exp(-\lambda_+ 2Ky)$, with increasing distance y from the interface, and thereby they oscillate with frequency $2\pi/K$. In the absence of an electric field and for $\rho_x = 0$ at the boundary, then inside the sample ρ_y and ρ_z do not vanish, if at least one of them does not vanish at the boundary. If, at the boundary, $\rho_x \neq 0$, but both the other components of the magnetic moment vanish, then inside the sample only the component ρ_x remains, and it decreases exponentially, as $\exp(-2Ky)$, with increasing distance y from the boundary ($1/2K$ - diffusion length of spin packet). It is worth noticing that the characteristic diffusion length for itinerant electrons, in the absence of an electric field and for weak spin-orbit coupling $\nu_F K\tau \ll 1$ (ν_F - velocity at the Fermi surface, τ - relaxation time for elastic scattering), obtained in [84], completely agrees with (139), if $\zeta = 0$ is put in the latter formula.

Analogously, a finite spin density can also be expected to occur near the boundary if the boundary conditions require zero spin-Hall current across the boundary. In this case an inhomogeneous stationary magnetic moment near the boundary arises from two competing processes — spin current flow towards the boundary owing to spin-Hall current, and relaxation of the occurring spin magnetic moment. Note that in this case the spin-Hall current does not agree with the above defined total spin-Hall current, and it does not include relaxation processes of the spin magnetic moment. If the stationary total spin current tends to zero (cf. (120)), then without relaxation it does not vanish identically. The condition of zero current across the boundary (at $\xi = 0$), without relaxation, is, according to (133) to (135) given by

$$\rho' + 2\Lambda\rho_x + 2\Lambda\zeta\rho_z = 0, \tag{140}$$

$$\rho'_x - 2\Lambda\rho = 0, \tag{141}$$

$$\rho'_y + 2\rho'_z = 0, \tag{142}$$

$$\rho'_z - 2\rho_y + 2\Lambda\zeta\rho = 0. \tag{143}$$

For such boundary conditions, all three components of the spin magnetic moment are finite near the boundary and decay exponentially with increasing distance from the boundary. A more detailed theoretical inspection of the distribution of the magnetic moment for such conditions is given in [17]. Note that just such an inhomogeneous magnetic moment was experimentally observed in GaAs layers [99, 100].

It is worth noticing that usage of the conditions (140), (141) of zero current across the boundary causes inhomogeneous electron and spin concentrations even in the absence of an electric field (for $\zeta = 0$). In [104, 105] arguments are put forward that, in the absence of an electric field, hard-wall boundary conditions give rise to a homogeneous electron concentration and a zero spin magnetic moment. Compliance with this result would require omission of the term $u_H(\mathbf{K} \times i\boldsymbol{\kappa})\rho/e\tau$ on the right hand side of (93), so that the boundary condition (141) becomes

$$\rho'_x = 0. \tag{144}$$

The above adopted approximation of weak spin-orbit coupling ($\Lambda \ll 1$) leaves the electronic subsystem in equilibrium, so that the concentration of electrons remains homogeneous in space. Account of corrections nonlinear in Λ gives rise to a spatially inhomogeneous contribution to the electron density, which induces an inhomogeneous spin magnetic moment. To study this effect analytically, we use equations (132) to (135), in the absence of an electric field ($\zeta = 0$), but for arbitrary values of the parameter Λ.

For $\zeta = 0$, the four equations (132) to (135) fall into two pairs of independent equations for ρ, ρ_x and ρ_y, ρ_z, respectively. This decomposition into two independent sets of equations is characteristic of Rashba spin-orbit coupling [106]. Recall that we already met this decomposition above while studying particle and spin diffusion in zero electrical field. The general solution for the pair of components ρ, ρ_x is given by

$$\rho = n + \rho_x^{(0)} \frac{2\Lambda}{\sqrt{1 - (2\Lambda)^2}} \exp\left(-\sqrt{1 - (2\Lambda)^2}\xi\right), \tag{145}$$

$$\rho_x = \rho_x^{(0)} \exp\left(-\sqrt{1 - (2\Lambda)^2}\xi\right), \tag{146}$$

where $\rho_x^{(0)}$ denotes the magnetic moment at the boundary. This solution holds in the case of $2\Lambda < 1$. In the small polaron model, this condition is undoubtedly fulfilled, as $u_K \approx 10^3 \mathrm{cm}^2/(\mathrm{Vs})$ for $K^{-1} \approx 10^{-6}\mathrm{cm}$, whereas $u_H < 1\mathrm{cm}^2/(\mathrm{Vs})$. Note that the virtue of relations (145), (146) is limited. These relations only indicate the occurrence of a spatially inhomogeneous electron density close to the boundary. Such a phenomenon causes space charge and thus induces an intrinsic electric field, which needs to be determined self-consistently by means of the Poisson equation.

Consider now the pair of equations for ρ_y and ρ_z, the solution of which is given by

$$\rho_y = \mathrm{Re}\left[c\frac{1-\lambda^2}{2\lambda}\exp(-\lambda\xi)\right], \tag{147}$$

$$\rho_z = \mathrm{Re}\left[c\exp(-\lambda\xi)\right], \tag{148}$$

where $\lambda = \left(\sqrt{\sqrt{8}-1}+i\sqrt{\sqrt{8}+1}\right)/2$, and c is a complex constant determined by the boundary conditions. Thus, near the boundary, a stratified magnetic moment occurs, which possesses components ρ_y and ρ_z, if at the boundary at least one of these components is non-zero. It is worth noticing that in [107], by means of numerical methods, such stratified spin magnetic moment (spin coherent standing wave) with components ρ_y and ρ_z was obtained, and it was pointed out that stratification of the magnetic moment is accompanied by long spin relaxation time.

Concluding this section, note that the basic physical phenomena of small polarons in the presence of spin-orbit coupling are governed by equations (92), (93), which in principle describe all transport properties: currents of particles and spins, spin accumulation, spin relaxation, diffusion processes, behavior patterns of spin magnetic moment near a boundary, etc. These equations, derived on assuming hopping transport, are in many respects reminiscent of corresponding equations for itinerant electrons. Accordingly, for both the transport regimes, similar facts and phenomena are predicted, such as spin accumulation (102), relation between spin current and spin accumulation (118), rotation of the vector of the spin magnetic moment in the plane defined by the vector of the electrical field and the normal of the surface of the sample, absence of coupling between the pairs of equations for ρ, $(\mathrm{rot}\boldsymbol{\rho})_z$ and ρ_z, $\mathrm{div}\boldsymbol{\rho}$, respectively, in the absence of an electrical field, and various other phenomena. This allows corresponding equations of motion to be formulated on a phenomenological basis, which are valid for hopping as well as itinerant transport. However, this is true only if $\nu_F K\tau \ll 1$, where ν_F is the velocity at the Fermi surface and τ is the relaxation time for elastic scattering. In the opposite case, there are additional degrees of freedom for itinerant transport, which complicate the equations of motion.

4.2 Spin Transport in Disorded Hopping Systems

Let us now turn to the study of spin transport in a spatially disordered hopping system. Specifically, we consider a two-dimensional structure with Rashba spin-orbit interaction. Due to the disorder, the transition to wave-vector space (as, e.g., in (85)) cannot be applied and we have to investigate the rate equations (83) directly in space representation. The increased complexity furthermore forces us to restrict the consideration to two-site hopping probabilites.

The rate equations corresponding to (83) read [17, 19]

$$\frac{d}{dt}\rho_m = \sum_{m_1} \left\{ \rho_{m_1} W_{m_1 m} - \rho_m W_{m m_1} \right\} \tag{149}$$

for the occupation numbers and

$$\frac{d}{dt}\boldsymbol{\rho}_m = \sum_{m_1} \left\{ \widehat{D}_{m_1 m} \cdot \boldsymbol{\rho}_{m_1} W_{m_1 m} - \boldsymbol{\rho}_m W_{m m_1} \right\} \tag{150}$$

for the spin orientation. Here, $W_{m_1 m}$ is the transition rate between sites m_1 and m, and $\widehat{D}_{m_1 m}$ is a 3×3-matrix, which describes a rotation of the spin during the transition. Note, that the two sets of equations are decoupled in the chosen approximation (due to the restriction to two-site hopping probabilities). This means, that the charge (or particle) transport has no influence on the spin transport and vice versa. Thus, the phenomenon of the spin Hall effect lies outside the scope of the theory presented here.

The aim is to derive macroscopic transport equations starting from the microscopic equations (149), the master equation for particle (respectively charge) transport, and (150), the equation governing the spin evolution. It can be shown [19], that under the approximation of a small Rashba interaction $KR_t \ll 1$, where R_t is a typical hopping length, an (as yet unspecified) disorder average of (149) also gives an averaged equation for spin.

The "generic" averaging procedure follows the procedure used in [108] which is employed to study the connection between the continuous time random walk approach and the exact (generalized) master equation. We assume that the spatially disordered system can be adequately represented using an ordered host lattice with very small lattice constant. Then only some sites of the host lattice correspond to the original sites and are available for transport, whereas all other host lattice sites are unavailable in a specific disorder realization. Using this ordered representation of the originally disordered problem, one can construct a formal solution and thereafter perform the disorder average (see [108]). In this way, one obtains a generalized master equation (GME) for the (averaged or macroscopic) disordered system, which is formally exact. Approximations (as, e.g., the continuous time random walk considered in [108]) are only needed thereafter, in order to obtain explicit expressions for the transition rates of the GME and thus be able to calculate solutions of the GME.

Under the assumption that a product of several rotation matrices $\widehat{D}$ obeys the relation

$$\widehat{D}_{m m_1} \cdot \widehat{D}_{m_1 m_2} \cdot \ldots \cdot \widehat{D}_{m_k n} \approx \widehat{D}_{mn}, \tag{151}$$

the same averaging procedure can be applied to (150). This equation is valid to first order in K. Therefore, the condition $KR_t \ll 1$, assures the applicability of the approximation. The fact, that (151) is valid only to first order in K, gives the reason for the restriction to two-site hopping probabilities.

Since the disorder averaged system is homogeneous, it is now convenient to work in wave-vector space. The long wavelength limit of the disorder averaged rate equations in Fourier-Laplace space are obtained as ($\widehat{I}_3$ is the 3×3 identity matrix and $\circ$ is the dyadic product) [19]

$$\left[s + D(s)\kappa^2 + iu(s)\boldsymbol{\kappa} \cdot \mathbf{E} \right] \rho(s|\boldsymbol{\kappa}) = \rho_0(\boldsymbol{\kappa}) \tag{152}$$

and

$$\left[\left(s + D(s)\kappa^2 + iu(s)\boldsymbol{\kappa} \cdot \mathbf{E} + 4D(s)K^2 \right) \widehat{I}_3 \right.$$

$$\left. + 4D(s)\mathbf{K} \circ \mathbf{K} \right] \cdot \boldsymbol{\rho}(s|\boldsymbol{\kappa})$$

$$+ \left[\mathbf{K} \times (4\mathrm{i}D(s)\boldsymbol{\kappa} - 2u(s)\mathbf{E}) \right] \times \boldsymbol{\rho}(s|\boldsymbol{\kappa}) = \boldsymbol{\rho}_0(\boldsymbol{\kappa}). \tag{153}$$

These evolution equations agree exactly with the corresponding equations of an ordered hopping system (the two-site hopping probability part of (92) and (93)), except that here the diffusion constant D and the mobility u are frequency dependent. This frequency dependence is entirely determined by charge transport and does not depend on spin-orbit interaction within the approximations considered here. Thus, one finds that under the condition $KR_t \ll 1$, the spin dynamics is affected by disorder only through the change of D and u. Note, that the above derivation is independent of the approximations used to derive concrete expressions for $D(s)$ and $u(s)$. In particular, it is *not* restricted to continuous time random walk, the procedure further considered in [108], even though we followed the general approach used there.

Let us consider the temporal evolution of the total spin magnetization. In this case we have to set $\boldsymbol{\kappa} = \mathbf{0}$ in (152) and (153). Then, (152) immediately yields particle number conservation. Writing the corresponding equation for spin polarization in matrix form, while fixing the coordinate system such that $\mathbf{E} \parallel \mathbf{e}_x$ and $\mathbf{K} \parallel \mathbf{e}_z$, one obtains

$$\begin{bmatrix} s + 4DK^2 & 0 & -2uKE \\ 0 & s + 4DK^2 & 0 \\ 2uKE & 0 & s + 8DK^2 \end{bmatrix} \cdot \boldsymbol{\rho}(s) = \boldsymbol{\rho}_0. \tag{154}$$

One can see that in zero electric field $E = 0$, the spin components decay with two different time constants (taking D for the moment as frequency independent): The z-component with $\tau_1 = 1/8DK^2$, and the in-plane components with twice this value $\tau_2 = 1/4DK^2$, i.e. the spin component perpendicular to the plane decays two times faster than the in-plane components, a fact which has also been found for ordered hopping systems [17] and itinerant electrons [85, 86]. Note, in particular, that the spin life-time is inversely proportional to the diffusion constant. Since D substantially decreases with increasing disorder in hopping systems, the spin life-time will strongly increase with increasing disorder.

The frequency dependence of D complicates the matter, but its overall effect is to further increase the decay time constant for later times. Thus, the spin decay slows down further in the progress of time.

Even in a finite electric field, the in-plane component of ρ perpendicular to the field follows a decay law [ρ_y in (154)], whereas the other two components are coupled and have the solution

$$\begin{pmatrix} \rho_x(s) \\ \rho_z(s) \end{pmatrix} = \frac{1}{\det(\Pi)} \Pi \cdot \begin{pmatrix} \rho_{x0} \\ \rho_{z0} \end{pmatrix}, \tag{155}$$

with the matrix

$$\Pi = \begin{bmatrix} s + 8DK^2 & 2uEK \\ -2uEK & s + 4DK^2 \end{bmatrix}. \tag{156}$$

When D and u do not depend on frequency, this corresponds to a sum of exponential functions in time, so that the spin components are either hyperbolic or trigonometric functions of time, times an exponential decay factor [see (105),(106)]. Here, in the disordered case, D and u are frequency dependent, and the transformation of the solution of (154) into t-space is impossible without choosing a specific model for the frequency dependence $D(s)$ and $u(s)$ beforehand. For large times, D and u approach their respective dc-values, i.e., the asymptotic behavior of ρ for large times is easily obtained. On the other hand, due to the exponential decay, the large time behavior is only relevant, if the time constant for dc-behavior is smaller than the time constant for spin decay $1/4DK^2$.

The dimensionless electric field $\zeta = uE/(DK)$ discriminates between two different behaviors of the total spin polarization: exponential decay in small electric fields and an additional oscillation in large electric fields [see (105), (106)]. The occurrence of one or the other regime can be tuned by varying (one or several of) the electric field, the temperature, and the Rashba length. The dimensionless electric field ζ is time-independent if the Einstein relation between D and u is valid, even when D and u themselves depend on time. In this case, ζ furthermore does not depend on the disorder, whereas the quantities D and u are strongly affected. Note, that the conclusions remain the same, whether one considers a single polarized spin initially placed at the origin, or a homogeneously polarized system.

As a second example, we determine the stationary state of a system with an in-plane electric field and spin injection. Taking, as before, $\mathbf{E} \propto \mathbf{e}_x$, and boundaries parallel to the y-axis, the charge and spin densities can only depend on the x-coordinate. Denoting the derivative with respect to x by a prime, one obtains

$$0 = D_0 \rho''(x) - u_0 E \rho'(x) \tag{157}$$

$$0 = D_0 \boldsymbol{\rho}''(x) - u_0 E \boldsymbol{\rho}'(x) - 4D_0 K^2 \boldsymbol{\rho}(x) - 4D_0 K^2 \mathbf{e}_z \rho_z(x)$$
$$- 4D_0 K \left[\mathbf{e}_x \rho_z'(x) - \mathbf{e}_z \rho_x'(x) \right]$$
$$- 2u_0 K E \left[\mathbf{e}_z \rho_x(x) - \mathbf{e}_x \rho_z(x) \right]. \tag{158}$$

Here, $D_0 = \int_0^\infty dt\, D(t)$ and $u_0 = \int_0^\infty dt\, u(t)$ are the dc-values of the corresponding quantities.

Specifically, if one considers a half-plane, and takes the boundary conditions at $x = 0$ and $x = \infty$ to be $\rho(0) = \rho(\infty) = \rho_0$, $\boldsymbol{\rho}(0) = \mathbf{e}_z$, and $\boldsymbol{\rho}(\infty) = 0$, the solution then reads $\rho(x) = \rho_0$,

$$
\begin{aligned}
\rho_x(x) &= e^{-x/\lambda} A_x \sin(x/\Lambda), \\
\rho_y(x) &= 0, \\
\rho_z(x) &= e^{-x/\lambda} \left[\cos(x/\Lambda) - A_z \sin(x/\Lambda)\right],
\end{aligned}
\tag{159}
$$

where the dimensionless electric field $\zeta = \mu_0 E/(D_0 K)$ assumes the dc-value and is the parameter determining the quantities

$$
\omega_\pm = \sqrt{\pm\frac{\zeta^2 - 8}{2} + \frac{1}{2}\sqrt{\zeta^4 + 48\zeta^2 + 512}},
\tag{160}
$$

the amplitudes $A_x = \dfrac{\zeta^2 + 32}{5\omega_- + \sqrt{\zeta^2 + 7\omega_+}}$ and $A_z = \dfrac{5\omega_+ - \sqrt{\zeta^2 + 7\omega_-}}{5\omega_- + \sqrt{\zeta^2 + 7\omega_+}}$, the decay length $\lambda = 2(\omega_+ - \zeta)^{-1}/K$ and the oscillation length $\Lambda = 2/(\omega_- K)$. Note, that the disorder enters only through the ratio u_0/D_0. Thus, provided the Einstein relation between D_0 and u_0 is valid, the spatial behavior of $\boldsymbol{\rho}(x)$ does not depend on the disorder, the relevant length scale being determined mainly by K, since the (dimensionless) quantity ω_- only very weakly depends on its sole parameter ζ (specifically, $3.95 < \omega_- \le 4$).

In conclusion, we have derived macroscopic spin transport equations for spatially disordered hopping systems with Rashba spin-orbit interaction. It is found that the introduction of disorder leaves the vectorial structure of these equations intact, the only effect being that diffusion constant and mobility become frequency dependent (or, expressed differently, that the transport equations obtain memory). This frequency dependence is already determined by the charge transport behavior and does not depend on the specifics of spin transport. The derivation of this relation between ordered and disordered hopping spin transport is subject to the approximation, that the Rashba length (the length scale of spin precession) is large against a typical hopping length. Furthermore, only two-site hopping processes can be dealt with in the present treatment, thus excluding, e.g., the discussion of a possible spin Hall effect.

Other effects, previously predicted for ordered spin hopping, also occur for disordered spin hopping. For instance, the spin decay is exponential in small electric fields, whereas it obtains an oscillatory component in large electric fields. Thus, there is a finite critical field, dividing both regimes, also in the disordered case. It is also conceivable that the behavior "oscillates" between exponential and vibrational behavior. But this can be excluded, so far as $D(t)$ and $u(t)$ change monotonically with time.

In comparison to the ordered case, the spin life-time is strongly increased by the introduction of disorder. This is explained by the significantly reduced diffusion constant, which enters the life-time reciprocally and which sharply decreases with increasing disorder. For the stationary state of spins injected through a boundary into a topologically disorder hopping system, it is found that the spatial behavior of the spin polarization is largely unaffected by varying the disorder.

References

[1]   T. Holstein, Phys. Rev. **121**, 1329 (1961).
[2]   L. Friedman, T. Holstein, Ann. Phys. (N.Y.) **21**, 494 (1963).
[3]   V. V. Bryksin, Yu. A. Firsov, Sov. Phys. Solid State **12**, 627 (1970).
[4]   D. Emin, Ann. Phys. **64**, 336 (1971).
[5]   H. Böttger, V. V. Bryksin, F. Schulz, Phys. Rev. B **48**, 161 (1993).
[6]   H. Böttger, V. V. Bryksin, Phys. Stat. Sol. (b) **80**, 569 (1977); **81**, 97, 433 (1977).
[7]   H. Böttger, V. V. Bryksin, Sol. St. Comm. **23**, 227 (1977).
[8]   L. Friedman, M. Pollak, Phil. Mag. B **38**, 173 (1978).
[9]   O. Bleibaum, H. Böttger, V. V. Bryksin, F. Schulz, Phys. Rev. B **51**, 14020 (1995).
[10] H. Böttger, V. V. Bryksin, F. Schulz, Phys. Rev. B **52**, 16305 (1995).
[11] O. Bleibaum, H. Böttger, V. V. Bryksin, Phys. Rev. B **56**, 6698 (1997).
[12] V. V. Bryksin, Fiz. Tverd. Tela **22**, 2441 (1980) [Sov. Phys. Solid State **22**, 1421 (1980)].
[13] Concepts in Spin Electronics (Series on Semiconductor Science and Technology) ed. by S. Maekawa, Oxford University Press, 2006.
[14] G. Dresselhaus, Phys. Rev. **100**, 580 (1955).
[15] Yu. Bychkov, E. I. Rashba, JETP Lett. **39**, 78 (1984).
[16] T. Damker, V. V. Bryksin, H. Böttger, Phys. Stat. Sol. (c) **1**, 84 (2004).
[17] T. Damker, H. Böttger, V. V. Bryksin, Phys. Rev. B **69**, 205324 (2004).
[18] O. Entin-Wohlman, A. Aharony, Y. M. Galperin, V. I. Kozub, V. Vinokur, Phys. Rev. Lett. **95**, 086603 (2005).
[19] U. Beckmann, T. Damker, H. Böttger, Phys. Rev. B **71**, 205311 (2005).
[20] I. G. Lang, Yu. A. Firsov, Zh. Eksp. Teor. Fiz. **43**, 1843 (1962).
[21] A. Alexandrov, J. Ranninger, Phys. Rev. B **23**, 1796 (1981); B **24**, 1164 (1981).
[22] H. Böttger, V. V. Bryksin, Hopping conduction in solids, Akademie-Verlag, Berlin (1985).
[23] T. Holstein, Ann. Phys. (N.Y.) **8**, 325, 343 (1959).
[24] Yu. A. Firsov in "Polarons", Nauka, Moskwa (1975) (in Russian).
[25] T. Holstein, Phil. Mag. **27**, 255 (1973).
[26] D. Emin, Phil. Mag. B **35**, 1189 (1977).
[27] V. V. Bryksin, Yu. A. Firsov, Sov. Phys. Sol. State **16**, 1941 (1974).
[28] V. N. Bogomolov, E. K. Kudinov, Yu. A. Firsov, Sov. Phys. Sol. State **9**, 3175 (1967).
[29] E. K. Kudinov, D. N. Mirlin, Yu. A. Firsov, Sov. Phys. Sol. State **11**, 2789 (1969).
[30] K. Schotte, Z. Physik **196**, 393 (1966).
[31] G. Pfister, H. Scher, Adv. Phys. **27**, 747 (1978).

[32] C. M. Hurd, J. Phys. C **18**, 6487 (1985).

[33] V. V. Bryksin, JETP **100**, 1556 (1991).

[34] P. Lagnel, B. Poumellec, C. Picard, phys. stat. sol. (b) **151**, 531 (1989).

[35] V. V. Bryksin, S. D. Khanin, Fiz. Tverd. Tela **35**, 2266 (1993).

[36] V. N. Andreev, F. A. Chudnovskiy, S. Perooly, J. M. Honig, phys. stat. sol (b) **234**, 623 (2002).

[37] V. N. Andreev, F. A. Chudnovskij, J. M. Honig, P. A. Metcalf, Phys. Rev. B **70**, 235124 (2004).

[38] D. Emin, Phil. Mag. B **52**, L71 (1985).

[39] V. V. Bryksin, JETP **101**, 1282 (1992).

[40] V. V. Bryksin, JETP **101**, 180 (1992).

[41] A. Miller, B. Abrahams, Phys. Rev. **120**, 745 (1960).

[42] S. Kirkpatrick, Rev. Mod. Phys. **45**, 574 (1973).

[43] V. Ambegaokar, B. I. Halperin, J. S. Langer, Phys. Rev. B **4**, 2612 (1971).

[44] B. I. Shklovskii, A. L. Efros, JETP **60**, 867 (1971).

[45] P. N. Butcher, J. A. McInnes, Phil Mag. B **44**, 595 (1981).

[46] P. N. Butcher, A. A. Kumar, Phil. Mag. B **42**, 201 (1980).

[47] B. Movaghar, B. Pohlmann, W. Schirmacher, Phil. Mag. B **41**, 49 (1980).

[48] N. F. Mott, J. non-crystall. Solids **1**, 1 (1967).

[49] M. Grünewald, H. Müller, P. Thomas, D. Würtz, Sol. St. Comm. **38**, 1011 (1981).

[50] L. Friedman, M. Pollak, Phil. Mag. B **44**, 487 (1981).

[51] R. Ne'meth, B. Mühlschlegel, Sol. St. Comm. **66**, 999 (1988).

[52] R. Ne'meth, Physica B **167**, 1 (1990).

[53] Yu. M. Gal'perin, E. P. German, V. G. Karpov, JETP **99**, 343 (1991) [Sov. Phys. JETP **72**, 193 (1991)].

[54] B. Movaghar, B. Pohlmann, D. Wuertz, J. Phys. C **16**, 3755 (1983).

[55] O. Bleibaum, H. Böttger, V. V. Bryksin, Phys. Rev. B **54**, 5444 (1996).

[56] J. R. Macdonald, Phys. Rev. B **49**, 9428 (1994).

[57] M. Amitay, M. Pollak, J. Phys. Soc. Japan Suppl. **21**, 549 (1966).

[58] V. Vomvas, M. Roilos, Phil. Mag. B **49**, 143 (1984).

[59] H. Böttger, V. V. Bryksin, F. Schulz, Phys. Rev. B **49**, 2447 (1994).

[60] V. L. Nguen, B. Z. Spivak, B. I. Shklovskii, Zh. Eksp. Teor. Fiz. **89**, 1770 (1985) [Sov. Phys. JETP **62**, 1021 (1985)].

[61] W. Schirmacher, Phys. Rev. B **41**, 2461 (1990).

[62] O. Entin-Wohlman, Y. Levinson, A. G. Aronov, Phys. Rev. B **49**, 5165 (1994).

[63] S. Kirkpatrick, Rev. Mod. Phys. **45**, 574 (1973).

[64] V. V. Bryksin, Fiz. Tverd. Tela **26**, 1362 (1984) [Sov. Phys. Solid State **26**, 827 (1984)].

[65] O. Bleibaum, H. Böttger, V. V. Bryksin, Phys. Rev. B **62**, 11450 (2000).

[66] O. Bleibaum, H. Böttger, V. V. Bryksin, Phys. Rev. B **64**, 104204 (2001).

[67] F. Tremblay, M. Pepper, R. Newburry, R. Ritchie, D. C. Peacock, J. E. F. Frost, G. A. C. Jones, Phys. Rev. B **40**, 10052 (1989).

[68] O. Faran, Z. Ovadyahu, Phys. Rev. B **38**, 5457 (1988).

[69] Y. Zhang, M. P. Sarachik, Phys. Rev. B **43**, 7212 (1991).

[70] L. Essaleh, G. Galibert, S. M. Wasim, H. Hernandez, J. Leotin, Phys. Rev. B **50**, 18040 (1994).

[71] M. Pollak, T. H. Geballe, Phys. Rev. B **122**, 1742 (1961).

[72] W. Schirmacher, H. T. Fritzsche, R. Kemper, Sol. St. Comm. **88**, 125 (1993).

[73] H. Fritzsche, W. Schirmacher, Europhys. Lett. **21**, 67 (1993).
[74] O. Entin-Wohlman, Y. Imry, U. Sivan, Phys. Rev. B **40**, 8342 (1989).
[75] E. Medina, M. Kardar, R. Rangel, Phys. Rev. B **53**, 7663 (1996).
[76] H. J. Zhu, M. Ramsteiner, H. Kostial, et al., Phys. Rev. Lett. **87**, 016601 (2001).
[77] E. I. Rashba, A. L. Efros, Phys. Rev. Lett. **91**, 126405 (2003).
[78] J. M. Kikkawa, D. D. Awschalom, Phys. Rev. Lett. **80**, 4313 (1998).
[79] S. D. Ganichev, E. L. Ivchenko, V. V. Belkov et al., Nature **417**, 153 (2002).
[80] P. R. Hammar, B. R. Bennett, M. J. Yang, M. Johnson, Phys. Rev. Lett. **83**, 203 (1999).
[81] A. Brataas, Y. Tserkovnyak, G. E. W. Bauer, B. I. Halperin, Phys, Rev. B **66**, 060404 (2002).
[82] I. Zutić, J. Fabian, S. Das Sarma, Rev. Mod. Phys. **76**, 323 (2004).
[83] P. Zhang, J. Shi, D. Xiao, Q. Niu, cond-mat/0503505 (2005).
[84] E. G. Mishchenko, A. V. Shitov, B. I. Halperin, Phys. Rev. Lett. **93**, 226602 (2004).
[85] O. Bleibaum, Phys. Rev. B **69**, 205202 (2004).
[86] A. A. Burkov, A. S. Nu´nez, A. H. MacDonald, Phys. Rev, B **70**, 155308 (2004).
[87] K. Nomura, A. H. Sinova, N. A. Sinitsyn, B. Kaestner, A. H. MacDonald, T. Jungwirth, Phys. Rev. B **72**, 245330 (2005).
[88] N. Suginoto, S. Onada, S. Murakami, N. Nagaosa, cond-mat/0503475 (2005).
[89] V. V. Bryksin, JETP **127**, 353 (2005).
[90] I. Adagideli, G. E. W. Bauer, cond-mat/0506531 (2005).
[91] O. Bleibaum, Phys. Rev. B **71**, 195329 (2005).
[92] V. M. Edelstein, Sol. St. Comm. **73**, 233 (1990).
[93] V. V. Bryksin, P. Kleinert, cond-mat/0512576 (2005); Phys. Rev. B **73**, 165313 (2006).
[94] J. Inoue, G. E. W. Bauer, L. W. Molenkamp, Phys. Rev. B **67**, 033104 (2003).
[95] E. N. Bulgakov, K. N. Pichugin, A. F. Sadreev, P. Str´eda, P. S´eba, Phys. Rev. Lett. **83**, 376 (1999).
[96] A. Cre´pieux, P. Bruno, Phys. Rev. B **64**, 014416 (2001).
[97] J. E. Hirsch, Phys. Rev. Lett. **83**, 1834 (1999).
[98] P. Zhang, J. Shi, D. Xiao, Q. Niu, cond-mat/0503505 (2005).
[99] Y. Kato, R. C. Myers. A. C. Gossard, D. D. Awschalom, Science **306**, 1910 (2004).
[100] J. Wunderlich, B. Kaestner, J. Sinova, T. Jungwirth, Phys. Rev. Lett. **94**, 047204 (2005).
[101] J. Sinova, D. Culcer, Q. Niu, N. A. Sinitsyn, T. Jungwirth, A. H. MacDonald, Phys. Rev. Lett. **92**, 126603 (2004).
[102] S. Zhang, Z. Yang, Phys. Rev. Lett. **94**, 066602 (2005).
[103] S. I. Erlingsson, J. Schliemann, D. Loss, Phys. Rev. B **71**, 035319 (2005).
[104] V. M. Galitski, A. A. Burkov, S. Das Sarma, cond-mat/0601677v2 (2006).
[105] O. Bleibaum, cond-mat/0606351.
[106] B. Andrei, Bernevig, Xiaowei Yu, Shou-Cheng Zhang, Phys. Rev. Lett. **95**, 076602 (2005).
[107] Yu. V. Pershin, Phys. Rev. B **71**, 155317 (2005).
[108] J. Klafter, R. Silbey, Phys. Rev. Lett. **44**, 55 (1980).
[109] V. V. Bryksin, H. Böttger, P. Kleinert, Phys. Rev. B **74**, 235302 (2006).

Single Polaron Properties in Different Electron Phonon Models

V. Cataudella[1], G. De Filippis[2], and C.A. Perroni[3]

[1] CNR-INFM Coherentia and University of Napoli, V. Cintia 80126 Napoli, Italy
`vit@na.infn.it`
[2] CNR-INFM Coherentia and University of Napoli, V. Cintia 80126 Napoli, Italy
`giuliode@na.infn.it`
[3] Institut für Festkörperforschung (IFF), Forschungszentrum Jülich, 52425 Jülich,
Germany `a.perroni@fz-juelich.de`

1 Introduction

One of the most studied problems in condensed matter physics is the behavior of an electron coupled to a quantum bosonic field. It is well known that, under specific conditions, the electron can form a composite quasi-particle consisting of the bare electron dressed by a cloud of field excitations. In the case of lattice excitations the quasi-particle takes the name of "polaron". The idea that an electron can bind itself to the lattice excitations (phonons) goes back to the pioneering work by Landau [1] who first introduced the concept of the polaron in a condensed matter context. This concept has since been used extensively to describe the behavior of electrons in ionic solids (KCl, KBr), III-V semiconductors ($PbTe$), doped oxides (TiO_2) and, more recently, perovskites, to mention only a few examples. One of the most interesting aspects of this problem, that has attracted the attention of many researchers, is the behavior of the system in the so-called intermediate coupling regime where the asymptotic (strong and weak couplings) perturbative descriptions are no longer able to describe the system and non perturbative methods of quantum field theory or numerical approaches have to be used. This was, for instance, the motivation behind the famous all coupling polaron theory by Feynman [2] who restarted, after Landau's intuition, the interest in the polaron problem. Furthermore, in recent years, a large amount of experimental evidence has been accumulating that shows the important role played by polarons in new materials of possible technological impact as manganites [3], cuprates [4], nichelates [5] and one-dimensional organic compounds [6]. More interestingly the electron-phonon (e-ph) intermediate coupling regime seems to be the relevant regime for many of these materials. Indeed the e-ph effects in such materials do not follow the traditional solid state paradigms: Migdal approximation in metallic compounds and polaronic self-trapping in ionic insulators.

Perovskites such as high-Tc superconductors and colossal magneto-resistance manganites are examples of those compounds where the intermediate coupling dominates their behavior.

In this chapter we will address the problem of polaron formation in different e-ph coupling models from a unifying variational point of view. This approach has the advantage of giving direct access to the ground state wavefunction making the understanding of the physical properties in the different regimes more transparent. Furthermore it allows us to study in a single framework the crossover between the weak coupling, where the electron moves coherently dragging a phonon cloud characterized by small lattice deformations involving a large area around the electron itself, and the strong coupling regime where the electron is self-trapped in the potential well created by the lattice deformations. It is worth noting that the use of strong and weak coupling is somehow not rigorous in the sense that, in different models, it can acquire different meanings. However, in this context, it is used in order to individuate the two asymptotic regimes that characterize the polaron physics. All the e-ph models discussed in this chapter are characterized by a linear coupling between the electron and the lattice displacements that are described by dispersion-less longitudinal optical phonons of frequency ω_0. We will study models where the lattice displacement is coupled either to the electron density (Fröhlich and Holstein-like models) or to the electron hopping (SSH-like models) trying to emphasize the common points and investigating the role of the coupling range on the polaron properties.

The first part is mainly dedicated to the ground state properties of these models. We will systematically compare our results with the best results available in literature with the aim to show that the variational approach can indeed reproduce at best more accurate numerical results giving, at the same time, a more physical view of the basic mechanisms involved in the polaron formation. In this way we give also an overview, necessarily incomplete, of some of the approaches used.

The second part of the chapter will be devoted to the calculation of polaron properties involving excited states. In particular we will focus our attention on the optical conductivity and the spectral function in two of the most studied polaron models: Holstein and Fröhlich models. This effort is quite important since it can allow a systematic comparison with experimental measurements leading to a validation of different e-ph models.

2 Ground State Properties

2.1 The Fröhlich Model.

The Fröhlich Hamiltonian [7] was introduced a long time ago to describe ionic solids and it has the form

$$H_F = \frac{p^2}{2m} + \sum_q \omega_0 a_q^\dagger a_q + \sum_q (M_q e^{i\boldsymbol{q}\cdot\boldsymbol{r}} a_q + h.c.). \tag{1}$$

In (1) m is the band mass of the electron, ω_0 is the longitudinal optical phonon energy, $\boldsymbol{r}$ and $\boldsymbol{p}$ are the position and momentum operators of the electron, $a_q^\dagger$ represents the creation operator for phonons with wave number $\boldsymbol{q}$ and M_q indicates the e-ph matrix element that takes the form:

$$M_q = i\omega_0 \frac{R_p^{1/2}}{q} \sqrt{\frac{4\pi\alpha}{V}} \,,$$

where $\alpha = e^2/(2R_p\omega_0\tilde{\varepsilon})$ is the dimensionless e-ph coupling constant, $R_p = \sqrt{\frac{1}{2m\omega_0}}$ is the typical polaron length, $\frac{1}{\tilde{\varepsilon}} = (\varepsilon_0 - \varepsilon_\infty)/\varepsilon_0\varepsilon_\infty$ is the inverse of the effective dielectric constant and V is the system's volume. The units are such that $\hbar = 1$ as throughout the chapter. The electron is treated in the effective mass approximation that is reasonable for ionic solids and polar semiconductors. The coupling function contains a q^{-1} term that is related to the electrostatic long range nature of the coupling in these materials. It is worth noting that, if we measure energy in units of ω_0 and lengths in units of R_p, the Fröhlich Hamiltonian depends on a single dimensionless parameter, α, that controls both the e-ph coupling strength and the adiabaticity regime.

The problem of finding the ground state energy of the Fröhlich Hamiltonian in all the coupling regimes attracted the interest of a lot of researchers mainly in the period 1950-1955, even if numerical approaches have only been developed recently. Numerous mathematical techniques have been used to solve this problem: from the perturbation theory in the weak coupling regime [8] to the strong coupling theory [9], from the linked cluster theory [10] to variational [11] and Monte Carlo approaches [12–14]. Among these approaches the Feynman approach [2] plays a special role for two reasons: first of all it gave the first understanding of the polaron properties in the crossover regime between weak and strong couplings and, secondly, it represents a very beautiful proof of how the path-integral formulation of the field theory can provide powerful non perturbative solutions [15]. In view of its importance and since, in the following, we will use some of the ideas contained in this formulation, we will recover some of Feynman's results from a Hamiltonian point of view.

Feynman Polaron Model Revisited

The main ideas behind this approach are to use the Feynman-Jensen inequality and the identification of a variational trial action that is able to catch the polaron properties both in the weak and strong coupling regimes. The trial action introduced by Feynman corresponds to a simple model made by an electron bound to an effective mass by means of a spring, the so-called Feynman polaron model (FPM). Both the effective mass and the spring constant are variational parameters. This simple model has the advantage of preserving the translational invariance of the system and provides a very appealing description of the polaron concept even if, somehow, oversimplified. In order to incorporate these ideas in a Hamiltonian formalism we add to the Fröhlich Hamiltonian an extra degree of freedom (the effective mass of Feynman's model) coupled harmonically with the electron:

$$H^+ = H_F + \frac{\boldsymbol{P}^2}{2M} + \frac{1}{2}k(\boldsymbol{r} - \boldsymbol{R})^2 - A \tag{2}$$

where $\mathbf{P}$ and $\mathbf{R}$ are the momentum and the position of the effective mass, M, and k is the spring constant. The constant A takes the form $A = \frac{3}{2}\sqrt{\frac{k}{M}}$. By introducing creation and annihilation operators for the effective mass, M, the Hamiltonian (2) becomes

$$H^+ = H_F + w\boldsymbol{b}^\dagger \cdot \boldsymbol{b} + \frac{2c}{w}r^2 - \sqrt{2c}\,\boldsymbol{r} \cdot (\boldsymbol{b}^\dagger + \boldsymbol{b}) \tag{3}$$

where $\boldsymbol{b} \equiv (b_x, b_y, b_z)$, $w = \sqrt{\frac{k}{M}}$ and $c = \frac{1}{4}\sqrt{\frac{k^3}{M}}$. Of course, the ground state (GS) energy of the Fröhlich model is related to the GS energy, $E(c, \alpha)$, of this more general problem by:

$$E_{Fr} = E(0, \alpha) = E(c, \alpha) - \int_0^c \frac{dE(c', \alpha)}{dc'} dc'$$

At this point we can use the Ritz principle to get an upper bound for the GS energy of the Hamiltonian (2). Guided by Feynman's approach we choose the following trial function:

$$|\Psi\rangle = \frac{1}{\sqrt{V}} \exp\left[-i\sum_q (\boldsymbol{q} \cdot \boldsymbol{R}_{CM}) a_q^\dagger a_q\right] \exp\left[\sum_q (h_q(\boldsymbol{r} - \boldsymbol{R})a_q - h.c.)\right] |0\rangle_a |0\rangle_b |\varphi_0\rangle \tag{4}$$

where $|\varphi_0\rangle$ is the GS of the FPM (electron + effective mass), $\boldsymbol{R}_{CM} = (m\boldsymbol{r} + M\boldsymbol{R})/(m + M)$ is its center of mass and $h_q(\boldsymbol{r} - \boldsymbol{R})$ takes the form

$$h_q(\boldsymbol{r} - \boldsymbol{R}) = \int_0^\infty \left\{ M_q \exp\left[-i\boldsymbol{q} \cdot (\boldsymbol{R} - \boldsymbol{r})\frac{v^2 - w^2}{v^2} e^{-v\tau}\right] \right.$$

$$\left. \exp\left[-\tau\left(\omega_0 + \frac{q^2}{2(m + M)}\right)\right] \exp\left[-\frac{q^2}{4mv}\left(1 - e^{2v\tau}\right)\frac{v^2 - w^2}{v^2}\right] \right\} d\tau,$$

where we have introduced $v = \sqrt{w^2 + \frac{4c}{mw}}$.

It is useful to give a physical interpretation of the wave-function (4). If we expand the second exponential in (4) to the first order we get the GS wave-function of Feynman's model corrected by the scattering with the phonons taken into account at the first order. In other words we first solve exactly the Feynman polaron model and we then consider the effects due to the phonon scattering on the composite-particle made by the electron and the effective mass M neglecting the correlations between the emission of successive virtual phonons. The expectation value of the extended Hamiltonian (2) on the wave-function (4) can be calculated in a closed form and gives the following upper bound:

$$E(c,\alpha) \leq \frac{3}{2}(v-w) - \alpha\omega_0^2 \sqrt{\frac{v}{\pi\omega_0}} \int_0^\infty \frac{e^{-\omega_0\tau}\,d\tau}{\sqrt{\frac{w^2}{v}\tau + (1-e^{-v\tau})\left(\frac{v^2-w^2}{v^2}\right)}}.$$

Denoting the previous expression as $\widetilde{E}$ and using the previous inequality, we get:

$$E_{Fr} \leq \widetilde{E} - \int_0^c \frac{dE(c',\alpha)}{dc'}dc' \leq \widetilde{E} - c\,\frac{dE(c',\alpha)}{dc'}\bigg|_{c'=c} \tag{5}$$

where the last inequality follows by direct inspection after eliminating both the phonon degree of freedom and the effective mass in the Hamiltonian (3). Unfortunately, the inequality (5) cannot be used to give an upper bound for E_{Fr} since it still requires the true GS energy of the extended model at any c. However, Feynman showed that if we calculate the last term in (5) by using the Feynman-Helmann theorem and replacing the exact GS function with the ansatz of (4) the inequality is still valid. The present derivation of the Feynman result allows us to identify a wave-function that is somehow related to the Feynman approach and has the advantage that it can be easily extended to study the excited states.

The Feynman approximation provides a very good GS energy both at weak and strong coupling (see Table I) and gives a very satisfactory description of the polaron GS energy at all couplings.

Table 1. Comparison between the GS Energy in the Feynman approach and asymptotic perturbation theory: weak and strong coupling limits.

Feynman approach	Best asymptotic approaches
$\alpha \ll 1$ $E_{Fr} = \omega_0(-\alpha - 0.0123\alpha^2)$	$E_{Fr} = \omega_0(-\alpha - 0.0159\alpha^2)$[22]
$\alpha \gg 1$ $E_{Fr} = \omega_0(-2.83 - 0.106\alpha^2)$	$E_{Fr} = \omega_0(-2.836 - 0.1085\alpha^2)$[18]

All Coupling Variational Approach

In this section we will report a variational approach that we have recently introduced [16]. This approach is based on a simple idea. We start from the best trial wave-functions available for strong and weak couplings and we then construct, keeping the translational invariance, a new trial function that is a linear superposition of the asymptotic wave functions. This method gives for all e-ph couplings a ground state energy better than that obtained within the Feynman approach presented in the previous section and it allows us to discuss and compare the many variational approaches proposed so far.

Strong Coupling

When the value of α is very large ($\alpha \gg 1$) the electron can follow adiabatically the lattice polarization and it becomes self-trapped in the induced polarization field. This idea goes back to the pioneering work by Landau and Pekar [17], where they propose a trial wave function, valid in the strong coupling limit, made as a product of normalized variational wave functions depending, respectively, on the electron and phonon coordinates:

$$|\psi\rangle = |\varphi\rangle|f\rangle. \tag{6}$$

The expectation value of the Hamiltonian (1) on the state (6) gives:

$$\langle\psi|H|\psi\rangle = \langle\varphi|\frac{p^2}{2m}|\varphi\rangle + \langle f|\sum_{q}\left[\omega_0 a_q^\dagger a_q + \rho_q a_q + \rho_q^* a_q^\dagger\right]|f\rangle \tag{7}$$

with

$$\rho_q = M_q\langle\varphi|e^{iq\cdot r}|\varphi\rangle. \tag{8}$$

The variational problem with respect to $|f>$ leads to the following lowest energy phonon state:

$$|f> = \exp\left[\sum_{q}\frac{\rho_q}{\omega_0}a_q - h.c.\right]|0>. \tag{9}$$

The minimization of the corresponding energy with respect to $|\varphi\rangle$ leads to a non-linear integro-differential equation that has been solved numerically by Miyake [18]. The result for the polaron ground state energy in the strong coupling limit is:

$$E = -0.108513\alpha^2\omega_0. \tag{10}$$

The simpler Landau-Pekar [17] Gaussian ansatz for $|\varphi\rangle$:

$$|\varphi_{lp}\rangle = e^{-(m\omega)^2\frac{r^2}{2}}\left(\frac{m\omega}{\pi}\right)^{3/4}, \tag{11}$$

provides a slightly higher estimate of the ground state energy:

$$E = -\frac{\alpha^2}{3\pi}\omega_0 \simeq -0.106103\alpha^2\omega_0 \tag{12}$$

that is very close to the exact result (10). The best value for ω turns out as:

$$\omega = \frac{4\alpha^2}{9\pi}\omega_0. \tag{13}$$

An excellent approximation for the true energy is obtained by using a trial wave function similar to that one introduced by Pekar [17]:

$$|\varphi_p\rangle = Ne^{-\gamma r}\left[1 + b\,(2\gamma r) + c\,(2\gamma r)^2\right], \tag{14}$$

with N the normalization constant and b, c and γ variational parameters. The minimization of $\langle\varphi_p|H|\varphi_p\rangle$ leads to:

$$E = -0.108507\alpha^2\omega_0. \tag{15}$$

This upper bound for the energy differs from the exact value by less than 0.01%.

Following this suggestion and exploiting what we learned from the discussion of the Feynman approach, we have proposed as trial wave-function a coherent state of this type:

$$|\psi\rangle = \exp\left[\sum_{\boldsymbol{q}}\left[\left(s_{\boldsymbol{q}}e^{i\boldsymbol{q}\cdot\boldsymbol{r}} + l_{\boldsymbol{q}}e^{i\boldsymbol{q}\cdot\boldsymbol{r}\eta}\right)a_{\boldsymbol{q}} - h.c.\right]\right]|0\rangle|\varphi_p\rangle. \tag{16}$$

where the variational parameters $(b,\ c,\ \gamma,\ \eta)$ and the functions $l_{\boldsymbol{q}}$ and $s_{\boldsymbol{q}}$ have to be determined by minimizing the expectation value of the Fröhlich Hamiltonian on this state. The function $s_{\boldsymbol{q}}e^{i\boldsymbol{q}\cdot\boldsymbol{r}} + l_{\boldsymbol{q}}e^{i\boldsymbol{q}\cdot\boldsymbol{r}\eta}$ at the exponent of the coherent state controls the lattice deformation. It is worth noting that for $s_{\boldsymbol{q}} = 0$ and $\eta = 0$ (16) returns the Landau-Pekar suggestion that takes into account correctly the adiabatic contributions. On the contrary, in the general case, the choice (16) introduces a dependence on the electron position, $\mathbf{r}$, in the function that controls the lattice deformation. This behavior allows us to include a non-adiabatic contribution where the lattice deformation tends to follow the electron position. This is, indeed, what we can learn from the analysis of the Feynman ansatz (4). The expectation value of (1) on the wave-function (16) gives:

$$\langle\psi|H|\psi\rangle = \langle\varphi_p|\frac{p^2}{2m}|\varphi_p\rangle + \sum_{\boldsymbol{q}}\left[\omega_0\left(|l_{\boldsymbol{q}}|^2 + |s_{\boldsymbol{q}}|^2\right) + \frac{q^2}{2m}\left(\eta^2|l_{\boldsymbol{q}}|^2 + |s_{\boldsymbol{q}}|^2\right)\right] +$$

$$+ \sum_{\boldsymbol{q}}\left[\left(\omega_0 + \frac{q^2}{2m}\eta\right)\left(r_{\boldsymbol{q}}s_{\boldsymbol{q}}l_{\boldsymbol{q}}^* + h.c.\right) - \left(M_{\boldsymbol{q}}s_{\boldsymbol{q}}^* + M_{\boldsymbol{q}}r_{\boldsymbol{q}}l_{\boldsymbol{q}}^* + h.c.\right)\right]$$

$$\tag{17}$$

with

$$r_{\boldsymbol{q}} = \langle \varphi_p | e^{i\boldsymbol{q} \cdot \boldsymbol{r}(1-\eta)} | \varphi_p \rangle. \tag{18}$$

Making $\langle \psi | H | \psi \rangle$ stationary with respect to arbitrary variations of the functions $l_{\boldsymbol{q}}$ and $s_{\boldsymbol{q}}$, we obtain two, easily solvable, algebraic equations. The minimization and the asymptotic expansion of the ground state energy provide, for $\alpha \to \infty$,

$$E = \left[-0.108507\alpha^2 - 1.89 \right] \omega_0. \tag{19}$$

The electron self-energy shows the exact dependence on α^2 typical of the strong coupling regime [18] together with a good estimate of the constant term due to the lattice fluctuations. This allows us to obtain, for $\alpha \geq 8.7$, an upper bound for the polaron ground state energy better than that obtained in the Feynman approach. However, for lower α the method gets worse and worse showing a non-physical discontinuity in the transition from strong to weak coupling regimes. This behavior is due to the lack of translational invariance in the proposed trial wave-function. To overcome this difficulty we construct an eigenstate of the total wave number by taking a superposition of the localized states (16):

$$|\psi_{(sc)}\rangle = \int \psi(\boldsymbol{r} - \boldsymbol{R}) d^3 R \ . \tag{20}$$

The minimization, with respect to the variational parameters, of the expectation value of the Fröhlich Hamiltonian on this state, that accounts for the translational symmetry, provides in the $\alpha \to \infty$ limit:

$$E = \left[-0.108507\alpha^2 - 2.67 \right] \omega_0. \tag{21}$$

This upper bound is lower than the best variational Feynman estimate which for large values of α assumes the form [2]:

$$E = \left[-\frac{\alpha^2}{3\pi} - 3\log 2 - \frac{3}{4} \right] \omega_0. \tag{22}$$

It is worth noting that the proposed ansatz, (16), collects together both the old proposal by Landau-Pekar and the Feynman approximation [2, 16]. In this sense our ansatz contains all the wave-functions proposed to describe the strong coupling regime and generalizes them.

Weak Coupling

A similar procedure can be adopted for the opposite weak-coupling regime. In this case the reference wave-function is given by the Lee-Low-Pines (LLP) variational coherent state [19]. Starting from this suggestion we choose a wave-function with the following structure:

$$|\psi_{wc}\rangle = \exp\left[\sum_q -i(\boldsymbol{q}\cdot\boldsymbol{r})a_q^\dagger a_q\right]\cdot\exp\left[\sum_q (g_q a_q - h.c.)\right]$$

$$\cdot\left[1 + \sum_{\boldsymbol{q}_1,\boldsymbol{q}_2} d_{\boldsymbol{q}_1,\boldsymbol{q}_2} a_{\boldsymbol{q}_1}^\dagger a_{\boldsymbol{q}_2}^\dagger\right]|0\rangle. \tag{23}$$

where the first exponential takes into account the translational invariance, the second one is related to the LLP ansatz and controls the lattice deformation and the third term introduces the correlation between the emission of pairs of virtual phonons [20] that are completely neglected in the coherent states. In particular, by following the suggestion contained in the LLP approach, we will choose:

$$g_q = \frac{M_q}{\left(\omega_0 + \frac{q^2}{2m}\epsilon^2\right)} \tag{24}$$

and

$$d_{\boldsymbol{q}_1,\boldsymbol{q}_2} = \frac{\gamma\omega_0}{2m}\boldsymbol{q}_1\cdot\boldsymbol{q}_2 \frac{M_{q_1}}{\left(\omega_0 + \frac{q_1^2}{2m}\delta^2\right)} \frac{M_{q_2}}{\left(\omega_0 + \frac{q_2^2}{2m}\delta^2\right)}. \tag{25}$$

As discussed in [16], the presence of a dependence on the electron spectrum in the energy denominators is able to take into account the recoil effects on the electron at least on average. In (24) and (25) γ, δ and ϵ are three variational parameters that have to be determined by minimizing the expectation value of the Hamiltonian (1) on the state (23). This procedure provides as upper bound for the polaron ground state energy at small values of α:

$$E = -\alpha\omega_0 - 0.0123\alpha^2\omega_0, \quad \alpha \to 0, \tag{26}$$

i.e. the same result, at α^2 order, of the Feynman approach [2]. We stress that, at the same order, the correct result for the electron self-energy is:

$$E = -\alpha\omega_0 - 0.0159\alpha^2\omega_0 \tag{27}$$

as found by Larsen [20], Höhler [21], and Roseler [22].

Intermediate Coupling

A careful inspection of the wave function (20) shows that, even if it is able to interpolate between strong and weak coupling regimes, the approximation is not very satisfying for small values of α. In this regime a much better description of the polaron ground state features is provided by the wave function (23). Moreover, in the weak and intermediate e-ph coupling, $\alpha \leq 7$, these two solutions are not orthogonal and have non-zero off diagonal matrix elements. This suggests that a better description of the lowest state of the system is made of a mixture of the two wave functions. Then, our best ansatz is a linear superposition of the two previously discussed wave functions. The minimization procedure can be performed in two steps. First, the expectation values

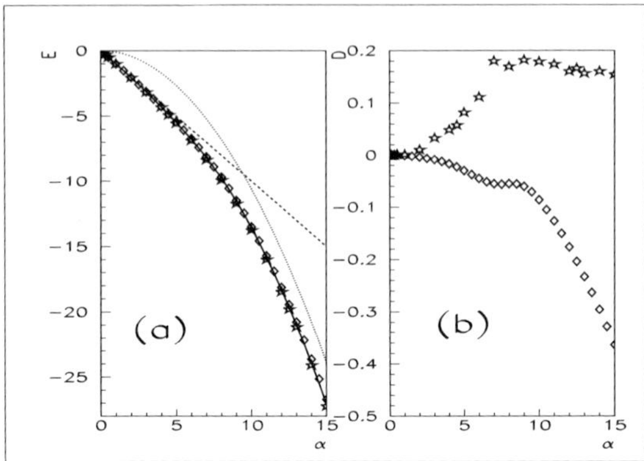

Fig. 1. (a) The polaron ground state energy, E, is reported as function of α in units of ω_0. The data (solid line), obtained within the approach discussed in this section, are compared with the results (diamonds) of the Feynman approach, E_F, and the results (stars) of the diagrammatic Quantum Monte-Carlo method, E_{MC}, kindly provided by A.S. Mishchenko [13]. For comparison we also report the weak (dashed) and strong (dotted) coupling GS energies. (b) differences: $E - E_F$ (diamonds) and $E - E_{MC}$ (stars) are reported as function of α.

of the Fröhlich Hamiltonian on the two trial wave functions in (20) and (23) are minimized and the variational parameters are determined. Then, the two constants A and B that provide the relative weight of the two components in the ground state of the system are obtained with a further minimization. This way to proceed simplifies the computational effort and makes all the described calculations accessible on a personal computer.

In Fig. 1 we plot the polaron ground state energy, obtained within our approach, as a function of the e-ph coupling constant α. The data are compared with the results of the variational treatments due to Lee, Low and Pines [19], Pekar [17], Feynman [2] and with the energies calculated within a diagrammatic Quantum Monte-Carlo method [12]. As is clear from the plots, our variational proposal recovers the asymptotic result of the Feynman approach in the weak coupling regime, improves the Feynman data particularly in the opposite regime, characterized by values of the e-ph coupling constant $\alpha \gg 1$, and is in very good agreement with the best available results in the literature, obtained with the Quantum Monte Carlo calculation [12].

In order to get a better understanding of the wave-function associated to the GS we show in Fig.2 its spectral weight:

$$Z = |\langle \psi | c_{\boldsymbol{k}=0}^{\dagger} | 0 \rangle|^2, \tag{28}$$

where $|0\rangle$ is the electronic vacuum state containing no phonons and $c_{\boldsymbol{k}}^{\dagger}$ is the electron creator operator in the momentum space. Z represents the renormalization coefficient of the one-electron Green's function and gives the fraction of the bare electron state in the polaron trial wave function. This quantity

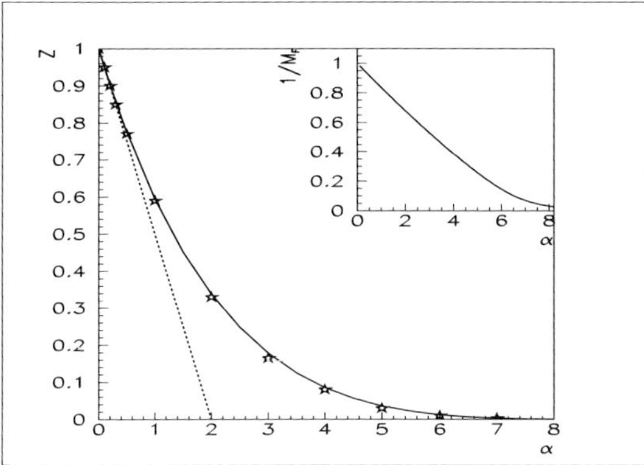

Fig. 2. The ground state spectral weight, Z, is plotted as a function of α. The data (solid line), obtained within the approach discussed in this paper, are compared with the results (stars) of the diagrammatic Quantum Monte-Carlo method [13]. The result of the weak coupling perturbation theory (dashed line) is also indicated. In the inset is reported the inverse of the polaron mass in the Feynman approach.

is compared with the one obtained in the diagrammatic Quantum Monte Carlo method [12, 13]. The agreement is again very good confirming that the proposed variational wave-function represents a very good approximation of the Fröhlich GS. The result of the weak coupling perturbation theory is also indicated: $Z = 1 - \alpha/2$. For small values of α the main part of the spectral weight is located at energies that correspond approximatively to the bare electronic levels. Increasing the e-ph interaction, the spectral weight decreases very fast and becomes very small in the strong coupling regime. Here most of the spectral weight is transferred to excited states. At the same time the polaron effective mass (see inset of Fig.2) increases with a similar behavior. As soon as the quasi-particle peak loses its spectral weight the polaron acquires a larger and larger mass and, eventually, gets trapped. A more detailed discussion on the link between the GS spectral weight and effective polaron mass will be given in Sect. (2.4) where we will stress the importance of the range of the e-ph interactions.

The diagrammatic Quantum Monte Carlo study [12, 13] of the Fröhlich polaron has pointed out that there are no stable excited states in the energy gap between the ground state energy and the continuum. There are, instead, several many–phonon unstable states at fixed energies: $E_f - E_0 \simeq 1, 3.5$ and $8.5\,\omega_0$. The nature of the excited states and the optical absorption of polarons in the Fröhlich model require further study and we postpone this discussion to the second part of this chapter.

We conclude this section emphasizing that we can think of the Fröhlich GS as a linear combination of two different components characterized by different deformations. The strong coupling component is dominated by adiabatic contributions and the non-adiabatic terms enter as corrections while, in the second component, the anti-adiabatic terms are very important and the corrections are due to correlations in the virtual phonon emission.

2.2 The Holstein Model

The large amount of experimental data on oxide perovskites has renewed the interest in studying the Holstein molecular crystal model that, for its relative simplicity, is the most considered model for the interaction of a single tight-binding electron coupled to an optical local phonon mode [23]. The Holstein Hamiltonian takes the following form:

$$H = -t \sum_{<i,j>} c_i^\dagger c_j + \omega_0 \sum_{q} a_q^\dagger a_q + \sum_{i,q} c_i^\dagger c_i \left[M e^{i q \cdot R_i} a_q + h.c. \right] \qquad (29)$$

where the electron operators are in the site representation while the phonon ones describe excitations in Fourier space. In (29) R_i indicates the position of the lattice site i, M indicates the e-ph matrix element and lengths are measured in units of a, as in the rest of the chapter. In the Holstein model the e-ph interaction is considered short range and, in fact, the local phonon mode at site i is coupled to the electron density at the same site and M does not depend on q:

$$M = \frac{g}{\sqrt{N}} \omega_0. \qquad (30)$$

Here N is the number of lattice sites. It is worth emphasizing that the Holstein model is controlled by two dimensionless parameters: the adiabaticity parameter $\gamma = \frac{\omega_0}{t}$ and the e-ph coupling constant g (we measure the energy in units of ω_0). This makes the Holstein model richer when compared to the Fröhlich model where only one parameter controls the different model regimes. For instance, in the Holstein model, the strong coupling region, where the electron gets self-trapped, is not uniformly reached with increasing g but it depends, crucially, on the value of γ. For this reason, in the adiabatic regime ($\gamma < 1$), it is better to use, instead of g, the effective e-ph coupling constant $\lambda = \frac{g^2 \omega_0}{2dt}$ (d is the system dimensionality) that represents the ratio between the small polaron binding energy and the energy gain of an itinerant electron on a rigid lattice and that, as we will see, roughly signals the electron self-trapping.

The Holstein model has been studied by many techniques. Beside the weak-coupling perturbative theory [24] an analytical approach is known for the strong coupling limit in the nonadiabatic regime (small polaron)[25, 26]. It is based on the Lang-Firsov canonical transformation and on expansion in powers of $1/\lambda$. It is well known that both these analytical techniques fail to describe the region, of greatest physical interest, characterized by intermediate couplings and by electronic and phononic energy scales not well separated. This regime has been analyzed in several works based on Monte Carlo simulations [27], numerical exact diagonalization of small clusters [28, 29], dynamical mean field theory [30], density matrix renormalization group [31] and variational approaches [32–35]. The general conclusion is that the ground state energy and the effective mass in the Holstein model are continuous functions of the e-ph coupling and that there is no phase transition in this one-body system [36]. In particular when the interaction strength is greater than a critical value the ground state properties change significantly but without breaking the translational symmetry.

A Variational Approach

Recently we have proposed a powerful variational approach [37], valid for all couplings, that is able to provide very accurate results without resorting to heavy numerical calculations. The approach is based on the explicit construction of the GS wave-function and allows more transparent access to the polaron properties without losing significant details. Interestingly the approach is very close to what we did for the Fröhlich model signaling that these two models have much more in common than believed.

Again the best way to proceed is to start from very good ansatz for the wave-functions in the two asymptotic regimes: localized and itinerant. Compared to the Fröhlich case, the two limit regimes are now more difficult to identify since we cannot associate them simply to strong and weak coupling regimes, respectively, but we have to take into account adiabatic and non-adiabatic contributions. However, once we have constructed these two asymptotic components (in the following we will refer to them as strong and weak coupling components), we can access the most interesting intermediate coupling regime choosing a linear combination of the two asymptotic wave-functions.

In both asymptotic regimes a very good variational wave-function, that takes into account the translational invariance, is given by

$$|\psi_{\boldsymbol{k}}> = \frac{1}{\sqrt{N}} \sum_{\boldsymbol{R}_i} e^{i\boldsymbol{k}\cdot\boldsymbol{R}_i} \sum_{\boldsymbol{R}_n,|\boldsymbol{R}_i-\boldsymbol{R}_n|\leq d_0} \eta_{\boldsymbol{k}}(\boldsymbol{R}_i - \boldsymbol{R}_n)|\phi_{\boldsymbol{k}}(\boldsymbol{R}_i, \boldsymbol{R}_n)> \qquad (31)$$

where $\eta_{\boldsymbol{k}}(\boldsymbol{R}_n)$ are variational functions to be determined and the $|\phi_{\boldsymbol{k}}(\boldsymbol{R}_i, \boldsymbol{R}_n)>$ are localized wave-functions that assume a different form in weak and strong coupling regimes. In (31) the first sum implements the translational invariance while the internal sum takes into account the retardation effects expected to play a significant role in the adiabatic regime. The sum is restricted to a sphere of radius d_0 around the site $\boldsymbol{R}_i$. Therefore, d_0 controls how strong the adiabatic contribution in (31) is. In the strong coupling case we take

$$|\phi_{\boldsymbol{k}}(\boldsymbol{R}_i, \boldsymbol{R}_n)> = c_{n+i}^{\dagger} e^{\sum_{\boldsymbol{q}}\left[f_{\boldsymbol{q}}(\boldsymbol{k})a_{\boldsymbol{q}}e^{i\boldsymbol{q}\cdot\boldsymbol{R}_i}+h.c.\right]}|0>_{ph}|0>_{el} \qquad (32)$$

and

$$f_{\boldsymbol{q}}(\boldsymbol{k}) = \frac{\rho_{\boldsymbol{q}}(\boldsymbol{k})}{\omega_0} = g\sum_{\boldsymbol{R}_m}|\eta_{\boldsymbol{k}}(\boldsymbol{R}_m)|^2 e^{i\boldsymbol{q}\cdot\boldsymbol{R}_m} \ . \qquad (33)$$

It is worth noting that the wave-function (31) is made by a linear combination of phonon coherent states (32) that, in general, describe a lattice deformation that is not centered at the electron position. Only when $i = n$ in (32) the center of the lattice deformation and the electron position coincide. These contributions are needed to describe the electron fluctuations in the potential well created by the lattice deformation. As already mentioned, this freedom in the wave-function is able to handle the adiabatic regime ($\gamma < 1$).

The wave-function (31) has been first introduced in the pioneering work by Toyozawa [38] and then applied to the Holstein model in [32, 37]. If one is not interested in the very adiabatic regime ($\gamma \ll 1$), it is sufficient to take three coherent states corresponding to on site, nearest and next nearest neighbors ($d_0 = 2a$, a being the lattice constant) to get a very good estimation of the GS energy. Finally we note that the form chosen for $f_q(\boldsymbol{k})$ (33) has the same structure used in the Pekar approach for the Fröhlich model [17](strong coupling adiabatic limit). This makes the analogies between the two models stronger.

In the weak coupling regime we make a different choice for $|\phi_{\boldsymbol{k}}(\boldsymbol{R}_i, \boldsymbol{R}_n) >$. We still start from a coherent state but we only consider deformations centered where the electron is sitting ($d_0 = 0$). In this regime the polaron formation is controlled by the non adiabatic term. Since recoil effects are important and have to be taken into account, we use a coherent state similar to that used in the LLP approach for the Fröhlich model. Summarizing, our choice for the weak coupling limit is:

$$|\phi_{\boldsymbol{k}}(\boldsymbol{R}_i, \boldsymbol{R}_i)\rangle = c_i^\dagger \left[\exp \sum_q \left(h_q(\boldsymbol{k}) a_q e^{i\boldsymbol{q}\cdot\boldsymbol{R}_i} - h.c. \right) \right]$$
$$\cdot \left[1 + \sum_q d_q^*(\boldsymbol{k}) e^{-i\boldsymbol{q}\cdot\boldsymbol{R}_i} a_q^\dagger \right] |0>_{ph} |0>_{el} \qquad (34)$$

and

$$h_q(\boldsymbol{k}) = \frac{M_q}{\omega_0 + E_b(\boldsymbol{q}) - E_b(\boldsymbol{q}=0)} . \qquad (35)$$

Here $E_b(\boldsymbol{q})$ is the free electron band energy:

$$E_b(\boldsymbol{q}) = -2t \sum_{i=1}^{d} \cos(q_i a) \qquad (36)$$

where $d_q(\boldsymbol{k})$ is a variational function that has to be determined by minimizing the expectation value of the Hamiltonian (29). We note that the term in the square brackets of (34) allows a considerable advantage over the independent phonon LLP approximation. In the LLP ansatz an important physical ingredient is missing: it does not take into account the fact that the polaron energy

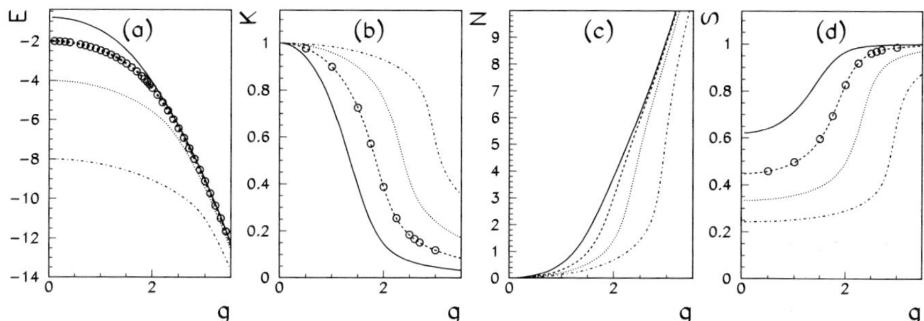

Fig. 3. The polaron ground-state energy (E), the polaron kinetic energy in units of the bare kinetic energy (K), the average number of phonons (N) and the e-ph local correlation function (S) are plotted as a function of g for different values of the adiabatic parameter ω_0/t: $\omega_0/t = 2.5$ (solid line), $\omega_0/t = 1$ (dashed line), $\omega_0/t = 0.5$ (dotted line), $\omega_0/t = 0.25$ (dashed-dotted line). The data reported are for the one dimensional case. In (a) the circles indicate the "global local variational method" data, kindly provided by A. Romero [32], and, in (b) and (d) the circles represent the DMRG data, kindly provided by E. Jeckelmann [31]. The energies are given in units of ω_0.

can approach ω_0. On the contrary the wave function (34) contains this physical information [39]. In particular when the polaron excitation energy becomes equal to the energy of the longitudinal optical phonon, the band dispersion flattens. For these values of k the polaron band has the bare phonon-like behavior with very small spectral weight.

For any particular value of t there is a value of the e-ph coupling constant (g_c) where the ground state energies of the two previously discussed solutions become equal. Nevertheless the two solutions exhibit very different polaron features. In particular when the coupling constant is smaller than g_c the stable solution (the one with lowest energy) is characterized by small lattice deformations that involve many lattice sites around the electron (large polaron) while for $g > g_c$ it is characterized by very strong and localized lattice deformations that are able to trap the electron (small polaron). Crossing g_c the mass of the polaronic quasi-particle increases in a discontinuous way. A more careful inspection shows that in this range of g values the wave functions describing the two solutions of large and small polarons are not orthogonal and have non-zero off diagonal matrix elements. This suggests that the lowest state of the system is made of a mixture of the large and small polaron solutions [40].

Then the idea is to use a variational method to determine the ground state energy of the Hamiltonian (29) by considering as trial state a linear superposition of the wave functions describing the two types of previously discussed polarons.

As in the case of the Fröhlich polaron the agreement with the most advanced numerical approaches is excellent (see Fig. 3).

2.3 The Su–Schrieffer–Heeger Model

In order to explain the anomalous transport properties of non-local polarons in various 1D systems [41–44] many models have been introduced. In particular the tight-binding Su–Schrieffer–Heeger (SSH) model [41] was introduced to explain the transport properties of quasi one-dimensional polymers such as polyacetylene where the CH monomers form chains of alternating double and single bonds. In this case the localization is due to large shrinkage of two particular bonds and the corresponding large hopping integral between the sites. As a result, the hopping between the two occupied sites and the surrounding ones is reduced resulting in a tendency towards localization. This class of models has been successfully applied to a very large number of 1D systems ranging from carbon nanotubes [45] to DNA [46].

Our purpose here is to examine the single polaron formation in a model where non-local e-ph interactions are present (SSH model) and the phonon spectrum is dispersionless. Also for this model we will use all the variational machinery that we discussed for the Fröhlich and Holstein model showing that, starting from a good description of the lattice deformations on the left and right bonds of the polaron, it is possible to get a very accurate variational GS wave-function.

If we consider only the 1D case, which is the most interesting for this class of Hamiltonians, the model takes the form

$$\mathcal{H} = -t \sum_i (c_i^\dagger c_{i+1} + c_{i+1}^\dagger c_i) + \omega_0 \sum_i a_i^\dagger a_i + H_{int}, \tag{37}$$

where H_{int} is

$$H_{int} = g\omega_0 \sum_i (c_i^\dagger c_{i+1} + c_{i+1}^\dagger c_i)(a_{i+1}^\dagger + a_{i+1} - a_i^\dagger - a_i), \tag{38}$$

with $a_i^\dagger$ (a_i) the site phonon creation (destruction) operator. The quantity g is the dimensionless SSH coupling constant that is proportional to $\sqrt[4]{M}$ where M is the ion mass. The difference with the model discussed so far is clear. The lattice deformation $(x_i \sim a_i^\dagger + a_i)$ is not coupled to the electron density but to the electron hopping $[c_i^\dagger c_{i+1}(x_{i+1} - x_i)]$ such that the electron hopping is directly influenced by the lattice dynamics. The non-local nature of the interaction is also clear. We study the coupling of a single electron to lattice deformations.

Variational Approach vs. Exact Diagonalization

In this section we discuss a variational approach based on the same ideas introduced for the previous models and compare our results to exact diagonalization of small clusters. First we introduce the variational wave function. We consider translation-invariant Bloch states obtained by superposition of localized states centered on different lattice sites of the same type introduced in the previous section for the Holstein model. Here we extend those kind of wave-functions to the SSH interaction model. Due to the asymmetry of the SSH coupling (shrinking of the bond on which the electron is localized and stretching of the neighboring bonds), we have to define two wave-functions that provide the correct description of the lattice deformations on the left and right bonds of the polaron. Naturally the left and right directions are relative to the site where the presence of the electron is more probable. We assume:

$$|\psi_k^{(s)}\rangle = \frac{1}{\sqrt{N}} \sum_i e^{ik \cdot i} \sum_n \eta_k^{(s)}(n-i)|\phi_k^{(s)}(i,n)\rangle. \tag{39}$$

where the apex s can assume the values L and R indicating the $Left$ (L) and $Right$ (R) polaron wave-function, respectively. In (39) we have introduced

$$|\phi_k^{(s)}(i,n)\rangle = c_{i+n}^\dagger \exp\left[U_k^{(s)}(i) + U_k^{(s)}(i-1) + U_k^{(s)}(i+1)\right]|0\rangle_{ph}|0\rangle_{el}, \tag{40}$$

with the quantity $U_k^{(s)}(j)$ given by

$$U_k^{(s)}(j) = \frac{g}{\sqrt{N}} \sum_q [f_{k,j}^{(s)}(q)a_q e^{iq \cdot R_j} - h.c.]. \tag{41}$$

The phonon distribution function $f_{k,j}^{(s)}(q)$ is chosen as

$$f_{k,j}^{(s)}(q) = \frac{\alpha_{k,j}^{(s)}}{1 + \frac{2}{\gamma}\beta_{k,j}^{(s)}[cos(k) - cos(k+q)]}, \tag{42}$$

with $\alpha_{k,j}^{(s)}$ and $\beta_{k,j}^{(s)}$ variational parameters. In (40), $|0\rangle_{ph}$ and $|0\rangle_{el}$ denote the phonon and electron vacuum state, respectively, and the variational functions $\phi_k^{(s)}(i,n)$ are assumed to be not zero up to the fifth neighbors $(|i-n|=5)$. It is worth noting that traditional variational approaches for the Holstein polaron problem use the localized state (40) where only the on-site operator $U_k^{(s)}(i)$ is applied. Thus we introduce in the expression of the trial wave-function the nearest-neighbor displacement operators $U_k^{(s)}(i+1)$ and $U_k^{(s)}(i-1)$, in order to take into account the dependence of the hopping integral on the relative distance between two adjacent ions.

The wave-functions L and R are related as follows

$$\begin{aligned}
f_{k,n}^{(R)}(q) &= -f_{k,n}^{(L)}(q) < 0 \\
f_{k,n-1}^{(R)}(q) &= -f_{k,n-1}^{(L)}(q) > 0 \\
f_{k,n+1}^{(R)}(q) &= -f_{k,n+1}^{(L)}(q) > 0 \\
\phi_k^{(R)}(m) &= \phi_k^{(L)}(-m).
\end{aligned} \tag{43}$$

All the variational parameters are determined by minimizing the expectation value of the Hamiltonian (37) on the states (40). Even though the wave-functions L and R describe the different lattice deformations of the left and right side of the polaron, respectively, the mean values of the Hamiltonian on these states are equal.

Of course the set of approximations proposed can be systematically improved by increasing the extension of the phonon contributions in (40) and the number of neighbors. Furthermore, they are not orthogonal and the off-diagonal matrix elements of the Hamiltonian between these two states are not zero. This allows us to determine the ground-state energy by considering as trial state the linear superposition of the wave-functions R and L

$$|\psi_k\rangle = \frac{A_k|\Phi_k^{(R)}\rangle + B_k|\Phi_k^{(L)}\rangle}{\sqrt{A_k^2 + B_k^2 + 2A_kB_kS_k}}. \qquad (44)$$

In (44) $|\Phi_k^{(L)}\rangle$ and $|\Phi_k^{(R)}\rangle$ are the wave-functions of (39) after normalization, the coefficients A_k and B_k are the weights to be found variationally and

$$S_k = \langle\Phi_k^{(L)}|\Phi_k^{(R)}\rangle \qquad (45)$$

is the overlap factor. The wave-function (44) correctly describes the properties of the lattice deformations on both sides of the polaron and we will find that it is in very good agreement with the results derived by the exact diagonalizations on a chain of 6 sites (see Fig. 4). Furthermore the variational approach involves a number of variational parameters that do not depend on the chain length, so it allows study of the thermodynamic limit of the system.

The minimization procedure is performed in two steps. First the energies of the left and right wave-functions are separately minimized, then these wave-functions are used in the minimization procedure of the quantity $E_k = \langle\psi_k|H|\psi_k\rangle/\langle\psi_k|\psi_k\rangle$ with respect to A_k and B_k defined in (44). Exploiting the equality

$$\langle\psi_k^{(L)}|H|\psi_k^{(L)}\rangle = \langle\psi_k^{(R)}|H|\psi_k^{(R)}\rangle = \varepsilon_k, \qquad (46)$$

we obtain

$$E_k = \frac{\varepsilon_k - S_kE_{kc} - |E_{kc} - S_k\varepsilon_k|}{1 - S_k^2}, \qquad (47)$$

where $E_{kc} = \langle\Phi_k^{(L)}|H|\Phi_k^{(R)}\rangle$ is the off-diagonal matrix element, and $|A_k| = |B_k|$. The matrix elements between the states $\psi_k^{(R)}$ and $\psi_k^{(L)}$ contained in (47) are reported in [47].

The results of the minimization are reported in Fig. 4 for a six-site lattice and $\gamma = 0.4$. We also study the thermodynamic limit and find energy curves very close to those of the finite system. In order to test the validity of our variational approach, exact numerical calculations on small clusters are performed by means of the Lanczos algorithm. The agreement between numerical

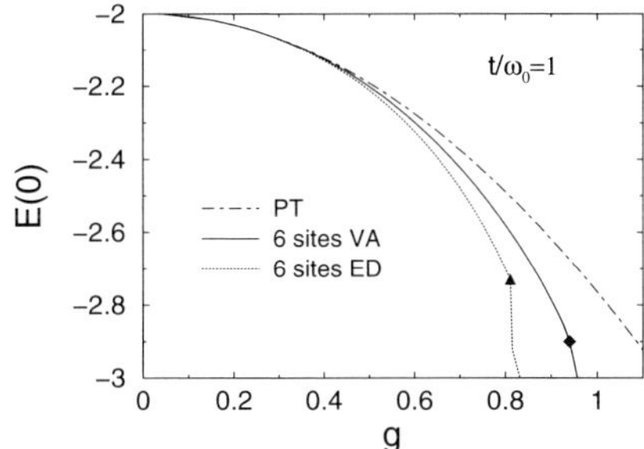

Fig. 4. Ground state energy $E(0)$ as a function of the SSH e-ph coupling g for $\gamma = 0.4$. Solid and dotted lines are obtained from the variational approach and the Lanczos data for a six-site lattice, respectively; perturbative curves (dot–dashed lines) are plotted for comparison. Symbols mark the kink values of the energy.

data and variational approach is very good up to g values close to the so-called unphysical transition. The presence of this "unphysical" transition, which is signaled by a kink in the GS energy as a function of g, deserves a brief comment. Indeed it has been shown [48] that the strong-coupling solution is characterized by an unphysical sign change of the effective next-nearest-neighbor hopping which is missing when acoustical phonons are considered [43]. For this reason a real strong coupling regime is never reached in this model where we find evidence of a crossover from the weak to the intermediate coupling regime. Consequently the wave-function does not contain a strong coupling component as in the models previously discussed.

As for the Fröhlich model it is useful to investigate the behavior of the quasiparticle spectral weight $Z(0)$ that signals the crossover from weak to intermediate coupling regime. We find that increasing the e-ph coupling for fixed values of γ, the spectral weight starts to drop but it never reaches a really small value before the unphysical sign change of the hopping occurs. Nevertheless we observe distinct signatures of the tendency towards localization. On the basis of these calculations we are able to build up a phase diagram that summarizes information on the weak to intermediate coupling and on the location of the "unphysical region" (see Fig. 5). The latter is calculated from the position of the kink in the ground state energy obtained by means of the variational approach (diamonds) and exact diagonalization (triangles). The agreement between the two methods improves moving towards the adiabatic limit. In analogy with the phase diagram obtained for the Holstein polaron [37], we mark a crossover region defined as the range of parameters for which $Z(0)$ is less than 0.9. As shown in Fig. 5, we find that the considered SSH model does not present any marked mixing of electronic and phononic degrees of freedom, being the strongly coupled state prevented from the unphysical behavior of the model. As far as the fully adiabatic limit, $\omega_0 = 0$, is concerned, we verify that the crossover line joins onto the line for

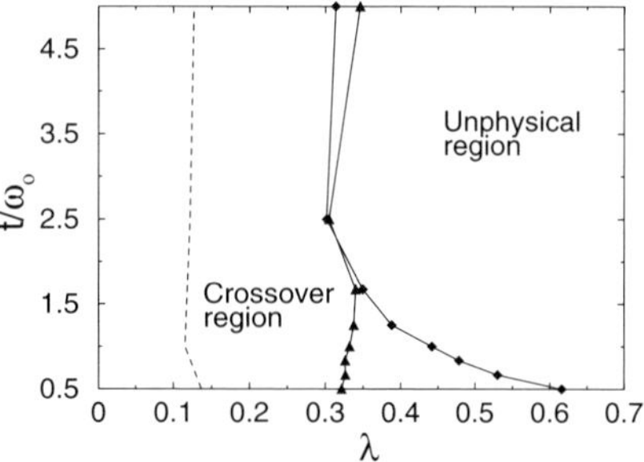

Fig. 5. Phase diagram for one electron in a six-site lattice. Triangles and diamonds correspond, respectively, to the couplings where the exact numerical ground state energy and the variational result have a kink. The dashed line indicates the boundary of the crossover region, where the spectral weight $Z(0)$ is less than 0.9.

the transition to the unphysical region at the critical value $\lambda = 0.25$, confirming the discussion in [48]. We finally notice that, as discussed in [48], both the crossover region boundary, and the instability line obtained by exact diagonalization are only weakly dependent on the adiabatic ratio, and that λ is the relevant e-ph coupling regardless of the value of γ. This is a peculiarity of the SSH coupling with respect to the Holstein one, where the polaron crossover moves to large values of λ as the phonon frequency increases [30, 48–50].

2.4 Intermediate and Long Range Models

The increasing interest in the effect of e-ph interaction in new materials of potential technological impact has not only renewed interest in studying simplified e-ph coupled systems such as Holstein or Fröhlich models but has also pushed researchers to propose more realistic interaction models [51, 52].

Recently a quite general e-ph lattice Hamiltonian with a "density displacement" type interaction has been introduced in order to understand the role of long-range (LR) coupling on polaron formation [51, 53]. The model is described by the Hamiltonian

$$H = -t \sum_{<i,j>} c_i^\dagger c_j + \omega_0 \sum_i \left(a_i^\dagger a_i + \frac{1}{2} \right) + g\omega_0 \sum_{i,j} f(|\boldsymbol{R}_i - \boldsymbol{R}_j|) c_i^\dagger c_i \left(a_j + a_j^\dagger \right).$$

$$(48)$$

where $f(|\boldsymbol{R}_i - \boldsymbol{R}_j|)$ is the interacting force between an electron on the site i and an ion displacement on the site j and the symbol $<>$ denotes nearest neighbors (nn) linked through the transfer integral t.

The Hamiltonian (48) reduces to the short range (SR) Holstein model if $f(|\boldsymbol{R}_i - \boldsymbol{R}_j|) = \delta_{\boldsymbol{R}_i, \boldsymbol{R}_j}$, while in general it contains longer range interaction. In particular when one attempts to mimic the non-screened coupling between doped holes and apical oxygen in some cuprates [51], a reasonable LR expression for the interaction force is given by

$$f(|\boldsymbol{R}_i - \boldsymbol{R}_j|) = \left(|\boldsymbol{R}_i - \boldsymbol{R}_j|^2 + 1\right)^{-\frac{3}{2}}, \tag{49}$$

if the distance $|\boldsymbol{R}_i - \boldsymbol{R}_j|$ is measured in units of lattice constant. The expression (49) can be viewed as the natural extension of the Fröhlich model to the tight-binding approximation for the electron. In addition to the SR and LR cases, we wish to analyze also intermediate coupling regimes (IR) where the electron couples with local and nn lattice displacements. In this case the interaction function becomes

$$f(|\boldsymbol{R}_i - \boldsymbol{R}_j|) = \delta_{\boldsymbol{R}_i,\boldsymbol{R}_j} + \frac{g_1}{g}\sum_{\boldsymbol{\delta}} \delta_{\boldsymbol{R}_i+\boldsymbol{\delta},\boldsymbol{R}_j}, \tag{50}$$

where $\boldsymbol{\delta}$ indicates the nn sites and g_1 controls the corresponding coupling strength. For all the couplings of (49,50) it is useful to define the e-ph matrix element in the momentum space $M_{\boldsymbol{q}}$ as

$$M_{\boldsymbol{q}} = \frac{g\omega_0}{\sqrt{N}}\sum_m f(|\boldsymbol{R}_m|)e^{i\boldsymbol{q}\cdot\boldsymbol{R}_m}, \tag{51}$$

and the polaronic shift E_p

$$E_p = \sum_{\boldsymbol{q}} \frac{M_{\boldsymbol{q}}^2}{\omega_0}. \tag{52}$$

Then the coupling constant $\lambda = E_p/zt$, with z lattice coordination number, represents a natural measure of the strength of the e-ph interaction in both SR and LR cases. Clearly, for LR interactions, the matrix element $M_{\boldsymbol{q}}$ peaks around $\boldsymbol{q} = 0$. Since it has been claimed that the enhancement of the forward direction in the e-ph scattering could play a role in explaining several anomalous properties of cuprates as the linear temperature behavior of the resistivity and the d-wave symmetry of the superconducting gap [54, 55], the study of lattice polaron features for LR interactions is important in order to clarify the role of the e-ph coupling in complex systems.

When the interaction force is given by (49), the model has been first investigated applying a path-integral Monte-Carlo (PIMC) algorithm [51, 53] that is able to reach the thermodynamic limit. The first investigations have been mainly limited to the determination of the polaron effective mass pointing out that, due to the LR coupling, the polaron is much lighter than in the Holstein model with the same binding energy in the strong coupling regime. Furthermore it has been found that this effect, due to the weaker band renormalization, becomes smaller in the antiadiabatic regime. The quasi-particle properties have been studied by an exact Lanczos diagonalization method [56] on finite one-dimensional lattices (up to 10 sites) making a close comparison with the corresponding properties of the Holstein model. As a result of the

LR interaction, the lattice deformation induced by the electron is spread over many lattice sites in the strong coupling region giving rise to the formation of a large polaron (LP) as in the weak coupling regime. All numerical and analytical results have been mainly obtained in the antiadiabatic and non-adiabatic regime. The behavior of the effective mass of a two-site system [57] in the adiabatic regime has been studied within the nearest-neighbor approximation for the e-ph interaction confirming that the LP is lighter than in the Holstein model at strong coupling. Recently we have introduced a variational solution [58] showing that there is a range of intermediate values of the e-ph coupling constant, in the adiabatic regime, where the GS has lost spectral weight but the polaron mass is only weakly renormalized. In the same regime, a further increase of the e-ph coupling leads to a smooth increase of the effective mass associated to an average kinetic energy not strongly reduced. The peculiar properties of this LR model suggest investigation of the crossover between short and long range coupling (intermediate coupling (IR)). This has been discussed in [59] where it has been shown that for large values of the coupling with nearest neighbor sites, most physical quantities show a strong resemblance with those obtained for the long range e-ph interaction. However, this limit is reached in a non monotonic way and, at intermediate values of the interaction strength, the correlation function between electron and nearest neighbor lattice displacements is characterized by an upturn as a function of the e-ph coupling constant.

Variational Approach

Also for this class of models (LR and IR) we can show that the variational scheme adopted in the previous cases can be applied successfully. We consider as trial wave functions translational invariant Bloch states of the same type of those used for the Holstein model (31). Both the weak and strong coupling components can be chosen as follows:

$$|\phi_{\boldsymbol{k}}^{(a)}(\boldsymbol{R}_i, \boldsymbol{R}_n) >= c_{n+i}^{\dagger} e^{\sum_{\boldsymbol{q}}\left[f_{\boldsymbol{q}}^{(a)}(\boldsymbol{k})a_{\boldsymbol{q}}e^{i\boldsymbol{q}\cdot\boldsymbol{R}_i}+h.c.\right]}|0>_{ph}|0>_{el}, \qquad (53)$$

where the apex $a = w, s$ indicates the weak and strong coupling wave function, respectively. Following the procedure discussed for the Holstein case, the phonon distribution functions $h_{\boldsymbol{q}}^{(a)}(\boldsymbol{k})$ are chosen in order to describe polaron features in the two asymptotic limits [37]:

$$h_{\boldsymbol{q}}^{(w)}(\boldsymbol{k}) = \frac{M_{\boldsymbol{q}}}{\omega_0 + E_b(\boldsymbol{k}+\boldsymbol{q}) - E_b(\boldsymbol{k})}, \qquad (54)$$

where $E_b(\boldsymbol{k})$ is the free electron band energy for the weak coupling case and phonon distribution function $h_{\boldsymbol{q}}^{(s)}(\boldsymbol{k})$ as

$$h_{\boldsymbol{q}}^{(s)}(\boldsymbol{k}) = \frac{M_{\boldsymbol{q}}}{\omega_0} \sum_{\boldsymbol{R}_m} |\eta_{\boldsymbol{k}}^{(s)}(\boldsymbol{R}_m)|^2 e^{i\boldsymbol{q}\cdot\boldsymbol{R}_m} \qquad (55)$$

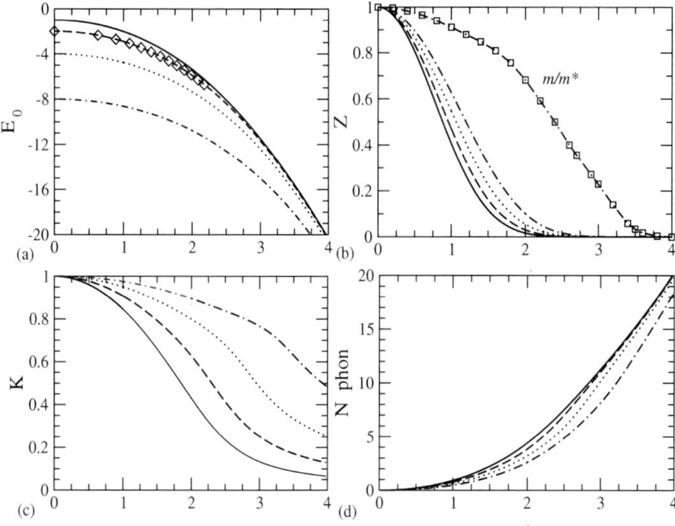

Fig. 6. The ground state energy E_0 in units of ω_0 (a), the spectral weight Z (b), the average kinetic energy K (c) and the average phonon number N (d) as a function of the coupling constant g for different values of the adiabatic ratio: $\omega_0/t = 2$ (solid line), $\omega_0/t = 1$ (dashed line), $\omega_0/t = 0.5$ (dotted line) and $\omega_0/t = 0.25$ (dash-dotted line). The diamonds in (a) indicate the $PIMC$ data for the energy kindly provided by P. E. Kornilovitch at $\omega_0/t = 1$, and the squares on the dash-dotted line in (b) denote the ratio m/m^* obtained within the variational approach at $\omega_0/t = 0.25$.

for the strong coupling case.

Again the complete variational wave function is chosen as a superposition of weak and strong coupling components, $A_{\boldsymbol{k}}$ and $B_{\boldsymbol{k}}$ being the relative weights.

We perform the minimization procedure with respect to the parameters $\eta_{\boldsymbol{k}}^{(w)}(\boldsymbol{R}_m)$, $\eta_{\boldsymbol{k}}^{(s)}(\boldsymbol{R}_m)$, $A_{\boldsymbol{k}}$ and $B_{\boldsymbol{k}}$, limiting the sum in (31) to third neighbors. The ground state energies obtained with this choice are slightly higher than PIMC mean energies, the difference being less than 0.5% in the worst case of the intermediate regime. We note that these wave functions can, in principle, be improved extending the sum in (31) further.

In Figs. 6, 7 and 8 we show some of the polaron GS properties in the one-dimensional case for LR and IR cases, respectively. We have checked that our variational proposal recovers the asymptotic perturbative results and improves significantly these asymptotic estimates in the intermediate region. In particular, in the LR case our data for the ground-state energy in the intermediate region are successfully compared with the results of the PIMC approach [51] shown as diamonds in Fig. 6a. The consistency of the results with a numerically more sophisticated approach indicates that the true wave function is very close to a superposition of weak and strong coupling states.

Long Range Case

A specific property of the long range model is the fact that the reduction of the GS spectral weight is not always accompanied by an equal increase of the polaron effective mass. This is a peculiar behavior of the LR coupling that is completely absent in the SR Holstein model. In Fig. 6b we show that the increase of the e-ph coupling strength induces a decrease of the spectral weight that is smooth also in the adiabatic regime. The reduction of Z is closely related to the decrease of the Drude weight obtained by exact diagonalizations [56] pointing out a gradual suppression of coherent motion. We note that the behavior of Z is different from that of the local Holstein model. In fact, for the latter, Z results to be very close to the ratio m/m^*, with m and m^* bare electron and effective polaron mass [56], respectively, while for LR couplings $Z < m/m^*$ in the intermediate to strong coupling adiabatic regime. This relation is confirmed by the results shown in Fig. 6b, where the dash-dotted line and the squares on a similar line indicate the spectral weight Z and the ratio m/m^* obtained within the variational approach, respectively, as a function of the coupling constant g at $\omega_0/t = 0.25$ [53]. Actually there is a large region of the parameters in the adiabatic regime where the ground state is well described by a particle with a weakly renormalized mass but a spectral weight Z much smaller than unity. In the adiabatic case, with increasing e-ph coupling, a band collapse occurs in the SR case, while the particle undergoes a weaker band renormalization in the case of LR interactions. Therefore in the LR case the polaron results lighter than in the SR Holstein model both in the intermediate and strong coupling adiabatic regimes.

Insight about the electron state is obtained by calculating its kinetic energy K in units of the bare kinetic energy. Since the average kinetic energy gives the total weight of the optical conductivity, K includes both coherent and incoherent transport processes [56]. At the same time, in the strong coupling adiabatic region before the electron self-trapping ($K \ll 1$), the average kinetic energy and the ratio m/m^* are weakly renormalized (Figs. 6b and 6c) and, therefore, the optical conductivity is dominated by the coherent motion of the polaron.

Another quantity associated to the polaron formation is the correlation function $S(R_l)$

$$S(R_l) = S_{k=0}(R_l) = \frac{\sum_n < \psi_{k=0}|c_n^\dagger c_n \left(a_{n+l}^\dagger + a_{n+l} \right) |\psi_{k=0} >}{< \psi_{k=0}|\psi_{k=0} >} \tag{56}$$

or equivalently the normalized correlation function $\chi(R_l) = S(R_l)/N$, with $N = \sum_l S(R_l)$. In Fig. 7a we report the correlation function $S(R_l)$ at $\omega_0/t = 1$ for several values of the e-ph interaction. The lattice deformation is spread over many lattice sites giving rise to the formation of LP also in the strong coupling regime where really the correlation function assumes the largest values. In the inset of Fig. 7a the normalized electron-lattice correlation function χ shows

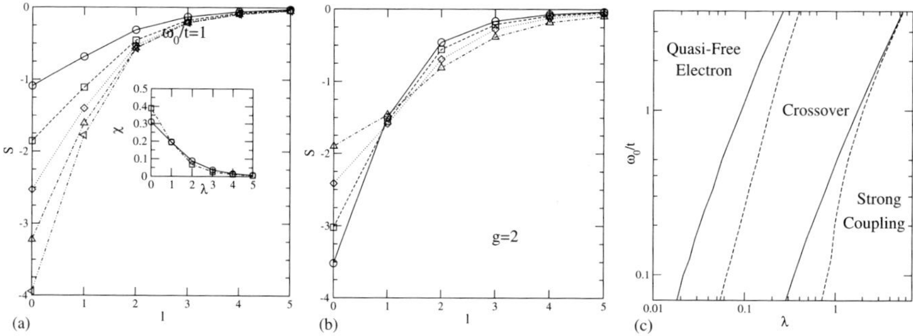

Fig. 7. (a) The electron–lattice correlation function $S(R_l)$ at $\omega_0/t = 1$ for different values of the coupling: $\lambda = 0.5$ (circles), $\lambda = 1.25$ (squares), $\lambda = 2.0$ (diamonds), $\lambda = 2.75$ (triangles up), and $\lambda = 3.5$ (triangles down). In the inset the normalized correlation function $\chi(R_l)$ at $\omega_0/t = 1$ for $\lambda = 0.5$ (circles) and $\lambda = 2.75$ (squares). (b) The electron-lattice correlation function $S(R_l)$ at g=2 for several values of the adiabatic parameter$\gamma = \omega_0$: $\gamma = 2$ (circles), $\gamma = 1$ (squares), $\gamma = 0.5$ (diamonds), and $\gamma = 0.25$ (triangles up). (c) Polaron "phase diagram" for long-range (solid line) and Holstein (dashed line) $e - ph$ interaction

consistency with the corresponding quantity calculated in a previous work [56]. While in the weak coupling regime the amplitude χ is smaller than the quantum lattice fluctuations, increasing the strength of the interaction, it becomes stronger and the lattice deformation is able to generate an attractive potential that can trap the charge carrier. Of course, even if the correlations between electron and lattice are large, the resulting polaron is delocalized over the lattice due to the translational invariance. Finally the variation of the lattice deformation as a function of ω_0/t shown in Fig. 7b can be understood as a retardation effect. In fact, for small ω_0/t, the fewer phonons excited by the passage of the electron take a long time to relax, therefore the lattice deformation increases far away from the current position of the electron.

On the basis of the previous discussion, in Fig. 7c, we propose a "phase diagram" based on the values assumed by the spectral weight in analogy with the Holstein polaron. Analyzing the behavior of Z it is possible to distinguish three different regimes: (1) quasi-free-electron regime ($0.9 < Z < 1$) where the electron has a weakly renormalized mass and the motion is coherent; (2) crossover regime ($0.1 < Z < 0.9$) characterized by intermediate values of spectral weight and a mass not strongly enhanced; (3) strong coupling regime ($Z < 0.1$) where the spectral weight is negligible and the mass is large but not enormous. We note that for LR interactions in the adiabatic case there is strong mixing of electronic and phononic degrees of freedom for values of the coupling constant λ (solid lines) smaller than those characteristic of local Holstein interaction (dashed lines). Furthermore in this case, entering the strong coupling regime, the charge carrier does not undergo any abrupt localization, on the contrary, as indicated also by the behavior of the average kinetic energy K, it is quite mobile.

Intermediate Range Coupling

In Fig. 8 we report some properties of the polaron ground state as a function of the e-ph constant coupling λ for different g_1 values that control the coupling range. One interesting property that gives a qualitative difference among SR and LR models is reported in Fig. 8b. With increasing range of coupling, the overlapping of the two components (weak and strong) becomes more important. Actually, as shown in Fig. 8b, there are marked differences in the ratio B/A, that is the weight of the strong coupling solution with respect to the weak coupling one. In the SR case the strong coupling solution provides almost all of the contribution. However, by increasing the range of the interaction, the weight of the weak coupling function increases and the polaronic crossover becomes smoother. Another quantity that gives insight about the properties of the electron state is the average kinetic energy K reported in Fig. 8c (in units of the bare electron energy). While in the SR case K is strongly reduced, in the LR case it is only weakly renormalized stressing that the self-trapping of the electron occurs for larger couplings with increasing the range of the interaction. Finally the mean number of phonons is plotted in Fig. 8d. In the weak-coupling regime the interaction of the electron with displacements on different sites is able to excite more phonons. However, in the strong coupling regime there is an inversion in the roles played by SR and IR interaction. Indeed the Holstein small polaron is strongly localized on the site allowing a larger number of local phonons to be excited.

In addition to the quantities discussed in Fig. 8, other properties change remarkably with increase of the ratio g_1/g. An interesting property is the ground state spectral weight Z. The increase of the e-ph coupling strength induces a decrease of the spectral weight that is more evident with increasing range of e-ph coupling. Only in the strong coupling regime the spectral weights calculated for different ranges assume similar small values. While for the local Holstein model $Z = m/m^*$, as the IR case is considered, Z becomes progressively smaller than m/m^* in analogy with the behavior identified for the LR interaction. We have found that for the ratio $g_1/g = 0.3$ there is a region of intermediate values of λ where the ground state is described by a particle with a weakly renormalized mass but a spectral weight Z much smaller than unity. Finally, we mention that, following the same criteria discussed in the previous section it is possible to build up a phase diagram that identifies three different regimes as in the case of LR coupling [59].

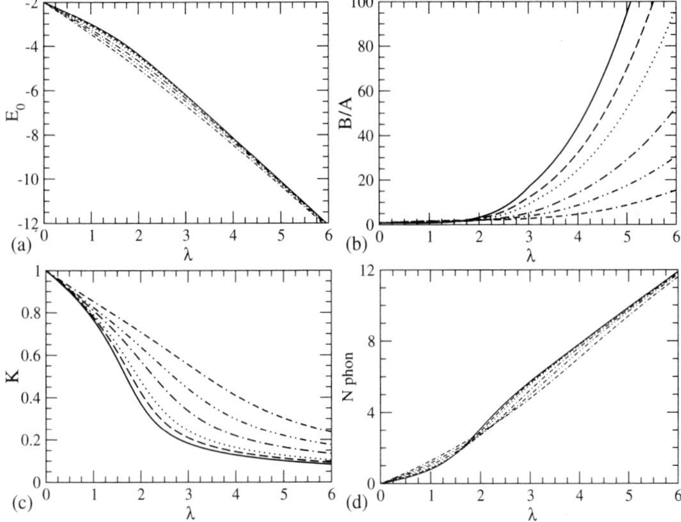

Fig. 8. The ground state energy E_0 in units of ω_0 (a), the ratio B/A at k=0 (b), the average kinetic energy K in units of the bare kinetic energy (c) and the average phonon number N (d) for $t = \omega_0$ as a function of the coupling constant λ for different ranges of the $e - ph$ interaction: SR (solid line), IR with $g_1/g = 0.05$ (dash line), IR with $g_1/g - 0.1$ (dot line), IR with $g_1/g = 0.2$ (dash-dot line), IR with $g_1/g = 0.3$ (dash-double dot line), LR (double dash-dot line).

3 Excited States: Optical Conductivity and Spectral Functions

It is well known that infrared spectroscopy is an excellent probe to investigate e-ph effects and, indeed, it was the attempt to explain the optical features of alkali halides by polaron absorption that motivated Landau [1] in 1933 to introduce the notion of the self-trapped carrier. More recently, the effort in understanding the non-conventional properties of complex materials such as transition metal oxides and the role played by e-ph interaction, has favored a growing interest in experimental data in such materials [60] and reliable calculations on the optical conductivity of e-ph models.

3.1 The Fröhlich Model

Although this model has attracted over the years the interest of many researchers [61], a complete understanding of its optical absorption has been obtained only recently [62] by integrating many different analytical and numerical approaches. For a long time the understanding of this problem has been mainly based on the response formalism developed by Feynman for path integrals [63] that, successively, has been shown to be equivalent to the memory function formalism associated with the Feynman polaron model that we

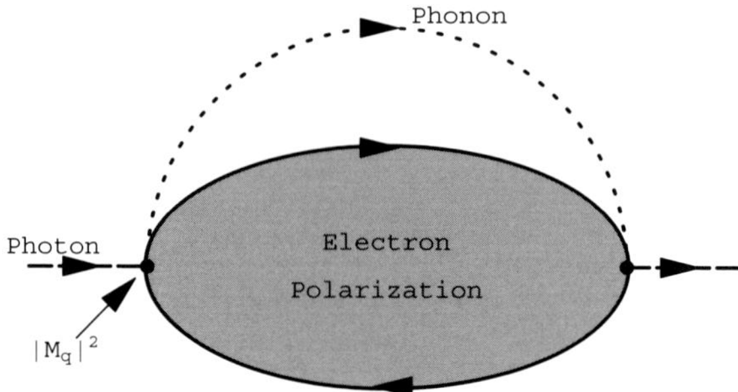

Fig. 9. The Feynman diagram used in the MFF approach eq.(58)

have discussed in Sect. (2.1)[64]. For a review of all the results obtained within these formalisms and, more generally, for a comprehensive account of the literature on optical conductivity we refer to [65]. This approach predicts in the strong coupling regime a sharp peak in the optical conductivity. The idea is that, for large values of α, there is a quasi-stable state, the relaxed excited state (RES), characterized by lattice distortion adapted to the excited electronic state. In addition to this narrow peak, a band at higher energies was predicted. Following the interpretation of [63, 64, 66], it is due to the transitions to the unstable FC state. Recently, results based on the Diagrammatic Quantum Monte Carlo (DQMC) methods [62, 67] have shown that there is no indication of any sharp resonance in the infrared spectra. From this analysis it is clear that going beyond the approximations used so far is crucial in order to clarify the different aspects of the problem [62].

From Weak to Intermediate Coupling Optical Conductivity

A description of the infrared absorption, able to reproduce the main structure of DQMC data up to $\alpha \simeq 8$, can be obtained by using the memory function formalism (MFF)[68]. Within this approach the real part of the conductivity, in the limit of a vanishing electron density n, and treating the interaction between the charge carriers and the phonons at the lowest order, can be written as [63, 64]

$$\text{Re } [\sigma_{xx}(\omega)] = -\frac{ne^2}{m} \frac{\text{Im } [\Sigma_{xx}(\omega)]}{(\omega - \text{Re } [\Sigma_{xx}(\omega)])^2 + (\text{Im } [\Sigma_{xx}(\omega)])^2} \tag{57}$$

where

$$\text{Im } [\Sigma_{xx}(\omega)] = \frac{1}{m\omega} \sum_{q} |M_q|^2 \, q_x^2 \text{Im } [\chi(\boldsymbol{q}, \omega - \omega_0)] \tag{58}$$

and

$$\chi(\boldsymbol{q},t) = -i\theta(t)\left\langle e^{i\boldsymbol{q}\cdot\boldsymbol{r}(t)}e^{-i\boldsymbol{q}\cdot\boldsymbol{r}(0)}\right\rangle. \tag{59}$$

In term of Feynman diagrams (58) corresponds to the phononic renormalization of the electron polarization bubble (Fig. 9). Of course, the key role in the MFF is played by the calculation of the Fourier transform of the electron density-density correlation function: $\chi(\boldsymbol{q},t)$. At the lowest level it can be calculated assuming that the electron does not interact with phonons and gives $\chi(\boldsymbol{q},t) = -i\theta(t)\exp\left[-iq^2 t/(2m)\right]$. This approximation returns the perturbative result ($\alpha \ll 1$)

$$\mathrm{Im}\ [\Sigma_{xx}(\omega)] = -\frac{2}{3}\alpha\frac{\omega_0^{3/2}}{\omega}\sqrt{\omega - \omega_0}\theta(\omega - \omega_0).$$

This quantity has been traditionally evaluated (see [63]) by using the FPM in which the electron is coupled via a harmonic force to a fictitious particle simulating the phonon degrees of freedom. The result is

$$\chi(\boldsymbol{q},t) = -i\theta(t)e^{-i\frac{q^2 t}{2M_e}}e^{-\frac{q^2 R}{2M_e}(1-e^{-ivt})} \tag{60}$$

where $M_e = m\frac{v^2}{w^2}$ is the total mass of FPM and $R = \frac{v^2 - w^2}{vw^2}$. v and w are related to the mass and the elastic constant of the model and are determined variationally (see Sect. (2.1)). Consequently $\mathrm{Im}\ [\Sigma_{xx}(\omega)]$ becomes:

$$\mathrm{Im}\ [\Sigma_{xx}(\omega)] = -\frac{2\alpha v^3\omega_0^{3/2}}{3\omega w^3}\sum_{n=0}^{\infty}\frac{\theta(\omega_n)}{n!}\exp\left[-\frac{\omega_n\left(v^2 - w^2\right)}{vw^2}\right]\frac{\left(v^2 - w^2\right)^n}{v^n w^{2n}}\omega_n^{n+1/2}$$

$$\tag{61}$$

where $\omega_n = \omega - \omega_0 - nv$. As already mentioned, the majority of the work on optical conductivity, given by (57), is based on (61). However, it contains two important limitations. The first one is that the exact sum rule

$$\int_{-\infty}^{\infty}\frac{d\omega}{\pi}\omega^2\mathrm{Im}\ [\chi(\boldsymbol{q},\omega)] = -\left(\frac{q^2}{2m}\right)^2 - \frac{2}{3}\frac{q^2}{m}\left\langle\frac{p^2}{2m}\right\rangle \tag{62}$$

is not satisfied and the other one is that this approximation does not contain dissipative effects. Both limitations can be removed introducing a finite lifetime in (60) in such a way that the sum rule (62) is fulfilled. Lifetime effects are expected to play a strong role since the FPM is only an effective model that only partially takes into account the phonon bath and a residual coupling with the original bath should still have important effects. If we replace the term e^{-ivt} in (60) with $(1 + it/\tau)^{-v\tau}$ we allow the FPM to scatter with the phonon bath attributing finite lifetimes to the states of the FPM. The quantity τ is determined making use of the sum rule (62) where $\left\langle\frac{p^2}{2m}\right\rangle$ is estimated by using Feynman approach for the polaron ground state. We stress

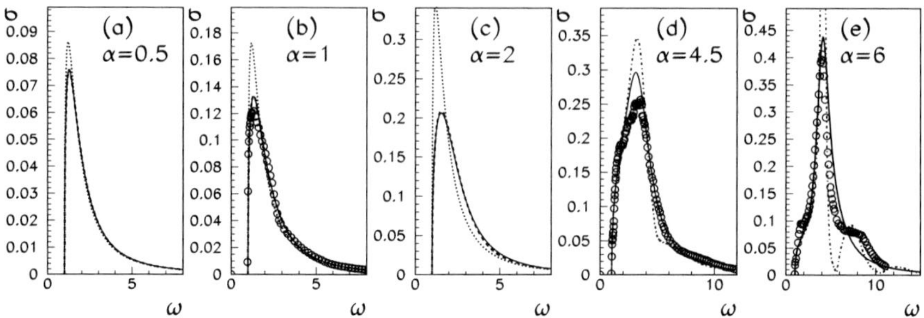

Fig. 10. The OC for different values of α calculated within the EMFF (solid line). Our EMFF data are compared with weak coupling perturbation theory (dotted line), MFF (dashed line) and DQMC data (open circles).

here the difference between Feynman approach and FPM. In the former the scattering among the FPM and the residual e-ph interaction is included. It is worthwhile noting that for $\tau \longrightarrow \infty$ we recover the expression (60). We will call this approximation extended memory function formalism (EMFF). As expected, τ turns out to be of the order of ω_0^{-1}. In Fig. 10 we compare the results with DQMC and with the theory based on (61). In the weak e-ph coupling regime, the memory function formalism, with or without damping τ, is in good agreement with DQMC data, indicating that the phonon assisted intra-band transitions are dominant in this regime (Fig. 10a). In any case the results shown (Figs. 10b,10c) exhibit a significant improvement with respect to weak coupling perturbation approach [69]. The latter of these, which is recovered within the memory function formalism, provides a good description of the infrared spectra only for very small values of α. As the plots in Fig. 10d show, the introduction of the damping becomes crucial already at $\alpha = 4.5$. Here the agreement with DQMC data is still complete while the results obtained neglecting the damping provide a sharper peak in the optical conductivity that is not present in the DQMC data. On the other hand, at $\alpha = 6$ the optical conductivity calculated within DQMC starts to present a small shoulder that is not present in our EMFF. In the next section we will try to understand this new structure starting from the opposite limit of strong adiabatic coupling.

Strong Coupling Optical Conductivity: Franck-Condon Regime

For $\alpha >> 1$, as discussed in Sect. 2.1, the GS of the Fröhlich model is well described by the adiabatic approximation and, therefore, we expect that in this regime, the OC is dominated by the Franck-Condon (FC) principle [70] that rests on the adiabatic decoupling of electron and lattice degrees of freedom. We present, here, a derivation of OC in the FC limit that includes phonon replica. As we will see this approach is able to reproduce correctly the approximation-free results of DQMC for large enough α.

Following the Landau-Pekar approach (6)-(15), we first apply the canonical transformation $U = \exp\left[-\sum_{\boldsymbol{q}}((\rho_q/\omega_0)\,a_{\boldsymbol{q}} - h.c.)\right]$ to the Frohlich Hamiltonian and obtain

$$\tilde{H} = \frac{p^2}{2m} + \sum_{\boldsymbol{q}} \omega_0 a_{\boldsymbol{q}}^{\dagger} a_{\boldsymbol{q}} + \sum_{\boldsymbol{q}} \left[(M_q e^{i\boldsymbol{q}\cdot\boldsymbol{r}} - \rho_q)a_{\boldsymbol{q}} + h.c.\right] \tag{63}$$

$$- \sum_{\boldsymbol{q}} \left[\frac{\rho_q^*}{\omega_0}(M_q e^{i\boldsymbol{q}\cdot\boldsymbol{r}} - \rho_q) + h.c\right] - \sum_{\boldsymbol{q}} \frac{|\rho_q|^2}{\omega_0}$$

where $\rho_q = M_q\langle\varphi|e^{i\boldsymbol{q}\cdot\boldsymbol{r}}|\varphi\rangle$ (8) is related to the GS electron wave-function $|\varphi\rangle$. It has to be emphasized that the use of the canonical transformation, U, breaks the translational invariance, but this is not a severe limitation in the strong coupling limit. The Hamiltonian (63) can be viewed as a Fröhlich Hamiltonian with a modified e-ph coupling function in an external potential given by the last two terms. In order to proceed further in the OC calculation we have to make a further approximation replacing in (63) the true potential with an harmonic one (see (11)). This allows us to identify in (63) a part that can be solved exactly (the sum of a harmonic oscillator and free phonons) and the part that we wish to treat perturbatively. Summarizing, we obtain $\tilde{H} \to \tilde{\mathcal{H}} = \tilde{\mathcal{H}}_0 + \sum_{\boldsymbol{q}} \omega_0 a_{\boldsymbol{q}}^{\dagger} a_{\boldsymbol{q}} + \mathcal{W}$ where·

$$\tilde{\mathcal{H}}_0 = \frac{p^2}{2m} + \frac{m\tilde{\omega}^2}{2}r^2 - \frac{3}{4}\tilde{\omega}; \text{ and } \mathcal{W} = \sum_{\boldsymbol{q}} \left[(M_q e^{i\boldsymbol{q}\cdot\boldsymbol{r}} - \rho_q)a_{\boldsymbol{q}} + h.c.\right]. \tag{64}$$

$\tilde{\omega} = (4\alpha^2\omega_0)/(9\pi)$ is the Landau-Pekar variational estimation (13). It is worth noting that the estimation of the energy difference between the GS and the first adiabatic excited state can be improved exploiting the derivation of the Feynman approach presented in Sect. 2.1. In that formalism we can choose a trial wave-function like that of (4) where the GS of FPM is replaced with the first electronic excited state. Actually we keep the phonon potential well fixed at the value found for the GS, but we put the electron in the first excited state. This gives the following value for $\tilde{\omega} = (4\alpha^2\omega_0)/(9\pi) - 3.8\omega_0$.

Since the real part of the OC in the limit of vanishing electron density can be written in terms of the position operator

$$\text{Re}\,[\sigma_{xx}(\omega)] = ne^2\omega\text{Re}\int_0^{\infty} dt e^{i(\omega+i\delta)t}\,\langle x(t)x(0)\rangle$$

we have to evaluate $\langle x(t)x(0)\rangle$ by using the Hamiltonian of (64). Within this set of approximations it is possible to eliminate exactly the phonon degrees of freedom showing that

$$\langle x(t)x(0)\rangle = \frac{1}{2m\widetilde{\omega}}e^{-i\widetilde{\omega}t}\,\langle f_{1,0,0}f\,|Te^{\chi}|\,f_{1,0,0}\rangle$$

where $|f_{1,0,0}\rangle$ is the first excited state of $\widetilde{H}_0$, T is the time ordering operator and

$$\chi(t) = -\int\limits_0^t d\tau \int\limits_0^\tau d\tau' e^{-i\omega_0(\tau-\tau')}\sum_q w_q(\tau)w_q^*(\tau'). \tag{65}$$

In (65) $w_q(s) = \exp\left[i\widetilde{H}_0 s\right](M_q e^{iq\cdot r} - \rho_q)\exp\left[-i\widetilde{H}_0 s\right]$. At this stage of the calculation, if we completely neglect χ, we recover the extreme FC regime in the Landau-Pekar approximation:

$$\text{Re}\,[\sigma_{xx}(\omega)] = \frac{ne^2\pi}{2m}\delta(\omega - \widetilde{\omega}).$$

In order to include higher order contributions we make two further approximations. First we neglect the time correlations:

$$\langle f_{1,0,0}f\,|Te^{\chi}|\,f_{1,0,0}\rangle \rightarrow \exp\langle f_{1,0,0}\,|\chi|\,f_{1,0,0}\rangle$$

and then retain in $\langle f_{1,0,0}\,|w_q(s)w_q^*(s')|\,f_{1,0,0}\rangle$ only those terms that do not depend on $(s - s\prime)$. While the first approximation is motivated by a real computational difficulty (even if we do not expect that time correlations are important at $\alpha >> 1$) , the second one is based on the fact that $\omega_0 \ll \widetilde{\omega}$. The inclusion of the terms proportional to $(s - s\prime)$ can be done exactly and should take into account the first non-adiabatic contributions. We get

$$\text{Re}\,[\sigma_{xx}(\omega)] = \frac{ne^2\pi}{2m}\frac{\omega}{\widetilde{\omega}}\sum_{n=0}^{\infty}\exp\left[-\frac{\omega_s}{\omega_0}\right]\left(\frac{\omega_s}{\omega_0}\right)^n\frac{1}{n!}\delta\left(\omega - \widetilde{\omega} + \omega_s - n\omega_0\right),$$

where $\omega_s = \alpha\sqrt{\widetilde{\omega}\omega_0^3}/(4\sqrt{\pi})$. This expression is very close to that expected for the exactly solvable independent oscillators model [71]. The OC contains a set of delta function peaks starting from $(\widetilde{\omega} - \omega_s)$ and separated by ω_0. However, for $\alpha >> 1$, the maximum spectral weight is attributed to the peak closest to $\widetilde{\omega}$, the FC frequency. The effect of the residual e-ph scattering does not change the peak maximum, but introduces phonon sidebands that concur to define an effective peak width. This becomes clearer for large enough coupling strengths (or, equivalently, for $\omega_0 \rightarrow 0$) when the envelope of the Poisson distribution can be described by the Gaussian

$$\text{Re}\,[\sigma_{xx}(\omega)] = \frac{ne^2\pi}{2m}\frac{\omega}{\widetilde{\omega}}\frac{1}{\sqrt{2\pi\omega_0\omega_s}}\exp\left\{-\frac{(\omega - \widetilde{\omega})^2}{2\omega_0\omega_s}\right\}. \tag{66}$$

The frequency $(\widetilde{\omega} - \omega_s)$ deserves a comment. Indeed, it can be interpreted as the RES introduced by Devreese [61], but, at the strong coupling regime

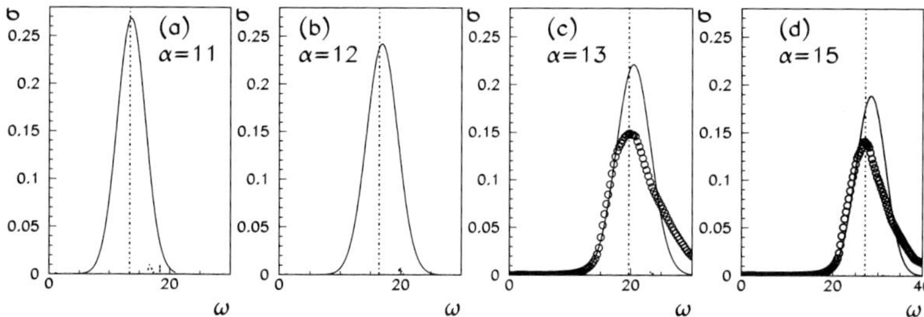

Fig. 11. The OC for different values of α calculated within the SC approach (solid line). Our SC data are compared with MFF (dashed line) and DQMC data (open circles).

where the present derivation is valid it has only a marginal spectral weight. Within this approximate scheme we cannot judge if in the intermediate coupling regime RES could play some role.

In Fig. 11 the very good agreement between the adiabatic SC approach and the DQMC data clearly shows that the broad peak observed in the DQMC data at $\alpha \gg 1$ is due to FC processes. We have also reported the MFF result for comparison. In this regime, it reduces to a delta function located at the FC energy and to a very small structure at higher energy. Both these structures follow the α^2 dependence characteristic of the FC transition. This behavior suggests that the MFF, even if it completely misses the correct OC peak shape, is able to reproduce the FC energy.

We end this section mentioning that the results obtained both in the weak to intermediate coupling and in the strong coupling regime are in very good agreement with DQMC data [62]. We refer to the chapter by Mishchenko and Nagaosa [72] for a specific account of those data. Here, we wish to stress that from the analysis outlined here and from DQMC data a clear scenario comes out. We find that different regimes, characterized by different excitations, can be recognized. i) For large values of α, $\alpha \geq 11$, the main contribution to the real part of the optical conductivity comes from the transition to the first excited state in the potential well characteristic of the polaron ground state. In this regime the spectral weight of the RES turns out to be negligible. Furthermore, the maximum of the optical conductivity follows the characteristic dependence predicted for the energy gap between the ground and FC states, i.e. it is proportional to $4\alpha^2/9\pi$; ii) in the opposite regime ($\alpha \leq 1$), as shown by the perturbation theory, the maximum of the optical conductivity is related to the phonon assisted intra-band transitions. The spectral weight of these optical transitions decreases with increasing strength of the e-ph interaction; iii) for $1 \leq \alpha \leq 6$, in addition to the phonon assisted intra-band transitions, which still survive, the main contribution to the real part of the conductivity

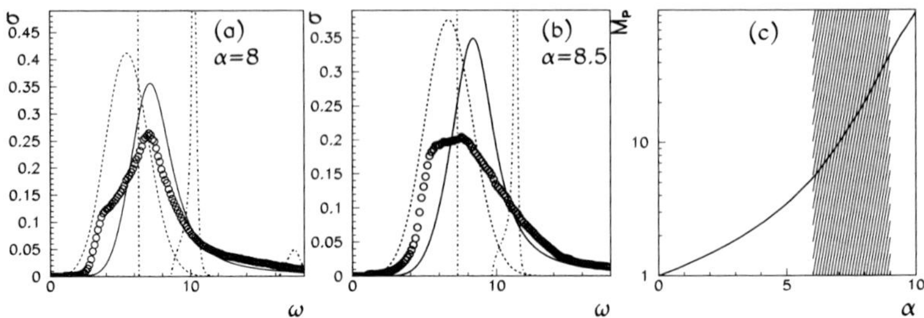

Fig. 12. (a) and (b). The OC for different values of α calculated within the EMFF (solid line) and SC approach (dotted line). Our data are compared with MFF (dashed line) and DQMC data (open circles). (c) The polaron effective mass as function of α [13]

comes from the electron transitions between the damped levels of the FPM. Here the interaction with free lattice oscillations is treated at the lowest order and the dynamic lattice effects are dominant. In any case we can say that, for these values of α, the FPM captures also the physics of the excited states of the Fröhlich model; iv) at $\alpha = 7$ a new excitation is present in the optical absorption. The DQMC data [67] exhibit a shoulder at $\omega \simeq 4\omega_0$. At $\alpha = 8.5$ two structures appear in the infrared spectra with almost the same spectral weight. This new excitation is due to the FC transitions, i.e. the lattice configuration does not change during the radiation absorption process. At this value of α static and dynamic lattice effects coexist (see Fig. 12). We end this section showing that the previous scenario is confirmed by the effective polaron mass behavior (see Fig. 12). Also for this quantity we can identify three regimes that correspond to the different OC regions: i) $\alpha < 6$ where the polaron mass enhancement is not very strong and the lattice is still able to partially follow the electrons; ii) $\alpha > 9$ where the lattice is completely static and the mass is huge; iii) $6 < \alpha < 9$ where the mass makes its rapid crossover between small and very large values and static and dynamic features coexists in the OC.

3.2 The Holstein Model

Spectral Function in the Coherent State Basis

Knowledge of the spectral function is crucial to get information on the excited states of any system and is directly related to photoemission experiments. In particular, the great importance played by this class of experiments in cuprates has renewed the interest in non perturbative calculation for the spectral function of the Holstein model. The open question is whether or not some spectroscopic features can be attributed to e-ph interaction [4, 73].

Recently a large number of numerically exact methods have been introduced. These approaches are based on different algorithms for exact diagonalization [28], but all need some kind of truncation in the phonon space. The introduction of an optimized phonon approach based on the analysis of the density matrix [31, 74] has produced an important improvement in this field. The idea is to exploit the knowledge of the largest eigenvalues and eigenvectors of the site density matrix to select the phonon linear combinations that, at the best, can describe the system: the so-called optimized phonon basis (OPB). Unfortunately, the density matrix of the target states is not available *a priori*: it must be calculated in a self-consistent way together with the OPB. To this aim different strategies have been discussed. Recently, we have introduced a variational technique, based on an expansion of coherent states (CS) that very much simplifies the selection of an optimized phonon basis and that does not require any truncation in the number of phonons [75, 76]. This method is able to provide accurate results for any e-ph coupling regime and for any value of the adiabatic ratio. In the Holstein model, the proposed expansion provides states surprisingly close to the eigenvectors of the site density matrix corresponding to the highest probabilities. Although the method requires a smaller phonon basis with respect to previous proposals, it has been applied so far to relatively small clusters. The idea is to replace the "natural" phonon basis at each site i $|\mu_i\rangle$ with a coherent state $|h, i\rangle$:

$$|\mu_i\rangle \rightarrow |h, i\rangle = e^{gh(b_i - b_i^\dagger)} |0\rangle_i^{(ph)} = e^{-\frac{g^2 h^2}{2}} \sum_{n=0}^{\infty} \frac{(-gh)^n}{\sqrt{n!}} |n\rangle_i. \tag{67}$$

This substitution does not introduce any truncation in the Hilbert space since, varying h in the complex plane, the local basis (67) is over-complete. In particular, for $h = 0$, the CS $|h = 0, i\rangle$ is the phonon vacuum at site i, i.e. the phonon ground state at site i when $g = 0$ (absence of e-ph interaction). On the other hand, for $h = 1$, the CS $|h = 1, i\rangle$, multiplied by the product of the phonon vacuum on the remainder of the sites, $\prod_{j \neq i} |0\rangle_j^{(ph)}$, gives the ground state of the Holstein Hamiltonian when $t = 0$, i being the site which the electron occupies.

However, in order to perform a numerical diagonalization of the Holstein Hamiltonian, we need to work with a finite number of coherent states. To this aim the second step is to choose a finite number M of CS, characterized by the corresponding values h_α ($\alpha = 0, ..., M - 1$), that we choose to be real and assume to be independent on the site index i. Then we consider only the phonon states $\prod_{i=1}^{N} |h_{\alpha_i}, i\rangle$, with $\sum_{i=1}^{N} \alpha_i \leq (M - 1)$. Taking into account that the coherent state $|h, i\rangle$ is a linear superposition of states with fixed number of phonons (see (67)) distributed according to the Poisson function with a maximum at $n \simeq (hg)^2$ and variance $\sigma^2 = (hg)^2$, it easy to recognize that the previous procedure does not introduce any truncation in energy and in the phonon number. In particular the mean value and the variance increase by increasing the values of h and g. Then, when the e-ph interaction is strong

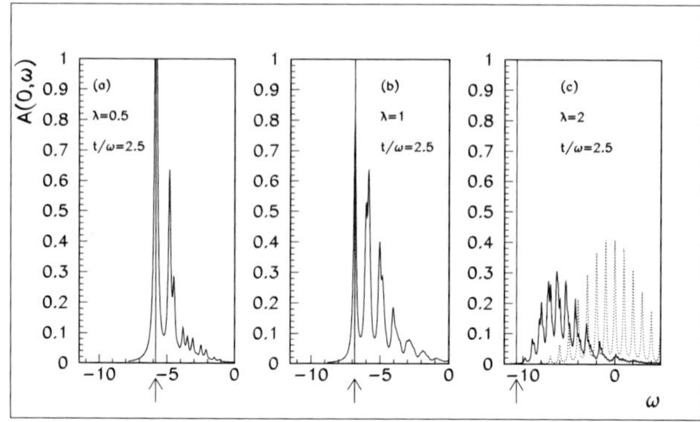

Fig. 13. The spectral weight function $A(k,\omega)/(2\pi)$ at $t/\omega_0 = 2.5$ and $k = 0$ for three different values of the e-ph coupling constant. In Fig. 3c, for comparison, the Lang-Firsov [26] spectral function is reported (dotted line). The arrows indicate the ground state energy. The energies are in units of ω_0. The value of the cutoff is $M = 16$ and the number of sites is N=8.

and h is of the order of one, the CS $|h, i\rangle$ represents a state with a large energy and phonon number dispersion and then it is the ideal candidate to describe small–polaron physics. In this case the presence, in the chosen basis, of coherent states with a small value of h allows us to take into account also the quantum fluctuations and the retardation effects of the e-ph interaction.

The ground state is obtained by using the modified Lanczos method [77]. We obtain a systematic improvement of the approximation adding more and more CS and this allows us to estimate the error introduced by a specific truncation. The variational calculation always provides values of h such that $0 \leq h \leq 1$. In particular, when the value of the cutoff M is about or greater than 10 the variationally chosen values of h turn out to be equally distributed within the interval $[0, 1]$. The main advantage of this approach is that the convergence of the ground state energy is very rapid so that we obtain the best subspace in which to find the target state with a little numerical effort, independently of the value of the e-ph coupling constant. This behavior can be understood by studying the site density matrix. In fact, we have shown in [75] that for a two-site one-electron system the eigenvectors corresponding to the largest eigenvalues of the exact site density matrix are in excellent agreement with those obtained in our approach, already with two CS per site.

In order to appreciate the computational improvement that can be obtained by using the coherent state basis we show in Fig. 13 the spectral weight for different values of the coupling constant in the moderate adiabatic regime.

Optical Conductivity

Following the analysis presented for the Fröhlich model we can calculate, also for the Holstein model, the OC in two limiting cases by using the MFF and the adiabatic FC approximation, respectively. Once again it is worth noting that that Fröhlich and Holstein models have many more similarities than expected. As discussed for the Fröhlich model, for the Holstein model we can also individuate a weak coupling regime dominated by intra-band excitations (Sect. 3.1) where dynamic phonon processes are relevant and an adiabatic strong coupling limit where only Franck-Condon processes between localized electron states contribute to the OC [78]. In the former case the MFF provides an excellent description of the relevant processes. Following closely the procedure outlined in Sect. 3.1 it is possible to obtain the OC that, in the 1D case, takes the closed form:

$$\text{Re } [\sigma(\omega)] = -\frac{2te^2}{N} \frac{\text{Im } [\Sigma(\omega)]}{(\omega - \text{Re } [\Sigma(\omega)])^2 + (\text{Im } [\Sigma(\omega)])^2},$$

where

$$\text{Im } [\Sigma(\omega)] = -\frac{g^2\omega_0^2}{\omega} \sqrt{1 - \frac{(\omega - \omega_0 - 2t)^2}{4t^2}} \theta(\omega - \omega_0)\theta(\omega_0 - \omega + 4t).$$

The real and imaginary parts of $\Sigma(\omega)$ are related by the usual Kramers-Kroning relations. In the adiabatic strong coupling limit we can proceed as in Sect. 2.2. Within the same set of approximations we get the following expression for the OC:

$$\text{Re } [\sigma_{xx}(\omega)] = \frac{e^2\pi}{N}\omega\beta^2 \sum_{n=0}^{\infty} \exp\left[-\frac{\omega_s}{\omega_0}\right] \left(\frac{\omega_s}{\omega_0}\right)^n \frac{1}{n!}\delta\left(\omega - \widetilde{\omega} + \omega_s - n\omega_0\right)$$

where $\omega_s = 3g^2\omega_0\alpha^4/2$, and $\widetilde{\omega} = E_1 - E_0$. The quantity $E_1 - E_0$ is the energy difference between ground and first adiabatic excited states that can be calculated, for instance, with the variational approaches discussed in Sect. 2.2 and $\beta = \left\langle f_{el}^{(0)} \middle| P_x \middle| f_{el}^{(1)} \right\rangle$ is the matrix element of the x component of the electron position in the tight-binding approximation between the ground state $(\left|f_{el}^{(0)}\right\rangle)$ and the first adiabatic excited state $(\left|f_{el}^{(1)}\right\rangle)$. Finally α is related to β via the normalization condition $|\alpha|^2 + d|\beta|^2 = 1$, d being the system dimensionality.

While those two limiting cases are well established in the literature the question of the Holstein OC in the intermediate coupling is still open. How does the crossover proceed between these limits? One possibility is that in the Holstein case the crossover regime is also characterized by the presence of both types of processes making the Fröhlich and Holstein models even more similar. Another possibility is that the crossover regime is characterized by new processes that are not relevant in the asymptotic limits. One proposal in this sense is given in [79] where adiabatic photoemission processes are considered responsible for the OC in the regime characterized by $g\omega_0 \ll 4dt$. We end this section mentioning that such process has been already proposed by Emin [80] for the OC of Fröhlich model, even if there is no evidence for them in the error free DQMC data [62].

4 Conclusion

In this chapter we have discussed the polaron formation and some of its properties in a number of e-ph models by using a unifying variational point of view. We have shown how variational semi-analytical approaches can give a very accurate description of these systems without resorting to heavy numerical calculations. These are crucial in order to have a reference point in the non perturbative regimes, but semi-analytical approaches have the advantage of giving direct access to the ground state wave-function making the understanding of the physical properties more transparent. A paradigmatic example of this statement is the case of Fröhlich OC where the joint effort of DQMC and semi-analytical results has been crucial in disentangling the complex physics of that model. We firmly believe that the merger of these different approaches will produce new and important improvements in the polaron field.

Acknowledgement. Let us finish by thanking G. Iadonisi, V. Marigliano Ramaglia, M. Capone, E. Piegari, S.A. Mishchenko and J. Devreese, for discussions and collaborations.

References

1. L.D. Landau: Phys. Z. Sowjetunion **3**, 664 (1933) [English translation: *Collected Papers* (Gordon and Breach New York, 1965), pp.67-68];
2. R.P. Feynman: Phys. Rev. **97**,660 (1955).
3. D.M. Edwards: Adv. Phys. **51**, 1259 (2002); E. Dagotto, *Nanoscale phase separation and colossal magnetoresistance: the physics of manganites and related compounds* (Springer, Heidelberg New York 2003).
4. *Lattice effects in High Temperature Superconductors* ed by Y. Bar.Yam, J. Mustre de Leon, and A. R. Bishop (World Scientific, Singapore, 1992); A. Damascelli, Z. Hussain, and Zhi-Xun Shen, Rev. Mod. Phys. **75**, 473 (2003); P.A. Lee, N. Nagaosa, and X.-G. Wen: Rev. Mod. Phys. **78**, 17 (2006).
5. J. Zaanen and P.B. Littlewood: Phys. Rev. B **50**, 7222 (1994); G. Wu and J.J. Neumeier: Phys. Rev. B **67**, 125116 (2003); Hotta Takashi and E. Dagotto: Phys. Rev. Lett. **92**, 227201 (2004). .
6. J.W. Bray, L.V. Interrante, L.S. Jacobs, and J.C. Bonner in *Extended linear Chain Compounds*, ed by J.S. Miller (Plenum, New York, 1985), p. 353;

S. Kagoshima, H. Nagasawa, and T. Sambongi *One-Dimensional Conductors* (Springer-Verlag, New York, 1988).

7. H. Fröhlich, H. Pelzer, and S. Zienau: Philos. Mag. **41**, 221 (1950).

8. G. Höhler and A.M. Mullensiefen: Z. Physik **157**, 159 (1959); D.M. Larsen: Phys. Rev. **144**, 697 (1966); G.D. Mahan: in *Polarons in Ionic Crystals and Polar Semiconductors* (North-Holland, Amsterdam, 1972), p. 553; M.A. Smondyrev: Teor. Mat. Fiz. **68**, 29 (1986).

9. N.N. Bogoliubov and S.V. Tiablikov: Zh. Eksp. i Theor. Fiz., **19**, 256 (1949); S.V. Tiablikov: Zh. Eksp. i Theor. Fiz., **21**, 377 (1951); S.V. Tiablikov Abhandl. Sowj. Phys., **4**, 54 (1954); G.R. Allcock: Advances in Physics **5**, 412 (1956); G. Höhler in *Field Theory and the Many Body Problem*, ed. E.R. Caianiello (Accademic Press), pp.285; E.P. Solodovnikova, A.N. Tavkhelidze, O.A. Khrustalev: Teor. Mat. Fiz. **11**, 217 (1972); V.D. Lakhno and G.N. Chuev: Physics-Uspekhi **38**, 273 (1995).

10. D. Dunn: Can. J. Phys., **53**, 321 (1975).

11. T.D. Lee, F. Low, and D. Pines: Phys. Rev. **90**, 297 (1953); M. Gurari: Phil. Mag., **44**, 329 (1953); E.P. Gross: Phys. Rev., **100**, 1571 (1955); G. Höhler: Z. Physik, **140**, 192 (1955); G. Höhler: Z. Physik, **146**, 372 (1956); Y. Osaka: Prog. Theor. Phys., **22**, 437 (1959); T.D. Schultz: Phys. Rev., **116**, 526 (1959); D.M. Larsen: Phys. Rev., **172**, 967 (1968); J. Roseler: Phys. Stat. Sol., **25**, 311 (1968).

12. N.V. Prokof'ev and B. V. Svistunov: Phys. Rev. Lett. **81**, 2514 (1998).

13. A.S. Mishchenko, N.V. Prokof'ev, A. Sakamoto, and B.V. Svistunov: Phys. Rev. B **62**,6317 (2000).

14. John T. Titantah, Carlo Pierleoni, and Sergio Ciuchi: Phys. Rev. Lett., **87**, 206406 (2001).

15. An interesting discussion on the accuracy of Feynman's polaron model can be found in J.T. Devreese and R. Evrard: Phys. Lett. **11**, 278 (1964).

16. G. De Filippis, V. Cataudella, V.M. Ramaglia, C.A. Perroni and D. Bercioux: Eur. Phys. J B **36**, 65 (2003).

17. L.D. Landau and S.I. Pekar: J. Exptl. Theor. Phys. **18**, 419 (1948); S.I. Pekar: Research in Electron Theory of Crystals, Moscow, Gostekhizdat (1951) [English translation: Research in Electron Theory of Crystals, US AEC Report AEC-tr-5575 (1963)].

18. S.J. Miyake: J. Phys. Soc. Jap. **38**, 181 (1975); **41**, 747 (1976).

19. T.D. Lee, F. Low, and D. Pines: Phys. Rev. **90**, 297 (1953).

20. D.N. Larsen: Phys. Rev. **172**,967 (1968).

21. G. Höhler and A.M. Mullensiefen: Z. Physik **157**,159 (1959).

22. J. Roseler: Phys. Stat. Sol. **25**, 311 (1968).

23. T. Holstein: Ann. Phys. (Leipzig) **8**, 325 (1959).

24. A. B. Migdal: Sov. Phys. JETP **7**, 996 (1958).

25. A. A. Gogolin: Phys. Status Solidi **109**, 95 (1982); F. Marsiglio: Physica C **244**, 21 (1995); W. Stephan: Phys. Rev. B **54**, 8981 (1996).

26. I.J. Lang, and Yu. Firsov: Sov. Phys. JEPT **16**, 1301 (1963).

27. H. de Raedt and Ad Lagendijk: Phys. Rev. B **27**, 6097 (1983) and Phys. Rev. B **30**, 1671 (1984); P. E. Kornilovitch: Phys. Rev. Lett. **81**, 5382 (1998); M. Hohenadler, H.G. Evertz, and W. von Linden: Phys Rev. **69**, 24301 (2004).

28. For a review see: E. Jeckelmann and H. Fehske:Exact numerical methods for electron phonon problems. In: *Polarons in Bulk Materials and Systems with*

Reduced Dimensionality, ed. by G. Iadonisi, J. Ranninger and G. De Filippis (IOS Press Amsterdam, Oxford, Tokio, Washington DC, 2006), p.247.

29. A. Kongeter and M. Wagner: J. Chem. Phys. **92**, 4003 (1990); J. Ranninger and U. Thibblin: Phys. Rev. B **45**, 7730 (1992); A. S. Alexandrov, V. V. Kabanov and D. K. Ray: Phys. Rev. B **49**, 9915 (1994); E. de Mello and J. Ranninger: Phys. Rev. B **55**, 14872 (1997); M. Capone, W. Stephan, and M. Grilli: Phys. Rev. B **56**, 4484 (1997); G. Wellein and H. Fehske: Phys. Rev. B **56**, 4513 (1997); H. Fehske, J. Loos, and G. Wellein: Z. Phys. B **104**, 619 (1997).

30. S. Ciuchi, F. de Pasquale, S. Fratini, and D. Feinberg: Phys. Rev. B **56**, 4494 (1997); S. Ciuchi, F. de Pasquale, and D. Feinberg: Physica C **235-240**, 2389 (1994).

31. S. R. White: Phys. Rev. B **48**, 10345 (1993); E. Jeckelmann and S. R. White: Phys. Rev. B **57**, 6376 (1998).

32. A. H. Romero, D. W. Brown, and K. Lindenberg: Phys. Rev. B **59**, 13728 (1999); **60**, 14080 (1999); **62**, 1496 2000; J. Comp. Phys. **109**, 6540 (1998).

33. G. Iadonisi, V. Cataudella, G. De Filippis, and D. Ninno: Europhys. Lett. **41**, 309 (1998); G. Iadonisi, V. Cataudella and D. Ninno: Phys. Sta. Sol. (b) **203**, 411 (1997); Y. Lepine and Y. Frongillo: Phys. Rev. B **46**, 14510 (1992).

34. J. Bonĉa, S.A. Trugman and I. Batistiĉ: Phys. Rev. B **60**, 1633 (1999); L.C. Ku, S.A. Trugman, and J. Bonĉa; Phys. Rev. B **65**, 174306 (2002).

35. O.S. Barisic: Phys. Rev B **65**, 144301 (2002).

36. H. Lowen: Phys. Rev. B **37**, 8661 (1988); B. Gerlach and H. Lowen: Rev. Mod. Phys. **63**, 63 (1991).

37. V. Cataudella, G. De Filippis and G. Iadonisi: Phys. Rev. B **60**, 15163 (1999); **62**, 1496 (2000).

38. Y. Toyozawa: Prog. Theor. Phys. **26**, 29 (1961).

39. D.N. Larsen: Phys. Rev. **144**, 697 (1966).

40. D.N. Eagles: Phys. Rev. B, **145**,645 (1966).

41. W. P. Su, J. R. Schrieffer, and A. J. Heeger: Phys. Rev. Lett. **42**, 1698 (1979); Phys. Rev. B **22**, 2099 (1980).

42. *Solitons and Polarons in Conducting Polymers*, ed by Y. Lu (World Scientific, Singapore 1988).

43. A. La Magna and R. Pucci: Phys. Rev. B, **55**, 6296 (1997).

44. M. Zoli: Phys. Rev. B **66**, 012303 (2002); **67**, 195102 (2003).

45. M. Verissimo-Alves, R. B. Capaz, B. Koiller, E. Artacho, and H. Chacham: Phys. Rev. Lett. **86**, 3372 (2001); E. Piegari, V. Cataudella, V. Marigliano, and G. Iadonisi: Phys. Rev. Lett. **89**, 49701 (2002); Yu N. Gartstein, A. A. Zakhidov, and R. H. Baughman: Phys. Rev. Lett. **89**, 045503 (2002).

46. S. S. Alexandre, E. Artacho, J. M. Soler, and H. Chacham: Phys. Rev. Lett. **91**, 108105 (2003).

47. CA Perroni, E. Piegari, M. Capone and V. Cataudella: Phys. Rev. B **69**, 174301 (2004).

48. M. Capone, W. Stephan, and M. Grilli: Phys. Rev. B **56**, 4484 (1997).

49. M. Capone, S. Ciuchi, and C. Grimaldi: Europhys. Lett. **42**, 523 (1998).

50. M. Capone and S. Ciuchi: Phys. Rev. Lett. **91**, 186405 (2003).

51. A. S. Alexandrov and P. E. Kornilovitch: Phys. Rev. Lett. **82**, 807 (1999).

52. A. W. Sandvik, D. J. Scalapino, and N. E. Bickers: Phys. Rev. **69**, 94523 (2004).

53. A. S. Alexandrov: Phys. Rev. B **61**, 12315 (2000); A. S. Alexandrov and C. Sricheewin: Europhys. Lett. **51**, 188 (2000).

54. E. Cappelluti and L. Pietronero: Europhys. Lett. **36**, 619 (1996).

55. M. L. Kulic: Phys. Rep. **338**, 1 (2000).

56. H. Fehske, J. Loos, and G. Wellein: Phys. Rev. B **61**, 8016 (2000).

57. A. S. Alexandrov and B. Ya. Yavidov: Phys. Rev. B **69**, 73101 (2004).

58. C.A. Perroni, V. Cataudella, G. De Filippis: J. Phys.: Condens. Matter **16**, 1593 (2004).

59. C.A. Perroni, V. Cataudella, G. De Filippis: Phys. Rev. B **71**, 54301 (2005).

60. For a recent review see P. Calvani: Effects of strong charge-lattice coupling on the optical conductivity of transition-metal oxides. In *Polarons in Bulk Materials and Systems with Reduced Dimensionality*, ed. by G. Iadonisi, J. Ranninger and G. De Filippis (IOS Press Amsterdam, Oxford, Tokio, Washington DC, 2006), p.53.

61. For a review see J.T. Devreese: Polarons. In *Encyclopedia of Applied physics*, edited by G.L. Trigg (VCH, New Jork, 1996), Vol. 14, p. 383.

62. G. De Filippis, V. Cataudella, A. Mishchenko, C.A. Perroni, and J. Devreese: Phys.Rev. Lett. **96**, 136405 (2006).

63. J.T. Devreese, J. De Sitter, and M. Govaerts: Phys. Rev. B **5**, 2367 (1972); see also J.T. Devreese: Internal structure of the free Fröhlich Polaron, optical absorption and cyclotronic resonance. In *Polarons in ionic crystals and polar semiconductors* (North-Holland, Amsterdam 1972) p. 83

64. F.M. Peeters and J.T. Devreese: Phys. Rev. B **28**, 6051 (1983).

65. J.T. Devreese, this volume

66. E. Kartheuser, R. Evrard, and J. Devreese: Phys. Rev. Lett. **22**, 94 (1969).

67. A.S. Mishchenko, N. Nagaosa, N.V. Prokofev, A. Sakamoto, and B.V. Svistunov: Phys. Rev. Lett. **91**, 236401 (2003).

68. H. Mori: Prog. Theor. Phys **33**, 423 (1965); **34**, 399 (1965); W. Götze, and P. Wölfle: Phys. Rev. B **6**, 1226 (1972).

69. V.L. Gurevich, I.E. Lang, and Yu A. Firsov: Fiz. Tverd. Tela: **4**, 1252 (1962) [Sov.Phys. Solid State **4**, 918 (1962)]; G.D. Mahan: *Many-Particle Physics*, (Plenum, New Jork, 1981), Chap. 8.

70. Y. Toyozawa, *Optical Processes in Solids*, (Cambridge: University Press, 2003).

71. G.D. Mahan: *Many-Particle Physics*, (Plenum, New Jork, 1981), Chap. 4, pag. 282.

72. A.S. Mishchenko and N. Nagaosa, this volume.

73. K.M. Shen, F. Ronning, D.H. Lu, W.S. Lee, N.J.C. Ingle, W. Meevasana, F. Baumberger, A. Damascelli, N.P. Armitage, L.L. Milelr, Y. Kohaska, M. Azuma, M. Takano, H. Takagi, and Z.-X. Shen: Phys. Rev. Lett. **93**, 267002 (2004).

74. G. Wellein and H. Fehske: Phys. Rev. B **58**,6280 (1998); A. Weisse, H. Fehske, G. Wellein, and A. R. Bishop: Phys. Rev. B **62** R747 (2000).

75. V. Cataudella, G. De Filippis, F. Martone and C.A. Perroni: Phys. Rev. B **70**, 193105 (2004).

76. G. De Filippis, V. Cataudella, V.M. Ramaglia and C.A. Perroni: Phys. Rev. B **72**, 14307 (2005).

77. E.R. Gagliano, E. Dagotto, A. Moreo and F.C. Alcaraz: Phys. Rev. B **70**, 1677 (1986).

78. H.G. Reik: Phys. Lett. **5**, 5 236 (1963); Z. Phys. **203**, 346 (1967).

79. S. Fratini and S. Ciuchi: Cond-Mat/0512202.

80. D. Emin: Phys. Rev. B **48**, 13691 (1993).

Path Integrals in the Physics of Lattice Polarons

Pavel Kornilovitch

Hewlett-Packard, Corvallis, Oregon, 97330, USA
pavel.kornilovich@hp.com

Summary. A path-integral approach to lattice polarons is developed. The method is based on exact analytical elimination of phonons and subsequent Monte Carlo simulation of self-interacting fermions. The analytical basis of the method is presented with emphasis on visualization of polaron effects, which path integrals provide. Numerical results on the polaron energy, mass, spectrum and density of states are given for short-range and long-range electron-phonon interactions. It is shown that certain long-range interactions significantly reduce the polaron mass, and anisotropic interactions enhance polaron anisotropy. The isotope effect on the polaron mass and spectrum is discussed. A path-integral approach to the Jahn-Teller polaron is developed. Extensions of the method to lattice bipolarons and to more complex polaron models are outlined.

1 Introduction

The polaron problem was one of the first applications of path integrals (PI). Just a few years after the introduction of the new technique into quantum mechanics [1, 2] and quantum statistics [3] Feynman published his seminal paper [4] on the theory of Fröhlich polarons [5], which set the stage for the development of polaron physics for the next half a century. The main feature of the statistical PI is an extra dimension that transforms each point-like particle into a one-dimensional line, or *path*. The extra dimension, to be called here the imaginary time τ, extends between 0 and $\beta = (k_B T)^{-1}$ where T is the absolute temperature at equilibrium. As a result, a quantum-mechanical system is mapped onto a purely classical system in one extra dimension. This enables visualization of quantum-mechanical objects, which is an instructive and useful feature. The quantum mechanical properties are fully recovered by considering an ensemble of paths, each contributing its own statistical weight into the system's partition function.

Having formulated the polaron problem in PI terms, Feynman made another key advance. He showed that if the ionic coordinates entered the model only quadratically (a free phonon field) and linearly (electron-phonon interaction in linear approximation) then the infinite-dimensional integration over ionic configurations could be performed analytically and exactly. As a result, *all* bosonic degrees of freedom (an infinite number of them) are eliminated in favour of just one fermionic degree of freedom. Phonons remain in the system as a retarded *self-interaction* of the electron. In the PI language, the statistical weight of each path is exponential in its Euclidian action and the latter contains a double integral over imaginary time as opposed to a single integral in cases of ordinary instantaneous interactions. Different segments of the fermion path "feel" each other if they are separated in imaginary time by less than the inverse phonon frequency. As a result, the electron path stiffens which leads to an enhanced effective mass and other characteristic effects as detailed later in the chapter.

The resulting electron path integral could not be calculated analytically due to the complex nature of the retarded self-interaction. Feynman resolved the difficulty by employing an original variational principle, in which the exact polaron action was replaced by an approximate, but exactly solvable, quadratic action. That led to a remarkably accurate top bound for the energy of the Fröhlich polaron (see [6–8]), which was only marginally improved by subsequent generalizations of the method [9–13]. The PI method became a workhorse of Fröhlich polaron research for decades and inspired numerous extensions which included polaron mobility [14–16] and optical conductivity [17], polaron in a magnetic field [18, 19], large bipolaron [20], and others. Feynman's method also inspired a Fourier Monte Carlo method, in which only the *difference* between the exact and variational energy was calculated numerically [21, 22]. Recently, the variational PI treatment was extended to a many-polaron system [23].

In the parallel development of small-polaron physics, Feynman's remarkable reduction had not been utilized for almost thirty years. Although the phonon integration could be performed as well, the self-interacting electron PI could not be calculated. The problem was that on the lattice even a free particle possessed a non-quadratic action. Because such a path integral could not be done analytically, it could not serve as a trial variational action. The situation changed in 1982 when De Raedt and Lagendijk (DRL) observed [24] that the electron PI could instead be evaluated numerically. Metropolis Monte Carlo [26] was ideally suited to the task because the polaron action was purely real, and as such resulted in a positive-definite statistical weight of the path. Using this approach, DRL obtained a number of nice results on the Holstein polaron: confirmed formation of a self-trapped state with increasing strength of the electron-phonon interaction, observed that the crossover gets sharper with decreasing phonon frequency and increasing lattice dimensionality [24], and that the critical coupling goes down as a dispersive phonon mode softens [27]. These results were reviewed in [25]. DRL also provided the first Monte Carlo analysis of the Holstein bipolaron [28].

The works of DRL were an important step forward in the application of PI to lattice polarons. Their method could handle infinite crystal lattices of arbitrary symmetry and dimensionality, dispersive phonons, and long-range electron-phonon interactions. At the same time, it was still limited to thermodynamical properties such as energy, specific heat, and static correlation functions. In addition, it suffered from a systematic error caused by finite discretization of imaginary time (the Trotter slicing). These limitations were removed in 1997–1998. Firstly, it was shown how open boundary conditions in imaginary time could enable direct calculation of the polaron effective mass [29] and even the entire spectrum and density of states [30]. The open boundary conditions allow projection of the partition function on states with definite polaron momenta. This is a particular manifestation of the general projection technique in the presence of a global symmetry. Systematic application of this principle makes it possible to separate states of different symmetries in many interesting situations, most notably bipolarons of different parity and orbital symmetry. More about the projection technique is in Sect. 2. Secondly, the polaron Monte Carlo method was formulated in continuous imaginary time [31], which completely eliminated the Trotter slicing and tremendously improved the computational efficiency of the method. This will be described in detail in Sect. 3. As a result, the versatility of DRL's method was enhanced by better computational efficiency and by a number of new polaron and bipolaron properties that could be computed with path integrals. The method was further developed in [32–39], the content of which will be described later in the chapter.

The path-integral Quantum Monte Carlo (PIQMC) with phonon integration is only one from an impressive list of numerical methods developed for polaron models in the last three decades. There exist several other QMC techniques, in particular the path-integral QMC without phonon integration [40, 41], Fourier QMC [21, 22], diagrammatic QMC [7, 8, 42–45] and Lang-Firsov QMC [46–48]. Non-QMC *classes* of methods include exact diagonalization [49–54], variational calculations [13, 55–63] and the density-matrix renormalization group [64–66]. Despite proliferation of methods, most of them have been applied to the two major polaron models: the ionic crystal model of Fröhlich [5] and molecular crystal model of Holstein [67]. (The notable exceptions are the Jahn-Teller (bi)polaron [34, 60] and Peierls instabilities [41, 68, 69].) Due to its versatility, PIQMC is uniquely positioned to study many other models, for example long-range and anisotropic electron-phonon interactions. In addition, it provides some (bi)polaron properties that are difficult to obtain by other methods such as the density of states. In this way, several physically interesting results have been obtained that include, in particular, the light polaron mass in the case of long-range electron-phonon interactions [32], the enhancement of polaron anisotropy by electron-phonon interaction [33], formation of a peak in the polaron density of states [30, 34, 37], the isotope effect on the polaron spectrum and density of states [35], and the "superlight" crablike bipolaron [39, 70, 71]. These and other results obtained by the PIQMC

method in the last decade will be reviewed in Sect. 4. Various extensions of
PIQMC and its prospects are given in Sect. 5.

2 Projected Partition Functions

It is commonly believed that the statistical path integral can provide information only on the ground state of a quantum mechanical system, and in general
this is true. However, when the system possesses a global symmetry suitable
projection operators can project the configurations on sectors of the Hilbert
space corresponding to different irreducible representations of the symmetry
group. This enables access to the lowest states within each sector and provides
valuable information about the system's excitations. This strategy proves very
useful in the path-integral studies of polaron models. It enables calculation of
the polaron mass, spectrum, density of states, bipolaron singlet-triplet splitting and other properties. The above considerations are formalized as follows.
The full thermodynamic partition function is a trace of the density matrix:

$$Z = \sum_{\mathbf{R}} \langle \mathbf{R} | e^{-\beta H} | \mathbf{R} \rangle \,, \tag{1}$$

where H is the Hamiltonian and $|\mathbf{R}\rangle$ is a complete set of basis states in
the real-space representation. If there is a global symmetry group G with a
set of irreducible representations U, it is meaningful to compose a projected
partition function Z_U which is a trace over the U-sector of the Hilbert space
only:

$$Z_U = \sum_{\mathbf{R}_U} \langle \mathbf{R}_U | e^{-\beta H} | \mathbf{R}_U \rangle = \sum_{\mathbf{R}} \langle \mathbf{R} | O_U^\dagger e^{-\beta H} O_U | \mathbf{R} \rangle \,, \tag{2}$$

where operator O_U generates a basis state $|\mathbf{R}_U\rangle$ from an arbitrary state $|\mathbf{R}\rangle$.
In the low-temperature limit $\beta \to \infty$, the partition function is dominated by
the lowest U-eigenvalue, $Z_U \to \exp\left(-\beta E_U\right)$. Thus E_U can be found by taking
the logarithm of Z_U in the low-temperature limit. This is particularly efficient
if the first excited state in U is separated from E_U by a finite gap.

The most important application of this technique is the formula for the
(bi)polaron effective mass. In this case, irreducible representations U are
labeled by the total momentum $\mathbf{K}$ and the projection operator is $O_{\mathbf{K}} =
\sum_{\mathbf{r}} e^{i\mathbf{K}\mathbf{r}} T_{\mathbf{r}}$, where $T_{\mathbf{r}}$ is the shift operator by a lattice vector $\mathbf{r}$. The $\mathbf{K}$-
projected partition function is then [30]

$$Z_{\mathbf{K}} = \sum_{\Delta\mathbf{r}} e^{i\mathbf{K}\Delta\mathbf{r}} \langle \mathbf{R} + \Delta\mathbf{r} | e^{-\beta H} | \mathbf{R} \rangle = \sum_{\Delta\mathbf{r}} e^{i\mathbf{K}\Delta\mathbf{r}} Z_{\Delta\mathbf{r}} \,. \tag{3}$$

Here $\mathbf{R} + \Delta\mathbf{r}$ denotes a many-particle configuration $\mathbf{R}$, which is parallel-shifted
as a whole by a lattice vector $\Delta\mathbf{r}$. The partition function $Z_{\Delta\mathbf{r}}$ is a "shifted
trace" of the density matrix: it connects $\mathbf{R}$ not with $\mathbf{R}$ but with $\mathbf{R}$ shifted by

$\Delta\mathbf{r}$. It follows from the above equation that $Z_{\mathbf{K}}$ and $Z_{\Delta\mathbf{r}}$ satisfy a Fourier-type relationship (for a more detailed derivation, cf. Refs. [30, 72]). The partition function $Z_{\Delta\mathbf{r}}$ is formulated in real space and as such can be represented by a real-space path integral. The partition function $Z_{\mathbf{K}}$ is diagonal in momentum space and therefore contains information about the variation of the system's properties with $\mathbf{K}$. The Fourier theorem (3) is key to the ability to infer the (bi)polaron spectrum and mass from the path integral.

Dividing $Z_{\mathbf{K}}$ by $Z_{\mathbf{K}=0}$ one obtains in the zero temperature limit

$$\frac{Z_{\mathbf{K}}}{Z_{\mathbf{K}=0}} = e^{-\beta(E_{\mathbf{K}}-E_0)} = \frac{\sum_{\Delta\mathbf{r}} e^{i\mathbf{K}\Delta\mathbf{r}}\langle\mathbf{R}+\Delta\mathbf{r}|e^{-\beta H}|\mathbf{R}\rangle}{\sum_{\Delta\mathbf{r}}\langle\mathbf{R}+\Delta\mathbf{r}|e^{-\beta H}|\mathbf{R}\rangle} = \langle\cos\mathbf{K}\Delta\mathbf{r}\rangle_{\text{shift}} . \quad (4)$$

The ratio of two sums over $\Delta\mathbf{r}$ represents the mean value of $\exp(i\mathbf{K}\Delta\mathbf{r})$ over an ensemble of paths in which the two end-point many-body configurations are identical up to an arbitrary parallel shift. In the QMC language, simulations need to be performed with *open boundary conditions in imaginary time*. There is a certain parallel with the uncertainty principle: fixing the boundary conditions in imaginary time (by making them periodic) results in a mix of all possible momenta. Conversely, relaxing the boundary conditions by mixing in all possible shifts $\Delta\mathbf{r}$ allows extraction of information on the given $\mathbf{K}$-sector of the Hilbert space. Resolving the last equation with respect to $E_{\mathbf{K}}$, one obtains an estimator for the (bi)polaron spectrum

$$E_{\mathbf{K}} - E_0 = -\lim_{\beta\to\infty}\frac{1}{\beta}\ln\langle\cos\mathbf{K}\Delta\mathbf{r}\rangle_{\text{shift}} . \quad (5)$$

The left-hand-side of this equation is a constant that depends on the parameters of the model being simulated. Therefore, the average cosine on the opposite side must decrease exponentially with β to compensate the growing denominator. At some β it becomes too small to be measured reliably with the available statistics of a Monte Carlo run. This is an incarnation of the infamous "sign problem" that plagues many QMC algorithms. In physical terms, at very low temperatures the probability to access an excited state is exponentially low due to the Boltzmann weight factor. As a result, the QMC process samples relevant configurations exponentially rarely. These considerations put the low limit on the temperature of the simulation (high limit on β). At the same time, the condition that $Z_{\mathbf{K}}$ is dominated by a single eigenstate in the $\mathbf{K}$ sector, puts a high limit on temperature. Thus there is an interval of temperatures at which a meaningful calculation of the (bi)polaron spectrum can be performed. If the expected bandwidth is too large the temperature interval shrinks to zero, which renders the calculation impossible. In general, the (bi)polaron spectrum can be computed by this method if the total bandwidth is smaller than the phonon frequency.

Expanding (5) for small momenta, one derives an estimator for the μ-th component of the *inverse* effective mass

$$\frac{1}{m_\mu^*} = \lim_{\beta\to\infty}\frac{\langle(\Delta r_\mu)^2\rangle_{\text{shift}}}{\hbar^2\beta} . \quad (6)$$

Since in the present formulation the end points of the path are not tied together path evolution between $\tau = 0$ and $\tau = \beta$ can be regarded as *diffusion of the top end* with respect to the bottom end over a diffusion time β. According to the Einstein relation, the mean square displacement of a diffusing particle is proportional to time. Then (6) implies that the *inverse effective mass is proportional to the diffusion coefficient D of the path in imaginary time* [29, 31, 72]. The combination $\hbar\beta$ has the dimensionality of time, so the inverse mass can be written as

$$\frac{1}{m_\mu^*} = \frac{2D}{\hbar} \, . \tag{7}$$

This formula works for both polarons and bipolarons [39, 44]. It can be viewed as a fluctuation-dissipation relation because it equates a dynamical characteristic, the mass, with an equilibrium thermodynamic property D. In addition, it provides a nice visualization of the main polaron effect: effective mass increase caused by an electron-phonon interaction. As will be shown in the next section, phonon integration results in correlations between distant parts of the imaginary-time polaron path. This increases the statistical weight of paths with straight segments. The average polaron path *stiffens*, which translates in a reduced diffusion coefficient D and, according to (7), enhanced effective mass. Note that these considerations apply equally well to other types of composite particles whose mass originates from interaction. The most notable examples are defects in quantum liquids [73] and the hadrons of quantum chromodynamics.

Another important application of the projection technique is the singlet-triplet splitting of the bipolaron, or, more generally, of a two-fermion bound state. Coming back to the projected partition function Z_U, (2), the relevant projection operators are identified as $O_S = I + X$ for singlet states and $O_T = I - X$ for triplet states. Here I is the identity operator and X is the fermion exchange operator, $X|\mathbf{r}_1\mathbf{r}_2\{\xi\}\rangle = |\mathbf{r}_2\mathbf{r}_1\{\xi\}\rangle$, where $\{\xi\}$ are the ionic displacements. Upon substitution in (2) one finds

$$Z^{S,T} = \sum_{\mathbf{r}_1\mathbf{r}_2\{\xi\}} \langle\mathbf{r}_1\mathbf{r}_2\{\xi\}|e^{-\beta H}|\mathbf{r}_1\mathbf{r}_2\{\xi\}\rangle \pm \langle\mathbf{r}_2\mathbf{r}_1\{\xi\}|e^{-\beta H}|\mathbf{r}_1\mathbf{r}_2\{\xi\}\rangle \, . \tag{8}$$

Monte Carlo simulation proceeds over an ensemble of paths that have identical end configurations *up to exchange of the top ends of the two fermion trajectories*. For the singlet, both types of configurations contribute a phase factor $(+1)$ whereas for the triplet the direct paths contribute $(+1)$ and exchanged paths contribute (-1). Forming the ratio of the triplet and singlet partition functions one obtains an estimator for the $S - T$ energy split

$$E_0^T - E_0^S = -\lim_{\beta\to\infty}\frac{1}{\beta}\ln\langle(-1)^P\rangle_{\text{ex}} \, . \tag{9}$$

Here $(-1)^P = \pm 1$ for the direct and exchanged two-fermion paths. If the ends of the two paths coincide then $(-1)^P = 0$. All said above about the sign problem and the role of temperature in calculating the average cosine applies to this expression as well.

Equation (9) provides the energy difference between the lowest triplet and singlet states. It is possible to compute the effective mass and the entire spectrum of the triplet bipolaron. To this end, one needs to combine the two symmetries considered above: the translational and exchange. This naturally leads to a path ensemble with the end configurations differing by any combination of parallel shifts and exchanges. Without repeating the derivation, the final expressions are

$$E_{\mathbf{K}}^T - E_0^T = -\frac{1}{\beta} \ln \left(\frac{Z_{\mathbf{K}}^T}{Z_0^S} \frac{Z_0^S}{Z_0^T} \right) = -\frac{1}{\beta} \ln \frac{\langle (-1)^P \cos(\mathbf{K}\Delta\mathbf{r}) \rangle_{\text{shift,ex}}}{\langle (-1)^P \rangle_{\text{shift,ex}}}, \qquad (10)$$

for the triplet spectrum and

$$\frac{1}{m_i^T} = \frac{1}{\beta\hbar^2} \frac{\langle (-1)^P (\Delta r_i)^2 \rangle_{\text{shift,ex}}}{\langle (-1)^P \rangle_{\text{shift,ex}}}, \qquad (11)$$

for the triplet mass. For the singlet spectrum and mass, (5) and (6) are still valid, except that the boundary conditions must be changed to "shift and exchange". The same applies to the singlet-triplet estimator (9).

3 Continuous-Time Path-Integral Quantum Monte Carlo Method

The main appeal of quantum Monte Carlo (QMC) methods is the ability to calculate physical properties without approximations. A quantum mechanical system is directly *simulated* taking into account all the details of particle dynamics and inter-particle interaction. Physically interesting observables can be calculated without bias while the statistical errors in general decrease with increasing simulation time. It is said therefore that QMC provides "numerically exact" values for the observables. A number of comprehensive reviews on QMC exist [25, 74–78]. Many problems of the early QMC methods, such as the finite-size effects, finite time-step effects, and critical slowing down, have been resolved with the development of novel algorithms and increasing computing power. The *sign problem*, that is the non-positive-definiteness of the statistical weight of basis configurations, remains the only fundamental problem. In real systems and models without a sign problem (for example liquid and solid He^4 [77, 79–82]), QMC works beautifully and provides us with accurate and valuable information about the thermodynamics and sometimes real-time dynamics.

One polaron is free from the sign-problem difficulties. Phonons are bosons and as such do not lead to a fermion sign problem. And as long as there is only one polaron in the system statistics does not matter. This is the main reason behind the success of the QMC approach to the polaron. The ground state energy, the effective mass, isotope exponents on mass, the number of excited phonons, static correlation functions and other quantities can be calculated without any approximations. The bipolaron ground state can also be investigated without a sign problem as long as the bipolaron is a spin singlet with a symmetric spatial wave function. Thus the power of QMC can be (and has been) applied to bipolarons of various kinds and to pairs of distinguishable particles such as the exciton. For many-polaron systems the fermion sign problem fully manifests itself, which limits the applicability of QMC methods. More about this will be said in Sect. 5. In this section the basics of the continuous-time path-integral quantum Monte Carlo (PIQMC) method for lattice polarons are described.

The starting point is the shift partition function $Z_{\Delta \mathbf{r}}$, see (3), and the electron-phonon (e-ph) Hamiltonian

$$H = -t \sum_{\mathbf{nn}'} c_{\mathbf{n}'}^{\dagger} c_{\mathbf{n}} - \sum_{\mathbf{nm}} f_{\mathbf{m}}(\mathbf{n}) c_{\mathbf{n}'}^{\dagger} c_{\mathbf{n}} \xi_{\mathbf{m}} + \sum_{\mathbf{m}} \left(-\frac{\hbar^2}{2M} \frac{\partial^2}{\partial \xi_{\mathbf{m}}^2} + \frac{M\omega^2}{2} \xi_{\mathbf{m}}^2 \right) . \quad (12)$$

The Hamiltonian is written in mixed representation, $c_{\mathbf{n}}$ being fermionic annihilation operators and $\xi_{\mathbf{m}}$ ion displacements. Index $\mathbf{n}$ denotes the spatial location of an electron Wannier orbital whereas $\mathbf{m}$ denotes that of an ion displacement. In general, $\mathbf{n} \neq \mathbf{m}$ even if they belong to the same lattice unit cell. For simplicity, the electron kinetic energy (the first term of H) is taken in the nearest-neighbor-hopping approximation with amplitude $-t$, and the lattice energy (third term in H) in the independent Einstein oscillator approximation with mass M and frequency ω. Neither approximation is necessary. Section 5 explains how PIQMC deals with more complex forms of kinetic energy and phonon spectrum. The most interesting part is the e-ph interaction (second term in H) which is written in the "density-displacement" form. The quantity $f_{\mathbf{m}}(\mathbf{n})$ is the *force* with which an electron $\mathbf{n}$ acts on the ion coordinate $\mathbf{m}$. Asymmetric notation emphasizes that the force $f_{\mathbf{m}}$ is an attribute of a given oscillator, while the argument $\mathbf{n}$ is a dynamic variable indicating the current position of the electrons interacting with $\mathbf{m}$. No constraints are imposed on the functional form of f, thus allowing studies of long-range e-ph interactions. The commonly used Holstein model [67] corresponds to a localized force function $f_{\mathbf{m}}(\mathbf{n}) = \kappa \delta_{\mathbf{nm}}$.

3.1 Handling Kinetic Energy in Continuous Time

Handling the kinetic energy on a lattice possesses a challenge for Monte Carlo. To see this consider just the kinetic part of H and one free particle. By introducing multiple resolutions of identity, $\widehat{I} = \sum_{\mathbf{r}} |\mathbf{r}\rangle\langle\mathbf{r}|$, the shift partition function is developed into a multidimensional sum

$$Z_{\Delta \mathbf{r}} = \sum_{\mathbf{r}_1 \mathbf{r}_2 \ldots \mathbf{r}_M} \langle \mathbf{r}_1 + \Delta \mathbf{r} | e^{-\Delta \tau H_{\mathrm{el}}} | \mathbf{r}_M \rangle \ldots \langle \mathbf{r}_3 | e^{-\Delta \tau H_{\mathrm{el}}} | \mathbf{r}_2 \rangle \langle \mathbf{r}_2 | e^{-\Delta \tau H_{\mathrm{el}}} | \mathbf{r}_1 \rangle, \quad (13)$$

where $\Delta\tau = \beta/L$ is the time step and $L+1$ is the number of time slices. Each matrix element from the product is expanded to the first power in $\Delta\tau$:

$$\langle \mathbf{r}_{j+1}|e^{-\Delta\tau H_{\mathrm{el}}}|\mathbf{r}_j\rangle = \delta_{\mathbf{r}_{j+1},\mathbf{r}_j} + t\Delta\tau \sum_{\mathbf{l}} \delta_{\mathbf{r}_{j+1},\mathbf{r}_j+\mathbf{l}} + \mathcal{O}(\Delta\tau^2)\,, \qquad (14)$$

where $\mathbf{l}$ runs over the nearest neighbors. Expanding the product, the partition function becomes a sum of a large number of terms, each of which represents a particular path of the electron in imaginary time $\mathbf{r}(\tau)$. The paths consist of two different building blocks: straight segments and kinks, which originate from the first and second terms in (14), respectively. On straight segments the electron position does not change, whereas each kink changes the electron coordinates by vector $\mathbf{l}$ depending on the kink type. The statistical weight of a path with N_k kinks is

$$W_{N_k} = \underbrace{(t\Delta\tau)(t\Delta\tau)\ldots(t\Delta\tau)}_{N_k \text{ times}} = (t\Delta\tau)^{N_k}\,. \qquad (15)$$

Now consider two paths, D and D', that differ by D' having one extra kink of type $\mathbf{l}$ at imaginary time τ. At small $\Delta\tau$, D' has a vanishingly small statistical weight compared to D. As a result, no meaningful detailed balance between *individual* paths with different number of kinks seems to be possible in the continuous-time limit. The resolution of this difficulty comes from the observation that there are *infinitely many* paths with N_k+1 kinks than with N_k kinks. The small weight of individual path is compensated by the large number of paths. The combined weight of all paths with N_k+1 kinks is of the same order as the combined weight of all paths with N_k kinks. Instead of detailed balance between individual paths one should seek detailed balance between path groups of similar combined weight. The paths D and D' are regarded as representatives of such groups.

In accordance with the general ideas of the Metropolis Monte Carlo [26], one has to compose an equation that equates the transition rate from D to D' to the reciprocal transition rate from D' to D. Both rates are products of three factors: the probability to have the original path, which is proportional to the path's weight W, the probability Q to propose the necessary change to the path, and the probability P to accept the change. The detailed balance reads:

$$W(D)Q(\text{add }\mathbf{l}\text{ at }\tau)P(D \to D') = W(D')Q(\text{remove }\mathbf{l}\text{ at }\tau)P(D' \to D)\,. \quad (16)$$

The probabilities to propose the addition and removal of the kink turn out to be two quite different quantities. In the direction $D' \to D$, the kink to be removed is chosen from a *finite* number of all other kinks. Therefore the probability that it is selected for removal is a finite number $Q_{\mathrm{remove}}(\mathbf{l};\tau)$. In contrast, in going from D to D' the kink does not yet exist and the probability to propose its creation *exactly at time τ* is proportional to the time interval

$\Delta\tau$. The probability to propose the kink's addition is an infinitesimal quantity: $Q(\text{add } \mathbf{l} \text{ at } \tau) = q_{\text{add}}(\mathbf{l}; \tau)(\Delta\tau)$, where q_{add} is the corresponding probability *density*. Substitution of these two results in the balance equation yields

$$(t\Delta\tau)^{N_k} q_{\text{add}}(\mathbf{l}; \tau)(\Delta\tau) P(D \to D') = (t\Delta\tau)^{N_k+1} Q_{\text{remove}}(\mathbf{l}; \tau) P(D' \to D) \,. \tag{17}$$

The extra $(\Delta\tau)$ from the addition probability on the left side exactly balances the smaller weight of the right side. The time step cancels out of the equation leaving only finite factors which are well defined in the continuous-time limit. According to the standard Metropolis recipe the solution is given by the following acceptance rules

$$P(D \to D') = \min\left\{1 \; ; \; \frac{t \cdot Q_{\text{remove}}(\mathbf{l}; \tau)}{q_{\text{add}}(\mathbf{l}; \tau)}\right\} \,, \tag{18}$$

$$P(D' \to D) = \min\left\{1 \; ; \; \frac{q_{\text{add}}(\mathbf{l}; \tau)}{t \cdot Q_{\text{remove}}(\mathbf{l}; \tau)}\right\} \,. \tag{19}$$

These expressions still leave a lot of freedom in specifying functions q_{add} and Q_{remove}. As an example, consider the simplest choice when both functions are constant. For the addition process this means that the time for the new kink of type $\mathbf{l}$, on top of the existing $N_{\mathbf{l}}$, is chosen with equal probability within the $[0, \beta]$ time interval, hence $q_{\text{add}}(\mathbf{l}; \tau) = \beta^{-1}$. For the removal process, the candidate kink for elimination is chosen from the existing $N_{\mathbf{l}}$ with equal probability, that is $Q_{\text{remove}}(\mathbf{l}; \tau) = N_{\mathbf{l}}^{-1}$. One must also take into account the fact that kinks can only be added to the path if $N_{\mathbf{l}} = 0$, while added or removed if $N_{\mathbf{l}} \geq 1$. The full set of update rules could be as follows. (i) Select the kink type $\mathbf{l}$, that is the hopping direction of the particle. The probability with which different $\mathbf{l}$ are selected may not be equal but must be non-zero. (ii) If the current path already has kinks of type $\mathbf{l}$, whether to add or remove a kink is proposed with, for example, equal probability of $\frac{1}{2}$. If the path possesses no kinks, addition is the only option, and it must be chosen with probability 1. (iii) If addition is selected, the time for the new kink is chosen with probability density $\frac{1}{\beta}$. The update is accepted with probability $\min\{1; (t\beta)/(N_{\mathbf{l}} + 1)\}$ for $N_{\mathbf{l}} \geq 1$ and with probability $\min\{1; (t\beta)/2\}$ for $N_{\mathbf{l}} = 1$. (iv) If removal is selected, the candidate kink is chosen with equal probability from the existing $N_{\mathbf{l}}$. The update is accepted with probability $\min\{1; N_{\mathbf{l}}/(t\beta)\}$ for $N_{\mathbf{l}} \geq 2$ and with probability $\min\{1; 2/(t\beta)\}$ for $N_{\mathbf{l}} = 1$.

The described method can be readily generalized to next-nearest neighbor hopping and anisotropic hopping. It can also be generalized to multi-kink updates, as needed for example in simulating the bipolaron. To close this section one should note that a path can be interpreted as a time-space *diagram*, where kinks play the role of vertices and straight segments the role of propagators. Integration over the ensemble of fluctuating paths can then be understood within the general concept of Diagrammatic Monte Carlo [7, 83].

3.2 Integration Over Phonons

Analytic integration over phonons is a key ingredient of the polaron PIQMC method. Although an electron-phonon (e-ph) system can be simulated without phonon integration [40, 41], it is the integration that makes the method powerful. Indeed, the integration reduces $N+1$ degrees of freedom to just one degree of freedom, which increases accuracy and reduces simulation time. The integration effectively eliminates all the distant parts of the system by folding the infinite number of ion coordinates into one retarded self-interaction function. The simulation then proceeds on an infinite lattice with no finite-size effects. In addition, the integration makes it possible to simulate long-range e-ph interactions with the same efficiency as local interactions, which extends QMC capabilities towards more realistic models.

Technical details of the phonon integration are somewhat cumbersome but well-documented in the literature. Here we will describe the main steps of the derivation and point out associated subtleties.

Upon insertion of H into (3), a path-integral representation for $Z_{\Delta \mathbf{r}}$ is developed by introducing L time intervals and $L-1$ intermediate sets of basis states. In the continuous time limit, $L \to \infty$, the kinetic energy of the electron is treated as described in the previous section. Integration over the electron coordinates is converted to stochastic summation over an ensemble of paths which are sampled by adding and removing kinks. For each electron path, there is an internal path integration over the ionic coordinates characterized by the electron-phonon action $A_{\mathrm{e-ph}}$

$$
Z_{\Delta \mathbf{r}} = \int_{(\mathbf{r},0)}^{(\mathbf{r}+\Delta\mathbf{r},\beta)} \mathcal{D}\mathbf{r}(\tau) \int_{(\{\xi_{\mathbf{m}}\},0)}^{(\{\xi_{\mathbf{m}-\Delta\mathbf{r}}\},\beta)} \mathcal{D}\xi_{\mathbf{m}}(\tau)\, e^{A_{\mathrm{e-ph}}[\mathbf{r}(\tau),\,\xi_{\mathbf{m}}(\tau)]} , \tag{20}
$$

$$
A_{\mathrm{e-ph}} = -\sum_{\mathbf{m}} \int_0^\beta \left[\frac{M\dot{\xi}_{\mathbf{m}}^2(\tau)}{2\hbar^2} + \frac{M\omega^2}{2}\xi_{\mathbf{m}}^2(\tau) \right] d\tau + \sum_{\mathbf{m}} \int_0^\beta f_{\mathbf{m}}[\mathbf{r}(\tau)]\xi_{\mathbf{m}}(\tau)d\tau . \tag{21}
$$

The path integral over $\xi_{\mathbf{m}}(\tau)$ is Gaussian and therefore can be calculated analytically. Actually, the integration is performed in two steps. Firstly, *path* integration is done with fixed but arbitrary boundary conditions at the two ends of the path. Secondly, a certain relationship between $\xi_{\mathbf{m}}(0)$ and $\xi_{\mathbf{m}}(\beta)$ is assumed followed by an additional one-dimensional *ordinary* (but still Gaussian) integration. In most PI studies, periodic boundary conditions $\xi_{\mathbf{m}}(\beta) = \xi_{\mathbf{m}}(0)$ are assumed. This leads to the classic Feynman formula for the oscillator action in the presence of an arbitrary time-dependent external force [84]. However, we have seen in the previous section that mass and spectrum calculation requires a partition function with the end configurations parallel-shifted with respect to each other. The shift of the ionic configuration must be the same as that of the fermion configuration, which implies $\xi_{\mathbf{m}}(\beta) = \xi_{\mathbf{m}-\Delta\mathbf{r}}(0)$, as indicated in the last equation. The correlation between the electron and phonon boundary conditions is illustrated in Fig. 1(a).

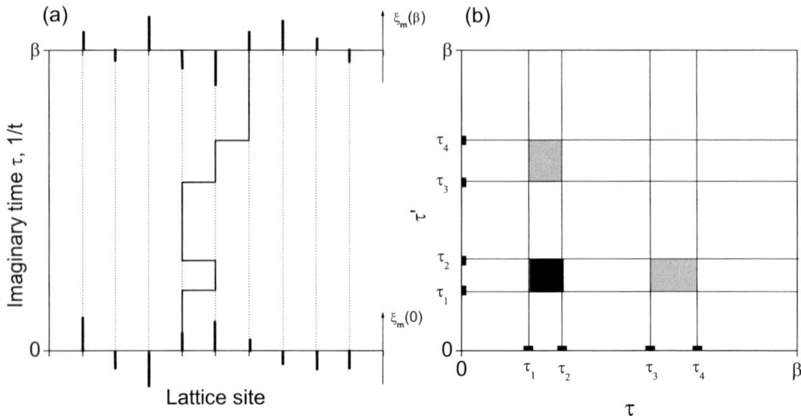

Fig. 1. (a) Integration over ionic paths under shifted boundary conditions. Ionic displacements are shown as vertical bars at $\tau = 0$ and $\tau = \beta$. Notice how the pattern of displacements is correlated with the shift of the electron path. The end displacements of individual oscillators are not equal, $\xi_{\mathbf{m}}(0) \neq \xi_{\mathbf{m}}(\beta)$. (b) Calculation of the polaron action as a double integral over imaginary time. The kink times τ_i break the $(\tau\tau')$ plane into a finite number of rectangles. Within each rectangle the electron coordinates are constant and the double integral can be calculated analytically for arbitrary τ_i [36]. After that the double integral reduces to a sum over the rectangles. Since the number of rectangles decreases at strong coupling, the algorithm gets faster at *strong* coupling.

Due to an additional shift, the second-phase integration results in a correction to the conventional "periodic" action. The full polaron action, upon summation over all oscillators, reads [31, 38]

$$A_{\mathrm{pol}}[\mathbf{r}(\tau)] = A_0 + \Delta A \,, \tag{22}$$

$$A_0 = \sum_{\mathbf{m}} \frac{\hbar}{4\omega M} \int_0^\beta\!\!\int_0^\beta d\tau d\tau' \frac{\cosh \hbar\omega(\frac{\beta}{2} - |\tau - \tau'|)}{\sinh \hbar\omega \frac{\beta}{2}} f_{\mathbf{m}}[\mathbf{r}(\tau)] f_{\mathbf{m}}[\mathbf{r}(\tau')] \,, \tag{23}$$

$$\Delta A = \sum_{\mathbf{m}} \frac{\hbar}{2\omega M} B_{\mathbf{m}} \left(C_{\mathbf{m}+\Delta\mathbf{r}} - C_{\mathbf{m}} \right) \,, \tag{24}$$

$$B_{\mathbf{m}} \equiv \int_0^\beta d\tau' \frac{\sinh \hbar\omega(\beta - \tau')}{\sinh \hbar\omega\beta} f_{\mathbf{m}}[\mathbf{r}(\tau')] \,, \tag{25}$$

$$C_{\mathbf{m}} \equiv \int_0^{\beta} d\tau' \, \frac{\sinh \hbar\omega\tau'}{\sinh \hbar\omega\beta} \, f_{\mathbf{m}}[\mathbf{r}(\tau')] \,. \tag{26}$$

The polaron action is a functional of the electron path and contains all the information about the ionic subsystem. The denominator of (4) finally becomes

$$Z_{\mathbf{K}=0} = \sum_{\Delta\tau} Z_{\Delta\mathbf{r}} = Z_{\mathrm{ph}} \sum_{N_k=0,1,\dots}^{\infty} \int_0^{\beta} \cdots \int_0^{\beta} (d\tau)^{N_k} \, t^{N_k} \, e^{A_{\mathrm{pol}}[\mathbf{r}(\tau)]} \,. \tag{27}$$

Here $Z_{\mathrm{ph}} = [2\sinh(\frac{1}{2}\hbar\beta\omega)]^{-N}$ (N is the number of lattice sites) is the partition function of free phonons. It is a multiplicative constant that cancels out from all statistical averages. The factor $e^{A_{\mathrm{pol}}}$ adds to the weight of each path on top of the "kinetic" contribution $(t\Delta\tau)^{N_k}$. As a result, the acceptance rules (18) and (19) are modified by the ratio of the respective factors for the paths D and D'

$$P(D \to D') = \min \left\{ 1 \;;\; \frac{t \cdot Q_{\mathrm{remove}}(\mathbf{l};\tau)}{q_{\mathrm{add}}(\mathbf{l};\tau)} \, e^{A_{\mathrm{pol}}(D') - A_{\mathrm{pol}}(D)} \right\} \,, \tag{28}$$

$$P(D' \to D) = \min \left\{ 1 \;;\; \frac{q_{\mathrm{add}}(\mathbf{l};\tau)}{t \cdot Q_{\mathrm{remove}}(\mathbf{l};\tau)} \, e^{A_{\mathrm{pol}}(D) - A_{\mathrm{pol}}(D')} \right\} \,. \tag{29}$$

These acceptance rules together with the formula for the polaron action and path update rules described in the previous section fully define the Monte Carlo update process. It is illustrated in Fig. 2. Numerical efficiency of the algorithm critically depends on how easily the polaron action can be computed at every update. The task is greatly aided by the fact that the electron path is defined on a discrete lattice. In other words, it consists of a finite number of straight segments. This has two important consequences. First, because within each segment the electron coordinate is independent of imaginary time τ, the forces $f_{\mathbf{m}}(\mathbf{r})$ can be taken outside the time integration. The rest of the integrand is an explicit function of τ and τ'. This time integral can be calculated analytically for arbitrary end times of the segment. The double integral over time becomes a double sum over the segments, as illustrated in Fig. 1(b). The summand is a simple function of kink times; the explicit expressions are given in [36]. The smaller the number of segments, that is the straighter the path, the less time is required to compute A_{pol}. The electron-phonon interaction increases the statistical weight of straight paths and reduces the mean number of segments. Thus, in contrast to most other numerical methods, PIQMC becomes *faster* at large electron-phonon couplings.

Secondly, the (ij)-th term of the sum over segments contains the electron-ion forces in the combination

$$\Phi(\mathbf{r}_i - \mathbf{r}_j) = \sum_{\mathbf{m}} f_{\mathbf{m}}(\mathbf{r}_i) f_{\mathbf{m}}(\mathbf{r}_j) \,, \tag{30}$$

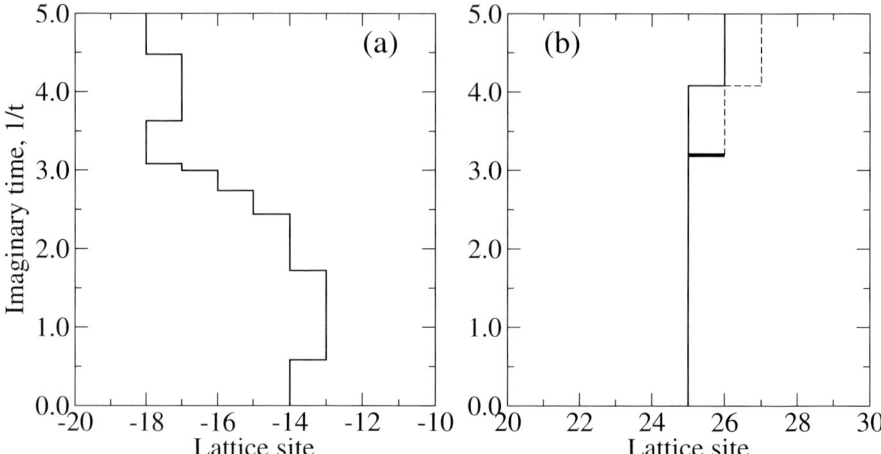

Fig. 2. (a) A typical fluctuating path of a free particle on a one-dimensional chain ($E_0 = -2.00\,t$, $m^*/m_0 = 1$). The path has drifted to the left by four lattice sites, which contributes to the effective mass according to (6). (b) A typical path of a one-dimensional Holstein polaron at $\hbar\omega/t = 2.0$ and $\lambda = 2.0$ ($E_0 = -4.66(2)\,t$, $m^*/m_0 = 3.06(3)$. The path has drifted to the right by just one site reflecting a larger mass. The thick bar is a kink update proposed by Monte Carlo. If accepted, the top part of the path is changed to the dashed line (*top* shift). Alternatively, the bottom part of the path can be shifted by one site to the left (*bottom* shift).

where $\mathbf{r}_i$ is the electron coordinate on i-th segment. The coefficients Φ are defined on a discrete set of points and can be pre-computed for a sufficiently large range of coordinate separation. After that the simulation takes essentially the same amount of time for *any* form of electron-phonon interaction. This is why the PIQMC polaron method is as efficient for long-range e-ph interactions as for the short-range Holstein model.

We close this section by noting that the requirement of phonon integration with shifted boundary conditions equally applies to the continuous case. This does not seem to be the case with Feynman's original calculation of the mass of the Fröhlich polaron and its subsequent generalizations. This issue was thoroughly investigated in [85]. It was shown that the periodic boundary conditions in the phonon integration, while absolutely correct for the energy calculation, lead to infinite terms in the polaron action in the mass calculation. Feynman's mass formula is obtained only if the terms are regarded as unphysical and thrown out by hand. In contrast, the shifted boundary conditions result in a self-consistent action and Feynman's result is obtained without any complications.

3.3 Observables

The preceding sections explained how to organize an efficient Monte Carlo sampling process that simulates the electron kinetic energy while fully taking into account the ion dynamics and electron-phonon interaction. Note that *no* approximation has been made so far except for the quadratic form of the elastic energy of the crystal. In this section it will be explained how to extract useful physical information from the one-to-one correspondence between the quantum mechanical-polaron and an ensemble of self-interacting paths. A number of important estimators have already been derived in Sect. 2. In particular, the effective mass is obtained from the mean-square end-to-end distribution, see (6). The polaron spectrum *relative to the ground state* is given by (5). It is essential that statistics for different $E_{\mathbf{K}}$ are collected in parallel so that the entire polaron spectrum is obtained in a single run. One consequence of this remarkable property is PIQMC's ability to compute the polaron density of states by simply histogramming $E_{\mathbf{K}}$ at the end. At present, PIQMC is the only method capable of efficiently calculating the polaron density of states.

The next important characteristic is the absolute polaron energy E_0. One way to obtain E_0 is from the low-temperature limit of the internal energy U. The latter can be computed from the thermodynamic relation $U = -\partial \ln Z / \partial \beta$. Since we are interested in the ground state, the corresponding partition function must be $Z_{\mathbf{K}=0} = \sum_{\Delta \mathbf{r}} Z_{\Delta \mathbf{r}} = \sum_{\text{shifted paths}} W[\mathbf{r}(\tau)]$. Returning momentarily to discrete time, $\beta = L\Delta\tau$,

$$U = -\frac{1}{L}\frac{\sum_{\text{shift}} \frac{\partial W}{\partial \Delta \tau}}{\sum_{\text{shift}} W} = -\frac{1}{L}\frac{\sum_{\text{shift}} \left[\frac{1}{W}\frac{\partial W}{\partial \Delta \tau}\right] W}{\sum_{\text{shift}} W} = -\frac{1}{M}\left\langle \frac{1}{W}\frac{\partial W}{\partial \Delta \tau}\right\rangle_{\text{shift}}. \quad (31)$$

The weight W is a product of the kinetic contribution $(t\Delta\tau)^{N_k}$ and the phonon term $e^{A_{\text{pol}}}$. Substituting in the above expression, one obtains a ground-state energy estimator

$$E_0 = -\frac{1}{L}\left\langle \frac{N_k}{\Delta \tau} + \frac{\partial A_{\text{pol}}}{\partial \Delta \tau}\right\rangle_{\text{shift}} = \left\langle -\frac{N_k}{\beta} - \frac{\partial A_{\text{pol}}}{\partial \beta}\right\rangle_{\text{shift}}. \quad (32)$$

The two terms here represent the kinetic and potential energy of the polaron, respectively. The kinetic energy is simply the mean number of kinks on the path. One can regard the kinks as "quanta" of the kinetic energy, each contributing the same amount $-\beta^{-1}$. With increasing coupling the paths become "stiffer", that is the mean number of kinks goes down. As a result, the polaron kinetic energy decreases by absolute value. Measuring the potential energy is harder for it is derived from the polaron action. Similarly to the latter, the potential energy estimator can be reduced to a double sum over the path's segments. Explicit expressions for the double sum terms are given in [36].

The number of excited phonons N_{ph} in the polaron cloud is another interesting characteristic. It measures the amount of energy stored in the lattice deformation around the electron. The polaron mass scales exponentially with N_{ph} so in addition the latter gives a rough estimate of the polaron mass. An estimator for N_{ph} can be derived by noting that the last term in the Hamiltonian (12) can be rewritten as $\hbar\omega(\widehat{N}_{\mathrm{ph}} + \frac{1}{2})$, where $\widehat{N}_{\mathrm{ph}}$ is the operator of the total number of phonons. The mean value of $\widehat{N}_{\mathrm{ph}}$ is obtained by differentiating the free energy with respect to $\hbar\omega$. To make sure that no contribution is received from the interaction term, the derivative must be taken with the combination $f^2/(M\omega)$ kept constant. Using $F = -\frac{1}{\beta}\ln Z$, the estimator is obtained as follows

$$N_{\mathrm{ph}} = -\frac{1}{\beta}\left.\frac{\partial F}{\partial(\hbar\omega)}\right|_{\frac{f^2}{M\omega}} = -\frac{1}{\beta}\left\langle \left.\frac{\partial A_{\mathrm{pol}}}{\partial(\hbar\omega)}\right|_{\frac{f^2}{M\omega}} \right\rangle_{\mathrm{shift}}. \tag{33}$$

The subscript "shift" implies that the number of phonons corresponds to the ground state of the polaron ($\mathbf{K} = 0$). It is also possible to derive estimators for non-zero total momenta, which are omitted here.

Finally, one can derive an estimator for the isotope exponent of the polaron effective mass α_μ. In the (bi)polaron mechanism of superconductivity, α_μ is related to the isotope effect on the critical temperature [86]. The mass isotope exponent is defined as $m_\mu^* = M^{\alpha_\mu}$, where M is the ion mass. Since the present method calculates the inverse polaron mass, α_μ is more conveniently expressed via the inverse effective mass

$$\alpha_\mu = -\frac{M}{\frac{1}{m_\mu^*}}\frac{\partial}{\partial M}\left(\frac{1}{m_\mu^*}\right) = \frac{\omega}{2\left(\frac{1}{m_\mu^*}\right)}\frac{\partial}{\partial\omega}\left(\frac{1}{m_\mu^*}\right). \tag{34}$$

The last transformation follows from the scaling $M \propto \omega^{-2}$. The estimator for α_μ is derived directly from the estimator for the inverse effective mass (6). A path's weight depends on the phonon frequency only via the polaron action $e^{A_{\mathrm{pol}}}$. (One should recall that the definition of a Monte Carlo average $\langle(\Delta r_\mu)^2\rangle$ contains A_{pol} in the numerator and denominator.) Upon differentiation one obtains

$$\alpha_\mu = \frac{\omega}{2\langle(\Delta r_\mu)^2\rangle}\left[\left\langle(\Delta r_\mu)^2\left.\frac{\partial A_{\mathrm{pol}}}{\partial\omega}\right|_{M\omega^2}\right\rangle - \langle(\Delta r_\mu)^2\rangle\left\langle\left.\frac{\partial A_{\mathrm{pol}}}{\partial\omega}\right|_{M\omega^2}\right\rangle\right]. \tag{35}$$

In taking the frequency derivative, one should keep the combination $M\omega^2$ constant, as it is independent of the ion mass. As with the phonon action and potential energy, the estimators for the number of phonons and isotope exponent can be split into a double sum over path's segments. The expressions for the respective summands can be found in [36].

4 Polaron Properties

The focus of this section is going to be the polaron properties that have been obtained with the continuous-time path-integral Quantum Monte Carlo method. For each topic, additional details of the algorithm will be provided, if necessary, on top of the general description presented above.

4.1 Long-Range Interaction Produces Light Polarons

The Holstein model [67] describes polarons in systems with well-screened short-range electron-phonon interaction, such as molecular crystals. It is in the short list of "main" polaron models. Due to its simplicity, the model has been extremely popular with the numerical community. Almost any new numerical method was first tried on the Holstein model, for which a lot of accurate results are available. Also, for a long time the Holstein model was regarded as a "generic" model for all narrow-band systems where the e-ph and lattice effects were equally important. The locality of the interaction was considered simply a matter of convenience, and the Holstein model itself was believed to contain all the qualitative features of lattice polarons. This turns out not to be the case. The main reason lies in the very sharp, exponential, dependence on the model parameters of the main polaron characteristic, the effective mass. Some properties of the Holstein polaron are presented in Fig. 3. Any approximation or model simplification that have little effect on the polaron energy can have dramatic consequences for the mass, leading to false qualitative conclusions. In particular, the Holstein model is *not* adequate for complex oxides of transition metals such as the cuprates or manganites. In those materials the bare bands are narrow, $\sim 1 - 2$ eV, so the lattice effects are important. At the same time, a low density of free carriers cannot fully screen the Coulomb interaction between electrons and distant anions and cations, leading to a long-range e-ph interaction.

A model of this kind was considered a long time ago by Eagles [87]. (The corresponding polaron was called there a "nearly-small polaron".) More recently, a long-range model with linearly polarized phonons was put forward in relation to high-temperature superconductors in [32]. The model is depicted in Fig. 4(a). An electron hops within a rigid chain (in the one-dimensional case) or plane (in the two-dimensional case) and interacts with ions that are positioned above the electron chain (or plane) and vibrate perpendicular to the chain (plane). Such a polarization was chosen to represent the c-polarized phonon modes that feature prominently in the cuprates [88]. The force function was taken to be the Coulomb force projected on the z-axis

$$f_{\mathbf{m}}(\mathbf{n}) = \kappa a^2 \frac{(\mathbf{n} - \mathbf{m})_z}{|\mathbf{n} - \mathbf{m}|^3} = \frac{\kappa a^2 d}{[(\mathbf{n} - \mathbf{n_m})^2 + d^2]^{3/2}} \, . \tag{36}$$

Here a is the lattice constant, d is the distance between the vibrating ions and the electron plane, and $\mathbf{n_m}$ is the vector of the electron Wannier orbital in the same unit cell as ion $\mathbf{m}$. The interaction strength is characterized by the parameter κ. Alternatively, one can use a dimensionless parameter

$$\lambda = \frac{E_p}{zt} = \frac{1}{2M\omega^2 zt} \sum_{\mathbf{m}} f_{\mathbf{m}}^2(0) \, , \tag{37}$$

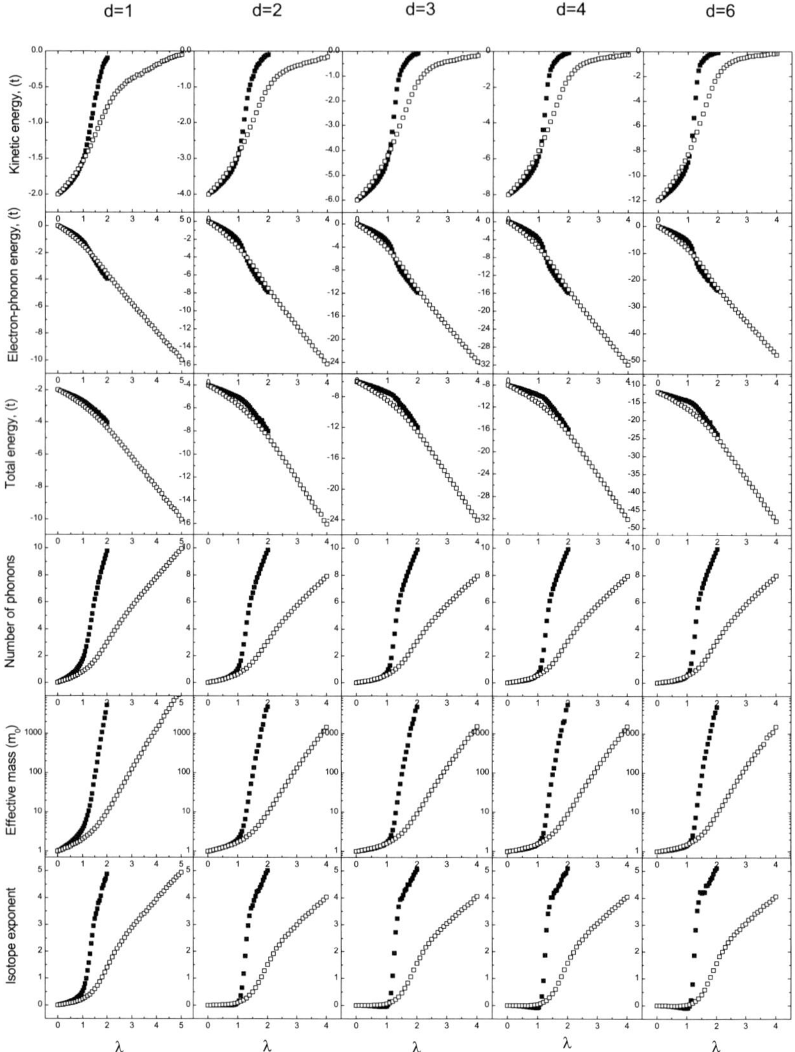

Fig. 3. Properties of the Holstein polaron on $d = 1, 2, 3, 4, 6$ hypercubic lattices. The filled and open symbols show data for the phonon frequencies $\hbar\omega/t = 0.4$ and 1.0 ($d = 1$), 0.8 and 2.0 ($d = 2$), 1.2 and 3.0 ($d = 3$), 1.6 and 4.0 ($d = 4$), and 2.4 and 6.0 ($d = 6$), respectively. The coupling constant is defined as $\lambda = E_p/(zt) = \kappa^2/(2M\omega^2 zt)$, where $z = 2d$ is the number of nearest neighbours. The effective mass is measured in units of $m_0 = \hbar^2/(2ta^2)$. Note that because the phonon frequency increases with d, the curves in different dimensions have similar shapes except for small differences around the λ of polaron formation.

which is a ratio of the polaron energy in the atomic limit E_p (at $t = 0$) to the kinetic energy of a free electron zt (at $\kappa = 0$). Here z is the number of nearest lattice sites. The confinement of the polaron path is determined by the function Φ, (30), whose profile is shown in Fig. 4(b). (It should be compared with the Holstein confining function $\kappa^2 \delta_{\mathbf{nn'}}$.) Due to loose confinement, the polaron path diffused farther than in the Holstein case, which is equivalent to a much lighter mass. Such a polaron was named the "small Fröhlich polaron" (SFP) in [32].

The masses of SFP and SHP calculated with PIQMC are compared in Fig. 5. SFP is slightly heavier at small λ but much lighter at large λ. The ratio of the two masses is a non-monotonic function. This observation was later confirmed by exact diagonalization [52] and variational [58, 62] methods. The most interesting property of SFP is exponential reduction of mass relative to SHP, see Fig. 5(c). This effect is independent of the dimensionality because it originates solely from the long-ranginess of e-ph interaction. The mass reduction can be very large. For example, in two dimensions at $\hbar\omega/t = 0.5$ and $\lambda = 1.2$, $m_{\mathrm{SHP}} \approx 220$ while m_{SFP} is only ≈ 9. In [36], an additional screening factor was added to the force function (36). All polaron properties smoothly interpolated between those of SHP and SFP as the screening radius was changed from zero to infinity.

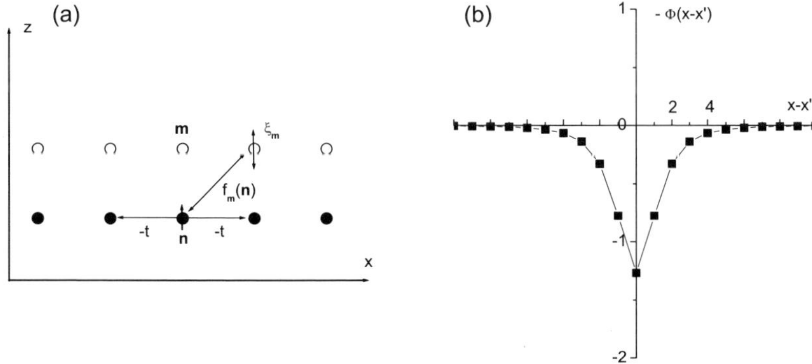

Fig. 4. (a) One-dimensional model with a long-range electron-phonon interaction. The electron moves along the bottom chain of sites shown by full circles (denoted by $\mathbf{n}$). The vibrating ions shown by open circles oscillate along the z-axis. The interaction is characterized by the z-projection of the Coulombic force (36). (b) Spatial profile of function Φ, (30), that characterizes the retarded interaction of different parts of the polaron path. The parameters are $\kappa = 1$ and $d = a$. *Negative* Φ is shown in order to draw analogy with a potential well. This shape of Φ should be contrasted with that of the Holstein model $\Phi_{\mathrm{Hol}} = \kappa^2 \delta_{xx'}$. A smooth Φ cannot localize the path as well as a sharp one, which results in a smaller effective mass in the case of a long-range model.

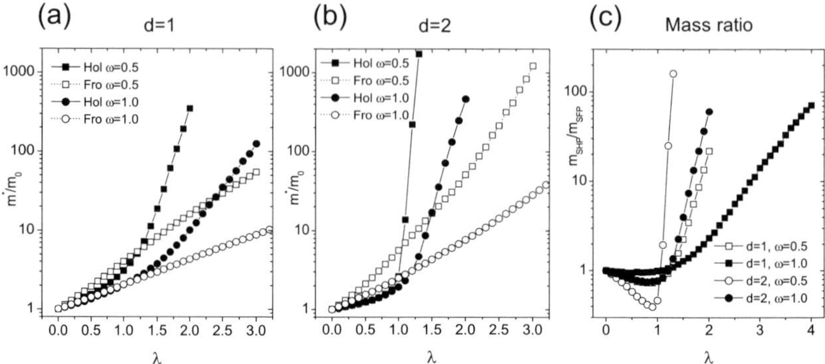

Fig. 5. Effective masses of small Holstein and Fröhlich polarons in units of $m_0 = \hbar^2/(2ta^2)$. (a) $d = 1$. (b) $d = 2$. (c) The mass ratio of the small Holstein and Fröhlich polarons for several model parameters. The ratio scales exponentially with the coupling, and could exceed 100.

The physical reason for the small mass of SFP is easy to understand. The mass is determined by the overlap of the ionic wave functions before and after the electron hop. In the case of short-range e-ph interaction, the lattice deformation must relax all the way back to the equilibrium state after the electron leaves the site. Likewise, the lattice deformation at the new electron's location must form anew from the equilibrium state. As a result, the overlap integral is exponentially small. In the case of a long-range interaction, the deformation at the new location is partially pre-existent. The ions must move by a smaller distance, which results in a much larger overlap of the wave functions. That also produces an exponentially large mass, but with a smaller exponent. (The fact that the mass is still exponential in coupling justifies the term "small" in the polaron's name.) This effect can also be understood within the Lang-Firsov (LF) approximation for the polaron mass [89, 90] that becomes exact in the limit of infinitely fast ions, $\hbar\omega \gg t$. According to LF, the polaron mass is given by $m^* = m_0\, e^{\gamma(E_p/\hbar\omega)}$, where E_p is the polaron shift from (37). The dimensionless parameter

$$\gamma = 1 - \frac{\sum_{\mathbf{m}} f_{\mathbf{m}}(0) f_{\mathbf{m}}(\mathbf{a})}{\sum_{\mathbf{m}} f_{\mathbf{m}}^2(0)}, \tag{38}$$

depends on the shape of the e-ph interaction. For the force function (36) at $d = a$, $\gamma = 0.387$ for the linear chain, 0.334 for the square lattice and 0.320 for the triangular lattice. Note also that a long-range e-ph interaction smoothens the polaron crossover, cf. Fig. 5, which makes the single exponential form be more applicable for the entire coupling range [32].

These results demonstrate that the Holstein model is an *extreme* model that predicts the largest mass for the given polaron energy. As such, the

Holstein model is not adequate for ionic systems with poor screening where the e-ph interaction is long range.

4.2 Enhancement of Anisotropy by Electron-Phonon Interaction

Two different behaviours of the lattice polaron can lead to an interesting situation when the same particle is Holstein-like and Fröhlich-like along two different lattice directions. A concrete model that features such properties was put forward in [33]. The model is an extension of the force function (36) to three dimensions, see Fig. 6(a). Clearly, the e-ph interaction is anisotropic, which translates to an even higher anisotropy of polaron spectrum. Upon hops along the z direction, the electron must reverse the sign of deformation. Thus the ion overlap integrals are even smaller than in the Holstein case. In contrast, the movement in the xy plane is Fröhlich-like, as described in the previous section. The deformation is partially prepared, the overlap integrals are large, and the effective mass is small. This reasoning is corroborated by the spatial profiles of the path confinement function Φ, see Fig. 6(b). When the path diffuses along the x direction, Φ is smooth which translates into easy "escape". Along the z direction, Φ is much steeper. In fact it even changes sign upon migration to the nearest plane, which reflects the sign reversal of the lattice deformation after the electron hop. Since Φ affects the path's weight exponentially, one expects an exponentially large difference between the polaron masses in z and x directions: $m_z^*/m_x^* \propto e^{(\gamma_z - \gamma_x)\lambda/\bar{\omega}}$, where $\lambda = 2.93\kappa^2/(12M\omega^2 t_x)$ and $\bar{\omega} = \hbar\omega/t_x$ is the dimensionless phonon frequency. As a result, the mass anisotropy grows exponentially with λ and can reach very big values. In addition, the mass anisotropy is a sharp function of the phonon frequency, which should lead, for example, to an isotope effect on anisotropy.

The results of PIQMC analysis of the model are summarized in Fig. 7 [33]. The inset shows a typical behaviour of the polaron masses at $\hbar\omega = 1.0\,t_x$ and $t_z = 0.25\,t_x$. (This choice of hoppings ensures the isotropy of the bare spectrum, since the lattice constant in the z direction is assumed to be twice that in the x direction.) As expected, m_x^* grows exponentially with coupling. Interestingly, the z mass grows *super*-exponentially with a large quadratic component: $m_z^* \approx m_{x0}\,e^{1.26\lambda+0.88\lambda^2}$. It is very possible though that this is still a transient regime, and m_z^* approaches pure exponential growth at still larger λ (at which m_z becomes so large it is difficult to calculate). The mass anisotropy for several sets of model parameters is shown in the main panel of Fig. 7. Due to the super-exponential growth of m_z^*, the anisotropy is also super-exponential, e.g. $m_z^*/m_{xy}^* \approx e^{0.42\lambda+0.71\lambda^2}$ for $\hbar\omega = 1.0\,t_x$ (circles). At a smaller frequency $\hbar\omega = 0.5\,t_x$ (squares) the anisotropy is exponentially larger, as expected from the reasoning given above. This implies the existence of an *isotope effect on the mass anisotropy.* The third model parameter, the bare hopping anisotropy, turns out to have little effect on the anisotropy of the polaron spectrum. The mass anisotropy for $t_z = 0.1\,t_x$ and the same phonon frequency $\hbar\omega = 0.5\,t_x$ is shown by triangles in Fig. 7. While being $2.0 - 2.5$

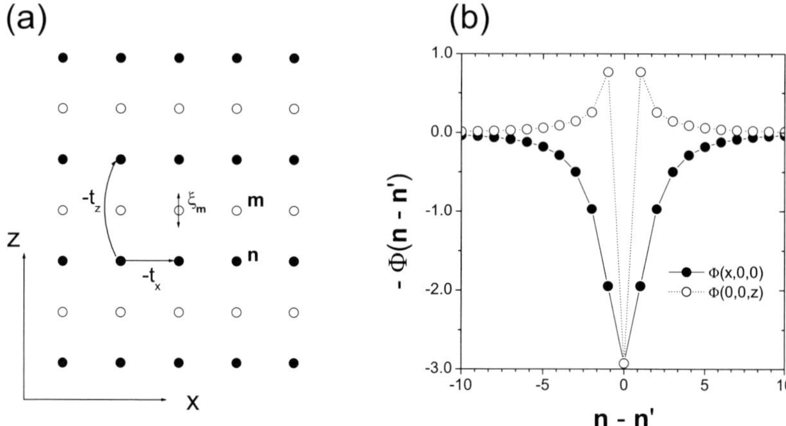

Fig. 6. (a) A three-dimensional polaron model with anisotropic e-ph interaction. The electron moves within and between the planes of filled circles with hopping integrals $-t_x$ and $-t_z$, respectively. It interacts with vibrating ions shown by open circles. The vibrations are polarized along the z-axis. In the case of a long-range interaction the polaron is Fröhlich-like in the xy plane and Holstein-like along the z direction. (b) The (negative) polaron path confinement function $-\kappa^{-2}\Phi(x,y,z)$. An additional factor $e^{-|\mathbf{n}-\mathbf{m}|/R}$ with $R = 10$ lattice constants in the x direction, has been added to the force function (36) to help the lattice sum to converge. The confining profile along z is much steeper than along x.

times larger at small λ, the anisotropy approaches that of the $t_z = 0.25\,t_x$ case at large λ. It shows that it is primarily the e-ph interaction that governs the polaron anisotropy. This conclusion is further supported by studies of the Holstein model with anisotropic bare hopping [33] where *no* enhancement of the polaron anisotropy was observed.

In relation to the effect described, it is interesting to note a well-documented discrepancy between the theoretical and experimental anisotropy of the cuprates [91]. According to band structure calculations, the anisotropy of resistivity of LSCO and YBCO compounds should be about $10 - 30$ [92, 93]. At the same time, the experimental anisotropy of resistivity is between 10^2 and 10^3 depending on the level of doping. The anisotropy of bismuthates is even higher, $(5 - 80) \cdot 10^4$ [91], which is difficult to explain with the conventional Bloch-Boltzmann theory. According to the present results, anisotropic interaction with z-polarized phonons is a sufficient condition for a very large z effective mass. Of course, the anisotropy of mass and resistivity are two different things, but it would be fair to assume that the former is at least partially responsible for the latter (see, e.g., [91] page 85). This idea still awaits proper

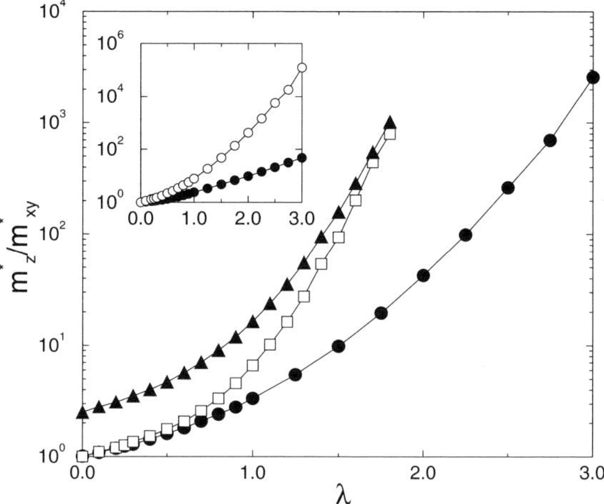

Fig. 7. Mass anisotropy of the three-dimensional model (36). Circles: $\hbar\omega = 1.0\,t_x$, $t_z = 0.25\,t_x$; squares: $\hbar\omega = 0.5\,t_x$, $t_z = 0.25\,t_x$; triangles: $\hbar\omega = 0.5\,t_x$, $t_z = 0.1\,t_x$. Inset: masses m_{xy}^* (filled circles) and m_z^* for $\hbar\omega = 1.0\,t_x$, $t_z = 0.25\,t_x$. From [33].

development. Alternative explanations of the anomalous z-transport exist as well [94].

4.3 Spectrum Flattening and Polaron Density of States

A great power of the PIQMC method is the ability to calculate an entire polaron spectrum in any dimensionality in a single run, as was explained in Sect. 3.3. This opens up an exciting possibility to calculate the polaron density of states (DOS) $\rho(E) = N^{-1}\sum_{\mathbf{K}}\delta(E - E_{\mathbf{K}} + E_0)$ by discretizing the energy interval and histogramming $E_{\mathbf{K}}$ values at the end of simulations. The polaron spectrum in the adiabatic limit $(t \gg \hbar\omega)$ possesses an interesting property of flattening at large lattice momenta, which has been well documented in the literature [50, 51, 55, 95]. In the weak-coupling limit, the flattening can be readily understood as hybridization between the bare electron spectrum and a momentum-independent phonon mode. The resulting polaron dispersion is cosine-like at small $\mathbf{K}$ and flat at large $\mathbf{K}$. At larger couplings this picture becomes less and less intuitive but the flattening remains as a matter of fact [96]. It is noteworthy that in large spatial dimensions the phase space is dominated by states with large momenta. If all those states have close energies, they should form a peak in the DOS close to the *top* of the polaron band. Exact PIQMC calculations have confirmed that this is indeed the case [30]. The evolution of the DOS of the isotropic Holstein model with phonon frequency ω in two and three dimensions is shown in Fig. 8(a) and (b). In all the cases presented the polaron is fully developed, with the bandwidth much smaller than ω.

At small ω, DOS develops a massive peak at the top of the band. The peak is more pronounced in $d = 3$ than in $d = 2$. The van Hove singularities are absorbed in the peak and as such cannot be seen. With increasing ω, the polaron spectrum approached the cosine-like shape in full accordance with the Lang-Firsov non-adiabatic formula. The respective DOS gradually assume the familiar shapes of tight-binding zones with renormalized hopping integrals. The van Hove singularities are clearly visible. These results have an interesting corollary for the Holstein model. At small-to-moderate ω in two and three dimensions, the bottom half of the polaron band contains a tiny minority of the total number of states (measured in low percentage points). In most real systems those states will be localized and hence irrelevant. All the system's responses will be dominated by the states in the peak. The physical properties of these physically relevant states are going to be quite different from those of the ground state that are usually analyzed theoretically.

It is interesting to look at the DOS of the long-range model (36). The two dimensional DOS is shown in Fig. 8(c). It is much closer to the tight-binding shape than the Holstein DOS at the same parameters. The polaron spectrum and density of states is another manifestation of the extremity of the Holstein model. Long-range e-ph interactions remove those peculiarities and make the polaron bands more "normal".

Some densities of states of the Holstein model with anisotropic hopping were presented in [30].

4.4 Isotope Exponents

Isotope substitution is a powerful tool in determining if a particular phenomenon or feature is of phonon origin. Some polaron properties, such as mass, are sharp functions of the lattice parameters, therefore isotope effects in (bi)polarons are strongly enhanced. In particular, anything that depends on the (bi)polaron mass should exhibit a large isotope exponent [86]. A large isotope effect was observed, for example, on the magnetic penetration depth in cuprates [97–99], which was interpreted as evidence for bipolaronic carriers in the superconducting state.

As was explained in Sect. 3.3, the PIQMC method enables approximation-free calculation of the isotope exponent on the (bi)polaron effective mass for a large class of e-ph models. A good feel for α in the strong-coupling limit can be obtained from the anti-adiabatic expression for the polaron mass, $m^* = m_0\, e^{\gamma(E_p/\hbar\omega)}$. Since the polaron shift E_p is ion-mass-independent, $\alpha = \frac{\gamma}{2}\frac{E_p}{\hbar\omega}$. Thus α is proportional to the coupling constant λ. In the weak coupling regime, α can be computed perturbatively. For example, for the one-dimensional Holstein model, the second order yields

$$\alpha^{(2)}_{\text{1d Holstein}} = \lambda\frac{\bar\omega^2(\bar\omega^2 + 2\bar\omega + 4)}{(\bar\omega^2 + 4\bar\omega)^{\frac{5}{2}}}, \tag{39}$$

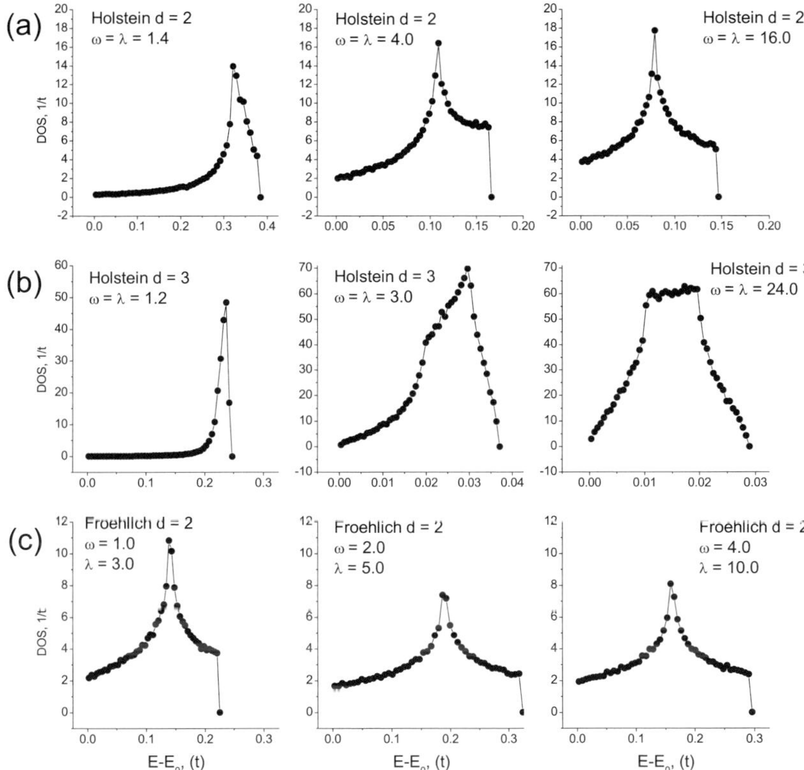

Fig. 8. (a) and (b) The evolution of the density of states of the Holstein polaron with phonon frequency ω in $d = 2$ and 3, respectively. Each graph was obtained by calculating the polaron spectrum at 100,000 **K** points randomly distributed in the Brillouin zone, and histogramming the results between 50 energy bins. Each spectrum point was calculated by averaging 250,000 values of $\cos \mathbf{K}\Delta\mathbf{r}$ taken every 10th path update. Every 5000 measurements the path was reset and then equilibrated for 1000 updates. (c) The same for the small Fröhlich polaron in $d = 2$.

where $\bar{\omega} = \hbar\omega/t$. (The second-order coefficients of other polaron properties of the one-dimensional Holstein model can be found, e.g., in [36]. The second order coefficients for higher-dimensional lattices were given in [37].) The isotope exponent is again proportional to λ but with a different coefficient. The two linear dependencies should be smoothly connected over the polaron crossover.

The mass isotope exponents for the small Holstein polaron are shown in the top row of Fig. 9. Before the polaron transition α is small reflecting the non-polaronic character of the carrier. Note that in $d = 2$ and 3, α is *negative* at small frequencies [37]. After the polaron crossover, which always begins at $\lambda \sim 1$ but ends at λ that increases with ω, the isotope exponent quickly reaches the strong-coupling asymptotic behaviour with $\gamma = 1$. Notice that the beginning and the end of the transition are clearly identifiable on the plots. Thus the mass isotope exponent can be useful in analyzing the Holstein polaron transition.

The case of the small Fröhlich polaron is somewhat different, see the bottom row of Fig. 9. At small coupling, the exponent grows linearly with a γ close to the strong coupling limit, but then deviates to smaller values. The exponent returns back to the strong-coupling limit at much larger λ, whose value decreases with increasing ω. This final approach happens when the entire path is mostly confined to one lattice site, and only rarely deviates to the first nearest neighbour. This interesting behaviour is not yet fully understood.

As was shown in the previous section, in the adiabatic regime the properties of the polaron mass do not fully represent those of the vast majority of polaron states. Additional insight can be gained from an isotope effect on the polaron bandwidth or even on individual spectrum points [35]. The isotope effect on polaron spectrum and density of states in $d = 2$ is illustrated in Fig 10. The ratio of the two phonon frequencies, $\hbar\omega = 0.80\,t$ and $0.75\,t$, has been chosen to roughly correspond to the substitution of ^{16}O for ^{18}O in complex oxides. One can see that the polaron band shrinks significantly, by 20-30%, for both polaron types. The middle panels show the isotope exponents

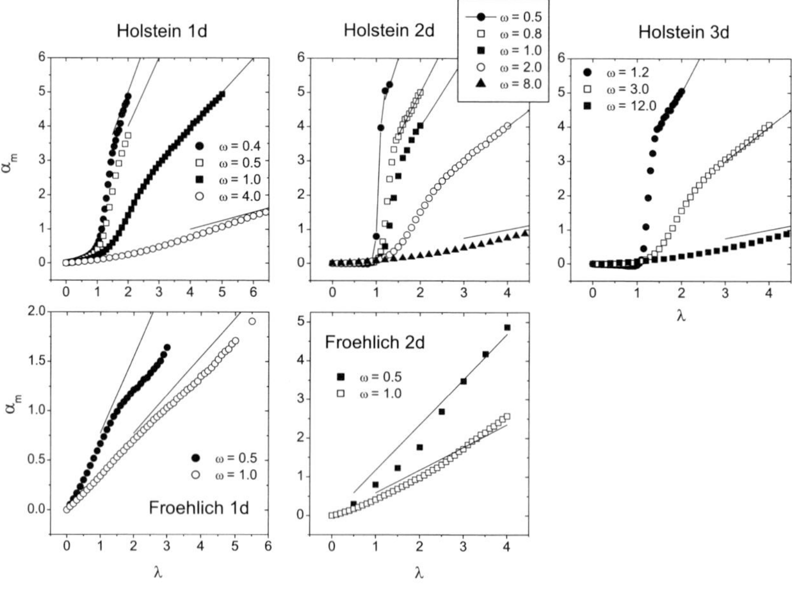

Fig. 9. Mass isotope exponents of the small Holstein and small Fröhlich polarons in different dimensions and for different phonon frequencies. The thin solid lines indicate the strong-coupling limit $\alpha = \frac{\gamma}{2}\frac{\lambda zt}{\hbar\omega}$. Note that α of the Holstein polaron in $d = 2$ and 3 is *negative* at small ω.

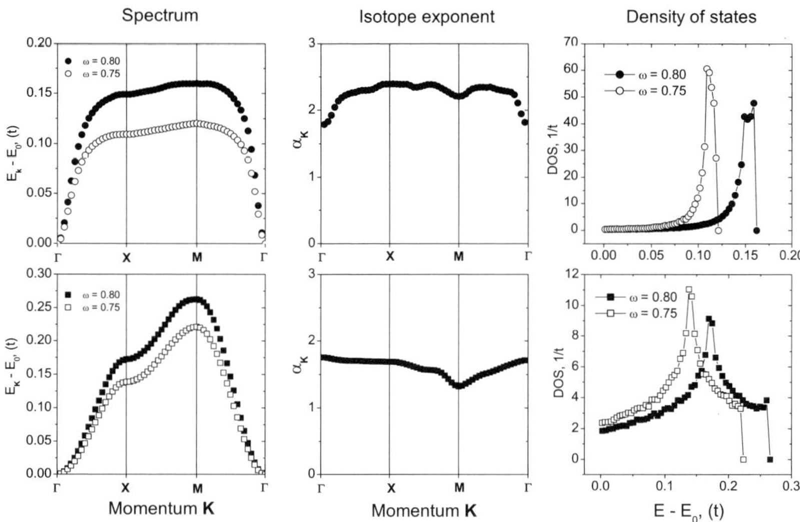

Fig. 10. Isotope effect on polaron spectrum and density of states. *Top row*: the $2d$ Holstein polaron at $\lambda = 1.2$. Left: polaron spectrum at $\bar{\omega} = 0.80$ and 0.75. Middle: the isotope exponent for each **K** point. Right: the density of states for the two frequencies. *Bottom row*: the same for the small $2d$ Fröhlich polaron at $\lambda = 2.4$.

on spectrum points calculated as

$$\alpha_{\mathbf{K}} = \frac{1}{2} \frac{\langle \omega \rangle}{\langle E_{\mathbf{K}} \rangle} \frac{\Delta E_{\mathbf{K}}}{\Delta \omega} \,, \tag{40}$$

where the angular brackets denote the mean value of either the two frequencies or of the two energy values. An interesting observation is that $\alpha_{\mathbf{K}}$ of the Fröhlich polaron is roughly independent of **K** ($\pm 10\%$). In the Holstein case $\alpha_{\mathbf{K}}$ dips in the vicinity of the Γ point. (Small fluctuations at intermediate momenta are due to statistical errors in averaging $\langle \cos \mathbf{K} \Delta \mathbf{r} \rangle$.)

4.5 Jahn-Teller Polaron

Density-displacement is only one possible type of e-ph interaction. Examples of other types are the deformation potential and Su-Schrieffer-Heeger (SSH) interaction. In the former, the electron density interacts with a gradient of lattice displacement whereas in the SSH case the lattice deformation is coupled to the electron's kinetic energy. The Jahn-Teller (JT) interaction is one of the most complex types because it usually involves a multidimensional electron basis and a multidimensional representation of the deformation group. The JT interaction is active in some molecules and crystals of high point symmetry.

The JT effect was also a guiding principle in the search for high-temperature superconductivity [100]. More recently, a JT theory of the cuprates was developed in [101–103]. There exist different flavors of the JT interaction. The simplest one is the $E \otimes e$ interaction [104] that describes a short-range coupling between twice-degenerate e_g electronic levels (c_1, c_2) and a local double-degenerate vibron mode (ζ, η). The Hamiltonian reads

$$
\begin{aligned}
H_{\mathrm{JT}} = & -t \sum_{\langle \mathbf{nn'} \rangle} \left(c^{\dagger}_{\mathbf{n'}1} c_{\mathbf{n}1} + c^{\dagger}_{\mathbf{n'}2} c_{\mathbf{n}2} \right) \\
& - \kappa \sum_{\mathbf{n}} \left[\left(c^{\dagger}_{\mathbf{n}1} c_{\mathbf{n}2} + c^{\dagger}_{\mathbf{n}2} c_{\mathbf{n}1} \right) \eta_{\mathbf{n}} + \left(c^{\dagger}_{\mathbf{n}1} c_{\mathbf{n}1} - c^{\dagger}_{\mathbf{n}2} c_{\mathbf{n}2} \right) \zeta_{\mathbf{n}} \right] \\
& + \sum_{\mathbf{n}} \left[-\frac{\hbar^2}{2M} \left(\frac{\partial^2}{\partial \zeta^2_{\mathbf{n}}} + \frac{\partial^2}{\partial \eta^2_{\mathbf{n}}} \right) + \frac{M\omega^2}{2} \left(\zeta^2_{\mathbf{n}} + \eta^2_{\mathbf{n}} \right) \right] .
\end{aligned} \tag{41}
$$

The symmetry of the interaction ensures the same coupling parameter κ for the two phonons. The kinetic energy is chosen to connect like electron orbitals of the nearest neighbours. This choice is somewhat arbitrary (makes it more "Holstein-like") and not dictated by symmetry. Because the ionic coordinates of different cells are not coupled, the model describes a collection of separate clusters that are linked only by electron hopping. To relate the Hamiltonian to more realistic situations, phonon dispersion must be added [101, 105].

An important property of the $E \otimes e$ interaction is the absence of an exact analytical solution in the atomic limit $t = 0$. In the density-displacement case the dynamics of each ζ is described by a one-dimensional differential equation under a constant force. That has a shifted oscillator solution, which serves as a convenient starting point for various strong-coupling expansions. Here, in contrast, the atomic limit is described by two coupled *partial* differential equations for the electron doublet $\psi_{1,2}(\zeta, \eta)$. Although it is possible to separate variables and reduce the system to two ordinary second-order equations, they seem to be too complex to admit an analytical solution. At large couplings, however, the elastic energy assumes the Mexican hat shape and the phonon dynamics separates into radial oscillatory motion and azimythal rotary motion. This results in an additional pre-exponential factor $\propto \kappa$ in the ion overlap integral, leading to the effective mass $m^*_{\mathrm{JT}} = m_0 \sqrt{\frac{2}{\pi g}} \, e^{g^2}$, where $g^2 = \frac{\kappa^2}{2M\hbar\omega^3}$ [106].

A path integral approach to Hamiltonian (41) was developed in [34]. Because there are two electron orbitals, the electron path must be assigned an additional orbital index (or *colour*) $a = 1, 2$. Colour 1 (or 2) of a given path segment means that it resides in the first (second) atomic orbital of the electron doublet. The short-term density matrix is

$$
\begin{aligned}
\rho(\Delta\tau) = & \langle \mathbf{r}'a'; \{\zeta'_{\mathbf{n}}\}\{\eta'_{\mathbf{n}}\} | e^{-\Delta\tau H} | \mathbf{r}a; \{\zeta_{\mathbf{n}}\}\{\eta_{\mathbf{n}}\} \rangle \\
= & \left[\delta_{\mathbf{rr}'}\delta_{aa'} + (\kappa\Delta\tau)\delta_{\mathbf{rr}'}\delta_{a\bar{a}'}\eta_{\mathbf{n}} + (t\Delta\tau)\delta_{aa'} \sum_{\mathbf{l}} \delta_{\mathbf{r}',\mathbf{r}+\mathbf{l}} \right] e^{A_{\mathrm{ph}}(\Delta\tau)} . \tag{42}
\end{aligned}
$$

$$A_{\rm ph}(\Delta\tau) = (\kappa\Delta\tau)(\delta_{a1} - \delta_{a2})\zeta_{\mathbf{n}}$$

$$-\sum_{\mathbf{n}}\left\{\frac{M}{2\hbar^2(\Delta\tau)}[(\zeta_{\mathbf{n}} - \zeta'_{\mathbf{n}})^2 + (\eta_{\mathbf{n}} - \eta'_{\mathbf{n}})^2] + (\Delta\tau)\frac{M\omega^2}{2}(\zeta_{\mathbf{n}}^2 + \eta_{\mathbf{n}}^2)\right\}. \quad (43)$$

Here $\bar{a} = 1(2)$ when $a = 2(1)$, that is $\bar{a}$ is "not" a. This expression reveals a difference between the two phonons. Phonon ζ is coupled to electron density, like in the Holstein case. The difference from the Holstein is that the direction of the force changes to the opposite when the electron changes orbitals. In contrast, phonon η is coupled to orbital changes themselves: the more often the electron changes orbitals, the more "active" is η. (Discrete orbital changes are analogous to electron hops between discrete lattice sites, and as such are associated with "kinetic orbital energy". Phonon coupling to orbital changes is analogous to phonon coupling to electron hopping in the Su-Schrieffer-Heeger interaction [107].) Multiplication of a large number of the short-time propagators generates multiple terms, each of which contains a finite number of lattice hops and orbital changes. On top of that, the exponents combine in a total phonon action $A_{\rm ph}(\beta)$ that comprises the free η action $A_{\eta 0}$ and the action of ζ phonons under an external force [A_ζ is similar to (21)]. The next step is integration over the paths $\eta(\tau)$ and $\zeta(\tau)$. Integration over $\zeta(\tau)$ is standard and performed as described as for the density-displacement interaction. The result is a factor e^{A_ζ} in the path's statistical weight, where

$$A_\zeta[\mathbf{r}(\tau), a(\tau)] = \kappa^2 \int_0^\beta\int_0^\beta \mathrm{d}\tau\mathrm{d}\tau' G(\tau - \tau')\left[\delta_{a(\tau),a(\tau')} - \delta_{a(\tau),\bar{a}(\tau')}\right], \quad (44)$$

$$G(\tau-\tau') = \frac{\hbar}{2M\omega}\left[e^{-\hbar\omega|\tau-\tau'|}\delta_{\mathbf{r}(\tau),\mathbf{r}(\tau')} + e^{-\hbar\omega(\beta-|\tau-\tau'|)}\delta_{\mathbf{r}(\tau),\mathbf{r}(\tau')+\mathrm{sgn}(\tau-\tau')\Delta\mathbf{r}}\right].$$
$$(45)$$

The last expression is valid under the condition $e^{\beta\hbar\omega} \gg 1$. As expected, action A_ζ is an explicit functional of the spatial path $\mathbf{r}(\tau)$ and of the orbital path $a(\tau)$. Note that the ζ phonon favours like orbitals and disfavours orbital changes. Integration over $\eta(\tau)$ is trickier for it contains η as pre-exponential factors. A typical term with N_s orbital changes has the following form:

$$(\kappa\Delta\tau)^n\eta_{\mathbf{r}(\tau_1)}(\tau_1)\eta_{\mathbf{r}(\tau_2)}(\tau_2)\ldots\eta_{\mathbf{r}(\tau_{N_s})}(\tau_{N_s})\,e^{A_{\eta 0}[\eta_{\mathbf{n}}(\tau)]}. \quad (46)$$

Here, τ_s is the time of sth orbital change ($s = 1, \ldots, N_s$), $\mathbf{r}(\tau_s)$ is the electron position at this time, and $\eta_{\mathbf{r}(\tau_s)}(\tau_s)$ is the η-displacement at this site at this time. For odd N_s, path integration over $\eta_{\mathbf{n}}(\tau)$ produces zero by parity. For even m, the integration can be performed by introducing fictitious sources, calculating the generating functional and differentiating it m times. The result is a sum of all possible pairings of τ_s. Within each term, each pair contributes the factor $\kappa^2 G(\tau_s-\tau_{s'})$, with G given by (45). Since G is positive, the statistical weight increases with the number of orbital flips. Thus, the η-phonon *favours* orbital flips, in complete contrast with the ζ-phonon. Since both phonons are governed by the same Green's function (which should be no surprise since the two phonons are related by symmetry) one should expect a dynamical equilibrium between the two phonons and some finite mean value of orbital flips for given coupling. The shift partition function is

$$
Z_{\Delta \mathbf{r}} = Z_{\mathrm{ph}} \sum_{N_k=0,1,\dots} \sum_{N_s=0,2,\dots} \int_0^\beta \cdots \int_0^\beta (\mathrm{d}\tau)^{N_k} (\mathrm{d}\tau)^{N_s} W_{N_k N_s} , \tag{47}
$$

$$
W_{N_k N_s} = t^{N_k} \kappa^{N_s} \left[\prod_{(\text{ pairs of } \tau_s)} G(\tau_s - \tau_{s'}) \right] e^{A_\zeta [\mathbf{r}(\tau), a(\tau)]} . \tag{48}
$$

Compared to (27), the above expression involves additional multiple integration over the times of orbital flips. Each term is associated with a spatial-orbital path with N_k kinks and $(N_s/2)$ η-phonon pairings, as shown in Fig. 11(a). A Markov process is organized by inserting/removing spatial kinks and by attaching/removing the pairing lines. The acceptance rules for the pairing lines can be derived by extending the method of Sect. 3.1 to two-time updates. For details see [34]. The path shown in Fig. 11(a) can be considered a space-time *diagram*. In fact, the update process just described is a version of the Diagrammatic Monte Carlo method [83].

After the update rules are established, the JT polaron properties can be calculated with no approximations. The mass, spectrum, and density of states are obtained as for the conventional lattice polarons. For the JT energy the thermodynamic estimator yields:

$$
E_0 = \left\langle -\frac{N_k}{\beta} - \frac{\partial A_\zeta}{\partial \beta} - \frac{N_s}{\beta} - \frac{\hbar\omega}{\beta} \sum_G \frac{1}{G(\tau_s - \tau_{s'})} \frac{\partial G(\tau_s - \tau_{s'})}{\partial(\hbar\omega)} \right\rangle_{\mathrm{shift}} . \tag{49}
$$

The first two terms are familiar from the Holstein case. The first is the kinetic energy of the electron. The second is the potential energy of interaction with the ζ plus the (positive) elastic energy of ζ. The other two terms are new. The third one represents the negative "orbital energy" caused by orbital flips. These processes excite phonons η, which results in the positive fourth term.

The number of excited phonons is an important characteristic because it provides an internal consistency check for the algorithm: one expects *equal* mean number of ζ and η phonons. The phonon number estimator is derived as the derivative of the free energy with respect to $(\hbar\omega)$ under a fixed combination $\kappa^2/(M\omega)$. One obtains

$$
N_{\mathrm{ph},\zeta} = -\frac{1}{\beta} \left\langle \frac{\partial A_\zeta}{\partial(\hbar\omega)} \bigg|_{\frac{\kappa^2}{M\omega}} \right\rangle_{\mathrm{shift}} , \tag{50}
$$

$$
N_{\mathrm{ph},\eta} = -\frac{1}{\beta} \left\langle \sum_G \frac{1}{G(\tau_s - \tau_{s'})} \frac{\partial G(\tau_s - \tau_{s'})}{\partial(\hbar\omega)} \right\rangle_{\mathrm{shift}} . \tag{51}
$$

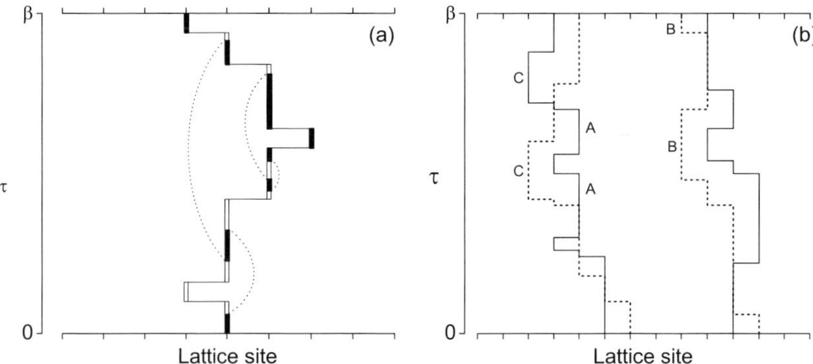

Fig. 11. Extensions of the basic polaron PIQMC method. (a) In the Jahn-Teller polaron, the path carries an internal orbital index shown as black or white sections of the path. The η-phonon parings are represented by dashed lines. Each vertex changes the index. (b) In the bipolaron case, the system is represented by two imaginary-time paths. A direct configuration is shown on the left and an exchange one on the right. The interaction of segments marked A or B contribute to the energy of individual polarons. The segments that belong to different paths, such as those marked C, contribute to the interaction between the polarons.

Results of PIQMC calculations are shown in Fig. 12 [34]. Most properties behave similarly to those of the $d = 3$ Holstein polaron at the same phonon frequency, $\hbar\omega = 1.0\,t$. For example, the kinetic energy, Fig. 12(a), sharply decreases by absolute value between $\lambda = 1.2$ and 1.4. The JT polaron mass is slightly larger at the small to intermediate coupling, but several times smaller at the strong coupling. This non-monotonic behaviour of the ratio of the JT and Holstein masses was later confirmed by accurate variational calculations [60], although in that work the JT polaron (and bipolaron) was investigated in one spatial dimension. The relative lightness of the JT polaron is consistent with Takada's result mentioned above [106]. The number of excited phonons of both types is shown in Fig. 12(c). As expected, numerical values coincide within the statistical error, which validates the numerical algorithm. Interestingly, the shape of the phonon curves is similar to that of the logarithm of the effective mass. This suggests an intimate relationship between the two quantities, again similarly to the Holstein case. Finally, the density of JT polaron states features the same peak at the top of the band, caused by the spectrum flattening at large polaron momenta.

In summary, the locality of the JT interaction and the independence of vibrating clusters result in the same extremity of polaron properties as in the $3d$ Holstein model. One should expect that either a long-range JT or phonon dispersion will soften the sharp polaron features and make them more "normal". This is an interesting research opportunity for the PIQMC method.

5 Prospects

In this chapter, a theoretical approach to lattice polarons based on the statistical path integral has been developed. A characteristic feature of the method is a tight integration of analytical and numerical elements within one technique. The numerical part of the algorithm is greatly aided by the preliminary analytical steps: path integration over the ionic coordinates and reduction of the double integral over imaginary time to a discrete sum over path segments. Thus, on the way from the Hamiltonian to observables, exact analytical transformations are carried as far as possible increasing the accuracy and speed of the simulations.

The PIQMC polaron method is still very much under development. Due to the space constraints, only key elements have been presented here and many technical details have been left out. This concluding section will be used to

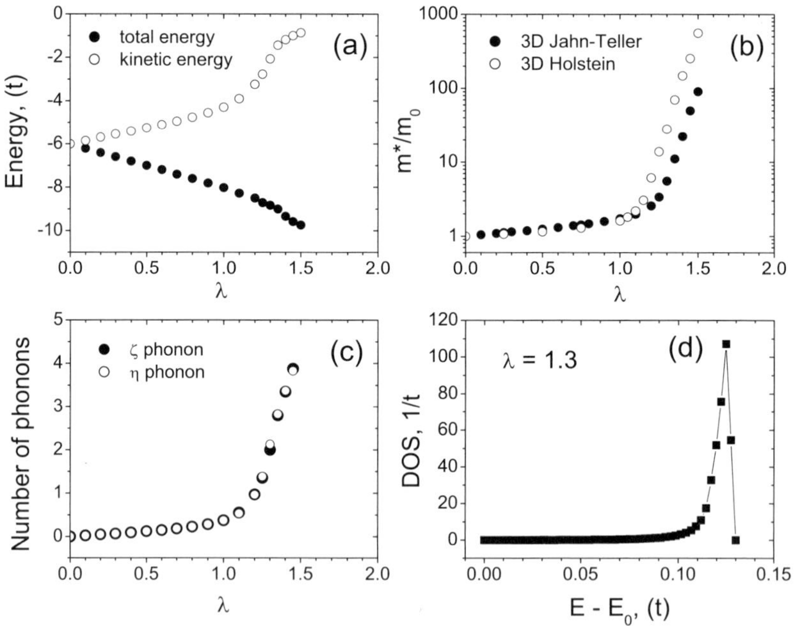

Fig. 12. Physical properties of the $d = 3$ Jahn-Teller polaron at $\hbar\omega = 1.0\,t$ [34]. (a) The total and kinetic energy. (b) The effective mass compared with the $d = 3$ Holstein polaron at the same phonon frequency. (c) The number of excited phonons of both types. (d) The density of states of the JT polaron at $\lambda = 1.3$.

outline various extensions of the method. Some of them have already been done, others *can* be done but await proper implementation. Certain problems have not yet been solved even on the conceptual level, and those will be discussed too.

5.1 Bipolaron

The possibility of two polarons forming bound pairs, or bipolarons, opens a whole new dimension in polaron physics [108, 109]. The two major competing factors are the phonon-mediated attraction between the polarons that favours paring and direct Coulombic repulsion that prevents pairing. Kinetic energy, exchange effects and short-range lattice effects play important supporting roles that can tip the balance toward or against paring. All of this may lead to a rich and complex phase diagram. The bipolaron concept has been well-developed in the literature. The large continuous-space (Fröhlich) bipolaron in ionic media was considered in [20, 110, 111], and the small, localized, lattice bipolaron in [112]. Since the bipolarons are charged bosons at low density n, they can superconduct below the Bose-Einstein condensation temperature

$$T_{\mathrm{BE}} = 3.31 \frac{h^2}{k_B} \frac{n^{2/3}}{m^{**}} \, . \tag{52}$$

The concept of *mobile* lattice bipolarons and bipolaronic superconductivity was put forward by Alexandrov and Ranninger in 1981 [113] and since then thoroughly developed by Alexandrov and co-workers [70, 71, 86, 94, 114–117]. Note that T_{BE} is inversely proportional to the bipolaron mass. This suggests the following recipe to increase the critical temperature according to the bipolaronic mechanism: *reduce the effective mass while keeping the bipolarons non-overlapping.* (Clearly, simply reducing the mass will increase the bipolaron radius and reduce the overlap density. There is a trade-off between the $n^{2/3}$ and $(m^{**})^{-1}$ factors in (52), which creates a challenging optimization problem.)

As with polarons, the first application of path integrals to the bipolaron was to the continuum-space Fröhlich bipolaron [20, 110]. Until very recently, the only application of path integrals to the lattice bipolaron was the work by de Raedt and Lagendijk [28] on the discrete-time version of their path-integral QMC method. The authors limited the consideration to the extreme adiabatic limit and computed the phase diagram of the Holstein bipolaron with an additional Hubbard repulsion. A continuous-time version of PIQMC was developed very recently and the first results presented in [39].

The lattice bipolaron, mostly with Holstein interaction, has been studied by a variety of other numerical methods: exact diagonalization [50, 118], variational [57, 58, 60], density matrix renormalization group [66], diagrammatic quantum Monte Carlo [44], and Lang-Firsov quantum Monte Carlo [47, 119].

Conceptually, generalizing the PIQMC method to the bipolaron is straightforward. There are two paths that visualize the imaginary-time evolution of the two fermions. To enable calculation of the effective mass, open boundary conditions in imaginary time must be used: the top ends of the paths must be the same as the bottom ones up to an arbitrary lattice translation $\Delta\mathbf{r}$. Singlet and triplet states can be separated by allowing the paths to exchange, see Fig. 11(b). Then the bipolaron spectrum, mass, and singlet-triplet splitting is calculated as explained in Sect. 2. Phonon integration is performed as before with the same resulting action, (22)-(26). The difference is the force that is acting on the oscillator $\mathbf{m}$ at time τ. In a bipolaron system it receives contribution from both electrons, $f_{\mathbf{m}}(\tau) = f_{\mathbf{m}}[\mathbf{r}_1(\tau)] + f_{\mathbf{m}}[\mathbf{r}_2(\tau)]$. Since the action is quadratic in f, it becomes a sum of three terms that contain the products $f[\mathbf{r}_1(\tau)]f[\mathbf{r}_1(\tau')]$, $f[\mathbf{r}_2(\tau)]f[\mathbf{r}_2(\tau')]$, and $f[\mathbf{r}_1(\tau)]f[\mathbf{r}_2(\tau')]$, respectively. The first two terms describe the retarded self-interaction of the two paths and are responsible for polaron formation. The third cross-term describes the interaction between the paths. It is responsible for the interaction between the polarons and for bipolaron formation. This is illustrated in Fig. 11(b). The estimators for various observables and specific expressions for the segment-to-segment contributions are derived analogously to the polaron case. Technical details of the derivation have not yet been published, and will be published elsewhere.

One interesting effect that has already been considered by the PIQMC method is the "crab-like" bipolaron that can exist on certain lattices [39]. Usually, the bipolaron is much heavier that the constituent polarons, its mass scaling as the second power of the polaron mass in the non-adiabatic regime and as the fourth power in the adiabatic limit [120]. (Path integrals provide a useful visualization of this effect, see Fig. 11(b). At a small phonon frequency, the two paths interact over their entire length. Therefore they are much more difficult to separate, which results in slower imaginary-time diffusion, and hence a heavier particle.) However, on the triangular, face-centered cubic and some other lattices the *intersite* bipolaron can move without breaking. This results in an effective mass that scales only *linearly* with the polaron mass. This effect was predicted some time ago by Alexandrov [70], and recently confirmed by exact PIQMC simulations in [39].

5.2 Further Extensions

Recall that the basic Hamiltonian (12) contains the simplest possible form of the electron kinetic energy and the simplest possible model of the crystal lattice. It will now be explained how to relax these restrictions. First of all, it is straightforward to add electron hopping beyond the nearest neighbours. Because the path is defined in real space, this only requires the introduction of additional kink types that move the path in the respective directions. The values of the distant hopping integrals can be absolutely independent of the nearest-neighbour values as long as they are negative. The methods of Sect. 3.1 are readily generalized to paths containing kinks of different magnitude. The statistical weight (15) is now a product of factors $(t_i \Delta\tau)$ along the path. The stochastic acceptance rules now contain the factor $(t_i\beta)$ where t_i is the hopping amplitude of the proposed kink. It is interesting though that contribution to the kinetic energy is the *same* (equal to β^{-1}) for all kinks independently of the t_i value. It is the mean number of kinks that is affected by t_i, leading to different partial kinetic energies along different lattice directions. In exactly the same manner one can study models with anisotropic (but nearest-neighbour) electron hopping. This was done in [30, 33], cf. Sect. 4.2. There remains one important restriction on the values of the hopping amplitudes: they must all be negative to ensure the positivity of the path's statistical weight. If even a single t_i is positive, some paths will continue being positive but *some* will have negative weights. That will introduce a sign-problem, which will eventually render the simulation numerically unstable. Another extension related to the kinetic energy is lattices of different symmetries. Again, all that is changed is the table of kink types that specifies which two spatial points (lattice sites) each particular kink connects. In this way, PIQMC has been applied to the triangular, face-centred-cubic, and hexagonal Bravais lattices in [37].

It should be mentioned that PIQMC can also simulate the (bi)polaron in dimensionalities larger than 3, see Fig. 3. (The $d = 4$ Holstein polaron was previously investigated in [59].) Although such calculations have little practical value, they can still be useful in assessing the accuracy of approximate methods that rely on the number of spatial dimensions being large. (An example of such a method would be the Dynamical Mean Field Theory [121].)

Consider now the phonon subsystem. It is well-known that a bosonic path integral can be calculated analytically for any quadratic action. This means that the ionic coordinates can be eliminated even when they are coupled, that is in the case of dispersive phonons. Moreover, integration can be done for any number of degrees of freedom per unit cell. De Raedt and Lagendijk studied a one-dimensional Holstein polaron with phonon dispersion as early as in 1984 [27]. They found that the critical coupling of polaron formation decreases as the phonon mode becomes soft. To our knowledge, this remains the only exact analysis of a model with dispersive phonons to-date. A general phonon integration for the purposes of the PIQMC method was performed in [38]. Conceptually, the result is similar to (22)-(27). The action is a sum of a periodic contribution and a correction due to the shifted boundary conditions. The action is a double integral over the imaginary time. However, there is an additional sum over the phonon momentum and branches. In addition, the action involves the eigenvalues and eigenvectors of the dynamical matrix. Thus, the polaron action comprises full information about the phonon spectrum and polarizations.

An important advantage of the PIQMC method is the ability to study temperature effects. At a finite temperature T, the projected partition function $Z_{\mathbf{K}}$ receives contributions from high-energy states with momentum $\mathbf{K}$. It is therefore not meaningful to calculate the polaron mass and spectrum. Instead, the conventional periodic boundary conditions in imaginary time must be used, which makes all states contribute to the full thermodynamic partition function Z. The parameter $\Delta \mathbf{r}$ is set to zero. The remaining periodic action (23) is valid at any temperature, large or small. This enables approximation free calculation of thermodynamic properties of the polaron: the internal energy, specific heat, number of excited phonons, and static e-ph correlation functions.

The grand challenge for PIQMC, as for almost any Quantum Monte Carlo method is efficient simulation of many-body systems. If the number of particles is three or more, the *fermionic* sign-problem is in general unavoidable. It is possible that the sign-problem in many-fermion e-ph models will be less severe that in the case of repulsive electron-electron models. Consider the limit of low electron density and strong e-ph interaction. The polarons will be bound in bipolarons, and the paths will fluctuate in pairs. To account for statistics, the top path ends must be constantly permuted, with even/odd permutations contributing a $+1/-1$ phase factor to the partition function. However, any odd permutation will have to separate a path pair. This will increase the energy of the configuration and such an update will likely be rejected. In contrast, *some* even permutations will exchange whole bipolarons, which will keep the paths together with no significant energy increase. Such updates will likely be accepted. As a result, there will be many more even than odd permutations, and the average sign will be well defined. The stronger the coupling the more pronounced this effect will be, and the more the system will approach a purely bosonic limit (which is sign-problem-free).

The temperature-control capability offers another unique opportunity: calculation of the Meissner fraction and superconducting critical temperature. The method was devised by Scalapino et al [122] and is based on the sum rule for the off-diagonal current-current correlation function. By comparing its static limit with the kinetic energy, one could determine a temperature at which the two quantities start to differ. This signifies the appearance of a Meissner fraction and determines the T_c.

Finally, we comment on the calculation of dynamical (bi)polaron properties such as the spectral function or optical conductivity. This is a classic and difficult problem for most existing Monte Carlo methods. Derivation of real-time correlators from imaginary-time correlators amounts to solving an ill-posed inversion problem, which is usually regularized by methods such as maximum entropy [123], Pade approximants, or stochastic optimization [8]. Any of this techniques can be used in conjunction with the PIQMC method.

5.3 Conclusions

In summary, path integrals have played and continue to play an important role in the development of polaron physics. The path-integral quantum Monte Carlo method has proven to be a powerful and versatile tool. It has some unique advantages, such as system-size independence, the ability to calculate the density of states, mass isotope exponents, and temperature effects. Compared to the exact diagonalization, density-matrix renormalization group, or variational methods, PIQMC has a larger statistical error, but provides unbiased estimates for the (bi)polaron properties. Several novel qualitative results have been obtained with the PIQMC method: small polaron mass in models with long-range electron-phonon interaction, enhancement of the anisotropy of the polaron spectrum, a peak in the polaron density of states in the adiabatic limit, and the isotope effect on the polaron spectrum. These and some other effects have been discussed in this chapter in detail.

Acknowledgement. It is my pleasure to thank Sasha Alexandrov for long-standing collaboration and motivation for this research. I also thank James Hague and John Samson for useful discussions on the subject of this Chapter, and Natalia Kornilovich for help with organizing numerical data. I am grateful to James Hague for help with Fig. 9. This work has been supported by EPSRC (UK), grant EP/C518365/1.

References

1. P.A.M. Dirac: Physikalische Zeitschrift der Sowjetunion **3**, 64 (1933)
2. R.P. Feynman: Rev. Mod. Phys. **20**, 367 (1948)
3. M. Kac. Trans. Am. Math. Soc. **65**, 1 (1949)
4. R.P. Feynman: Phys. Rev. **97**, 660 (1955)
5. H. Fröhlich, H. Pelzer, S. Zienau: Philos. Mag. **41**, 221 (1950); H. Fröhlich: Advan. Phys. **3**, 325 (1954); H. Fröhlich: Introduction to the theory of the polaron. In: *Polarons and Excitons*, ed by C.G. Kuper, G.D. Whitfield (Plenum Press, New York 1963) pp 1–32, and references therein
6. J.T. Devreese, R. Evrard: Phys. Lett. **23**, 196 (1996)
7. N.V. Prokof'ev, B.V. Svistunov: Phys. Rev. Lett. **81**, 2514 (1998)
8. A.S. Mishchenko, N.V. Prokof'ev, A. Sakamoto, B.V. Svistunov: Phys. Rev. B **62**, 6317 (2000)
9. Y. Ōsaka: Prog. Theor. Phys. **22**, 437 (1959)
10. R. Abe, K. Okamoto: J. Phys. Soc. Jpn. **31**, 1337 (1971); **33**, 343 (1972)
11. M. Saitoh: J. Phys. Soc. Jpn. **49**, 878 (1980)
12. V.K. Fedyanin, C. Rodriguez: Physica A **112**, 615 (1982)
13. G. De Filippis, V. Cataudella, V. Marigliano Ramaglia, C.A. Perroni and D. Bercioux: Eur. Phys. J. B **36**, 65 (2003)
14. R.P. Feynman, R.W. Hellwarth, C.K. Iddings, P.M. Platzman: Phys. Rev. **127**, 1004 (1962)
15. K.K. Thornber, R.P. Feynman: Phys. Rev. B **1**, 4099 (1970)
16. F.M. Peeters, J.T. Devreese: Theory of polaron mobility. In: *Solid State Physics*, vol 38, ed by F. Seitz, D. Turnbull (Academic Press, New York 1984) pp 81–133
17. J.T. Devreese: Optical properties of few and many Fröhlich polarons from 3D to 0D. In: this volume; and references therein
18. F.M. Peeters, J.T. Devreese: Phys. Rev. B **25**, 7281 (1982)
19. J.T. Devreese, F. Brosens: Phys. Rev. B **45**, 6459 (1992)
20. G. Verbist, F.M. Peeters, J.T. Devreese: Solid State Commun. **76**, 1005 (1990); Phys. Rev. B **43**, 2712 (1991)
21. C. Alexandrou, W. Fleischer, R. Rosenfelder: Phys. Rev. Lett. **65**, 2615 (1990)

22. C. Alexandrou, R. Rosenfelder: Phys. Rep. **215**, 1 (1992)
23. F. Brosens, S.N. Klimin, J.T. Devreese: Phys. Rev. B **71**, 214301 (2005)
24. H. de Raedt, A. Lagendijk: Phys. Rev. Lett. **49**, 1522 (1982); Phys. Rev. B **27**, 6097 (1983)
25. H. de Raedt, A. Lagendijk: Phys. Rep. **127**, 234 (1985)
26. N. Metropolis, A.W. Rosenbluth, M.N. Rosenbluth, A.H. Teller, E. Teller: J. Chem. Phys. **21**, 1987 (1953)
27. H. de Raedt, A. Lagendijk: Phys. Rev. B **30**, 1671 (1984)
28. H. de Raedt, A. Lagendijk: Z. Phys. B: Condens. Matter **65**, 43 (1986)
29. P.E. Kornilovitch, E.R. Pike: Phys. Rev. B **55**, 8634 (1997); **69**, 059902 (2004)
30. P.E. Kornilovitch: Phys. Rev. B **60**, 3237 (1999)
31. P.E. Kornilovitch: Phys. Rev. Lett. **81**, 5382 (1998)
32. A.S. Alexandrov, P.E. Kornilovitch: Phys. Rev. Lett. **82**, 807 (1999)
33. P.E. Kornilovitch: Phys. Rev. B **59**, 13531 (1999)
34. P.E. Kornilovitch: Phys. Rev. Lett. **84**, 1551 (2000)
35. P.E. Kornilovitch, A.S. Alexandrov: Phys. Rev. B **70**, 224511 (2004)
36. P.E. Spencer, J.H. Samson, P.E. Kornilovitch, A.S. Alexandrov: Phys. Rev. B **71**, 184310 (2005)
37. J.P. Hague, P.E. Kornilovitch, A.S. Alexandrov, J.H. Samson: Phys. Rev. B **73**, 054303 (2006)
38. P.E. Kornilovitch: Phys. Rev. B **73**, 094305 (2006)
39. J.P. Hague, P.E. Kornilovitch, J.H. Samson, A.S. Alexandrov: Phys. Rev. Lett. **98**, 037002 (2007)
40. J.E. Hirsch, E. Fradkin: Phys. Rev. Lett. **49**, 402 (1982); Phys. Rev. B **27**, 4302 (1983)
41. J.E. Hirsch, E. Fradkin: Phys. Rev. B **27**, 1680 (1983)
42. A.S. Mishchenko, N. Nagaosa, N.V. Prokof'ev, A. Sakamoto, B.V. Svistunov: Phys. Rev. Lett. **91**, 236401 (2003)
43. A.S. Mishchenko, N. Nagaosa: Phys. Rev. Lett. **93**, 036402 (2004)
44. A. Macridin, G.A. Sawatzky, M. Jarrell: Phys. Rev. B **69**, 245111 (2004)
45. A.S. Mishchenko, N. Nagaosa: Spectroscopic properties of polarons in strongly correlated systems by exact diagrammatic Monte Carlo method. In: this volume
46. M. Hohenadler, H.G. Evertz, W. von der Linden: Phys. Rev. B **69**, 024301 (2004)
47. M. Hohenadler, W. von der Linden: Phys. Rev. B **71**, 184309 (2005)
48. M. Hohenadler, W. von der Linden: Lang-Firsov approaches to polaron physics: From variational methods to unbiased quantum Monte Carlo simulations. In: this volume
49. A.S. Alexandrov, V.V. Kabanov, D.K. Ray: Phys. Rev. B **49**, 9915 (1994)
50. G. Wellein, H. Röder, H. Fehske: Phys. Rev. B **53**, 9666 (1996)
51. G. Wellein, H. Fehske: Phys. Rev. B **56**, 4513 (1997)
52. H. Fehske, J. Loos, G. Wellein: Phys. Rev. B **61**, 8016 (2000)
53. F.X. Bronold, H. Fehske: Phys. Rev. B **66**, 073102 (2002)
54. H. Fehske, S.A. Trugman: Numerical solution of the Holstein polaron problem. In: this volume
55. A.H. Romero, D.W. Brown, K. Lindenberg: J. Chem. Phys. **109**, 6540 (1998); Phys. Rev. B **59**, 13728 (1999)
56. J. Bonča, S.A. Trugman, I. Batistić: Phys. Rev. B **60**, 1633 (1999)

57. J. Bonča, T. Katrašnik, S.A. Trugman: Phys. Rev. Lett. **84**, 3153 (2000)
58. J. Bonča, S.A. Trugman: Phys. Rev. B **64**, 094507 (2001)
59. Li-Chung Ku, S.A. Trugman, J. Bonča: Phys. Rev. B **65**, 174306 (2002)
60. E. El Shawish, J. Bonča, Li-Chung Ku, S.A. Trugman: Phys. Rev. B **67**, 014301 (2003)
61. V. Cataudella, G. De Filippis, G. Iadonisi: Phys. Rev. B **62**, 1496 (2000)
62. C.A. Perroni, V. Cataudella, G. De Filippis: J. Phys.: Condensed Matter **16**, 1593 (2004)
63. V. Cataudella, G. De Filippis, C.A. Perroni: Single polaron properties in different electron phonon models. In: this volume.
64. E. Jeckelmann, S.R. White: Phys. Rev. B **57**, 6376 (1998)
65. E. Jeckelmann, C. Zhang, S.R. White: Phys. Rev. B **60**, 7950 (1999)
66. C. Zhang, E. Jeckelmann, S.R. White: Phys. Rev. B **60**, 14092 (1999)
67. T. Holstein: Ann. Phys. **8**, 325 (1959); Ann. Phys. **8**, 343 (1959)
68. G. Wellein, H. Fehske, A.P. Kampf: Phys. Rev. Lett. **81**, 3956 (1998)
69. A. Weiße, H. Fehske: Phys. Rev. B **58**, 13526 (1998)
70. A.S. Alexandrov: Phys. Rev. B **53**, 2863 (1996); Phys. Rev. Lett. **82**, 2620 (1999)
71. A.S. Alexandrov, P.E. Kornilovitch: J. Phys.: Condensed Matter **14**, 5337 (2002)
72. P.E. Kornilovitch: Phys. Rev. B **71**, 094301 (2005)
73. M. Boninsegni, D.M. Ceperley: Phys. Rev. Lett. **74**, 2288 (1995)
74. K. Binder, ed: *Monte Carlo Methods in Statistical Physics*, Topics in Current Physics No. 7 (Springer, New York 1978)
75. K. Binder, ed: *Applications of the Monte Carlo Method in Statistical Physics*, Topics in Current Physics No. 36 (Springer, New York 1984)
76. K. Binder, ed: *The Monte Carlo Method in Condensed Matter Physics*, Topics in Applied Physics No. 71 (Springer, New York 1995)
77. D.M. Ceperley: Rev. Mod. Phys. **67**, 279 (1995)
78. W.M.C. Foulkes, L. Mitas, R.J. Needs, G. Rajagopal: Rev. Mod. Phys. **73**, 33 (2001)
79. M.C. Gordillo, D.M. Ceperley: Phys. Rev. B **58**, 6447 (1998)
80. G.H. Bauer, D.M. Ceperley, N. Goldenfeld: Phys. Rev. B **61**, 9055 (2000)
81. D.M. Ceperley, B. Bernu: Phys. Rev. Lett. **93**, 155303 (2004)
82. B.K. Clark, D.M. Ceperley: Phys. Rev. Lett. **96**, 105302 (2006)
83. N.V. Prokof'ev, B.V. Svistunov, I.S. Tupitsyn: Zh. Eksp. Teor. Fiz. **114**, 570 (1998) [Sov. Phys. JETP **87**, 310 (1998)]
84. R.P. Feynman: *Statistical mechanics* (Benjamin, Reading 1972) pp 78–84 (chapter 3)
85. P.E. Kornilovitch: Phys. Rev. B **71**, 094301 (2005)
86. A.S. Alexandrov: Phys. Rev. B **46**, 14932 (1992)
87. D.M. Eagles: Phys. Rev. **145**, 645 (1966); Phys. Rev. **181**, 1278 (1969)
88. T. Timusk et al. In: *Anharmonic Properties of High-T_c Cuprates*, ed by D. Mihailović et al (World Scientific, Singapore, 1995) pp 171
89. S.V. Tjablikov: Zh. Eksp. Teor. Fiz. **23**, 381 (1952)
90. I.G. Lang, Yu.A. Firsov: Zh. Eksp. Teor. Fiz. **43**, 1843 (1962) [Sov. Phys. JETP **16**, 1301 (1963)]
91. S.L. Cooper, K.E. Gray: Anisotropy and interlayer coupling in the high-T_c cuprates. In: *Physical Properties of High-Temperature Superconductors*, vol IV, ed by D.M. Ginsberg (World Scientific, Singapore 1994) pp 61–188

92. P.B. Allen, W.E. Pickett, H. Krakauer: Phys. Rev. B **36**, 3926 (1987)
93. W.E. Pickett: Rev. Mod. Phys. **61**, 433 (1989)
94. A.S. Alexandrov, V.V. Kabanov, N.F. Mott: Phys. Rev. Lett. **77**, 4796 (1996)
95. W. Stephan: Phys. Rev. B **54**, 8981 (1996)
96. Y.B. Levinson, E.I. Rashba: Rep. Prog. Phys. **36**, 1499 (1973)
97. G. Zhao, M.B. Hunt, H. Keller, K.A. Müller: Nature **385**, 236 (1997)
98. R. Khasanov, D.G. Eshchenko, H. Luetkens, E. Morenzoni, T. Prokscha, A. Suter, N. Garifianov, M. Mali, J. Roos, K. Conder, H. Keller: Phys. Rev. Lett. **92**, 057602 (2004)
99. Guo-meng Zhao: Polarons in colossal magnetoresistive and high-temperature superconducting materials. In: this volume
100. J.G. Bednorz, K.A. Müller: Rev. Mod. Phys. **60**, 585 (1988)
101. D. Mihailovic, V.V. Kabanov: Phys. Rev. B **63**, 054505 (2001)
102. V.V. Kabanov, D. Mihailovic: Phys. Rev. B **65**, 212508 (2002)
103. T. Mertelj, V.V. Kabanov, D. Mihailovic: Phys. Rev. Lett. **94**, 147003 (2005)
104. J. Kanamori: J. Appl. Phys. **31**, S14 (1960)
105. P.B. Allen, V. Perebeinos: Phys. Rev. B **60**, 10747 (1999)
106. Y. Takada: Phys. Rev. B **61**, 8631 (2000)
107. W.P. Su, J.R. Schrieffer, A.J. Heeger: Phys. Rev. Lett. **42**, 1698 (1979)
108. A.S. Alexandrov, N.F. Mott: *Polarons and Bipolarons* (World Scientific, Singapore 1995)
109. J.T. Devreese: Polarons. In: *Encyclopedia of Applied Physics*, vol 14 (VCH Publishers 1996) pp 383–413
110. H. Hiramoto, Y. Toyozawa: J. Phys. Soc. Japan **54**, 245 (1985)
111. D. Emin, J. Ye, C.L. Beckel: Phys. Rev. B **46**, 10710 (1992)
112. P.W. Anderson: Phys. Rev. Lett. **34** 953 (1975)
113. A.S. Alexandrov, J. Ranninger: Phys. Rev. B **23**, 1796 (1981); Phys. Rev. B **24**, 1164 (1981)
114. A.S. Alexandrov: Phys. Rev. B **48**, 10571 (1993)
115. A.S. Alexandrov, N.F. Mott: *High-Temperature Superconductors and Other Superfluids* (Taylor & Fransis Springer, London 1994)
116. A.S. Alexandrov: *Theory of Superconductivity: From Weak to Strong Coupling* (IoP Publishing, Bristol and Philadelphia, 2003)
117. A.S. Alexandrov: Superconducting polarons and bipolarons. In: this volume
118. A. Weiße, H. Fehske, G. Wellein, A.R. Bishop: Phys. Rev. B **62**, 747 (2000)
119. M. Hohenadler, M. Aichhorn, W. von der Linden: Phys. Rev. B **71**, 014302 (2005)
120. A.S. Alexandrov, V.V. Kabanov: Sov. Phys. Solid State **28**, 631 (1986)
121. M. Capone, S. Ciuchi: Phys. Rev. Lett. **91**, 186405 (2003)
122. D.J. Scalapino, S.R. White, S. Zhang: Phys. Rev. B **47**, 7995 (1993)
123. M. Jarrell, J. Gubernatis: Phys. Rep. **269**, 133 (1996)

Path Integral Methods in the Su–Schrieffer–Heeger Polaron Problem

Marco Zoli

Istituto Nazionale Fisica della Materia - Dipartimento di Fisica, Universitá di Camerino, 62032 Camerino, Italy `marco.zoli@unicam.it`

Summary. I propose a path integral description of the Su–Schrieffer–Heeger Hamiltonian, both in one and two dimensions, after mapping the real space model onto the time scale. While the lattice degrees of freedom are classical functions of time and are integrated out exactly, the electron particle paths are treated quantum mechanically. The method accounts for the variable range of the electronic hopping processes. The free energy of the system and its temperature derivatives are computed by summing at any T over the ensemble of relevant particle paths which mainly contribute to the total partition function. In the low T regime, the *heat capacity over T* ratio shows an upturn peculiar to a glass-like behavior. This feature is more sizeable in the square lattice than in the linear chain as the overall hopping potential contribution to the total action is larger in higher dimensionality. The effects of the electron-phonon anharmonic interactions on the phonon subsystem are studied by the path integral cumulant expansion method.

1 Introduction

There has been growing interest towards polarons over recent years particularly in view of the technological potential of polymers, organic molecules [1–7], carbon nanotubes [8] and high T_c superconductors [9] in which polaronic properties, essentially dependent on the type and strength of the electron-phonon coupling [10–12], have been widely discussed.

One dimensional (1D) systems with a half-filled band undergo a structural distortion [13] which increases the elastic energy and opens a gap at the Fermi surface thus lowering the electronic energy. The competition between lattice and electronic subsystems stabilizes the 1D structure which accordingly acquires semiconducting properties whereas the behavior of the 3D system would be metallic-like.

Conjugated polymers, for example polyacetylene, show anisotropic electrical and optical properties [14] due to intrinsic delocalization of π electrons along the chain of CH units. As the intrachain bonding between adjacent CH monomers is much stronger than the interchain coupling the lattice is quasi-1D. Hence, as a result of the Peierls instability, *trans*-polyacetylene shows an alternation of short and long neighboring carbon bonds, a dimerization, accompanied by a two fold degenerate ground state energy. Charge injection in polymers induces a lattice distortion with the associated formation of localized excitations, polarons and/or charged solitons. While the latter can exist only in *trans*-polyacetylene, the former solutions are more general since they do not require such degeneracy.

The Su–Schrieffer–Heeger (SSH) model Hamiltonian [15] has become a successful tool in polymer physics as it hosts the peculiar ground state excitations of the 1D conjugated structure and it accounts for a broad range of polymer properties [16–18]. In the weak coupling regime a continuum version [19] of the SSH model has been developed and polaronic solutions have been obtained analytically [20]. Although the 1D properties of the SSH model have been mainly investigated so far [21–28], extensions to two dimensions were considered in the late eighties in connection with the Fermi surface nesting effect on quasi 2D high T_c superconductors [29]. Recent numerical analysis [30] have revealed the rich physics of 2D-SSH polarons whose mass seems to be larger than in the 1D case, at least in the intermediate regime of the adiabatic parameter [31].

As a fundamental feature of the SSH Hamiltonian the electronic hopping integral linearly depends on the relative displacement between adjacent atomic sites thus leading to a nonlocal *e-ph* coupling [32, 33] with vertex function depending both on the electronic and the phononic wave vector. The latter property induces, in the Matsubara formalism [34], an electron hopping associated with a time dependent lattice displacement. As a consequence time retarded electron-phonon interactions arise in the system yielding a source current which depends both on time and on the electron path coordinates. This causes large *e-ph* anharmonicities in the equilibrium thermodynamics of the SSH model [35]. Hopping of electrons from site to site accompanied by a coupling to the lattice vibration modes is a fundamental process determining the transport [36] and equilibrium properties of many body systems [37]. A variable range hopping may introduce some degree of disorder thus affecting the charge mobility [38, 39] and the thermodynamic functions.

These issues are hereafter analyzed by means of the path integral formalism [40–45] which fully accounts for the time retarded *e-ph* interactions as a retarded potential naturally emerges in the exact integral action [46]. Using the main property of the *e-ph* coupling model, the Hamiltonian linearly dependent on the phonon displacement field, I attack the SSH model after introducing a generalized version of the semiclassical model which treats only the electrons quantum mechanically. Being valid for any *e-ph* coupling value, the path integral method allows one to derive the partition function and the related temperature derivatives without those limitations which affect the perturbative studies. The general formalism, both for the 1D and 2D system, is described in Sect. 2 while the path integral approach used to derive the full partition function of the interacting system is developed in Sect. 3. In Sect. 4, I outline the main features of the computational method and derive the thermo-dynamical properties for a particle described by a SSH Hamiltonian in a bath of harmonic oscillators. In Sect. 5, the model is generalized in order to include the electron-phonon effects on the oscillator bath: the anharmonic corrections to the total heat capacity are evaluated by a path integral cumulant expansion in terms of the source current of the Hamiltonian model. Some conclusions are drawn in Sect. 6.

2 The Hamiltonian Model

In a square lattice with isotropic nearest neighbors hopping integral J, the SSH Hamiltonian for electrons plus *e-ph* interactions reads:

$$
\begin{aligned}
H = \quad & \sum_{r,s}\Big[(J_{r,s})_x \big(f^{\dagger}_{r+1,s}f_{r,s} + h.c.\big) \\
& + (J_{r,s})_y \big(f^{\dagger}_{r,s+1}f_{r,s} + h.c.\big)\Big] \\
(J_{r,s})_x = \quad & -\frac{1}{2}\big[J + \alpha \Delta u_x\big] \\
(J_{r,s})_y = \quad & -\frac{1}{2}\big[J + \alpha \Delta u_y\big] \\
\Delta u_x = \quad & u_x(r+1,s) - u_x(r,s) \\
\Delta u_y = \quad & u_y(r,s+1) - u_y(r,s) \quad ,
\end{aligned}
\tag{1}
$$

where α is the electron-phonon coupling, $\mathbf{u}(r,s)$ is the dimerization coordi-nate indicating the displacement of the monomer group on the (r,s)–lattice site, $f^{\dagger}_{r,s}$ and $f_{r,s}$ create and destroy electrons (i.e., π band electrons in poly-acetylene). The phonon Hamiltonian is given by a set of 2D classical harmonic oscillators. The two addenda in (1) deal with one dimensional *e-ph* couplings along the x and y axis respectively, with first neighbors electron hopping. Second neighbors hopping processes (with overlap integral $J^{(2)}$) may be ac-counted for by adding to the Hamiltonian the term $H^{(2)}$ such that

$$
\begin{aligned}
H^{(2)} = \quad & (J_{r,s})_{x,y}\big(f^{\dagger}_{r+1,s+1}f_{r,s} + h.c.\big) \\
(J_{r,s})_{x,y} = \quad & -\frac{1}{2}\Big[J^{(2)} + \alpha\sqrt{(\Delta u_x)^2 + (\Delta u_y)^2}\Big] \quad .
\end{aligned}
\tag{2}
$$

The 1D SSH Hamiltonian is obtained by (1) just dropping the second term depending on the y coordinate. The real space Hamiltonian in (1) can be transformed into a time dependent Hamiltonian [47] by introducing the electron coordinates: i) $\big(x(\tau'), y(\tau')\big)$ at the (r, s) lattice site, ii) $\big(x(\tau), y(\tau')\big)$ at the $(r + 1, s)$ lattice site and iii) $\big(x(\tau'), y(\tau)\big)$ at the $(r, s + 1)$ lattice site, respectively. τ and τ' vary on the scale of the inverse temperature β. The spatial *e-ph* correlations contained in (1) are mapped onto the time axis by changing: $u_{x,(y)}(r, s) \to u_{x,(y)}(\tau')$, $u_x(r + 1, s) \to u_x(\tau)$ and $u_y(r, s + 1) \to u_y(\tau)$. Now we set $\tau' = 0$, $\big(x(\tau'), y(\tau')\big) \equiv (0, 0)$, $\big(u_x(\tau'), u_y(\tau')\big) \equiv (0, 0)$. Accordingly, (1) transforms into the time dependent Hamiltonian:

$$
\begin{aligned}
H(\tau) = {} & J_x(\tau)\Big(f^\dagger(x(\tau), 0) f(0, 0) + h.c. \Big) \\
& + J_y(\tau)\Big(f^\dagger(0, y(\tau)) f(0, 0) + h.c. \Big) \\
J_x(\tau) = {} & -\frac{1}{2}\big[J + \alpha u_x(\tau) \big] \\
J_y(\tau) = {} & -\frac{1}{2}\big[J + \alpha u_y(\tau) \big] \quad .
\end{aligned}
$$

$$(3)$$

While the ground state of the 1D SSH Hamiltonian is twofold degenerate, the degree of phase degeneracy is believed to be much higher in 2D [48] as many lattice distortion modes contribute to open the gap at the Fermi surface. However, as in 1D, these phases are connected by localized and nonlinear excitations, the soliton solutions. Thus, also in 2D both electron hopping between solitons [49] and thermal excitation of electrons to band states may take place within the model. These features are accounted for by the time dependent version of the Hamiltonian. As τ varies continuously on the β scale and the τ-dependent displacement fields are continuous variables (whose amplitudes are in principle unbound in the path integral), long range hopping processes are automatically included in $H(\tau)$ which is therefore more general than the real space SSH Hamiltonian in (1) (and (2)). Thus, by means of the path integral formalism we look at the low temperature thermodynamical behavior both in 1D and 2D searching for those features which may be ascribable to some local disorder related to the variable range of the hopping processes.

The semiclassical nature of the model is evident from (3) in which quantum mechanical degrees of freedom interact with the classical variables $u_{x(y)}(\tau)$. Averaging the electron operators over the ground state we obtain the time dependent semiclassical energy per lattice site N which is linear in the atomic displacements:

$$
\begin{aligned}
\frac{< H(\tau) >}{N} = {} & J_x(\tau) P\big(J, \tau, x(\tau) \big) + J_y(\tau) P\big(J, \tau, y(\tau) \big) \\
P\big(J, \tau, \mathbf{v}(\tau) \big) = {} & \frac{1}{\pi^2} \int d\mathbf{k} \cos[\mathbf{k} \cdot \mathbf{v}(\tau)] \cosh(\epsilon_\mathbf{k} \tau) n_F(\epsilon_\mathbf{k}) \quad ,
\end{aligned}
$$

$$(4)$$

with $\mathbf{v}(\tau) = (x(\tau), 0)$ and $\mathbf{v}(\tau) = (0, y(\tau))$ in the first and second term respectively. $\epsilon_{\mathbf{k}} = -J\sum_{i=x,y}\cos(k_i)$ is the electron dispersion relation and n_F is the Fermi function. The chemical potential has been pinned to the zero energy level. Equation (4) can be rewritten in a way suitable to the path integral approach by defining

$$
\begin{aligned}
\frac{<H(\tau)>}{N} &= V\big(x(\tau)\big) + V\big(y(\tau)\big) + \mathbf{u}(\tau) \cdot \mathbf{j}(\mathbf{v}(\tau)) \\
V\big(x(\tau)\big) &= -JP\big(J, \tau, x(\tau)\big) \\
V\big(y(\tau)\big) &= -JP\big(J, \tau, y(\tau)\big) \\
\mathbf{j}(\mathbf{v}(\tau)) &= -\alpha P\big(J, \tau, \mathbf{v}(\tau)\big) \\
\mathbf{u}(\tau) &= \big(u_x(\tau), u_y(\tau)\big) \quad .
\end{aligned}
$$

$$(5)$$

$V\big(x(\tau)\big)$ and $V\big(y(\tau)\big)$ are the effective terms accounting for the τ dependent electronic hopping while $\mathbf{j}(\mathbf{v}(\tau))$ is interpreted as the external source [46] current for the oscillator field $\mathbf{u}(\tau)$. Averaging the electrons over the ground state we neglect the fermion-fermion correlations [50] which lead to effective polaron-polaron interactions in non-perturbative analysis of the model [51]. This approximation however is not expected to substantially affect the following calculations.

3 The Path Integral Formalism

Taking a bath of $\bar{N}$ 2D oscillators, we write the SSH electron path integral, $\zeta(\tau) \equiv (x(\tau), y(\tau))$, as:

$$
\begin{aligned}
<\zeta(\beta)\mid\zeta(0)> &= \prod_{i=1}^{\bar{N}} \int D\mathbf{u}_i(\tau) \int D\zeta(\tau) \\
&\quad \cdot \exp\left[-\int_0^\beta d\tau \sum_{i=1}^{\bar{N}} \frac{M}{2}\big(\dot{\mathbf{u}}_i^2(\tau) + \omega_i^2 \mathbf{u}_i^2(\tau)\big)\right] \\
&\quad \cdot \exp\left[-\int_0^\beta d\tau\left(\frac{m}{2}\dot{\zeta}^2(\tau) + V\big(x(\tau)\big) + V\big(y(\tau)\big)\right)\right. \\
&\quad \left.+ \sum_{i=1}^{\bar{N}} \mathbf{u}_i(\tau) \cdot \mathbf{j}(\mathbf{v}(\tau))\right] \quad ,
\end{aligned}
$$

$$(6)$$

where m is the electron mass, M is the atomic mass and ω_i is the oscillator frequency. As a main feature we notice that the interacting energy is linear in the atomic displacement field. Then, the electronic path integral can be derived after integrating out the oscillator degrees of freedom which are decoupled along the x and y axis. Accordingly, we get:

$$< \zeta(\beta)|\zeta(0) >= \prod_{i=1}^{\bar{N}} Z_i \left[\int Dx(\tau) \right.$$

$$\cdot \, exp\left[- \int_0^\beta d\tau \left(\frac{m}{2} \dot{x}^2(\tau) + V(x(\tau)) \right) - \frac{1}{\hbar} A(x(\tau)) \right]^2 ,$$

$$A(x(\tau)) = - \frac{\hbar^2}{4M} \sum_{i=1}^{\bar{N}} \frac{1}{\hbar\omega_i \sinh(\hbar\omega_i\beta/2)}$$

$$\cdot \int_0^\beta d\tau j(x(\tau)) \int_0^\beta d\tau'' \cosh\left(\hbar\omega_i \left(|\tau - \tau''| - \beta/2 \right) \right) j(x(\tau'')),$$

$$Z_i = \left[\frac{1}{2\sinh(\hbar\omega_i\beta/2)} \right]^2 .$$

$$(7)$$

Thus, the 2D electron path integral is obtained after squaring the sum over one dimensional electron paths. This permits us to reduce the computational problem which is nonetheless highly time consuming particularly in the low temperature limit. Note in fact that the source current $j(x(\tau))$ requires integration over the 2D Brillouin Zone (BZ) according to (4) and (5) and this occurs for any choice of the electron path coordinates. The quadratic (in the coupling α) source action $A(x(\tau))$ is time retarded as the particle moving through the lattice drags the excitations of the oscillator fields which take time to readjust to the electron motion. When the interaction is sufficiently strong the conditions for polaron formation may be fulfilled in the system according to the degree of adiabaticity [52]. However the present path integral description is valid independent of the existence of polarons as it applies also to the weak coupling regime. Assuming periodic conditions $x(\tau) = x(\tau + \beta)$, the particle paths can be expanded in Fourier components

$$x(\tau) = x_o + \sum_{n=1}^{\infty} 2\left(\mathrm{Re}\, x_n \cos(\omega_n\tau) - \mathrm{Im}\, x_n \sin(\omega_n\tau) \right)$$

$$\omega_n = 2\pi n/\beta \qquad\qquad (8)$$

and the open ends integral over the paths $\int Dx(\tau)$ transforms into the measure of integration $\oint Dx(\tau)$. Taking:

$$\oint Dx(\tau) \equiv \int\limits_{-\infty}^{\infty} \frac{dx_o}{\left(2\pi\hbar^2/mK_BT\right)^{(1/2)}} \prod_{n=1}^{\infty} \left[\frac{\int_{-\infty}^{\infty} d\mathrm{Re}\; x_n \int_{-\infty}^{\infty} d\mathrm{Im}\; x_n}{\left(\pi\hbar^2 K_BT/m\omega_n^2\right)} \right] \quad , \quad (9)$$

we proceed to integrate (7) in order to derive the full partition function of the system versus temperature.

Let us point out that, by mapping the electronic hopping motion onto the time scale, a continuum version of the interacting Hamiltonian (3) has been de facto introduced. However, unlike previous [14] approaches, our path integral method is not constrained to the weak *e-ph* coupling regime and it can be applied to any range of physical parameters.

4 Computational Method and Thermodynamical Results

As a preliminary step we determine, for a given path and at a given temperature: i) the minimum number $(N_{\mathbf{k}})$ of $\mathbf{k}$–points in the BZ to accurately estimate the average interacting energy per lattice site and, ii) the minimum number (N_τ) of points in the double time integration to get a numerically stable source action in (7). The momentum integrations required by (4) converge by summing over 1600 and 70 points in the reduced 2D and 1D BZ, respectively. Moreover, $N_\tau = 300$ at $T = 1K$.

Computation of (4) - (7) requires fixing two sets of input parameters. The first set contains the physical quantities characterizing the system: the bare hopping integral J, the oscillator frequencies ω_i and the effective coupling $\chi = \alpha^2\hbar^2/M$ (in units meV^3). The second set defines the paths for the particle motion which mainly contribute to the partition function through: the number of pairs $(\mathrm{Re}\; x_n, \mathrm{Im}\; x_n)$ in the Fourier expansion of (8), the cutoff (Λ) on the integration range of the expansion coefficients in (9) and the related number of points (N_Λ) in the measure of integration which ensures numerical convergence.

After introducing a dimensionless path

$$x(\tau)/a = \bar{x}_o + \sum_{n=1}^{N_p} \left(\bar{a}_n \cos(\omega_n\tau) + \bar{b}_n \sin(\omega_n\tau) \right) \quad , \quad (10)$$

with: $\bar{x}_o \equiv x_o/a$, $\bar{a}_n \equiv 2\mathrm{Re}\; x_n/a$ and $\bar{b}_n \equiv -2\mathrm{Im}\; x_n/a$, the functional measure of (9) can be rewritten for computational purposes as:

$$\oint Dx\,(\tau) \approx \frac{a}{\sqrt{2}} \left(\frac{a}{2}\right)^{2N_p} \frac{(2\pi \cdot 4\pi \cdots 2N_p\pi)^2}{(\pi\hbar^2/mK_BT)^{N_p+1/2}} \int\limits_{-\Lambda/a}^{\Lambda/a} d\bar{x}_o \cdot$$

$$\int\limits_{-2\Lambda/a}^{2\Lambda/a} d\bar{a}_1 \int\limits_{-2\Lambda/a}^{2\Lambda/a} d\bar{b}_1 \cdots \int\limits_{-2\Lambda/a}^{2\Lambda/a} d\bar{a}_{N_p} \int\limits_{-2\Lambda/a}^{2\Lambda/a} d\bar{b}_{N_p} \quad .$$

$$(11)$$

In the following we take the lattice constant $a = 1\mathring{A}$. As a criterion to set the cutoff Λ on the integration range, we notice that the functional measure normalizes the kinetic term in (7):

$$\oint Dx(\tau)exp\left[-\frac{m}{2} \int\limits_0^\beta d\tau \dot{x}^2(\tau)\right] \equiv 1 \tag{12}$$

and this condition holds for any number of pairs N_p truncating the Fourier expansion in (9). Then, taking $N_p = 1$, the left hand side of (12) transforms as:

$$\oint Dx(\tau)exp\left[-\frac{m}{2} \int\limits_0^\beta d\tau \dot{x}^2(\tau)\right] \simeq \frac{4}{\pi}\left[\int\limits_0^U dyexp(-y^2)\right]^2$$

$$U \equiv \sqrt{2\pi^3}\frac{\Lambda}{\lambda}$$

$$\lambda = \sqrt{\frac{2\pi\hbar^2}{mK_BT}} \quad . \tag{13}$$

Using the series representation [53]

$$\int\limits_0^U dyexp(-y^2) = \sum_{k=0}^\infty \frac{(-1)^k U^{2k+1}}{k!(2k+1)} \quad , \tag{14}$$

one determines U (after setting the series cutoff k_{max} which ensures convergence) by fitting the Poisson integral value $\sqrt{\pi}/2$.

Thus, we find that the cutoff Λ can be expressed in terms of the thermal wavelength λ as $\Lambda \sim \lambda/\sqrt{2\pi^3}$ and hence it scales versus temperature as $\Lambda \propto 1/\sqrt{T}$. This means physically that, at low temperatures, Λ is large since many paths are required to yield the correct normalization. For example, at $T = 1K$, we get $\Lambda \sim 284\mathring{A}$. Numerical investigation of (7) shows however that a much shorter cutoff suffices to guarantee convergence in the path integral, while the cutoff temperature dependence implied by (12) holds also in the computation of the interacting partition function. The thermodynamical results hereafter presented have been obtained by taking $\Lambda \sim \lambda/(10\sqrt{2\pi^3})$. Summing in the 1D system over $N_\Lambda \sim 20/\sqrt{T}$ points for each integration range and taking $N_p = 2$, we are then evaluating the contribution of $(N_\Lambda + 1)^{2N_p+1}$ paths (the integer part of N_Λ is obviously selected at any temperature). Thus, at $T = 1K$ and in 1D, we are considering $\sim 4 \cdot 10^6$ different paths for the particle motion while, at $T = 100K$ the number of paths drops to 243 [54]. In the 2D problem, $N_\Lambda \sim 35/\sqrt{T}$ points for each Fourier coefficient are required [55]. Computation of the second order derivatives of the free energy in the range $T \in [1, 301]K$, with a spacing of 3K, takes 55 hours and 15 minutes on a Pentium 4. Note that larger N_p in the path Fourier expansion would further increase the computing time without introducing any substantial improvement in the thermodynamical output of our calculation.

Although the history of the SSH model is mainly related to wide band polymers, we take here a narrow band system ($J = 100meV$) [28] with the caveat that electron-electron correlations may become relevant in narrow bands.

Free energy and heat capacity have been first computed up to room temperature, both in 1D and 2D, assuming a bath of $\bar{N} = 10$ low energy oscillators separated by $2meV$: $\hbar\omega_1 = 2meV, ..., \hbar\omega_{10} = 20meV$. The lowest energy oscillator yields the largest contribution to the phonon partition function mainly in the low temperature regime while the ω_{10} oscillator essentially sets the phonon energy scale which determines the size of the e-ph coupling. A larger number $\bar{N}$ of oscillators in the aforegiven range would not significantly modify the calculation whereas lower ω_i values would yield a larger contribution to the phonon partition function mainly at low T.

In the discrete SSH model, the value $\bar{\alpha} \equiv 4\alpha^2/(\pi\kappa J) \sim 1$ marks the crossover between weak and strong e-ph coupling, with κ being the effective spring constant. In our continuum and semiclassical model the effective coupling is χ. Although in principle, discrete and continuum models may feature non coincident crossover parameters, we assume that the relation between α and J obtained by the discrete model crossover condition still holds in our model. Hence, at the crossover we get: $\chi_c \sim \pi J\hbar^2\omega_{10}^2/64$. This means that, in Figs. 1-3, the crossover value is set at $\chi_c \sim 2000meV^3$.

In Fig. 1, a comparison between the 1D and the 2D free energies is presented for two values of χ, one lying in the weak and one in the strong e-ph coupling regime. The oscillator free energies F_{ph} are plotted separately while the free energies arising from the total action in Eq. (7), shortly termed F_{source}, results from the competition between the free path action (kinetic term plus hopping potential in the exponential integrand) and the source action depending on the e-ph coupling. While the former enhances the free energy, the latter becomes dominant at increasing temperatures thus reducing the total free energy. In general, the 2D free energies have a larger gradient (versus temperature) than the corresponding 1D terms. The F_{ph} lie above the F_{source} both in 1D and 2D because of the choice of the $\hbar\omega_i$: in general one may expect a crossing point between F_{ph} and F_{source} with temperature

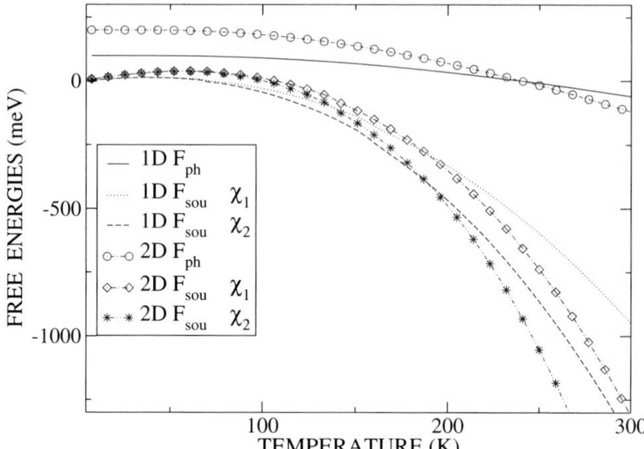

Fig. 1. Phonon (F_{ph}) and Source Term (F_{sou}) contributions to the 1D and 2D free energies for two values of the effective coupling χ: $\chi_1 = 1440meV^3$ (weak *e-ph* coupling) $\chi_2 = 2560meV^3$ (strong *e-ph* coupling). A bath of ten phonon oscillators is considered, the largest phonon energy being $\hbar\omega_{10} = 20meV$. Reproduced from [55], ©(2005) by the American Physical Society.

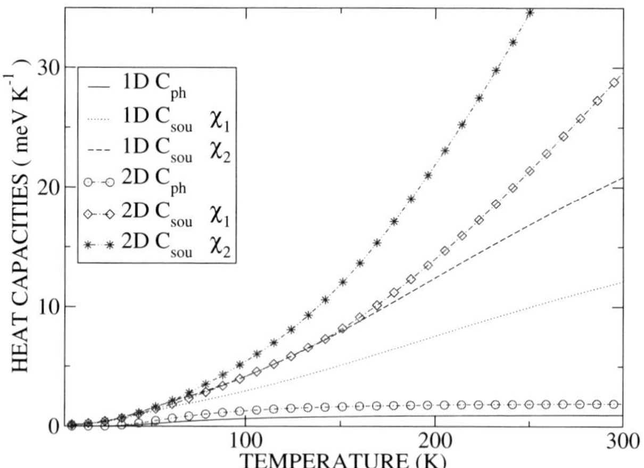

Fig. 2. Phonon and Source Term contributions (normalized over the number of oscillators) to the 1D and 2D heat capacities for the same parameters as in Fig. 1. The oscillator heat capacities are also plotted. Reproduced from [55], ©(2005) by the American Physical Society.

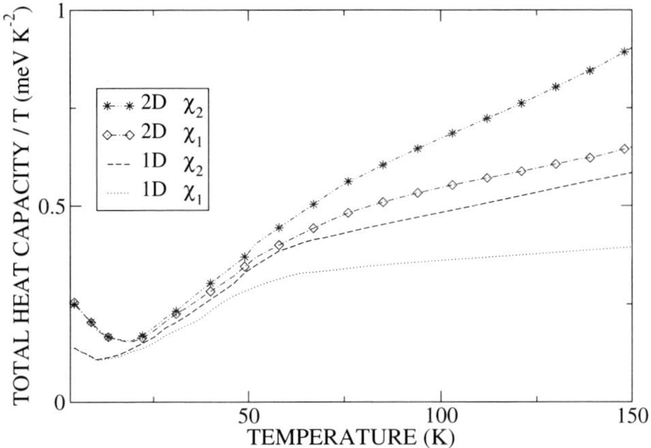

Fig. 3. Total heat capacity over temperature for the same parameters as in Fig. 1. Reproduced from [55], ©(2005) by the American Physical Society.

location depending on the value of χ. Note that the F_{source} plots have a positive temperature derivative in the low temperature regime and this feature is more pronounced in 2D. In fact, at low T, the source action is dominated by the hopping potential $V(x(\tau))$ while, at increasing T, the *e-ph* effects become progressively more important (as the bifurcation between the χ_1 and χ_2 curves shows) and the F_{source} have a negative derivative. In 2D, the weight of the $V(x(\tau))$ term is larger because there is a higher hopping probability. This physical expectation is taken into account by the path integral method. At any temperature, we monitor the ensemble of relevant particle paths over which the hopping potential is evaluated. For a selected set of Fourier components in (10) the hopping decreases by lowering T but its value is still significant at low T. Considered that: i) the total action is obtained after a $d\tau$ integration of $V(x(\tau))$ and ii) the $d\tau$ integration range is larger at lower temperatures, we explain why the overall hopping potential contribution to the total action is responsible for the anomalous free energy behavior at low T.

In Fig. 2, the heat capacity contributions due to the oscillators (C_{ph}) and electrons plus *e-ph* coupling (C_{sou}) are reported. The values are normalized over $\bar{N}$. The previously described summation over a large number of paths turns out to be essential to recover the correct thermodynamical behavior in the zero temperature limit. The dimensionality effects are seen to be large and, for a given dimensionality, the role of the *e-ph* interactions is magnified at increasing T. The total heat capacity ($C_{ph} + C_{sou}$) *over* T ratios are plotted in Fig. 3 in the low T regime to emphasize the presence of an anomalous upturn which appears at $T \preceq 10K$ in 1D and $T \preceq 20K$ in 2D. This feature in the heat capacity linear coefficient is ultimately related to the sizable effective hopping integral term $V(x(\tau))$. The strength of the *e-ph* coupling has a minor role in the low T limit although it determines the shape of the anomaly versus T.

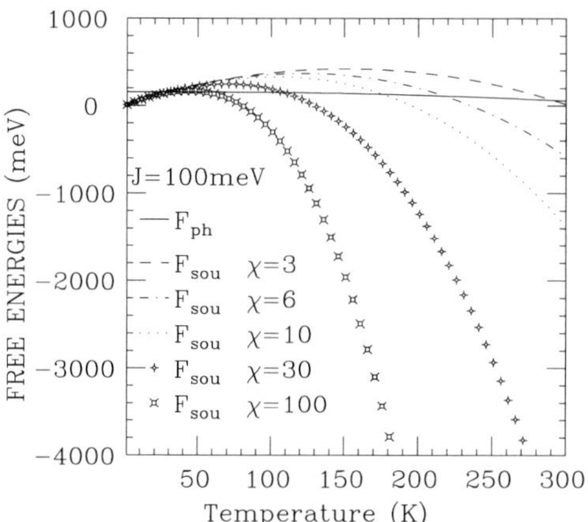

Fig. 4. 1D Phonon and Source Term contributions to the free energy for five values of the effective coupling χ (in units $10^3 meV^3$) and a narrow electron band. A bath of ten phonon oscillators has been taken, the largest phonon energy is $\hbar\omega_{10} = 40meV$. Reproduced from [54], ©(2003) by the American Physical Society.

Let us focus now on the 1D system and consider the effect of the oscillator bath on the thermodynamical properties: in Figs. 4-6 the ten phonon energies are: $\hbar\omega_1 = 22meV, ..., \hbar\omega_{10} = 40meV$. Accordingly the crossover is set at $\chi_c \sim 8000meV^3$ and three plots out of five lie in the strong *e-ph* coupling regime. As shown in Fig. 4, large χ values are required to get strongly decreasing free energies versus temperature while the $\chi = 3000meV^3$ curve now hardly intersects the phonon free energy at room temperature. Figure 5 shows the rapid growth of the source heat capacity versus temperature at strong couplings whereas the presence of the low T upturn in the *total heat capacity over T* ratio is confirmed in Fig. 6. Note that, due to the enhanced oscillators energies, the phonon heat capacity saturates here at $T \sim 400K$ (in Fig. 2, for the 1D case, $T \sim 200K$).

5 Electron-Phonon Anharmonicity

So far we have considered a bath of harmonic phonons. Now we face the following question: what is the effect of the particle-phonon interaction on the phonon subsystem?

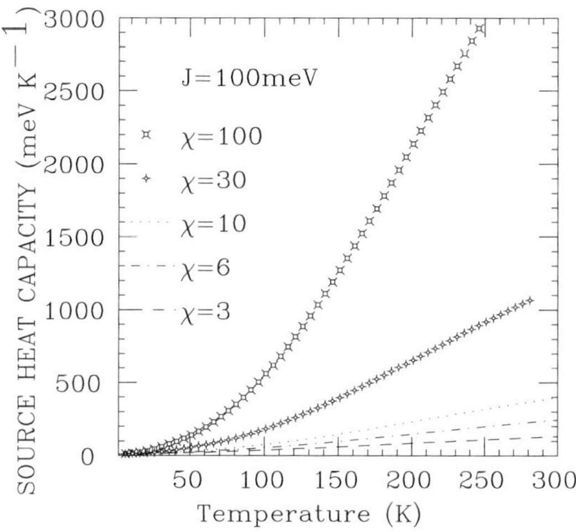

Fig. 5. Source Term contributions to the heat capacity for the same parameters as in Fig. 4. Reproduced from [54], ©(2003) by the American Physical Society.

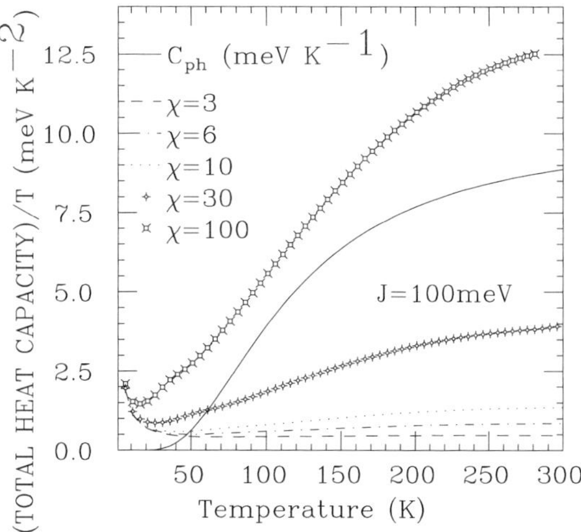

Fig. 6. Total heat capacity over temperature for the same parameters as in Fig.4. The phonon heat capacity is also plotted. Reproduced from [54], ©(2003) by the American Physical Society.

In general, the phonon partition function perturbed by a source current $j(\tau)$ can be expanded in anharmonic series as:

$$Z_{ph}[j(\tau)] \simeq Z_{ph}\left(1 + \sum_{l=1}^{k}(-1)^l < C^l >_{j(\tau)}\right) \quad ,$$

(15)

where the cumulant terms $< C^l >_{j(\tau)}$ are expectation values of powers of correlation functions of the perturbing potential. The averages are meant over the ensemble of the harmonic oscillators whose partition function is Z_{ph}.

5.1 The Holstein-Type Current

First we consider the general problem of an electron path linearly coupled to a single oscillator with energy ω and displacement $u(\tau)$, through the current $j_x(\tau) = -\alpha x(\tau)$. This type of current models the Holstein interaction [56]. In this case odd k cumulant terms vanish and the lowest order even k cumulants can be straightforwardly derived [57]. To obtain a closed analytical expression for the cumulants to any order we approximate the electron path by its τ averaged value: $< x(\tau) >\equiv \frac{1}{\beta} \int_0^\beta d\tau x(\tau) = x_0/a$ and expand the oscillator path in N_F Fourier components:

$$u(\tau) = u_o + \sum_{n=1}^{N_F} 2\Big(\mathrm{Re}\, u_n \cos(\omega_n \tau) - \mathrm{Im}\, x_n \sin(\omega_n \tau)\Big)$$

$$\omega_n = 2\pi n/\beta \quad .$$

(16)

Next we choose the measure of integration

$$\oint Du(\tau) \equiv \left(\frac{1}{2}\right)^{2N_F} \frac{\left(2\pi \cdot \cdot 2N_F \pi\right)^2}{\sqrt{2}\lambda_M^{(2N_F+1)}} \int_{-\infty}^{\infty} du_o$$

$$\times \prod_{n=1}^{N_F} \int_{-\infty}^{\infty} d\mathrm{Re}\, u_n \int_{-\infty}^{\infty} d\mathrm{Im}\, u_n \quad ,$$

(17)

being $\lambda_M = \sqrt{\pi\hbar^2\beta/M}$. Such a measure normalizes the kinetic term in the oscillator field action

$$\oint Du(\tau)exp\left[-\frac{M}{2}\int_0^\beta d\tau\dot{u}^2(\tau)\right] \equiv 1 \quad .$$

(18)

Then, using (15) - (17), we obtain for the $k - th$ cumulant

$$< C^k >_{N_F} = Z_{ph}^{-1} \frac{(\alpha_R \beta \lambda_M)^k (k-1)!!}{k! \pi^{k/2} (\omega\beta)^{k+1}} \prod_{n=1}^{N_F} \frac{(2n\pi)^2}{(2n\pi)^2 + (\omega\beta)^2}$$

$$\alpha_R = \alpha x_0/a \quad . \tag{19}$$

Let us set $x_0/a = 0.1$ in the following calculations thus reducing the effective coupling α_R by one order of magnitude with respect to the bare value. However, the trend shown by the results hereafter presented does not depend on this choice since x_0/a and α can be varied independently. As the cumulants should be stable against the number of Fourier components in the oscillator path expansion, using Eq. (19) we set the minimum N_F through the condition $2N_F\pi \gg \omega\beta$. The thermodynamics of the anharmonic oscillator can be computed by the cumulant corrections to the harmonic phonon free energy:

$$F^{(k)}(T) = -\frac{1}{\beta} \ln\left[1 + \sum_{l=1}^{k} < C^{2l} >_{N_F} \right] \quad . \tag{20}$$

To proceed one needs a criterion to find the temperature dependent cutoff k^* in the cumulant series. We feel that, in the low T limit, the third law of thermodynamics may offer the suitable constraint to determine k^*. Then, given α and ω, the program searches for the cumulant order such that the phonon heat capacity and the entropy tend to zero in the zero temperature limit. At any finite temperature T, the constant volume heat capacity is computed as

$$C_V^{(k)}(T) = -\left[F^{(k)}(T + 2\Delta) - 2F^{(k)}(T + \Delta) + F^{(k)}(T) \right]$$

$$\times \left(\frac{1}{\Delta} + \frac{T}{\Delta^2} \right) \quad , \tag{21}$$

Δ being the incremental step and k^* is determined as the minimum value for which the heat capacity converges with an accuracy of 10^{-4}. Figs. 7(a) and 7(b) show phonon heat capacity and free energy respectively in the case of a low energy oscillator for an intermediate value of e-ph coupling. Harmonic functions, anharmonic functions with second order cumulant and anharmonic functions with k^* corrections are reported on in each figure. The second order cumulant is clearly inadequate to account for the low temperature trend yielding a negative phonon heat capacity below $\sim 40K$ while at high T the second order cumulant contribution tends to vanish. Instead, the inclusion of k^* terms in (20), (21) leads to the correct zero temperature limit although there is no visible anharmonic effect on the phonon heat capacity throughout the whole temperature range being $C_V^{(k^*)}$ perfectly superimposed on the harmonic C_V^h. Note in Fig. 7(b) that the k^* corrections simply shift the free energy downwards without changing its slope versus temperature. By increasing α_R, the low T range with wrong (negative) $C_V^{(2)}$ broadens whereas the k^*

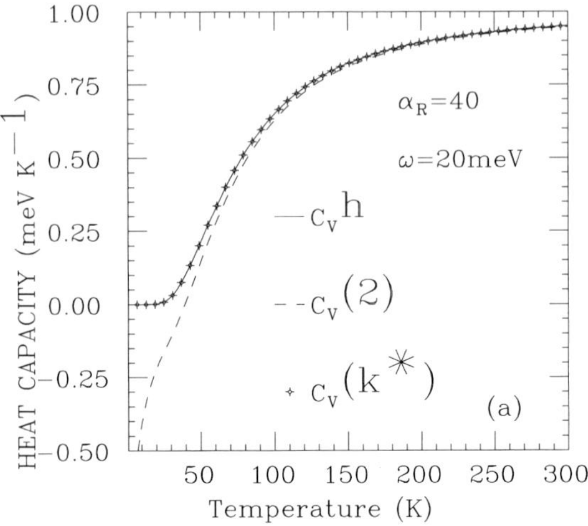

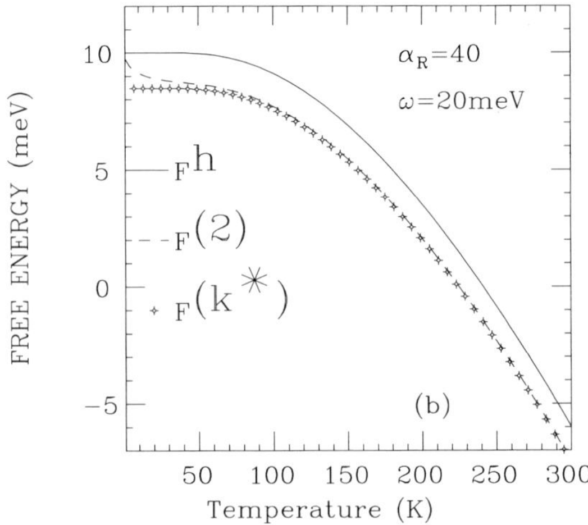

Fig. 7. (a) 1D Phonon heat capacity and (b) 1D Phonon free energy calculated in i) the harmonic model, ii) anharmonic model with second order cumulant, iii) anharmonic model with k^* cumulants (see text). α_R is the effective *e-ph* coupling in units $meV \, \mathring{A}^{-1}$ and ω is the phonon energy.

contributions permit us to fulfill the zero temperature constraint and substantially lower the phonon free energy. Thus, for the particular choice of constant (in τ) source current we find that the *e-ph* anharmonicity renormalizes the phonon partition function although no change occurs in the thermodynamical behavior of the free energy derivatives. Anharmonicity is essential to stabilize the system but it leaves no trace in the heat capacity [58]. Figure 8(a) displays the k^* temperature dependence for three choices of *e-ph* coupling in the case of a low energy oscillator: while, at high T, the number of required cumulants ranges between six and ten according to the coupling, k^* strongly grows at low temperatures reaching the value 100 at $T = 1K$ for $\alpha_R = 60 meV\,\mathring{A}^{-1}$. The k^* versus α_R behavior is depicted in Fig. 8(b) for three selected temperatures: at low T the cutoff varies strongly with the strength of the coupling while, by enhancing T, the number of cumulant terms in the series is smaller and becomes much less dependent on α_R.

5.2 The SSH-Type Current

Next we turn to the computation of the equilibrium thermodynamics of the phonon subsystem perturbed by the source current of the semiclassical SSH model described in Sects. 2 and 3. Assuming that the electron particle path interacts with each of the $\bar{N}$ oscillators through the coupling α (taken independent of i), we write the kth cumulant term as

$$
< C^k >_{j(\tau)} = Z_{ph}^{-1} \prod_{i=1}^{\bar{N}} \oint Du_i(\tau) \frac{1}{k!} \prod_{l=1}^{k} \left[\int_0^\beta d\tau_l u_i(\tau_l) j(\tau_l) \right]^l
$$

$$
\times\ exp\left[-\int_0^\beta d\tau \sum_{i=1}^{\bar{N}} \frac{M_i}{2}\left(\dot{u_i}^2(\tau) + \omega_i^2 u_i^2(\tau) \right) \right] ,
$$

$$
\tag{22}
$$

where $j(\tau)$ is given in (5). We take here the 1D system. Since the oscillators are in fact decoupled in our model (and anharmonic effects mediated by the electron particle path are neglected) the behavior of the cumulant terms $< C^k >_{j(\tau)}$ can be studied by selecting a single oscillator having energy ω and displacement $u(\tau)$.

As the electron propagator depends on the bare hopping integral we set as before $J = 100 meV$. Any electron path yields in principle a different cumulant contribution. Numerical investigation shows however that convergent k–order cumulants are achieved by taking $N_p = 2$ Fourier components in the electron path expansion and summing over $\sim 5^{2N_p+1}$ electron paths.

As in the case of Sect. 5.1, we truncate the cumulant series by invoking the third law of thermodynamics to determine the cutoff k^* in the low temperature limit and by searching for numerical convergence on the first and second free

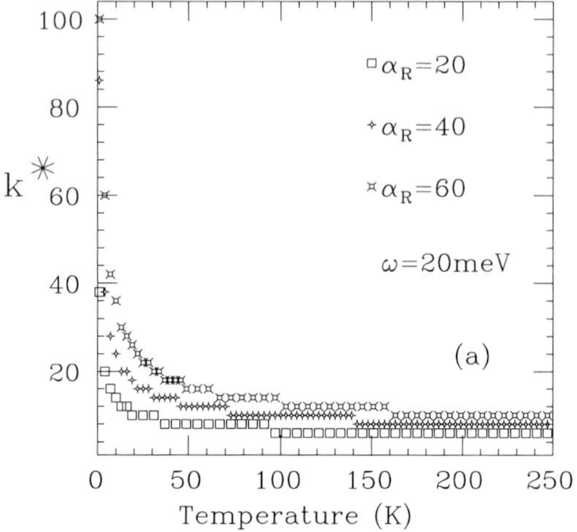

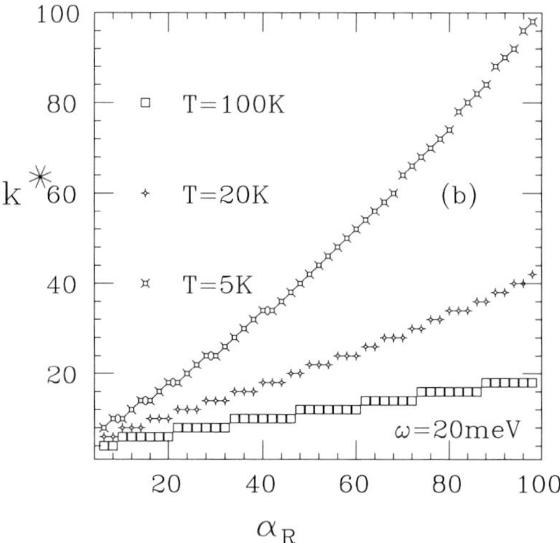

Fig. 8. (a) Number of cumulants required to obtain a convergent phonon heat capacity at any temperature for different choices of *e-ph* couplings. (b) Number of cumulants yielding a convergent phonon heat capacity at any *e-ph* coupling for three selected temperatures.

energy derivatives at any finite temperature. Again, we can start our analysis from (20) after checking that odd k cumulants yield vanishing contributions. Now however the picture of the anharmonic effects changes drastically. The

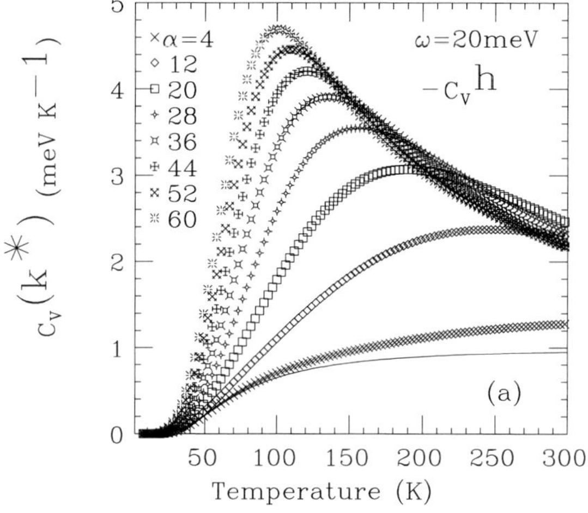

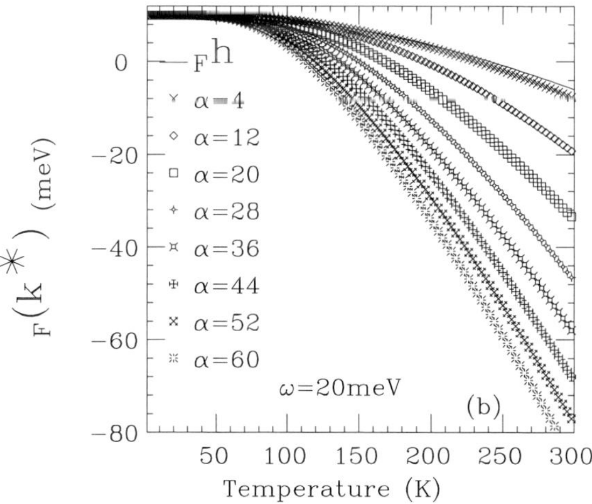

Fig. 9. 1D Anharmonic (a) phonon heat capacity and (b) free energy versus temperature for eight values of *e-ph* coupling. The harmonic plots are also reported. A low energy oscillator is assumed.

e-ph coupling strongly modifies the shape of the heat capacity and free energy plots with respect to the harmonic result as it is seen in Figs. 9(a) and 9(b) respectively. The heat capacity versus temperature curves show a peculiar peak above a threshold value $\alpha \sim 10 meV \mathring{A}^{-1}$ which clearly varies according to the energy of the harmonic oscillator. Here we set $\omega = 20 meV$ to emphasize the size of the anharmonic effects on a low energy oscillator. By enhancing α the height of the peak grows and the bulk of the anharmonic effects on the heat capacity is shifted towards lower T. At $\alpha \sim 60 meV \mathring{A}^{-1}$ the crossover temperature is around 100K. Note that the size of the anharmonic enhancement is ~ 10 times the value of the harmonic oscillator heat capacity at $T = 100K$. It is worth noting that previous numerical studies of a classical one dimensional anharmonic model undergoing a Peierls instability [59] also found a specific heat peak as a signature of anharmonicity. However such a large anharmonic effect on the phonon subsystem is partly covered in the total heat capacity by the source action $A(j(\tau))$ and mainly by the hopping potential $V(x(\tau))$ contributions analysed in the previous section.

Taking for instance a bath of ten low energy oscillators with $\omega = 20 meV$, setting $\alpha \sim 60 meV \mathring{A}^{-1}$ which implies an effective coupling $\chi \sim 700 meV^3$ (last equation of (5)) we get a source heat capacity a factor two larger than the harmonic phonon heat capacity at temperatures of order $100K$. Thus the anharmonic peak, although substantially smeared by the electronic contributions to the total heat capacity, should still appear in systems with low energy phonons and sizeable *e-ph* coupling to which the SSH Hamiltonian applies. Let us focus on this point.

The total heat capacity is given by the phonon contribution plus a *source* heat capacity which includes both the electronic contribution (related to the electron hopping integral) and the contribution due to the source action (the latter being $\propto \alpha^2$). Fig. 10(a) shows the comparison between the anharmonic phonon heat capacity (C_V^{anh}) and the source heat capacity, here termed C_V^{e-p} to emphasize the dependence both on the electronic and on the *e-ph* coupling terms. C_V^{e-p} is computed as described above setting $\alpha = 21.74 meV \mathring{A}^{-1}$ [35]. Also the total heat capacity ($C_V^{tot} = C_V^{anh} + C_V^{e-p}$) is shown in Fig. 10(a). At low temperatures, C_V^{e-p} yields the largest effect mainly due to the electronic hopping while at high T, C_V^{e-p} prevails as the source action becomes dominant. In the intermediate range ($T \in [90, 210]$) the anharmonic phonons provide the highest contribution although their characteristic peak is substantially smeared in the total heat capacity by the source term background. Fig. 10(b) compares C_V^{anh} and C_V^{e-p} for two increasing values of α: while the anharmonic peak shifts downwards (along the T axis) by enhancing α, C_V^{anh} remains larger than C_V^{e-p} in a temperature range which progressively shrinks due to the strong dependence of the source action on the strength of the *e-ph* coupling. Finally, we observe that the low temperature upturn displayed in the *total heat capacity over T* ratio discussed above is not affected by the

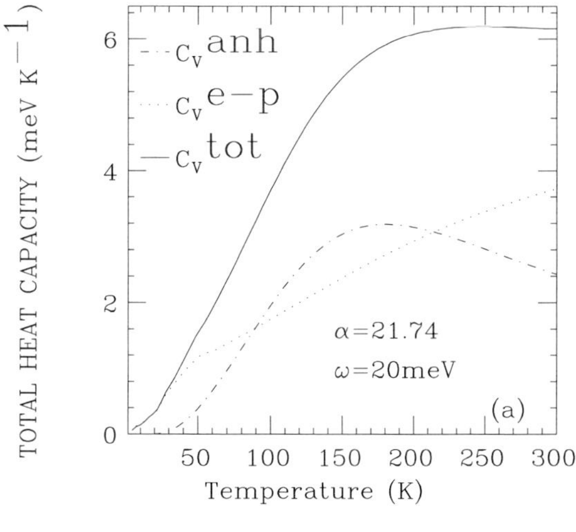

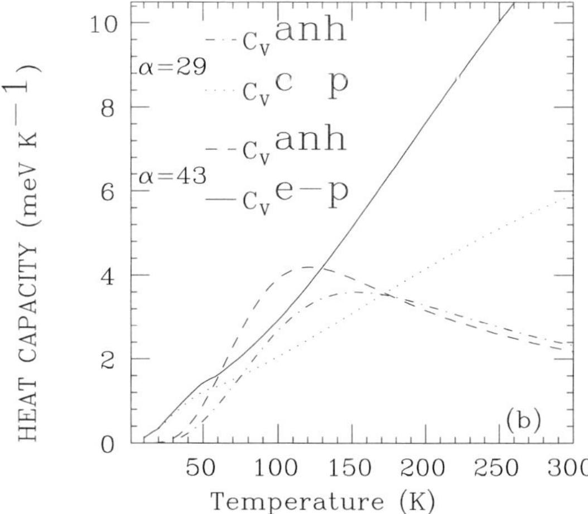

Fig. 10. (a) Total Heat Capacity versus temperature in the 1D Su–Schrieffer–Heeger model. The contributions due to anharmonic phonons (C_V^{anh}) and electrons *plus* electron-phonon interactions (C_V^{e-p}) are plotted separately. The largest α of Fig. 1 is assumed. (b) C_V^{anh} and C_V^{e-p} for two values of α in units $meV\,\mathring{A}^{-1}$. $\omega = 20meV$. Reproduced from [35], ©(2004) by the American Physical Society.

inclusion of phonon anharmonic effects which tend to become negligible at low temperatures.

6 Conclusions

Mapping the real space Su–Schrieffer–Heeger model onto the time scale I have developed a semiclassical version of the interacting model Hamiltonian in one and two dimensions suitable to be attacked by path integral methods. The acoustical phonons of the standard SSH model have been replaced by a set of oscillators providing a bath for the electron interacting with the displacements field. Time retarded interactions are naturally introduced in the formalism through the source action $A(x(\tau))$ which depends quadratically on the bare *e-ph* coupling strength α. Via calculation of the electronic motion path integral, the partition function can be derived in principle for any value of α thus avoiding those limitations on the *e-ph* coupling range which burden any perturbative method. Particular attention has been paid to establish a reliable and general procedure which allows one to determine those input parameters intrinsic to the path integral formalism. It turns out that a large number of paths is required to carry out low temperature calculations which therefore become highly time consuming. The physical parameters have been specified to a narrow band system and the behavior of some thermodynamical properties, free energy and heat capacity, has been analysed for some values of the effective coupling strength lying both in the weak and in the strong coupling regime. We find, both in 1D and 2D, a peculiar upturn in the low temperature plots of the heat capacity over temperature ratio indicating that a glass-like behavior can arise in the linear chain as a consequence of a time dependent electronic hopping with variable range.

According to our integration method (7), at any temperature, a specific set of Fourier coefficients defines the ensemble of relevant particle paths over which the hopping potential $V(x(\tau))$ is evaluated. This ensemble is therefore T dependent. However, given a single set of path parameters one can monitor the $V(x(\tau))$ behavior versus T. I find that the hopping decreases (as expected) by lowering T but its value remains appreciable also at low temperatures ($\precsim 20K$ in 2D and $\precsim 10K$ in 1D). Since the $d\tau$ integration range is larger at lower temperatures, the overall hopping potential contribution to the total action is relevant also at low T. It is precisely this property which is responsible for the anomalous upturn in the heat capacity linear coefficient. Further investigation also reveals that the upturn persists both in the extremely narrow ($J \sim 10meV$) and in the wide band ($J \sim 1eV$) regimes. Moreover, the upturn is not modified by the inclusion of electron-phonon anharmonicity in the phonon subsystem.

The presented computational method accounts for variable range hopping on the τ scale which corresponds physically to the introduction of some degree of disorder along the linear chain. This feature makes my model more general than the standard SSH Hamiltonian (1) with only real space nearest neighbor hops. While hopping type mechanisms have been suggested [49] to explain the striking conducting properties of doped polyacetylene at low temperatures I am not aware of any other direct calculation of the specific heat in the SSH model. Since the latter quantity directly probes the density of states and integrating over T the *specific heat over* T ratio one can have access to the experimental entropy, this method may provide a new approach to analyse the transition to a disordered state which indeed exists in polymers. In this connection it is also worth noting that the low T upturn in the specific heat over T ratio is a peculiar property of glasses [60, 61] in which tunneling states for atoms (or group of atoms) provide a non magnetic internal degree of freedom in the potential structure [62, 63].

References

1. N.Tessler, G.J.Denton, R.H.Friend, Nature (London) **397**, 121 (1996)
2. F.Capasso, C.Gmachl, D.L.Sivco, A.Y.Cho, Phys.Today **55** (5), 34 (2002)
3. C.Q.Wu, Y.Qiu, Z.An, K.Nasu, Phys. Rev. B **68**, 125416 (2003)
4. S.Dallakyan, M.Chandross, S.Mazumdar, Phys. Rev. B **68**, 075204 (2003)
5. M.Krishnan, S.Balasubramanian, Phys. Rev. B **68**, 064304 (2003)
6. K.Hannewald, V.M.Stojanović, J.M.T.Schellekens, P.A.Bobbert, G.Kresse, J.Hafner, Phys. Rov. B **60**, 075211 (2004)
7. A.M.Bratkovsky, this volume
8. M.Verissimo-Alves, R.B.Capaz, B.Koiller, E.Artacho, H.Chacham, Phys. Rev. Lett. **86**, 3372 (2001)
9. A.S.Alexandrov, A.B.Krebs, Sov. Phys. Usp. **35**, 345 (1992)
10. A.S.Alexandrov, P.E.Kornilovitch, Phys. Rev. Lett. **82**, 807 (1999)
11. A.S.Alexandrov, this volume
12. M.Zoli, A.N.Das, J.Phys.: Cond. Matter **16**, 3597 (2004)
13. R.E.Peierls, *Quantum Theory of Solids* (Clarendon, Oxford 1955).
14. Yu Lu, *Solitons and Polarons in Conducting Polymers* (World Scientific, Singapore 1988)
15. W.P.Su, J.R.Schrieffer, A.J.Heeger, Phys. Rev. Lett. **42**, 1698 (1979)
16. C.K.Chiang, C.R.Finker Jr., Y.W.Park, A.J.Heeger, H.Shirakawa, E.J.Louis, S.C.Gau, A.G.MacDiarmid, Phys. Rev. Lett. **39**, 1098 (1977)
17. V.Cataudella, G.De Filippis, C.A.Perroni, this volume
18. A.J. Heeger, S.Kivelson, J.R.Schrieffer, W.-P.Su, Rev. Mod. Phys. **60**, 781 (1988)
19. H.Takayama, Y.R.Lin-Liu, K.Maki, Phys. Rev. B **21**, 2388 (1980)
20. D.K.Campbell, A.R.Bishop, Phys. Rev. B **24**, 4859 (1981)
21. H.J.Schulz, Phys. Rev. B **18**, 5756 (1978)
22. J.E.Hirsch, E.Fradkin, Phys. Rev. Lett. **49**, 402 (1982)
23. S.Stafström, K.A.Chao, Phys. Rev. B **30**, 2098 (1984)
24. K.Michielsen, H.De Raedt, Z.Phys.B **103**, 391 (1997)
25. H.Zheng, Phys. Rev. B **56**, 14414 (1997)
26. A.La Magna, R.Pucci, Phys. Rev. B **55**, 6296 (1997)
27. M.Capone, W.Stephan, M.Grilli, Phys. Rev. B **56**, 4484 (1997)

28. M.Zoli, Phys. Rev. B **66**, 012303 (2002)

29. S.Tang, J.E.Hirsch, Phys. Rev. B **37**, 9546 (1988)

30. N.Miyasaka,Y.Ono, J.Phys.Soc.Jpn. **70**, 2968 (2001)

31. M.Zoli, Physica C **384**, 274 (2003)

32. V.M.Stojanović, P.A.Bobbert, M.A.J.Michels, Phys. Rev. B **69**, 144302 (2004)

33. C.A.Perroni, E.Piegari, M.Capone, V.Cataudella, Phys. Rev. B **69**, 174301 (2004)

34. G.D.Mahan, *Many Particle Physics*, (Plenum Press, NY 1981)

35. M.Zoli, Phys.Rev.B **70**, 184301 (2004)

36. I.J.Lang, Yu.A.Firsov, Sov. Phys. JETP **16**, 1301 (1963)

37. Th.M.Nieuwenhuizen, J.Mod.Optic **50**, 2433 (2003); cond-mat/9701044

38. I.I.Fishchuk, A.Kadashchuk, H.Bässler, S.Nešpurek, Phys. Rev. B **67**, 224303 (2003)

39. A.V.Plyukhin, Europhys.Lett. **71**, 716 (2005)

40. R.P.Feynman, Phys. Rev. **97**, 660 (1955)

41. J.T.Devreese, *Polarons* in Encyclopedia of Applied Physics (VCH Publishers, NY 1996) **14**, 383; J.T.Devreese, this volume

42. G.A.Farias, W.B.da Costa, F.M.Peeters, Phys. Rev. B **54**, 12835 (1996)

43. G.Ganbold, G.V.Efimov, in *Proceedings of the 6th International Conference on: Path Integrals from peV to Tev - 50 Years after Feynman's paper*, (World Scientific Publishing 1999) pp.387

44. H.De Raedt, A.Lagendijk, Phys. Rev. B **27**, 6097 (1983); ibid., **30**, 1671 (1984)

45. P.Kornilovitch, this volume.

46. H.Kleinert, *Path Integrals in Quantum Mechanics, Statistics and Polymer Physycs*, (World Scientific Publishing, Singapore 1995).

47. D.R.Hamann, Phys. Rev. B **2**, 1373 (1970)

48. Y.Ono, T.Hamano, J.Phys.Soc.Jpn. **69**, 1769 (2000)

49. S.Kivelson, Phys. Rev. Lett. **46**, 1344 (1981)

50. J.E.Hirsch, Phys. Rev. Lett. **51**, 296 (1983)

51. M.Cococcioni, M.Acquarone, Int. J. Mod. Phys. B **14**, 2956 (2000)

52. D.W.Brown, K.Lindenberg, Y.Zhao, J.Chem.Phys. **107**, 3179 (1997)

53. I.S.Gradshteyn, I.M.Ryzhik, *Tables of Integrals, Series and Products*, (Academic Press NY 1965).

54. M.Zoli, Phys.Rev.B **67**, 195102 (2003)

55. M.Zoli, Phys.Rev.B **71**, 205111 (2005)

56. M.Zoli, Phys. Rev. B **71**, 184308 (2005); Phys. Rev. B **72**, 214302 (2005)

57. M.Zoli, Eur.Phys.J. B **40**, 79 (2004)

58. V.L. Gurevich, D.A. Parshin, H.R. Schober, Phys. Rev. B **67**, 094203 (2003)

59. V.Perebeinos, P.B.Allen, J.Napolitano, Solid State Commun. **118**, 215 (2001)

60. R.C.Zeller, R.O.Pohl, Phys.Rev.B **4**, 2029 (1971)

61. P.W.Anderson, B.I.Halperin, C.H.Varma, Philos.Mag. **25**, 1 (1971)

62. K.Vladar, A.Zawadowski, Phys. Rev. B **28**, 1564 (1983)

63. M.Zoli, Phys. Rev. B **44**, 7163 (1991); ibid, Acta Physica Polonica **A77**, 639 (1990)

Part II

Bipolarons in Multi-Polaron Systems

Superconducting Polarons and Bipolarons

A. S. Alexandrov

Department of Physics, Loughborough University, Loughborough LE11 3TU,
United Kingdom
a.s.alexandrov@lboro.ac.uk

Summary. The seminal work by Bardeen, Cooper and Schrieffer (BCS) extended
further by Eliashberg to the intermediate coupling regime solved one of the major
scientific problems of Condensed Matter Physics in the last century. The BCS theory
provides qualitative and in many cases quantitative descriptions of low-temperature
superconducting metals and their alloys, and some novel high-temperature superconductors like magnesium diboride. The theory has been extended by us to the
strong-coupling regime where carriers are small lattice polarons and bipolarons.
Here I review the multi-polaron strong-coupling theory of superconductivity. Attractive electron correlations, prerequisite to any superconductivity, are caused by
an almost unretarded electron-phonon (e-ph) interaction sufficient to overcome the
direct Coulomb repulsion in this regime. Low energy physics is that of small polarons and bipolarons, which are real-space electron (hole) pairs dressed by phonons.
They are itinerant quasiparticles existing in the Bloch states at temperatures below the characteristic phonon frequency. Since there is almost no retardation (i.e.
no Tolmachev–Morel–Anderson logarithm) reducing the Coulomb repulsion, e-ph
interactions should be relatively strong to overcome the direct Coulomb repulsion,
so carriers *must* be polaronic to form pairs in novel superconductors. I identify the
long-range Fröhlich electron-phonon interaction as the most essential for pairing in
superconducting cuprates. A number of key observations have been predicted or
explained with polarons and bipolarons including unusual isotope effects and upper
critical fields, normal state (pseudo)gaps and kinetic properties, normal state diamagnetism, and giant proximity effects. These and many other observations provide
strong evidence for a novel state of electronic matter in layered cuprates, which is a
charged Bose-liquid of small mobile bipolarons.

1 Introduction

While the single polaron problem has been actively researched for a long
time (for reviews see [1–9] and the present volume), multi-polaron physics
has gained particular attention in the last two decades. For weak electron-
phonon coupling, $\lambda < 1$, and the adiabatic limit, $\omega/E_F \ll 1$, Migdal theory

describes electron dynamics in the normal Fermi-liquid state [10], and BCS-Eliashberg theory in the superconducting state [11, 12] (here and further I use $\hbar = c = k_B = 1$). While the electron-phonon (e-ph) interaction is weak Migdal's theorem is perfectly applied. The theorem proves that the contribution of diagrams with "crossing" phonon lines (so called "vertex" corrections) is small if the parameter $\lambda\omega/E_F$ is small, where λ is the dimensionless (BCS) e-ph coupling constant, ω is the characteristic phonon frequency, and E_F is the Fermi energy. Neglecting the vertex corrections, Migdal [10] calculated the renormalized electron mass as $m^* = m_0(1 + \lambda)$ (near the Fermi level), where m_0 is the band mass in the absence of e-ph interaction, and Eliashberg [12] extended Migdal's theory to describe the BCS superconducting state at intermediate values of $\lambda < 1$. Later on many researchers applied Migdal–Eliashberg theory with λ even larger than 1 (see, for example, [13], and references therein). With increasing strength of interaction and increasing phonon frequency, ω, finite bandwidth [14, 15] and vertex corrections [16] become increasingly important. But unexpectedly for many researchers who applied the non-crossing approximation even at $\lambda > 1$ we have found that the Migdal–BCS–Eliashberg theory (with or without vertex corrections) breaks down entirely at $\lambda \sim 1$ for any value of the adiabatic ratio ω/E_F since the bandwidth is narrowed and the Fermi energy, E_F is renormalised down exponentially so the effective parameter $\lambda\omega/E_F$ becomes large [17].

The electron-phonon coupling constant λ is about the ratio of the electron-phonon interaction energy E_p to the half bandwidth $D \approx N(E_F)^{-1}$, where $N(E)$ is the density of electron states in a rigid lattice. One expects [17] that when the coupling is strong, $\lambda > 1$, all electrons in the bare Bloch band are "dressed" by phonons since their kinetic energy $(< D)$ is small compared with the potential energy due to the local lattice deformation, E_p, caused by electrons. In this strong coupling regime the canonical Lang–Firsov transformation [18, 19] can be used to determine the properties of the system. Under certain conditions [20–23], the multi-polaron system is metallic but with polaronic carriers rather than bare electrons. This regime is beyond Migdal–Eliashberg theory, where the effective mass approximation is used and the electron bandwidth is infinite. In particular, the small polaron regime cannot be reached by summation of the standard Feynman–Dyson perturbation diagrams using a translation-invariant Green's function $G(\mathbf{r}, \mathbf{r}', \tau) = G(\mathbf{r} - \mathbf{r}', \tau)$ with the Fourier transform $G(\mathbf{k}, \Omega)$ prior to solving the Dyson equations on a discrete lattice. This assumption excludes the possibility of local violation of the translational symmetry [24] due to the lattice deformation in any order of the Feynman–Dyson perturbation theory similar to the absence of the anomalous (Bogoliubov) averages in any order of perturbation theory [10]. To enable electrons to relax into the lowest polaronic band, one has to introduce an infinitesimal translation-noninvariant potential, which should be set to zero only in the final solution obtained by the summation of Feynman diagrams for the Fourier transform $G(\mathbf{k}, \mathbf{k}', \Omega)$ of $G(\mathbf{r}, \mathbf{r}', \tau)$ rather than for $G(\mathbf{k}, \Omega)$ [25]. As in the case of the off-diagonal superconducting order parameter, the off-diagonal terms of the Green function, in particular the Umklapp terms with $\mathbf{k}' = \mathbf{k} + \mathbf{G}$, drive the system into a small polaron ground state at sufficiently large coupling. Setting the translation-noninvariant potential to zero in the solution of the equations of motion restores the translation symmetry but in a polaron band rather than in the bare electron band, which turns out to be an excited state.

Alternatively, one can work with momentum eigenstates throughout the whole coupling region, but taking into account the finite electron bandwidth from the very beginning. In recent years many numerical and analytical studies have confirmed the conclusion [17] that the Migdal–Eliashberg theory breaks down at $\lambda \geq 1$ (see, for example [26–39] and contributions to this book). With increasing phonon frequency the range of validity of the $1/\lambda$ polaron expansion extends to smaller values of λ [40]. As a result, the region of applicability of the Migdal–Eliashberg approach (even with vertex corrections) shrinks to smaller values of the coupling, $\lambda < 1$, with increasing ω. Strong correlations between carriers reduce this region further (see [31]).

In advanced materials such as high-temperature superconductors, colossal magnetoresistance oxides and organic molecules (see Part IV), carriers are strongly coupled with high-frequency optical phonons, which makes small polarons and non-adiabatic effects relevant. Indeed the characteristic phonon energies $0.05 - 0.2$ eV in cuprates, manganites and in many organic materials are of the same order as generally accepted values of the hopping integrals $0.1 - 0.3$ eV.

As reviewed in this book lattice polarons are different from ordinary electrons in many aspects, but perhaps one of the most remarkable differences is found in their superconducting properties. By extending the BCS theory towards the strong interaction between electrons and ion vibrations, a charged Bose gas (CBG) of tightly bound small bipolarons was predicted by us [42] with a further prediction that high critical temperature T_c is found in the crossover region of the e-ph interaction strength from the BCS-like *polaronic* to *bipolaronic* superconductivity [17]. This contribution describes what happens to the conventional BCS theory when the electron-phonon coupling becomes strong. The author's particular view of cuprates is also presented.

2 Electron–Phonon and Coulomb Interactions in Wannier Representation

For doped semiconductors and metals with a strong electron-phonon interaction it is convenient to transform the Bloch states $|\mathbf{k}\rangle$ to the site (Wannier) states $|\mathbf{m}\rangle$ using the canonical linear transformation of the electron operators,

$$c_i = \frac{1}{\sqrt{N}} \sum_{\mathbf{k}} e^{i\mathbf{k}\cdot\mathbf{m}} c_{\mathbf{k}s}, \tag{1}$$

where $i = (\mathbf{m}, s)$ includes both site $\mathbf{m}$ and spin s quantum numbers, and N is the number of sites in a crystal. In the site representation the electron kinetic energy takes the following form

$$H_e = \sum_{i,j} \left[T(\mathbf{m} - \mathbf{m}')\delta_{ss'} - \mu\delta_{ij} \right] c_i^\dagger c_j, \tag{2}$$

where

$$T(\mathbf{m}) = \frac{1}{N} \sum_{\mathbf{k}} E_{\mathbf{k}} e^{i\mathbf{k}\cdot\mathbf{m}}$$

is the bare hopping integral in the rigid lattice, μ is the chemical potential, $j = (\mathbf{n}, s')$, and $E_{\mathbf{k}}$ is the bare Bloch band dispersion.

The electron–phonon and Coulomb interactions acquire simple forms in the Wannier representation, if their matrix elements in the momentum representation $\gamma(\mathbf{q},\nu)$ and $V_c(\mathbf{q})$ depend only on the momentum transfer $\mathbf{q}$,

$$H_{e-ph} = \sum_{\mathbf{q},\nu,i} \omega_{\mathbf{q}\nu} \hat{n}_i \left[u_i(\mathbf{q},\nu)d_{\mathbf{q}\nu} + H.c. \right], \tag{3}$$

and

$$H_{e-e} = \frac{1}{2} \sum_{i \neq j} V_c(\mathbf{m} - \mathbf{n})\hat{n}_i\hat{n}_j. \tag{4}$$

Here

$$u_i(\mathbf{q},\nu) = \frac{1}{\sqrt{2N}} \gamma(\mathbf{q},\nu)e^{i\mathbf{q}\cdot\mathbf{m}} \tag{5}$$

and

$$V_c(\mathbf{m}) = \frac{1}{N} \sum_{\mathbf{q}} V_c(\mathbf{q})e^{i\mathbf{q}\cdot\mathbf{m}}, \tag{6}$$

are the matrix elements of electron–phonon and Coulomb interactions, respectively, in the Wannier representation for electrons, $\hat{n}_i = c_i^\dagger c_i$ is the electron density operator, and $d_{\mathbf{q}\nu}$ annihilates the ν-branch phonon with the wave vector $\mathbf{q}$ and frequency $\omega_{\mathbf{q}\nu}$. Taking the interaction matrix elements depending only on the momentum transfer one neglects terms in the electron-phonon and Coulomb interactions, which are proportional to the overlap integrals of the Wannier orbitals on different sites. This approximation is justified for narrow band materials with the bandwidth $2D$ less than the characteristic value of the crystal field. As a result, the generic Hamiltonian takes the following form in the Wannier representation,

$$H = \sum_{i,j} \left[T(\mathbf{m} - \mathbf{m}')\delta_{ss'} - \mu\delta_{ij} \right] c_i^\dagger c_j + \sum_{\mathbf{q},\nu,i} \omega_{\mathbf{q}\nu} \hat{n}_i \left[u_i(\mathbf{q},\nu)d_{\mathbf{q}\nu} + H.c. \right]$$
$$+ \frac{1}{2} \sum_{i \neq j} V_c(\mathbf{m} - \mathbf{n})\hat{n}_i\hat{n}_j + \sum_{\mathbf{q}} \omega_{\mathbf{q}\nu}(d_{\mathbf{q}\nu}^\dagger d_{\mathbf{q}\nu} + 1/2). \tag{7}$$

Here we confine our discussions to a single electron band and the electron–phonon matrix element depending only on the momentum transfer $\mathbf{q}$. This approximation allows for qualitative and in many cases quantitative descriptions of essential polaronic effects in advanced materials. There are might be degenerate atomic orbitals in solids coupled to local molecular-type Jahn–Teller distortions, where one has to consider multi-band electron energy structures (see [43]).

The quantitative calculation of the matrix element in the whole region of momenta has to be performed from pseudopotentials [44–46]. On the other hand one can parameterize the e-ph interaction rather than computing it from first principles in many physically important cases [47]. The three most important interactions in doped semiconductors are polar coupling to optical phonons (i.e the Fröhlich e-ph interaction), deformation potential coupling to acoustical phonons, and the local (Holstein) e-ph interaction with molecular type vibrations in complex lattices. While the matrix element is ill defined in metals, it is well defined in doped semiconductors, which have their parent dielectric compounds, together with bare phonons $\omega_{\mathbf{q}\nu}$ and the electron band structure $E_{\mathbf{k}}$. Here the effect of carriers on the crystal field and on the dynamic matrix is small while the carrier density is much less than the atomic one (for phonon self-energies and frequency renormalizations in polaronic systems see [20, 48]). Hence one can use the band structure and the crystal field of parent insulators to calculate the matrix element in doped semiconductors.

The e-ph matrix element $\gamma(\mathbf{q})$ has different q-dependence for different phonon branches. In the long wavelength limit ($q \ll \pi/a$, a is the lattice constant), $\gamma(\mathbf{q}) \propto q^n$, where $n = -1, 0$ and $n = -1/2$ for polar optical, molecular ($\omega_{\mathbf{q}} = \omega_0$) and acoustic ($\omega_{\mathbf{q}} \propto q$) phonons, respectively. Not only is q dependence known but the absolute values of $\gamma(\mathbf{q})$ are also well parameterized in this limit. For example in polar semiconductors the interaction of two slow electrons at some distance r is found as

$$v(r) = V_c(r) - \frac{1}{N}\sum_{\mathbf{q}}|\gamma(\mathbf{q})|^2 \omega_{\mathbf{q}} e^{i\mathbf{q}\cdot\mathbf{r}}. \tag{8}$$

The Coulomb repulsion in a rigid lattice is $V_c(r) = e^2/\epsilon_\infty r$, and the second term represents the difference between the Coulomb repulsion screened with the core electrons and the repulsion screened with both core electrons and ions. Hence the matrix element of the Fröhlich interaction depends only on the dielectric constants and the optical phonon frequency ω_0 as

$$|\gamma(\mathbf{q})|^2 = \frac{4\pi e^2}{\kappa \omega_0}, \tag{9}$$

where $\kappa = (\epsilon_\infty^{-1} - \epsilon_0^{-1})^{-1}$.

One can transform the e-ph interaction further using the site-representation also for phonons. The site representation of phonons is particularly convenient for the interaction with dispersionless local modes, whose $\omega_{\mathbf{q}\nu} = \omega_\nu$ and the polarization vectors $\mathbf{e}_{\mathbf{q}\nu} = \mathbf{e}_\nu$ are $\mathbf{q}$ independent. Introducing the phonon site-operators

$$d_{\mathbf{n}\nu} = \frac{1}{\sqrt{N}} \sum_{\mathbf{k}} e^{i\mathbf{q}\cdot\mathbf{n}} d_{\mathbf{q}\nu} \tag{10}$$

one transforms the deformation energy and the e-ph interaction as [23]

$$H_{ph} = \sum_{\mathbf{n},\nu} \omega_\nu (d_{\mathbf{n}\nu}^\dagger d_{\mathbf{n}\nu} + 1/2), \tag{11}$$

and

$$H_{e-ph} = \sum_{\mathbf{n},\mathbf{m},\nu} \omega_\nu g_\nu (\mathbf{m} - \mathbf{n})(\mathbf{e}_\nu \cdot \mathbf{e}_{\mathbf{m}-\mathbf{n}}) \widehat{n}_{\mathbf{m}s} (d_{\mathbf{n}\nu}^\dagger + d_{\mathbf{n}\nu}), \tag{12}$$

respectively.

Here $g_\nu(\mathbf{m})$ is a dimensionless *force* acting between the electron on site $\mathbf{m}$ and the displacement of ion $\mathbf{n}$, and $\mathbf{e}_{\mathbf{m}-\mathbf{n}} \equiv (\mathbf{m} - \mathbf{n})/|\mathbf{m} - \mathbf{n}|$ is the unit vector in the direction from the electron $\mathbf{m}$ to the ion $\mathbf{n}$. This real space representation is convenient in modeling the electron-phonon interaction in complex lattices. Atomic orbitals of an ion adiabatically follow its motion. Therefore the electron does not interact with the displacement of the ion, whose orbital it occupies, that is $g_\nu(0) = 0$.

3 Breakdown of Migdal–Eliashberg Theory in the Strong-Coupling Regime

Obviously a perturbative approach to the e-ph interaction fails when $\lambda > 1$. However one might expect that the self-consistent Migdal–Eliashberg (ME) theory is still valid in the strong-coupling regime because it sums the infinite set of particular non-crossing diagrams in the electron self-energy. One of the problems with such an extension of the ME theory is a lattice instability. The same theory applied to phonons yields the renormalised phonon frequency $\widetilde{\omega} = \omega(1 - 2\lambda)^{1/2}$ [10] (here ω is the bare acoustic phonon frequency). The frequency turns out to be zero at $\lambda = 0.5$. Because of this lattice instability Migdal [10] and Eliashberg [12] restricted the applicability of their approach to $\lambda < 1$. However, it was shown later that there was no lattice instability, but only a small renormalisation of the phonon frequencies of the order of the adiabatic ratio, $\omega/\mu \ll 1$, for *any* value of λ, if the adiabatic Born–Oppenheimer approach was properly applied [49]. The conclusion was that the Fröhlich Hamiltonian correctly describes the electron self-energy for any value of λ, but it should not be applied to further renormalise phonons.

In fact, ME theory cannot be applied at $\lambda > 1$ for a reason which has nothing to do with the lattice instability. Actually the $1/\lambda$ multi-polaron expansion technique [20] shows that the many-electron system collapses into the small polaron (or bipolaron) regime at $\lambda \approx 1$ for any adiabatic ratio.

To illustrate the point let us compare the Migdal solution of the simple molecular-chain Holstein model [50] with the exact solution [22] in the adiabatic limit, $\omega_0/t \to 0$, where $t = T(a)$ is the nearest neighbour hopping integral. The Hamiltonian of the model is

$$H = -t \sum_{<ij>} c_i^\dagger c_j + H.c. + 2(\lambda kt)^{1/2} \sum_i x_i c_i^\dagger c_i \qquad (13)$$

$$+ \sum_i \left(-\frac{1}{2M}\frac{\partial^2}{\partial x_i^2} + \frac{kx_i^2}{2} \right),$$

where x_i is the normal coordinate of the molecule (site) i, and $k = M\omega_0^2$. The Migdal theorem is exact in this limit. Hence in the framework of the Migdal–Eliashberg theory one would expect the Fermi-liquid behaviour above T_c and the BCS ground state below T_c at any value of λ. In fact, the exact ground state is a self-trapped insulator at any filling of the band, if $\lambda \geq 1$.

First we consider a two-site case (zero dimensional limit), $i, j = 1, 2$ with one electron, and than generalise the result for an infinite lattice with many electrons. The transformation $X = (x_1+x_2)$, $\xi = x_1-x_2$ allows us to eliminate the coordinate X, which is coupled only with the total density $(n_1 + n_2 = 1)$. That leaves the following Hamiltonian to be solved in the extreme adiabatic limit $M \to \infty$:

$$H = -t(c_1^\dagger c_2 + c_2^\dagger c_1) + (\lambda kt)^{1/2}\xi(c_1^\dagger c_1 - c_2^\dagger c_2) + \frac{k\xi^2}{4}. \qquad (14)$$

The solution is

$$\psi = (\alpha c_1^\dagger + \beta c_2^\dagger) |0\rangle , \qquad (15)$$

where

$$\alpha = \frac{t}{[t^2 + ((\lambda kt)^{1/2}\xi + (t^2 + \lambda kt\xi^2)^{1/2})^2]^{1/2}}, \qquad (16)$$

$$\beta = -\frac{(\lambda kt)^{1/2}\xi + (t^2 + \lambda kt\xi^2)^{1/2}}{[t^2 + ((\lambda kt)^{1/2}\xi + (t^2 + \lambda kt\xi^2)^{1/2})^2]^{1/2}}, \qquad (17)$$

and the energy is

$$E = \frac{k\xi^2}{4} - (t^2 + \lambda kt\xi^2)^{1/2}. \qquad (18)$$

In the extreme adiabatic limit the displacement ξ is classical, so the ground state energy E_0 and the ground state displacement ξ_0 are obtained by minimising (18) with respect to ξ. If $\lambda \geq 0.5$ we obtain

$$E_0 = -t(\lambda + \frac{1}{4\lambda}), \qquad (19)$$

and

$$\xi_0 = \left[\frac{t(4\lambda^2 - 1)}{\lambda k}\right]^{1/2}. \qquad (20)$$

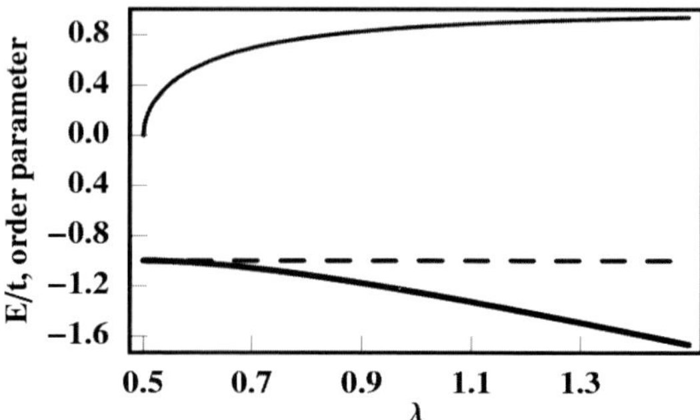

Fig. 1. The ground-state energy (in units of t, solid line) and the order parameter (thin solid line) of the adiabatic Holstein model. The Migdal solution is shown as the dashed line.

The symmetry-breaking "order" parameter is

$$\Delta \equiv \beta^2 - \alpha^2 = \frac{[2\lambda + (4\lambda^2 - 1)^{1/2}]^2 - 1}{[2\lambda + (4\lambda^2 - 1)^{1/2}]^2 + 1}. \tag{21}$$

If $\lambda < 0.5$, the ground state is translation invariant, $\Delta = 0$, and $E_0 = -t$, $\xi = 0$, $\beta = -\alpha$. This state is the "Migdal" solution of the Holstein model, which is symmetric (translation invariant) with $|\alpha| = |\beta|$. When $\lambda < 0.5$, the Migdal solution is the *only* solution. However, when $\lambda > 0.5$ this solution is *not* the ground state of the system, Fig. 1. The system collapses into a localised adiabatic polaron trapped on the "right-hand" or on the "left-hand" site due to the finite lattice deformation $\xi_0 \neq 0$.

The generalisation for a multi-polaron system on the infinite lattice of any dimension is straightforward in the extreme adiabatic regime. The adiabatic solution of the infinite one-dimensional (1D) chain with one electron was obtained by Rashba [51] in a continuum (i.e. effective mass) approximation, and by Holstein [50] and Kabanov and Mashtakov [26, 48] for a discrete lattice. The latter authors studied the Holstein two-dimensional (2D) and three-dimensional (3D) lattices in the adiabatic limit. According to [26] the self-trapping of a single electron occurs at $\lambda \geq 0.875$ and at $\lambda \geq 0.92$ in 2D and 3D, respectively. The radius of the self-trapped adiabatic polaron, r_p, is readily derived from its continuous wave function [51]

$$\psi(r) \sim 1/\cosh(\lambda r/a). \tag{22}$$

It becomes smaller than the lattice constant, $r_p = a/\lambda$ for $\lambda \geq 1$. Hence the multi-polaron system remains in the self-trapped insulating state in the strong-coupling adiabatic regime, no matter how many polarons it has. The only instability which might occur in this regime is the formation of self-trapped bipolarons, if the on-site attractive interaction, $2\lambda zt$, is larger than the repulsive Hubbard U [52]. Self-trapped on-site bipolarons form a charge ordered state due to a weak repulsion between them [37, 42] (see also [53]).

The transition into the self-trapped state due to a broken translational symmetry is expected at $0.5 < \lambda < 1.3$ (depending on the lattice dimensionality) for any electron-phonon interaction conserving the on-site electron occupation numbers. For example, Hiramoto and Toyozawa [54] calculated the strength of the deformation potential, which transforms electrons into small polarons and bipolarons. They found that the transition of two electrons into a self-trapped small bipolaron occurs at the electron-acoustic phonon coupling $\lambda \simeq 0.5$, that is half of the critical value of λ at which the transition of the electron into the small acoustic polaron takes place in the extreme adiabatic limit, $\omega << zt$ (here z is the coordination lattice number). The effect of the adiabatic ratio ω/zt on the critical value of λ was found to be negligible. The radius of the acoustic polaron and bipolaron is roughly the lattice constant, so that the critical value of λ does not depend very much on the number of electrons in this case either. As discussed below the non-adiabatic corrections (phonons) allow polarons and bipolarons to propagate as the Bloch states in narrow bands.

4 Polaron Dynamics

4.1 Polaron Band

A self-consistent approach to the multi-polaron problem is possible with the "$1/\lambda$" expansion technique [20], which treats the kinetic energy as a perturbation. The technique is based on the fact, known for a long time, that there is an analytical exact solution of a *single* polaron problem in the strong-coupling limit $\lambda \to \infty$ [18]. Following Lang and Firsov we apply the canonical transformation e^S to diagonalise the Hamiltonian. The diagonalisation is exact, if $T(\mathbf{m}) = 0$ (or $\lambda = \infty$):

$$\widetilde{H} = e^S H e^{-S}, \tag{23}$$

where

$$S = -\sum_{\mathbf{q},\nu,i} \widehat{n}_i \left[u_i(\mathbf{q},\nu) d_{\mathbf{q}\nu} - H.c. \right] \tag{24}$$

is such that $S^\dagger = -S$. The electron and phonon operators are transformed as

$$\widetilde{c}_i = c_i \exp \left[\sum_{\mathbf{q}} u_i(\mathbf{q},\nu) d_{\mathbf{q}\nu} - H.c. \right], \tag{25}$$

and

$$\tilde{d}_{\mathbf{q}\nu} = d_{\mathbf{q}\nu} - \sum_i \widehat{n}_i u_i^*(\mathbf{q},\nu), \tag{26}$$

respectively. It follows from (26) that the Lang–Firsov canonical transformation shifts the ions to new equilibrium positions. In a more general sense it changes the boson vacuum. As a result, the transformed Hamiltonian takes the following form

$$\tilde{H} = \sum_{i,j}[\widehat{\sigma}_{ij} - \mu\delta_{ij}]c_i^\dagger c_j - E_p \sum_i \widehat{n}_i + \sum_{\mathbf{q},\nu} \omega_{\mathbf{q}\nu}(d_{\mathbf{q}\nu}^\dagger d_{\mathbf{q}\nu} + 1/2) + \frac{1}{2}\sum_{i\neq j} v_{ij}\widehat{n}_i\widehat{n}_j, \tag{27}$$

where

$$\widehat{\sigma}_{ij} = T(\mathbf{m} - \mathbf{n})\delta_{ss'} \exp\left(\sum_{\mathbf{q},\nu}[u_j(\mathbf{q},\nu) - u_i(\mathbf{q},\nu)]d_{\mathbf{q}\nu} - H.c.\right) \tag{28}$$

is the renormalised hopping integral depending on the phonon operators, and

$$v_{ij} \equiv v(\mathbf{m} - \mathbf{n}) = \tag{29}$$

$$V_c(\mathbf{m} - \mathbf{n}) - \frac{1}{N}\sum_{\mathbf{q},\nu} |\gamma(\mathbf{q},\nu)|^2 \omega_{\mathbf{q}\nu} \cos[\mathbf{q} \cdot (\mathbf{m} - \mathbf{n})]$$

is the interaction of polarons comprising their Coulomb repulsion and the interaction via a local lattice deformation. In the extreme infinite-coupling limit, $\lambda \to \infty$, we can neglect the hopping term of the transformed Hamiltonian. The remaining Hamiltonian has analytically determined eigenstates and eigenvalues. The eigenstates $|\tilde{N}\rangle = |n_i, n_{\mathbf{q}\nu}\rangle$ are sorted by the polaron $n_{\mathbf{ms}}$ and phonon $n_{\mathbf{q}\nu}$ occupation numbers. The energy levels are

$$E = -(\mu + E_p)\sum_i n_i + \frac{1}{2}\sum_{i\neq j} v_{ij}n_i n_j + \sum_{\mathbf{q}} \omega_{\mathbf{q}\nu}(n_{\mathbf{q}\nu} + 1/2), \tag{30}$$

where $n_i = 0, 1$ and $n_{\mathbf{q}\nu} = 0, 1, 2, 3, ...\infty$.

The Hamiltonian, (27), in zero order with respect to the hopping describes localised polarons and independent phonons, which are vibrations of ions relative to new equilibrium positions depending on the polaron occupation numbers. The phonon frequencies remain unchanged in this limit. The middle of the electron band falls by the polaron level shift E_p due to a potential well created by lattice deformation, Fig. 2,

$$E_p = \frac{1}{2N}\sum_{\mathbf{q},\nu} |\gamma(\mathbf{q},\nu)|^2 \omega_{\mathbf{q}\nu}. \tag{31}$$

Now let us discuss the $1/\lambda$ expansion. First we restrict the discussion to a single-polaron problem with no polaron-polaron interaction and $\mu = 0$. The finite hopping term leads to polaron tunnelling because of degeneracy of the zero order Hamiltonian with respect to the site position of the polaron. To see how the tunnelling occurs we apply the perturbation theory using $1/\lambda$ as a small parameter, where

$$\lambda \equiv \frac{E_p}{D}, \tag{32}$$

and $D = zt$. The proper Bloch set of N-degenerate zero order eigenstates with the lowest energy $(-E_p)$ of the unperturbed Hamiltonian is

$$|\mathbf{k}, 0\rangle = \frac{1}{\sqrt{N}} \sum_{\mathbf{m}} c^{\dagger}_{\mathbf{m}s} \exp(i\mathbf{k} \cdot \mathbf{m})|0\rangle, \tag{33}$$

where $|0\rangle$ is the vacuum. By applying textbook perturbation theory one readily calculates the perturbed energy levels. Up to the second order in the hopping integral they are given by

$$E(\mathbf{k}) = -E_p + \epsilon_{\mathbf{k}} \tag{34}$$

$$- \sum_{\mathbf{k}', n_{\mathbf{q}\nu}} \frac{|\langle \mathbf{k}, 0| \sum_{i,j} \hat{\sigma}_{ij} c^{\dagger}_i c_j |\mathbf{k}', n_{\mathbf{q}\nu}\rangle|^2}{\sum_{\mathbf{q},\nu} \omega_{\mathbf{q}\nu} n_{\mathbf{q}\nu}},$$

where $|\mathbf{k}', n_{\mathbf{q}\nu}\rangle$ are the exited states of the unperturbed Hamiltonian with one electron and at least one real phonon. The second term in (34), which is linear with respect to the bare hopping $T(\mathbf{m})$, describes the polaron-band dispersion,

$$\epsilon_{\mathbf{k}} = \sum_{\mathbf{m}} T(\mathbf{m}) \, e^{-g^2(\mathbf{m})} \exp(-i\mathbf{k} \cdot \mathbf{m}), \tag{35}$$

where

$$g^2(\mathbf{m}) = \frac{1}{2N} \sum_{\mathbf{q},\nu} |\gamma(\mathbf{q},\nu)|^2 [1 - \cos(\mathbf{q} \cdot \mathbf{m})] \tag{36}$$

is the *band-narrowing factor* at zero temperature. The third term in (34), quadratic in $T(\mathbf{m})$, yields a negative almost **k**-*independent* correction to the polaron level shift of the order of $1/\lambda^2$. The origin of this correction, which could be much larger than the first-order contribution ((35) contains a small exponent), is understood in Fig. 2. The polaron localised in a potential well of depth E_p on the site $\mathbf{m}$, hops onto a neighbouring site $\mathbf{n}$ with no deformation around it and then returns. As for any second order correction this transition shifts the energy down by an amount of about $-t^2/E_p$. It has little to do with the polaron effective mass and the polaron tunneling mobility because the lattice deformation around $\mathbf{m}$ does not follow the electron. The electron hops back and forth many times (about e^{g^2}) waiting for a sufficient lattice deformation to appear around the site $\mathbf{n}$. Only after the deformation around $\mathbf{n}$ is created does the polaron tunnel onto the next site together with the deformation.

Oddly enough, analysing the Holstein two-site model some authors took the second-order correction in t in (34) as a measure of the polaron motion and

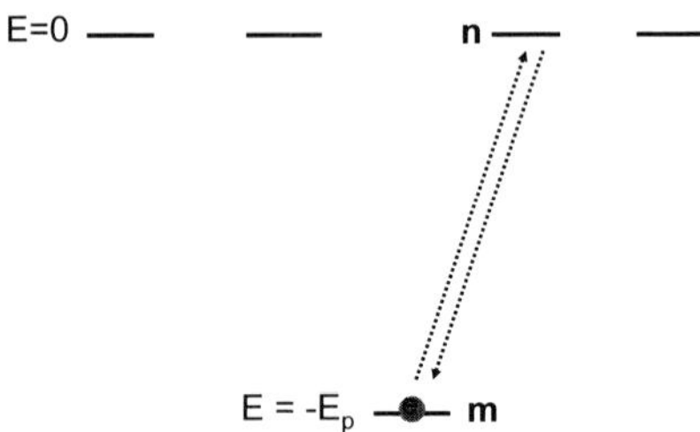

Fig. 2. "Back and forth" virtual transitions of the polaron without any transfer of the lattice deformation from one site to another. These transitions shift the middle of the band further down without any real charge delocalization

arrived at the erroneous conclusion that "polarons are no longer describable in terms of quasiparticles having a well-defined dispersion" [55] and "the Lang–Firsov approach, which is generally believed to become exact in the limit of antiadiabaticity and an electron-phonon coupling going to infinity, actually diverges most from the exact results precisely in this limit..." [56, 57]. Later on it became clear [58–60] that this controversy is the result of the erroneous identification of the polaron kinetic energy by the authors of [55–57].

4.2 Polaron Spectral and Green's Functions

The multi-polaron problem has an exact solution in the extreme infinite-coupling limit, $\lambda = \infty$, for any type of e-ph interaction conserving the on-site occupation numbers of electrons. For the finite coupling $1/\lambda$ perturbation expansion is applied. The expansion parameter is actually [8, 20, 61, 62]

$$\frac{1}{2z\lambda^2} \ll 1,$$

so that the analytical perturbation theory has a wider region of applicability than one can expect using a semiclassical estimate $E_p > D$. However, the expansion convergency is different for different e-ph interactions. Exact numerical diagonalisations of vibrating clusters, variational calculations (see [27–29, 33, 36, 38] and [63, 64]), dynamical mean-field approach in infinite dimensions [35], and Quantum-Monte-Carlo simulations (see [39, 65–71] and [43, 72]) simulations revealed that the ground state energy ($\approx - E_p$) is not very sensitive to the parameters. On the contrary, the effective mass, the bandwidth, and the polaron density of states strongly depend on the adiabatic ratio ω_0/t and on the radius of the interaction. The first-order in $1/\lambda$ perturbation theory is practically exact in the non-adiabatic regime $\omega_0 > t$ *for any value* of the coupling constant and any type of e-ph interaction. However, it *overestimates* the polaron mass by a few orders of magnitude in the adiabatic case, $\omega_0 \ll t$, if the interaction is short-ranged [27].

A much lower effective mass of the adiabatic Holstein polaron compared with that estimated using the first order perturbation theory is the result of poor convergency of the perturbation expansion owing to the double-well potential [50] in the adiabatic limit. The tunnelling probability is extremely sensitive to the shape of this potential and also to the phonon frequency dispersion. The latter leads to a much lighter Holstein polaron compared with the nondispersive approximation [73]. Importantly, the analytical perturbation theory becomes practically exact for a wider range of the adiabatic parameter for the long-range Fröhlich interaction [39].

Keeping this in mind, let us calculate the one-particle GF in the first order in $1/\lambda$. Applying the canonical transformation we write the transformed Hamiltonian as

$$\widetilde{H} = H_p + H_{ph} + H_{int}, \tag{37}$$

where

$$H_p - \sum_{\mathbf{k}} \zeta(\mathbf{k}) c_{\mathbf{k}}^\dagger c_{\mathbf{k}} \tag{38}$$

is the "free" *polaron* contribution,

$$H_{ph} = \sum_{\mathbf{q}} \omega_{\mathbf{q}} (d_{\mathbf{q}}^\dagger d_{\mathbf{q}} + 1/2) \tag{39}$$

is the free phonon part (spin and phonon branch quantum numbers are dropped here), and $\xi_{\mathbf{k}} = Z' E_{\mathbf{k}} - \mu$ is the renormalised polaron-band dispersion. The chemical potential μ includes the polaron level shift $-E_p$, and it could also include all higher orders in $1/\lambda$ corrections to the polaron spectrum, which are independent of $\mathbf{k}$. The band-narrowing factor Z' is defined as

$$Z' = \frac{\sum_{\mathbf{m}} T(\mathbf{m}) e^{-g^2(\mathbf{m})} \exp(-i\mathbf{k} \cdot \mathbf{m})}{\sum_{\mathbf{m}} T(\mathbf{m}) \exp(-i\mathbf{k} \cdot \mathbf{m})}, \tag{40}$$

which is $Z' = \exp(-\gamma E_p/\omega))$ with $\gamma \leq 1$ depending on the range of the e-ph interaction and phonon frequency dispersions. The interaction term H_{int} comprises the polaron-polaron interaction, (29), and the residual polaron-phonon interaction

$$H_{p-ph} \equiv \sum_{i \neq j} [\widehat{\sigma}_{ij} - \langle \widehat{\sigma}_{ij} \rangle_{ph}] c_i^\dagger c_j, \tag{41}$$

where $\langle \widehat{\sigma}_{ij} \rangle_{ph}$ means averaging with respect to the bare phonon distribution. We can neglect H_{p-ph} in the first-order of $1/\lambda \ll 1$. To understand the spectral properties of a single polaron we also neglect the polaron-polaron interaction. Then the energy levels are

$$E_{\widetilde{m}} = \sum_{\mathbf{k}} \xi_{\mathbf{k}} n_{\mathbf{k}} + \sum_{\mathbf{q}} \omega_{\mathbf{q}}[n_{\mathbf{q}} + 1/2], \tag{42}$$

and the transformed eigenstates $|\widetilde{m}\rangle$ are sorted by the polaron Bloch-state occupation numbers, $n_{\mathbf{k}} = 0, 1$, and the phonon occupation numbers, $n_{\mathbf{q}} = 0, 1, 2, ..., \infty$.

The spectral function of any system described by quantum numbers m, n and eigenvalues E_n, E_m is defined as (see, for example [23])

$$A(\mathbf{k}, \omega) \equiv \pi(1 + e^{-\omega/T})e^{\Omega/T} \sum_{n,m} e^{-E_n/T} |\langle n| c_{\mathbf{k}} |m\rangle|^2 \delta(\omega_{nm} + \omega). \tag{43}$$

It is real and positive, $A(\mathbf{k}, \omega) > 0$, and obeys the important sum rule

$$\frac{1}{\pi} \int_{-\infty}^{\infty} d\omega A(\mathbf{k}, \omega) = 1. \tag{44}$$

Here Ω is the thermodynamic potential. The matrix elements of the electron operators can be written as

$$\langle n| c_{\mathbf{k}} |m\rangle = \frac{1}{\sqrt{N}} \sum_{\mathbf{m}} e^{-i\mathbf{k}\cdot\mathbf{m}} \langle \widetilde{n}| c_i \widehat{X}_i |\widetilde{m}\rangle \tag{45}$$

by the use of the Wannier representation and the Lang–Firsov transformation. Here

$$\widehat{X}_i = \exp\left[\sum_{\mathbf{q}} u_i(\mathbf{q})d_{\mathbf{q}} - H.c.\right].$$

Now, applying the Fourier transform of the δ-function in (43),

$$\delta(\omega_{nm} + \omega) = \frac{1}{2\pi} \int_{-\infty}^{\infty} dt e^{i(\omega_{nm}+\omega)t},$$

the spectral function is expressed as

$$A(\mathbf{k}, \omega) = \frac{1}{2} \int_{-\infty}^{\infty} dt e^{i\omega t} \frac{1}{N} \sum_{\mathbf{m},\mathbf{n}} e^{i\mathbf{k}\cdot(\mathbf{n}-\mathbf{m})} \times \tag{46}$$

$$\left\{\left\langle\left\langle c_i(t)\widehat{X}_i(t)c_j^\dagger \widehat{X}_j^\dagger\right\rangle\right\rangle + \left\langle\left\langle c_j^\dagger \widehat{X}_j^\dagger c_i(t)\widehat{X}_i(t)\right\rangle\right\rangle\right\}.$$

Here the quantum and statistical averages are performed for independent polarons and phonons, therefore

$$\left\langle\!\!\left\langle c_i(t)\widehat{X}_i(t)\widehat{X}_j^\dagger c_i^\dagger\right\rangle\!\!\right\rangle = \left\langle\!\!\left\langle c_i(t)c_j^\dagger\right\rangle\!\!\right\rangle\left\langle\!\left\langle \widehat{X}_i(t)\widehat{X}_j^\dagger\right\rangle\!\right\rangle. \tag{47}$$

The Heisenberg free-polaron operator evolves with time as

$$c_{\mathbf{k}}(t) = c_{\mathbf{k}}e^{-i\xi_{\mathbf{k}}t}, \tag{48}$$

and

$$\left\langle\!\!\left\langle c_i(t)c_i^\dagger\right\rangle\!\!\right\rangle = \frac{1}{N}\sum_{\mathbf{k}',\mathbf{k}''} e^{i(\mathbf{k}'\cdot\mathbf{m}-\mathbf{k}''\cdot\mathbf{n})}\left\langle\!\!\left\langle c_{\mathbf{k}'}(t)c_{\mathbf{k}''}^\dagger\right\rangle\!\!\right\rangle = \tag{49}$$

$$\frac{1}{N}\sum_{\mathbf{k}'}[1-\bar{n}(\mathbf{k}')]e^{i\mathbf{k}'\cdot(\mathbf{m}-\mathbf{n})-i\xi_{\mathbf{k}'}t},$$

$$\left\langle\!\!\left\langle c_i^\dagger c_i(t)\right\rangle\!\!\right\rangle = \frac{1}{N}\sum_{\mathbf{k}'} \bar{n}(\mathbf{k}')e^{i\mathbf{k}'\cdot(\mathbf{m}-\mathbf{n})-i\xi_{\mathbf{k}'}t} \tag{50}$$

where $\bar{n}(\mathbf{k}) = [1 + \exp\xi_{\mathbf{k}}/T]^{-1}$ is the Fermi–Dirac distribution function of polarons. The Heisenberg free-phonon operator evolves in a similar way,

$$d_{\mathbf{q}}(t) = d_{\mathbf{q}}e^{-i\omega_{\mathbf{q}}t},$$

and

$$\left\langle\!\left\langle \widehat{X}_i(t)\widehat{X}_j^\dagger\right\rangle\!\right\rangle = \prod_{\mathbf{q}} \langle\!\langle\exp[u_i(\mathbf{q},t)d_{\mathbf{q}} - H.c.]\exp[-u_j(\mathbf{q})d_{\mathbf{q}} - H.c.]\rangle\!\rangle, \tag{51}$$

where $u_{i,j}(\mathbf{q},t) = u_{i,j}(\mathbf{q})e^{-i\omega_{\mathbf{q}}t}$. This average is calculated using the operator identity

$$e^{\widehat{A}+\widehat{B}} = e^{\widehat{A}}e^{\widehat{B}}e^{-[\widehat{A},\widehat{B}]/2}, \tag{52}$$

which is applied for any two operators $\widehat{A}$ and $\widehat{B}$, whose commutator $[\widehat{A},\widehat{B}]$ is a number. Because $[d_{\mathbf{q}},d_{\mathbf{q}}^\dagger] = 1$, we can apply this identity in (51) to obtain

$$e^{[u_i(\mathbf{q},t)d_{\mathbf{q}}-H.c.]}e^{[-u_j(\mathbf{q})d_{\mathbf{q}}-H.c.]} = e^{(\alpha^* d_{\mathbf{q}}^\dagger-\alpha d_{\mathbf{q}})}\times$$
$$e^{[u_i(\mathbf{q},t)u_j^*(\mathbf{q})-u_i^*(\mathbf{q},t)u_j(\mathbf{q})]/2},$$

where $\alpha \equiv u_j(\mathbf{q},t) - u_i(\mathbf{q})$. Applying once again the same identity yields

$$e^{[u_i(\mathbf{q},t)d_{\mathbf{q}}-H.c.]}e^{[-u_j(\mathbf{q})d_{\mathbf{q}}-H.c.]} = e^{\alpha^* d_{\mathbf{q}}^\dagger}e^{-\alpha d_{\mathbf{q}}}e^{-|\alpha|^2/2}\times \tag{53}$$
$$e^{[u_i(\mathbf{q},t)u_j^*(\mathbf{q})-u_i^*(\mathbf{q},t)u_j(\mathbf{q})]/2}.$$

Now the quantum and statistical averages are calculated by the use of

$$\left\langle\!\!\left\langle e^{\alpha^* d_{\mathbf{q}}^\dagger}e^{-\alpha d_{\mathbf{q}}}\right\rangle\!\!\right\rangle = e^{-|\alpha|^2 n_\omega}, \tag{54}$$

where $n_\omega = [\exp(\omega_{\mathbf{q}}/T) - 1]^{-1}$ is the Bose–Einstein distribution function of phonons. Collecting all multiplies in (51) we arrive at

$$\left\langle\left\langle \widehat{X}_i(t)\widehat{X}_j^\dagger\right\rangle\right\rangle = \exp\left\{-\frac{1}{2N}\sum_{\mathbf{q}}|\gamma(\mathbf{q})|^2 f_{\mathbf{q}}(\mathbf{m}-\mathbf{n},t)\right\}, \tag{55}$$

where

$$f_{\mathbf{q}}(\mathbf{m},t) = [1-\cos(\mathbf{q}\cdot\mathbf{m})\cos(\omega_{\mathbf{q}}t)]\coth\frac{\omega_{\mathbf{q}}}{2T} + i\cos(\mathbf{q}\cdot\mathbf{m})\sin(\omega_{\mathbf{q}}t). \tag{56}$$

Here we used the symmetry of $\gamma(-\mathbf{q}) = \gamma(\mathbf{q})$, because of which terms containing $\sin(\mathbf{q}\cdot\mathbf{m})$ disappeared. The average $\left\langle\left\langle \widehat{X}_j^\dagger \widehat{X}_i(t)\right\rangle\right\rangle$, which is a multiplier in the second term in the brackets of (46), is obtained as

$$\left\langle\left\langle \widehat{X}_j^\dagger \widehat{X}_i(t)\right\rangle\right\rangle = \left\langle\left\langle \widehat{X}_i(t)\widehat{X}_j^\dagger\right\rangle\right\rangle^*. \tag{57}$$

To proceed with the analytical results we consider low temperatures, $T \ll \omega_{\mathbf{q}}$, when $\coth(\omega_{\mathbf{q}}/2T) \approx 1$. Then expanding the exponent in (55) yields

$$\left\langle\left\langle \widehat{X}_i(t)\widehat{X}_j^\dagger\right\rangle\right\rangle = Z\sum_{l=0}^{\infty}\frac{\left\{\sum_{\mathbf{q}}|\gamma(\mathbf{q})|^2 e^{i[\mathbf{q}\cdot(\mathbf{m}-\mathbf{n})-\omega_{\mathbf{q}}t]}\right\}^l}{(2N)^l l!}, \tag{58}$$

where

$$Z = \exp\left[-\frac{1}{2N}\sum_{\mathbf{q}}|\gamma(\mathbf{q})|^2\right]. \tag{59}$$

Then performing summation over $\mathbf{m}, \mathbf{n}, \mathbf{k}'$ and integration over time in (46) we arrive at [74]

$$A(\mathbf{k},\omega) = \sum_{l=0}^{\infty}\left[A_l^{(-)}(\mathbf{k},\omega) + A_l^{(+)}(\mathbf{k},\omega)\right], \tag{60}$$

where

$$A_l^{(-)}(\mathbf{k},\omega) = \pi Z\sum_{\mathbf{q}_1,\ldots\mathbf{q}_l}\frac{\prod_{r=1}^l|\gamma(\mathbf{q}_r)|^2}{(2N)^l l!}\times \tag{61}$$
$$\left[1-\bar{n}\left(\mathbf{k}-\sum_{r=1}^l\mathbf{q}_r\right)\right]\delta\left(\omega-\sum_{r=1}^l\omega_{\mathbf{q}_r}-\xi_{\mathbf{k}-\sum_{r=1}^l\mathbf{q}_r}\right),$$

and

$$A_l^{(+)}(\mathbf{k},\omega) = \pi Z\sum_{\mathbf{q}_1,\ldots\mathbf{q}_l}\frac{\prod_{r=1}^l|\gamma(\mathbf{q}_r)|^2}{(2N)^l l!}\times \tag{62}$$
$$\bar{n}\left(\mathbf{k}+\sum_{r=1}^l\mathbf{q}_r\right)\delta\left(\omega+\sum_{r=1}^l\omega_{\mathbf{q}_r}-\xi_{\mathbf{k}+\sum_{r=1}^l\mathbf{q}_r}\right).$$

Clearly (60) is in the form of a perturbative multi-phonon expansion. Each contribution $A_l^{(\pm)}(\mathbf{k}, \omega)$ to the spectral function describes the transition from the initial state $\mathbf{k}$ of the polaron band to the final state $\mathbf{k} \pm \sum_{r=1}^{l} \mathbf{q}_r$ with the emission (or absorption) of l phonons.

The $1/\lambda$ expansion result, (60), is applied to *low-energy* polaron excitations in the strong-coupling limit. In the case of the long-range Fröhlich interaction with high-frequency phonons it is also applied in the intermediate regime [39, 70]. Different from the conventional ME spectral function there is no imaginary part of the self-energy since the exponentially small at low temperatures polaronic damping [23] is neglected. Instead the e-ph coupling leads to the coherent dressing of electrons by phonons. The dressing can be seen as the phonon "side-bands" with $l \geq 1$.

While the major sum rule, (44) is satisfied,

$$\frac{1}{\pi} \int_{-\infty}^{\infty} d\omega\, A(\mathbf{k}, \omega) = Z \sum_{l=0}^{\infty} \sum_{\mathbf{q}_1, \ldots \mathbf{q}_l} \frac{\prod_{r=1}^{l} |\gamma(\mathbf{q}_r)|^2}{(2N)^l l!} = \tag{63}$$

$$Z \sum_{l=0}^{\infty} \frac{1}{l!} \left\{ \frac{1}{2N} \sum_{\mathbf{q}} |\gamma(\mathbf{q})|^2 \right\}^l = Z \exp\left[\frac{1}{2N} \sum_{\mathbf{q}} |\gamma(\mathbf{q})|^2 \right] = 1,$$

the higher-momentum integrals, $\int_{-\infty}^{\infty} d\omega \omega^p A(\mathbf{k}, \omega)$ with $p > 0$, calculated using (60), differ from the exact values (see Part III) by an amount proportional to $1/\lambda$. The difference is due to a partial "undressing" of high-energy excitations in the side-bands, which is beyond the first order $1/\lambda$ expansion.

The spectral function of the polaronic carriers comprises two different parts. The first ($l = 0$) $\mathbf{k}$-dependent *coherent* term arises from the polaron band tunnelling,

$$A_{coh}(\mathbf{k}, \omega) = \left[A_0^{(-)}(\mathbf{k}, \omega) + A_0^{(+)}(\mathbf{k}, \omega) \right] = \pi Z \delta(\omega - \xi_{\mathbf{k}}). \tag{64}$$

The spectral weight of the coherent part is suppressed as $Z \ll 1$. However in the case of the Fröhlich interaction the effective mass is less enhanced, $\xi_{\mathbf{k}} = Z' E_{\mathbf{k}} - \mu$, because $Z << Z' < 1$.

The second *incoherent* part $A_{incoh}(\mathbf{k}, \omega)$ comprises all the terms with $l \geq 1$. It describes the excitations accompanied by emission and absorption of phonons. We notice that its spectral density spreads over a wide energy range of about twice the polaron level shift E_p, which might be larger than the unrenormalised bandwidth $2D$ in the rigid lattice without phonons. On the contrary, the coherent part shows a dispersion only in the energy window of the order of the polaron bandwidth, $w = Z'D$. It is interesting that there is some $\mathbf{k}$ dependence of the *incoherent* background as well [74], if the matrix element of the e-ph interaction and/or phonon frequencies depend on $\mathbf{q}$. Only in the Holstein model with the short-range dispersionless e-ph interaction $\gamma(\mathbf{q}) = \gamma_0$ and $\omega_{\mathbf{q}} = \omega_0$ the incoherent part is momentum independent (see also [47]),

$$A_{incoh}(\mathbf{k}, \omega) = \pi \frac{Z}{N} \sum_{l=1}^{\infty} \frac{\gamma_0^{2l}}{2^l l!} \times \tag{65}$$

$$\sum_{\mathbf{k}'} \left\{ [1 - \bar{n}(\mathbf{k}')] \delta(\omega - l\omega_0 - \xi_{\mathbf{k}'}) + \bar{n}(\mathbf{k}') \delta(\omega + l\omega_0 - \xi_{\mathbf{k}'}) \right\}.$$

As soon as we know the spectral function, different Green's functions (GF) are readily obtained using their analytical properties. For example, the temperature GF is given by the integral [23]

$$\mathcal{G}(\mathbf{k}, \omega_k) = \frac{1}{\pi} \int_{-\infty}^{\infty} d\omega' \frac{A(\mathbf{k}, \omega')}{i\omega_k - \omega'}. \tag{66}$$

where $\omega_k = \pi T(2k + 1), k = 0, \pm 1, \pm 2, \dots$ Calculating the integral with the spectral density (65) we find in the Holstein model [75]

$$\mathcal{G}(\mathbf{k}, \omega_n) = \frac{Z}{i\omega_n - \xi_{\mathbf{k}}} + \frac{Z}{N} \sum_{l=1}^{\infty} \frac{\gamma_0^{2l}}{2^l l!} \sum_{\mathbf{k}'} \left\{ \frac{1 - \bar{n}(\mathbf{k}')}{i\omega_n - l\omega_0 - \xi_{\mathbf{k}'}} + \frac{\bar{n}(\mathbf{k}')}{i\omega_n + l\omega_0 - \xi_{\mathbf{k}'}} \right\}. \tag{67}$$

Here the first term describes the coherent tunnelling in the narrow polaron band while the second $\mathbf{k}$-independent sum describes the phonon-side bands.

5 Polaron–Polaron Interaction and Small Bipolaron

Polarons interact with each other, (29). The range of the deformation surrounding Fröhlich polarons is quite large, and their deformation fields overlap at finite density. Taking into account both the long-range attraction of polarons owing to the lattice deformations *and* their direct Coulomb repulsion, the residual *long-range* interaction turns out to be rather weak and repulsive in ionic crystals [1]. The Fourier component of the polaron-polaron interaction, $v(\mathbf{q})$, comprising the direct Coulomb repulsion and the attraction mediated by phonons is

$$v(\mathbf{q}) = \frac{4\pi e^2}{\epsilon_\infty q^2} - |\gamma(\mathbf{q})|^2 \omega_{\mathbf{q}}. \tag{68}$$

In the long-wave limit ($q \ll \pi/a$) the Fröhlich interaction [76] dominates in the attractive part, which is described by (9). Fourier transforming (68) yields the repulsive interaction in real space,

$$v(\mathbf{m} - \mathbf{n}) = \frac{e^2}{\epsilon_0 |\mathbf{m} - \mathbf{n}|} > 0. \tag{69}$$

We see that optical phonons nearly nullify the bare Coulomb repulsion in ionic solids, where $\epsilon_0 \gg 1$, but cannot overscreen it at large distances.

Considering the polaron–phonon interaction in the multi-polaron system we have to take into account dynamic properties of the polaron response function [60]. One may erroneously believe that the long-range Fröhlich interaction becomes a short-range (Holstein) one due to the screening of ions by heavy polaronic carriers. In fact, small polarons cannot screen high-frequency optical vibrations because their renormalised plasma frequency is comparable with or even less than the phonon frequency. In the absence of bipolarons (see below) we can apply the ordinary bubble approximation to calculate the dielectric response function of polarons at the frequency Ω as

$$\epsilon(\mathbf{q}, \Omega) = 1 - 2v(\mathbf{q}) \sum_{\mathbf{k}} \frac{\bar{n}(\mathbf{k}+\mathbf{q}) - \bar{n}(\mathbf{k})}{\Omega - \epsilon_{\mathbf{k}} + \epsilon_{\mathbf{k}+\mathbf{q}}}. \tag{70}$$

This expression describes the response of small polarons to any external field of the frequency $\Omega \leq \omega_0$, when phonons in the polaron cloud follow the polaron motion. In the static limit we obtain the usual Debye screening at large distances ($q \to 0$). For a temperature larger than the polaron half-bandwidth, $T > w$, we can approximate the polaron distribution function as

$$\bar{n}(\mathbf{k}) \approx \frac{n}{2a^3} \left(1 - \frac{(2-n)\epsilon_{\mathbf{k}}}{2T} \right), \tag{71}$$

and obtain

$$\epsilon(q, 0) = 1 + \frac{q_s^2}{q^2}, \tag{72}$$

where

$$q_s = \left[\frac{2\pi e^2 n(2-n)}{\epsilon_0 T a^3} \right]^{1/2},$$

and n is the number of polarons per unit cell. For a finite but rather low frequency, $\omega_0 > \Omega \gg w$, the polaron response becomes dynamic,

$$\epsilon(\mathbf{q}, \Omega) = 1 - \frac{\omega_p^2(\mathbf{q})}{\Omega^2} \tag{73}$$

where

$$\omega_p^2(\mathbf{q}) = 2v(\mathbf{q}) \sum_{\mathbf{k}} n(\mathbf{k})(\epsilon_{\mathbf{k}+\mathbf{q}} - \epsilon_{\mathbf{k}}) \tag{74}$$

is the temperature-dependent polaron plasma frequency squared. The polaron plasma frequency is rather low due to the large static dielectric constant, $\epsilon_0 \gg 1$, and the enhanced polaron mass $m^* \gg m_e$.

Now replacing the bare electron-phonon interaction vertex $\gamma(\mathbf{q})$ by a screened one, $\gamma_{sc}(\mathbf{q}, \omega_0)$, as shown in Fig. 3, we obtain

$$\gamma_{sc}(\mathbf{q}, \omega_0) = \frac{\gamma(\mathbf{q})}{\epsilon(\mathbf{q}, \omega_0)} \approx \gamma(\mathbf{q}) \tag{75}$$

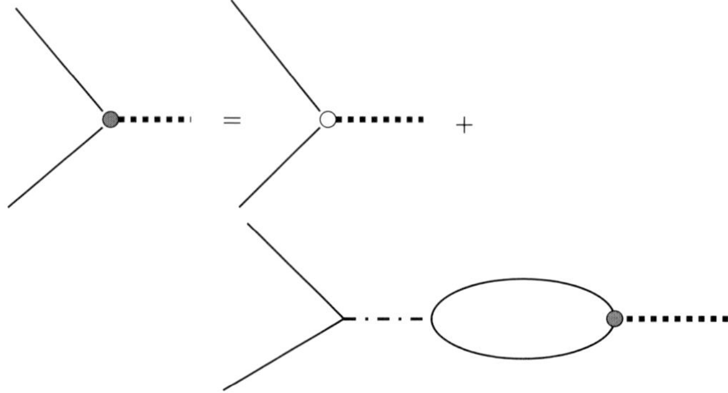

Fig. 3. E-ph vertex, $\gamma(\mathbf{q})$ screened by the polaron–polaron interaction, $v(\mathbf{q})$ (dashed–dotted line). Solid and dotted lines are polaron and phonon propagators, respectively.

because $\omega_0 > \omega_p$. Therefore, the singular behaviour of $\gamma(\mathbf{q}) \sim 1/q$ is unaffected by screening. Polarons are too slow to screen high-frequency crystal field oscillations. As a result, the strong interaction with high-frequency optical phonons in ionic solids remains unscreened at any density of small polarons.

Another important point is the possibility of Wigner crystallization of the polaronic liquid. Because the net long-range repulsion is relatively weak, the relevant dimensionless parameter $r_s = m^* e^2 / \epsilon_0 (4\pi n/3)^{1/3}$ is not very large in ionic semiconductors. The Wigner crystallization appears around $r_s \simeq 100$ or larger, which corresponds to the atomic density of polarons $n \leq 10^{-6}$ with $\epsilon_0 = 30$ and $m^* = 5m_e$. This estimate tells us that polaronic carriers are usually in the liquid state at relevant doping levels.

At large distances polarons repel each other. Nevertheless two *large* polarons can be bound into a *large* bipolaron by an exchange interaction even with no additional e-ph interaction but the Fröhlich one [77, 78]. When a short-range deformation potential and molecular-type (i.e. Holstein) e-ph interactions are taken into account together with the Fröhlich interaction, they overcome the Coulomb repulsion at a short distance of roughly the lattice constant. Then, owing to the narrow band, two small polarons easily form a bound state, i.e. a *small* bipolaron. Let us estimate the coupling constant λ and the adiabatic ratio ω_0/t, at which the "bipolaronic" instability occurs. The characteristic attractive potential is $|v| = D(\lambda - \mu_c)$, where μ_c is the dimensionless Coulomb repulsion, and λ includes the interaction with all phonon branches. The radius of the potential is roughly a. In three dimensions a bound state of two attractive particles appears, if

$$|v| \geq \frac{\pi^2}{8m^* a^2}. \tag{76}$$

Substituting the polaron mass, $m^* = [2a^2 t]^{-1} \exp(\gamma \lambda D/\omega_0)$, we find

$$\frac{t}{\omega_0} \leq (\gamma z \lambda)^{-1} \ln \left[\frac{\pi^2}{4z(\lambda - \mu_c)} \right]. \tag{77}$$

As a result, small bipolarons form at $\lambda \geq \mu_c + \pi^2/4z$ almost independent of the adiabatic ratio. In the case of the Fröhlich interaction there is no sharp transition between small and large polarons, and the first-order $1/\lambda$ expansion is accurate in the whole region of the e-ph coupling, if the adiabatic parameter is not too small. Hence we can say that in the antiadiabatic and intermediate regime the carriers are small polarons *independent* of the value of λ if the e-ph interaction is long-ranged. It means that they tunnel together with the entire phonon cloud no matter how "thin" the cloud is.

6 Polaronic Superconductivity

The polaron-polaron interaction is the sum of two large contributions of the opposite sign, (29). It is generally larger than the polaron bandwidth and the polaron Fermi-energy, $\epsilon_F = Z' E_F$. This condition is opposite to the weak-coupling BCS regime, where the Fermi energy is the largest. However, there is still a narrow window of parameters, where bipolarons are extended enough, and pairs of two small polarons overlap in a similar way to Cooper pairs. Here the BCS approach is applied to nonadiabatic carriers with a *nonretarded* attraction, so that bipolarons are Cooper pairs formed by two *small polarons* [17]. The size of the bipolaron is estimated as

$$r_b \approx \frac{1}{(m^* \Delta)^{1/2}}, \tag{78}$$

where Δ is the binding energy of the order of an attraction potential v. The BCS approach is applied if $r_b \gg n^{-1/3}$, which puts a severe constraint on the value of the attraction

$$|v| \ll \epsilon_F. \tag{79}$$

There is no "Tolmachev" logarithm in the case of nonadiabatic carriers, because the attraction is nonretarded as soon as $\epsilon_F \leq \omega_0$. Hence a superconducting state of small polarons is possible only if $\lambda > \mu_c$. This consideration leaves a rather narrow *crossover* region from the normal polaron Fermi liquid to a superconductor, where one can still apply the BCS mean-field approach,

$$0 < \lambda - \mu_c \ll Z' < 1. \tag{80}$$

In the case of the Fröhlich interaction Z' is about 0.1 for typical values of λ. Hence the region, (80), is on the borderline between a polaronic normal metal and a bipolaronic superconductor (Sect. 8).

   In the crossover region polarons behave like fermions in a narrow band with a weak nonretarded attraction. As long as $\lambda \gg 1/\sqrt{2z}$, we can neglect their residual interaction with phonons in the transformed Hamiltonian,

$$\tilde{H} \approx \sum_{i,j} \left[((\langle \hat{\sigma}_{ij} \rangle_{ph} - \mu \delta_{ij}) c_i^\dagger c_j + \frac{1}{2} v_{ij} c_i^\dagger c_j^\dagger c_j c_i \right] \tag{81}$$

written in the Wannier representation. If the condition (80) is satisfied, we can treat the polaron-polaron interaction approximately by the use of the BCS theory. For simplicity let us keep only the on-site v_0 and the nearest-neighbour v_1 interactions. At least one of them should be attractive to ensure that the ground state is superconducting. Introducing two order parameters

$$\Delta_0 = -v_0 \langle c_{\mathbf{m},\uparrow} c_{\mathbf{m},\downarrow} \rangle, \tag{82}$$

$$\Delta_1 = -v_1 \langle c_{\mathbf{m},\uparrow} c_{\mathbf{m}+\mathbf{a},\downarrow} \rangle \tag{83}$$

and transforming to the $\mathbf{k}$-space results in the familiar BCS Hamiltonian,

$$H_p = \sum_{\mathbf{k},s} \xi_{\mathbf{k}} c_{\mathbf{k}s}^\dagger c_{\mathbf{k}s} + \sum_{\mathbf{k}} [\Delta_{\mathbf{k}} c_{\mathbf{k}\uparrow}^\dagger c_{-\mathbf{k}\downarrow}^\dagger + H.c.], \tag{84}$$

where $\xi_{\mathbf{k}} = \epsilon_{\mathbf{k}} - \mu$ is the renormalised kinetic energy and

$$\Delta_{\mathbf{k}} = \Delta_0 - \Delta_1 \frac{\xi_{\mathbf{k}} + \mu}{w} \tag{85}$$

is the order parameter.

The Bogoliubov anomalous averages are found as

$$\langle c_{\mathbf{k},\uparrow} c_{-\mathbf{k},\downarrow} \rangle = \frac{\Delta_{\mathbf{k}}}{2\sqrt{\xi_{\mathbf{k}}^2 + \Delta_{\mathbf{k}}^2}} \tanh \frac{\sqrt{\xi_{\mathbf{k}}^2 + \Delta_{\mathbf{k}}^2}}{2T}, \tag{86}$$

and two coupled equations for the on-site and inter-site order parameters are

$$\Delta_0 = -\frac{v_0}{N} \sum_{\mathbf{k}} \frac{\Delta_{\mathbf{k}}}{2\sqrt{\xi_{\mathbf{k}}^2 + \Delta_{\mathbf{k}}^2}} \tanh \frac{\sqrt{\xi_{\mathbf{k}}^2 + \Delta_{\mathbf{k}}^2}}{2T}, \tag{87}$$

$$\Delta_1 = -\frac{v_1}{Nw} \sum_{\mathbf{k}} \frac{\Delta_{\mathbf{k}}(\xi_{\mathbf{k}} + \mu)}{2\sqrt{\xi_{\mathbf{k}}^2 + \Delta_{\mathbf{k}}^2}} \tanh \frac{\sqrt{\xi_{\mathbf{k}}^2 + \Delta_{\mathbf{k}}^2}}{2T}. \tag{88}$$

These equations are equivalent to a single BCS equation for $\Delta_{\mathbf{k}} = \Delta(\xi_{\mathbf{k}})$, but with the half polaron bandwidth w cutting the integral, rather than the Debye temperature,

$$\Delta(\xi) = \int_{-w-\mu}^{w-\mu} d\eta \, N_p(\eta) V(\xi, \eta) \frac{\Delta(\eta)}{2\sqrt{\eta^2 + \Delta(\eta)^2}} \tanh \frac{\sqrt{\eta^2 + \Delta(\eta)^2}}{2T}. \tag{89}$$

Here $V(\xi, \eta) = -v_0 - zv_1(\xi + \mu)(\eta + \mu)/w^2$.

The critical temperature T_c of the polaronic superconductor is determined by two linearised equations in the limit $\Delta_{0,1} \to 0$,

$$\left[1 + A\left(\frac{v_0}{zv_1} + \frac{\mu^2}{w^2}\right)\right]\Delta - \frac{B\mu}{w}\Delta_1 = 0, \tag{90}$$

$$-\frac{A\mu}{w}\Delta + (1 + B)\Delta_1 = 0, \tag{91}$$

where $\Delta = \Delta_0 - \Delta_1\mu/w$, and

$$A = \frac{zv_1}{2w}\int\limits_{-w-\mu}^{w-\mu} d\eta\,\frac{\tanh\frac{\eta}{2T_c}}{\eta},$$

$$B = \frac{zv_1}{2w}\int\limits_{-w-\mu}^{w-\mu} d\eta\,\frac{\eta\tanh\frac{\eta}{2T_c}}{w^2}.$$

These equations are applied only if the polaron-polaron coupling is weak, $|v_{0,1}| < w$. A nontrivial solution is found at

$$T_c \approx 1.14w\sqrt{1 - \frac{\mu^2}{w^2}}\,\exp\left(\frac{2w}{v_0 + zv_1\mu^2/w^2}\right), \tag{92}$$

if $v_0 + zv_1\mu^2/w^2 < 0$, so that superconductivity exists even in the case of the on-site repulsion, $v_0 > 0$, if this repulsion is less than the total intersite attraction, $z|v_1|$. There is a nontrivial dependence of T_c on doping. With a constant density of states in the polaron band, the Fermi energy $\epsilon_F \approx \mu$ is expressed via the number of polarons per atom n as

$$\mu = w(n - 1), \tag{93}$$

so that

$$T_c \simeq 1.14w\sqrt{n(2 - n)}\,\exp\left(\frac{2w}{v_0 + zv_1[n - 1]^2}\right), \tag{94}$$

which has two maxima as a function of n separated by a deep minimum in the half-filled band ($n = 1$), where the nearest-neighbour contributions to pairing are canceled.

7 Mobile Small Bipolarons

The attractive energy of two small polarons is generally larger than the polaron bandwidth, $\lambda - \mu_c \gg Z'$. When this condition is satisfied, small bipolarons are not overlapped. Consideration of particular lattice structures shows that small bipolarons are mobile even when the electron-phonon coupling is strong and the bipolaron binding energy is large [41, 71]. Here we encounter a novel electronic state of matter, a charged Bose liquid, qualitatively different from the normal Fermi-liquid and the BCS superfluid. The Bose-liquid is stable because bipolarons repel each other (see below).

7.1 On-Site Bipolarons and Bipolaronic Hamiltonian

The small parameter, $Z'/(\lambda - \mu_c) \ll 1$, allows for a consistent treatment of bipolaronic systems [17, 42]. Under this condition the hopping term in the transformed Hamiltonian $\widetilde{H}$ is a small perturbation of the ground state of immobile bipolarons and free phonons,

$$\widetilde{H} = H_0 + H_{pert}, \tag{95}$$

where

$$H_0 = \frac{1}{2} \sum_{i,j} v_{ij} c_i^\dagger c_j^\dagger c_j c_i + \sum_{\mathbf{q},\nu} \omega_{\mathbf{q}\nu} [d_{\mathbf{q}\nu}^\dagger d_{\mathbf{q}\nu} + 1/2] \tag{96}$$

and

$$H_{pert} = \sum_{i,j} \widehat{\sigma}_{ij} c_i^\dagger c_j. \tag{97}$$

Let us first discuss the dynamics of *onsite* bipolarons, which are the ground state of the system with the Holstein non-dispersive e-ph interaction [42, 52]. The onsite bipolaron is formed if

$$2E_p > U, \tag{98}$$

where U is the onsite Coulomb correlation energy (the Hubbard U). The intersite polaron-polaron interaction is just the Coulomb repulsion since the phonon mediated attraction between two polarons on different sites is zero in the Holstein model. Two or more onsite bipolarons as well as three or more polarons cannot occupy the same site because of the Pauli exclusion principle. Hence, bipolarons repel single polarons and each other. Their binding energy, $\Delta = 2E_p - U$, is larger than the polaron half-bandwidth, $\Delta \gg w$, so that there are no unbound polarons in the ground state. H_{pert}, (97), destroys bipolarons in the first order. Hence it has no diagonal matrix elements. Then the bipolaron dynamics, including superconductivity, is described by the use of a new canonical transformation $\exp(S_2)$ [42], which eliminates the first order of H_{pert},

$$(S_2)_{fp} = \sum_{i,j} \frac{\langle f|\widehat{\sigma}_{ij} c_i^\dagger c_j|p\rangle}{E_f - E_p}. \tag{99}$$

Here $E_{f,p}$ and $|f\rangle, |p\rangle$ are the energy levels and the eigenstates of H_0. Neglecting the terms of the order higher than $(w/\Delta)^2$ we obtain

$$(H_b)_{ff'} \equiv \left(e^{S_2} \widetilde{H} e^{-S_2}\right)_{ff'}, \tag{100}$$

$$(H_b)_{ff'} \approx (H_0)_{ff'} - \frac{1}{2} \sum_{\nu} \sum_{i\neq i', j\neq j'} \langle f|\widehat{\sigma}_{ii'} c_i^\dagger c_{i'}|p\rangle \langle p|\widehat{\sigma}_{jj'} c_j^\dagger c_{j'}|f'\rangle \times$$
$$\left(\frac{1}{E_p - E_{f'}} + \frac{1}{E_p - E_f}\right).$$

S_2 couples a localised onsite bipolaron and a state of two unbound polarons on different sites. The expression (100) determines the matrix elements of the transformed *bipolaronic* Hamiltonian H_b in the subspace $|f\rangle, |f'\rangle$ with no single (unbound) polarons. On the other hand, the intermediate *bra* $\langle p|$ and *ket* $|p\rangle$ refer to configurations involving two unpaired polarons and any number of phonons. Hence we have

$$E_p - E_f = \Delta + \sum_{\mathbf{q},\nu} \omega_{\mathbf{q}\nu} \left(n^p_{\mathbf{q}\nu} - n^f_{\mathbf{q}\nu}\right), \tag{101}$$

where $n^{f,p}_{\mathbf{q}\nu}$ are phonon occupation numbers $(0, 1, 2, 3...\infty)$. This equation is an explicit definition of the bipolaron binding energy Δ which takes into account the residual inter-site repulsion between bipolarons and between two unpaired polarons. The lowest eigenstates of H_b are in the subspace, which has only doubly occupied $c^\dagger_{\mathbf{m}s} c^\dagger_{\mathbf{m}s'}|0\rangle$ or empty $|0\rangle$ sites. On-site bipolaron tunnelling is a two-step transition. It takes place via a single polaron tunneling to a neighbouring site. The subsequent tunnelling of its "partner" to the same site restores the initial energy state of the system. There are no *real* phonons emitted or absorbed because the (bi)polaron band is narrow. Hence we can average H_b with respect to phonons. Replacing the energy denominators in the second term in (100) by the integrals with respect to time,

$$\frac{1}{E_p - E_f} = i \int_0^\infty dt\, e^{i(E_f - E_p + i\delta)t},$$

we obtain

$$H_b = H_0 - i \sum_{\mathbf{m}\neq\mathbf{m}',s} \sum_{\mathbf{n}\neq\mathbf{n}',s'} T(\mathbf{m} - \mathbf{m}')T(\mathbf{n} - \mathbf{n}') \times \tag{102}$$

$$c^\dagger_{\mathbf{m}s} c_{\mathbf{m}'s} c^\dagger_{\mathbf{n}s'} c_{\mathbf{n}'s'} \int_0^\infty dt\, e^{-i\Delta t} \Phi^{\mathbf{nn}'}_{\mathbf{mm}'}(t).$$

Here $\Phi^{\mathbf{nn}'}_{\mathbf{mm}'}(t)$ is a multiphonon correlator,

$$\Phi^{\mathbf{nn}'}_{\mathbf{mm}'}(t) \equiv \left\langle\left\langle \widehat{X}^\dagger_i(t)\widehat{X}_{i'}(t)\widehat{X}^\dagger_j\widehat{X}_{j'} \right\rangle\right\rangle. \tag{103}$$

$\widehat{X}^\dagger_i(t)$ and $\widehat{X}_{i'}(t)$ commute for any $\gamma(\mathbf{q},\nu) = \gamma(-\mathbf{q},\nu)$. $\widehat{X}^\dagger_j$ and $\widehat{X}_{j'}$ commute as well, so that we can write

$$\widehat{X}^\dagger_i(t)\widehat{X}_{i'}(t) = \prod_{\mathbf{q}} e^{[u_{i'}(\mathbf{q},t) - u_i(\mathbf{q},t)]d_{\mathbf{q}} - H.c.}, \tag{104}$$

$$\widehat{X}^\dagger_j\widehat{X}_{j'} = \prod_{\mathbf{q}} e^{[u_{j'}(\mathbf{q}) - u_j(\mathbf{q})]d_{\mathbf{q}} - H.c.}, \tag{105}$$

where the phonon branch index ν is dropped for transparency. Applying twice the identity (52) yields

$$\hat{X}_i^\dagger(t)\hat{X}_{i'}(t)\hat{X}_j^\dagger\hat{X}_{j'} = \prod_{\mathbf{q}} e^{\beta^* d_{\mathbf{q}}^\dagger} e^{-\beta d_{\mathbf{q}}} e^{-|\beta|^2/2} \times \tag{106}$$

$$e^{[u_{i'}(\mathbf{q},t)-u_i(\mathbf{q},t)][u_{j'}^*(\mathbf{q})-u_j^*(\mathbf{q})]/2-H.c.},$$

where

$$\beta = u_i(\mathbf{q},t) - u_{i'}(\mathbf{q},t) + u_j(\mathbf{q}) - u_j'(\mathbf{q}).$$

Finally using the average (54) we find

$$\Phi_{\mathbf{mm'}}^{\mathbf{nn'}}(t) = e^{-g^2(\mathbf{m}-\mathbf{m'})} e^{-g^2(\mathbf{n}-\mathbf{n'})} \times \tag{107}$$

$$\exp\left\{ \frac{1}{2N} \sum_{\mathbf{q},\nu} |\gamma(\mathbf{q},\nu)|^2 F_{\mathbf{q}}(\mathbf{m},\mathbf{m'},\mathbf{n},\mathbf{n'}) \frac{\cosh\left[\omega_{\mathbf{q}\nu}\left(\frac{1}{2T} - it\right)\right]}{\sinh\left[\frac{\omega_{\mathbf{q}\nu}}{2T}\right]} \right\},$$

where

$$F_{\mathbf{q}}(\mathbf{m},\mathbf{m'},\mathbf{n},\mathbf{n'}) = \cos[\mathbf{q}\cdot(\mathbf{n'}-\mathbf{m})] + \cos[\mathbf{q}\cdot(\mathbf{n}-\mathbf{m'})] - \tag{108}$$

$$\cos[\mathbf{q}\cdot(\mathbf{n'}-\mathbf{m'})] - \cos[\mathbf{q}\cdot(\mathbf{n}-\mathbf{m})].$$

Taking into account that there are only bipolarons in the subspace, where H_b operates, we finally rewrite the Hamiltonian in terms of the creation $b_{\mathbf{m}}^\dagger = c_{\mathbf{m}\uparrow}^\dagger c_{\mathbf{m}\downarrow}^\dagger$ and annihilation $b_{\mathbf{m}} = c_{\mathbf{m}\downarrow}c_{\mathbf{m}\uparrow}$ operators of singlet pairs as

$$H_b = -\sum_{\mathbf{m}}\left[\Delta + \frac{1}{2}\sum_{\mathbf{m'}} v^{(2)}(\mathbf{m}-\mathbf{m'})\right] n_{\mathbf{m}} + \tag{109}$$

$$\sum_{\mathbf{m}\neq\mathbf{m'}}\left[t(\mathbf{m}-\mathbf{m'})b_{\mathbf{m}}^\dagger b_{\mathbf{m'}} + \frac{1}{2}\bar{v}(\mathbf{m}-\mathbf{m'})n_{\mathbf{m}}n_{\mathbf{m'}}\right].$$

There are no triplet pairs in the Holstein model, because the Pauli exclusion principle does not allow two electrons with the same spin to occupy the same site. Here $n_{\mathbf{m}} = b_{\mathbf{m}}^\dagger b_{\mathbf{m}}$ is the bipolaron site-occupation operator,

$$\bar{v}(\mathbf{m}-\mathbf{m'}) = 4v(\mathbf{m}-\mathbf{m'}) + v^{(2)}(\mathbf{m}-\mathbf{m'}), \tag{110}$$

is the bipolaron-bipolaron interaction including the direct polaron-polaron interaction $v(\mathbf{m}-\mathbf{m'})$ and a second order in $T(\mathbf{m})$ repulsive correlation

$$v^{(2)}(\mathbf{m}-\mathbf{m'}) = 2i \int_0^\infty dt\, e^{-i\Delta t} \Phi_{\mathbf{mm'}}^{\mathbf{m'm}}(t). \tag{111}$$

This additional repulsion appears because a virtual hop of the pair polarons is forbidden, if the neighbouring site is occupied by another pair. The bipolaron transfer integral is of the second order in $T(\mathbf{m})$

$$t(\mathbf{m} - \mathbf{m}') = -2iT^2(\mathbf{m} - \mathbf{m}') \int_0^\infty dt\, e^{-i\Delta t} \Phi_{\mathbf{mm}'}^{\mathbf{mm}'}(t). \qquad (112)$$

The *bipolaronic* Hamiltonian, (109) describes the low-energy physics of strongly coupled electrons and phonons. Using the explicit form of the multiphonon correlator, (107), we obtain for dispersionless phonons at $T \ll \omega_0$,

$$\Phi_{\mathbf{mm}'}^{\mathbf{mm}'}(t) = e^{-2g^2(\mathbf{m}-\mathbf{m}')} \exp\left[-2g^2(\mathbf{m} - \mathbf{m}')e^{-i\omega_0 t}\right],$$

$$\Phi_{\mathbf{mm}'}^{\mathbf{m}'\mathbf{m}}(t) = e^{-2g^2(\mathbf{m}-\mathbf{m}')} \exp\left[2g^2(\mathbf{m} - \mathbf{m}')e^{-i\omega_0 t}\right].$$

Expanding the time-dependent exponents in the Fourier series and calculating the integrals in (112) and (111) yields [79]

$$t(\mathbf{m}) = -\frac{2T^2(\mathbf{m})}{\Delta} e^{-2g^2(\mathbf{m})} \sum_{l=0}^\infty \frac{[-2g^2(\mathbf{m})]^l}{l!(1 + l\omega_0/\Delta)} \qquad (113)$$

and

$$v^{(2)}(\mathbf{m}) = \frac{2T^2(\mathbf{m})}{\Delta} e^{-2g^2(\mathbf{m})} \sum_{l=0}^\infty \frac{[2g^2(\mathbf{m})]^l}{l!(1 + l\omega_0/\Delta)}. \qquad (114)$$

When $\Delta \ll \omega_0$, we can keep the first term only with $l = 0$ in the bipolaron hopping integral, (113). In this case the bipolaron half-bandwidth $zt(\mathbf{a})$ is of the order of $2w^2/(z\Delta)$. However, if the bipolaron binding energy is large, $\Delta \gg \omega_0$, the bipolaron bandwidth dramatically decreases proportionally to e^{-4g^2} in the limit $\Delta \to \infty$. However, this limit is not realistic because $\Delta = 2E_p - V_c < 2g^2\omega_0$. In a more realistic regime, $\omega_0 < \Delta < 2g^2\omega_0$, (113) yields

$$t(\mathbf{m}) \approx \frac{2\sqrt{2\pi}T^2(\mathbf{m})}{\sqrt{\omega_0\Delta}} \exp\left[-2g^2 - \frac{\Delta}{\omega_0}\left(1 + \ln\frac{2g^2(\mathbf{m})\omega_0}{\Delta}\right)\right]. \qquad (115)$$

On the contrary, the bipolaron-bipolaron repulsion, (114) has no small exponent in the limit $\Delta \to \infty$, $v^{(2)} \propto D^2/\Delta$. Together with the direct Coulomb repulsion the second order $v^{(2)}$ ensures stability of the bipolaronic liquid against clustering.

The high temperature behavior of the bipolaron bandwidth is just the opposite to that of the small polaron bandwidth. While the polaron band collapses with increasing temperature [8], the bipolaron band becomes wider [80],

$$t(\mathbf{m}) \propto \frac{1}{\sqrt{T}} \exp\left[-\frac{E_p + \Delta}{2T}\right] \qquad (116)$$

for $T > \omega_0$.

7.2 Superlight Intersite Bipolarons in the Fröhlich–Coulomb Model (FCM)

Any realistic theory of doped ionic insulators must include both the long-range Coulomb repulsion between carriers and the strong long-range electron-phonon interaction. From a theoretical standpoint, the inclusion of the long-range Coulomb repulsion is critical in ensuring that the carriers would not form clusters. Indeed, in order to form stable bipolarons, the e-ph interaction has to be strong enough to overcome the Coulomb repulsion at short distances. Since the realistic e-ph interaction is long-ranged, there is a potential possibility for clustering. The inclusion of the Coulomb repulsion V_c makes the clusters unstable. More precisely, there is a certain window of V_c/E_p inside which the clusters are unstable but mobile bipolarons form nonetheless. In this parameter window bipolarons repel each other and propagate in a narrow band.

Let us consider a generic "Fröhlich–Coulomb" Hamiltonian, which explicitly includes the infinite-range Coulomb and electron-phonon interactions, in a particular lattice structure [81]. The implicitly present infinite Hubbard U prohibits double occupancy and removes the need to distinguish the fermionic spin, as soon as we are interested in the charge excitations alone. Introducing spinless fermion operators $c_{\mathbf{n}}$ and phonon operators $d_{\mathbf{m}\nu}$, the Hamiltonian is written as

$$H = \sum_{\mathbf{n}\neq\mathbf{n}'} T(\mathbf{n} - \mathbf{n}')c_{\mathbf{n}}^{\dagger}c_{\mathbf{n}'} + \sum_{\mathbf{n}\neq\mathbf{n}'} V_c(\mathbf{n} - \mathbf{n}')c_{\mathbf{n}}^{\dagger}c_{\mathbf{n}}c_{\mathbf{n}'}^{\dagger}c_{\mathbf{n}'} + \tag{117}$$

$$\omega_0 \sum_{\mathbf{n}\neq\mathbf{m},\nu} g_\nu(\mathbf{m} - \mathbf{n})(\mathbf{e}_\nu \cdot \mathbf{e}_{\mathbf{m}-\mathbf{n}})c_{\mathbf{n}}^{\dagger}c_{\mathbf{n}}(d_{\mathbf{m}\nu}^{\dagger} + d_{\mathbf{m}\nu}) +$$

$$\omega_0 \sum_{\mathbf{m},\nu} \left(d_{\mathbf{m}\nu}^{\dagger}d_{\mathbf{m}\nu} + \frac{1}{2}\right).$$

The e-ph term is written in the real space representation (Sect. 2), which is more convenient in working with complex lattices.

In general, the many-body model (117) is of considerable complexity. However, we are interested in the non/near adiabatic limit of the strong e-ph interaction. In this case, the kinetic energy is a perturbation and the model can be grossly simplified using the Lang–Firsov canonical transformation in the Wannier representation for electrons and phonons,

$$S = \sum_{\mathbf{m}\neq\mathbf{n},\nu} g_\nu(\mathbf{m} - \mathbf{n})(\mathbf{e}_\nu \cdot \mathbf{e}_{\mathbf{m}-\mathbf{n}})c_{\mathbf{n}}^{\dagger}c_{\mathbf{n}}(d_{\mathbf{m}\nu}^{\dagger} - d_{\mathbf{m}\nu}).$$

The transformed Hamiltonian is

$$\tilde{H} = e^{-S}He^{S} = \sum_{\mathbf{n} \neq \mathbf{n}'} \hat{\sigma}_{\mathbf{nn}'} c_{\mathbf{n}}^{\dagger} c_{\mathbf{n}'} + \omega_0 \sum_{\mathbf{m}\alpha} \left(d_{\mathbf{m}\nu}^{\dagger} d_{\mathbf{m}\nu} + \frac{1}{2} \right) + \quad (118)$$

$$\frac{1}{2} \sum_{\mathbf{n} \neq \mathbf{n}'} v(\mathbf{n} - \mathbf{n}') c_{\mathbf{n}}^{\dagger} c_{\mathbf{n}} c_{\mathbf{n}'}^{\dagger} c_{\mathbf{n}'} - E_p \sum_{\mathbf{n}} c_{\mathbf{n}}^{\dagger} c_{\mathbf{n}}.$$

The last term describes the energy gained by polarons due to e-ph interactions. E_p is the familiar polaron level shift,

$$E_p = \omega_0 \sum_{\mathbf{m}\nu} g_{\nu}^2(\mathbf{m} - \mathbf{n})(\mathbf{e}_{\nu} \cdot \mathbf{e}_{\mathbf{m}-\mathbf{n}})^2, \quad (119)$$

which is independent of $\mathbf{n}$. The third term on the right-hand side in (118) is the polaron-polaron interaction:

$$v(\mathbf{n} - \mathbf{n}') = V_c(\mathbf{n} - \mathbf{n}') - V_{ph}(\mathbf{n} - \mathbf{n}'), \quad (120)$$

where

$$V_{ph}(\mathbf{n} - \mathbf{n}') = 2\omega_0 \sum_{\mathbf{m},\nu} g_{\nu}(\mathbf{m} - \mathbf{n}) g_{\nu}(\mathbf{m} - \mathbf{n}') \times$$

$$(\mathbf{e}_{\nu} \cdot \mathbf{e}_{\mathbf{m}-\mathbf{n}})(\mathbf{e}_{\nu} \cdot \mathbf{e}_{\mathbf{m}-\mathbf{n}'}).$$

The phonon-induced interaction V_{ph} is due to displacements of common ions by two electrons. The transformed hopping operator $\hat{\sigma}_{\mathbf{nn}'}$ in the first term in (118) is given by

$$\hat{\sigma}_{\mathbf{nn}'} = T(\mathbf{n} - \mathbf{n}') \exp \left[\sum_{\mathbf{m},\nu} [g_{\nu}(\mathbf{m} - \mathbf{n})(\mathbf{e}_{\nu} \cdot \mathbf{e}_{\mathbf{m}-\mathbf{n}}) \right. \quad (121)$$

$$\left. - g_{\nu}(\mathbf{m} - \mathbf{n}')(\mathbf{e}_{\nu} \cdot \mathbf{e}_{\mathbf{m}-\mathbf{n}'})] (d_{\mathbf{m}\alpha}^{\dagger} - d_{\mathbf{m}\alpha}) \right].$$

This term is a perturbation at large λ. Here we consider a particular lattice structure (ladder), where bipolarons tunnel already in the first order in $T(\mathbf{n})$, so that $\hat{\sigma}_{\mathbf{nn}'}$ can be averaged over phonons. When $T \ll \omega_0$ the result is

$$t(\mathbf{n} - \mathbf{n}') \equiv \langle\langle \hat{\sigma}_{\mathbf{nn}'} \rangle\rangle_{ph} = T(\mathbf{n} - \mathbf{n}') \exp[-g^2(\mathbf{n} - \mathbf{n}')], \quad (122)$$

$$g^2(\mathbf{n} - \mathbf{n}') = \sum_{\mathbf{m},\nu} g_{\nu}(\mathbf{m} - \mathbf{n})(\mathbf{e}_{\nu} \cdot \mathbf{e}_{\mathbf{m}-\mathbf{n}}) \times$$

$$[g_{\nu}(\mathbf{m} - \mathbf{n})(\mathbf{e}_{\nu} \cdot \mathbf{e}_{\mathbf{m}-\mathbf{n}}) - g_{\nu}(\mathbf{m} - \mathbf{n}')(\mathbf{e}_{\nu} \cdot \mathbf{e}_{\mathbf{m}-\mathbf{n}'})].$$

The mass renormalization exponent can be expressed via E_p and V_{ph} as

$$g^2(\mathbf{n} - \mathbf{n}') = \frac{1}{\omega_0} \left[E_p - \frac{1}{2} V_{ph}(\mathbf{n} - \mathbf{n}') \right]. \quad (123)$$

Now phonons are "integrated out" and the polaronic Hamiltonian is given by

$$H_p = H_0 + H_{pert}, \tag{124}$$

$$H_0 = -E_p \sum_{\mathbf{n}} c_{\mathbf{n}}^\dagger c_{\mathbf{n}} + \frac{1}{2} \sum_{\mathbf{n} \neq \mathbf{n}'} v(\mathbf{n} - \mathbf{n}') c_{\mathbf{n}}^\dagger c_{\mathbf{n}} c_{\mathbf{n}'}^\dagger c_{\mathbf{n}'},$$

$$H_{pert} = \sum_{\mathbf{n} \neq \mathbf{n}'} t(\mathbf{n} - \mathbf{n}') c_{\mathbf{n}}^\dagger c_{\mathbf{n}'}.$$

When V_{ph} exceeds V_c the full interaction becomes negative and polarons form pairs. The real space representation allows us to elaborate more physics behind the lattice sums in (119) and (120) [81]. When a carrier (electron or hole) acts on an ion with a force $\mathbf{f}$, it displaces the ion by some vector $\mathbf{x} = \mathbf{f}/k$. Here k is the ion's force constant. The total energy of the carrier-ion pair is $-\mathbf{f}^2/(2k)$. This is precisely the summand in (119) expressed via dimensionless coupling constants. Now consider two carriers interacting with the *same* ion, Fig. 4a. The ion displacement is $\mathbf{x} = (\mathbf{f}_1 + \mathbf{f}_2)/k$ and the energy is $-\mathbf{f}_1^2/(2k) - \mathbf{f}_2^2/(2k) - (\mathbf{f}_1 \cdot \mathbf{f}_2)/k$. Here the last term should be interpreted as an ion-mediated interaction between the two carriers. It depends on the scalar product of $\mathbf{f}_1$ and $\mathbf{f}_2$ and consequently on the relative positions of the carriers with respect to the ion. If the ion is an isotropic harmonic oscillator, as we assume here, then the following simple rule applies. If the angle ϕ between $\mathbf{f}_1$ and $\mathbf{f}_2$ is less than $\pi/2$ the polaron-polaron interaction will be attractive, if otherwise it will be repulsive. In general, some ions will generate attraction between polarons, and some will generate repulsion, Fig. 4b.

The overall sign and magnitude of the interaction is given by the lattice sum in (120), the evaluation of which is elementary. One should also note that according to (123) an attractive e-ph interaction reduces the polaron mass (and consequently the bipolaron mass), while repulsive e-ph interaction enhances the mass. Thus, the long-range nature of the e-ph interaction serves a double purpose. Firstly, it generates an additional inter-polaron attraction because the distant ions have small angle ϕ. This additional attraction helps to overcome the direct Coulomb repulsion between polarons. And secondly, the Fröhlich interaction makes the bipolarons lighter.

The many-particle ground state of H_0 depends on the sign of the polaron-polaron interaction, the carrier density, and the lattice geometry. Here we consider the zig-zag ladder, Fig. 5a, assuming that all sites are isotropic two-dimensional harmonic oscillators. For simplicity, we also adopt the nearest-neighbour approximation for both interactions, $g_\nu(\mathbf{l}) \equiv g$, $V_c(\mathbf{n}) \equiv V_c$, and for the hopping integrals, $T(\mathbf{m}) = T_{NN}$ for $l = n = m = a$, and zero otherwise. Hereafter we set the lattice period $a = 1$. There are four nearest neighbours in the ladder, $z = 4$. Then the *one-particle* polaronic Hamiltonian takes the form

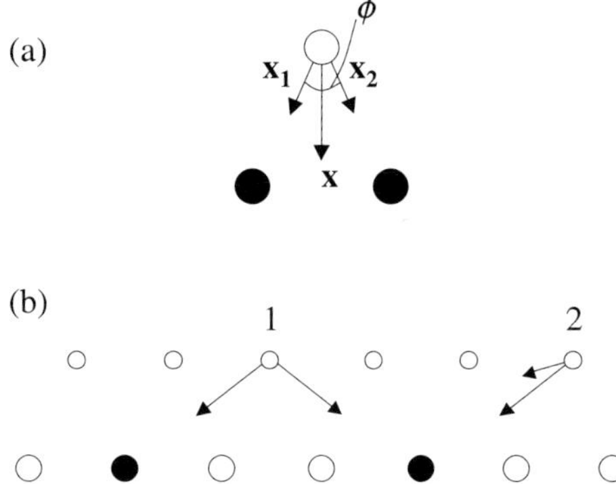

Fig. 4. Mechanism of the polaron-polaron interaction. (a) Together, two polarons (solid circles) deform the lattice more effectively than they would separately. An effective attraction occurs when the angle ϕ between $\mathbf{x}_1$ and $\mathbf{x}_2$ is less than $\pi/2$. (b) A mixed situation: ion 1 results in repulsion between two polarons while ion 2 results in attraction. Reproduced from [81], ©(2002) IOP Publishing Ltd.

$$H_p = -E_p \sum_n (c_n^\dagger c_n + p_n^\dagger p_n) + \tag{125}$$

$$\sum_n [t'(c_{n+1}^\dagger c_n + p_{n+1}^\dagger p_n) + t(p_n^\dagger c_n + p_{n-1}^\dagger c_n) + H.c.],$$

where c_n and p_n are polaron annihilation operators on the lower and upper legs of the ladder, respectively, Fig. 5b. Using (119), (120) and (123) we find

$$E_p = 4g^2\omega_0, \tag{126}$$

$$t' = T_{NN} \exp\left(-\frac{7E_p}{8\omega_0}\right),$$

$$t = T_{NN} \exp\left(-\frac{3E_p}{4\omega_0}\right).$$

The Fourier transform of (125) into momentum space yields

$$H_p = \sum_k (2t' \cos k - E_p)(c_k^\dagger c_k + p_k^\dagger p_k) + \tag{127}$$

$$t \sum_k [(1 + e^{ik})p_k^\dagger c_k + H.c.].$$

A linear transformation of c_k and p_k diagonalises the Hamiltonian. There are two overlapping polaronic bands,

$$E_1(k) = -E_p + 2t' \cos(k) \pm 2t \cos(k/2) \tag{128}$$

with the effective mass $m^* = 2/|4t' \pm t|$ near their edges.

Let us now place two polarons on the ladder. The nearest neighbour interaction is $v = V_c - E_p/2$, if two polarons are on different legs of the ladder, and $v = V_c - E_p/4$, if both polarons are on the same leg. The attractive interaction is provided via the displacement of the lattice sites, which are the common nearest neighbours to both polarons. There are two such nearest neighbours for the intersite bipolaron of type A or B, Fig. 5c, but there is only one common nearest neighbour for bipolaron C, Fig. 5d. When $V_c > E_p/2$, there are no bound states and the multi-polaron system is a one-dimensional Luttinger liquid. However, when $V_c < E_p/2$ and consequently $v < 0$, the two polarons are bound into an inter-site bipolaron of types A or B.

It is quite remarkable that the bipolaron tunnelling in the ladder already appears in the first order with respect to a single-electron tunnelling. This case is different from both onsite bipolarons discussed above, and from intersite chain bipolarons of [82], where the bipolaron tunnelling was of the second order in $T(a)$. Indeed, the lowest energy degenerate configurations A and B are degenerate. They are coupled by H_{pert}. Neglecting all higher-energy configurations, we can project the Hamiltonian onto the subspace containing A, B, and empty sites.

The result of such a projection is the bipolaronic Hamiltonian

$$H_b = \left(V_c - \frac{5}{2}E_p \right) \sum_n [A_n^\dagger A_n + B_n^\dagger B_n] - t' \sum_n [B_n^\dagger A_n + B_{n-1}^\dagger A_n + H.c.], \quad (129)$$

where $A_n = c_n p_n$ and $B_n = p_n c_{n+1}$ are intersite bipolaron annihilation operators, and the bipolaron-bipolaron interaction is dropped (see below). The Fourier transform of (129) yields two *bipolaron* bands,

$$E_2(k) = V_c - \frac{5}{2}E_p \pm 2t' \cos(k/2). \quad (130)$$

with a combined width $4|t'|$. The bipolaron binding energy in zero order with respect to t, t' is

$$\Delta \equiv 2E_1(0) - E_2(0) = \frac{E_p}{2} - V_c. \quad (131)$$

The bipolaron mass near the bottom of the lowest band, $m^{**} = 2/t'$, is

$$m^{**} = 4m^* \left[1 + 0.25 \exp\left(\frac{E_p}{8\omega_0} \right) \right]. \quad (132)$$

The numerical coefficient $1/8$ in the exponent ensures that m^{**} remains of the order of m^* even at large E_p. This fact combines with a weaker renormalization of m^* providing a *superlight* bipolaron.

In models with strong intersite attraction there is a possibility of clustering. Similar to the two-particle case above, the lowest energy of n polarons placed on the nearest neighbours of the ladder is found as

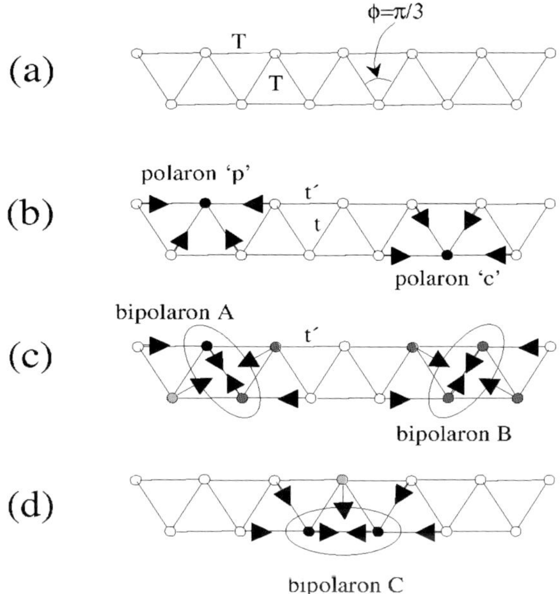

Fig. 5. One-dimensional zig-zag ladder. (a) Initial ladder with the bare hopping amplitude $T(a)$. (b) Two types of polarons with their respective deformations. (c) Two degenerate bipolaron configurations A and B. (d) A different bipolaron configuration, C, whose energy is higher than that of A and B. Reproduced from [81], ©(2002) IOP Publishing Ltd.

$$E_n = (2n - 3)V_c - \frac{6n - 1}{4}E_p \tag{133}$$

for any $n \geq 3$. There are *no* resonating states for an n-polaron configuration if $n \geq 3$. Therefore there is no first-order kinetic energy contribution to their energy. E_n should be compared with the energy $E_1 + (n - 1)E_2/2$ of widely separated $(n - 1)/2$ bipolarons and a single polaron for odd $n \geq 3$, or with the energy of widely separated n bipolarons for even $n \geq 4$. "Odd" clusters are stable if

$$V_c < \frac{n}{6n - 10}E_p, \tag{134}$$

and "even" clusters are stable if

$$V_c < \frac{n - 1}{6n - 12}E_p. \tag{135}$$

As a result we find that bipolarons repel each other and single polarons at $V_c > 3E_p/8$. If V_c is less than $3E_p/8$ then immobile bound clusters of three and more polarons could form. One should notice that at distances much larger than the lattice constant the polaron-polaron interaction is always repulsive, and the formation of infinite clusters, stripes or strings is impossible (see also [48]). Combining the condition of bipolaron formation and that of the instability of larger clusters we obtain a window of parameters

$$\frac{3}{8}E_p < V_c < \frac{1}{2}E_p, \tag{136}$$

where the ladder is a bipolaronic conductor. Outside the window the ladder is either charge-segregated into finite-size clusters (small V_c), or it is a liquid of repulsive polarons (large V_c).

There is strong experimental evidence for superlight intersite bipolarons in cuprate superconductors (see below), where they form in-plane oxygen - apex oxygen pairs (so called apex bipolarons) and/or in-plane oxygen-oxygen pairs [1, 21, 23, 41, 83]. While the long-range Fröhlich interaction combined with Coulomb repulsion might cause clustering of polarons into finite-size quasi-metallic mesoscopic textures, the analytical [84] and QMC [85] studies of mesoscopic textures with lattice deformations and Coulomb repulsion show that pairs dominate over phase separation since bipolarons effectively repel each other (see also [48].)

8 Bipolaronic Superconductivity

In the subspace with no single polarons, the Hamiltonian of electrons strongly-coupled with phonons is reduced to the bipolaronic Hamiltonian written in terms of creation, $b_{\mathbf{m}}^\dagger = c_{\mathbf{m}\uparrow}^\dagger c_{\mathbf{m}\downarrow}^\dagger$ and annihilation, $b_{\mathbf{m}}$, bipolaron operators as

$$H_b = \sum_{\mathbf{m}\neq\mathbf{m}'} \left[t(\mathbf{m} - \mathbf{m}')b_{\mathbf{m}}^\dagger b_{\mathbf{m}'} + \frac{1}{2}\bar{v}(\mathbf{m} - \mathbf{m}')n_{\mathbf{m}}n_{\mathbf{m}'} \right], \tag{137}$$

where $\bar{v}(\mathbf{m} - \mathbf{m}')$ is the bipolaron-bipolaron interaction, $n_{\mathbf{m}} = b_{\mathbf{m}}^\dagger b_{\mathbf{m}}$, and the position of the middle of the bipolaron band is taken as zero. There are additional spin quantum numbers $S = 0, 1; S_z = 0, \pm 1$, which should be added to the definition of $b_{\mathbf{m}}$ in the case of intersite bipolarons, which tunnel via one-particle hopping. This "crab-like" tunnelling, Fig. 5, results in a bipolaron bandwidth of the same order as the polaron one. Keeping this in mind we can apply H_b, (137) to both on-site and/or inter-site bipolarons, and even to more extended non-overlapping pairs, implying that the site index $\mathbf{m}$ is the position of the center of mass of a pair.

Bipolarons are not perfect bosons. In the subspace of pairs and empty sites their operators commute as

$$b_{\mathbf{m}}b_{\mathbf{m}}^\dagger + b_{\mathbf{m}}^\dagger b_{\mathbf{m}} = 1, \tag{138}$$

$$b_{\mathbf{m}}b_{\mathbf{m}'}^\dagger - b_{\mathbf{m}'}^\dagger b_{\mathbf{m}} = 0 \tag{139}$$

for $\mathbf{m} \neq \mathbf{m}'$. This makes useful the pseudospin analogy [42],

$$b_{\mathbf{m}}^{\dagger} = S_{\mathbf{m}}^{x} - iS_{\mathbf{m}}^{y} \tag{140}$$

and

$$b_{\mathbf{m}}^{\dagger}b_{\mathbf{m}} = \frac{1}{2} - S_{\mathbf{m}}^{z} \tag{141}$$

with the pseudospin 1/2 operators $S^{x,y,z} = \frac{1}{2}\tau_{1,2,3}$. $S_{\mathbf{m}}^{z} = 1/2$ corresponds to an empty site $\mathbf{m}$ and $S_{\mathbf{m}}^{z} = -1/2$ to a site occupied by the bipolaron. Spin operators preserve the bosonic nature of bipolarons, when they are on different sites, and their fermionic internal structure. Replacing bipolarons by spin operators we transform the bipolaronic Hamiltonian into the anisotropic Heisenberg Hamiltonian,

$$H_b = \sum_{\mathbf{m}\neq\mathbf{m}'}\left[\frac{1}{2}\bar{v}_{\mathbf{m}\mathbf{m}'}S_{\mathbf{m}}^{z}S_{\mathbf{m}'}^{z} + t_{\mathbf{m}\mathbf{m}'}(S_{\mathbf{m}}^{x}S_{\mathbf{m}'}^{x} + S_{\mathbf{m}}^{y}S_{\mathbf{m}'}^{y})\right]. \tag{142}$$

This Hamiltonian has been investigated in detail as a relevant form for magnetism and also for quantum solids like a lattice model of ^{4}He. However, while in those cases the magnetic field is an independent thermodynamic variable, in our case the total "magnetization" is fixed,

$$\frac{1}{N}\sum_{\mathbf{m}}\langle\langle S_{\mathbf{m}}^{z}\rangle\rangle = \frac{1}{2} - n_b, \tag{143}$$

if the bipolaron density n_b is conserved. The spin 1/2 Heisenberg Hamiltonian, (142) cannot be solved analytically. Complicated commutation rules for bipolaron operators make the problem hard, but not in the limit of low atomic density of bipolarons, $n_b \ll 1$ (for a complete phase diagram of bipolarons on a lattice see [42, 86]). In this limit we can reduce the problem to a charged Bose gas on a lattice [87]. Let us transform the bipolaronic Hamiltonian to a representation containing only the Bose operators $a_{\mathbf{m}}$ and $a_{\mathbf{m}}^{\dagger}$ defined as

$$b_{\mathbf{m}} = \sum_{k=0}^{\infty}\beta_k(a_{\mathbf{m}}^{\dagger})^k a_{\mathbf{m}}^{k+1}, \tag{144}$$

$$b_{\mathbf{m}}^{\dagger} = \sum_{k=0}^{\infty}\beta_k(a_{\mathbf{m}}^{\dagger})^{k+1} a_{\mathbf{m}}^{k}, \tag{145}$$

where

$$a_{\mathbf{m}}a_{\mathbf{m}'}^{\dagger} - a_{\mathbf{m}'}^{\dagger}a_{\mathbf{m}} = \delta_{\mathbf{m},\mathbf{m}'}. \tag{146}$$

The first few coefficients β_k are found by substituting (144) and (145) into (138) and (139),

$$\beta_0 = 1, \beta_1 = -1, \beta_2 = \frac{1}{2} + \frac{\sqrt{3}}{6}. \tag{147}$$

We also introduce bipolaron and boson Ψ-operators as

$$\Phi(\mathbf{r}) = \frac{1}{\sqrt{N}} \sum_{\mathbf{m}} \delta(\mathbf{r} - \mathbf{m}) b_{\mathbf{m}}, \tag{148}$$

$$\Psi(\mathbf{r}) = \frac{1}{\sqrt{N}} \sum_{\mathbf{m}} \delta(\mathbf{r} - \mathbf{m}) a_{\mathbf{m}}. \tag{149}$$

The transformation of the field operators takes the form

$$\Phi(\mathbf{r}) = \left[1 - \frac{\Psi^{\dagger}(\mathbf{r})\Psi(\mathbf{r})}{N} + \frac{(1/2 + \sqrt{3}/6)\Psi^{\dagger}(\mathbf{r})\Psi(\mathbf{r})\Psi(\mathbf{r})}{N^2} + ... \right] \Psi(\mathbf{r}). \tag{150}$$

Then we write the bipolaronic Hamiltonian as

$$H_b = \int d\mathbf{r} \int d\mathbf{r}' \Psi^{\dagger}(\mathbf{r}) t(\mathbf{r} - \mathbf{r}') \Psi(\mathbf{r}') + H_d + H_k + H^{(3)}, \tag{151}$$

where

$$H_d = \frac{1}{2} \int d\mathbf{r} \int d\mathbf{r}' \bar{v}(\mathbf{r} - \mathbf{r}') \Psi^{\dagger}(\mathbf{r}) \Psi^{\dagger}(\mathbf{r}') \Psi(\mathbf{r}') \Psi(\mathbf{r}), \tag{152}$$

is the dynamic part,

$$H_k = \frac{2}{N} \int d\mathbf{r} \int d\mathbf{r}' t(\mathbf{r} - \mathbf{r}') \times \tag{153}$$

$$\left[\Psi^{\dagger}(\mathbf{r}) \Psi^{\dagger}(\mathbf{r}') \Psi(\mathbf{r}') \Psi(\mathbf{r}') + \Psi^{\dagger}(\mathbf{r}) \Psi^{\dagger}(\mathbf{r}) \Psi(\mathbf{r}) \Psi(\mathbf{r}') \right].$$

is the kinematic (hard-core) part due to the "imperfect" commutation rules, and $H^{(3)}$ includes three- and higher-body collisions. Here

$$t(\mathbf{r} - \mathbf{r}') = \sum_{\mathbf{k}} \epsilon_{\mathbf{k}}^{**} e^{i\mathbf{k}\cdot(\mathbf{r}-\mathbf{r}')},$$

$$\bar{v}(\mathbf{r} - \mathbf{r}') = \frac{1}{N} \sum_{\mathbf{k}} \bar{v}_{\mathbf{k}} e^{i\mathbf{k}\cdot(\mathbf{r}-\mathbf{r}')},$$

$\bar{v}_{\mathbf{k}} = \sum_{\mathbf{m}\neq 0} \bar{v}(\mathbf{m}) \exp(i\mathbf{k} \cdot \mathbf{m})$ is the Fourier component of the dynamic interaction and

$$\epsilon_{\mathbf{k}}^{**} = \sum_{\mathbf{m}\neq 0} t(\mathbf{m}) \exp(-i\mathbf{k} \cdot \mathbf{m}) \tag{154}$$

is the bipolaron band dispersion. $H^{(3)}$ contains powers of the field operator higher than four. In the dilute limit, $n_b \ll 1$, only two-particle interactions are essential which include the short-range kinematic and direct density-density repulsions. Because $\bar{v}$ already has the short range part $v^{(2)}$, (111), the kinematic contribution can be included in the definition of $\bar{v}$. As a result H_b is reduced to the Hamiltonian of interacting hard-core charged bosons tunnelling in the narrow band.

To describe electrodynamics of bipolarons we introduce the vector potential $\mathbf{A}(\mathbf{r})$ using the so-called Peierls substitution [88],

$$t(\mathbf{m} - \mathbf{m}') \rightarrow t(\mathbf{m} - \mathbf{m}')e^{i2e\mathbf{A}(\mathbf{m})\cdot(\mathbf{m}-\mathbf{m}')},$$

which is a fair approximation when the magnetic field is weak compared with the atomic field, $eHa^2 << 1$. It has the following form,

$$t(\mathbf{r} - \mathbf{r}') \rightarrow t(\mathbf{r}, \mathbf{r}') = \sum_{\mathbf{k}} \epsilon^{**}_{\mathbf{k}-2e\mathbf{A}} e^{i\mathbf{k}\cdot(\mathbf{r}-\mathbf{r}')} \tag{155}$$

in real space. If the magnetic field is weak, we can expand $\epsilon^{**}_{\mathbf{k}}$ in the vicinity of $\mathbf{k} = 0$ to obtain

$$t(\mathbf{r}, \mathbf{r}') \approx -\frac{[\nabla + 2ie\mathbf{A}(\mathbf{r})]^2}{2m^{**}}\delta(\mathbf{r} - \mathbf{r}'), \tag{156}$$

where

$$\frac{1}{m^{**}} = \left(\frac{d^2 \epsilon^{**}_{\mathbf{k}}}{dk^2}\right)_{k \rightarrow 0} \tag{157}$$

is the inverse bipolaron mass. Here we assume a parabolic dispersion near the bottom of the band, $\epsilon^{**}_{\mathbf{k}} \sim k^2$, so that

$$H_b \approx -\int d\mathbf{r}\Psi^\dagger(\mathbf{r})\left\{\frac{[\nabla + 2ie\mathbf{A}(\mathbf{r})]^2}{2m^{**}} + \mu\right\}\Psi(\mathbf{r}) + \tag{158}$$

$$\frac{1}{2}\int d\mathbf{r}d\mathbf{r}'\bar{v}(\mathbf{r} - \mathbf{r}')\Psi^\dagger(\mathbf{r})\Psi^\dagger(\mathbf{r}')\Psi(\mathbf{r})\Psi(\mathbf{r}'),$$

where we add the *bipolaron* chemical potential μ. We note that the bipolaron-bipolaron interaction is the Coulomb repulsion, $\bar{v}(\mathbf{r}) \sim 1/(\epsilon_0 r)$ at large distances, and the hard-core repulsion is not important in the dilute limit. The Hamiltonian (158) describes the charged Bose gas with the effective boson mass m^{**} and charge $2e$, which is a superconductor [89].

9 Bipolaronic Superconductivity in Cuprates

The fact that Helium-4 and its isotope Helium-3 are well known Bose and Fermi superfluids, respectively, with very different superfluid transition temperatures ($T_c = 2.17K$ in ^4He and $T_c = 0.0026K$ in ^3He) already kindles the view that high-temperature superconductivity might derive from preformed real-space charged bosons rather than in the BCS state with Cooper pairs, which are correlations in momentum space. The possibility of real-space pairing of carriers in cuprates, as opposed to Cooper pairing, has been the subject of much discussion. Some authors dismissed any real-space pairs even in underdoped cuprates, where a low density of carriers appears to favor individual pairing rather than Cooper pairing. But on the other hand real-space pairing is strongly supported by our strong-coupling extension of the BCS theory since the e-ph interaction is very strong in cuprates. Also on the experimental side a growing number of other independent observations point to the possibility that high-T_c superconductors may not be conventional Bardeen–Cooper–Schrieffer (BCS) superconductors, but rather derive from the Bose–Einstein condensation (BEC) of real-space superlight small bipolarons. There is strong evidence for real-space pairing and the three-dimensional BEC in cuprates from unusual upper critical fields [90] and isotope effects [91] predicted by us and the λ-like electronic specific heat [92], parameter-free fitting of experimental T_c with BEC T_c [93], normal state pseudogaps [94, 95] and anisotropy [96], and more recently from normal state diamagnetism [97], the Hall–Lorenz numbers [98, 99], the normal state Nernst effect [100, 101], and the giant proximity effect (GPE) [102]. Here I briefly discuss a few of these remarkable observations (for more details see [23] and Part IV).

9.1 Upper Critical Field, the Hall–Lorenz Number, and Isotope Effects

Magnetotransport [103] and thermal magnetotransport [98, 104] data strongly support preformed pairs in cuprates. In particular, many high magnetic field studies revealed a non-BCS upward curvature of the upper critical field, $H_{c2}(T)$ and its non-linear temperature dependence in the vicinity of T_c in a number of cuprates as well as in a few other unconventional superconductors, Fig. 6. If unconventional superconductors are in the "bosonic" limit of preformed real-space pairs, such unusual critical fields are expected in accordance with the theoretical prediction for the Bose–Einstein condensation of charged bosons in an external magnetic field [90].

Notwithstanding, some "direct" evidence for the existence of a charge $2e$ Bose liquid in the normal state of cuprates is highly desirable. Alexandrov and Mott [105] discussed the thermal conductivity κ; the contribution from the carriers given by the Wiedemann–Franz ratio depends strongly on the elementary charge as $\sim (e^*)^{-2}$ and should be significantly suppressed in the case of $e^* = 2e$ compared with the Fermi-liquid contribution. As a result, the Lorenz number, $L = (e/k_B)^2 \kappa_e/(T\sigma)$ differs significantly from the Sommerfeld value $L_e = \pi^2/3$ of the standard Fermi-liquid theory, if carriers are double-charged bosons. Here κ_e, σ, and e are the electronic thermal conductivity, the electrical conductivity, and the elementary charge, respectively. Reference [105] predicted a rather low Lorenz number for bipolarons, $L = 6L_e/(4\pi^2) \approx 0.15L_e$, due to the double charge of carriers, and also due to their nearly classical distribution function above T_c.

The extraction of the electron thermal conductivity has proven difficult since both the electron term, κ_e and the phonon term, κ_{ph} are comparable to each other in the cuprates. A new way to determine the Lorenz number has been realized by Zhang et al. [104], based on the thermal Hall conductivity. The thermal Hall effect allowed for an efficient way to separate the phonon heat current even when it is dominant. As a result, the "Hall" Lorenz number,

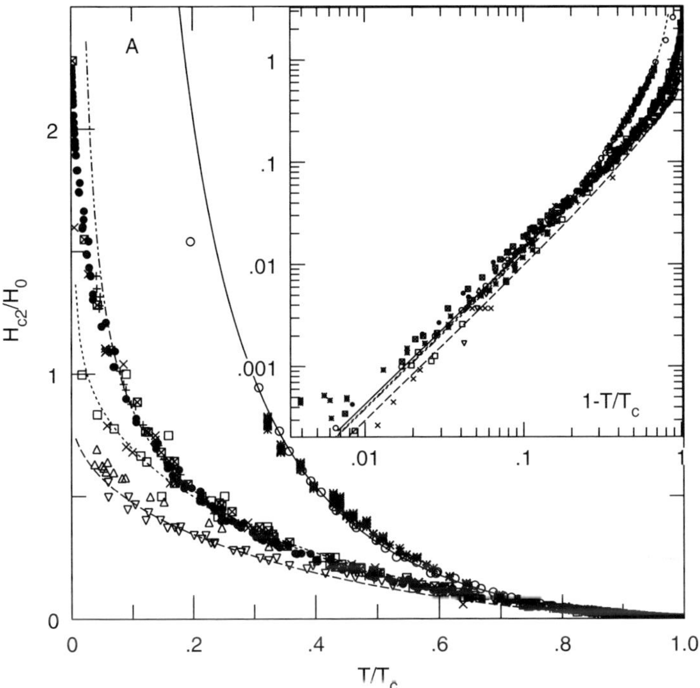

Fig. 6. Resistive upper critical field [103] (determined at 50% of the transition) of cuprates, spin-ladders and organic superconductors scaled according to the Bose–Einstein condensation field of charged bosons [90], $H_{c2}(T) \propto \left[b(1-t)/t + 1 - t^{1/2} \right]^{3/2}$ with $t = T/T_c$. The parameter b is proportional to the number of delocalised bosons at zero temperature, b is 1 (solid line), 0.02 (dashed-dotted line), 0.0012 (dotted line), and 0 (dashed line). The inset shows a universal scaling of the same data near T_c on the logarithmic scale. Symbols correspond to $Tl-2201(\bullet)$, $La_{1.85}Sr_{0.15}CuO_4(\triangle)$, $Bi-2201(\times)$, $Bi-2212(*)$, $YBa_2Cu_3O_{6+x}(\circ)$, $La_{2-x}Ce_xCuO_{4-y}$ (squares), $Sr_2Ca_{12}Cu_{24}O_{41}(+)$, and Bechgaard salt organic superconductor (∇) .

$L_H = (e/k_B)^2 \kappa_{xy}/(T\sigma_{xy})$, has been directly measured in $YBa_2Cu_3O_{6.95}$ because transverse thermal κ_{xy} and electrical σ_{xy} conductivities involve only the electrons. Remarkably, the measured value of L_H just above T_c is about the same as predicted by the bipolaron model, $L_H \approx 0.15 L_e$. The experimental L_H showed a strong temperature dependence, which violates the Wiedemann–Franz law. This experimental observation has been accounted for by taking into account thermally excited polarons and also triplet pairs in the bipolaron model [98], Fig. 7.

Further compelling evidence for (bi)polaronic carries in novel superconductors was provided by the discovery of substantial isotope effects on T_c

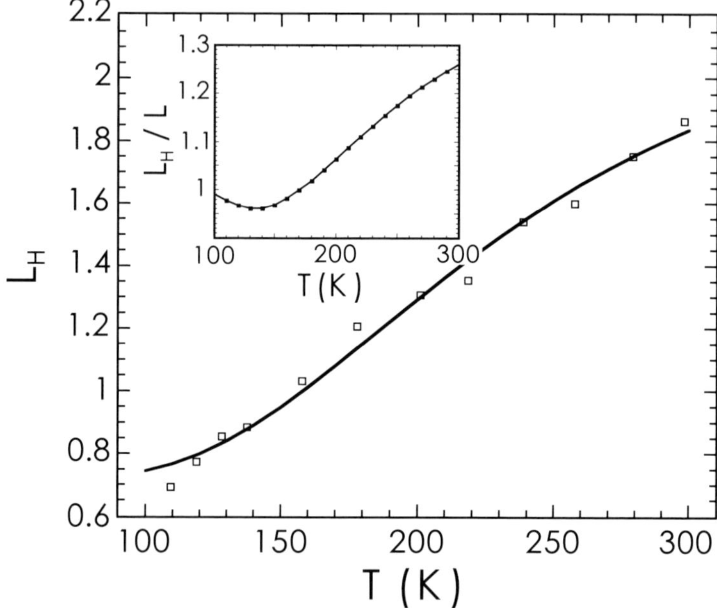

Fig. 7. The Hall Lorenz number L_H [98] of charged bosons fits the experiment in $YBa_2Cu_3O_{6.95}$ [104]. The pseudogap is taken as 675 K. The inset gives the ratio of the Hall Lorenz number to the Lorenz number in the model.

and on the carrier mass [91, 106]. The advances in the fabrication of the isotope substituted samples made it possible to measure a sizable isotope effect, $\alpha = -d \ln T_c / d \ln M$ in many high-T_c oxides. This led to a general conclusion that phonons are relevant for high T_c. Moreover the isotope effect in cuprates was found to be quite different from the BCS prediction, $\alpha = 0.5$ (or less). Several compounds showed $\alpha > 0.5$, and sometimes negative values of α were observed.

Essential features of the isotope effect, in particular large values in low T_c cuprates, an overall trend to lower value as T_c increases, and a small or even negative α in some high T_c cuprates can be understood in the framework of the bipolaron theory [107]. With increasing ion mass the bipolaron mass increases and the Bose–Einstein condensation temperature $T_c \propto 1/m^{**}$ decreases in the bipolaronic superconductor (Sect. 8). On the contrary in polaronic superconductors (Sect. 6) an increase of the ion mass leads to a band narrowing enhancing the polaron density of states and increasing T_c. Hence the isotope exponent of T_c can distinguish the BCS–like polaronic superconductivity with $\alpha < 0$, and the Bose–Einstein condensation of small bipolarons with $\alpha > 0$. Moreover, underdoped cuprates, which are certainly in the BEC regime, could have $\alpha > 0.5$, as observed.

The isotope effect on T_c is linked with the isotope effect on the carrier mass, α_{m^*}, as [107]

$$\alpha = -d\ln T_c/d\ln M = \alpha_{m^*}[1 - Z/(\lambda - \mu_c)], \qquad (159)$$

where $\alpha_{m^*} = d\ln m^*/d\ln M$ and $Z = m/m^* \ll 1$. In ordinary metals, where the Migdal approximation is believed to be valid, the renormalized effective mass of electrons is independent of the ion mass M because the electron-phonon interaction constant λ does not depend on M. However, when the e-ph interaction is sufficiently strong, the electrons form polarons dressed by lattice distortions, with an effective mass $m^* = m\exp(\gamma E_p/\omega)$. While E_p in the above expression does not depend on the ion mass, the phonon frequency does. As a result, there is a large isotope effect on the carrier mass in polaronic conductors, $\alpha_{m^*} = (1/2)\ln(m^*/m)$ [107], in contrast to the zero isotope effect in ordinary metals. Such an effect was observed in cuprates in the London penetration depth λ_H of isotope-substituted samples [106]. The carrier density is unchanged with the isotope substitution of O^{16} by O^{18}, so that the isotope effect on λ_H measures directly the isotope effect on the carrier mass. In particular, the carrier mass isotope exponent α_{m^*} was found as large as $\alpha_{m^*} = 0.8$ in $La_{1.895}Sr_{0.105}CuO_4$.

More recent high resolution angle resolved photoemission spectroscopy [108] provided further compelling evidence for strong e-ph interaction in cuprates. It revealed a fine phonon structure in the electron self-energy of the underdoped $La_{2-x}Sr_xCuO_4$ samples. Remarkably, an isotope effect on the electron spectral function in Bi-2212 [109] has been discovered. These experiments together with a number of earlier optical [110–115] and neutron-scattering [116] experimental and theoretical studies firmly established the strong coupling of carriers with optical phonons in cuprates (see also Part IV).

9.2 Normal State Diamagnetism: BEC Versus Phase Fluctuations

Above T_c the charged bipolaronic Bose liquid is non-degenerate and below T_c phase coherence (ODLRO) of the preformed bosons sets in. The state above T_c is perfectly "normal" in the sense that the off-diagonal order parameter (i.e. the Bogoliubov–Gor'kov anomalous average $\mathcal{F}(\mathbf{r}, \mathbf{r}') = \langle \psi_\downarrow(\mathbf{r})\psi_\uparrow(\mathbf{r}') \rangle$ is zero above the resistive transition temperature T_c as in the BCS theory. Here $\psi_{\downarrow,\uparrow}(\mathbf{r})$ annihilates electrons with spin $\downarrow, \uparrow$ at point $\mathbf{r}$.

However in contrast with the bipolaron and BCS theories a significant fraction of research in the field of cuprate superconductors suggests a so-called phase fluctuation scenario [117–119], where $\mathcal{F}(\mathbf{r}, \mathbf{r}')$ remains nonzero well above T_c. I believe that the phase fluctuation scenario is impossible to reconcile with the extremely sharp resistive transitions at T_c in high-quality underdoped, optimally doped and overdoped cuprates. For example, the in-plane and out-of-plane resistivity of Bi-2212, where the anomalous Nernst signal has been measured [118], is perfectly "normal" above T_c, (Fig. 8), showing

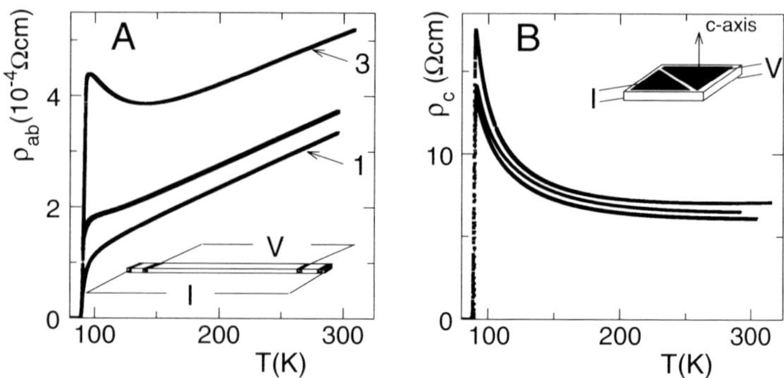

Fig. 8. In-plane (A) and out-of-plane (B) resistivity of 3 single crystals of $Bi_2Sr_2CaCu_2O_8$ [120] showing no signature of phase fluctuations well above the resistive transition. Reproduced from [120], ©(2005) American Physical Society.

only a few percent positive or negative magnetoresistance [120], explained with bipolarons [121].

Both in-plane [122–126] and out-of-plane [127–129] resistive transitions of high-quality samples remain sharp in the magnetic field providing a reliable determination of the genuine $H_{c2}(T)$. The preformed Cooper-pair (or phase fluctuation) model [117] is incompatible with a great number of thermodynamic, magnetic, and kinetic measurements, which show that only holes (density x), doped into a parent insulator are carriers *both* in the normal and the superconducting states of cuprates. The assumption [117] that the superfluid density x is small compared with the normal-state carrier density is also inconsistent with the theorem [130], which proves that the number of supercarriers at $T = 0\,K$ should be the same as the number of normal-state carriers in any clean superfluid.

The normal state diamagnetism of cuprates provides further clear evidence for BEC rather than for the phase fluctuation scenario. A number of experiments (see, for example, [119, 131–135] and references therein), including torque magnetometries, showed enhanced diamagnetism above T_c, which has been explained as the fluctuation diamagnetism in quasi-2D superconducting cuprates (see, for example [133]). The data taken at relatively low magnetic fields (typically below 5 Tesla) revealed a crossing point in the magnetization $M(T, B)$ of most anisotropic cuprates (e.g. $Bi - 2212$), or in $M(T, B)/B^{1/2}$ of less anisotropic $YBCO$ [132]. The dependence of magnetization (or $M/B^{1/2}$) on the magnetic field has been shown to vanish at some characteristic temperature below T_c. However the data taken in high magnetic fields (up to 30 Tesla) have shown that the crossing point, anticipated for low-dimensional superconductors and associated with superconducting fluctuations, does not explicitly exist in magnetic fields above 5 Tesla [134].

Most surprisingly the torque magnetometery [131, 134] uncovered a diamagnetic signal somewhat above T_c which increases in magnitude with applied magnetic field. It has been linked with the Nernst signal and mobile vortexes

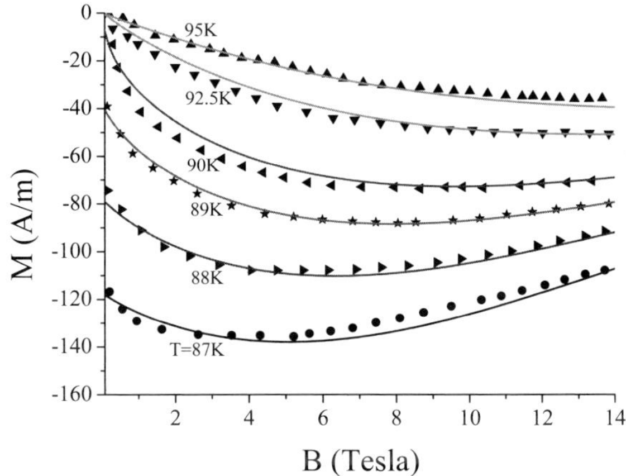

Fig. 9. Diamagnetism of optimally doped Bi-2212 (symbols)[119] compared with magnetization of CBG [97] near and above T_c (lines). Reproduced from [97], ©(2006) American Physical Society.

in the normal state of cuprates [119]. However, apart from the inconsistences mentioned above, the vortex scenario of the normal-state diamagnetism is internally inconsistent. Accepting the vortex scenario and fitting the magnetization data in $Bi-2212$ with the conventional logarithmic field dependence [119], one obtains surprisingly high upper critical fields $H_{c2} > 120$ Tesla and a very large Ginzburg–Landau parameter, $\kappa = \lambda/\xi > 450$ even at temperatures close to T_c. The in-plane low-temperature magnetic field penetration depth is $\lambda = 200$ nm in optimally doped $Bi-2212$ (see, for example [136]). Hence the zero temperature coherence length ξ turns out to be roughly the lattice constant, $\xi = 0.45$nm, or even smaller. Such a small coherence length rules out the "preformed Cooper pairs" [117], since the pairs are well separated at any size of the Fermi surface in $Bi-2212$. Moreover the magnetic field dependence of $M(T,B)$ at and above T_c is entirely inconsistent with what one expects from a vortex liquid. While $-M(B)$ decreases logarithmically at temperatures well below T_c, the experimental curves [119, 131, 134] clearly show that $-M(B)$ increases with the field at and above T_c , just opposite to what one could expect in the vortex liquid. This significant departure from the London liquid behavior clearly indicates that the vortex liquid does not appear above the resistive phase transition [131].

Some time ago we explained the anomalous diamagnetism in cuprates as the Landau normal-state diamagnetism of preformed bosons [137]. More recently the model has been extended to high magnetic fields taking into account the magnetic pair-breaking of singlet bipolarons and the anisotropy of the energy spectrum [97]. When the magnetic field is applied perpendicular to the copper-oxygen planes the quasi-2D bipolaron energy spectrum is quantized as $E_\alpha = \omega(n + 1/2) + 2t_c[1 - \cos(K_z d)]$, where α comprises $n = 0, 1, 2, \ldots$ and in-plane K_x and out-of-plane K_z center-of-mass quasi-momenta, $\omega = 2eB/\sqrt{m_x^{**}m_y^{**}}$, t_c and d are the hopping integral and the lattice period perpendicular to the planes. We assume here that the spectrum consists of two degenerate branches, so-called "x" and "y" bipolarons as in the case of apex intersite pairs [41] with anisotropic in-plane bipolaron masses $m_x^{**} \equiv m$ and $m_y^{**} \approx 4m$. Expanding the Bose–Einstein distribution function in powers of $\exp[(\mu - E)/T]$ with the negative chemical potential μ one can after summation over n readily obtain the boson density

$$n_b = \frac{2eB}{\pi d} \sum_{r=1}^{\infty} I_0(2t_c r/T) \frac{\exp[(\mu - \omega/2 - 2t_c)r/T]}{1 - \exp(-\omega r/T)}, \tag{160}$$

and the magnetization,

$$M(T, B) = -n_b\mu_b + \frac{2eT}{\pi d} \sum_{r=1}^{\infty} I_0\left(\frac{2t_c r}{T}\right) \times \tag{161}$$

$$\frac{\exp[(\mu - \omega/2 - 2t_c)r/T]}{1 - \exp(-\omega r/T)} \left(\frac{1}{r} - \frac{\omega \exp(-\omega r/T)}{k_B T[1 - \exp(-\omega r/T)]}\right).$$

Here $\mu_b = e/\sqrt{m_x^{**}m_y^{**}}$ and $I_0(x)$ is the modified Bessel function. At low temperatures $T \to 0$ Schafroth's result [89] is recovered, $M(0, B) = -n_b\mu_b$. The magnetization of charged bosons is field-independent at low temperatures. At high temperatures, $T \gg T_c$ the chemical potential has a large magnitude, and we can keep only the terms with $r = 1$ in (160,161) to obtain $M(T, B) = -n_b\mu_b\omega/(6T)$ at $T \gg T_c \gg \omega$, which is the familiar Landau orbital diamagnetism of nondegenerate carriers. Here T_c is the Bose–Einstein condensation temperature $T_c = 3.31(n_b/2)^{2/3}/(m_x^{**}m_y^{**}m_c^{**})^{1/3}$, with $m_c = 1/2|t_c|d^2$.

Comparing with experimental data one has to take into account the temperature and field depletion of singlets due to their thermal excitations into spin-split triplet states, $n_b(T, B) = n_c[1 - \alpha\tau - (B/B^*)^2]$. Here $\alpha = 3(2n_c t)^{-1}[J(e^{J/T_c} - 1)^{-1} - T_c \ln(1 - e^{-J/T_c})]$, $\mu_B B^* = (2T_c n_c t)^{1/2} \sinh(J/2T_c)$, $\mu_B \approx 0.93 \times 10^{-23}$ Am^2 is the Bohr magneton, n_c is the density of singlets at $T = T_c$ in zero field, $\tau = T/T_c - 1$, J is the singlet-triplet exchange energy, and $2t$ is the triplet bandwidth. As a result, (161) fits remarkably well the experimental curves in the critical region of optimally doped Bi-2212, Fig. 9, with $n_c\mu_b = 2100$A/m, $T_c = 90$K, $\alpha = 0.62$ and $B^* = 56$ Tesla, which corresponds to the singlet-triplet exchange energy $J \approx 20$K.

On the other hand the experimental data, Fig. 9, contradict BCS and the phase-fluctuation scenarios [117, 119]. Indeed, if we define a critical exponent as $\delta = \ln B / \ln |M(T, B)|$ for $B \to 0$, the T dependence of $\delta(T)$ in the charged Bose gas (CBG) is dramatically different from the Berezinski–Kosterlitz–Thouless (BKT) transition critical exponents (as proposed in the phase fluctuation scenario), but it is very close to the experimental [119] $\delta(T)$, Fig. 10.

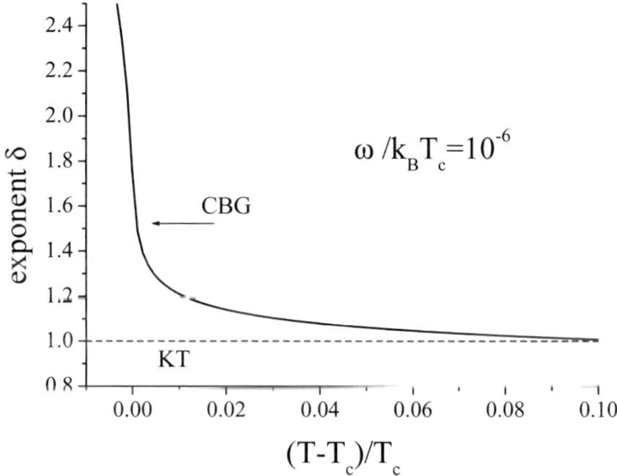

Fig. 10. Critical exponents of the low-field magnetization in CBG and in BKT transition. Reproduced from [97], ©(2006) American Physical Society.

Also the large Nernst signal, allegedly supporting vortex liquid in the normal state of cuprates [119], has been explained as the normal state phenomenon owing to a partial localization of charge carriers in a random potential inevitable in cuprates [100]. The coexistence of the large Nernst signal and the insulating-like resistivity in slightly doped cuprates sharply disagrees with the vortex scenario, but agrees remarkably well with our theory [101].

9.3 Giant Proximity Effect

Several groups reported that in the Josephson cuprate SNS junctions supercurrent can run through normal N-barriers as thick as 100 nm in strong conflict with the standard theoretical picture, if the barrier is made from non-superconducting cuprates. Using advanced molecular beam epitaxy, Bozovic et al. [138] proved that this giant proximity effect (GPE) is intrinsic, rather than extrinsic caused by any inhomogeneity of the barrier. Hence GPE defies the conventional explanation, which predicts that the critical current should decay exponentially with characteristic length of about the coherence length, which is $\xi \leq 1$ nm in the cuprates.

This effect can be broadly understood as the Bose–Einstein condensate tunnelling into a cuprate *semiconductor* [102]. Indeed the chemical potential μ remains in the charge-transfer gap of doped cuprates like $La_{2-x}Sr_xCuO_4$ [139] because of the bipolaron formation. The condensate wave function, $\psi(Z)$, is described by the Gross–Pitaevskii (GP) equation. In the superconducting region, $Z < 0$, near the SN boundary, Fig. 11, the equation is

$$\frac{1}{2m_c^{**}}\frac{d^2\psi(Z)}{dZ^2} = [V|\psi(Z)|^2 - \mu]\psi(Z), \tag{162}$$

where V is a short-range repulsion of bosons, and m_c^{**} is the boson mass along Z. Deep inside the superconductor $|\psi(Z)|^2 = n_s$ and $\mu = Vn_s$, where the condensate density n_s is about $x/2$, if the temperature is well below T_c of the superconducting electrode (the in-plane lattice constant a and the unit cell volume are taken as unity).

The normal barrier at $Z > 0$ is an underdoped cuprate semiconductor above its transition temperature, where the chemical potential μ lies below the bosonic band by some energy ϵ, Fig. 11. For quasi-two dimensional bosons one readily obtains [23]

$$\epsilon(T) = -T\ln(1 - e^{-T_0/T}), \tag{163}$$

where $T_0 = \pi x'/m^{**}$, m^{**} is the in-plane boson mass, and $x' < x$ is the doping level of the barrier. Then the GP equation in the barrier can be written as

$$\frac{1}{2m_c^{**}}\frac{d^2\psi(Z)}{dZ^2} = [V|\psi(Z)|^2 + \epsilon]\psi(Z). \tag{164}$$

Introducing the bulk coherence length, $\xi = 1/(2m_c^{**}n_sV)^{1/2}$ and dimensionless $f(z) = \psi(Z)/n_s^{1/2}$, $\tilde{\mu} = \epsilon/n_sV$, and $z = Z/\xi$, one obtains for a real $f(z)$

$$\frac{d^2f}{dz^2} = f^3 - f, \tag{165}$$

if $z < 0$, and

$$\frac{d^2f}{dz^2} = f^3 + \tilde{\mu}f, \tag{166}$$

if $z > 0$. These equations can be readily solved using first integrals of motion respecting the boundary conditions, $f(-\infty) = 1$, and $f(\infty) = 0$,

$$\frac{df}{dz} = -(1/2 + f^4/2 - f^2)^{1/2}, \tag{167}$$

and

$$\frac{df}{dz} = -(\tilde{\mu}f^2 + f^4/2)^{1/2}, \tag{168}$$

for $z < 0$ and $z > 0$, respectively. The solution in the superconducting electrode is given by

$$f(z) = \tanh\left[-2^{-1/2}z + 0.5\ln\frac{2^{1/2}(1+\widetilde{\mu})^{1/2}+1}{2^{1/2}(1+\widetilde{\mu})^{1/2}-1}\right]. \tag{169}$$

It decays in the close vicinity of the barrier from 1 to $f(0) = [2(1+\widetilde{\mu})]^{-1/2}$ in the interval about the coherence length ξ. On the other side of the boundary, $z > 0$, it is given by

$$f(z) = \frac{(2\widetilde{\mu})^{1/2}}{\sinh\{z\widetilde{\mu}^{1/2} + \ln[2(\widetilde{\mu}(1+\widetilde{\mu}))^{1/2} + (1+4\widetilde{\mu}(1+\widetilde{\mu}))^{1/2}]\}}. \tag{170}$$

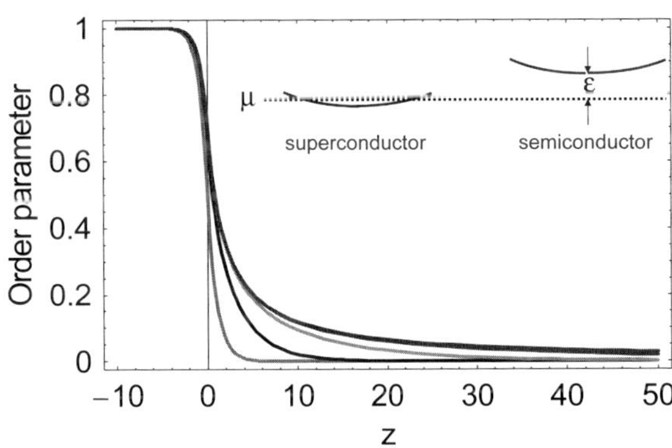

Fig. 11. BEC order parameter at the SN boundary for $\widetilde{\mu} = 1.0, 0.1, 0.01$ and ≤ 0.001 (upper curve).

Its profile is shown in Fig. 11. Remarkably, the order parameter penetrates into the normal layer up to the length $Z^* \approx (\widetilde{\mu})^{-1/2}\xi$, which could be larger than ξ by many orders of magnitude, if $\widetilde{\mu}$ is small. It is indeed the case, if the barrier layer is sufficiently doped. For example, taking $x' = 0.1$, c-axis $m_c^{**} = 2000m_e$, in-plane $m^{**} = 10m_e$ [23], $a = 0.4$ nm, and $\xi = 0.6$ nm, yields $T_0 \approx 140$ K and $(\widetilde{\mu})^{-1/2} \approx 50$ at $T = 25$K. Hence the order parameter could penetrate into the normal cuprate semiconductor up to more than a hundred coherence lengths as observed [138]. If the thickness of the barrier L is small compared with Z^*, and $(\widetilde{\mu})^{1/2} \ll 1$, the order parameter decays following the power law, rather than exponentially,

$$f(z) = \frac{\sqrt{2}}{z+2}. \tag{171}$$

Hence, for $L \leq Z^*$, the critical current should also decay following the power law [102]. On the other hand, for an *undoped* barrier $\tilde{\mu}$ becomes larger than unity, $\tilde{\mu} \propto \ln(m^{**}T/\pi x') \to \infty$ for any finite temperature T when $x' \to 0$, and the current should decay exponentially with a characteristic length smaller than ξ, as is also experimentally observed [139]. As a result the bipolaron theory accounts for the giant and nil proximity effects in slightly doped semiconducting and undoped insulating cuprates, respectively. It predicts the occurrence of a new length scale, $\hbar/\sqrt{2m_c^{**}\epsilon(T)}$, and explains the temperature dependence of the critical current of SNS junctions [102].

10 Conclusion

Extending the BCS theory towards the strong interaction between electrons and ion vibrations, a charged Bose gas of tightly bound small bipolarons was predicted by us [42] with a further prediction that high T_c should exist in the crossover region of the e-ph interaction strength from the BCS-like to bipolaronic superconductivity [17].

For very strong electron-phonon coupling, polarons become self-trapped on a single lattice site. The energy of the resulting small polaron is given as $-E_p = -\lambda zt$. Expanding about the atomic limit in hopping integrals t (which is small compared to E_p in the small polaron regime, $\lambda > 1$) the polaron mass is computed as $m^* = m_0 \exp(\gamma E_p/\omega_0)$, where ω_0 is the frequency of Einstein phonons, m_0 is the rigid band mass on a cubic lattice, and γ is a numerical constant. For the Holstein model, which is purely site local, $\gamma = 1$. Bipolarons are on-site singlets in the Holstein model and their mass m_H^{**} appears only in the second order of t [42] scaling as $m_H^{**} \propto (m^*)^2$ for $\omega_0 \gg \Delta$, and as $m_H^{**} \propto (m^*)^4$ in a more realistic regime $\omega_0 \ll \Delta$ (Sect. 7). Here $\Delta = 2E_p - U$ is the bipolaron binding energy, and U is the on-site (Hubbard) repulsion. Since the Hubbard U is about 1 eV or larger in strongly correlated materials, the electron-phonon coupling must be large to stabilize on-site bipolarons and the Holstein bipolaron mass appears very large, $m_H^{**}/m_0 > 1000$, for realistic values of the phonon frequency.

This estimate led some authors to the conclusion that the formation of itinerant small polarons and bipolarons in real materials is unlikely [141], and high-temperature bipolaronic superconductivity is impossible [142]. However, one should note that the Holstein model is an extreme polaron model, and typically yields the highest possible value of the (bi)polaron mass in the strong coupling limit. Many advanced materials with low density of free carriers and poor mobility (at least in one direction) are characterized by poor screening of high-frequency optical phonons and are more appropriately described by the long-range Fröhlich electron-phonon interaction [41]. For this interaction the parameter γ is less than 1 ($\gamma \approx 0.3$ or less), reflecting the fact that in a hopping event the lattice deformation is partially pre-existent. Hence the unscreened Fröhlich electron-phonon interaction provides relatively light small polarons, which are several orders of magnitude lighter than small Holstein polarons.

As shown above FCM is reduced to an extended Hubbard model with inter-site attraction and suppressed double-occupancy in the limit of high phonon frequency $\omega_0 \geq t$ and large on-site Coulomb repulsion. Then the Hamiltonian can be projected onto the subspace of nearest neighbor intersite bipolarons. In contrast with the "crawler" motion of the on-site bipolaron, the intersite bipolaron tunnelling is "crab-like", so that its mass scales linearly with the polaron mass ($m^{**} \approx 4m^*$ on the staggered chain [81]) as confirmed numerically using CTQMC algorithm by Kornilovitch [43]. As a result, the crab bipolarons could Bose–condense already at room temperature [71].

We believe that the following recipe is worth investigating to look for room-temperature superconductivity [71]: (a) The parent compound should be an ionic insulator with light ions to form high-frequency optical phonons, (b) The structure should be quasi two-dimensional to ensure poor screening of high-frequency c-axis polarized phonons, (c) A triangular lattice is desirable in combination with strong, on-site Coulomb repulsion to form the superlight crab bipolaron, and (d) Moderate carrier densities are required to keep the system of small bipolarons close to the dilute regime. I believe that most of these conditions are already met in cuprate superconductors. As discussed above there is strong evidence for 3D bipolaronic BEC in cuprates from unusual upper critical fields and the electronic specific heat, normal state pseudogaps and anisotropy, normal state diamagnetism, the Hall–Lorenz numbers, and the giant proximity effect.

Acknowledgement. I thank A. F. Andreev, J. P. Hague, V. V. Kabanov, P. E. Kornilovitch, and J. H. Samson for illuminating discussions and collaboration. The work was supported by EPSRC (UK) (grant no. EP/C518365/1).

References

1. A. S. Alexandrov and N. F. Mott, Rep. Prog. Phys. **57**, 1197 (1994); *'Polarons and Bipolarons'* (World Scientific, Singapore, 1995).
2. J. T. Devreese, in *Encyclopedia of Applied Physics* (VCH Publishers, 1996), vol. 14, p. 383; *Polaron* (In Encyclopedia of Physics, vol. 2, ed by R.G. Lerner, G. L. Trigg, Wiley-VCH, Weinheim, 2005) p. 2004.
3. E. I. Rashba, in *Encyclopedia of Condensed Natter Physics*, eds. G. F. Bassani, G. L. Liedl, and P. Wyder (Elsevier, 2005), p. 347.
4. *Anharmonic Properties of High-T_c Cuprates*, eds. D. Mihailović *et al.*, (World Scientific, Singapore, 1995).
5. *Polarons and Bipolarons in High-T_c Superconductors and Related Materials*, eds E. K. H. Salje, A. S. Alexandrov and W. Y. Liang (Cambridge University Press, Cambridge, 1995).

6. N. Itoh and A. M. Stoneham, *Materials Modification by Electronic Excitation* (Cambridge University Press, Cambridge, 2001).

7. J. Appel, in *Solid State Physics* **21**(eds. F. Seitz, D. Turnbull, and H. Ehrenreich, Academic Press, 1968).

8. Yu. A. Firsov (ed) *Polarons* (Nauka, Moscow, 1975).

9. H. Boettger and V. V. Bryksin, *Hopping Conduction in Solids* (Academie-Verlag, Berlin, 1985).

10. A. B. Migdal, Zh. Eksp. Teor. Fiz. **34**, 1438 (1958) [Sov. Phys. JETP **7**, 996 (1958)].

11. J. Bardeen, L. N. Cooper, and J. R. Schrieffer, Phys. Rev. **108**, 1175 (1957).

12. G. M. Eliashberg, Zh. Eksp. Teor. Fiz. **38**, 966 (1960); **39**, 1437 (1960) [Sov. Phys. JETP **11**, 696; **12**, 1000 (1960)].

13. D. J. Scalapino, in *Superconductivity*, ed. R. D. Parks (Marcel Dekker, NY, 1069), p. 449.

14. A. S. Alexandrov, V. N. Grebenev, and E. A. Mazur, Pis'ma Zh. Eksp. Teor. Fiz. **45**, 357 (1987)[JETP Lett. **45**, 455 (1987)].

15. F. Dogan and F. Marsiglio, Phys. Rev. B**68**, 165102 (2003).

16. J. P. Hague, J. Phys.: Condens. Mat. **15**, 2535 (2003), and references therein.

17. A. S. Alexandrov, Zh. Fiz. Khim. **57**, 273 (1983) (Russ. J. Phys. Chem.**57**, 167 (1983)).

18. I. G. Lang and Yu. A. Firsov, Zh. Eksp. Teor. Fiz. **43**, 1843 (1962) [Sov. Phys. JETP **16**, 1301 (1963)].

19. Yu. A. Firsov, in the present volume.

20. A. S. Alexandrov, Phys. Rev. B **46**, 2838 (1992).

21. A. S. Alexandrov, in *Models and Phenomenology for Conventional and High-Temperature Superconductivity*, Course CXXXVI of the Intenational School of Physics 'Enrico Fermi', eds. G. Iadonisi, J. R. Schrieffer, and M. L. Chiofalo, (IOS Press, Amsterdam, 1998) p 309.

22. A. S. Alexandrov, Europhys. Lett., **56** , 92 (2001).

23. A. S. Alexandrov, *Theory of Superconductivity: From Weak to Strong Coupling* (IoP Publishing, Bristol, 2003).

24. L. D. Landau, Physikalische Zeitschrift der Sowjetunion, **3**, 664 (1933).

25. A. S. Alexandrov and E. A. Mazur, Zh. Eksp. Teor. Fiz. **96**, 1773 (1989)

26. V. V. Kabanov and O. Yu. Mashtakov, Phys. Rev. B **47**, 6060 (1993).

27. A. S. Alexandrov, V. V. Kabanov, and D. K. Ray, Phys. Rev. B **49**, 9915 (1994).

28. H. Fehske, H. Röder, G. Wellein, and A. Mistriotis, Phys. Rev. B **51**, 16582 (1995); G. Wellein, H. Röder, and H. Fehske, Phys. Rev. B **53**, 9666 (1996).

29. F. Marsiglio, Physica C **244**, 21 (1995).

30. Y. Takada and T. Higuchi, Phys. Rev. B **52**, 12720 (1995).

31. H. Fehske, J. Loos, and G. Wellein, Z. Phys. B**104**, 619 (1997).

32. T. Hotta and Y. Takada, Phys. Rev. B **56**, 13 916 (1997).

33. A. H. Romero, D. W. Brown and K. Lindenberg, J. Chem. Phys. **109**, 6504 (1998); Phys. Rev. B **59**, 13728 (1999); Phys. Rev. B **60**, 4618 (1999); Phys. Rev. B **60**, 14080 (1999).

34. A. La Magna and R. Pucci, Phys. Rev. B **53**, 8449 (1996).

35. P. Benedetti and R. Zeyher, Phys. Rev. B **58**, 14320 (1998).

36. T. Frank and M. Wagner, Phys. Rev. B **60**, 3252 (1999).

37. L. Proville and S. Aubry, S. Eur. Phys. J. B **11**, 41 (1999).

38. J. Bonča, S. A. Trugman, and I. Batistić, Phys. Rev. B **60**, 1633 (1999); J. Bonča, T. Katrasnic, and S. A. Trugman, Phys. Rev. Lett. **84**, 3153 (2000); L.-C. Ku, S. A. Trugman, and J. Bonča, Phys. Rev. B **65**, 174306 (2002); S. El Shawish, J. Bonča, L.-C. Ku, and S. A. Trugman, Phys. Rev. B **67**, 014301 (2003).

39. A. S. Alexandrov and P. E. Kornilovitch, Phys. Rev. Lett. **82**, 807 (1999).

40. A. S. Alexandrov, Phys. Rev. B **61**, 12315 (2000).

41. A. S. Alexandrov, Phys. Rev. B **53**, 2863 (1996).

42. A. S. Alexandrov and J. Ranninger, Phys. Rev. B **23**, 1796 (1981), ibid **24**, 1164 (1981).

43. P. Kornilovitch, in the present volume.

44. R. T. Shuey, Phys. Rev. A, **139**, 1675 (1994).

45. E. G. Maximov, D. Yu. Savrasov, and S. Yu. Savrasov, Uspechi Fiz. Nauk., **167**, 353 (1997).

46. V. G. Baryakhtar, E. V. Zarochentsev, and E. P. Troitskaya *Theory of Adiabatic Potential and Atomic Properties of Simple Metals* (Gordon and Breach, Amsterdam, 1999).

47. G. D. Mahan, *Many-Particle Physics* (Plenum Press, New York, 1990).

48. V.V. Kabanov, in the present volume.

49. B.T. Geilikman, Usp. Fiz. Nauk **115**, 403 (1975) [Sov. Phys. Usp. **18**, 190 (1975)].

50. T. Holstein, Ann. Phys. **8**, 325; 343 (1959).

51. E. I. Rashba, Opt. Spectr. **2**, 75 (1957); *Excitons*, eds. E. I. Rashba and D. M. Struge (Nauka, Moscow, 1985).

52. P. W. Anderson, Phys. Rev. Lett. **34**, 953 (1975).

53. S. Aubry, in the present volume.

54. H. Hiramoto and Y. Toyozawa, J. Phys. Soc. Jpn. **54**, 245 (1985).

55. J. Ranninger and U. Thibblin, Phys. Rev. B **45**, 7730 (1992).

56. E. V. L. de Mello and J. Ranninger, Phys. Rev. B **55**, 14872 (1997).

57. E. V. L. de Mello and J. Ranninger, Phys. Rev. B **58**, 9098 (1998).

58. E. K. Kudinov and Yu. A. Firsov, Fiz. Tverd. Tela (S.-Peterburg) **39**, 2159 (1997) [Phys. Solid State **39**, 1930 (1997)].

59. Yu. A. Firsov, E. K. Kudinov, V. V. Kabanov, and A. S. Alexandrov, Phys. Rev. B **59**, 12132 (1999).

60. A. S. Alexandrov, Phys. Rev. B **61**, 315 (2000).

61. D. M. Eagles, Phys. Rev. **130**, 1381 (1963); Phys. Rev. **181**, 1278 (1969); Phys. Rev. **186**, 456 (1969).

62. A. A. Gogolin, Phys. Status Solidi B **109**, 95 (1982).

63. V. Cataudella, G. De Filippis, and C.A. Perroni, in the present volume.

64. H. Fehske and S. A. Trugman, in the present volume.

65. H. de Raedt and A. Lagendijk, Phys. Rev. Lett. **49**, 1522 (1982); Phys. Rev. B **27**, 6097 (1983); Phys. Rev. B **30**, 1671 (1984); Phys. Rep. **127**, 234 (1985).

66. M. Hohenadler, H. G. Evertz, and W. von der Linden, Phys. Rev. B **69**, 024301 (2004).

67. A. S. Mishchenko, N.V. Prokof'ev, A. Sakamoto, and B.V. Svistunov, Phys. Rev. B **62**, 6317 (2000); A. S. Mishchenko, N. Nagaosa, N. V. Prokof'ev, A. Sakamoto, and B. V. Svistunov, Phys. Rev. Lett. **91**, 236401 (2003).

68. A. Macridin, G. A. Sawatzky, and M. Jarrell, Phys. Rev. B **69**, 245111 (2004).

69. P. E. Spencer, J. H. Samson, P. E. Kornilovitch, and A. S. Alexandrov, Phys. Rev. B **71**, 184319 (2005).

70. J. P. Hague, P. E. Kornilovitch, A. S. Alexandrov, and J. H. Samson, Phys. Rev. B **73**, 054303 (2006).
71. J. P. Hague, P. E. Kornilovitch, J. H. Samson, and A. S. Alexandrov, Phys. Rev. Lett. **98**, 037002 (2007)
72. A. S. Mishchenko and N. Nagaosa, this volume.
73. M. Zoli, Phys. Rev. B **57**, 555 (1998); ibid **61**, 14523 (2000).
74. A. S. Alexandrov and C. Sricheewin, Europhys. Lett. **51**, 188 (2000).
75. A. S. Alexandrov and J. Ranninger, Phys. Rev. B **45**, 13109 (1992); Physica C **198**, 360 (1992).
76. H. Fröhlich, Adv. Phys. **3**, 325 (1954).
77. V. L. Vinetskii and M. Sh. Giterman, Zh. Eksp. Teor. Fiz. **33**, 730 (1957) [Sov. Phys. JETP **6**, 560 (1958)].
78. G. Verbist, F. M. Peeters, and J. T. Devreese, Phys. Rev. B **43**, 2712 (1991); Solid State Commun. **76**, 1005 (1990).
79. A. S. Alexandrov(Aleksandrov) and V. V. Kabanov, Fiz. Tverd. Tela **28**, 1129 (1986) [1986 Soviet Phys. Solid St. **28**, 631 (1986)].
80. V. V. Bryksin and A. V. Gol'tsev, Fiz. Tverd. Tela **30**, 1476 (1988) [Sov. Phys. Solid State **30**, 851 (1988)].
81. A. S. Alexandrov and P. E. Kornilovitch, J. Phys.: Condens. Matter **14**, 5337 (2002).
82. J. Bonča and S. A. Trugman, Phys. Rev B **64**, 094507 (2001).
83. C. R. A. Catlow, M. S. Islam and X. Zhang, J. Phys.: Condens. Matter **10**, L49 (1998).
84. A. S. Alexandrov and V. V. Kabanov, JETP Lett. **72**, 569 (2000).
85. T. Mertelj, V. V. Kabanov, and D. Mihailovic, Phys. Rev. Lett. **94**, 147003 (2005).
86. A. S. Alexandrov, J. Ranninger, and S. Robaszkiewicz, Phys. Rev. B **33**, 4526 (1986).
87. A. S. Alexandrov, D. A. Samarchenko, and S. V. Traven, Zh. Eksp. Teor. Fiz. **93**, 1007 (1987) [Sov. Phys. JETP **66**, 567 (1987)].
88. R. E. Peierls, Z. Phys. **80**, 763 (1933).
89. M. R. Schafroth, Phys. Rev. **100**, 463 (1955); J. M. Blatt and S. T. Butler, Phys. Rev. **100**, 476 (1955).
90. A. S. Alexandrov, Doctoral Thesis MEPHI (Moscow, 1984); Phys. Rev. B **48**, 10571 (1993).
91. See chapter by Guo-meng Zhao in the present volume.
92. A. S. Alexandrov, W. H. Beere, V. V. Kabanov, and W. Y. Liang, Phys. Rev. Lett. **79**, 1551 (1997).
93. A. S. Alexandrov, Phys. Rev. Lett. **82**, 2620 (1999); A. S. Alexandrov and V. V. Kabanov, Phys. Rev. B **59**, 13628 (1999).
94. A. S. Alexandrov, J. Low Temp. Phys. **87**, 721 (1992).
95. D. Mihailovic, V. V. Kabanov, K. Zagar, and J. Demsar, Phys. Rev. B **60**, 6995 (1999) and references therein.
96. A. S. Alexandrov, V. V. Kabanov, and N. F. Mott, Phys. Rev. Lett. **77**, 4796 (1996).
97. A. S. Alexandrov, Phys. Rev. Lett. **96**, 147003 (2006).
98. K. K. Lee, A. S. Alexandrov, and W. Y. Liang, Phys. Rev. Lett. **90**, 217001 (2003); Eur. Phys. J. B**30**, 459 (2004).
99. A. S. Alexandrov, Phys. Rev. B **73**, 100501 (2006).

100. A. S. Alexandrov and V. N. Zavaritsky, Phys. Rev. Lett. **93**, 217002 (2004).

101. A. S. Alexandrov, Phys. Rev. Lett. **95**, 259704 (2005).

102. A.S. Alexandrov, cond-mat/0607770.

103. V. N. Zavaritsky, V. V. Kabanov, A. S. Alexandrov, Europhys. Lett. **60**, 127 (2002).

104. Y. Zhang, N. P. Ong, Z. A. Xu, K. Krishana, R. Gagnon, L. Taillefer, Phys. Rev. Lett. **84**, 2219 (2000).

105. A. S. Alexandrov, N. F. Mott, Phys. Rev. Lett. **1993**, 71, 1075.

106. G. Zhao and D. E. Morris, Phys. Rev. B **51**, 16487 (1995); G.-M. Zhao, M. B. Hunt, H. Keller, and K. A. Müller, Nature **385**, 236 (1997); R. Khasanov, D. G. Eshchenko, H. Luetkens, E. Morenzoni, T. Prokscha, A. Suter, N. Garifianov, M. Mali, J. Roos, K. Conder, and H. Keller Phys. Rev. Lett. **92**, 057602 (2004).

107. A. S. Alexandrov, Phys. Rev. B **46**, 14932 (1992).

108. A. Lanzara, P. V. Bogdanov, X. J. Zhou, S. A. Kellar, D. L. Feng, E. D. Lu, T. Yoshida, H. Eisaki, A. Fujimori, K. Kishio, J. I. Shimoyana, T. Noda, S. Uchida, Z. Hussain, Z. X. Shen, Nature **2001**, 412, 510.

109. G. H. Gweon, T. Sasagawa, S. Y. Zhou, J. Craf, H. Takagi, D. H. Lee, A. Lanzara, Nature **2004**, 430, 187.

110. D. Mihailovic, C.M. Foster, K. Voss, and A.J. Heeger, Phys. Rev. B**42**, 7989 (1990).

111. P. Calvani, M. Capizzi, S. Lupi, P. Maselli, A. Paolone, P. Roy, S.W. Cheong, W. Sadowski, and E. Walker, Solid State Commun. **91**, 113 (1994).

112. R. Zamboni, G. Ruani, A.J. Pal, and C. Taliani, Solid St. Commun. **70**, 813 (1989).

113. T. Timusk, C.C. Homes, and W. Reichardt, in *Anharmonic properties of High Tc cuprates* (eds. D. Mihailovic, G. Ruani, E. Kaldis, and K.A. Müller, World Scientific, Singapore, 1995) p.171.

114. J. Tempere and J. T. Devreese, Phys. Rev. B **64**, 104504 (2001).

115. J. T. Devreese, in the present volume.

116. T.R. Sendyka, W. Dmowski, and T. Egami, Phys. Rev. B**51**, 6747 (1995); T. Egami, J. Low Temp. Phys. **105**, 791 (1996).

117. V. J. Emery and S. A. Kivelson, Nature (London), **374**, 434 (1995).

118. Z. A. Xu, N. P. Ong, Y. Wang, T. Kakeshita, and S. Uchida, Nature (London) **406**, 486 (2000); N. P. Ong and Y. Wang, Physica C **408**, 11 (2004).

119. Y. Wang, L. Li, M. J. Naughton, G.D. Gu, S. Uchida, and N. P. Ong, Phys. Rev. Lett. **95**, 247002 (2005); L. Li, Y. Wang, M. J. Naughton, S. Ono, Y. Ando, and N. P. Ong, Europhys. Lett. **72**, 451(2005); Y. Wang, L. Li, and N. P. Ong, Phys. Rev. B **73**, 024510 (2006).

120. V. N. Zavaritsky and A. S. Alexandrov, Phys. Rev. B **71**, 012502 (2005).

121. V. N. Zavaritsky, J. Vanacken, V. V. Moshchalkov, and A. S. Alexandrov, Eur. Phys. J. B **42**, 367 (2004).

122. B. Bucher, J. Karpinski, E. Kaldis, and P. Wachter, Physica C **167**, 324 (1990).

123. A. P. Mackenzie, S. R. Julian, G. G. Lonzarich, A. Carrington, S. D. Hughes, R. S. Liu, and D. C. Sinclair, Phys. Rev. Lett. **71**, 1238 (1993)

124. M. A. Osofsky, R. J. Soulen, A. A. Wolf, J. M. Broto, H. Rakoto, J. C. Ousset, G. Coffe, S. Askenazy, P. Pari, I. Bozovic, J. N. Eckstein, and G. F. Virshup, Phys. Rev. Lett. **71**, 2315 (1993); ibid **72**, 3292 (1994).

125. D. D. Lawrie, J. P. Franck, J. R. Beamish, E. B. Molz, W. M. Chen, and M. J. Graf, J. Low Temp. Phys. **107**, 491 (1997).

126. V. F. Gantmakher, G. E. Tsydynzhapov, L. P. Kozeeva, and A. N. Lavrov, Zh. Eksp. Teor. Fiz. **88**, 148 (1999).
127. A. S. Alexandrov, V. N. Zavaritsky, W. Y. Liang, and P. L. Nevsky, Phys. Rev. Lett. **76**, 983 (1996).
128. J. Hofer, J. Karpinski, M. Willemin, G. I. Meijer, E. M. Kopnin, R. Molinski, H. Schwer, C. Rossel, and H. Keller, Physica C **297**, 103 (1998).
129. V. N. Zverev and D. V. Shovkun, JETP Lett. **72**, 73 (2000).
130. A. J. Leggett, Physica Fennica **8**, 125 (1973); J Stat. Phys. **93**, 927 (1998); V. N. Popov, *Functional Integrals and Collective Excitations* (Cambridge University Press, Cambridge, 1987).
131. C. Bergemann, A. W. Tyler, A. P. Mackenzie, J. R. Cooper, S. R. Julian, and D. E. Farrel, Phys. Rev. B **57**, 14387 (1998).
132. A. Junod, J-Y. Genouda, G. Trisconea, and T. Schneider, Physica C **294**, 115 (1998).
133. J. Hofer, T. Schneider, J. M. Singer, M. Willemin, H. Keller, T. Sasagawa, K. Kishio, K. Conder, and J. Karpinski, Phys. Rev. B **62**, 631 (2000).
134. M. J. Naughton, Phys. Rev. B **61**, 1605 (2000).
135. I. Iguchi, A. Sugimoto, and H. Sato, J. Low Temp. Phys. **131**, 451 (2003).
136. J. L. Tallon, J. W. Loram, J. R. Cooper, C. Panagopoulos, and C. Bernhard, Phys. Rev. B **68**, 180501(R) (2003).
137. C. J. Dent, A. S. Alexandrov, and V. V. Kabanov, Physica C **341-348**, 153 (2000).
138. I. Bozovic, G. Logvenov, M. A. J. Verhoeven, P. Caputo, E. Goldobin, and M. R. Beasley, Phys. Rev. Lett. **93**, 157002 (2004).
139. I. Bozovic, G. Logvenov, M. A. J. Verhoeven, P. Caputo, E. Goldobin, and T. H. Geballe, Nature (London) **422**, 873 (2003)
140. R. A. Ogg Jr., Phys. Rev. **69**, 243 (1946).
141. E. V. L. de Mello and J. Ranninger, Phys. Rev. B **58**, 9098 (1998).
142. P. W. Anderson, *The Theory of Superconductivity in the Cuprates* (Princeton University Press, Princeton NY, 1997).

Small Adiabatic Polarons and Bipolarons

Serge Aubry

Laboratoire Léon Brillouin, CEA Saclay (CEA-CNRS), 91191 Gif-sur-Yvette, France `saubry@dsm-mail.saclay.cea.fr`

Summary. We review some results concerning small adiabatic polarons, bipolarons and many polaron-bipolaron structures which were described in a series of works of the present author during recent decades.

We first investigate the existence of the single small polaron in models with short range interactions as a function of dimensionality We show by the variational method that it always exists in one dimension (1D) while in two and more dimensions, it appears discontinuously only beyond a critical coupling as a small polaron. When a magnetic field is added in 2D and 3D, large polarons now exist at small coupling and there is a first order transition versus electron phonon coupling between large and small polarons which disappears beyond a critical magnetic field. We also mention commensurability effects generated when the number of quantum fluxes per plaquette is rational. We next discuss the existence of multipeaked polarons and show that they do not exist as ground-states in a variation of the Holstein model, in contradiction with early claims. We also mention the possible existence of polaro-breathers where an electron may become trapped by an anharmonic and localized lattice vibration.

Next we investigate the possible mobility of small polarons and emphasize the role of the Peierls-Nabarro (PN) energy barrier which could be strongly depressed near first order transitions of the polaron. These ideas are applied to the bipolaron in the adiabatic Hostein-Hubbard model. We show that, depending on the Hubbard repulsion, the bipolaron may be single site or multipeaked. Then, it consists of two polarons bonded by a magnetic interaction. In 2D, more complex bipolarons which are spin quadrisinglet may become ground-state.

Next we describe theorems obtained for the adiabatic Holstein-Hubbard model near the anticontinuous limit where the electronic transfer integral is zero. We prove that the bipolaronic and polaronic structures with many electrons which trivially exist at this limit, persist by continuity as insulating structures when the transfer integral becomes nonzero. We also prove that the system ground-state belongs to this family of solutions. It could be either a Bipolaronic Charge Density Wave when the Hubbard term is not too large or a polaronic charge density wave when it is large enough with a superposed spin ordering.

In the 1D adiabatic Holstein models with small transfer integral, the ground-state is an insulating bipolaronic charge density wave which may be commensurate or incommensurate whether the band filling is rational or irrational. When the transfer integral increases, the incommensurate bipolaronic CDWs undergo a reverse transition by breaking of analyticity where the CDW loses its bipolaronic character and becomes a conducting Peierls-Fröhlich CDW.

Finally, we briefly discuss the role of quantum fluctuations on the bipolaronic and polaronic structure. We argue that this role becomes essential when the PN energy barrier becomes small. Then the spatial ordering of the bipolaronic structure may disappear and instead we could expect Bose condensation of the bipolarons which are hard core bosons, that is bipolaronic superconductivity. We also briefly mention the role of adiabatic polarons in chemistry in chemical reactions which consist in electron transfer between molecules. Exceptional phenomena such as ultrafast electron transfer could be described with polaronic models which are nonadiabatic.

1 Introduction

The concept of the polaron was discovered more than 70 years ago by Landau [1] who suggested that because of its electric charge, the wave function of an extra electron in matter could localize in its polarisation field generated selfconsistently. Later Pekar [2] suggested that an electron could similarly self-localize in the potential well created by the lattice deformation generated by its interaction with the surrounding atoms. After half a century of subsequent works, the concept of the polaron is now ubiquitous in physics. It has been extended to any quantum excitation which localizes self-consistently (self-trapping) in a potential generated by its interaction with a deformable field (generally atomic displacements). Polarons form when the energy gain obtained by lowering the electronic eigenenergy in the self-consistent local potential is larger than the energy cost of creating the distortion which generates this potential (see Fig. 1). Polarons have been found to be involved in many materials and could interpret a wide variety of phenomena.

It is usual to distinguish between small and large polarons. Large polarons are relatively extended at the scale of the spacing between the atoms so that they can be described within continuous models where the discreteness of the system is neglected. As a counterpart, since the electronic wave function is large, the binding energy of the electron in its self-consistent potential is small. In contrast, small polarons localize at the atomic scale and should be described within discrete models. Unlike large polarons, small polarons are sensitive to the discreteness of the underlying lattice and are pinned by a PN energy barrier. We shall mostly focus here on small polarons in the adiabatic Holstein model and also on bipolarons in the adiabatic Holstein-Hubbard model.

Since the electron has a spin $1/2$, a polaron exhibits a spin and thus is magnetic. However two electrons with opposite spins may occupy the same quantum state if the electronic repulsion (Hubbard or electrostatic origin) is not too large. We then obtain a *bipolaron* where the pair of electrons is a non-magnetic singlet state. Two electrons in the same state cooperatively generate

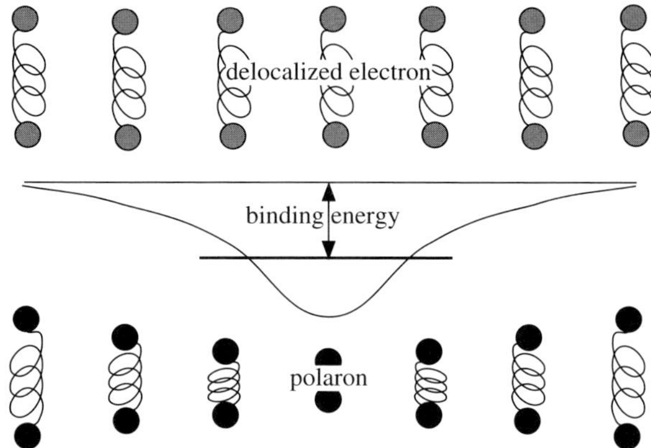

Fig. 1. Scheme of a polaron formation in the example of a molecular crystal of deformable diatomic molecules. The wave function of an extra electron in the system may remain an extended planewave and there is no molecule distortion (above) or a local distortion may spontaneously appears thus creating self-consistently a potential well in which the electronic wave function localizes (below).

a deeper self-consistent potential well which makes the binding energy for a bipolaron generally substantially larger than that for two independent single polarons. This is not true when there is a strong electronic repulsion. Then, bipolarons may break into two polarons but however these two polarons may remain bounded by magnetic spin interactions.

This self-trapping phenomena which produces a polaron (or a bipolaron) requires that the potential well which traps the electron is not erased by the quantum fluctuations of the deformable field. Thus, electron localization to be effective requires that the amplitude of distortion of the deformable field, which generates the local potential, is much larger than the zero point quantum fluctuations of this field. Polarons can only be strictly localized over a long lifetime when the amplitude of the zero point quantum fluctuations is negligible, that is in the limit of a strong interaction of the electron (for generating large distortion) and/or a very massive deformable field so that the classical approximation for the deformable field is valid.

Indeed it is well known that in principle the excitations of any fully quantum model which is translationnally invariant, and in particular polarons and bipolarons, form bands. As a consequence, when considering the deformable field as quantum, as it should be physically, polarons which should tunnel through the lattice can not remain localised and form polaron bands. The classical polaron limit is obtained when the band width vanishes which in other words means that the effective mass of this quantum particle is infinite. Thus it cannot tunnel and remains localized. This situation is met in many physical systems where the surrounding field of the electron consists of charged and heavy ions which strongly interact with the electron and displacements. However, there are also situations where small polarons or bipolarons become highly quantum with a relatively small effective mass. This situation occurs because the PN energy barrier becomes small. which also manifests by a low frequency pinning mode.

Although bipolarons and polarons are nonlinear excitations which cannot be linearly combined, it can be proved that they may persist with arbitrary random configurations when there are many electrons in the system. Their ground states correspond to specific ordered arrangements which are either bipolaronic charge density waves or polaronic charge density waves with a magnetic superstructure which is insulating. Some studies of these structures were performed only in 1D models. It was found that commensurate bipolaronic ground-states were always insulating while, when the electron-phonon coupling becomes small enough, incommensurate bipolaronic ground-states could undergo a second order transition (inverse transition by breaking of analyticity) and transform into Peierls-Fröhlich CDWs which are conducting. Moreover, the absence of the PN energy barrier and the strictly zero frequency pinning mode should make the role of the quantum lattice fluctuations essential. It is conjectured that Peierls-Fröhlich CDWs are unstable against quantum lattice fluctuations and could become superconducting.

Electrons in chemistry are responsible for the formation of chemical bonds between the atoms of the molecule. Thus, when electron transfer occurs between different molecules or sites, there is generally a very large reorganization of the environment not only because of chemical bond rearrangement but by the electrostatic interactions of the electron with the dipoles and charges of the molecules. Because of the relatively large nuclei displacements involved in electron transfer, a classical description of their displacement is generally valid. The standard theory of electron transfer has been developed on this basis [3] independent of polaron theory in physics. Actually, electron transfer in chemistry is nothing but polaron transfer.

The literature related to the concept of the adiabatic polaron is very abundant and cannot be discussed extensively. The aim of this review is only to focus on some aspects of the theory of polarons mostly related to my own contribution to this problem during the two last decades.

2 Single Adiabatic Polaron in Continuous Models

We first recall some basic results about polarons in the simplest models. Continuous models are appropriate for describing large polarons while discrete models are necessary for describing small polarons.

A simple continuous model is instructive for analyzing the conditions for the formation of large polarons. An adiabatic polaron can be obtained within the standard adiabatic approximation where the electronic state moves in the static potential generated by a field deformation. The Hamiltonian of a single electron in this continuous deformable medium may be written [4]

$$H = \frac{p^2}{2m_e} + V(\mathbf{r}) \tag{1}$$

where the momentum operator is $p = \frac{\hbar}{i}\nabla_{\mathbf{r}}$ and the electron potential

$$V(\mathbf{r}) = \int \chi(\mathbf{r} - \mathbf{r}')U(\mathbf{r}')d\mathbf{r}' \tag{2}$$

linearly depends on the strain field $U(\mathbf{r})$ through the kernel $\chi(\mathbf{r} - \mathbf{r}')$. The energy of the strain field is assumed to be quadratic

$$E_s = \frac{K}{2}\int U^2(\mathbf{r}')d\mathbf{r}' \tag{3}$$

Considering the lowest-energy eigenfunction $\Psi(\mathbf{r})$, the kinetic energy of the electron is

$$E_{el} = -\frac{\hbar^2}{2m_e}\int \Psi^\star(\mathbf{r})\Delta.\Psi(\mathbf{r})d\mathbf{r} \tag{4}$$

and the interaction energy is

$$E_{int} = \int |\Psi(\mathbf{r})|^2\chi(\mathbf{r} - \mathbf{r}')U(\mathbf{r}')d\mathbf{r}d\mathbf{r}' \tag{5}$$

The polaron solution is obtained by minimizing the total energy $E_P = E_{el} + E_{int} + E_s$ of the system with respect to $U(\mathbf{r})$ which yields

$$U(\mathbf{r}) = -\frac{1}{K}\int \chi(\mathbf{r}' - \mathbf{r})|\Psi(\mathbf{r}')|^2 d\mathbf{r}' \tag{6}$$

so that the total energy of the polaron is the minimum of $E_P = E_{el} - V_{int}$ where

$$V_{int} = \frac{1}{2K}\int \kappa(\mathbf{r}' - \mathbf{r}'')|\Psi(\mathbf{r}')|^2|\Psi(\mathbf{r}'')|^2 d\mathbf{r}'d\mathbf{r}'' \tag{7}$$

and

$$\kappa(\mathbf{r}' - \mathbf{r}'') = \int \chi(\mathbf{r}' - \mathbf{r})\chi(\mathbf{r}'' - \mathbf{r})d\mathbf{r}$$

In the limit of an infinite volume $\mathcal{V}$, the extended state $\Psi(\mathbf{r}) = 1/\sqrt{\mathcal{V}}$ minimizes and the electron kinetic energy E_{el} and V_{int} both vanish. Then, there is no strain $U(\mathbf{r}) = 0$ and the total energy $E = E_{el} - V_{int}$ is zero. For proving the existence of a polaron (where $\Psi(\mathbf{r})$ is not zero for $\mathcal{V} = +\infty$) it suffices to exhibit a trial function $\Psi(\mathbf{r})$ which has a total energy lower than the extended state that is strictly negative.

Emin and Holstein use scaling arguments to prove the existence of the polaron. Assuming that a nonvanishing electronic wave function $\Psi(\mathbf{r})$ is known for the ground-state, they consider the rescaled normalized wave-function

$\Psi_\xi(\mathbf{r}) = \xi^{-d/2}\Psi(\mathbf{r}/\xi)$ where d is the dimension of the space. The total energy of this wave function $< \Psi_\xi|H|\Psi_\xi >$ as a function of ξ should be a minimum for $\xi = 1$. It readily comes out that the electronic kinetic energy scales as $1/\xi^2$

$$E_{el}(\xi) = E_{el}\xi^{-2} > 0 \tag{8}$$

The behavior of V_{int} depends on the behavior of the kernel $\chi(\mathbf{r})$ describing the range of the interaction between the electron and the deformable field.

If one assumes for example a long range dipolar interaction $\chi(\mathbf{r}) = \chi_0\mathbf{r}^{-2}$, one obtains

$$V_{int}(\xi) = V_{int}\xi^{d-4} > 0 \tag{9}$$

Then, $E(\xi) = E_{el}\xi^{-2} - V_{int}\xi^{d-4}$ should have a negative minimum at $\xi = 1$. This is not possible for $d \geq 4$ since the minimum of $E(\xi)$ is at $\xi = +\infty$. Then, the electron has the lowest energy when it remains extended. When $d < 2$ the minimum of $E(\xi)$ is $-\infty$ at $\xi = 0$. This result means the polaron collapses at the microscopic scale and forms a small polaron.

When $2 < d < 4$, any non-vanishing trial function yields a negative minimum for $E(\xi)$ at some finite ξ (no answer can be obtained for $d = 2$ which is marginal). Consequently, we get large polarons in models with dipolar interactions only for 3D models. Other results would be obtained with different interactions $\chi(\mathbf{r}) = \chi_0\mathbf{r}^{-\alpha}$ for which large polarons exist when $2(\alpha - 1) < d < 2\alpha$.

If one assumes that the electric field due to the extra electron is totally screened by the background electrons of the material (at finite temperature), the interaction between the extra electron and the deformable field may be assumed to be purely local, that is

$$\chi(\mathbf{r}) = \chi_0\delta(\mathbf{r}) \tag{10}$$

is a Dirac function. Then the strain $U(\mathbf{r}) = -\frac{\chi_0}{K}|\Psi(\mathbf{r})|^2$ is proportional to the electron density at the same point and

$$V_{int} = \frac{\chi_0^2}{2K} \int |\Psi(\mathbf{r})|^4 d\mathbf{r} \tag{11}$$

Then, we get $V_{int}(\xi) = V_{int}\xi^{-d}$. Again the condition that

$$E(\xi) = \frac{E_{el}}{\xi^2} - \frac{V_{int}}{\xi^d} \tag{12}$$

is minimum for $\xi = 1$ requires $0 < d < 2$. We essentially get large polarons only in 1D (see Fig. 2). For $d > 2$, there are two minima for the energy which are at $\xi = +\infty$ and at $\xi = 0$ which thus cannot be at $\xi = 1$ (see Fig. 2). The absolute minima of $E(\xi)$ at $\xi = 0$ is negative and infinite which means that the polaron would collapse down to a vanishing extension. Then it should reach the atomic scale so that the discreteness of the system can no longer be neglected.

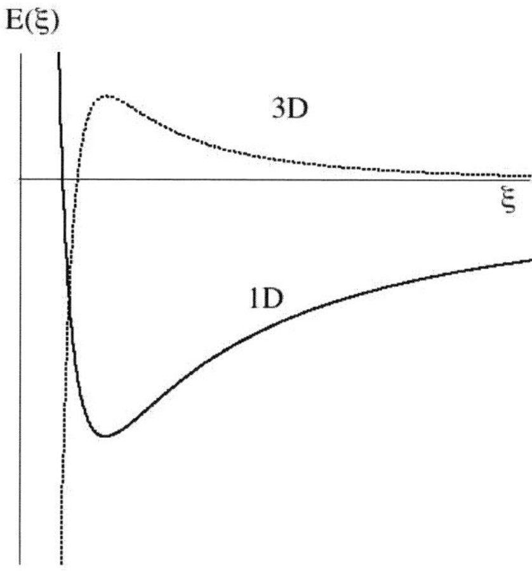

Fig. 2. Sketch of the variation of the polaron energy $E(\xi)$ versus the scale factor ξ in 1D (full line) and in 3D (dashed line).

The 2D case is marginal because a minimum at $\xi = 1$ requires $E_{el} = V_{int}$ but this minimum is degenerate since the total energy does not depend on the radius ξ. Actually, we have

$$E_{el} = \frac{\hbar^2}{2m_e} \int |\nabla \Psi(x,y)|^2 dxdy \tag{13}$$

and

$$V_{int} = \frac{\chi_0^2}{2K} \int |\Psi(x,y)|^4 dxdy \tag{14}$$

It was claimed in [5] that any normalized complex wavefunction in 2D $\Psi(x,y)$ fulfils the inequality

$$\kappa_c \int |\Psi(x,y)|^4 dxdy \leq \int |\nabla \Psi(x,y)|^2 dxdy \tag{15}$$

where $\pi \leq \kappa_c \leq 2\pi$ is a non-vanishing constant. Consequently, for

$$\frac{m_e \chi_0^2}{\hbar^2 K} < \kappa_c \tag{16}$$

we always have $E_{el} > V_{int}$ which implies the electron necessarily remains extended and there is no polaron. Conversely, when inequality (16) is reversed, Ψ can be always chosen in order that $E_{el} < V_{int}$ so that there is polaron collapse toward a negative infinite energy. This polaron requires description within a discrete model.

However, the rigorous proof of (15) which is quite non trivial, has never been published since this claim [5]. This review is an opportunity to present it (see appendix A).

3 Discrete Models for Small Polarons

In discrete models, the minimum size of a small polaron is a single site, which prevents it collapsing to a single point with a negative infinite energy as in the continuous 2D and 3D Holstein models. Only such discrete models allow a correct description of small polarons.

However, discrete models may involve microscopic details which disappear in the continuous limit and thus there are many discrete models with the same continuous limit. Actually, as we shall see later, these microscopic details may be important concerning the properties of small polarons. We first consider here one of the simplest discrete models which consists of an electron in a single band on a square lattice. The electronic wave function is now a function of the discrete sites i of a lattice. More precisely, it consists of the set of components Ψ_i of the real electronic wave function on the base of Wannier functions $\phi_i(\widetilde{r}) = \phi(\widetilde{r} - \widetilde{r}_i)$ which spans the space of electronic wave functions associated with a single band (tight binding representation). These complex amplitudes fulfil the normalization condition $\sum_i |\Psi_i|^2 = 1$.

Assuming nearest neighbor transfer integral only and that the atomic displacements are small, we expand at lowest significant order the energy of the distortion and the interaction of the electron density $|\Psi_i|^2$ with the discrete strain field u_i at site i of a d-dimensional square lattice, and we obtain a discrete version of the continuous model (1) with the Hamiltonian

$$H = -T \sum_{<i,j>} (\Psi_i^\star \Psi_j + \Psi_j^\star \Psi_i) + \sum_{i,n} \chi_n |\Psi_i|^2 u_{i+n} + \frac{K}{2} \sum_i u_i^2 + \frac{p_i^2}{2M} \quad (17)$$

The transfer integral T of the electrons between nearest neighbor sites $< i, j >$ determines the bare band width. χ_n couples the electron density operator $|\Psi_i|^2$ at site i with the strain u_{i+n} at the distant site $i + n$.

When χ_n is long range, we obtain a Fröhlich model [6] while when the interaction is purely local, that is only χ_0 is non-zero, we obtain at this opposite limit a Holstein model [7].

The ground-state of model (17) is first obtained by minimizing the energy of the system with respect to p_i and u_i which yields $p_i = 0$ and

$$Ku_i + \sum_n \chi_n |\Psi_{i-n}|^2 = 0 \quad (18)$$

Then, the polaron is obtained as the ground-state of the variational form

$$\Phi = -T \sum_{<i,j>} (\Psi_i^\star \Psi_j + \Psi_j^\star \Psi_i) - \frac{1}{2K} \sum_i (\sum_n \chi_n |\Psi_{i-n}|^2)^2 \tag{19}$$

which yields by minimization with respect to Ψ_i, with the normalization condition that the normalized electronic wave function fulfils the nonlinear discrete Schrödinger eigenequation

$$-T \sum_{j:i} \Psi_j - (\sum_n \kappa_n |\Psi_{i+n}|^2) \Psi_i = E \Psi_i \tag{20}$$

where $j : i$ represents the set of nearest neighbor sites j to site i and

$$\kappa_n = \kappa_{-n} = \frac{1}{K} \sum_p \chi_p \chi_{n+p} \tag{21}$$

The lowest energy of an extended electron is $-2Td$ where d is the dimension of the square lattice. Thus it suffices to exhibit a trial solution with a lower energy for proving the existence of a polaron. Actually, there are many other microscopic discrete models and variations exhibiting polarons which were designed for describing specific physical situations and that we shall not describe in detail. We shall mostly focus in this review on the Holstein model for which many essential features concerning polarons can be obtained and possibly extended with variations to more complex models.

For the Holstein model where $\chi_n = 0$ for $n \neq 0$, we used [9] the normalized trial solution

$$\Psi_{\{n_\alpha\}} = (\frac{1 - \eta^2}{1 + \eta^2})^{d/2} \eta^{|n|_s} \tag{22}$$

where $|n|_s = \sum_\alpha |n_\alpha|$ and n_α are the integer components of the coordinates $n = \{n_\alpha\}$ in the d dimensional square lattice. The parameter $\eta = e^{-1/\xi}$ determines the characteristic size of the polaron. The variational energy may be written as

$$\Phi = -T \left(\sum_{<i,j>} (\Psi_i^\star \Psi_j + \Psi_j^\star \Psi_i) + \frac{1}{2} k^2 \sum_i |\Psi_i|^4 \right) \tag{23}$$

where k is a dimensionless parameter

$$k^2 = \frac{\chi_0^2}{KT} \tag{24}$$

which characterizes the strength of the electron–phonon coupling. This variational energy can explicitly calculated [9] as a function of η

$$\Phi(\eta) = -T \left(4d \frac{\eta}{1 + \eta^2} + \frac{k^2}{2} \frac{(1 - \eta^2)^d (1 + \eta^4)^d}{(1 + \eta^2)^d} \right) \tag{25}$$

It has been shown [9] that the trial function which yields the minimum of $\Phi(\eta)$ yields a remarkably accurate approximation of the polaron when compared with precise numerical calculation of the adiabatic polaron (as done for example in [8]). The extrema of $\Phi(\eta)$ (25) are obtained when η fulfils the relation

$$k^2 = \frac{(1+\eta^2)^{3d-1}}{\eta(\eta^4 - \eta^2 + 1)(1+\eta^4)^{d-1}(1-\eta^2)^{d-2}} \tag{26}$$

The behavior of the energy $\Phi(\eta)/T$ as a function of the variational parameter

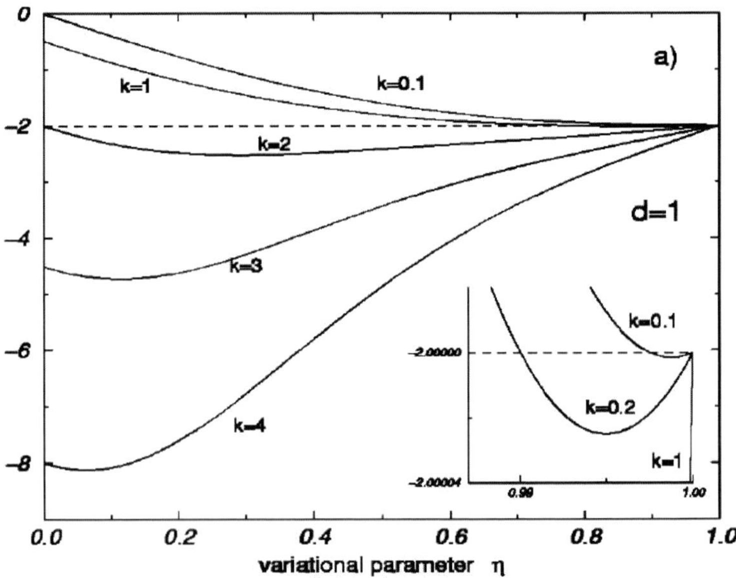

Fig. 3. Energy $\Phi(\eta)/T$ defined by (25) versus variational parameter η in the 1D Holstein model for different values of k. Inset shows a magnification for reduced small electron-phonon coupling k. Reproduced with permission from [9], ©(1998) by the American Physical Society.

η depends on the dimension of the model. Fig. 3 shows the variation of $\Phi(\eta)$ versus η in 1D. There is always a minimum in the interval $0 < \eta < 1$ which corresponds to the polaron solution. Consequently, the polaron exists at any coupling. However, the size of this polaron diverges ($\eta \to 1$) at small electron-phonon coupling k and shrinks to a single site at large k ($\eta \to 0$). Figure 4 shows the polaron energy obtained as the minimum of $\Phi(\eta)$ versus η which becomes asymptote to the lowest energy of the extended states at small k. Figure 5 shows that in 2D, $\Phi(\eta)$ exhibits a single minimum at large electron-phonon coupling k but for smaller k a second minimum at $\eta = 1$ appears. It does not correspond to a polaron but to an extended electron. For k still

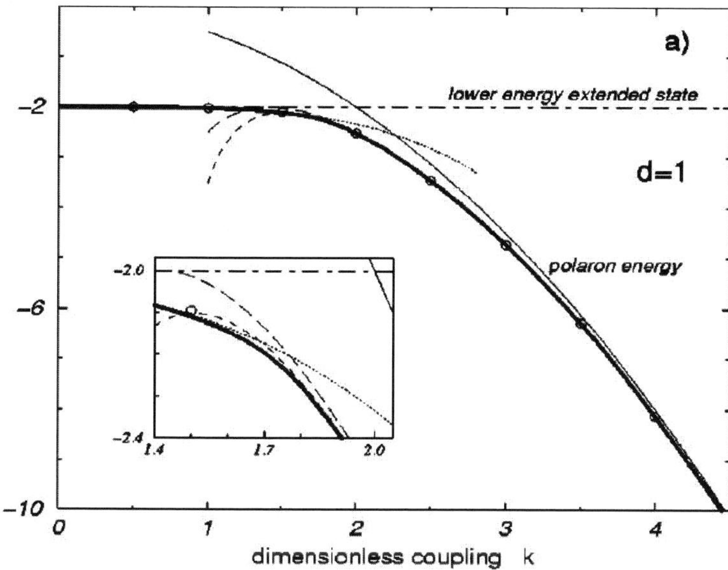

dimensionless coupling k

Fig. 4. Polaron energy versus the reduced electron–phonon coupling k (thick line) (obtained as $\min_\eta \Phi(\eta)$) for the 1D Holstein model. The open circle corresponds to exact numerical calculations. The other thin lines corresponds to other various approximations at small and large coupling (see [9]). Inset: magnification of the crossover region between large and small polarons. Reproduced with permission from [9], ©(1998) by the American Physical Society.

smaller, it becomes the lowest minimum and next the small polaron solution disappears. The consequence is that we now have a first order transition versus k between a localized small polaron and an extended electron and that there is no large polaron in the model.

Figure 6 shows the energy of the polaron versus k. The small polaron is the ground-state only for large enough k, that is $k > k_{c2}$. However, this solution remains metastable as a local minimum for $k_{c1} < k < k_{c2}$, but then it has a larger energy than the extended electron. This small polaron solution disappears for $k < k_{c1}$ beyond the critical point at $k = k_{c1}$, through a bifurcation with an unstable polaron solution (it is the minimax associated with the two local minima corresponding to the small polaron and the extended electron).

The behavior of the 3D Holstein model (see Figs. 7 and 8) is similar to that of the 2D Holstein model. There is also a first order transition between the extended electron and the small polaron at $k = k_{c2}$ and a critical point at $k = k_{c1} < k_{c2}$ where the small polaron disappears.

Thus, we have seen that in the 1D model, a local electron-phonon interaction always produces a polaron which interpolates continuously between large and small polarons. In contrast, in the 2D and 3D models, the strength of the

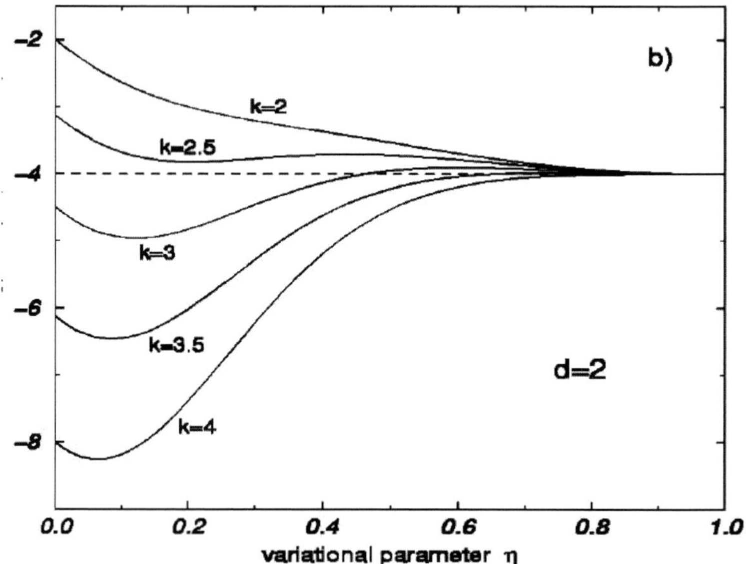

Fig. 5. Same as Fig. 3 but in the 2D Holstein model. Reproduced with permission from [9], ©(1998) by the American Physical Society.

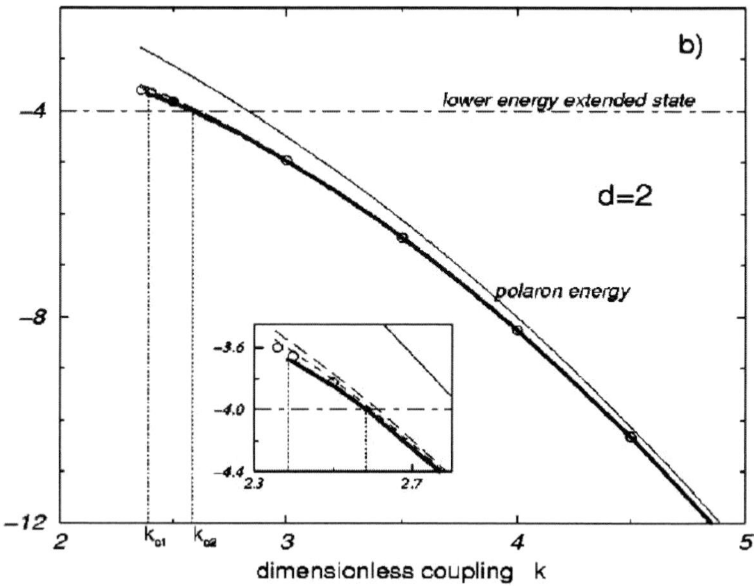

Fig. 6. Same as Fig. 4 but in the 2D Holstein model. Inset: Magnification of the region of first order transition. Reproduced with permission from [9], ©(1998) by the American Physical Society.

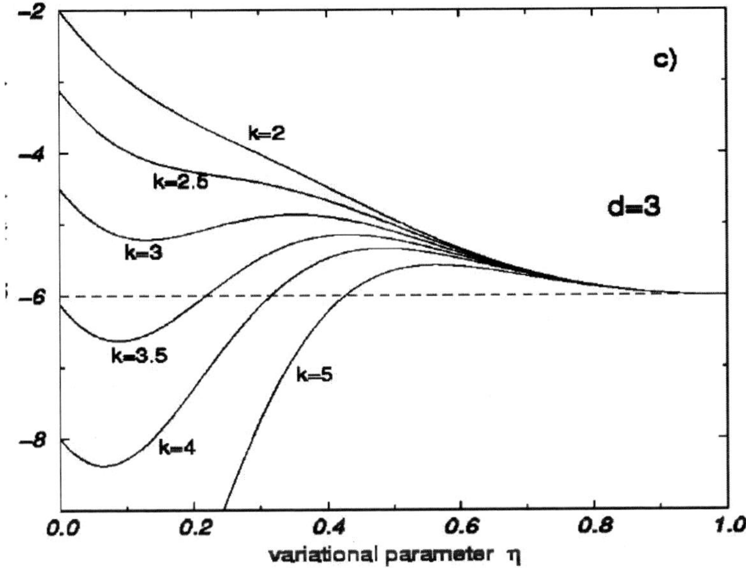

Fig. 7. Same as Fig. 3 but in the 3D Holstein model. Reproduced with permission from [9], ©(1998) by the American Physical Society.

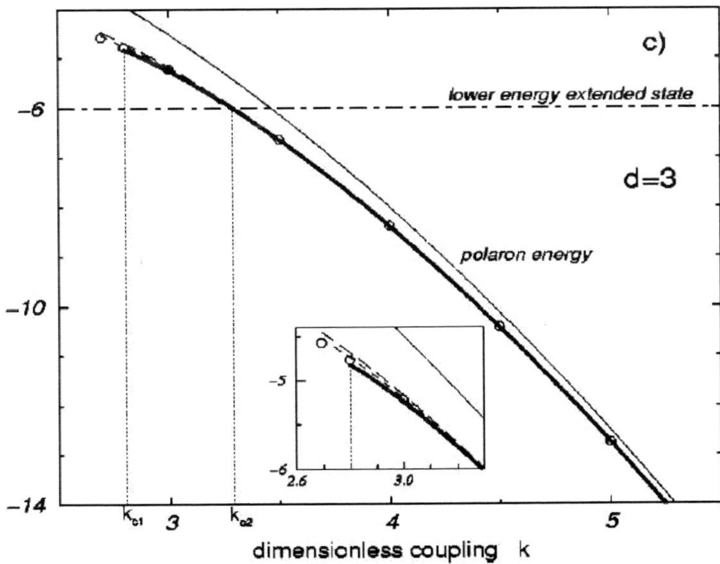

Fig. 8. Same as Fig. 4 but in the 3D Holstein model. Reproduced with permission from [9], ©(1998) by the American Physical Society.

local electron-phonon coupling should exceed a critical coupling in order to form small polarons and there is a first order transition between the extended electron and the polaronic state.

4 Polaron in a Uniform Magnetic Field

This problem has been studied previously to our investigations by Peeters and Devreese in 1982 [10, 11] for Fröhlich polarons in three dimensions treated with the Feynman approximation. First order phase transitions between different polaron ground states were found.

We consider here the adiabatic 2D and 3D Holstein models which have different behavior. At small enough electron-phonon coupling without magnetic field, these models do not sustain polarons. We show that when a magnetic field is applied in this case, a large polaron appears [5, 12]. Moreover, it is found that at intermediate electron-phonon coupling, there is a first order transition between a large and a small polaron. We first investigate these features in the continuous Holstein model with dimensionality arguments similar to those developed above and then we numerically investigate the discrete Holstein model.

4.1 Two-Dimensional Continuous Holstein Model

To prove this feature, we may consider continuous models (Sect. 2) which are 2D and 3D with a uniform magnetic field $\mathbf{H} = \left(0 \; 0 \; \mathcal{H} \right) = \mathrm{curl}\mathbf{A}$ in the z direction perpendicular to the $x - y$ plane. The vector potential $\mathbf{A}$ may then be chosen to be $\mathbf{A} = \frac{1}{2}\mathbf{H} \times \mathbf{r} = \frac{1}{2}\left(-\mathcal{H}y \; \mathcal{H}x \; 0 \right)$. The kinetic energy operator becomes (SI units)

$$H_K = \frac{1}{2m_e}(\mathbf{p} + e\mathbf{A})^2 \tag{27}$$

In the 2D case, the z component of the vector potential has to be simply dropped.

It is well known that a free electron in 2D and in a magnetic field (described by the kinetic energy operator H_K) has the same quantized eigenenergies $(n + \frac{1}{2})\hbar\omega_c$ (Landau levels) as a harmonic oscillator with the cyclotron frequency $\omega_c = \frac{e\mathcal{H}}{m_e}$. The eigenspace corresponding to each eigen energy is degenerate and is called a Bargmann space [13–15] which can be represented as a space of entire functions.

The eigenspace corresponding to the degenerate ground-state (lowest Landau level) is a Bargmann space $\mathcal{B}_n$. It is spanned by the set of localized wave functions

$$\Psi_g(\mathbf{r}) = \frac{1}{\xi_c\sqrt{\pi}} \exp{-\frac{(|\mathbf{r} - \mathbf{r}_i|)^2}{2\xi_c^2}} \tag{28}$$

where ξ_c is the cyclotron radius

$$\xi_c^2 = \frac{2\hbar}{e\mathcal{H}} = \frac{2\hbar^2}{m_e\hbar\omega_c}$$

and $\mathbf{r}_i$ is arbitrary. However note that this set of wave functions is not orthogonal and moreover are not linearly independent. There is a minimum density $\frac{1}{V}\int_V \delta(\mathbf{r} - \mathbf{r}_i)d\mathbf{r}$ of $\mathbf{r}_i$ required for spanning $\mathcal{B}_n$.

If we replace the kinetic energy operator in the continuous model without magnetic field (2) and with the Holstein coupling (10) by the kinetic energy operator (27) with magnetic field, the polaron solution in the presence of a uniform magnetic field is obtained by minimizing the variational energy $E_{el-h} - V_{int}$ as a function of the wave function $\Psi(\mathbf{r})$ where V_{int} is unchanged and defined by (7) but the electronic kinetic energy

$$E_{el-h} = \frac{1}{2m_e}\int \Psi^\star(\mathbf{r})(p + e\mathbf{A})^2\Psi(\mathbf{r})d\mathbf{r} = E_{el} + E_H + E_R \qquad (29)$$

where E_{el} is defined by (4) and

$$E_H = \frac{i}{2}\hbar\omega_c\int \Psi^\star(\mathbf{r})(y\frac{\partial.}{\partial x} - x\frac{\partial.}{\partial y})\Psi(\mathbf{r})d\mathbf{r} \qquad (30)$$

and

$$E_R = \frac{\hbar\omega_c}{4\xi_c^2}\int \Psi^\star(\mathbf{r})(x^2 + y^2)\Psi(\mathbf{r})d\mathbf{r} \qquad (31)$$

Assuming the ground state wave function is known, we may use the same scaling arguments [4] and consider the family of wavefunction in 2D models

$$\Psi_\xi(\mathbf{r}) = \xi^{-1}\Psi(\mathbf{r}/\xi) \qquad (32)$$

Then, we have $E_{el}(\xi) = E_{el}.\xi^{-2}$, $E_H(\xi) = E_H$, $E_R(\xi) = E_R\xi^2$ and $V_{int}(\xi) = V_{int}.\xi^{-2}$. The total energy should have a minimum at

$$E(\xi) = E_{el}.\xi^{-2} + E_H + E_R\xi^2 - V_{int}.\xi^{-2} \qquad (33)$$

At low enough electron phonon coupling when (16) is fulfilled, $E_{el} - V_{int} > 0$, the minimum of $E(\xi)$ given by (33) is unique, equal to $E_H + 2\sqrt{(E_{el} - V_{int})E_R}$. It is positive for any choice of Ψ and obtained for a finite non-vanishing value of $\xi^2 = \sqrt{(E_{el} - V_{int})/E_R}$ which can be fixed to 1 by choosing properly the initial scale for Ψ. Consequently, large polarons exist in a uniform magnetic field when (16) is fulfilled. Conversely, when this inequality is not fulfilled, it is possible to find Ψ such that $E_{el} - V_{int} < 0$. Then, the absolute minimum is $-\infty$ and the wave function collapses to a small polaron.

When the electron–phonon coupling is small, it can be treated as a perturbation which raises the degeneracy of the lowest Landau level. The wave function which yields the lowest energy for $E_{el-h} - V_{int}$ is a single Gaussian given by (28) for r_i arbitrary (which is not exponentially localized). However, a characteristic localization length $(< \int(x^2 + y^2)|\Psi(r)|^2dxdy)^{1/2}$ is nothing but the cyclotron radius ξ_c. When the electron–phonon coupling increases, we may choose a Gaussian $\Psi(x,y) = \frac{1}{\pi} e^{-(x^2+y^2)/2}$ as a trial function which has to be rescaled by the factor ξ as in (32) (which is then nothing but its characteristic size). This factor is obtained by minimizing the total energy (33) with respect to ξ where $E_{el} = \frac{\hbar^2}{2m_e\xi^2}$, $E_H = 0$, $E_R = \frac{\hbar\omega_c}{4\xi_c^2}\xi^2$ and $V_{int} = \frac{\chi_0^2}{2K\pi\xi^2}$ which yields

$$\xi \approx \xi_c\left(1 - \frac{m_e\chi_0^2}{\pi\hbar^2 K}\right)^{1/4} \tag{34}$$

This approximate formula yields that ξ vanishes when $\frac{m_e\chi_0^2}{\hbar^2 K} \approx \pi$. Actually, the exact value should be $\frac{m_e\chi_0^2}{K\hbar^2} = \kappa_c \approx 1.92\pi$ which corresponds to the transition between the extended electron and the small polaron in the continuous Hostein model without magnetic field.

4.2 3D Continuous Holstein Model

In 3D, a scaling argument assuming the ground state wave function $\Psi(x,y,z)$ is known may be also used. However, we should introduce two scale factors which are ξ_{xy} for the plane x,y and ξ_z for the direction z parallel to the uniform magnetic field. Then, we consider the family of normalized wave functions $\Psi_\xi(x,y,z) = \xi_{xy}^{-1}\xi_z^{-1/2}\Psi(x/\xi_{xy}, y/\xi_{xy}, z/\xi_z)$, the total energy of which has the form which extends (12)

$$E(\xi_{xy},\xi_z) = \frac{E_{el,xy}}{\xi_{x,y}^2} + \frac{E_{el,z}}{\xi_z^2} + E_H + E_R\xi_{xy}^2 - \frac{V_{int}}{\xi_{xy}^2\xi_z} \tag{35}$$

where

$$E_{el,xy} = \frac{\hbar^2}{2m_e}\int\left(|\frac{\partial\Psi}{\partial x}|^2 + |\frac{\partial\Psi}{\partial y}|^2\right)dxdydz$$

$$E_{el,z} = \frac{\hbar^2}{2m_e}\int|\frac{\partial\Psi}{\partial z}|^2 dxdydz$$

E_H, E_R and V_{int} are defined as above by (30), (31) and (11). The minimum of (35) with respect to ξ_z is obtained for $\xi_z = \frac{2E_{el,z}\xi_{xy}^2}{V_{int}}$ and then

$$E_m(\xi_{xy}) = \min_{\xi_z} E(\xi_{xy},\xi_z) = \frac{E_{el,xy}}{\xi_{x,y}^2} + E_H + E_R\xi_{xy}^2 - \frac{V_{int}^2}{4E_{el,z}\xi_{xy}^4} \tag{36}$$

The absolute minimum of $E_m(\xi_{xy})$ is obtained for $\xi_{xy} = 0$ and is $-\infty$. When the magnetic field is zero $E_R = 0$, E_m has a zero local minimum with respect to $\xi_{x,y}$ at $\xi_{xy} = +\infty$ and there is an intermediate maximum at finite ξ_{xy} (see Fig. 2). When the magnetic field is switched to a non-zero value, this minimum moves from $+\infty$ to a finite value. In that situation, we have a large polaron which is metastable. Next, there is a critical magnetic field obtained for $E_R = 16E_{el,z}^2 E_{el,xy}^3/(27V_{int}^4)$ where this minimum bifurcates with the intermediate maximum and where the large polaron collapses into a small polaron.

We may choose a normalized Gaussian for $\Psi(x,y,y) = \frac{1}{\pi^{3/2}} e^{-(x^2+y^2+z^2)/2}$ and then we have $E_{el,xy} = \frac{\hbar^2}{2m_e}, E_{el,z} = \frac{\hbar^2}{4m_e}, E_H = 0, E_R = \frac{\hbar\omega_c}{4\xi_c^2}$ and $V_{int} = \frac{\chi_0^2}{4\sqrt{2}K}$.

Setting the scaling parameters $\xi_{xy} = \alpha_{xy}\xi_c$ and $\xi_z = \alpha_z\xi_c$ in cyclotron radius units, we obtain

$$E(\alpha_{xy}, \alpha_z) = \frac{\hbar\omega_c}{4}\left(\frac{1}{\alpha_{xy}^2} + \frac{1}{2\alpha_z^2} + \alpha_{xy}^2 - \frac{C}{\alpha_{xy}^2\alpha_z}\right)$$

where

$$C = \frac{\chi_0^2}{4\sqrt{2}K\xi_c^3} = \frac{\chi_0^2}{4K}\left(\frac{m_e}{\hbar^2}\right)^{3/2}\sqrt{\hbar\omega_c}$$

which yields $\alpha_z = \alpha_{xy}^2/C$. When the uniform magnetic field is small $\alpha_{xy} \approx 1 - C^2/2$ and $\alpha_z \approx 1/C$ is large. The characteristic size ξ_{xy} of the polaron in the xy plane orthogonal to the magnetic field is close to the cyclotron radius while ξ_z is much larger in the z direction parallel to the magnetic field. We can say that the polaron is *cigar–shaped*. When C increases the large polaron becomes more spherical but it reaches a critical point where it has its minimum size when $C^2 = 2/3^{3/2}$ Then, $\alpha_{xy} = 1/\sqrt{3}$ and $\alpha_z = 1/12^{1/4}$. When C becomes large, the large polaron solution does not exist anymore and it collapses into a small polaron.

4.3 Polaron in a Uniform Magnetic Field in Discrete Models

Some numerical investigations of a polaron in the discrete Holstein model in a magnetic field were only done in 2D [5, 12]. The discrete Holstein model (17) in a magnetic field becomes

$$H = -T \sum_{<i,j>} (e^{i\theta_{i\to j}}\Psi_i^\star\Psi_j + e^{i\theta_{j\to i}}\Psi_j^\star\Psi_i) + \sum_i \chi_0|\Psi_i|^2 u_i + \frac{K}{2}\sum_i u_i^2 \quad (37)$$

where $\theta_{i\to j} = -\theta_{j\to i}$. The sum of the angles $\frac{1}{2\pi}\sum_{i\to j}\theta_{i\to j}$ over the oriented edges of a plaquette of the square lattice is the flux of the magnetic field $\mathcal{H}$ counted in magnetic flux quantum $2\varphi_0 = 2\pi\hbar/e = h/e$ through that plaquette. For a uniform magnetic field we may choose $\theta_{(i_x,i_y)\to(i_x+1,i_y)} = -i_y\varphi$ and $\theta_{(i_x,i_y)\to(i_x+1,i_y)} = 0$ where $i = (i_x, i_y)$ are the integer coordinates of the lattice site, a^2 the area of a plaquette, $\mathcal{H}$ is the magnetic field and where $\varphi = \frac{a^2 e\mathcal{H}}{\hbar} = \pi\frac{a^2\mathcal{H}}{\varphi_0}$. Then $\frac{1}{2\pi}\sum_{i\to j}\theta_{i\to j} = \varphi$ over the oriented edges of a plaquette. Let us note that there are many equivalent gauges for the same magnetic field, for example $\theta_{(i_x,i_y)\to(i_x+1,i_y)} = -i_y\varphi/2$ and $\theta_{(i_x,i_y)\to(i_x+1,i_y)} = i_x\varphi/2$ etc...

By rescaling the energy units to T and the displacement unit for u_i to $\sqrt{T/K}$, the electron-phonon coupling becomes a dimensionless parameter k defined as

$$k^2 = \frac{\chi_0^2}{KT} \tag{38}$$

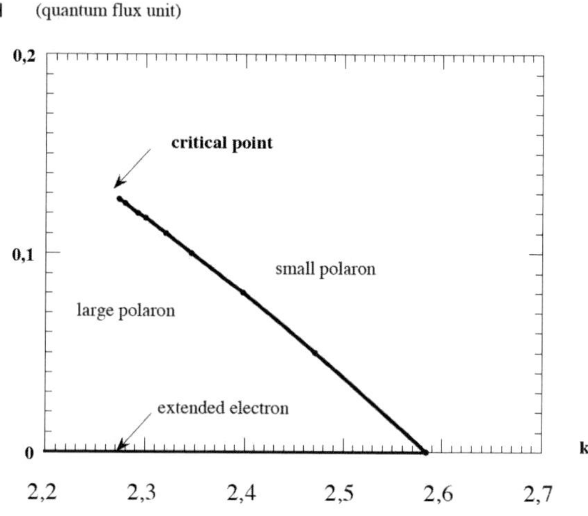

Fig. 9. First order transition line between the large and the small polaron in a magnetic field in the 2D Holstein model on a square lattice.

Polarons can be easily calculated numerically in this model [5, 12]. It is found that at small electron–phonon coupling k we have a large polaron instead of an extended solution in the absence of a magnetic field, while the small polaron at large coupling which is almost single site, persists. Moreover, it is found that the ground-state solution exhibits a first order transition line between the large and the small polaron when the magnetic field is not too large ($\varphi = \frac{a^2 e\mathcal{H}}{\hbar} < \varphi_c \approx 0.125.$) When the magnetic field increases beyond this critical value, this first order transition disappears and the polaron smoothly extends from small to large when the electron phonon–coupling decreases. Thus, the first order transition line ends at a critical point (see Fig. 9).

Figure 10 (left) shows for a magnetic field not too large that the energy curves versus electron–phonon coupling k (38) corresponding to the large polaron and the small polaron intersect at the first order transition point. When the magnetic flux per plaquette (in magnetic flux quantum units) increases beyond the critical value ≈ 0.125, there is a unique curve showing that the polaron continuously varies from the limit of large polaron to small polaron. Figure 10 shows the profile in the x direction of the large and small polaron at the critical point.

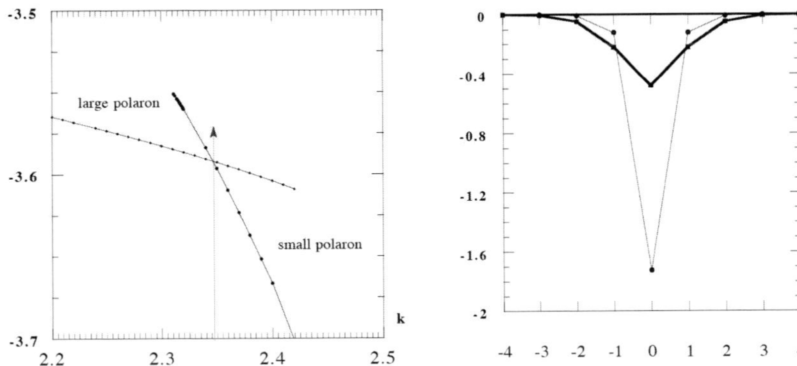

Fig. 10. Energy (in band width units T) versus k for the large and the small polaron at $\varphi = 0.1$ (left) and profile of the large (full line) and small (dashed line) polaron densities (central section in the axis direction) at the intersection at $k \approx 2.35$ and $\varphi = 0.1$.

We suggest that the same variational approaches as those used in [9], could reproduce analytically with a reasonable agreement this first order transition line ending at a critical point, but this work has not been done yet.

4.4 Commensurability Effects

We now mention some preliminary but uncompleted results mentioned in [12] concerning commensurability effects of the magnetic field which are due to the discreteness of the lattice. In the absence of electron–phonon coupling ($k = 0$), the problem of an electron on a discrete square lattice (37) in a magnetic field reduces to a 1D Harper equation [16] which is known to have a fractal spectrum (known as the *Hofstadter butterfly* [17]) when the number of quantum flux $2\varphi_0$ per plaquette in the discrete lattice is an irrational number, that is when $\frac{a^2\mathcal{H}}{2\varphi_0}$ is irrational. Moreover, this equation is just at the critical point of the self-dual transformation [18]. The fractal spectrum is a zero measure Cantor set (it is a singular continuous spectrum) and the corresponding eigenstates are marginally extended with fractal spatial structure.

The situation becomes much simpler when the number of quantum fluxes $2\varphi_0$ per plaquette in the discrete lattice is a rational number, $\frac{a^2\mathcal{H}}{2\varphi_0} = r/s$ where r and s are irreducible integers, then $\varphi = 2\pi r/s$. Then the corresponding Harper equation becomes spatially periodic with a supercell $\times s$ and its eigen spectrum is absolutely continuous. It simply consists of s non overlapping bands with gaps in between.

When the electron–phonon coupling is switched on but remains small enough, one may consider that the electron remains in the lowest single band. Then, the problem becomes similar to that of a 2D electron in a single band without magnetic field and with electron–phonon coupling. Thus, we should expect that the electron remains extended at small electron–phonon coupling. This is indeed what is numerically observed. For each commensurate magnetic flux, the ground state of the electron remains extended up to a certain critical $k_c(r/s)$. Beyond this value, the electron localizes as a small polaron. Figure 11 shows horizontal lines at some commensurate number of quantum fluxes per plaquette. The few numerical calculations of $k_c(r/s)$ shown suggests that at the incommensurate limit we have $\lim_{s\to+\infty} k_c(r/s) = 0$. In that limit, the size of the supercell diverges and there is no more spatial periodicity in the system. When the number of quantum fluxes per plaquette is close to a ratio-

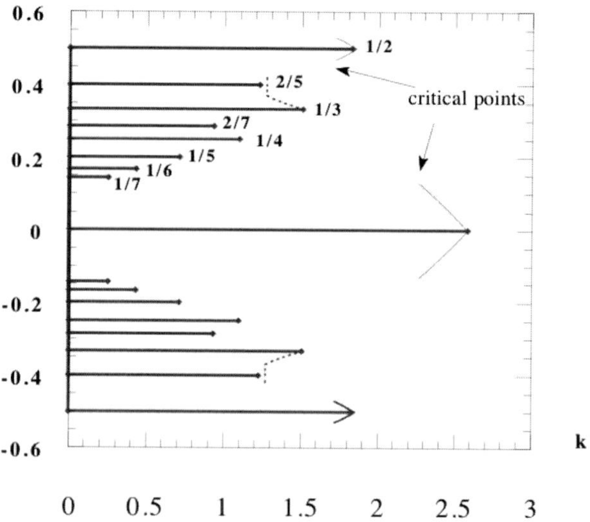

Fig. 11. The electronic ground state is extended on the thick horizontal lines which correspond to commensurate magnetic fluxes φ. First order transition line starting from the commensurate segments for $s = 1$ (zero magnetic field) as shown in Fig. 10 right [12].

nal number but not equal, we expect to be in a situation similar to that of a small magnetic field. Then we expect to have a large polaron at small electron phonon coupling with a first order transition to a small polaron as for the case $s = 1$ (corresponding to zero magnetic field or an integer number of quantum fluxes). There should exist two first order transition lines starting at the end point of the commensurate segment at $k = k_c r/s$ on both sides for larger and smaller magnetic fields. This feature has been confirmed numerically in the vicinity of few commensurate ratio where these first order transition lines were explicitly calculated for $r/s = 0$ and $r/s = 1/2$ (see Fig. 10)(thin lines). For higher commensurability (eg $r/s = 1/3$) the first order discontinuity was found to be very small and difficult to evaluate numerically. At the present stage, the first order transition line was calculated for magnetic fields above the magnetic field corresponding to $r/s = 1/3$ quantum fluxes per plaquette.

It might be possible to produce some analytic calculations based on a renormalization technique for finding the detailed behavior of the polaron hopefully with good accuracy. Each step of the renormalization would consist in determining a new supercell from which the polaron problem could be mapped on the initial problem. These supercells should be determined from the sequence of rational fields which are the best approximation of the irrational magnetic field and which correspond to the sequence of rational truncations of the continuous fraction expansion of the magnetic field (counted in quantum fluxes).

Although this numerical analysis is rather incomplete, it confirms the expected complexity of the polaron phase diagram and the existence of subtle commensurability effects in the discrete model with a magnetic field. No numerical studies for a polaron in a discrete 3D model in a magnetic field have been performed up to now. Let us also note that although up to now it is impossible to apply a magnetic field which corresponds to a substantial number of quantum fluxes per cell in crystals with a small unit cell, this could become more realistic for nanostructures which could be artificially built (or natural structures?) with large supercells.

5 Multipeaked Polarons

Extended polarons in discrete models are well described by continuous models where their spatial location is continuously degenerate. As a result, they can move without jumping any energy barrier. In contrast, small polarons in discrete model are sensitive to the microscopic details of the lattice (coordinance, nonlinear coupling with phonons, anharmonic phonons, etc...).

As we have already seen for polarons in a magnetic field, one may have a first order transition between different polaronic structures when the model parameters vary. These transition points may be of special physical interest because at that point the PN energy barrier which pins the polaron to the lattice may be strongly depressed which favors its mobility (see Sect. 7). Otherwise, when the quantum fluctuations are taken into account the effective mass of the polaron may become small which favors quantum structures (or superconductivity for bipolarons). The existence of a multipeaked polaron was mentioned first in an appendix of [19] for close to the anticontinuous model of the adiabatic Holstein model (at any dimension). Actually, it is a rather universal feature in the extreme discrete case (anticontinuous limit). A single polaron may be split into several smaller peaks which because of the lattice discreteness may be maintained in a metastable configuration. It was argued that multipeaked polarons may become ground-state in some conditions which was invoked [20, 21] to play a role in favoring polaron mobility. Actually, multipeaked polarons are not ground-states in the simple models which have been considered. Otherwise, multipeaked solutions of Discrete Nonlinear Schrödinger models were investigated in [22].

Because of lattice discreteness many metastable multipeaked polarons may be generated in (extended) Holstein models as we shall demonstrate now. We use two approaches which are complementary for understanding this problem.

5.1 Approach in 1D with Discrete Maps

The first approach used for proving the possible existence of multipeak polarons was related to the occurrence of chaotic trajectories in discrete maps $\mathcal{T}$ [19]. Because of this effect, it is now well-known that the discretized analogue of continuous models may exhibit many more local minima (and extrema) than the corresponding continuous model. In particular the stationary states of 1D models may be represented by discrete trajectories of discrete maps which are non integrable and exhibit chaotic trajectories [23, 24]. For example, let us consider the continuous 1D Holstein model with energy (in dimensionless units)

$$H_c = \int \left(\frac{1}{2}|\frac{\partial\psi}{\partial x}|^2 + ku(x)|\psi(x)|^2 + \frac{1}{2}u^2(x) \right) dx \tag{39}$$

and its energy in the discrete version

$$H_d = \sum_i \frac{1}{2}|\psi_{i+1} - \psi_i|^2 + ku_i|\psi_i|^2 + \frac{1}{2}u_i^2 \tag{40}$$

The equations for extremalization of (39) $u(x) = -k|\psi(x)|^2$ and $-\frac{\partial^2\psi}{\partial x^2}+ku\psi = E_0\psi$ yield the time independent DNLS equation

$$-\frac{d^2\psi}{dx^2} - k^2|\psi|^2\psi = E_0\psi \tag{41}$$

where E_0 is the Lagrange parameter associated with the normalization condition $\int |\psi|^2 dx = 1$. This equation is integrable since the first integral $|\frac{d\psi}{dx}|^2 + \frac{k^2}{2}|\psi|^4 + E_0|\psi|^2 = K$ is independent of x. All solutions of (41) can be formally obtained by integrations and moreover in this special case, all the solutions can be expressed in terms of Jacobi elliptic functions (which we do not exhibit here). The consequence is that all the solutions are smooth and do not exhibit any chaotic behavior.

Taking into account the normalization condition, the polaron solution is unique (up to an arbitrary translation and phase rotation) and is

$$\psi_p(x) = \frac{k}{2\sqrt{2}} \frac{1}{\cosh(k^2 x/4)} \tag{42}$$

Thus, in the continuous model, there is no multipeak polarons.

In the discrete case, (41) is replaced by the equation

$$-(\psi_{i+1} + \psi_{i-1} - 2\psi_i) - k^2|\psi_i|^2\psi_i = E_0\psi_i \tag{43}$$

In the discrete case, it is obvious that changing ψ_i into $|\psi_i|$ yields a lower energy (40) so that the minima of this energy can be searched for ψ_i real only. Then, (43) may be solved recursively using the cubic area preserving non linear map $\mathcal{T} : \mathrm{R}^2 \to \mathrm{R}^2$ which is defined as

$$\begin{pmatrix} \psi_{i+1} \\ \psi_i \end{pmatrix} = \mathcal{T}\begin{pmatrix} \psi_i \\ \psi_{i-1} \end{pmatrix} = \begin{pmatrix} 2\psi_i - \psi_{i-1} - k^2\psi_i^3 - E_0\psi_i \\ \psi_i \end{pmatrix}$$

Many representations of similar 2D maps can be found in the literature (see for example [23]). Such maps may exhibit many kinds of bounded trajectories which may be regular such as fixed point, periodic cycles and smooth Kolmogorov-Arnold-Moser tori but also irregular such as chaotic trajectories. Actually, the existence of chaotic trajectories for such maps was interpreted as due to the existence of Peierls Nabarro energy barriers allowing the pinning of single defects (corresponding to homoclinic trajectories) at random spatial locations [23]. We shall not detail the numerical analysis of $\mathcal{T}$ which is a bit technical and standard.

The polaron solutions, which have to be square summable, necessarily correspond to trajectories which are homoclinic to the fixed point $F_0 = (\psi_{i+1}, \psi_i) = (0,0)$. Such homoclinic trajectories exist only when $E_0 < 0$ and then $\mathcal{T}$ has three fixed points, F_0 , $F_1 = (\psi_{i+1}, \psi_i) = (\sqrt{E_0/k^2}, \sqrt{E_0/k^2})$ and $F_{-1} = (\psi_{i+1}, \psi_i) = (-\sqrt{E_0/k^2}, -\sqrt{E_0/k^2}$. F_0 is an unstable fixed point. Then, these homoclinic trajectories belong to the transverse intersections of the dilating and contracting manifold of the exponentially unstable fixed point F_0 which generate chaotic trajectories. There are also infinitely many homoclinic trajectories which correspond to multipeak polarons with a finite number of peaks. The location of these peaks may be chosen randomly. When their number is infinite they correspond to chaotic trajectories.

When the electron–phonon coupling k becomes small, most trajectories of the map become smooth KAM orbits. However, the homoclinic orbit persists although it corresponds to a rather extended defect with an envelope close to the solution obtained (42) in the continuous limit. However, although small, the energy barrier does not vanish, so that chaotic arrangements of such defects are still possible. These solutions manifest by the fact that weakly chaotic trajectories persist in the vicinity of the fixed point F_0.

We discarded for simplicity the normalization condition in the above arguments. Actually, for corresponding to a polaron, such homoclinic trajectories $\psi_i(E_0)$ should be normalized by the factor $\lambda = \sqrt{\sum_i \psi_i^2}$ but then it is a polaron for a different model where the electron-phonon coupling is also renormalized to λk. When E_0 increases to $-\infty$, this homoclinic solution may be continued and λ varies monotonously up to $+\infty$. Then, this homoclinic trajectory corresponds to a normalized polaron provided k is large enough.

Among those solutions, many are stable, that is, they are local minima of the variational energy (40).

5.2 Approach from the Anticontinuous Limit

In [19] (see page 746), the principle of anticontinuity allows a better approach to this problem which also holds for models at any dimension. The anticontinuous limit is a limit where the variational energy is separable into a sum of variational energies which are only functions of the variables at each site (which are thus decoupled). When there are several possible minima (or extrema) for each unit, it is trivial that the minima (or extrema) of the total variational energy are obtained by a random choice of the minima (or extrema) at each site. Note that this limit was initially called the *anti-integrable* limit [19] and later renamed the *anticontinuous* limit. The anticontinuous limit is the situation opposite to the continuous limit corresponding to T large and/or equivalently small electron–phonon coupling which may be treated as a perturbation for free extended electrons (in that regime quantum lattice fluctuations cannot be neglected).

The interesting result is that each of the chaotic solutions which minimize (or extremalize) the variational energy at the anticontinuous limit, is continuable as a local minima (or extrema) of the variational form when the coupling between the local units is switched on. This continuation holds at least up to a nonvanishing coupling parameter and sometime quite far away from the anticontinuous limit.

We show now how we can apply this concept to the discrete Holstein model for proving the possible existence of multipeaked polarons as metastable states. For convenience, instead of choosing as a unit of energy the transfer integral T, we keep the energy of the discrete Holstein model (17) in the original form and write it in the form

$$\Phi = \Phi_{AC} - T \sum_{<i,j>} (\psi_i \psi_j^\star + \psi_j \psi_i^\star) \tag{44}$$

where

$$\Phi_{AC} = \sum_i \left(V(u_i) + F(u_i)|\Psi_i|^2 \right) \tag{45}$$

is a sum of uncoupled local potentials. This is the most general form when it is assumed that the local energy depends only linearly on the electronic density. Generally, those potentials are anharmonic although we could choose the original harmonic potential $V(u_i) = \frac{K}{2}u_i^2$ and $F(u) = \chi_0 u$ as a special case. We assume that the potential $V(X)$ has its absolute minimum at $X = 0$ where $V(0) = 0$. Then, we have $V'(X) > 0$ for $X > 0$ and $V'(X) < 0$ for $X < 0$. We also assume that $F(X)$ is a monotone function of X with $F(0) = 0$.

Models which are physically acceptable require that the total energy (44) is bounded from below. Since the electronic kinetic energy is bounded from below (and above) this condition requires that $V(X) + \rho F(X)$ has a lower bound for any $0 < \rho < 1$ and any X which implies that $V'(X) + \rho F'(X) = 0$ has solutions for any $0 < \rho < 1$.

The extrema of the energy (45) at $t = 0$ fulfil

$$V'(u_i) + F'(u_i)|\psi_i|^2 = 0 \qquad (46)$$

and the eigen equation

$$-t \sum_{j:i} \psi_j + F(u_i)\psi_i = E_0(\{u_i\})\psi_i \qquad (47)$$

where E_0 is the Lagrange multiplier associated with the constraint $\sum_i |\psi_i|^2 = 1$. It is also the eigen energy of the electron and we have $\Phi(\{u_i\}) = E_0(\{u_i\}) + \sum_i V(u_i)$.

Then, (47) for $t = 0$ yields either $\psi_i = 0$ or $E_0 = F(u_i)$. We now introduce the arbitrary sequence of variables $\{\sigma_i\}$ (coding sequences) which characterizes the solution which is chosen at each site. We set $\sigma_i = 0$ when $\psi_i = 0$ (and then $u_i = 0$ from (46)) and $\sigma_i = 1$ when $\psi_i \neq 0$, (47) which implies $F(u_i) = E_0$. When $\sigma_i = 1$, u_i is independent of i since $F(X)$ is assumed to be monotone, and $|\psi_i|^2 = -\frac{V'(u_i)}{F'(u_i)}$ is also independent of i. Since $\sum_i |\psi_i|^2 = 1$, we have $|\psi_i|^2 = \frac{1}{N}$ where $N = \sum_i \sigma_i$ is an integer which is the number of peaks of the polaron. We define X_N by the condition

$$N = -\frac{F'(X_N)}{V'(X_N)} \qquad (48)$$

Note that with the assumption that the model energy is bounded from below, this equation always has solutions which however may not be unique. X_N may have several determinations. Then, we may choose $u_i = \sigma_i X_N$ for all i and then

$$\Phi_{AC} = NV(X_N) + F(X_N) \qquad (49)$$

It can be proven that each of these extrema solutions at $T = 0$ can be continued by the implicit function theorem for $T \neq 0$ up to a nonvanishing critical value. Those which are local minima of the energy can be continued as local minima as a function of T at least in some interval around 0.

We do not explicitly state here this proof which is formally identical to those used for proving the existence of discrete multisite breathers [25]. However, let us briefly mention an essential point of the proof. Since only the modulus of ψ_i is determined, its phase is arbitrary when it is non zero. Although a priori, the continuation of this degenerate multipeaked polaron from the anticontinuous limit is not possible, the continuation technique consists in proving first that the polaron can be continued as a (pseudo) solution under the constraint of quenched phases at arbitrary values. Next, the polaron energy is extremalized (or minimized) with respect to the phases. The local minima yields exact polaron solutions. See [25] where this problem was solved for a very similar discrete breather problem. Note however that in the case $T > 0$, this continuation is trivial because all the local minima can be obtained for ψ real positive (then, all the phases are quenched identically at zero value). The continuation of the polaron with ψ real with the implicit function theorem is then possible at T non zero.

As a result, we prove that at $T \neq 0$ there are many multipeaked polaron solutions with peaks at arbitrary locations which are local minima of the energy. Moreover, each of these solutions is stable up to a nonvanishing bifurcation value of T. The ground-state polaron is one of these many solutions but it is necessarily obtained for the single peak solution ($N = 1$) which we prove now.

However, let us mention that multipeaked polarons were obtained numerically in [21] and claimed to be ground-state (in models built on purpose to find criteria which favor polaron mobility in conducting polymers or DNA chains). Their model belongs to the class presented above with a Morse potential $V(X) = D(e^{-\alpha X} - 1)^2$ and $F(X) = -\chi X$ with $\chi > 0$, but however was not physically acceptable because with that choice $V(X) + \rho F(X)$ is not bounded from below for any electronic density where $0 < \rho < 1$. Since when T is not zero, the kinetic energy is bounded, the ground-state energy is always $-\infty$ with infinite atomic displacements for any value of T. Thus the single and multipeaked polarons which have been calculated in [21] cannot be ground-state but are only metastable states (which undergo bifurcations). This fact obviously manifests in their numerical finding at $T \neq 0$ because the lower envelope of the energies of their polaron states (supposed to be ground-state) cannot be a discontinuous function of their model parameters as they show but in all cases should be a continuous function (though it might not be differentiable at first order transitions).

Moreover, their model cannot be amended while in the same class of models because it can be easily proven that for any choice of functions $V(X)$ and $F(X)$, that a multipeaked polaron cannot be ground state at the anticontinuous limit.

Indeed, the energy of the N-peak polaron (49) which has the minimum energy is $E_p(N) = \min_X (NV(X) + F(X))$ (as an energy minimum, it is obviously a stable polaron). Then we have

$$E_P(N) = \min_X \left(V(X) + ((N - 1)V(X) + F(X)) \right)$$
$$> \min_X V(X) + \min_X \left((N - 1)V(X) + F(X) \right) > E_P(N - 1)$$

Since by definition, $\min V(X) = 0$ is realized only at $X = 0$, we have recursively $E_p(N) > E - P(N - 1) > ..... > E_P(1)$ which implies that at the anticontinuous limit the polaron which has the minimum energy is always the single peak polaron obtained for $N = 1$

It is not clear, however, whether a multipeaked polaron could become ground-state when the model is not close to an anticontinuous limit. Actually, we shall see later that for similar bipolaron models, the presence of a Hubbard term may favor the two peak bipolaron or an even more complex structure as ground-state instead of the single peak.

6 Polarobreathers

Instead of being trapped by a local static distortion, an electron may also be trapped by a localized nonlinear vibration as proven in [25, 26]. These solutions cannot be obtained by continuation of linear polaron modes up to finite amplitudes but can only exist with anharmonic local potential $V(u)$. In addition the frequency of the Discrete Breather and its harmonics should avoid any resonance with the linear phonon spectrum. Actually, we prove that a Discrete Breather which is a time periodic and spatially localized solution of an anharmonic system (and cannot exist in harmonic systems) is able to self-consistently trap an electron in order to form a composite stable dynamical object we call a polarobreather.

We consider the extended Holstein model on a lattice with Hamiltonian (44) where $V(u_i)$ is the potential of a convex anharmonic oscillator with unit mass.

We now consider the class of exact time quasiperiodic and time reversible solutions which fulfill

$$u_i(t + t_b) = u_i(t) \qquad \text{and} \qquad u_i(t) = u_i(-t) \tag{50}$$
$$\psi_i(t + t_b) = e^{-iE_b t_b} \qquad \text{and} \qquad \psi_i(t) = \psi_i^{\star}(-t) \tag{51}$$

for a given period $t_b = 2\pi/\omega_b$ and quasieigen energy E_b.

At the anticontinous limit when $T = 0$, $|\psi(t)|^2$ is time independent. Then there are trivial quasi-periodic solutions which belong to such classes since the oscillators with potential $V(u_i) + |\psi_i(0)|^2 u_i$ are uncoupled and exhibit time reversible and time periodic solutions with frequency ω_b

$$u_i(t) = g(\omega_b t) \tag{52}$$

Two time dependent solutions are obtained, separated from each other by a time shift of half a period (then this site is coded $\sigma_i = \pm 1$) or the solution may be at rest (then $\sigma_i = 0$). Since the oscillator is anharmonic, its frequency determines the amplitude of the solution.

Then, we have time reversible solutions for

$$\psi_i(t) = \psi_i(0) \exp^{-i\chi_0 \int_0^t u_i(\tau)d\tau} \tag{53}$$

where $\psi_i(0)$ is real. We have $\psi(t + t_b) = \psi(t)e^{-iE_b t}$ where the average displacement over one period t_b, $E_b = \frac{1}{t_b}\int_0^{t_b} u_i(\tau)d\tau$ has to be independent of i. This condition requires either $\psi_i(0) = 0$ (then we set the new coding sequence $\sigma_i' = 0$) or $\psi_i \neq 0$ (then $\sigma_i' = \pm 1$ according to the sign of $\psi_i(0)$). In that case,

E_b should be independent of i which requires the same average displacement for $u_i(t)$ for all i where $\sigma'_i \neq 0$.

Consequently, when $\sigma'^2_i = 1$, we have either $\sigma^2_i = 1$ or $\sigma_i = 0$ that is either $\sigma'^2_i(1 - \sigma^2_i) = 0$ for all i or $\sigma'^2_i\sigma^2_i = 0$ for all i.

Actually, we get two families of solutions. For the class where $\sigma'^2_i\sigma^2_i = 0$, the location of the electron and of the moving oscillators are different which means the solution can be viewed as a polaron (or multipeak polaron) and a breather (or multisite breather) initially non interacting. The class where $\sigma'^2_i(1 - \sigma^2_i) = 0$ is more interesting because the electron is precisely located with equal densities at the sites which oscillate at the same frequency ω_b. They correspond to polarobreathers where the electron localizes specifically at the lattice oscillation.

It is proven by the implicit function theorem [25] that (generically) for a given ω_b and E_b any of the solutions described by coding sequences $\{\sigma_i, \sigma'_i\}$ fulfilling the above rules at the anticontinuous limit can be continued as a function of the transfer integral T at least up to non zero values. It is worthwhile mentioning that the strength of our theory is that it is not restricted to special models but is applicable to large classes of models. Polarobreathers at the the anticontinuous limit are robust against any kind of perturbations of the Hamiltonian (providing they only involve short range interactions). We have considered for simplicity in our model only one kind of perturbation, which is a nearest neighbor electronic transfer. It is possible to add other perturbations which could be, for example, a small extra coupling between nearest neighbors atoms (which generate phonon dispersion) or any more complex anharmonic coupling, even mixing electronic density and atomic positions.

To be more complete, let us mention that in a first step, we proved the continuation of the polarobreather solution by the implicit function theorem at fixed E_b but not at fixed norm $\sum_i |\psi_i|^2$. Next, we proved that E_b can be varied in order to maintain the norm equal to 1 [25] which proves in the end that the polarobreather can be continued at fixed frequency ω_b and fixed norm 1 up to the non vanishing perturbation parameter.

Consequently, we have proven the existence of quite a large number of exact dynamical solutions which generalize the static polaron solution. However, the most extended solutions might be stable only for very small perturbations. The simplest solution and likely the most stable, is a single peak polarobreather obtained when choosing $\sigma_i = \sigma'_i = 1$ at a unique site i and zero everywhere else. This polarobreather has been numerically and accurately calculated for the above model in 1D with an anharmonic cubic potential $V(u) = \frac{1}{2}u^2 + \frac{\kappa}{3}u^3$ (see Fig. 12) and has been found stable.

Up to now, very few studies have been devoted to these dynamical excitations, which may deserve deeper studies. We think that they may play an essential role in certain dynamical electronic processes. We suggest, as an example of potential application, that polarobreathers may be involved in the luminescence decay of an electronic excitation. When an electronic excitation is created in an insulating crystal by a photon, an electron of the valence

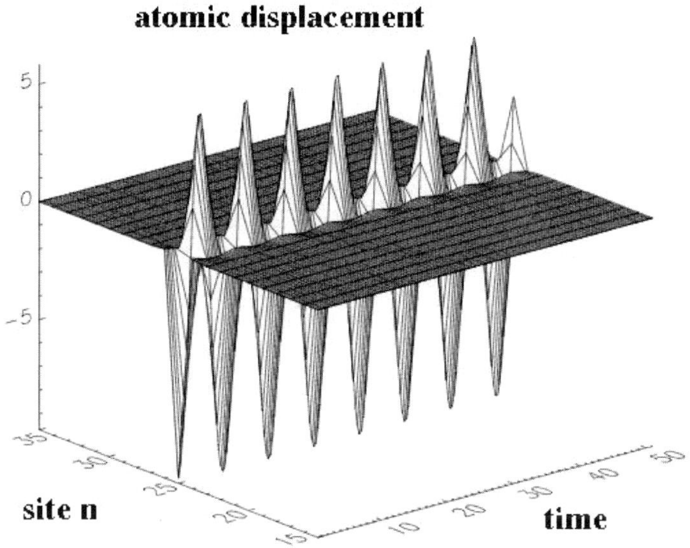

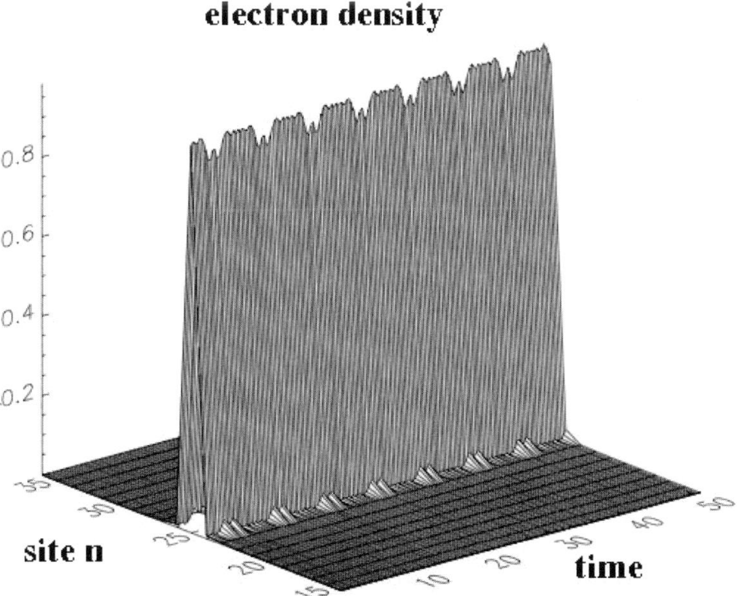

Fig. 12. Atomic displacements $u_n(t)$ (top) and electron density $|\psi_n(t)|^2$ (bottom) versus lattice site n and time t for a stable polarobreather at frequency $\omega_b = 0.908$ in the Holstein model with anharmonic potential $\kappa = 0 = 0.04$, $t = 0.14$ [26].

band is sent to an empty conduction band. According to the standard Franck-Condon theory, it should be expected that the lattice relaxes which implies that the excited electron in its conduction band self-localizes as a static polaron. It is believed that generally the lattice relaxation is very fast ($\sim 1\,\mathrm{ps}$). From that state, the electron relaxes to its initial lower energy state over a much longer time scale by emitting a photon which has a frequency smaller than the initial exciting photon (Franck-Condon shift). However, there are systems where the lattice relaxation may last an unusually long time (up to ms).

We believe that this long relaxation time may be due to the formation of a polarobreather. In systems which are anharmonic enough to sustain polarobreathers and when the temperature is low enough that the polarobreather is not rapidly destroyed by its interaction with the thermal phonons, polarobreathers may spontaneously form after the electron excitation and slow down the lattice relaxation. When the amplitude of the lattice distortion which is generated after the initial electronic excitation is large, polarobreathers might spontaneously form. Similarly, it is well-known that Discrete Breathers may spontaneously form when a large amount of vibrational energy is injected locally in an anharmonic system at low temperature (for a review see [27]). Since polarobreathers are exact solutions at zero temperature, their lifetime should be unusually long at low temperature but should decay when the temperature increases, down to the standard phonon relaxation time at some crossover temperature. The long lifetime of the polarobreather retarding the lattice relaxation should have experimental consequences such as a broadening of the line width of the re-emitted radiation (which should be time dependent).

There are puzzling experimental features in doped alkali halides mentioned for example in [28, 29] which could be related to the formation of polarobreathers, although in these systems the situation is more complex because of the existence of several almost degenerate electronic excited states which hybridize with each other.

7 Pinning Mode and Peierls-Nabarro Energy Barrier of Polarons

7.1 Peierls-Nabarro Energy Barrier

In general, a polaron is obtained as a local minima of a variational energy. In a discrete periodic lattice, this solution is degenerate only under discrete lattice translations. Because of that, there is generally an energy barrier for moving the polaron by one lattice spacing called the Peierls-Nabarro (PN) energy barrier. Actually, this concept was initially introduced for dislocation models [30, 31] but holds for any localized defect or excitation in discrete models. It is only a consequence of discreteness.

At the adiabatic limit, polarons (or bipolarons) when they exist are spatially well localized though their location is arbitrary and thus degenerate. When the quantum lattice fluctuations are taken into account, the polaron (or the bipolaron) can tunnel through the PN energy barrier between equivalent sites. However, at the adiabatic limit, the tunnelling energy of the polarons or bipolarons goes to zero and becomes negligible. When the quantum fluctuations are not negligible, it is intuitively obvious that the smaller the PN energy barrier, the larger the tunnelling energy will be, or equivalently, the smaller the effective mass of the polaron (or bipolaron) will be. Thus, the role of the quantum fluctuations will become important when this PN energy barrier becomes small. It is thus essential to study this PN energy barrier to understand the role of the quantum fluctuations. It is also important for studying the polaron mobility at zero or finite temperature.

The PN energy barrier is defined between two different sites by considering a variational energy $\Phi(\{u_i, \psi_i\}$ and the set of continuous paths $\mathcal{C}(\xi) = \{u_i(\xi), \psi_i(\xi)\}$ with $0 \leq \xi \leq 1$ which connect the two local minima corresponding to the polaron of either the first site or the second site. Then $\mathcal{C}(0) = \{u_i(0), \psi_i(0)\}$ is the configuration of a polaron at a given site n and $\mathcal{C}(1) = \{u_i(1), \psi_i(1)\}$ is the configuration of the same polaron configuration at another site m. Generally m and n are considered as nearest neighbor sites. We define for each path $\varphi_m(\mathcal{C}) = \max_{0 \leq \xi \leq 1} \Phi\{u_i(\xi), \psi_i(\xi)\})$ as the maximum of the variational energy along this path. Since the polaron is a local minimum of the variational energy, we have $E_b(\mathcal{C}) > E_p = \Phi(\mathcal{C}(0)) = \Phi(\mathcal{C}(1))$. Next, we consider $\varphi_{PN} = \min_{\mathcal{C}} \varphi(\mathcal{C})$ as the smallest maximum. This value is called the minimax. Then, the energy difference $E_{PN} = \varphi_{PN} - E_p \geq 0$ is the PN energy barrier, that is, the smallest amount of energy which must be provided to the polaron to move it from site n to m. Thus the minimax is nothing but the top of the energy barrier and corresponds to an unstable polaron.

This PN energy barrier becomes systematically negligible for large polarons for which the continuous approximation holds. The standard mechanism for producing highly mobile polarons or more generally mobile localized excitations is to make them extended (see for example [32]). On the contrary, this PN energy barrier is generally non-vanishing and is quite large for small polarons. It is however possible to get small polarons which are quite mobile according to the mechanism we describe. This mechanism can be transposed for Discrete Breathers [27].

7.2 Small Polaron with Small PN Energy Barrier

The minimax polaron often has simple symmetry. For example, if the ground-state polaron is single site at n, the minimax between n and a neighboring site m (the top of the energy barrier) is generally an unstable polaron localized on the two nearest neighbor sites n and m with equal densities. It may be the reverse, when the two site polaron is ground-state and the single site the minimax.

When the model parameters vary, it may happen in some models and for special values of the parameters, that the energy of the single-site polaron and the two-site polaron cross with each other. When they are equal, one may expect that this PN energy barrier becomes strictly zero with a degenerate polaron ground-state. Actually, this situation rarely occurs because the extrema bifurcate and new intermediate polaron extrema appear (see Fig. 13).

Let us assume for example that when model parameters vary there is a first order transition where the energy of the two-site polaron **2s** crosses that of the single-site polaron **1s** which is initially ground-state. Then, the two-site polaron **2s** becomes ground-state with a lower energy. Figure 13 shows the generic evolution of the energy profile versus parameters.

- **a)** The two-site polaron **2s** is initially the minimax for moving to the single site polaron **1s**.
- **b)** The minimax bifurcates into two new minimax **i1** and **i2** while the former minimax **2s** becomes a local minimum.
- **c)** The PN energy barrier reaches its minimum but does not vanish.
- **d)** The two-site polaron **2s** becomes the ground-state and polaron **1s** becomes the local minimum.
- **e)** The minimax **i1** and **i2** bifurcate with this local minimum **1s** which becomes the new minimax.

Further generic behavior is obtained when the ground-state polaron **1s** first bifurcates into two minima **i1** and **i2** and becomes an unstable maximum. The ground-state polarons then become two degenerate asymmetric polarons intermediate between single-site and two-site which later bifurcate with the two-site polaron **2s**. Although the PN energy barrier generally does not strictly vanish, it could become quite small at the first order transition on the ground-state between a single-site polaron and the two-site polaron. It is thus an interesting mechanism for obtaining polarons which are small and nevertheless rather mobile. Moreover, in that situation, considering now the phonons as quantum, the quantum tunnelling of the polaron through the lattice is greatly enhanced. In the case of bipolarons which are bosons, this phenomena could favor quantum superconducting ordering.

The vanishing of the PN energy barrier is related to a degenerate continuum of minima and implies the existence of a zero frequency mode for the polaron. When this energy barrier is small, there is nevertheless a low frequency (soft) pinning mode of the polaron which can be found from analysis of the linear modes of the polaron.

7.3 Linear Modes

The polarons as static solutions are local minima of the energy of the adiabatic system, for example (17) where the kinetic energy of the atoms vanishes. Actually, the analysis of the dynamics of the system requires us to consider the

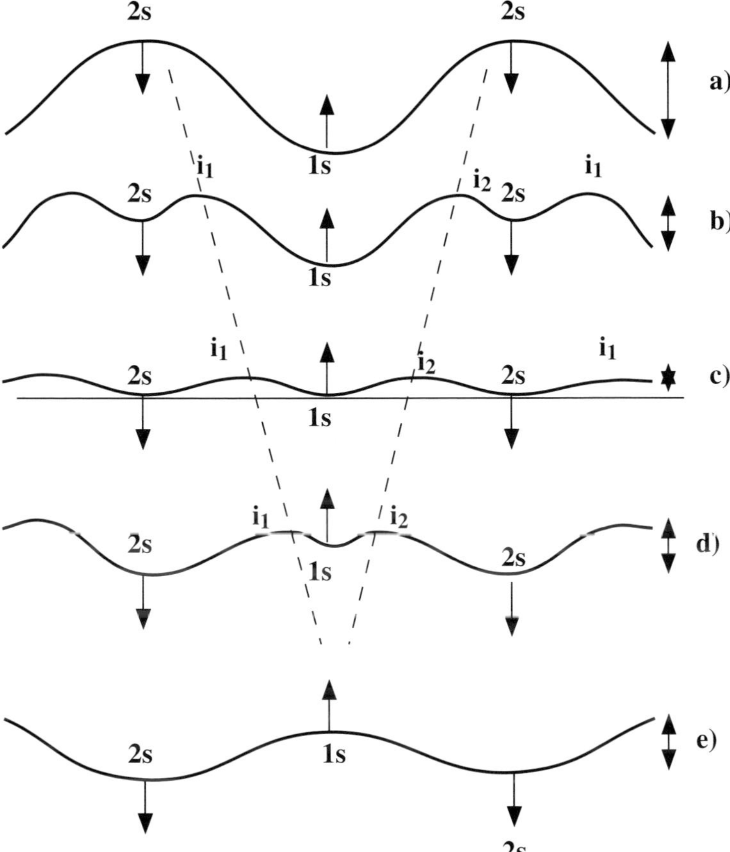

Fig. 13. An example of generic evolution of the energy profile versus parameters when the PN energy barrier almost vanishes. The abscissa represent the polaron location in the lattice.

complete Hamiltonian with its kinetic energy which could be for the Holstein model

$$H = -T \sum_{<i,j>} (\Psi_i^\star \Psi_j + \Psi_j^\star \Psi_i) + \sum_i \left(\chi_0 |\Psi_i|^2 u_i + V(u_i) + \frac{1}{2M} p_i^2 \right) \qquad (54)$$

where $V(u)$ is a local potential with its minimum at $u = 0$, for example $V(u) = \frac{K}{2} u^2$. This Hamiltonian is semiclassical in the sense that the atoms are classical particles while the electron is treated quantum mechanically in the potential created by the atoms. Then we get the dynamical equations

$$i\dot{\psi}_i = \frac{\partial H}{\partial \psi^\star} = \chi_0 u_i \psi_i - T \sum_{j:i} \psi_j \qquad (55)$$

$$\ddot{u}_i + V'(u_i) + \chi_0 |\psi_i|^2 = 0 \qquad (56)$$

For the static solutions, we have u_i time independent but $\psi_i(t) = \phi_i e^{-iEt}$ where E is the eigen energy of the electron and ϕ_i is time independent. The solutions $u_i + \epsilon_i(t), (\phi_i + \eta_i(t))e^{-Et}$ around the stable static solution u_i, ψ_i, where ϵ_i, η_i is a small perturbation, are obtained from the linearized (Hill) equations of (55) and (56) with time independent coefficients

$$i\dot{\eta}_i = (\chi_0 u_i - E)\eta_i + \chi_0 \phi_i \epsilon_i - T \sum_{j:i} \eta_j \qquad (57)$$

$$\ddot{\epsilon}_i + V''(u_i)\epsilon_i + \chi_0(\phi_i^\star \eta_i + \phi_i \eta_i^\star) = 0 \qquad (58)$$

Apart from the trivial static solution $\epsilon_i = 0, \eta_i = i\phi_i$ which just corresponds to a small phase rotation of ϕ, the solutions are linear combinations of time periodic eigen modes. For a stable polaron all these frequencies are real. If this polaron becomes unstable when varying model parameters, they become complex. The spatially localized and time periodic eigen modes are called the internal modes of the polaron.

When there is no PN energy barrier between two nearest neighbor sites, for example, the polaron $\{u_i, \psi_i\}$ is continuously degenerate and may be written as a continuous function $\{u_i(\xi), \phi_i(\xi)e^{-iEt}\}$ of a parameter ξ. Then, by differentiation of (55) and (56) with respect to ξ, it comes out that $\epsilon_i = \frac{du_i}{d\xi}$, $\eta_i = \frac{d\phi_i}{d\xi}$ are time independent solutions of (57) and (58). When E_{PN} is small, there is a low frequency mode called pinning mode.

Numerical calculations of polarons usually yield several internal modes (see Fig. 14) which detach from the band of extended phonons (in the case of the Holstein model, this band width is zero because there is no coupling between the atoms).

In many of the usual models, the polaron has spatial symmetries, so the eigen modes also have spatial symmetries. For example, in 1D, the modes are either symmetric or antisymmetric.

The antisymmetric mode with the lowest frequency may be recognized as the pinning mode because it corresponds to a small oscillation of the center of electronic density of the polaron. When its frequency is very low, the excitation of this mode may generate a mobile polaron as often observed. However, it should be excited with an appropriate phase which corresponds to an initial kick on the initial velocities of the atoms at their equilibrium positions (this excitation corresponds to the so-called marginal mode in the case of discrete breathers (see [27] and refs. therein). The polaron velocity depends on the amplitude of this kick. Although there is no exact polaron solution, the mobile polaron may persist over a very long time while it slows down gradually. This situation occurs in the example of Fig. 14 only when $k <\sim 1$, that is in

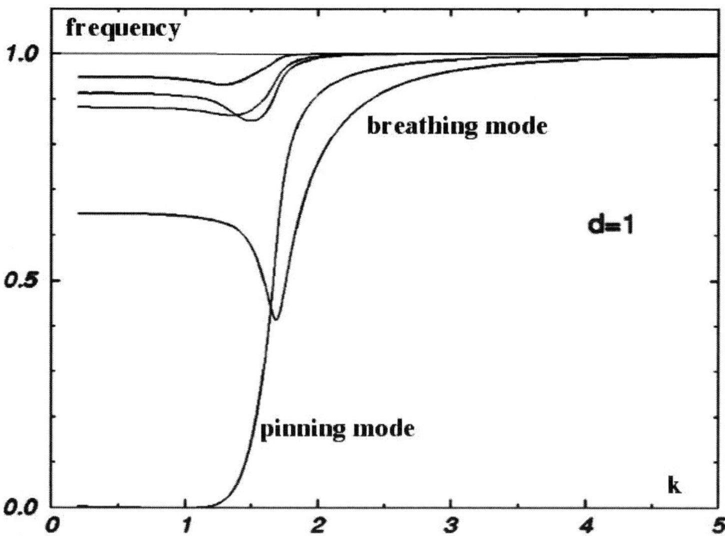

Fig. 14. Frequencies of the 5 internal modes of the single peak polaron in the 1D Holstein model (54) with harmonic potential $T = 1, K = 1, M = 1$ versus the reduced electron–phonon coupling k defined by (24). The phonon band is degenerate at the single frequency 1. The spatially antisymmetric (resp. symmetric) eigenmode with the lowest frequency is a pinning (resp. breathing) mode [9]. Reproduced with permission from [9], ©(1998) by the American Physical Society.

the regime where the polaron becomes spatially extended and the continuous approximation is valid. Note that the breathing mode also substantially softens in the cross-over region which separates the continuous regime (small k) from the anticontinuous regime (large k).

The situation is quite different for the Holstein model in higher dimensions. There is no cross-over but a first order transition between a small polaron (large k)and an extended electron (small k). When the stable polaron solution bifurcates with an unstable polaron solution, one of the internal modes softens to zero (as should be expected) but then it is no longer the ground-state. In our case, it is a spatially symmetric breathing mode which indicates polaron instability against dilation as suggested from the above calculation Sect. 3 (see figures in [9] for details).

8 Bipolarons in the Holstein-Hubbard Model

The self-trapping of a pair of electrons in a deformable lattice may generate a bound state which is called a bipolaron [8, 12, 33–37]. When the electrons are non-interacting, and since electrons are fermions with spin $1/2$, it is rather obvious that instead of forming two widely separated polarons, the self-trapped electrons gain the lattice deformation energy of the lattice in sharing the same potential well. Then the two electrons with opposite spins occupy the same lowest energy bound state in this potential well.

For example, in model (17), assuming that there are two electrons with opposite spins in the same state $\{\Psi_i\}$, the electronic energy (which is negative) is double. The total energy is

$$\Phi(\{u_i, \Psi_i\}) = 2\left(-T\sum_{<i,j>}(\Psi_i^\star \Psi_j + \Psi_j^\star \Psi_i) + \sum_{i,n}\chi_n|\Psi_i|^2 u_{i+n}\right) + \frac{K}{2}\sum_i u_i^2 \tag{59}$$

Setting $u_i = \sqrt{2}v_i$, it emerges that the bipolaron energy $E_b(K) = 2E_p(K/2) < 0$ is twice the polaron energy $E_p(K/2)$ in the same model but where the elastic constant K is twice as small. Since a variational argument readily proves that $E_p(K)$ is a monotonically increasing function of K, we obviously have $E_b(K) = 2E_p(K/2) < 2E_p(K)$. This inequality proves that the bipolaron energy is systematically smaller than the energy of two widely separated polarons.

The situation becomes more interesting when a repulsive Hubbard term is introduced, the bipolaron energy increases and may become larger than the energy of two widely separated polarons (which is independent of the Hubbard coefficient U). There is necessarily a first-order transition at some critical value of U at which the bipolaron breaks into two polarons and thus becomes multipeaked. We investigated numerically the phase diagram of the bipolaron ground-state in [35–37] in 1D and 2D models. Our motivation was to show that in the vicinity of the first-order transitions of the adiabatic bipolaron, the PN energy barrier may be drastically depressed which drastically enhances the mobility of the adiabatic bipolaron. Moreover, considering now the phonon field as quantum, this mobility still manifests as a drastic decrease of the bipolaron effective mass. We then argued that if the effective mass becomes small enough, when many bipolarons are present, a Bose condensate of these hard core bosons may become the superconducting ground-state instead of a bipolaron or polaron spatial ordering.

For considering this problem, it becomes more convenient to rewrite the same model with two electrons and a Hubbard interaction, within a full second quantization formalism. We also assume the electron-phonon interaction is on-site which yields the quantum Holstein-Hubbard model

$$H = -T\sum_{<i,j>,\sigma} c_{i,\sigma}^+ c_{j,\sigma} + g\sum_i (a_i^+ + a_i)n_i$$
$$+ v\sum_i n_{i,\uparrow}n_{i,\downarrow} + \hbar\omega_0 \sum_i (a_i^+ a_i + \frac{1}{2}) \tag{60}$$

where $g^2 = \hbar \frac{\chi_0^2}{2\sqrt{KM}}$ and $\omega_0^2 = \frac{K}{M}$. $c_{i,\sigma}^+$ and $c_{i,\sigma}$ are the standard creation and annihilation fermion operators of an electron at site i with spin $\sigma = \pm\frac{1}{2}$ and a_i^+ and a_i are standard creation and annihilation boson operators at site i, $n_{i,\uparrow} = c_{i,\uparrow}^+ c_{i,\uparrow}$, $n_{i,\downarrow} = c_{i,\downarrow}^+ c_{i,\downarrow}$ and $n_i = n_{i,\uparrow} + n_{i,\downarrow}$ are electron density operators. v is the constant of the on-site Hubbard interaction.

Defining new conjugate operators $u_i = \frac{\hbar\omega_0}{4g}(a_i^+ + a_i)$ and $p_i = \frac{2g}{\hbar\omega_0}(a_i^+ - a_i)$ we have $[u_i, p_i] = -i$. Then, considering $E_0 = \frac{8g^2}{\hbar\omega_0} = 4\frac{\chi_0^2}{K}$ as the energy unit, we obtain the reduced Hamiltonian

$$\widetilde{H} = \sum_i (\frac{1}{2}u_i^2 + \frac{1}{2}u_i n_i + U n_{i,\uparrow} n_{i,\downarrow}) - \frac{t}{2}\sum_{<i,j>,\sigma} c_{i,\sigma}^+ c_{j,\sigma} + \frac{\gamma}{2}\sum_i p_i^2 \qquad (61)$$

where $U = v/E_0$, $t = \frac{2T}{E_0}$ and $\gamma = \frac{1}{2}(\frac{\hbar\omega_0}{2g})^4$, When $g >> \hbar\omega_0$, γ is very small and the last term in (61) may be neglected. Then, the adiabatic approximation is valid since u_i commutes with the Hamiltonian. This condition means $\frac{\chi_0^2}{4K} >> \frac{\hbar\omega_0}{2}$, that is, the energy involved by the bipolaron distortion is much larger than the zero point energy of the phonons.

The electronic singlet state may be written very generally as $\sum_{i,j} \Psi_{i,j} c_{i,\uparrow}^+ c_{j,\downarrow}^+ |\emptyset>$ where $|\emptyset>$ is the vacuum where $\sum_{i,j}|\Psi_{i,j}|^2 = 1$ and $\Psi_{i,j} = \Psi_{j,i}$. Then, within the classical approximation of the phonon variable (u_i is scalar), the total energy may be written variationally as

$$\Phi(\{\Psi_{i,j}\}, \{u_i\}) = \frac{1}{2}\sum_i u_i^2 + \frac{1}{2}\sum_i u_i(\sum_j(|\Psi_{i,j}|^2 + |\Psi_{j,i}|^2)) + U(\sum_i |\Psi_{i,i}|^2)$$

$$- \frac{t}{2}\sum_{<(i,j),(k,l)>} (\Psi_{i,j}^\star \Psi_{k,l} + \Psi_{i,j}\Psi_{k,l}^\star) \qquad (62)$$

Extremalizing (62) with respect to u_i yields $u_i + \frac{1}{2}\sum_j(|\Psi_{i,j}|^2 + |\Psi_{j,i}|^2) = 0$ so that the problem is now to find a symmetric normalized wave function $\{\Psi_{i,j}\}$ which minimizes

$$F\{\Psi_{i,j}\}) = -\frac{1}{8}\sum_{i,j}(|\Psi_{i,j}|^2 + |\Psi_{j,i}|^2)^2 + U\sum_i |\Psi_{i,i}|^2 - \frac{t}{2} <\Psi|\Delta|\Psi> \qquad (63)$$

where $<\Psi|\Delta|\Psi> = \sum_{<(i,j),(k,l)>}(\Psi_{i,j}^\star \Psi_{k,l} + \Psi_{i,j}\Psi_{k,l}^\star)$.

This problem is trivial at the anticontinuous limit $t = 0$. As usual at the anticontinuous limit, there are many multipeaked solutions we shall not describe but it can be readily found that the ground state is either a single site or two site bipolaron and that the other solutions are not involved.

The electronic state of the single site bipolaron located at site i_0, is at the anticontinuous limit $|\Psi> = c_{i_0,\uparrow} c_{i_0,\downarrow}|\emptyset>$ and the corresponding bipolaron energy (in units E_0) is $F = -\frac{1}{2} + U$. The electronic state for a two-site bipolaron located at site i_0 and j_0, is $|\Psi> = \frac{1}{\sqrt{2}}(c_{i_0,\uparrow}c_{j_0,\downarrow} + c_{j_0,\uparrow}c_{i_0,\downarrow})|\emptyset >$, and the corresponding bipolaron energy is $F = -\frac{1}{4}$. Thus, at the anticontinuous limit, the ground-state is a single-site bipolaron, located at an arbitrary site for $U < \frac{1}{4} = U_c$. When $U > \frac{1}{4} = U_c$, the ground-state is a two-site bipolaron which is highly degenerate because

- the two sites i_0 and j_0 are arbitrary
- the spins of the electrons are also degenerate

The single site bipolaron (S0) solution can be continued according to the implicit function theorem when the transfer integral t is non vanishing. The two site bipolaron can be also continued at $t \neq 0$ but only providing its degeneracy is raised. Actually, this two-site bipolaron may be viewed as a pair of polarons with interacting spins. The ground-state is obtained in that model when the pair of electrons is in its singlet state (total spin zero) and when the distance $i_0 - j_0$ between the polarons is at a certain value.

These bipolaronic states were calculated numerically in [35] as a function of t and U. The ground-state was found to be the single-site bipolaron (S0) for small positive U. Increasing U, there is a first-order transition beyond which it becomes a two-site bipolaron at the nearest neighbor site (S1). Increasing U to larger values, there is a second first order transition at which it becomes a two-site bipolaron where the two peaks are at distance 2 and so on (see Fig. 15). Actually, according to the picture shown in Fig. 13, new intermediate bipolaron solutions which are spatially asymmetric should appear although they have not been calculated.

The bipolaron ground-state has been studied in this 1D Holstein-Hubbard model by perturbation theory close to the anticontinuous limit in [12, 38]. It was found that the spins of two polarons interact antiferromagnetically which creates an interaction potential with its minimum at a distance n. Then the ground-state is obtained for these two polarons at distance n. When U varies beyond $U_c(t) \approx 1/4$, there is a series of first-order transitions where the ground-state jumps from a two-site bipolaron at distance n to the distance $n+1$. When U reaches $U_{c2}(t) \approx 1/2$, these first order transition lines accumulates to a transition line where the polaron distance diverges at infinity. When U goes beyond this limit, the two polarons which form the bipolaron become unbounded with a ground-state obtained when their relative distance is infinite.

The phonon eigenmodes were also calculated. The lowest frequencies correspond to a pinning mode which is spatially antisymmetric and tends to move the bipolaron and a breathing mode which is spatially symmetric. Their variation versus U are shown in Fig. 16. As expected, the pinning mode substantially softens at the first-order transition although it does not strictly vanish. Actually it vanishes only at the edge of the region of metastability. The PN energy barrier does not strictly vanish, as depicted by Fig.13, but nevertheless it becomes much smaller at the first-order transition where bipolarons (S0) and (S1) have the same energy.

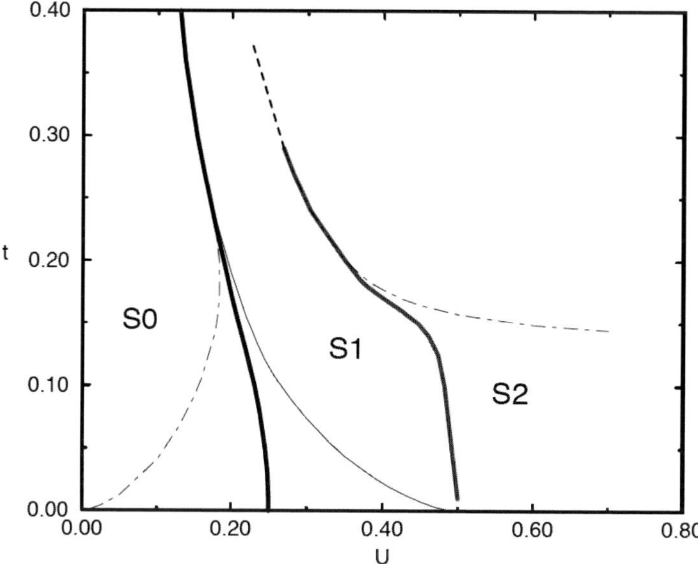

Fig. 15. Phase diagram versus U and t of the 1D Holstein model (62). The bipolaron ground-states where found to be (S0),(S1),(S2)   in the corresponding labelled
domains. Bipolaron (S0) has a larger domain of metastability delimited by the thin
line as well as bipolaron (S1), the domain of metastability which is delimited by the
two dot-dashed line. Reprinted from [35] with permission from Elsevier.

When the frequency of the pinning mode is small, a small perturbation of
the initial zero velocity field of the atoms in the direction of this pinning mode
may produce a mobile bipolaron (see Fig. 17). In the example which is shown,
one obtains a remarkably good mobility for a small bipolaron which extends
over two or three sites only and thus is quite far from the continuum limit.
This mobility only occurs in the vicinity of the first order transition line where
the pinning mode softens sufficiently and disappears when U varies away from
the transition line.

Similar investigations were done in the 2D Holstein-Hubbard model but
there are substantial differences. When increasing the transfer integral t there
is a first-order transition to a parameter domain where the ground-state is
not a bipolaron but is obtained for two extended electrons and no lattice
distortion. When t is small enough, there is a new phase where the bipolaronic
state is a quadrisinglet (QS).

This quadrisinglet bipolaron centered, for example, at site 0 of a square
lattice, can be obtained at the anticontinuous limit as an unstable solution
where the electronic wave function is obtained as the combination of four
singlet states along the bonds surrounding this site

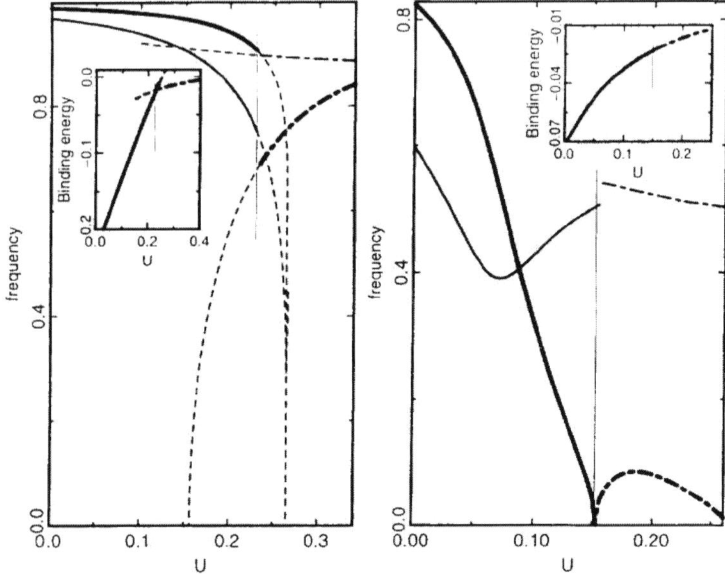

Fig. 16. Eigen mode frequencies ($\omega_0 = 1$) versus U at $t = 0.1$ (left) and $t = 0.3$ (right) for bipolarons (S0) and (S1). The pinning mode is represented by thick lines in the region where (S0) is ground-state (S0) (full line) and where (S1) is ground-state (dashed-dotted lines). These lines are continued as thin lines in the regions of metastability. The breathing modes are represented by thin lines. The insets represent the intersection of the bipolaron energies curves of (S1) and (S2). Reprinted from [35] with permission from Elsevier.

$$|\Psi> = \frac{1}{\sqrt{8}} \sum_{j:0} (c^+_{0,\uparrow} c^+_{j,\downarrow} + c^+_{j,\uparrow} c^+_{0,\downarrow})|\emptyset>$$

The electron density at site 0 is 1 and is $\frac{1}{4}$ at the four nearest neighboring sites.

This solution can be continued as an unstable solution when the transfer integral t increases from zero. This solution becomes stable in the domain delimited by the dashed lines of Fig. 18 (and then there is a bifurcation where new unstable bipolaronic solutions appear). Moreover, this solution becomes the ground-state in the smaller triangular-like domain in the middle of the phase diagram shown Fig. 18.

In the vicinity of the triple point between the phases of (S0), (S1) and (QS), the PN energy barrier for moving the bipolaron is sharply depressed while its binding energy is still non-negligible. This situation greatly favors the role of quantum lattice fluctuations which should be taken into account when the coefficient γ in (61) is not neglected.

The physical origin of this phenomenon which has been observed in the simple Holstein-Hubbard model, is essentially due to the competition between

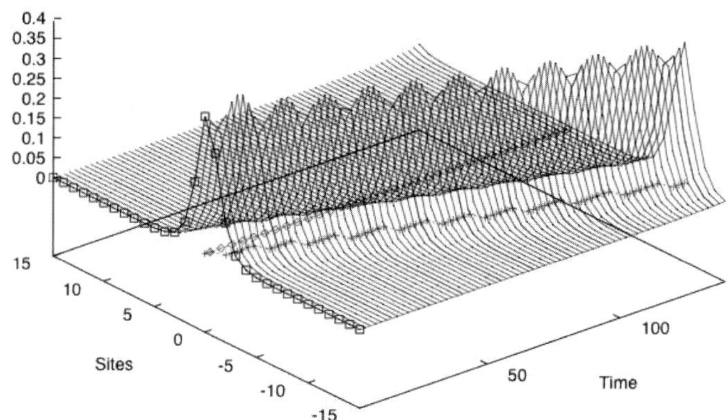

Fig. 17. 3D plot of the electronic density $\rho_i(t)$ versus i and time for a mobile bipolaron in the Holstein-Hubbard model at $t = 0.3$ and $U = 0.14$. Reprinted from [35] with permission from Elsevier.

the electron-phonon coupling which favors a single-site bipolaron (S0) and the Hubbard interaction which favors multipeaked bipolarons such as (S1) or (QS).

We suggested in [36] that other bipolaronic structures could appear in more complex models, where for example bipolarons with d-type symmetry could exist.

9 Many Polaron and Bipolaron Structures in the Holstein-Hubbard Model

Polarons and bipolarons are nonlinear solutions which cannot be linearly superposed. However, there exist infinitely many metastable states which correspond to many bipolaron or polaron structures.

9.1 General Theorem

Theorems about the existence and the properties of these solutions in the adiabatic Holstein model were given in [19, 39] for the adiabatic Holstein and Hostein-Hubbard models. Some of these results were improved with a simplified proof in [40, 41].

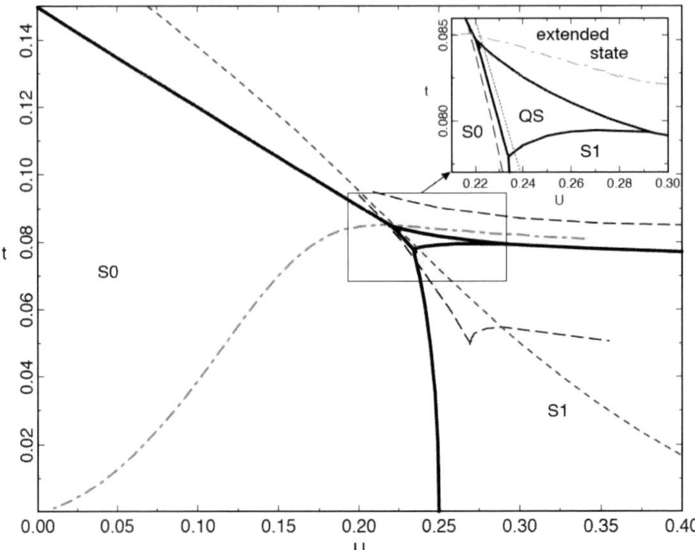

Fig. 18. Phase diagram versus U and t of the 2D Holstein model (62). The bipolaron ground-states were found to be (S0), (S1) in the corresponding labelled domains, with a larger domain of metastability delimited by the thin line for (S0) and two dot-dashed line for bipolaron (S1), the domain of metastability which is delimited by the two dot-dashed line. There is also a new domain where the bipolaron (SQ) is the ground-state with a larger domain of metastability. Insert: magnification of the domain where the QS bipolaron is the ground-state. Reprinted from [35] with permission from Elsevier.

We briefly explain the ideas of the proof which are based on the principle of anticontinuity (later extended to the existence proof of Discrete Breathers [25, 42]).

We consider the adiabatic Hamiltonian (61) of the Holstein-Hubbard model ($\gamma = 0$)

$$\widetilde{H} = \sum_i \left(\frac{1}{2}u_i^2 + (\frac{1}{2}u_i - \mu)n_i + Un_{i,\uparrow}n_{i,\downarrow} \right) - \frac{t}{2} \sum_{<i,j>,\sigma} c_{i,\sigma}^+ c_{j,\sigma} \qquad (64)$$

obtained from Hamiltonian (61) with $\gamma = 0$ with an arbitrary number of electrons. For convenience, we added an arbitrary chemical potential μ for fixing the density of electrons at a given value $0 < \rho =< n_i >< 2$. Then the electrons are supposed to be in their ground-state with respect to the atomic configuration $\{u_i\}$ (adiabatic approximation).

At the anticontinuous limit where $t = 0$, the model becomes separable and is the sum of independent Hamiltonians. Then the operator n_i commutes with the Hamiltonian so that $n_i = 0$, 1 or 2. Each atom u_i is then submitted to a local potential $V(u_i) = \min_{n_i} \frac{1}{2}u_i^2 + (\frac{1}{2}u_i - \mu)n_i + \frac{1}{2}Un_i(n_i - 1)$ which is the minimum of three parabola. Only the parabola obtained for $n_i = 2$ depends on U.

As μ only shifts the height of parabola $V(u)$ for $n_i = 1$ by $-\mu$ and the parabola $n_i = 2$ by -2μ, it can be chosen at convenience for what concerns the number of minima of $V(u)$ (providing there are at least two minima). A scheme is represented Fig. 19 where in order to fix the ideas, we chose $\mu = 1/8$ in order the parabola at $n_i = 0$ and $n_i = 1$ have the same minimum value.

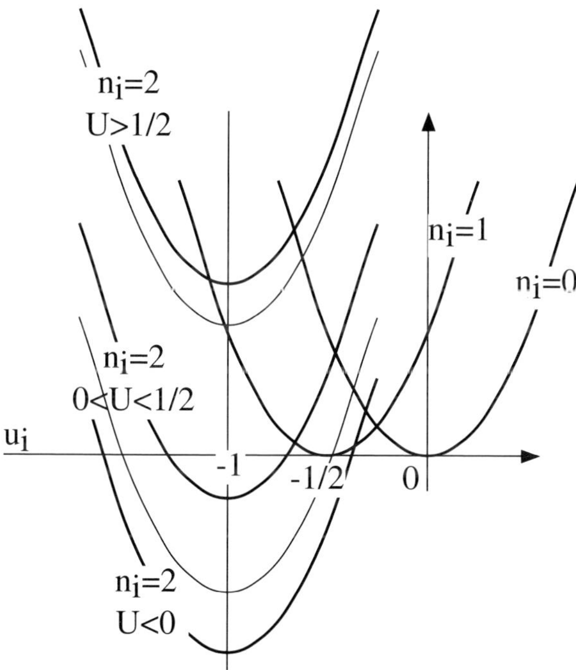

Fig. 19. Local atomic potential versus u_i at the anticontinuous limit. It is the minimum of three parabola obtained for $n_i = 0$, $n_i = 1$ and $n_i = 2$.

There are three possible situations (see Fig. 19)

- When $U \leq 0$, this local potential has only two minima obtained for $n_i = 2$ and $n_i = 0$. The minima of the anticontinuous adiabatic Hamiltonian for an electronic density strictly between 0 and 2 are arbitrary bipolaronic structures.
- When $0 < U < \frac{1}{2}$, this local potential has three minima obtained for $n_i = 2$, $n_i = 1$ and $n_i = 0$. The minima of the anticontinuous adiabatic Hamiltonian are arbitrary mixed polaronic and bipolaronic structures.
- When $U \geq \frac{1}{2}$, the local potential has only two minima obtained for $n_i = 0$ and $n_i = 1$ when the electronic density is between 0 and 1 or $n_i = 1$

and $n_i = 2$ when the electronic density is between 1 and 2 (the chemical potential then changes to negative values.).

Each local minimum of the adiabatic Hamiltonian can be coded by its electronic distribution at the anticontinuous limit $\{n_i\}$ where $n_i = 0, 1$ or 2. Such solutions can be interpreted as a many polaron-bipolaron structure (see Fig. 20). By the implicit function theorem [40], each local minimum can be

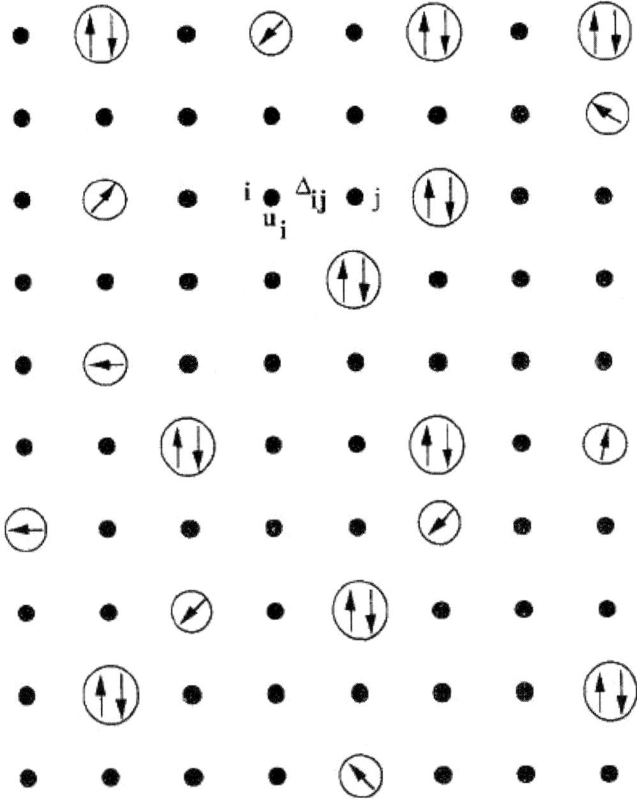

Fig. 20. Scheme of an arbitrary polaronic-bipolaronic configuration in the Holstein-Hubbard model. Any of the allowed configurations at the anticontinuous limit persist when the transfer integral is not zero, providing it is not too large .

continued as a local minimum when the transfer integral t is not zero (see Fig. 21). Thus, these configurations are stable. The number of states for the mixed polaronic-bipolaronic structures is 3^N for N lattice sites which yields an entropy per site $\ln 3$. It is $\ln 2$ when only bipolaronic (or polaronic) structures are allowed.

When $U \to 1/2$ the interval of continuation of the purely bipolaronic states goes to zero. When $U \to 0$, this interval of continuation goes to zero for the purely polaronic states.

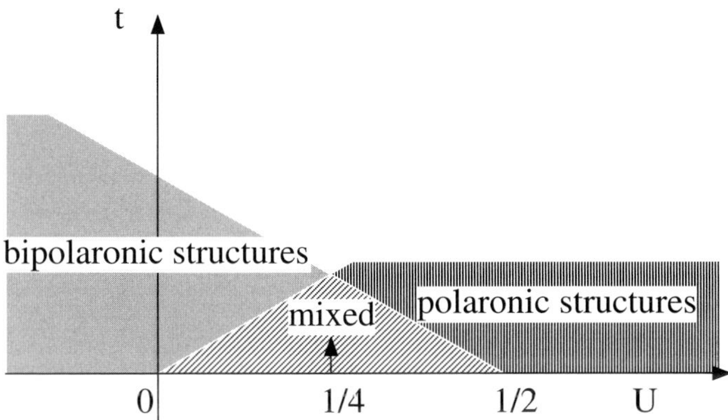

Fig. 21. Qualitative scheme of the domain in the parameter space U, t where the bipolaronic structures, the polaronic structures and the mixed polaronic–bipolaronic structures can be continued.

When the continued structure contains polarons (that is $n_i = 1$ at some sites i), the orientations of the polaron spins are degenerate at the anticontinuous limit. As soon as the transfer integral t is not zero, there are magnetic interactions between the spins which appear. We expect that they generally raise the degeneracy. In this specific Holstein–Hubbard model, the interaction between the spins of two polarons (calculated at lowest order) is antiferromagnetic [5]. We have seen above (8), that for a single pair of polarons the spin ground-state is a singlet state.

When the transfer integral t increases beyond a certain critical value, many bipolaronic or polaronic structures are expected to disappear through bifurcations with other unstable bipolaronic or polaronic structures. This is the phenomenon of pruning which has been mentioned for different variational forms [43, 44]. The entropy of the ensemble of polaronic–bipolaronic configurations then decreases since certain bipolaronic–polaronic configurations are no longer allowed. Otherwise, new polaronic or bipolaronic structures may appear. Bipolaron (SQ) described above is an example of a bipolaron which is not single-site at the anticontinuous limit and was not taken into account by our theorem.

It has also been proven, at least for bipolaronic states, that the adiabatic ground-state of the Holstein model for a specific band filling, when t is not too large, is obtained for a specific distribution of the bipolarons $n_i = 2$ and holes $n_i = 0$. The energy of the system at the anticontinuous limit is for N sites

$$\Phi(\{n_i\}) = \sum_i \left(\frac{1}{2} n_i (n_i - 1)(-\frac{1}{2} + U - 2\mu) + n_i(2 - n_i)(-\frac{1}{8} - \mu) \right)$$

$$= N\frac{1}{2} < n_i >^2 (U - \frac{1}{4}) - < n_i > (\frac{U}{2} + \mu) \tag{65}$$

Then, for a given electronic density $< n_i >= \rho$, the minimum of this energy at the anticontinuous limit is obtained;

- when $U < \frac{1}{4}$, the average $< n_i^2 >$ has to be maximum for $< n_i >$ fixed. Then, either $n_i = 0$ or $n_i = 2$ which means the ground state structure is bipolaronic (with bipolarons and holes). It is spatially degenerate because the bipolaron distribution is arbitrary.
- when $U > \frac{1}{4}$, the average $< n_i^2 >$ has to be minimum for fixed $< n_i >$ which implies either $n_i = 0$ or 1 when $0 \leq< n_i >\leq 1$, or $n_i = 1$ or 2 when $1 \leq< n_i >\leq 2$. Then the ground state structure is polaronic, either with polarons and holes or with polarons and bipolarons. It is degenerate not only because the polaron distribution is arbitrary, but also because the spins of the polarons are degenerate.

It can be noted that the energies of the configurations coded by $\{n_i\}$ continued as a function of t depend on $\{n_i\}$ and also on polaron spins S_i when $n_i = 1$. It formally defines an effective Hamiltonian $H(\{n_i\}, \{n_i(2 - n_i)S_i\})$. This Hamiltonian is constant and degenerate at $t = 0$ but could be expanded at the lowest orders as a function of t.

We have no general theorems which yield precise information about the ground-state of this Hamiltonian and how the (quantum) spin degeneracy will be raised in the general case when there are many polarons with spins. A spatial magnetic ordering might be expected to be superposed on the polaronic structure but, however, states like Resonating Valence Bond (RVB) states or others might occur when the system dimensionality is low enough. In the case of random distributions of polarons with frustration, spin glasses might also be expected.

Finding the ground-state of the adiabatic Holstein model and a fortiori of the Holstein-Hubbard model at arbitrary band filling and a dimension larger than 2 is an open problem. However, it can be found numerically in the 1D Holstein model where only bipolarons are present. We then obtain bipolaronic Charge Density Waves (CDW) which may be either commensurate or incommensurate depending on the band filling.

9.2 Charge Density Waves in 1D

It is well-known that 1D conductors exhibit a Peierls instability [45]. Considering the electron-phonon coupling as a perturbation, it is found that the unmodulated chain is necessarily unstable. There is always an energy gain by opening a gap at the Fermi surface which is generated by a spatially periodic lattice distortion (PLD) at wave vector $2k_F$ where $k_F = \pi\rho$ is the Fermi wavevector and $0 < \rho < 1$ is the density of electron pairs. This PLD is associated with a Charge Density Wave (CDW) where the electron density is modulated with the same period. When this density ρ is an irrational number, the period ρ of the CDW is incommensurate with the period of the lattice.

It has been commonly assumed without proof in the incommensurate case that since the phase of the CDW-PLD is continuously degenerate, there is a strictly zero frequency Goldstone mode corresponding to a uniform phase rotation or a sliding incommensurate CDW-PLD. Thus it was argued that incommensurate CDW-PLD were necessarily conducting which was an early tentative model proposed by Fröhlich as a theory for superconductivity before the standard BCS theory. Actually, we shall see that CDW-PLD may be also bipolaronic and insulating. Many real systems which exhibit CDW-PLD are now known [46]. Moreover, these systems often exhibit unusual nonlinear conductivity phenomena, the origin of which is still controversial.

We investigated [47–49, 51], this CDW-PLD in the adiabatic Holstein model with Hamiltonian and spinless electrons:

$$H = \sum_i \left(-(c_i^+ c_{i+1} + c_{i+1}^+ c_i) + k u_i c_i^+ c_i + \frac{1}{2} u_i^2 \right) \tag{66}$$

(this model is equivalent to a model with electron with spins where the electronic states are doubly occupied. Then the reduced electron-phonon coupling k should be replaced by $k\sqrt{2}$ in (66)). Note that $k^2 = 1/t$ is related to the reduced transfer integral t by (24).

The CDW and the PLD are obviously proportional to each other since $u_i = -k < c_i^+ c_i >$. Following the prediction of Peierls [45], numerical investigations confirm that the ground-state is a PLD described by

$$u_i = g(2k_F i + \alpha) \tag{67}$$

where $g(x)$ is a 2π periodic hull function and α is an arbitrary phase, Fig. 22.

When $\rho = \frac{r}{s}$ is a rational number, this function is piecewise constant and thus discontinuous with s constant steps in a period. When ρ is an irrational number, this function is sine-like for small k but exhibits a well defined transition at a critical value $k_c(\rho)$. At this transition, the hull function which was smooth and likely analytic becomes discontinuous with infinitely many discontinuities. Although the transition is very sharp, it is second order because it was checked that the hull function is a continuous function of k

We called this transition *Transition by Breaking of Analyticity* (TBA) because it exhibits the same physical characteristics of the TBA as those of the Frenkel-Kontorowa (FK) model [23] where, moreover, its existence has been rigorously proven [24, 50].

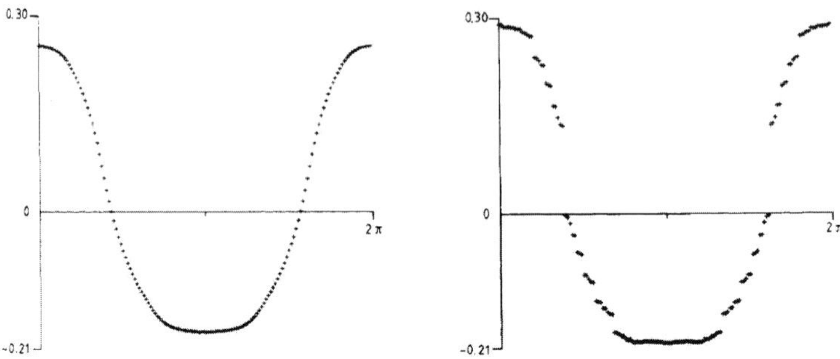

Fig. 22. Hull function of the Holstein model for the irrational $\rho = (3 - \sqrt{5})/2$ in the vicinity of the Transition by Breaking of Analyticity at $k = k_c \approx 1.58$ for $k = 1.57$ (left) and for $k = 1.59$ (right). Actually we used the rational approximate $\rho_r = \frac{34}{89}$ which has a practically identical behavior.

The analytic CDW-PLD has the characteristics of a Peierls-Fröhlich CDW which is conducting by phase sliding (Fröhlich mode). In contrast, the non-analytic CDW-PLD is insulating with properties which are different to the analytic CDW-PLD.

- In the nonanalytic phase, a local perturbation of the structure decays exponentially over a distance ξ (coherence length) which is finite. This coherence length diverges at the TBA and becomes infinite for the analytic CDW-PLD.
- The gap of the phonon frequencies spectrum is non-zero for the nonanalytic CDW-PLD but vanishes at the TBA. The gap is zero in the analytic CDW-PLD because of the existence of a strictly zero frequency sliding mode.
- The PN energy barrier for displacing nonanalytic CDW-PLD is non zero and vanishes at the TBA and is zero for the analytic CDW-PLD.
- The nonanalytic phase is defectible, that is, it can sustain many defects. In contrast, the analytic CDW-PLD is the unique stable configuration of the system (apart from an arbitrary phase shift) and thus it cannot sustain any metastable defects (as expected from a conducting material, since these defects would be charged).

The critical quantities at the TBA vanish or diverge with critical exponents which are related by scaling laws and depend on the incommensurability ratio ρ.

Actually, the TBA is nothing but a transition of the Peierls-Fröhlich CDW-PLD to a bipolaronic structure. The anticontinuous limit is reached when the electron coupling $k^2 = \frac{1}{t}$ becomes large. Then, according to our general results, the electronic density $< c_i^+ c_i >= \chi(2\pi\rho i + \alpha)$ can be described by a hull function which essentially takes two values 0 or 1 and is defined in the period $[0, 2\pi]$ by $\chi(x) = 1$ for $0 < x \leq 2\pi\rho$ and $\chi(x) = 0$ for $2\pi\rho < x \leq 2\pi$.

It is sometime convenient to describe the incommensurate bipolaronic CDWs as an array of periodic discommensurations (with fractional charges) referred to as a close arbitrary commensurate CDW. Then, the metastable defects which could be sustained correspond to extra discommensuration, either retarded or advanced, which could be randomly distributed.

Moreover, according to general theorems proven for the FK model [52] the CDW can be written as the convolution of the bipolaron distribution by a shape factor [19] which may be interpreted as the effective shape of the polaron, Fig. 23.

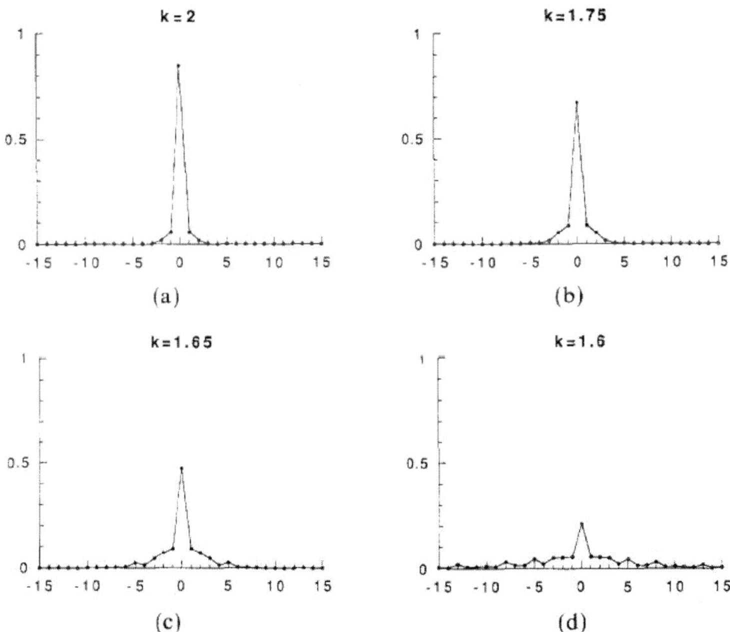

Fig. 23. Effective shape of polarons. Their spread and their size diverge at the TBA

When the electron density $\rho = \frac{r}{s}$ is rational (r and s are irreducible), the CDW-PLD is commensurate with an always nonanalytic hull function. It is thus a bipolaronic structure. However, as soon the commensurability order s has values larger than a few units (typically $5 - 8$), the TBA appears as a crossover where the effective size of the bipolarons almost diverges when the effective electron–phonon coupling decreases.

The existence of TBA appears to be ubiquitous when there is 1D incommensurate ordering. It is observed in other 1D models with Peierls instabilities, for example the SSH model [48, 51], where the electron-phonon coupling occurs within the transfer integral. It can also be found for other incommensurate structures involving both polarons and bipolarons. For example, the half-filled adiabatic Holstein model with electrons with spins submitted to a magnetic field was found to exhibit a TBA induced by a Zeeman effect [53]. The magnetic field breaks some of the bipolarons into polarons and the resulting structure order into a mixed polaron-bipolaron and incommensurate structure. The magnetic response is then Devil's staircase with plateaus at commensurate ordering. When the electron phonon coupling is not too large, the Devils's staircase is incomplete so that we get an insulating metal transition induced by a magnetic field at values where the ordering becomes incommensurate.

We also found that variations of the Holstein model may induce more complex bipolaronic phases [54] which corresponds to CDW which are not at the expected wavevector $2k_F$.

In models in more that 1D, we saw that a large enough electron–phonon coupling is necessary for generating bipolarons or polarons. Then we should generally obtain insulating bipolaron ordering. We may however reasonably expect from the results known in the 1D model, that commensurate ordering with a supercell favors bipolaronic structures. These could form for example regular stripes. When the electron density becomes irrational, the bipolaronic structure may disappear. However, up to now very few numerical studies were performed for understanding the bipolaronic ordering in more than 1D.

10 Further Prospects and Final Comments

We gave here a review of our early works on polarons and bipolarons. We first described the properties of the single polaron, mostly within the adiabatic Holstein model. In more than 1D, the polaron exists only at large enough electron-phonon coupling and is always a small polaron. There is a first order transition to the extended electron when the electron-phonon coupling decreases. When a magnetic field is added to the model, the polaron also exists at small electron–phonon coupling and the first order transition between large and small polarons ends at a triple point for a large enough magnetic field. We discussed the existence of multipeaked polarons in variations of the Holstein model. Next we also noted that the electron could self-localize on a local anharmonic vibration as a polarobreather.

We discussed the role of the PN barrier and the mechanism which could depress it. As an illustration, we investigated the bipolaron in the Holstein-Hubbard model, where the PN energy barrier could be depressed in specific regions of the phase diagram because of the Hubbard term which tends to break the bipolaron into two polarons. Then, we emphasize the role of the magnetic interactions which could bind the two polarons and form a more complex bipolaron. In 1D, we obtained highly mobile bipolarons which are however quite small. We found even more complex bipolarons in 2D which are quadrisinglets of spins. We did not obtain classical mobility in 2D as in 1D but nevertheless found a strong depression of this PN energy barrier.

It is conceivable that in a variation of the Holstein-Hubbard model with several sublattices, we could obtain ground-state bipolarons with an orbital momentum (that is which carry internal current) for example having p or d symmetry.

Next we studied many bipolaron and polaron structures and proved that at large enough electron-phonon coupling, the bipolaron and the polaron persist and may form disordered structures. Their ground-state is also a bipolaronic-polaronic structure which should be generally ordered as either a CDW when there are only bipolarons or Spin Density Waves (SQW) which superpose to a charge ordering when there are polarons. These structures are better understood in the 1D model where they form either commensurate or incommensurate CDWs whether the electron density is rational or not. The commensurate CDWs are always bipolaronic. The incommensurate CDW are purely bipolaronic at large electron-phonon coupling and become Peierls-Fröhlich CDW at smaller coupling. We also found that when the polarons or the bipolarons are embedded in a structure their interaction modifies their shape. They generally extend further than the single polaron or bipolaron. Their size may even diverge in case the of TBA. Nothing is known about the spatial ordering of (S1) or (SQ) bipolarons.

10.1 Quantum Lattice Fluctuations: CDW-SDW Versus Superconductivity

The main question about the existence and the role of polarons and bipolarons in real systems comes from the quantum fluctuations of the lattice, the study of which has been neglected. The quantum lattice fluctuations manifest in model (61) when γ is non-vanishing. Its main effect at the lowest order is to allow quantum tunnelling of the polarons or bipolarons from a local site to a nearest neighbor site. Thus single polarons or bipolarons should be extended and form bands.

In the case of many bipolaron structures, the energy of the bipolaronic structure is a function of its coding sequence $\{n_i\}$ where $n_i = 0$ or 2. A pseudospin with z component $\sigma_i^z = (n_i - 1)/2$ may be defined such that the bipolaronic configuration of the system may be viewed as that of an Ising model (note that this spin is not a magnetic spin). Then the tunnelling energy appears at lowest order as extra terms $\sum_{<i,j>} \Gamma(\sigma_i^x \sigma_i^x + \sigma_i^y \sigma_i^y)$ which tunnels a bipolaron from site i to site j or vice versa when one of the two sites is occupied.

Alexandrov et al. (ARR) already obtained this form many years ago [55] by an expansion of the quantum Holstein model at large electron-phonon coupling. They obtained a hardcore Boson model where the pseudospin σ_i may order along the z direction thus creating a bipolaron ordering (CDW). When the tunnelling term Γ becomes large enough, a transverse ordered component may appear in the xy plane. This quantum ordering corresponds to a bipolaronic superconductor. A uniform gradient of the average angle of the planar component ($< \sigma_i^x >, < \sigma_i^y >$) of these pseudospins generates a quantum state which carries a permanent electric current without dissipation. This quantum state is obtained by the application of a magnetic field (the Meissner effect).

Thus for this pseudospin model of bipolaronic superconductivity the electron pairs are assumed to preexist as bipolarons above the transition. They form a classical liquid of bipolarons when the planar phases are disordered and thus without quantum coherence. There is already an electronic gap above the superconducting transition corresponding to the energy required to break a bipolaron. The superconducting transition is obtained when the quantum phases (angle of the xy component) order coherently with long range order. This situation contrasts with standard BCS superconductivity which holds only at small electron–phonon coupling (when single bipolarons cannot exist). In that case, there are no Cooper pairs above the transition temperature. All the Cooper pairs appear self-consistently, simultaneously and coherently only at the superconducting transition.

Actually, bipolaronic and BCS superconductivity may be viewed as two limiting cases of the same phenomenon. This situation is similar to the order-disorder limit and the displacive limit of structural phase transitions. In the displacive case, the fluctuations of the order parameter only occur at large scale which could be characterized by a coherence length (wall thickness), while in the order-disorder case, these fluctuations occur at the scale of the lattice cell. Displacive transition and BCS superconductivity are thus well described by mean field theories while in contrast order-disorder transitions and bipolaronic superconductivity exhibit an extended critical behavior.

Unfortunately, in the case of the quantum Holstein model, realistic parameters in the ARR model generally yield a tunnelling parameter Γ much smaller than the coupling terms which favor spatial ordering. It is thus more reasonable to believe that the pseudospin ordering occurs mostly in the z direction and if there is any ordered component in the xy plane, it should order only at extremely low temperature. Although Γ could become much larger in the antiadiabatic regime, the bipolaron lattice distortion is then much smaller than the quantum zero point motion of the atoms and thus very small.

It is essential to realize that the tunnelling of the quantum bipolarons (or polarons) occurs through the PN energy barrier of the adiabatic bipolaron (or polaron). Then it is clear that the adiabatic approximation is valid when the PN energy barrier for moving the bipolaron is large (at the scale of quantum phonon energies) because the tunnelling energy Γ becomes negligible. These structures order as CDW or SDWs.

Thus the key point for realizing bipolaronic superconductivity is that we have to find situations where the PN energy barrier for displacing an adiabatic bipolaron through the lattice becomes small, which consequently drastically increases the tunnelling energy of the corresponding quantum bipolaron. Then, if the bipolaron becomes sufficiently light, the spatial ordering may be destroyed and a superconducting ordering should take place.

For small bipolarons with a restricted spatial extension, the PN energy barrier is generally very large but it could be reduced in the vicinity of first order transitions between different structures. For example in the Holstein-Hubbard model, it has been found [37] that the effective mass of the bipolaron drops sharply in the vicinity of the first order transition between different bipolaron structures like (S0)-(S1) or (SQ).

Actually, the vicinity of such first order transitions, is characterized by a quite well balanced competition between the electron–phonon coupling which favors compact single-site bipolarons and the repulsive Hubbard interaction which favors bipolarons with more complex structures consisting of two polarons bonded by magnetic interactions or the quadisinglet bipolaron. Despite the existence of several equivalent structures, the binding energy (of the order of the magnetic interactions between polarons) could remain substantially large.

If one wishes to extend these ideas to situations with many electrons, more complex models which extend those of ARR are necessary. These models should not involve only bipolarons but polarons which interact magnetically in order to allow different kinds of polaron ordering which could compete. For example, it is quite intuitive that in the Holstein–Hubbard model, a band fill ing $\rho = 1$ yields an insulating commensurate structure with one polaron per site which is antiferromagnetically ordered (this situation would correspond to undoped cuprates). More generally, other rational band filling could favor a commensurate ordering of polarons with a relatively small supercell which is magnetically ordered and insulating (this could correspond to some cuprates which at particular doping exhibiting stripes?). When no simple polaron ordering can be obtained at a given density of polarons, other ordering may appear where free polarons may bind into pairs as bipolarons with several degenerate structures favoring a light mass and condensing into a superconducting state.

In a different context, we considered incommensurate Peierls-Fröhlich CDW which have a strictly zero phonon gap and strictly zero PN energy barrier, the quantum lattice fluctuations become essential. We conjectured long ago [56, 57] that such CDWs should be unstable against any quantum lattice fluctuations and thus could not exist. We conjectured that instead of a CDW, those materials should become a BCS superconductor. As a consequence, all the existing CDWs should be bipolaronic. Although we still lack a proof, we still believe this conjecture. It was nevertheless shown by numerical means in the SSH model [58] that at low coupling, when the phonon gap becomes small enough, the dimerized chain becomes unstable because of the quantum lattice fluctuations. This result was obtained by showing that the energy of the quantum discommensuration of the dimerized SSH model becomes negative at low coupling. Otherwise, this conjecture could be a helpful assumption for understanding the puzzling properties of CDW materials for which many predictions of the Peierls-Fröhlich theory are not experimentally verified [46].

10.2 Dynamics of Polarons: Chemical Reactions by Polaron Transfer

Finally, concerning polaron theory it is worthwhile mentioning that electrons in chemistry are usually considered as polarons although they are not explicitly named so. The standard theory of chemical reactions, consisting of an electron transfer (ET) between two sites of different molecules or of the same molecule [3] takes into account large displacements of the environment of the electron (involving a large reorganization energy). Considering that the atomic displacements in the vicinity of the electron are large, the atomic coordinates are treated as classical variables (and called reaction coordinates). The rest of the system is considered as a thermal bath. Thus the problem of electron transfer becomes nothing but a problem of transfer of an adiabatic polaron between two nonequivalent states which are both local minima of the energy. The first state represents the reactants (R) and the second state the products (P). There is a PN energy barrier between these two states (R) and (P). Thus we are specifically interested in the dynamics of this polaron (Kramers model) under the effect of thermal fluctuations.

Actually, there are two situations which are schematized in Fig. 24. This figure represents the energy of the system as a function of the (many) reaction coordinates when the electron is in the initial state (R) and in the final state (P). The difference between these two curves is the energy required for a direct electron transfer at fixed reaction coordinates. In the normal regime, direct transfer of the electron from (R) to (P) at fixed reaction coordinates requires a positive energy while in the inverted regime this energy is negative.

The electron transfer occurs by tunnelling when the electronic levels at fixed reaction coordinates becomes resonant which requires that the thermal fluctuations bring the system to the top of the PN energy barrier. Considering the tunnelling time is negligible, the characteristic time for ET is exponentially related to the height of this barrier $\Delta G^{\star}$ and proportional to $\exp -\Delta G^{\star}/(k_B T)$ (Arrhenius law).

Since the PN energy barrier vanishes at the inversion point intermediate between the normal and the inverted regime, ultrafast electron transfer is expected precisely at this inversion point. This is the most common interpretation of ultrafast ET which occurs in real (mostly biological) systems such as, for example, the photosynthetic reaction center. In that situation, the characteristic time for ET is essentially the tunnelling time. Since in most real situations the transfer integral between the two states (R) and (P) is small (weak reactant limit), this tunnelling is relatively slow (of the order or longer than the characteristic time for atomic reorganization). Thus this process cannot be treated within the adiabatic approximation.

For this reason, we developed a nonadiabatic model which takes into account this essential feature. The (large amplitude) atomic displacements are treated classically but the electronic state is not assumed to be in equilibrium with the local environment with its own quantum dynamics. Thus, we obtain

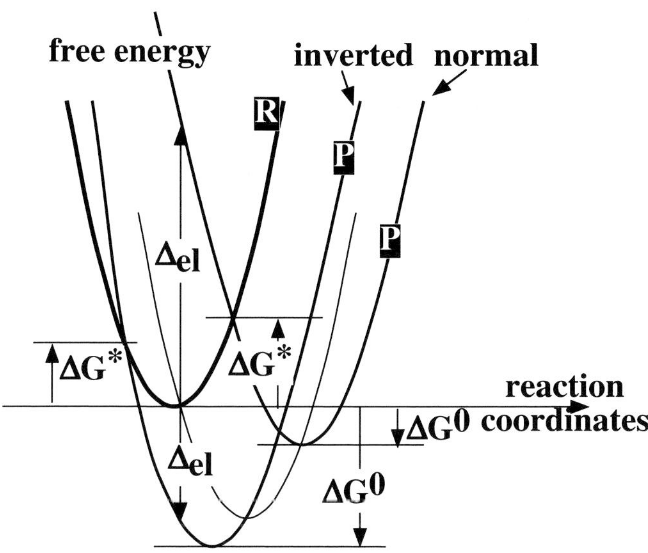

Fig. 24. Standard model for electron transfer. The energies of the system with the electron in the initial state (R) and the final state (P) are represented by a parabola. The excitation energy Δ_{el} for transferring the electron from state (R) to (P) at fixed reaction coordinates is positive in the normal regime and negative in the inverted regime.

a model where the quantum dynamics of the electron is coupled with the classical dynamics of the atoms and which is coupled to a thermal bath. We found that the most efficient way to have ultrafast ET is to involve a third site (called the catalytic site). This catalytic site is tuned with the donor site in order to have the phenomenon of Targeted Electron Transfer (TET). We then obtain coherent electron-phonon oscillations triggering ultrafast electron transfer toward the acceptor. For details see [59–61]. A model with mobile polarons along a chain of selected sites also based on TET was proposed recently [62]. Other applications for understanding biomotors, which convert ATP chemical energy into mechanical energy, are currently being investigated.

10.3 Final Comments

We have presented here a review of our past contribution concerning small adiabatic polarons and bipolarons. We have shown how rich the phenomenology associated with the existence of adiabatic polarons and bipolarons in physical systems could be. We have also shown the possibility of many kinds of transitions which could occur either on single polarons or bipolarons and also concerning many polaron-bipolaron structures. Up to now, these phenomena are still widely unexplored.

Understanding high-Tc cuprate superconductors is still a challenge nowadays. Most tentative theories proposed up to now discard the role of the electron-phonon coupling. We think however that models which are more realistic than the Holstein-Hubbard model will be necessary for producing a phenomenology which is similar to the observed facts, for example, for obtaining d-type bipolarons (with internal currents). Nevertheless, more studies on the polaronic-bipolaronic structures of the Holstein-Hubbard model and their quantum fluctuations would be useful and would greatly help future prospects on more sophisticated models.

Another field of application concerns biology, where highly selective and also ultrafast electron transfer may occur. Since we know that proteins and other biopolymers are both soft and heavily charged, the atomic reorganization is important when moving electrons and the effect of quantum lattice fluctuations could be discarded (excepts perhaps when protons are involved). Thus the atoms could be considered as classical particles but do not form a spatially periodic structure. Otherwise the interesting features to study will mostly concern the dynamics of polarons and other similar excitations for example excitons.

A Proof of inequality (15)

Let us define $\Psi(x, y)$ as an arbitrary complex differentiable function $\Psi : \mathcal{R} \times \mathcal{R} \to \mathcal{C}$ of two variables x and y which is square summable, as are its derivatives. Then we have the inequality

$$\pi \leq \frac{\int |\Psi(x, y)|^2 dx dy . \int |\nabla \Psi(x, y)|^2 dx dy}{\int |\Psi(x, y)|^4 dx dy} \tag{68}$$

Inequality (15) is the special case when $\Psi(x, y)$ is normalized ($\int |\Psi(x, y)|^2 dx dy = 1$).

Proof:

- We first prove that the minimum of the right term in (68) can be obtained for $\Psi(x, y)$ real only:

  $\Psi(x, y)$ complex can be written $\Psi = A e^{i\Xi}$ where $A(x, y)$ is the real positive amplitude and the phase $\Xi(x, y)$ is a real angle which is a function of x and y. Then $|\Psi(x, y)|^2 = A^2(x, y)$, $|\Psi(x, y)|^4 = A^4(x, y)$ and $\nabla \Psi = (\nabla A + iA\nabla \Xi)e^{i\Xi}$ which implies $|\nabla \Psi|^2 = |\nabla A|^2 + A^2|\nabla \Xi|^2 \geq |\nabla A|^2$. Consequently, we have

$$\int A^2(x, y) dx dy = \int |\Psi(x, y)|^2 dx dy \tag{69}$$

$$\int A^4(x, y) dx dy = \int |\Psi(x, y)|^4 dx dy \tag{70}$$

$$\int |\nabla A|^2 dx dy \leq \int |\nabla \Psi|^2 dx dy \tag{71}$$

- We consider the amplitude $A(x,y)$ of $\Psi(x,y)$ as a function of the area $S(x,y)$ delimited by the contour line of A passing at (x,y) and prove

$$\int_0^{+\infty} A^2(S)dS = \int |\Psi(x,y)|^2 dxdy \tag{72}$$

$$\int_0^{+\infty} A^4(S)dS = \int |\Psi(x,y)|^4 dxdy \tag{73}$$

$$4\pi \int_0^{+\infty} \left(\frac{dA}{dS}\right)^2 SdS \le \int |\nabla\Psi|^2 dxdy \tag{74}$$

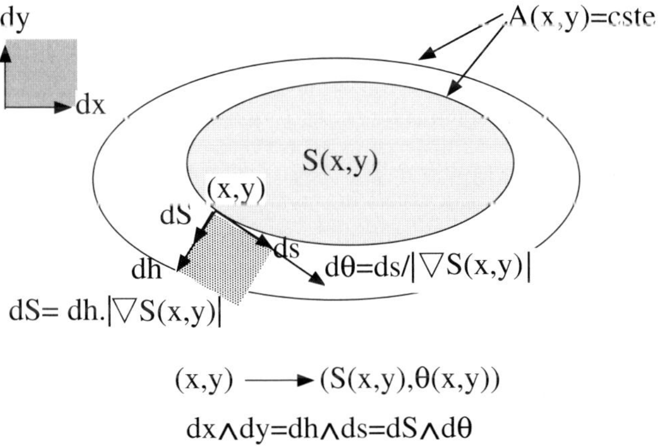

Fig. 25. The area of the contour line of the real positive amplitude A passing at (x,y) is $S(x,y)$. New variables $S(x,y)$ and $s(x,y)$ are defined such that $dS \wedge ds = dx \wedge dy$.

Actually, we define a change of variables (S,θ) which are functions of (x,y) and such that $dS \wedge d\theta = dx \wedge dy$ (see Fig. 25). Since $A(x,y)$ is differentiable with respect to (x,y), the contour lines defined by $A(x,y) = a$ where a is a constant, are continuous and differentiable closed curves (note that these closed curves may consist of several closed loops). $S(x,y)$ is the area of the contour line (counted algebraically) which passes by (x,y), that is the measure of the domain $\mathcal{D}(x,y)$ defined by $(x',y') \in \mathcal{D}(x,y)$ if $A(x',y') \ge A(x,y)$. $A(x,y)$ and $S(x,y)$ are constant on the same contour lines by definition.

The gradient flow $\nabla S(x,y) = \left(\frac{\partial S}{\partial x}, \frac{\partial S}{\partial y}\right)$ at (x,y) is orthogonal to the contour line of S passing at (x,y) (see Fig. 25). The curve abscissa on the contour line is s and $L(S) = \int ds$ is the total length of the contour line with area S. The length of the gradient $|\nabla S(x,y)| = (|\frac{\partial S}{\partial x}|^2 + |\frac{\partial S}{\partial y}|^2)^{1/2} = g(S,s)$ is a function of S and of the abscissa s on the contour line at S. Since the area of the domain $d\mathcal{D}$ delimited by two contour lines at S and $S + dS$, is by definition dS, we have $dS = g(S,s)dh$ where dh is the normal distance between the two contour lines at the curve abscissa s.

We have $dS = \int_{d\mathcal{D}} dxdy = \int dhds = dS \int \frac{ds}{g(S,s)}$ Since we must have $dS \wedge d\theta = dx \wedge dy$ we get $d\theta = \frac{ds}{g(S,s)}$ which defines θ (modulo an arbitrary origin). As a result, we have each contour line $\int d\theta = 1$ so that at fixed S, $\theta(S,s)$ may be viewed as an angle defined modulo 1.

It emerges that with the new variables S and θ that A only depends on S and is independent of θ. Eqations (72) and (73) are obviously fulfilled. We consider now

$$\int |\nabla A|^2 dxdy = \int (\frac{dA}{dS})^2 |\nabla S|^2 dxdy$$

$$= \int (\frac{dA}{dS})^2 g^2(S,\theta)dSd\theta = \int_0^{+\infty} (\frac{dA}{dS})^2 I(S)dS \qquad (75)$$

where $I(S) = \int g^2(S,\theta)d\theta$. Since $\int d\theta = 1$, we have the Schwarz inequality $(\int g(S,\theta)d\theta)^2 \leq \int g^2(S,\theta)d\theta = I(S)$. Since $d\theta = \frac{ds}{g(S,s)}$, we have $\int g(S,\theta)d\theta = \int ds = L(S)$ which is nothing but the total length of the contour line with area S. Finally, we have $I(S) \geq L^2(S)$.

It is well known (and easy to prove) that the length $L(S)$ of a loop (or a set of loops) surrounding a given area S, is minimum for a single loop which is a circle. The radius R of this disk fulfills $S = \pi R^2$ which yields the minimum perimeter $2\pi R = 2\sqrt{\pi S} \leq L(S)$. Consequently, we obtain the inequality $4\pi S \leq L^2(S) \leq I(S)$ which after substitution in (75) yields (74).

- Finally, we prove (68).

Because of (72) and (74), we have the inequality

$$\frac{1}{4\pi} \int |\nabla \Psi(x,y)|^2 dxdy. \int |\Psi(x,y)|^2 dxdy$$

$$\geq N(A) = \int_0^{+\infty} (\frac{dA}{dS})^2 SdS. \int_0^{+\infty} A^2(S)dS \qquad (76)$$

An integration by parts yields

$$\int_0^{+\infty} (\frac{dA}{dS})^2 SdS = \int_0^{+\infty} \left(\int_S^{+\infty} (\frac{dA}{d\sigma})^2 d\sigma \right) dS$$

and then we may write the product of the integral of the right term of (76) as

$$N(A) = \int\limits_{0}^{+\infty} \left(\int\limits_{S}^{+\infty} (\frac{dA}{d\sigma})^2 d\sigma . \int\limits_{0}^{\infty} A^2(\sigma)d\sigma \right) dS$$

$$\geq \int\limits_{0}^{+\infty} \left(\int\limits_{S}^{+\infty} (\frac{dA}{d\sigma})^2 d\sigma . \int\limits_{S}^{\infty} A^2(\sigma)d\sigma \right) dS$$

Because of the Schwarz inequality

$$\int\limits_{S}^{+\infty} (\frac{dA}{d\sigma})^2 d\sigma . \int\limits_{S}^{\infty} A^2(\sigma)d\sigma \geq \left(\int\limits_{S}^{+\infty} A(\sigma)\frac{dA}{d\sigma} d\sigma \right)^2 = \frac{1}{4}A^4(S)$$

we obtain

$$N(A) \geq \frac{1}{4} \int\limits_{0}^{+\infty} A^4(S)dS = \frac{1}{4} \int |\Psi(x,y)|^4 dxdy$$

which with (76) proves (68).

As a consequence of inequality (68), κ_c defined as

$$\kappa_c = \inf_{\Psi} \frac{\int |\Psi(x,y)|^2 dxdy . \int |\nabla\Psi(x,y)|^2 dxdy}{\int |\Psi(x,y)|^4 dxdy} \tag{77}$$

is non vanishing and we have $\kappa_c \geq \pi$. We may find an upper bound by choosing a special form for $\Psi(x,y)$, for example, a Gaussian $\Psi(x,y) = e^{-(x^2+y^2)}$. Then, we obtain $\kappa_c \leq 2\pi$. However, κ_c may be estimated numerically more precisely[5]. We found $\kappa_c \approx 1.92\pi$.

It is not known if there is a function $\Psi(x,y)$ which realizes the minimum in (77). However, this function is not unique but belongs to a family of functions $\lambda\psi(\alpha x, \alpha y)$ where λ and α are arbitrary parameters since they yield the same values in (77).

References

1. L. Landau, Phys. *Z. Sowjetunion* **33**, 664 (1933)
2. S.I. Pekar, *J. Phys. (Moscow)* **10**, 341 (1946); *Untersuchungen über die Electronentheorie der Kristalle* (Akademie-Verlag, Berlin, 1954)
3. R.A. Marcus, *Rev. Mod. Phys.* **65** 599-610 (1993)
4. D. Emin and T. Holstein, *Phys.Rev.Letts.* **36** 323-326 (1976)
5. S. Aubry, in *Fluctuation Phenomena: Nonlinearity and Disorder*, eds. A.R. Bishop, S. Jimenez and L. Vasquez, World Scientific p. 307-315 (1995)
6. H. Fröhlich, *Proc.Roy.Soc.London* Ser **A160** 230 (1937)
7. T. Holstein, *Ann.Phys. (N.Y)* **8** 325 (1959)

8. S. Aubry, in *Phase separation in Cuprate Superconductors* Edited K.A.Müller and G.Benedek, World Scientific Pub. (the Science and Culture Series-Physics) (1993) 304-334

9. G. Kalosakas, S.Aubry and G.P. Tsironis, *Phys. Rev.* **B58** (1998) 3094–3104

10. F.M. Peeters and J.T. Devreese, *Phys. Stat. Sol.* (b) **110** 631-635 (1982)

11. F.M. Peeters and J.T. Devreese, *Phys.Rev.* **B25** 7281-7301 (1982); ibid. *Phys.Rev.* **B25** 7302-7326 (1982)

12. S.Aubry, in *The Physics and Mathematical Physics of the Hubbard Model* NATO ASI SERIES B Physics, 1995 **343** page 125-144 (Eds. D.Baeriswyl, D.K. Campbell, J.M.P. Carmelo and F. Guinea)

13. S.M. Girvin and T. Jach, *Phys.Rev.* **B29** 5617-5625

14. Y. Avishai, R.M. Redheffer and Y.B. Band, *J.Phys.* **A25** 3883-3889 (1992)

15. N. Rohringer, J. Burgdörfer and N. Macris, **A36** 4173-4190 (2003)

16. P.G. Harper, *Proc.Phys.Soc.Lon* **A68** 874 (1955)

17. D. Hofstadter, *Phys. Rev.* **B14** 2239 (1976)

18. S. Aubry and G. André, Ann. of the Israël Phys.Soc. **3** 133-164 (1980) (reprinted in *The Physics of Quasicrystals* Ed. P.J. Steinhardt and S. Ostlund p.554-593 World Scientific Publishing Co. Singapore (1987))

19. S.Aubry, G.Abramovici and J.L. Raimbault, *J. Stat. Phys.* **67** 675-780 (1992)

20. Y. Zolotaryuk, P.L. Christiansen and J.J. Rasmussen, *Phys.Rev.* **B58** 14305 (1998)

21. M.A. Fuentes, P. Maniadis, G. Kalosakas, K.ØRasmussen, A.R. Bishop, V.M. Kenkre and Yu. Gaididei, *Phys.Rev.* **E70** (2004) 025601

22. G. Kalosakas, Physica **216D** (2006) 44-61

23. S. Aubry, in *Soliton and Condensed Matter* Sol.Stat.Sciences **8** 264-278 (1978) (Eds. Bishop A.R. and T.Schneider)

24. S. Aubry, in *Structure et Instabilité*(in french) Ed. C.Godrèche Edition de Physique (1986) p.73-194

25. S.Aubry, *Physica* **103D** (1997) 201-250

26. G. Kalosakas and S.Aubry, *Physica* **113D** 228-232 (1998)

27. S. Aubry, Physica **216D** (2006) 1-30

28. L. S. Schulman, E. Mihóková, A. Scardicchio, P. Facchi, M. Nikl, K. Polak, and B. Gaveau, *Phys. Rev. Lett.* **88** 224101 (2002)

29. E. Mihóková, L. S. Schulman, M. Nikl, B. Gaveau, K. Polak, K. Nitsch, and D. Zimmerman, *Phys. Rev.* **B 66** 155102 (2002).

30. R. Peierls, *Proc.Roy.Soc.Lon.* **52** 34 (1940)

31. F.R.N. Nabarro, *Proc.Phys.Soc.Lon.* **59** 256 (1947)

32. A.S. Alexandrov and P.E. Kornilovitch, Phys. Rev. Letts. **82** (1999) 807-810

33. S. Aubry, J.Physique IV colloque **C2 3** 349-355 (1993)

34. S.Aubry, in *Polarons and Bipolarons in High Tc Superconductors and related materials* Cambridge University Press (1995) p. 271-308 (Eds. E.K.H. Salje, A.S. Alexandrov and W.Y.Liang)

35. L. Proville and S. Aubry, *Physica* **113D** 307-317 (1998)

36. L. Proville and S. Aubry, European Physical Journal **B11** (1999) 41–58

37. L. Proville and S. Aubry, European Journal of Physics **B15** (2000) 405-421

38. J. Dorignac, J. Zhou and D.K. Campbell, Physica **D216** (2006) 207-213

39. L Proville and S. Aubry, J. Stat. Phys. **106** (2002) 1185–1195

40. C. Baesens and R.S. MacKay, Nonlinearity **7** (1994) 59-84

41. C. Baesens and R.S. MacKay, J. Phys A-Math and General, 31 (1998) 10065-10085

42. R.S. MacKay and S. Aubry, Nonlinearity **7** (1994) 1623-1643

43. S. Aubry, in *Twist Mappings and their Applications* eds. Richard McGehee and Kenneth R. Meyer, The IMA Volumes in Mathematics and Applications **44** (1992) 7-54 (Springer)

44. R. Schilling, Physica **D206** (2006) 157-166

45. R.E. Peierls, *Quantum theory of Solids* Oxford Unviversity Press (1955) p.108

46. P. Monceau, Physica **D206** (2006) 167-171

47. P.Y. Le Daëron and S. Aubry, J. Phys. **C16** 4827-4838 (1983)

48. P.Y. Le Daëron and S. Aubry, J.Physique (Paris) **C3** (1983) 1573-1577

49. S. Aubry, Solid State Sciences **47** p.126-143 (1984)

50. S. Aubry, Physica **7D** (1983) 240-258

51. S. Aubry and P. Quemerais, in *Low Dimensional Electronic Properties of Molybdenum Bronzes and Oxides* Ed. Claire SCHLENKER p. 295-405 Kluwer Academic Publishers Group (1989)

52. S. Aubry, J.P. Gosso, G. Abramovici, J.L. Raimbault, P. Quemerais, Physica **47D** 461-497 (1991)

53. C.Kuhn and S.Aubry, J.Phys. Cond.matter **6** 5891-5918 (1994)

54. J.L. Raimbault and S.Aubry, J. Phys.: Condensed Matter **7** (1995) 8287-8315

55. A.S. Alexandrov, J. Ranninger and S. Robaszkiewicz, Phys. Rev **B33** (1986) 4526-4542

56. S. Aubry, G. Abramovici, D. Feinberg, P. Quemerais and J.L. Raimbault, in *Non-Linear Coherent Structures in Physics, Mechanics and Biological Systems* Lectures Notes in Physics (Springer) **353** pp.103-116 (1989)

57. P. Quemerais, J.L. Raimbault and S. Aubry, Fisica **21** Supp.3, 106-108 (1990)

58. P.Quemerais, J.L. Raimbault, D.Campbell and S.Aubry, International Journal of Modern Physics **B7** (1993) 4289-4303

59. S. Aubry and G. Kopidakis, Int. J. OF Mod. Phys. **B 17** (2003): 3908-3921 (2003);ibid *Localization and Energy Transfer in Nonlinear Systems* Eds. L. Vázquez, R.S. MacKay and M.P. Zorzano, World Scientific Publishing (2003) pp.1-27

60. S. Aubry, in *Nonlinear Waves: Classical and Quantum Aspects* Eds. F. Kh. Abdullaev and V.V Konotop, NATO Science series II Mathematics, Physics and Chemistry Vol 153 (2004) pp.443–471

61. S. Aubry and G. Kopidakis, Journal of Biological Physics **31** (2005) 375 - 402

62. G. Hervé and S. Aubry, Physica **D216** 235-245 (2006)

From Single Polaron to Short Scale Phase Separation

V.V. Kabanov

J. Stefan Institute, Jamova 39, 1001, Ljubljana, Slovenia `viktor.kabanov@ijs.si`

1 Introduction

There is substantial evidence that the ground state in many oxides is inhomogeneous [1]. In cuprates, for example, neutron-scattering experiments suggest that phase segregation takes place in the form of stripes or short segments of stripes [2, 3]. There is some controversy whether this phase segregation is associated with magnetic interactions. On the other hand it is also generally accepted that the charge density in cuprates is not homogeneous.

The idea of charge segregation has quite a long history (see for example [4–6]). In charged systems phase separation is often accompanied by charge segregation. Breaking of the charge neutrality leads to the appearance of the electric field and substantial contribution of the electrostatic energy to the thermodynamic potential [7, 8]. Recently it was suggested that interplay of the short range lattice attraction and the long-range Coulomb repulsion between charge carriers could lead to the formation of short metallic [9, 10] or insulating [10, 11] stripes of polarons. If the attractive potential is isotropic, charged bubbles have a spherical shape. Khomskii and Kugel [12] suggested recently that the anisotropic attraction forces caused by Jahn-Teller centers could lead to phase segregation in the form of stripes. The long-range anisotropic attraction forces appear as the solution of the full set of elasticity equations (see [13]). An alternative approach to take into account elasticity potentials was proposed in [14] and is based on the proper consideration of compatibility constraints caused by the absence of a dislocation in the solid. Phenomenological aspects of the phase separation were discussed recently in the model of Coulomb frustrated phase transitions [15–17].

Here we consider some aspects of the phase separation associated with different types of ordering of *charged* polarons. The formation of polaronic droplets in this case is due to competition of two types of interactions: the long range Coulomb repulsion and attraction generated by the deformation field. It is important to underline that if we consider the system of neutral polarons (without Coulomb repulsion), it shows a first order phase transition at constant chemical potential, and is unstable with respect to global phase separation at fixed density [17]. Electron–phonon interaction may be short range or long range, depending on the type of phonons involved. In most cases we consider phonons of the molecular type, leading to short range forces. In some cases we consider long-range Fröhlich electron–phonon interaction and interaction with the strain.

2 Single Polaron in the Adiabatic Approximation

The adiabatic theory of polarons was formulated many years ago [18, 19] and we briefly formulate here the main principles of adiabatic theory for the particular case of interaction with molecular vibrations (Holstein model). The central equation in the adiabatic theory is the Schrödinger equation for the electron in the external potential of the deformation field. In the discrete version it has the following form [20]:

$$-\sum_{\mathbf{m}\neq 0} t(\mathbf{m})[\psi_{\mathbf{n}}^{k} - \psi_{\mathbf{n+m}}^{k}] + \sqrt{2}g\omega_0\varphi_{\mathbf{n}}\psi_{\mathbf{n}}^{k} = E_k\psi_{\mathbf{n}}^{k}. \tag{1}$$

Here $t(\mathbf{m})$ is the hopping integral, $\psi_{\mathbf{n}}$ is the electronic wave function on the site $\mathbf{n}$, $\varphi_{\mathbf{n}}$ is the deformation at the site $\mathbf{n}$, g is the electron–phonon coupling constant and ω_0 is the phonon frequency, k describes the quantum numbers of the problem. An important assumption of the adiabatic approximation is that the deformation field is very slow and we assume that $\varphi_{\mathbf{n}}$ is time independent $\partial\varphi/\partial t = 0$ when we substitute it into the Schrödinger equation for the electron. Therefore, $\omega_0 \to 0$ and $g^2 \to \infty$ but the product $g^2\omega_0 = E_p$ is finite and called the polaron shift. The equation for $\varphi_{\mathbf{n}}$ has the form [20]:

$$\varphi_{\mathbf{n}} = -\sqrt{2}g|\psi_{\mathbf{n}}^{0}|^2 \tag{2}$$

Here $\psi_{\mathbf{n}}^{0}$ corresponds to the ground state solution of (1). After substitution of (2) into (1) we obtain:

$$-\sum_{\mathbf{m}\neq 0} t(\mathbf{m})[\psi_{\mathbf{n}}^{k} - \psi_{\mathbf{n+m}}^{k}] - 2E_p|\psi_{\mathbf{n}}^{0}|^2\psi_{\mathbf{n}}^{k} = E_k\psi_{\mathbf{n}}^{k}. \tag{3}$$

As a result the nonlinear Schrödinger equation (3) describes the ground state of the polaron. All the excited states are the eigenvalues and eigenfunctions of the *linear* Schrödinger equation in the presence of the external field determined by the deformation field (2). The polaron energy is the sum of two contributions. The first contribution is the energy of the electron in the self-consistent potential well, determined by (3), and the second one is the energy of the strain field φ itself $E_{pol} = E_0 + \omega_0 \sum_{\mathbf{n}} \varphi_{\mathbf{n}}^2/2$. The polaron energy is presented in Fig. 1 as a function of the dimensionless coupling constant $2g^2\omega_0/t$ in 1D, 2D and 3D cases.

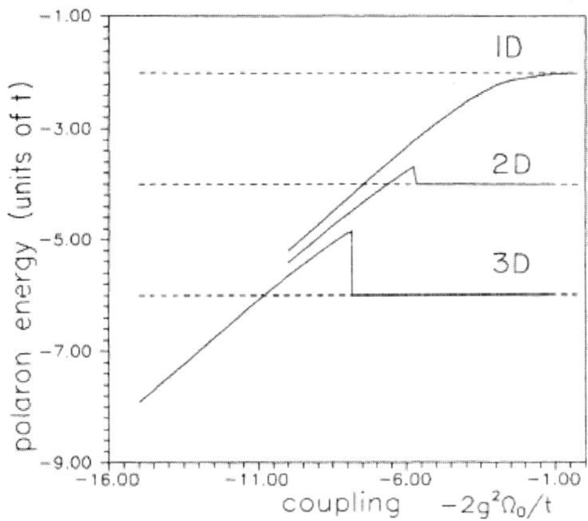

Fig. 1. Polaron energy as a function of the coupling constant in 1D,2D and 3D cases. Dashed lines represent the energy of the delocalized solution in 1D, 2D and 3D respectively [20].

There is a very important difference between the 1D, 2D and 3D cases. The polaron energy for the 1D case is always less then the energy of the delocalized state (dashed line). The polaron is always stable in 1D. In the 2D and 3D cases there is a critical value of the coupling constant where the first localized solution of (3) appears. It is interesting that the energy of the solution is higher then the energy of the delocalized state. Therefore, in the 2D and 3D cases there is a range of the coupling constant where the polaron is metastable. The delocalized solution is always stable in the 3D case. Therefore, the barrier which separates localized and delocalized states exists in the whole region of the coupling constants, where the self-trapped solution exists. In the 2D case the delocalized state is unstable at large values of the coupling constant. The barrier separating localized and delocalized states forms only in the restricted region of the coupling constant [20]. To demonstrate that, we have plotted polaron energy as a function of its radius in 2D in Fig. 2. As is clearly seen from this figure, the barrier has disappeared at $g > g_{c3} = \sqrt{2\pi t/\omega_0}$ [20].

There are two types of non-adiabatic corrections to the adiabatic polaron. The first is related to renormalization of local phonon modes. Fast motion of the electron within the polaronic potential well leads to the shift of the local vibrational frequency:

$$\omega = \omega_0[1 - zt^2/2(g^2\omega_0)^2]^{1/2} \tag{4}$$

here z is the number of nearest neighbors. This formula is valid in the strong coupling limit $g^2\omega_0 \gg t$ and when the tunneling frequency of the polaron is much smaller than the phonon frequency.

Another type of correction is related to the restoration of the translational symmetry (Goldstone mode) and describes polaron tunnelling and formation

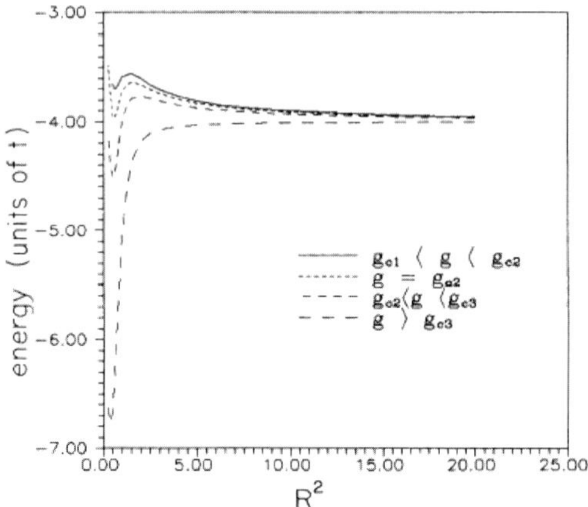

Fig. 2. Polaron energy as a function of radius in 2D. For $g > g_{c3}$ $\partial E_{pol}/\partial R < 0$. $g_{c1} = 1.69\sqrt{t/\omega_0}$, and $g_{c2} = 1.87\sqrt{t/\omega_0}$

of the polaron band. This correction was calculated in the original paper of Holstein [18]. A slightly improved formula was derived in [21]. In the adiabatic limit polaron tunnelling is exponentially suppressed $t_{eff} \propto \sqrt{E_p\omega_0}\exp\left(-g^2\right)$ (see (9) of [21]) and should be smaller than the phonon frequency ω_0.

In the following we will neglect all nonadiabatic corrections. We will consider the polaron as a pure localized state, and all corrections which contain the phonon frequency itself and the tunnelling amplitude for the polaron are neglected.

3 Strings in Charge-Transfer Mott Insulators: Effects of Lattice Vibrations and the Coulomb Interaction

Here we prove that the Fröhlich electron–phonon interaction, combined with the direct Coulomb repulsion, does not lead to string-like charge segregation in doped narrow-band insulators, either in the nonadiabatic or the adiabatic regime. However, this interaction significantly reduces the Coulomb repulsion, which might allow much weaker antiferromagnetic and/or short-range electron–phonon interactions to segregate charges in doped insulators, as suggested by previous studies [4, 5, 11].

To begin with we consider a generic Hamiltonian including, respectively, the kinetic energy of carriers, the Fröhlich electron-phonon interaction, the phonon energy, and the Coulomb repulsion as

$$H = \sum_{i\neq j} t(\mathbf{m} - \mathbf{n})\delta_{s,s'}c_i^\dagger c_j + \sum_{\mathbf{q},i} \omega_\mathbf{q} n_i \left[u_i(\mathbf{q})d_\mathbf{q} + H.c.\right]$$

$$+ \sum_{\mathbf{q}} \omega_{\mathbf{q}}(d_{\mathbf{q}}^{\dagger} d_{\mathbf{q}} + 1/2) + \frac{1}{2} \sum_{i \neq j} V(\mathbf{m} - \mathbf{n}) n_i n_j \tag{5}$$

with the bare hopping integral $t(\mathbf{m})$, and the matrix element of the electron–phonon interaction

$$u_i(\mathbf{q}) = \frac{1}{\sqrt{2N}} \gamma(\mathbf{q}) e^{i\mathbf{q} \cdot \mathbf{m}}. \tag{6}$$

Here $i = (\mathbf{m}, s)$, $j = (\mathbf{n}, s')$ include site $\mathbf{m}, \mathbf{n}$ and spin s, s' quantum numbers, $n_i = c_i^{\dagger} c_i$, $c_i, d_{\mathbf{q}}$ are the electron (hole) and phonon operators, respectively, and N is the number of sites. At large distances (or small q) one finds

$$\gamma(\mathbf{q})^2 \omega_{\mathbf{q}} = \frac{4\pi e^2}{\kappa q^2}, \tag{7}$$

and

$$V(\mathbf{m} - \mathbf{n}) = \frac{e^2}{\epsilon_{\infty} |\mathbf{m} - \mathbf{n}|}. \tag{8}$$

The phonon frequency $\omega_{\mathbf{q}}$ and the static and high-frequency dielectric constants in $\kappa^{-1} - \epsilon_{\infty}^{-1} - \epsilon_{0}^{-1}$ are those of the host insulator ($\hbar = c = 1$).

In the adiabatic limit one can apply a discrete version of the continuous nonlinear equation [22] proposed for the Holstein model (1), extended to the case of the deformation and Fröhlich interactions in [9–11]. Applying the Hartree approximation for the Coulomb repulsion, the single-particle wavefunction, $\psi_{\mathbf{n}}$ (the amplitude of the Wannier state $|\mathbf{n}\rangle$) obeys the following equation

$$- \sum_{\mathbf{m} \neq 0} t(\mathbf{m}) [\psi_{\mathbf{n}} - \psi_{\mathbf{n}+\mathbf{m}}] - e \phi_{\mathbf{n}} \psi_{\mathbf{n}} = E \psi_{\mathbf{n}}. \tag{9}$$

The potential $\phi_{\mathbf{n},k}$ acting on a fermion k at the site $\mathbf{n}$ is created by the polarization of the lattice $\phi_{\mathbf{n},k}^{l}$ and by the Coulomb repulsion from the other $M - 1$ fermions, $\phi_{\mathbf{n},k}^{c}$,

$$\phi_{\mathbf{n},k} = \phi_{\mathbf{n},k}^{l} + \phi_{\mathbf{n},k}^{c}. \tag{10}$$

Both potentials satisfy the discrete Poisson equation as

$$\kappa \Delta \phi_{\mathbf{n},k}^{l} = 4\pi e \sum_{p=1}^{M} |\psi_{\mathbf{n},p}|^2, \tag{11}$$

and

$$\epsilon_{\infty} \Delta \phi_{\mathbf{n},k}^{c} = -4\pi e \sum_{p=1, p \neq k}^{M} |\psi_{\mathbf{n},p}|^2, \tag{12}$$

with $\Delta \phi_{\mathbf{n}} = \sum_{\mathbf{m}}(\phi_{\mathbf{n}} - \phi_{\mathbf{n}+\mathbf{m}})$. In contrast to the approach used in [9], we include the Coulomb interaction in Pekar's functional J [22], describing the total energy in a self-consistent manner using the Hartree approximation, such that [10]

$$J = - \sum_{\mathbf{n},p,\mathbf{m}\neq 0} \psi_{\mathbf{n},p}^{*} t(\mathbf{m})[\psi_{\mathbf{n},p} - \psi_{\mathbf{n}+\mathbf{m},p}]$$
$$- \frac{2\pi e^2}{\kappa} \sum_{\mathbf{n},p,\mathbf{m},q} |\psi_{\mathbf{n},p}|^2 \Delta^{-1} |\psi_{\mathbf{m},q}|^2$$
$$+ \frac{2\pi e^2}{\epsilon_\infty} \sum_{\mathbf{n},p,\mathbf{m},q\neq p} |\psi_{\mathbf{n},p}|^2 \Delta^{-1} |\psi_{\mathbf{m},q}|^2. \tag{13}$$

If we assume, following [11] that the single-particle function of a fermion trapped in a string of length N is a simple exponent, $\psi_n = N^{-1/2}\exp(ikn)$ with periodic boundary conditions, then the functional J is expressed as $J = T + U$, where $T = -2t(N-1)\sin(\pi M/N)/[N\sin(\pi/N)]$ is the kinetic energy (for an *odd* number M of spinless fermions), proportional to t, and

$$U = -\frac{e^2}{\kappa}M^2 I_N + \frac{e^2}{\epsilon_\infty}M(M-1)I_N, \tag{14}$$

corresponds to the polarisation and the Coulomb energies. Here the integral I_N is given by

$$I_N = \frac{\pi}{(2\pi)^3} \int_{-\pi}^{\pi} dx \int_{-\pi}^{\pi} dy \int_{-\pi}^{\pi} dz \frac{\sin(Nx/2)^2}{N^2 \sin(x/2)^2}$$
$$\times \, (3 - \cos x - \cos y - \cos z)^{-1}. \tag{15}$$

I_N has the following asymptotic [10]:

$$I_N = \frac{1.31 + \ln N}{N}, \tag{16}$$

The asymptotic is derived also analytically at large N by the use of the fact that $\sin(Nx/2)^2/(2\pi N \sin(x/2)^2)$ can be replaced by a δ- function. If we split the first (attractive) term in (14) into two parts by replacing M^2 with $M + M(M-1)$, it becomes clear that the net interaction between polarons remains repulsive in the adiabatic regime because $\kappa > \epsilon_\infty$. Hence, there are no strings within the Hartree approximation for the Coulomb interaction. Strong correlations do not change this conclusion. Indeed, if we take the Coulomb energy of spinless one-dimensional fermions comprising both Hartree and exchange terms as

$$E_C = \frac{e^2 M(M-1)}{N\epsilon_\infty}[0.916 + \ln M], \tag{17}$$

the polarisation and Coulomb energy per particle becomes (for large $M >> 1$)

$$U/M = \frac{e^2 M}{N\epsilon_\infty}[0.916 + \ln M - \alpha(1.31 + \ln N)], \tag{18}$$

where $\alpha = 1 - \epsilon_\infty/\epsilon_0 < 1$. Minimising this energy with respect to the length of the string N we find

$$N = M^{1/\alpha} \exp(-0.31 + 0.916/\alpha), \qquad (19)$$

and

$$(U/M)_{min} = -\frac{e^2}{\kappa} M^{1-1/\alpha} \exp(0.31 - 0.916/\alpha). \qquad (20)$$

Hence, the potential energy per particle increases with the number of particles so that the energy of M well-separated polarons is lower than the energy of polarons trapped in a string, no matter whether they are correlated or not. The opposite conclusion of [9] originates in an incorrect approximation of the integral $I_N \propto N^{0.15}/N$. The correct asymptotic result is $I_N = \ln(N)/N$.

One can argue [9] that a finite kinetic energy (t) can stabilise a string of a finite length. Unfortunately, this is not correct either. We performed exact (numerical) calculations of the total energy $E(M, N)$ of M spinless fermions in a string of length N including both kinetic and potential energy with typical values of $\epsilon_\infty = 5$ and $\epsilon_0 = 30$. The local energy minima (per particle) in a string of length $1 \leq N \leq 69$ containing $M \leq N/2$ particles are presented in the Table 1. Strings with even fermion numbers carry a finite current and hence the local minima are found for odd M. In the extreme wide band regime with t as large as 1 eV the global string energy minimum is found at $M = 3, N = 25$ ($E = -2.1167$ eV), and at $M = 3, N = 13$ for $t = 0.5$ eV ($E = -1.2138$ eV). However, this is *not* the ground state energy in both cases. The energy of well-separated $d \geq 2$-dimensional polarons is well below this, less than $-2dt$ per particle (i.e. -6 eV in the first case and -3 eV in the second one in the three dimensional cubic lattice, and -4 eV and -2 eV, respectively, in the two-dimensional square lattice). This argument is applied for any values of $\epsilon_0, \epsilon_\infty$ and t. As a result we have proven that strings are impossible with the Fröhlich interaction alone, contrary to the erroneous conclusion of [9].

Table 1. $E(M, N)$ for $t = 1$ eV and $t = 0.5$ eV

	$t = 1$eV		$t = 0.5$eV	
M	N	E(M,N)	N	E(M,N)
1	11	-2.0328	3	-1.1919
3	25	-2.1167	13	-1.2138
5	42	-2.1166	25	-1.1840
7	61	-2.1127	40	-1.1661

The Fröhlich interaction is, of course, not the only electron-phonon interaction in ionic solids. As discussed in [23], any short range electron-phonon interaction, like, for example, the Jahn-Teller (JT) distortion can overcome the residual weak repulsion of Fröhlich polarons to form small bipolarons. At large distances small nonadibatic bipolarons weakly repel each other due to the long-range Coulomb interaction, with a strength four times of that of polarons. Hence, they form a liquid state [23], or bipolaronic-polaronic crystal-like structures [24] depending on their effective mass and density. The fact that the Fröhlich interaction almost nullifies the Coulomb repulsion in oxides justifies the use of the Holstein-Hubbard model [25, 26]. The ground state of the 1D Holstein-Hubbard model is a liquid of intersite bipolarons with a significantly reduced mass (compared with the on-site bipolaron) as shown recently [27]. The bound states of three or more polarons are not stable in this model, thus ruling out phase separation. However, the situation might be different if the antiferromagnetic [4, 5] and JT interaction [6] or any other short (but finite) range electron-phonon interactions are strong enough. Due to the long-range nature of the Coulomb repulsion the length of a string should be finite (see also [10, 11, 28]).

To summarize we conclude that there are no strings in ionic doped insulators with the Fröhlich interaction alone. Depending on their density and mass polarons remain in a liquid state or Wigner crystal. On the other hand the short-range electron-phonon and/or antiferromagnetic interactions might provide a liquid bipolaronic state and/or charge segregation (strings of a finite length) since the long-range Fröhlich interaction significantly reduces the Coulomb repulsion in highly polarizable ionic insulators.

4 Ordering of Charged Polarons: Lattice Gas Model

In this section we consider a macroscopic system of polarons in the thermodynamic limit. To underline the nontrivial geometry of the phase separation we consider two-fold degenerate electronic states which interact with a non-symmetric deformation field. In our derivation we follow a particular model for high-T_c superconductors. Nevertheless the results are general enough and are applicable to many Jahn-Teller systems.

Recently we formulated the model [29] where we suggested that interaction of a two-fold degenerate electronic state with fully symmetric phonon modes of the small group τ_1 at a finite wave-vector can lead to a local nonsymmetric deformation and short-length scale charge segregation in high-T_c materials. We reduced the proposed model to the lattice gas model [17] and showed that the model indeed displays phase separation, which may occur in the form of stripes or clusters depending on the anisotropy of the short range attraction between localized carriers [17]. We also generalized the model taking into account interaction of the Jahn-Teller centers via elasticity induced fields [17]. We showed that the model without Coulomb repulsion displays a first order phase transition at a constant chemical potential. When the number of particles is fixed, the system is unstable with respect to the global phase separation below a certain critical temperature. In the presence of the Coulomb repulsion the global phase separation becomes unfavorable due to a large contribution to the energy from long range Coulomb interaction. The system shows mesoscopic phase separation where the size of charged regions is determined by the competition between the energy gain due to ordering and energy cost due to breaking of the local charge neutrality. Since the short range attraction is anisotropic the phase separation may be in the form of short segments and/or stripes.

Let us start with the construction of a real-space Hamiltonian which couples 2-fold degenerate electronic states (or near-degenerate states) with optical phonons of τ_1 symmetry. Two-fold degeneracy is essential because in this case formation of the polaronic complexes leads to reduction not only of translational symmetry but also reduction of the point group symmetry. Since the Hamiltonian needs to describe a 2-fold degenerate system, the 2-fold degenerate states - for example the two E_u states corresponding to the planar hybridized Cu $d_{x^2-y^2}$, O p_x and p_y orbitals, or the E_u and E_g states of the apical O - are written in the form of Pauli matrices σ_i. Taking into account that the states are real, the Pauli matrices which describe transitions between the levels transform as A_{1g} $(k_x^2+k_y^2)$ for σ_0, B_{1g} $(k_x^2-k_y^2)$ for σ_3, B_{2g} $(k_x k_y)$ for σ_1, and A_{2g} (s_z) for σ_2 representations respectively. Collecting terms together by symmetry we can construct the *effective* electron-spin-lattice interaction Hamiltonian given in [29]. Here we consider a simplified version of the JT model Hamiltonian [29], taking only the deformation of the B_{1g} symmetry:

$$H_{JT} = g \sum_{\mathbf{n,l}} \sigma_{3,\mathbf{l}} f(\mathbf{n})(b^\dagger_{\mathbf{l+n}} + b_{\mathbf{l+n}}), \tag{21}$$

here the Pauli matrix $\sigma_{3,\mathbf{l}}$ describes two components of the electronic doublet, and $f(\mathbf{n}) = (n_x^2 - n_y^2)f_0(n)$ where $f_0(n)$ is a symmetric function describing the range of the interaction. For simplicity we omit the spin index in the sum. The model could be easily reduced to a lattice gas model [17]. Let us introduce the classical variable $\Phi_{\mathbf{i}} =< b^+_{\mathbf{i}} + b_{\mathbf{i}} > /\sqrt{2}$ and minimize the energy as a function of $\Phi_{\mathbf{i}}$ in the presence of the harmonic term $\omega_0 \sum_{\mathbf{i}} \Phi_{\mathbf{i}}^2/2$. We obtain the deformation, corresponding to the minimum of energy,

$$\Phi_{\mathbf{i}}^{(0)} = -\sqrt{2}g/\omega_0 \sum_{\mathbf{n}} \sigma_{3,\mathbf{i+n}} f(\mathbf{n}). \tag{22}$$

Substituting $\Phi_{\mathbf{i}}^{(0)}$ into the Hamiltonian (21) and taking into account that the carriers are charged we arrive at the lattice gas model. We use a pseudospin operator $S = 1$ to describe the occupancies of the two electronic levels n_1 and n_2. Here $S^z = 1$ corresponds to the state with $n_1 = 1$, $n_2 = 0$, $S_i^z = -1$ to $n_1 = 0$, $n_2 = 1$ and $S_i^z = 0$ to $n_1 = n_2 = 0$. Simultaneous occupancy of both levels is excluded due to a high on-site Coulomb repulsion energy. The Hamiltonian in terms of the pseudospin operator is given by[17]

$$H_{LG} = \sum_{\mathbf{i,j}}(-V_l(\mathbf{i} - \mathbf{j})S_{\mathbf{i}}^z S_{\mathbf{j}}^z + V_c(\mathbf{i} - \mathbf{j})Q_{\mathbf{i}}Q_{\mathbf{j}}), \tag{23}$$

where $Q_{\mathbf{i}} = (S_{\mathbf{i}}^z)^2$. $V_c(\mathbf{n}) = e^2/\epsilon_0 a(n_x^2 + n_y^2)^{1/2}$ is the Coulomb potential, e is the electron charge, ϵ_0 is the static dielectric constant and a is the effective unit cell period. The anisotropic short range attraction potential is given by $V_l(\mathbf{n}) = g^2/\omega_0 \sum_{\mathbf{m}} f(\mathbf{m}) f(\mathbf{n} + \mathbf{m})$. The attraction in this model is generated by the interaction of electrons with optical phonons. The radius of the attraction force is determined by the radius of the electron–phonon interaction and the dispersion of the optical phonons [10].

A similar model can be formulated in the limit of continuous media [17]. The deformation is characterized by the components of the strain tensor. For the two dimensional case we can define 3 components of the strain tensor: $e_1 = u_{xx} + u_{yy}$, $\epsilon = u_{xx} - u_{yy}$ and $e_2 = u_{xy}$ transforming as the A_{1g}, B_{1g} and B_{2g} representations of the D_{4h} group respectively. These components of the tensor are coupled linearly with the two-fold degenerate electronic state which transforms as the E_g or E_u representation of the point group. Similarly to the case of the previously considered interaction with optical phonons we keep the interaction with the deformation ϵ of the B_{1g} symmetry only. The Hamiltonian without the Coulomb repulsion term has the form:

$$H = \sum_{\mathbf{i}} \widetilde{g} S_{\mathbf{i}}^z \epsilon_{\mathbf{i}} + \frac{1}{2} \left(A_1 e_{1,\mathbf{i}}^2 + A_2 \epsilon_{\mathbf{i}}^2 + A_3 e_{3,\mathbf{i}}^2 \right), \qquad (24)$$

where A_j are the corresponding components of the elastic modulus tensor, $\widetilde{g} \propto g$. The components of the strain tensor are not independent [14, 30] and satisfy the compatibility condition:

$$\nabla^2 e_1(\mathbf{r}) - 4\partial^2 e_2(\mathbf{r})/\partial x \partial y = (\partial^2/\partial x^2 - \partial^2/\partial y^2)\epsilon(\mathbf{r})$$

The compatibility condition leads to a long range anisotropic interaction between polarons. The Hamiltonian in the reciprocal space has the form:

$$H = \sum_{\mathbf{k}} \widetilde{g} S_{\mathbf{k}}^z \epsilon_{\mathbf{k}} + (A_2 + A_1 U(\mathbf{k})) \frac{\epsilon_{\mathbf{k}}^2}{2}. \qquad (25)$$

The Fourier transform of the potential is given by:

$$U(\mathbf{k}) = \frac{(k_x^2 - k_y^2)^2}{k^4 + 16(A_1/A_3)k_x^2 k_y^2}. \qquad (26)$$

By minimizing the energy with respect to $\epsilon_{\mathbf{k}}$ and taking into account the long-range Coulomb repulsion we again derive (23). The anisotropic interaction potential $V_l(\mathbf{n}) = -\sum_{\mathbf{k}} \exp\left(i\mathbf{k} \cdot \mathbf{n}\right) \frac{\widetilde{g^2}}{2(A_2 + A_1 U(\mathbf{k}))}$ is determined by the interaction with the classical deformation and is long-range. It decays as $1/r^2$ at large distances in 2D. Since at large distances the attraction forces decay faster then the Coulomb repulsion forces the attraction can overcome the Coulomb repulsion at short distances, leading to a mesoscopic phase separation.

Irrespective of whether the resulting interaction between polarons is generated by acoustic or optical phonons the main physical picture remains the same. In both cases there is an anisotropic attraction between polarons at short distances. The interaction could be either ferromagnetic or antiferromagnetic in terms of the pseudospin operators, depending on the mutual orientation of the orbitals. Without losing generality we assume that $V(\mathbf{n})$ is non-zero only for the nearest neighbors.

Our aim is to study the model (23) at constant average density,

$$n = \frac{1}{N} \sum_{\mathbf{i}} Q_{\mathbf{i}}, \tag{27}$$

where N is the total number of sites. However, to clarify the physical picture it is more appropriate to perform calculations with a fixed chemical potential first by adding the term $-\mu \sum_{\mathbf{i}} Q_{\mathbf{i}}$ to the Hamiltonian (23).

Models such as (23) without long-range forces were studied many years ago on the basis of the molecular-field approximation in the Bragg-Williams formalism [31, 32]. The mean-field equations for the particle density n and the pseudospin magnetization $M = \frac{1}{N} \sum_{\mathbf{i}} S_{\mathbf{i}}^z$ have the form [31]:

$$M = \frac{2 \sinh\left(2zV_l M/k_B T\right)}{\exp\left(-\mu/k_B T\right) + 2 \cosh\left(2zV_l M/k_B T\right)}, \tag{28}$$

$$n = \frac{2 \cosh\left(2zV_l M/k_B T\right)}{\exp\left(-\mu/k_B T\right) + 2 \cosh\left(2zV_l M/k_B T\right)}. \tag{29}$$

Here $z = 4$ is the number of the nearest neighbours for a square lattice in 2D and k_B is the Boltzmann constant. A phase transition to an ordered state with finite M may be of either first or second order, depending on the value of μ. For the physically important case $-2zV_l < \mu < 0$, ordering occurs as a result of the first order phase transition. The two solutions of (28,29) with $M = 0$ and $M \neq 0$ correspond to two different minima of the free energy. The temperature of the phase transition T_{crit} is determined by the condition: $F(M = 0, \mu, T) = F(M, \mu, T)$ where M is the solution of (28). When the number of particles is fixed (29), the system is unstable with respect to global phase separation below T_{crit}. As a result, at fixed n two phases coexist with $n_0 = n(M = 0, \mu, T)$ and $n_M = n(M, \mu, T)$, resulting in a liquid-gas-like phase diagram (Fig. 3).

To investigate the effects of the long range-forces, we performed Monte Carlo simulations on the system (23)[17]. The simulations were performed on a square lattice with dimensions up to $L \times L$ sites with $10 \leq L \leq 100$ using a standard Metropolis algorithm [33] in combination with simulated annealing [34]. At constant n one Monte Carlo step included a single update for each site with nonzero Q_i, where the trial move consisted of setting $S_z = 0$ at the site with nonzero Q_i and $S_z = \pm 1$ at a randomly selected site with zero Q_i. A typical simulated annealing run consisted of a sequence of Monte Carlo simulations at different temperatures. At each temperature the equilibration phase ($10^3 - 10^6$ Monte Carlo steps) was followed by an averaging phase with the same or greater number of Monte Carlo steps. Observables were measured after each Monte Carlo step during the averaging phase only. For $L \gtrsim 20$ we observe virtually no dependence of the results on the system size.

Comparing the Monte Carlo results in the absence of Coulomb repulsion shown by t_{crit} in Fig. 3 with MF theory we find the usual reduction of t_{crit} due to fluctuations in 2D by a factor of ~ 2.

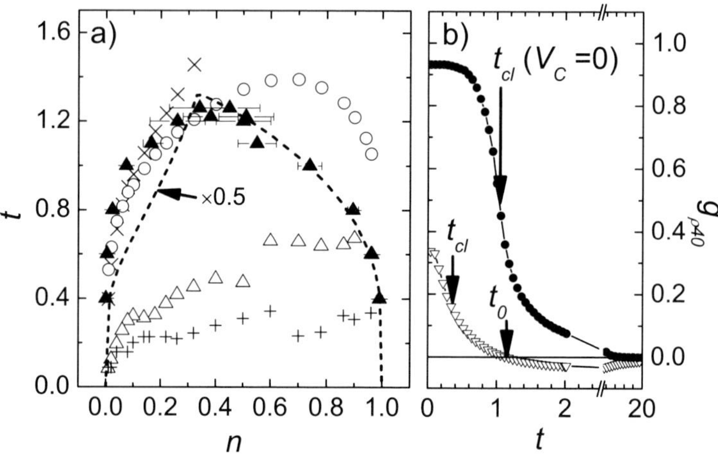

Fig. 3. a) The phase diagram generated by H_{JT} (23) with, and without the Coulomb repulsion (CR). The dashed line is the MF critical temperature, while the full triangles (▲) represent the Monte Carlo critical temperature, t_{crit}, *without CR*. The open circles (○) represent t_{cl}, *without CR*. The open triangles (△) represent t_{cl} while the diagonal crosses (×) represent the onset of clustering, t_0, *in the presence of* CR. The cluster-ordering temperature (see text), t_{co}, (also including CR) is shown as crosses (+). The size of the symbols corresponds to the error bars. b) Typical temperature dependencies of the nearest neighbor density correlation function $g_{\rho L}$ for $n = 0.18$ *in the absence of* CR (●) and *in the presence of* CR (▽). Arrows indicate the characteristic temperatures.

Next, we include the Coulomb interaction $V_c(r)$. We use open boundary conditions to avoid complications due to the long range Coulomb forces and ensure overall electroneutrality by adding a uniformly charged background electrostatic potential (jellium) to (23). The short range potential $v_l(\mathbf{i}) = V_l(\mathbf{i})\epsilon_0 a/e^2$ was taken to be non-zero only for $|\mathbf{i}| < 2$ and is therefore specified only for nearest and next-nearest neighbours as $v_l(1,0)$ and $v_l(1,1)$ respectively.

The anisotropy of the short range potential has a profound influence on the particle ordering. We can see this if we fix $v_l(1,0) = -1$, at a density $n = 0.2$ and vary the next-nearest neighbour potential $v_l(1,1)$ in the range from -1 to 1. When $v_l(1,1) < 0$, the attraction is "ferrodistortive" in all directions, while for positive $v_l(1,1) > 0$ the interaction is "antiferrodistortive" along the diagonals. The resulting clustering and ordering of clusters at $t = 0.04$ is shown in Fig. 4a. As expected, a more symmetric attraction potential leads to the formation of more symmetric clusters. On the other hand, for $v_l(1,1) = 1$, the "antiferrodistortive" interaction along diagonals prevails, resulting in diagonal stripes.

In the temperature region where clusters partially order the heat capacity ($c_L = \partial\langle E\rangle_L/\partial T$ where E is the total energy) displays a peak at t_{co}. The peak displays no scaling with L indicating that no long range ordering of clusters appears. Inspection of the particle distribution snapshots at low temperatures (Fig. 4a) reveals that finite size domains form. Within domains the clusters are perfectly ordered. The domain wall dynamics seems to be much slower than our Monte Carlo simulation timescale preventing domain growth. The effective L is therefore limited by the domain size. This explains the absence of the scaling and gives clear evidence for a phase transition near t_{co}.

We now focus on the shape of the short range potential which promotes the formation of stripes shown in Fig. 4a. We set $v_l(1,0) = -1$ and $v_l(1,1) = 0$ and study the density dependence. Since the inclusion of the Coulomb interaction completely suppresses the first order phase transition at t_{crit}, we measure the nearest neighbor density correlation function $g_{\rho L} = \frac{1}{4n(1-n)L^2} \sum_{|\mathbf{m}|=1} \langle \sum_{\mathbf{i}} (Q_{\mathbf{i+m}} - n)(Q_{\mathbf{i}} - n)\rangle_L$ to detect clustering. Here $\langle\rangle_L$ represents the Monte Carlo average. We define a dimensionless temperature $t_{cl} = k_B T_{cl}\epsilon_0 a/e^2$ as the characteristic crossover temperature related to the formation of clusters, at which $g_{\rho L}$ rises to 50% of its low temperature value. The dependence of t_{cl} on the density n is shown in the phase diagram in Fig. 3. Without Coulomb repulsion $V_c(r)$, t_{cl} follows t_{crit}, as expected. The addition of Coulomb repulsion $V_C(r)$ results in a significant decrease of t_{cl} and suppression of clustering. At low densities we can estimate the onset for cluster formation by the temperature, t_0, at which $g_{\rho L}$ becomes positive. It is interesting to note that t_0 almost coincides with the t_{crit} line at low n (Fig. 3).

To illustrate this behaviour, in Fig. 4b we show snapshots of the calculated Monte Carlo particle distributions at two different temperatures for different densities. The growth and ordering of clusters with decreasing temperature is clearly observed. At low n, the particles mostly form pairs with some short stripes. With further increasing density, quadruples gradually replace pairs, then longer stripes appear, mixed with quadruples, etc.. At the highest density, stripes prevail forming a labyrinth-like pattern. The density correlation function shows that the correlation length increases with doping, but long range order is never achieved (in contrast to the case without V_c). Note that while locally there is no four-fold symmetry the overall correlation function still retains 4-fold symmetry.

To get further insight into the cluster formation we measured the cluster-size distribution [17]. In Fig. 5 we present the temperature and density dependence of the cluster-size distribution function $x_L(j) = \langle N_p(j)\rangle_L/(nL^2)$, where $N_p(j)$ is the total number of particles within clusters of size j. At the highest temperature $x_L(j)$ is close to the distribution expected for random

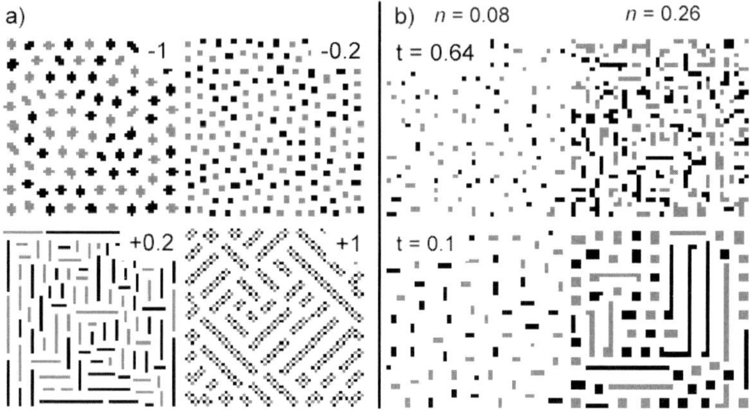

Fig. 4. a) Snapshots of clusters ordering at $t = 0.04$, $n = 0.2$ and $v_l(1,0) = -1$ for different diagonal $v_l(1,1)$ (given in each figure). Grey and black dots represent particles clusters in state $S_i^z = 1$ and state $S_i^z = -1$ respectively. The preference for even-particle-number clusters in certain cases is clearly observed, for example for $v_l(1,1) = -0.2$. b) Snapshots of the particle distribution for two densities at two different temperatures $t = 0.64$ and $t = 0.1$ respectively.

ordering. As the temperature decreases, the number of larger clusters starts to increase at the expense of single particles. Remarkably, as the temperature is further reduced, clusters of certain size start to prevail. This is clearly seen at higher densities (Fig. 3). Depending on the density, the prevailing clusters are pairs up to $n \approx 0.2$, quadruples for $0.1 \lesssim n \lesssim 0.3$ etc.. We note that for a large range of $v_l(1,0)$, the system prefers clusters with an even number of particles. Odd particle-number clusters can also form, but have a much narrower parameter range of stability. The preference to certain cluster sizes becomes clearly apparent only at temperatures lower than t_{cl}, and the transition is not abrupt but gradual with decreasing temperature. Similarly, with increasing density changes in textures also indicate a series of crossovers.

The results of the Monte Carlo simulation [17] presented above allow a quite general interpretation in terms of the kinetics of first order phase transitions [35]. Let us assume that a single cluster of ordered phase with radius R appears. As was discussed in [7, 29], the energy of the cluster is determined by three terms: $\epsilon = -F\pi R^2 + \alpha\pi R + \gamma R^3$. The first term is the energy gain due to the ordering phase transition where F is the energy difference between the two minima in the free energy density. The second term is the surface energy parameterized by α, and the third term is the Coulomb energy, parameterized by γ. If $\alpha < \pi F/3\gamma$, ϵ has a well defined minimum at $R = R_0$ corresponding to the optimal size of clusters in the system. Of course, these clusters are also interacting among themselves via Coulomb and strain forces, which leads to

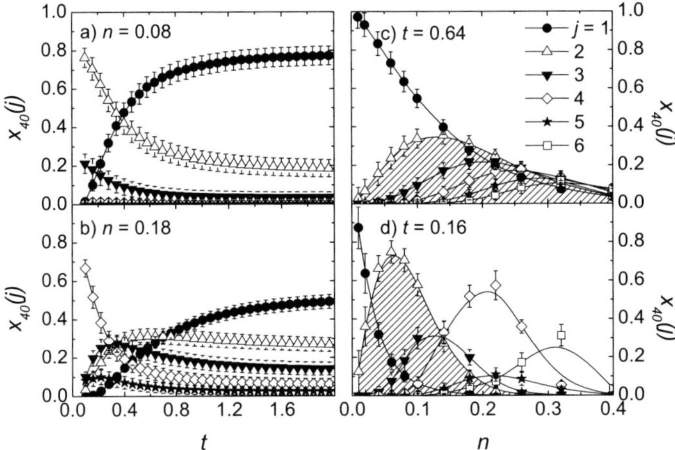

Fig. 5. The temperature dependence of the cluster-size distribution function $x_L(j)$ (for the smallest cluster sizes) at two different average densities $n = 0.08$ (a) and $n = 0.18$ (b). $x_L(j)$ as a function of n at the temperature between t_0 and t_{cl} (c), and near t_{co} (d). The ranges of the density where pairs prevail are very clearly seen in (d). Error bars represent the standard deviation.

cluster ordering or freezing of cluster motion at low temperatures as shown by the Monte Carlo simulations.

We conclude that a model with only anisotropic JT strain and long-range Coulomb interaction is indeed unstable with respect to the short scale phase separation and gives rise to a remarkably rich phase diagram including pairs, stripes and charge- and orbital- ordered phases, of clear relevance to oxides. The energy scale of the phenomena is defined by the parameters used in H_{JT} (23). For example, using the measured value $\epsilon_0 \simeq 40$ [36] for La_2CuO_4, we estimate $V_c(1,0) = 0.1$ eV, which is also the typical energy scale of the "pseudogap" in the cuprates. The robust prevalence of the paired state in a wide region of parameters (Fig. 5c,d) is particularly interesting from the point of view of superconductivity. A similar situation occurs in manganites and other oxides with the onset of a conductive state at the threshold of percolation, but different textures are expected to arise due to different magnitude (and anisotropy) of $V_l(\mathbf{n})$, and static dielectric constant ϵ_0 in the different materials [37].

5 Coulomb Frustrated First Order Phase Transition

As stated in the previous section, uncharged JT polarons have a tendency towards ordering. The ordering transition is a phase transition of the first order. At a fixed density of polarons the system is unstable with respect to the global phase separation. The global phase separation is frustrated by charging effects leading to short-scale phase separation. Therefore the results of the Monte-Carlo simulation of the model (23) allow general model independent interpretation. Let us consider the classical free energy density corresponding to the first order phase transition:

$$F_1 = ((t - 1) + (\eta^2 - 1)^2)\eta^2 \tag{30}$$

Here $t = (T - T_c)/(T_0 - T_c)$ is the dimensionless temperature. At $t = 4/3$ ($T = T_0 + (T_0 - T_c)/3$) the nontrivial minimum in the free energy appears. At $t = 1$ ($T = T_0$) the first order phase transition occurs, but the trivial solution $\eta = 0$ corresponds to the metastable phase. At $t = 0$ ($T = T_c$) the trivial solution becomes unstable. In order to study the case of the Coulomb frustrated phase transition we have to add coupling of the order parameter to the local charge density. Our order parameter describes sublattice magnetization and therefore only the square of the order parameter may be coupled to the local charge density ρ:,

$$F_{coupl} = -\alpha\eta^2\rho \tag{31}$$

The total free energy density should contain the gradient term and the electrostatic energy:

$$F_{grad} + F_{el} = C(\nabla\eta)^2 + \frac{K}{2}[\rho(\mathbf{r}) - \bar\rho]\int d\mathbf{r}'[\rho(\mathbf{r}') - \bar\rho]/|\mathbf{r} - \mathbf{r}'| \tag{32}$$

Here we write $\bar\rho$ explicitly to take into account global electroneutrality. The total free energy (30-32) should be minimized at fixed t and $\bar\rho$.

Let us demonstrate that the Coulomb term leads to phase separation in the 2D case. Minimization of F with respect to the charge density $\rho(\mathbf{r})$ leads to the following equation:

$$-\alpha\nabla_{3D}^2\eta^2 = 4\pi[\rho(\mathbf{r}) - \bar\rho]d\delta(z) \tag{33}$$

here we write explicitly that the density $\rho(\mathbf{r})$ depends only on the 2D vector $\mathbf{r}$ and introduce layer thickness d, to preserve the correct dimensionality. Solving this equation by applying the Fourier transform and substituting the solution back to the free energy density we obtain:

$$F = F_1 + \alpha\eta^2\bar\rho + C(\nabla\eta)^2 - \frac{\alpha^2}{8\pi^2 K}\int d\mathbf{r}'\frac{\nabla(\eta(\mathbf{r})^2)\nabla(\eta(\mathbf{r}')^2)}{|\mathbf{r} - \mathbf{r}'|} \tag{34}$$

As a results the free energy functional is similar to the case of the first order phase transition with shifted critical temperature due to the presence of the term $\alpha\eta^2\bar\rho$ and nonlocal gradient term of higher order.

To demonstrate that the uniform solution has higher energy than the non-homogeneous solution we make a Fourier transform of the gradient term:

$$F_{grad} \propto Ck^2|\eta_{\mathbf{k}}|^2 - \frac{\alpha^2 k|(\eta^2)_{\mathbf{k}}|^2}{4\pi K} \tag{35}$$

here $\eta_\mathbf{k}$ and $(\eta^2)_\mathbf{k}$ are Fourier components of the order parameter and square of the order parameter, respectively. If we assume that the solution is uniform i.e. $\eta_0 \neq 0$ and $(\eta^2)_0 \neq 0$, small nonuniform corrections to the solution reduces the free energy at small $\mathbf{k}$, where the second term dominates.

The proposed free energy functional is similar to that proposed in [16]. The important difference is that in our case the charge is coupled to the square of the order parameter and it plays the role of a local temperature, while in the case of [16] there is linear coupling of the charge to the order parameter. The charge in that case plays the role of the external field. Moreover, contrary to the case of [16] where charge is accumulated near *domain walls*, in our case charge is accumulated near *interphase boundaries*.

6 Conclusion

We have demonstrated that anisotropic interaction between Jahn-Teller centers generated by optical and/or acoustical phonons leads, in the presence of the long range Coulomb repulsion, to short scale phase separation. The topology of texturing differs from charged bubbles to oriented charged stripes depending on the anisotropy of the short range potential. On the phenomenological level, an inhomogeneous phase with charged regions appears due to the tendency of the system of polarons to global phase separation, while this phase separation is frustrated by long-range Coulomb forces. Effectively this system may be described by the standard Landau functional with a nonlocal long-range gradient term.

References

1. T. Egami et al., J Supercond. **13**, 709 (2000); J. Tranquada et al., Nature **375**, 561 (1995).
2. A. Bianconi et al., Phys. Rev. Lett. **76**, 3412 (1996).
3. H. A. Mook and F. Dogan, Nature **401**, 145 (1999)
4. J. Zaanen and O. Gunnarsson, Phys. Rev. B **40**, 7391 (1989)
5. V. J. Emery, S. Kivelson, and O. Zachar, Phys. Rev. B **56** 6120 (1997), and references therein.
6. L. P. Gorkov, A.V. Sokol, Pisma ZhETF, **46**, 333 (1987).
7. L. P. Gorkov, J. Supercond., **14**, 365, (2001).
8. U. Löw, V.J. Emery, K. Fabricious, and S. Kivelson Phys. Rev. Lett. **72**, 1918 (1994).
9. F.V. Kusmartsev, Phys. Rev. Lett., **84**, 530, (2000), *ibid* **84**, 5026 (2000).
10. A. S. Alexandrov and V. V. Kabanov, Pisma ZhETF **72**, 825 (2000) (JETP Lett. **72**, 569 (2000)).
11. F. V. Kusmartsev, J. Phys. IV **9**, 321 (1999).
12. D. I. Khomskii and K. I. Kugel, Europhys. Lett. **55**, 208 (2001); Phys. Rev. B **67**, 134401 (2003).

13. M. B. Eremin, A. Yu. Zavidonov, B. I. Kochelaev, ZhETF **90**, 537 (1986).
14. S. R. Shenoy, T. Lookman, A. Saxena, A. R. Bishop, Phys. Rev. B **60**, R12537 (1999); T. Lookman, S.R. Shenoy, K.O. Rasmussen, A. Saxena, and A. R. Bishop, Phys. Rev. B **67**, 024114 (2003).
15. C. Ortix, J. Lorenzana, and C. Di Castro, Phys. Rev. B **73**, 245117 (2006) and referece therein.
16. R. Jamei, S. Kivelson, and B. Spivak, Phys. Rev. Lett., **94**, 056805 (2005).
17. T. Mertelj, V.V. Kabanov, and D. Mihailovic, Phys. Rev. Lett. **94**, 147003 (2005)
18. T. Holstein, Ann. Phys.(N.Y.)**8**, 325, (1959); **8**, 343, (1959).
19. E. I. Rashba, Opt. Spektrosk. **2**, 78, (1957); **2**, 88, (1957)
20. V. V. Kabanov and O. Yu. Mashtakov, Phys. Rev. B **47**, 6060, (1993).
21. A. S. Alexandrov, V. V. Kabanov, and D. K. Ray, Phys. Rev. B **49**, 9915 (1994).
22. S. I. Pekar, Zh. Eksp. Teor. Fiz. **16**, 335 (1946).
23. A. S. Alexandrov and N. F. Mott, Rep. Progr. Phys. **57**, 1197, (1994); *Polarons and Bipolarons* (World Scientific, Singapore, 1995).
24. S. Aubry, in: 'Polarons and Bipolarons in High-T_c Superconductors and Related Materials', eds E. K. H. Salje, A. S. Alexandrov and W. Y. Liang, Cambridge University Press, Cambridge, 271 (1995).
25. A. R. Bishop and M. Salkola , in: 'Polarons and Bipolarons in High-T_c Superconductors and Related Materials', eds E. K. H. Salje, A. S. Alexandrov and W. Y. Liang, Cambridge University Press, Cambridge, 353 (1995).
26. H. Fehske *et al*, Phys. Rev. B **51**, 16582 (1995).
27. J. Bonca and S. A. Trugman, Phys. Rev. B **64**, 094507 (2001).
28. A. Bianconi, J. Phys. IV France **9**, 325 (1999), and references therein.
29. D. Mihailovic and V. V. Kabanov, Phys. Rev. B **63**, 054505, (2001); Phys. Rev., B **65**, 212508, (2002).
30. A. R. Bishop, T. Lookman, A. Saxena, S. R. Shenoy, Europhys. Lett. **63**, 289, (2003).
31. J. Lajzerovicz, J. Sivardiere, Phys. Rev. A **11**, 2079 (1975).
32. J. Sivardiere, J. Lajzerovicz, Phys. Rev. A **11**, 2090 (1975)
33. N. Metropolis, A. W. Rosenbluth, M. N. Rosenbluth, A.H. Teller and E. Teller, J. Chem. Phys. **21,** 1087 (1953).
34. S. Kirkpatrick, C. D. Gelatt and M. P. Vecchi, Science **220**, 671 (1983).
35. E. M. Lifshitz and L. P. Pitaevski, *Physical Kinetics*, ch.12 (Butterworth-Heinemann, 1980).
36. D. Reagor et al., Phys. Rev. Lett. **62**, 2048 (1989).
37. E. Dagotto, T. Hotta and A. Moreo, Phys. Rep. **344**, 1 (2001); A. S. Alexandrov, A. M. Bratkovsky, and V. V. Kabanov, Phys. Rev. Lett. **96**, 117003 (2006).

Part III

Strongly Correlated Polarons

Numerical Solution of the Holstein Polaron Problem

H. Fehske[1] and S. A. Trugman[2]

[1] Institut für Physik, Ernst-Moritz-Arndt-Universität Greifswald, D-17487 Greifswald, Germany holger.fehske@physik.uni-greifswald.de
[2] Theoretical Division, Los Alamos National Laboratory, Los Alamos, New Mexico 87545, U.S.A. sat@lanl.gov

1 Introduction

Noninteracting itinerant electrons in a solid occupy Bloch one-electron states. Phonons are collective vibrational excitations of the crystal lattice. The basic electron-phonon (EP) interaction process is the absorption or emission of a phonon by the electron with a simultaneous change of the electron state. From this it is clear that the motion of even a single electron in a deformable lattice constitutes a complex many-body problem, in that phonons are excited at various positions, with highly non-trivial dynamical correlations.

The mutual interaction between the charge carrier and the lattice deformations may lead to the formation of a new quasiparticle, an electron dressed by a phonon cloud. This composite entity is called a polaron [1, 2]. Since the induced distortion (polarisation) of the lattice will follow the electron when it is moving through the crystal, one of the most important ground-state properties of the polaron is an increased inertial mass. A polaronic quasiparticle is referred to as a "large polaron" if the spatial extent of the phonon cloud is large compared to the lattice parameter. By contrast, if the lattice deformation is basically confined to a single site, the polaron is designated as "small". Of course, depending on the strength, range and retardation of the electron-phonon interaction, the spectral properties of a polaron will also notably differ from those of a normal band carrier. Since there is only one electron in the problem, these findings are independent of the statistics of the particle, i.e. we can think of any fermion or boson, such as an electron-, hole-, exciton- or Jahn-Teller polarons (for details see [3–5]).

The microscopic structure of polarons is very diverse. The possible situations are determined by the type of particle-phonon coupling [4, 6]. Systems characterised by optical phonons with polar long-range interactions are usually described by the Fröhlich Hamiltonian [7–9]. If the optical phonons have nonpolar short-range EP interactions, Holstein's (molecular crystal) model applies [10, 11]. For a large class of Fröhlich- and Holstein-type models it has been proven that the ground-state energy of a polaron is an analytic function of the EP coupling parameter for all interaction strengths [12–15]. The dimensionality of space here has no qualitative influence. In this sense a (formal) abrupt (nonanalytical) polaron transition does not exist: The standard phase transition concept fails to describe polaron formation. It is, instead, a (possibly rapid) crossover. (We mention parenthetically that in contrast to the ground state, the polaron first excited state may be nonanalytic in the EP coupling.)

The fundamental theoretical question in the context of polaron physics concerns the possibility of a local lattice instability that traps the charge carrier upon increasing the EP coupling [1]. Such trapping is energetically favoured over wide-band Bloch states if the binding energy of the particle exceeds the strain energy required to produce the trap. Since the potential itself depends on the carrier's state, this highly non-linear feedback phenomenon is called "self-trapping" [3, 4, 16, 17]. Self-trapping does not imply a breaking of translational invariance. In a crystal the polaron ground state is still extended allowing, in principle, for coherent transport although with an extremely narrow band. One way to think of this is that a hypothetical self-trapped state can coherently tunnel with its phonon cloud to neighbouring locations, thus delocalising. The problem of self-trapping, i.e. the crossover from rather mobile large polarons to quasi-immobile small polarons, basically could not be addressed within the continuum approach. Self-trapping requires a physics which is related to particle and phonon dynamics on the scale of the unit cell [18]. On the experimental side, an increasing number of advanced materials show polaronic effects on such short length and time scales. Self-trapped polarons can be found, e.g., in (non-stoichiometric) uranium dioxide, alkaline earth halides, II-IV- and group-IV semiconductors, organic molecular crystals, high-T_c cuprates, charge-ordered nickelates and colossal magneto-resistance manganites [6, 19–26].

As stated above, the generic model to capture such a situation is the Holstein Hamiltonian, which is most simply written in real space [10]. Here the orbital states are identical on each site and the particle can move from site to site exactly as in a tight-binding model. The phonons are coupled to the particle at whichever site it is on. The dynamics of the lattice is treated purely locally with Einstein oscillators describing the intra-molecular oscillations.

Theoretical research on the Holstein model spans over five decades. As yet none of the various analytical treatments, based on variational approaches [27, 28] or on weak-coupling [29] and strong-coupling adiabatic [10] and non-adiabatic [30, 31] perturbation expansions, are suitable for the investigation of the physically most interesting crossover regime where the self-trapping crossover of the charge carrier takes place. That is because precisely in this situation the characteristic electronic and phononic energy scales are not well separated and non-adiabatic effects become increasingly important, implying a breakdown of the standard Migdal approximation [32]. The Holstein polaron can be solved in infinite dimensions ($D = \infty$) using dynamical mean-field theory [33, 34]. While this method treats the local dynamics exactly, it cannot account for the spatial correlations being of vital importance in finite-dimensional systems.

In principle, quasi approximation-free numerical methods like exact diagonalisation (ED) [35–39], quantum Monte Carlo (QMC) [40–45] and diagrammatic Monte Carlo [46] simulations, the global-local (GL) method [47] or the recently developed density-matrix renormalisation group (DMRG) technique [48, 49] can overcome all these difficulties. Although most of these methods give reliable results in a wide range of parameters, thereby closing the gap between the weak and strong EP coupling, low- and high-frequency limits, each suffers from different shortcomings. ED is probably the best controlled numerical method for the calculation of ground- and excited state properties. In practice, however, memory limitations have restricted brute force ED to small lattices (typically up to 20 sites). So results are limited to discrete momentum points. QMC can treat large system sizes (over 1000 sites) and provide accurate results for the thermodynamic properties. On the other hand, the calculation of spectral properties is less reliable mainly because of the ill-posed analytic continuation from imaginary time. The GL method is basically limited to the analysis of ground-state properties. DMRG and the recently developed dynamical DMRG [50] have proved to be extremely accurate for the investigation of 1D EP systems, where they can deal with sufficiently large system sizes (e.g., 128 sites and 40 phonons). The determination of spectral functions (in particular of the high-energy incoherent features), however, is computationally expensive and so far there exists no really efficient DMRG algorithm to tackle non-trivial problems in $D > 1$.

In this contribution we provide an exact numerical solution of the Holstein polaron problem by elaborate ED techniques, in the whole range of parameters and, at least concerning the properties of the ground state and low-lying excited states, in the thermodynamic limit. Combining Lanczos ED [51] with kernel polynomial [52, 53] and cluster perturbation [54, 55] expansion methods also allows the polaron's spectral and dynamical properties to be computed exactly. A numerical calculation is said to be exact if no approximations are involved aside from the restriction imposed by finite computational resources, the accuracy can be systematically improved with increasing computational effort, and actual numerical errors are quantifiable or completely negligible. In most numerical approaches to many-body problems, the numerical error decreases as $1/\log(\text{effort})$, where effort means either execution time or storage required. Thus even a large increase in computational power will not greatly improve the accuracy. Despite some progress by virtue of DMRG-based basis optimisation [56] or coherent-state variational approaches [57, 58], ED of EP systems remained inefficient. Recently ED methods have been developed that converge far more rapidly, with error $\sim 1/(\text{effort})^{\theta}$, where θ is a favourable power ($\theta \approx 3$ at intermediate coupling) [59]. Thus doubling the size of the Hilbert space results in almost an extra significant figure in the energy. The algorithm [60–62] we will apply in the following is based on the construction of a variational Hilbert space on a infinite lattice and can be expanded in a systematic way to easily achieve greater than 10-digit accuracy for static correlation functions and 20 digits for energies, with modest computational resources. The increased power makes it possible to solve the Holstein polaron problem at continuous wave vectors in dimensions D=1, 2, 3, 4, ...

The paper is organised as follows: In the remaining introductory part, Sect. 2 presents the Holstein model and outlines the numerical methods we will employ for its solution. The second, main part of this paper reviews our numerical results for the ground-state and spectral properties of the Holstein polaron. The polaron's effective mass and band structure, as well as static electron-lattice correlations, will be analysed in Sect. 3. Section 4 is devoted to the investigation of the excited states of the Holstein model. The dynamics of polaron formation is studied in Sect. 5. Characteristic results for electron and phonon spectral functions will be presented in Sect. 6. The optical response is examined in Sect. 7. Here also finite-temperature properties such as activated transport will be discussed. In the third part of this paper finite-density and correlation effects will be addressed. First we investigate the possibility of bipolaron formation and discuss the many-polaron problem (Sect. 8). Second we comment on the interplay of strong electronic correlations and EP interaction in advanced materials (Sect. 9). Some open problems are listed in the concluding Sect. 10.

2 Model and Methods

2.1 Holstein Hamiltonian

With our focus on polaron formation in systems with short-range non-polar EP interaction only, we consider the Holstein molecular crystal model on a D-dimensional hyper-cubic lattice,

$$H = -t \sum_{\langle i,j \rangle} c_i^\dagger c_j - \bar{g} \sum_i (b_i^\dagger + b_i) n_i + \omega_0 \sum_i b_i^\dagger b_i \,, \qquad (1)$$

where $c_i^\dagger$ (c_i) and $b_i^\dagger$ (b_i) are, respectively, creation (annihilation) operators for electrons and dispersionless optical phonons on site i, and $n_i = c_i^\dagger c_i$ is the corresponding particle number operator. The parameters of the model are the nearest-neighbour hopping integral t, the EP coupling strength $\bar{g}$, and the phonon frequency ω_0. The parameters t, ω_0, and $\bar{g}$ all have units of energy, and can be used to form two independent dimensionless ratios.

The first ratio is the so-called adiabaticity parameter,

$$\alpha = \omega_0/t \,, \qquad (2)$$

which determines which of the two subsystems, electrons or phonons, is the fast or the slow one. In the adiabatic limit $\alpha \ll 1$, the motion of the particle is affected by quasi-static lattice deformations (adiabatic potential surface). In contrast, in the anti-adiabatic limit $\alpha \gg 1$, the lattice deformation is presumed to adjust instantaneously to the position of the carrier. The particle is referred to as a "light" or "heavy" polaron in the adiabatic or anti-adiabatic regimes [4].

The second ratio is the dimensionless EP coupling constant. Here

$$g = \bar{g}/\omega_0 \tag{3}$$

appears in (small polaron) strong-coupling perturbation theory. Defining the polaron binding energy as $\varepsilon_p = \bar{g}^2/\omega_0 = g^2\omega_0$, the EP coupling can be parametrised alternatively as the ratio of polaron energy for an electron confined to a single site and the free electron half bandwidth $2Dt$:

$$\lambda = \varepsilon_p/2Dt\,. \tag{4}$$

In the limit of small particle density, a crossover from essentially free carriers to heavy quasiparticles is known to occur from early quantum Monte Carlo calculations [63], provided that two conditions, $g > 1$ and $\lambda > 1$, are fulfilled. So, while the first requirement is more restrictive if α is large, i.e. in the anti-adiabatic case, the formation of a small polaron state will be determined by the second criterion in the adiabatic regime [64, 65].

Perhaps it is not surprising that standard perturbative techniques are less able to describe the Holstein system close to the large- to small-polaron crossover, where $\varepsilon_p \sim 2Dt$ or $\varepsilon_p \sim \omega_0$. In principle, variational approaches, that give correct results in the weak- and strong-coupling limits, could provide an interpolation scheme. Most variational calculations, however, lead to a discontinuous transition in the wave function and the derivative of the ground-state energy, considered as a function of the coupling parameter. Clearly the analytical behaviour of an exact wave function may deviate considerably from that of a variational approximation [15]. With regard to the Holstein polaron problem the nonanalytic behaviour found for the adapted wave function in many variational approaches, see, e.g., [66] and references therein, is an artifact of the approximations, as we will demonstrate below for all dimensions [61].

Nevertheless, variational calculations are an indispensable tool for numerical work. In the next subsection we describe a variational exact diagonalisation (VED) scheme [60] that does not suffer from the above drawback of (ground-state) non-analyticities at the small-polaron transition. Above all, in contrast to finite-lattice ED, it yields a ground-state energy which is a variational bound for the exact energy in the thermodynamic limit. As yet the VED technique is fully worked out for the single polaron and bipolaron problem only. At finite particle densities the construction of the variational Hilbert space becomes delicate. On this account we will also outline some more general (robust) ED schemes, which can be applied for the calculation of ground-state and spectral properties of a larger class of strongly correlated EP systems.

2.2 Numerical Techniques

Hilbert Space and Basis Construction

The total Hilbert space of the Holstein model (1) can be written as the tensorial product space of electrons and phonons, spanned by the complete basis set $\{|b\rangle = |e\rangle \otimes |p\rangle\}$ with

$$|e\rangle = \prod_{i=1}^{N} \prod_{\sigma=\uparrow,\downarrow} (c_{i\sigma}^{\dagger})^{n_{i\sigma,e}} |0\rangle_e \quad \text{and} \quad |p\rangle = \prod_{i=1}^{N} \frac{1}{\sqrt{m_{i,p}!}} (b_i^{\dagger})^{m_{i,p}} |0\rangle_p . \tag{5}$$

Here $n_{i\sigma,e} \in \{0,1\}$, i.e. with respect to the electrons Wannier site i might be empty, singly or doubly occupied, whereas we have no such restriction for the phonon number, $m_{i,p} \in \{0, \ldots, \infty\}$. Consequently, $e = 1, \ldots, D_e$ and $p = 1, \ldots, D_p$ label basic states of the electronic and phononic subspaces having dimensions $D_e = \binom{N}{N_{e,\sigma}}\binom{N}{N_{e,-\sigma}}$ and $D_p = \infty$, respectively. $|0\rangle_{e/p}$ denote the corresponding vacua. This also holds including electron-electron (e.g. Hubbard-type) interaction terms [67]. For Holstein-t-J-type models, acting in a projected Hilbert space without doubly occupied sites, we have $D_e = \binom{N}{N_{e,\sigma}}\binom{N-N_{e,\sigma}}{N_{e,-\sigma}}$ only [39]. Since these model Hamiltonians commute with the total electron number operator $\widehat{N}_e = \sum_{\sigma} \widehat{N}_{e,\sigma}$, where $\widehat{N}_{e,\sigma} = \sum_{i=1}^{N} n_{i,\sigma}$, and the z-component of the total spin $S^z = \frac{1}{2} \sum_{i=1}^{N} (n_{i,\uparrow} - n_{i,\downarrow})$, the many-particle basis $\{|b\rangle\}$ can be constructed for fixed N_e and S^z.

Variational Approach

Let us first describe an efficient variational exact diagonalisation (VED) method to solve the Holstein model in the single-particle subspace. For generalisation of this method to the case of two particles (bipolaron) see [68].

A variational Hilbert space is constructed beginning with an initial root state, taken to be an electron at the origin with no phonon excitations, and acting repeatedly with the hopping (t) and EP coupling ($\bar{g}$) terms of the Holstein Hamiltonian (see Fig. 1). States in generation L are those obtained by acting L times with these "off-diagonal" terms. Only one copy of each state is retained. Importantly, all translations of these states on an infinite lattice are included. A translation moves the electron and all phonons j sites to the right. Then, according to Bloch's theorem, each eigenstate can be written as $\psi = e^{ikj} a_L$, where a_L is a set of complex amplitudes related to the states in the unit cell j, e.g. $L = 7$ for the small variational space shown in Fig. 1. For each momentum k the resulting numerical problem is then to diagonalise a Hermitian $L \times L$ matrix. The total number of states per unit cell (N_{st}) after L generations increases exponentially as $(D + 1)^L$ (note that the bipolaron has the same exponential dependence with only a larger prefactor). Most notably the error in the ground-state energy E_0 decreases exponentially, because states are added in a fairly efficient order. Thus in most cases $10^4 - 10^6$ basis

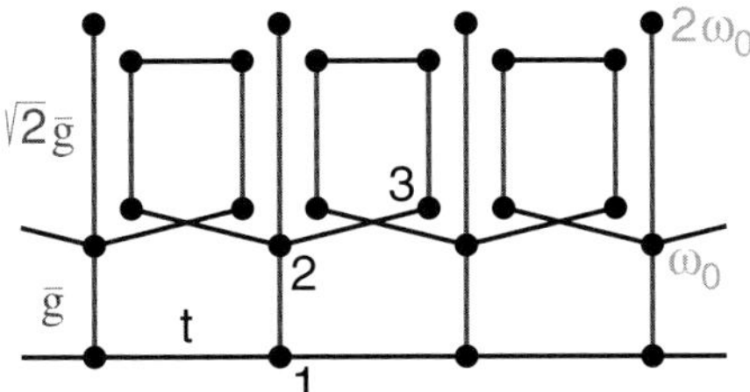

Fig. 1. Variational Hilbert space construction for the 1D polaron problem. Basis states are represented by dots, off-diagonal matrix elements by lines. Vertical bonds create or destroy phonons with frequency ω_0. Horizontal bonds correspond to electron hops ($\propto t$). Accordingly, state $|1\rangle$ describes an electron at the origin (0) and no phonon, state $|2\rangle$ is an electron and one phonon both at site 0, $|3\rangle$ is an electron at the nearest-neighbour site 1, and a phonon at site 0, and so on [60].

states are sufficient to obtain an 8-16 digit accuracy for E_0 (see Fig. 2). The ground-state energy calculated this way is variational for the infinite system.

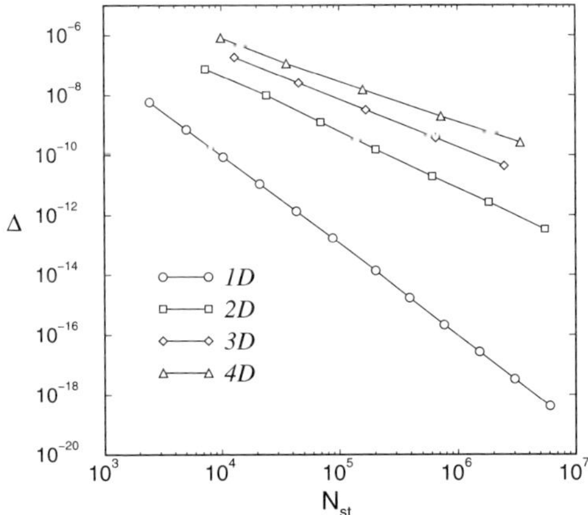

Fig. 2. Fractional error Δ in the ground-state energy of a D-dimensional polaron as a function of the number of basis states N_{st} retained. Parameters are $\lambda = 0.5$, $g = 1$, and $t = 1$. Figure is taken from [61], ©(2002) by the American Physical Society

Symmetrisation and Phonon Truncation

Treating more complex many-particle Hamilton operators on finite lattices, the dimension of the total Hilbert space can also be reduced. To this end we can exploit the space group symmetries [translations (G_T) and point group operations (G_L)] and the spin rotational invariance [(G_S); $S^z = 0$ subspace only]. Working, e.g., on finite 1D or 2D bipartite clusters with periodic boundary conditions (PBC), we do not have all the symmetry properties of the underlying 1D or 2D (square) lattices [39]. Restricting ourselves to the 1D non-equivalent irreducible representations of the group $G(\boldsymbol{K}) = G_T \times G_L(\boldsymbol{K}) \times G_S$, we can use the projection operator $\mathcal{P}_{\boldsymbol{K},rs} = [g(\boldsymbol{K})]^{-1} \sum_{\mathcal{G} \in G(\boldsymbol{K})} \chi_{\boldsymbol{K},rs}^{(\mathcal{G})} \, \mathcal{G}$ (with $[H, \mathcal{P}_{\boldsymbol{K},rs}] = 0$, $\mathcal{P}_{\boldsymbol{K},rs}^{\dagger} = \mathcal{P}_{\boldsymbol{K},rs}$ and $\mathcal{P}_{\boldsymbol{K},rs} \, \mathcal{P}_{\boldsymbol{K}',r's'} = \mathcal{P}_{\boldsymbol{K},rs} \, \delta_{\boldsymbol{K},\boldsymbol{K}'} \, \delta_{r,r'} \, \delta_{s,s'}$) in order to generate a new symmetrised basis set: $\{|b\rangle\} \xrightarrow{\mathcal{P}} \{|\tilde{b}\rangle\}$. $\mathcal{G}$ denotes the $g(\boldsymbol{K})$ elements of the group $G(\boldsymbol{K})$ and $\chi_{\boldsymbol{K},rs}^{(\mathcal{G})}$ is the (complex) character of $\mathcal{G}$ in the $[\boldsymbol{K},rs]$–representation, where $\boldsymbol{K}$ refers to one of the N allowed wave vectors in the first Brillouin zone, r labels the irreducible representations of the little group of $\boldsymbol{K}$, $G_L(\boldsymbol{K})$, and s parameterises G_S. For an efficient parallel implementation of the matrix vector multiplication (see below) it is extremely important that the symmetrised basis can be constructed preserving the tensor product structure of the Hilbert space, i.e.,

$$\{|\tilde{b}\rangle = N_{\tilde{b}}^{[\boldsymbol{K}rs]} \, \mathcal{P}_{\boldsymbol{K},rs} \, [|\tilde{e}\rangle \otimes |p\rangle]\} \tag{6}$$

with $\tilde{e} = 1, \ldots, \tilde{D}_e^{g(\boldsymbol{K})}$ $[\tilde{D}_e^{g(\boldsymbol{K})} \sim D_e/g(\boldsymbol{K})]$. The $N_{\tilde{b}}^{[\boldsymbol{K}rs]}$ are normalisation factors.

Since the Hilbert space associated to the phonons is infinite even for a finite system, we use a truncation procedure [38] retaining only basis states with at most M phonons:

$$\{|p\rangle \; ; \; m_p = \sum_{i=1}^{N} m_{i,p} \leq M\}. \tag{7}$$

Then the resulting Hilbert space has a total dimension $\tilde{D} = \tilde{D}_e^{g(\boldsymbol{K})} \times D_p^M$ with $D_p^M = (M+N)!/M!N!$, and a general state of the Holstein model is represented as

$$|\psi_{\boldsymbol{K},rs}\rangle = \sum_{\tilde{e}=1}^{\tilde{D}_e^{g(\boldsymbol{K})}} \sum_{p=1}^{D_p^M} c_{\tilde{e}p}^{\psi} \, |\tilde{b}\rangle. \tag{8}$$

The computational requirements can be further reduced if one separates the symmetric phonon mode, $B_0 = \frac{1}{\sqrt{N}} \sum_i b_i$, and calculates its contribution to H analytically [69].

Note that switching from a real space representation to a momentum space description the truncation scheme takes into account all dynamical phonon modes, which has to be contrasted with the frequently used single-mode approach [70]. In other words, depending on the model parameters and the band filling, the system "decides" by itself how the M phonons will be distributed among the independent Einstein oscillators related to the N Wannier sites or, alternatively, among the different phonon modes in $\boldsymbol{Q}$-space. Hence with the same accuracy phonon dynamical effects on lattice distortions being quasi-localised in real space (such as polarons, Frenkel excitons,. . .) or in momentum space (like charge-density-waves,. . .) can be studied.

Of course, one has to check carefully for the convergence of the above truncation procedure by calculating the ground-state energy as a function of the cut-off parameter M. In the numerical work below convergence is assumed to be achieved if E_0 is determined with a relative error less than 10^{-6}.

Phonon Basis Optimisation

In this section we outline an advanced phonon optimisation procedure based on controlled density-matrix basis truncation [56]. The method provides a natural way to dress the particles with phonons which allows the use of a very small optimal basis without significant loss of accuracy.

Starting with an arbitrary normalised quantum state,

$$|\psi\rangle = \sum_{e=0}^{D_e-1} \sum_{p=0}^{D_p-1} c_{ep}^{\psi}[|e\rangle \otimes |p\rangle] , \qquad (9)$$

expressed in terms of the basis of the direct product space, we wish to reduce the dimension D_p of the phonon space H_p by introducing a new basis,

$$|\widetilde{p}\rangle = \sum_{p=0}^{D_p-1} \alpha_{\widetilde{p}p}|p\rangle , \qquad (10)$$

with $\widetilde{p} = 0, \ldots, (D_{\widetilde{p}} - 1)$ and $D_{\widetilde{p}} < D_p$. We call $\{|\widetilde{p}\rangle\}$ an optimised basis, if the projection of $|\psi\rangle$ on the corresponding subspace $\widetilde{H} = H_e \otimes H_{\widetilde{p}} \subset H$ is as close as possible to the original state. Therefore we minimise

$$\||\psi\rangle - |\widetilde{\psi}\rangle\|^2 = 1 - \mathrm{Tr}\,(\alpha\rho\alpha^\dagger) \qquad (11)$$

with respect to the $\alpha_{\widetilde{p}p}$ under the condition $\langle \widetilde{p}'|\widetilde{p}\rangle = \delta_{\widetilde{p}'\widetilde{p}}$, where

$$|\widetilde{\psi}\rangle = \sum_{e=0}^{D_e-1} \sum_{\widetilde{p}=0}^{D_{\widetilde{p}}-1} \sum_{p,p'=0}^{D_p-1} \alpha_{\widetilde{p}p}\alpha_{\widetilde{p}p'}^{*} c_{p'e}^{\psi}[|e\rangle \otimes |p\rangle] \qquad (12)$$

is the projected state. $\rho = \sum_{e=0}^{D_e-1}(c_{ep}^{\psi})^* c_{ep}^{\psi}$ is called the density matrix of the state $|\psi\rangle$. Clearly the states $\{|\widetilde{p}\rangle\}$ are optimal if they are elements of the eigenspace of ρ corresponding to its $D_{\widetilde{p}}$ largest eigenvalues $w_{\widetilde{p}}$. If we interpret $w_{\widetilde{p}} \sim \exp(-a\widetilde{p})$ as the probability of the system to occupy the corresponding optimised state $|\widetilde{p}\rangle$, we immediately find that the probability for the complete phonon basis state $\otimes_{i=0}^{N-1}|\widetilde{p}_i\rangle$ is proportional to $\exp(-a\sum_{i=0}^{N-1}\widetilde{p}_i)$. This is reminiscent of an energy cut-off, and we therefore propose the following choice of a mixed phonon basis $\{|\mu_i\rangle\}$ at each site,

$$\forall\, i: \{|\mu_i\rangle\} = \mathrm{ON}(\{|\mu\rangle\}) \qquad (13)$$

$$|\mu\rangle = \begin{cases} \text{optimal state} & |\widetilde{p}\rangle,\ 0 \le \mu < M_{\mathrm{opt}} \\ \text{bare state} & |p\rangle,\ M_{\mathrm{opt}} \le \mu < M \end{cases}, \qquad (14)$$

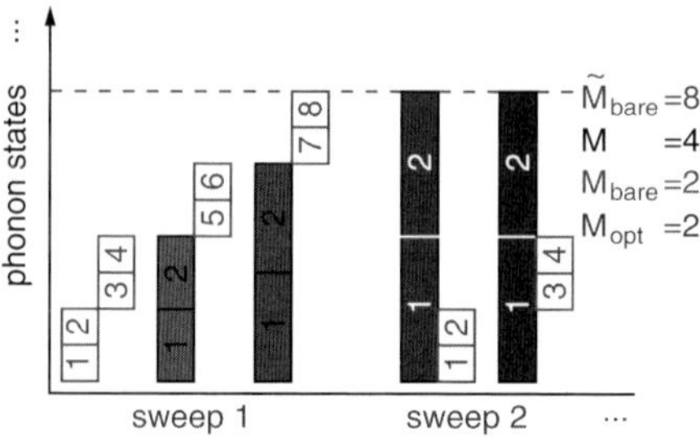

Fig. 3. Sweep technique in constructing optimised phonon states.

and for the complete phonon basis $\{\otimes_{\Sigma_i \mu_i < M} |\mu_i\rangle\}$, yielding $D_{\mathrm{ph}} = \binom{N+M-1}{N}$.

After a first initialisation the optimised states are improved iteratively through the following steps: (1) calculate the requested eigenstate $|\psi\rangle$ of the Hamiltonian H in terms of the actual basis, (2) replace $\{|\widetilde{p}\rangle\}$ with the most important (i.e. largest eigenvalues $w_{\widetilde{p}}$) eigenstates of the density matrix ρ; (3) change the additional states $\{|p\rangle\}$ in the set $\{|\mu\rangle\}$; (4) orthonormalise the set $\{|\mu\rangle\}$, and return to step (1).

A simple way to proceed in step (3) is to sweep the bare states $\{|p\rangle\}$ through a sufficiently large part of the infinite dimensional phonon Hilbert space. One can think of the algorithm as "feeding" the optimised states with bare phonons, thus allowing the optimised states to become increasingly perfect linear combinations of bare phonon states (see Fig. 3).

Solution of the Eigenvalue Problem

In all the above cases the numerical problem that remains is to find the eigenstates of a (sparse) Hermitian matrix. Here iterative (Krylov) subspace methods like Lanczos [51] and variants of Davidson [71] diagonalisation techniques are frequently applied. These algorithms contain basically three steps: (1) project the problem matrix $A \in R^n$ onto a subspace $\bar{A}^k \in V^k$ ($k \ll n$); (2) solve the eigenvalue problem in V^k using standard routines; (3) extend the subspace $V^k \to V^{k+1}$ by a vector $t \perp V^k$ and go back to (2). This way we obtain a sequence of approximate inverses of the original matrix A. A powerful and widely used technique is the Lanczos algorithm which recursively generates a set of orthogonal states (Lanczos vectors):

$$|\varphi_{l+1}\rangle = \mathrm{H}^{\tilde{D}}|\varphi_l\rangle - a_l|\varphi_l\rangle - b_l^2|\varphi_{l-1}\rangle, \qquad (15)$$

where

$$a_l = \frac{\langle\varphi_l|\mathrm{H}^{\tilde{D}}|\varphi_l\rangle}{\langle\varphi_l|\varphi_l\rangle}, \qquad b_l^2 = \frac{\langle\varphi_l|\varphi_l\rangle}{\langle\varphi_{l-1}|\varphi_{l-1}\rangle}, \qquad b_0^2 = 0, \qquad (16)$$

and $|\varphi_{-1}\rangle = 0$. Obviously, the representation matrix $[\mathrm{T}^L]_{l,l'} = \langle\varphi_l|\mathrm{H}^{\tilde{D}}|\varphi_{l'}\rangle$ of $\mathrm{H}^{\tilde{D}}$ is tridiagonal in the L-dimensional Hilbert space spanned by the

$\{|\varphi_l\rangle\}_{l=0,\ldots,L-1}$, where $L \ll \widetilde{D}$. The eigenvalue spectrum of T^L can be easily determined using standard routines from libraries such as EISPACK (see http://www.netlib.org). Note that the convergence of the Lanczos algorithm is excellent at the edges of the spectrum (the ground state for example is obtained with high precision after at most ~ 100 Lanczos iterations) but rapidly worsens inside the spectrum.

In general the computational requirements of these eigenvalue algorithms are determined by matrix-vector multiplications (MVM), which have to be implemented in a parallel, fast and memory saving way on modern supercomputers. The MVM step can be be done in parallel by using a parallel library such as PETSc (see http://www-unix.mcs.anl.gov/petsc/petsc-as/).

Our matrices are extremely sparse because the number of non-zero entries per row of our Hamilton matrix scales linearly with the number of electrons. Therefore a standard implementation of the MVM step uses a sparse storage format for the matrix, holding the non-zero elements only. The typical storage requirement per non-zero entry is 12-16 Byte, i.e. for a matrix dimension of $\widetilde{D} = 10^9$ about one TByte main memory is required to store only the matrix elements of the EP Hamiltonian. To extend our EP studies to even larger matrix sizes we no longer store the non-zero matrix elements but generate them in each MVM step. Of course, at that point standard libraries are no longer useful and a parallel code tailored to each specific class of Hamiltonians must be developed. Clearly the parallelisation approach follows the inherent natural parallelism of the Hilbert space. Assuming that the electronic dimension $(\widetilde{D}_e)$ is a multiple of the number of processors used we can easily distribute the electronic basis states among the processors. As a consequence of this choice only the electronic hopping term generates inter-processor communication in the MVM while all other (diagonal electronic) contributions can be computed locally on each processor. Using supercomputers with hundreds of processors and one TByte of main memory, such as the IBM p690 cluster, at the moment, one is able to run simulations up to a matrix dimension of 3×10^{10}.

Determination of Dynamical Correlation Functions

The numerical calculation of dynamical spectral functions,

$$A^{\mathcal{O}}(\omega) = -\lim_{\eta \to 0^+} \frac{1}{\pi} \mathrm{Im} \left[\langle \psi_0 | \mathrm{O}^\dagger \frac{1}{\omega - \mathrm{H} + E_0 + i\eta} \mathrm{O} | \psi_0 \rangle \right]$$

$$= \sum_{n=0}^{\widetilde{D}-1} |\langle \psi_n | \mathrm{O} | \psi_0 \rangle|^2 \delta[\omega - (E_n - E_0)] , \tag{17}$$

where O is the matrix representation of a certain operator $\mathcal{O}$ [e.g., the creation operator $c_{\boldsymbol{K}}^\dagger$ of an electron with wavevector $\boldsymbol{K}$ if one wants to calculate the single-particle spectral function; or the current operator $\widehat{\jmath} = -iet \sum_i (c_i^\dagger c_{i+1} - c_{i+1}^\dagger c_i)$ if one is interested in the optical conductivity], involves the resolvent of the Hamilton matrix H. Once we have obtained the eigenvalues and eigenvectors of H we can plug them into (17) and directly obtain the corresponding dynamical correlation or Green functions. For the typical EP problems under investigation we deal with Hilbert spaces having total dimensions $\widetilde{D}$ of 10^6 - 10^{11}. Finding all eigenvectors and eigenstates of such huge Hamilton matrices is impossible, because the CPU time required for exact diagonalisation of H scales as $\widetilde{D}^3$ and memory as $\widetilde{D}^2$. So in practice this "naive" approach is applicable for small Hilbert spaces only, where the complete diagonalisation of the Hamilton matrix is feasible. Fortunately, there exist very accurate and well-conditioned linear scaling algorithms for a direct approximate calculation of $A^{\mathcal{O}}(\omega)$.

Kernel Polynomial Method (KPM)

The idea of the KPM (for a review see [53]), is to expand $A^{\mathcal{O}}(\omega)$ in a finite series of $L + 1$ Chebyshev polynomials $T_m(x) = \cos[m \arccos(x)]$. Since the Chebyshev polynomials are defined on the real interval $[-1, 1]$, first a simple linear transformation to the Hamiltonian and all energy scales has to be applied: $X = (H - b)/a$, $x = (\omega - b)/a$, $a = (E_{max} - E_{min})/2(1 - \epsilon)$, and $b = (E_{max} + E_{min})/2$ (the small constant ϵ is introduced in order to avoid convergence problems at the endpoints of the interval – a typical choice is $\epsilon \sim 0.01$ which has only 1% impact on the energy resolution [72]). Then the expansion is

$$A^{\mathcal{O}}(x) = \frac{1}{\pi\sqrt{1 - x^2}} \left(\mu_0^{\mathcal{O}} + 2 \sum_{m=1}^{L} \mu_m^{\mathcal{O}} T_m(x) \right) , \tag{18}$$

with moments

$$\mu_m^{\mathcal{O}} = \int_{-1}^{1} dx\, T_m(x) A^{\mathcal{O}}(x) = \langle \psi_0 | O^\dagger T_m(X) O | \psi_0 \rangle . \tag{19}$$

(18) converges to the correct function for $L \to \infty$. The moments

$$\mu_{2m}^{\mathcal{O}} = 2\langle \phi_m | \phi_m \rangle - \mu_0^{\mathcal{O}} \quad \text{and} \quad \mu_{2m+1}^{\mathcal{O}} = 2\langle \phi_{m+1} | \phi_m \rangle - \mu_1^{\mathcal{O}} \tag{20}$$

can be efficiently obtained by repeated parallelised MVM, where $|\phi_{m+1}\rangle = 2X|\phi_m\rangle - |\phi_{m-1}\rangle$ but now $|\phi_1\rangle = X|O\phi_0\rangle$ with $|\phi_0\rangle \equiv |\psi_0\rangle$ determined by Lanczos ED.

As is well known from Fourier expansion, the series (18) with L finite suffers from rapid oscillations (Gibbs phenomenon) leading to a poor approximation to $A^{\mathcal{O}}(\omega)$. To improve the approximation the moments μ_n are modified $\mu_n \to g_n \mu_n$, where the damping factors g_n are chosen to give the "best" approximation for a given L. In more abstract terms this truncation of the infinite series to order L together with the corresponding modification of the coefficients is equivalent to a convolution of the spectral function with a smooth approximation kernel $K_L(x, y)$. The appropriate (optimal) choice of this kernel, that is of g_n, e.g. to guarantee positivity of $A^{\mathcal{O}}(\omega)$, lies at the heart of KPM. We mainly use the Jackson kernel which results in a uniform approximation whose resolution increases as $1/L$, but for the determination of the single-particle Green functions below we use a Lorentz kernel which mimics a finite imaginary part η in (17) [53].

In view of the uniform convergence of the expansion, the accuracy of the spectral functions can be made as good as required by just increasing L.

Cluster Perturbation Theory (CPT)

The spectrum of a finite system of N sites which we obtain through KPM differs in many respects from that in the thermodynamic limit $N \to \infty$, especially in that it is obtained for a finite number of momenta $K = \pi\, m/N$ only. While we cannot easily increase N without going beyond computationally accessible Hilbert spaces we can try to extrapolate from a finite to the infinite system.

For this purpose we first calculate the Green function $G^c_{ij}(\omega)$ for all sites $i, j = 1, \ldots, N_c$ of an N_c – size cluster with open boundary conditions, and then recover the infinite lattice by pasting identical copies of this cluster at their edges [54]. The "glue" is the hopping V between these clusters, where $V_{kl} = t$ for $|k - l| = 1$ and $k, l \equiv 0, 1 \,(\mathrm{mod}N)$, which is dealt with in first order perturbation theory. Then the Green function $G_{ij}(\omega)$ of the infinite lattice is given through a Dyson equation $G_{ij}(\omega) = G^c_{ij}(\omega) + \sum_{kl} G^c_{ik}(\omega)V_{kl}G_{lj}(\omega)$, where indices of $G^c(\omega)$ are counted modulo N_c. The Dyson equation is solved by Fourier transformation over momenta $K = kN_c$ corresponding to translations by N_c sites

$$G_{ij}(K, \omega) = \left[\frac{G^c(\omega)}{1 - V(K)G^c(\omega)} \right]_{ij}, \tag{21}$$

from which one finally obtains

$$G(k, \omega) = \frac{1}{N_c} \sum_{i,j=1}^{N_c} G^c_{ij}(N_c k, \omega) \exp(-\mathrm{i}k(i - j)) . \tag{22}$$

In this way we obtain a Green function $G(k, \omega)$ with continuous momenta k from the cluster Green function G^c. Two approximations are made, one by using first order perturbation theory in $V = t$, the second on assuming translational symmetry in $G_{ij}(\omega)$ which is only approximatively satisfied.

3 Ground State Results

The VED method can compute polaron properties in dimensions 1 through 4 and higher. The energies for 1D to 4D polarons at $\boldsymbol{k} = 0$ for intermediate

Table 1. Polaron ground state energies at $k = 0$ in 1D - 4D for $\lambda = 0.5$ and $g = 1.0$ In the following $t = 1$ unless specifically noted.

	1D	2D	3D	4D
E_0/t	-2.46968472393287071561	-4.814735778337	-7.1623948409	-9.513174069

to weak coupling on linear, square, cubic, and hypercubic lattices are listed in Table 1. The number of significant digits is determined by comparing the energy as the size of the Hilbert space is increased. The accuracy is high compared to that of other numerical methods, even when limited by the single-processor desktop workstations of several years ago, on which the results were obtained [61]. Correlation functions are generally less accurate than energies.

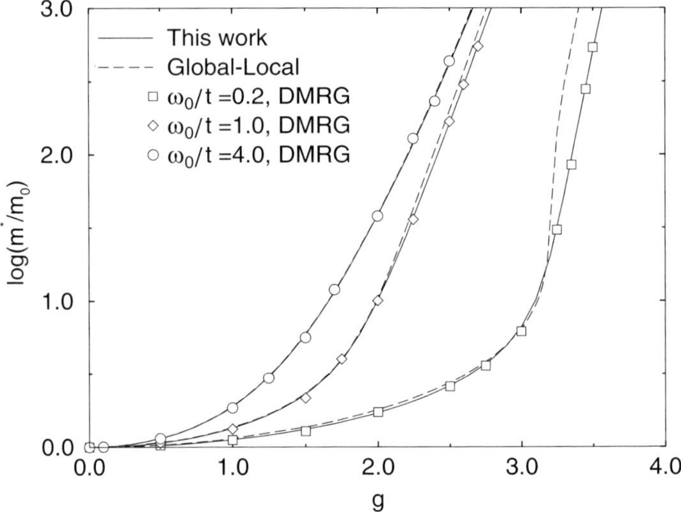

Fig. 4. Logarithm of the polaron effective mass in 1D as a function of g. VED results (full lines) were obtained operating repeatedly $L = 20$ times with the off-diagonal pieces of the Holstein Hamiltonian (cf. Sect. 2.2). For comparison GL (global-local) results (dashed lines) are included [47]. Open symbols, indicating the value of ω_0/t, are DMRG results [48].

Figure 4 shows the effective mass computed by VED [60] in comparison with GL and DMRG methods. m^* is obtained from

$$\frac{m_0}{m^*} = \frac{1}{2t} \frac{\partial^2 E_k}{\partial k^2}\bigg|_{k=0} , \tag{23}$$

where $m_0 = 1/(2t)$ is the effective mass of a free electron. The second derivative is evaluated by small finite differences in the neighbourhood of $\mathbf{k} = 0$. Note that although the calculated energy $E_{\mathbf{k}}$ is a variational bound for the exact energy, there is no such control on the mass, which may be either above or below the exact answer, and is expected to be more difficult to obtain accurately. Nevertheless, in the intermediate coupling regime where VED at $L = 20$ gives an energy accuracy of 12 decimal places, one can calculate the effective mass accurately (6-8 decimal places) by letting $\Delta k \to 0$.

In Fig. 4 the parameters span different physical regimes including weak and strong coupling, and low and high phonon frequency. We find good agreement between VED and GL away from strong coupling and excellent agreement in all regimes with DMRG results. DMRG calculations are not based on finite-k calculations due to a lack of periodic boundary conditions, so they extrapolate the effective mass from the ground state data using chains of different sizes, which leads to larger error bars and demands more computational effort. Notice that their discrete data is slightly scattered around the VED curves. Nevertheless, both methods agree well. We have compared the VED results for effective mass obtained on different systems from $L = 16$ with $N_{st} = 178617$ states to $L = 20$ with $N_{st} = 2975104$ states and obtained convergence of results to at least 4 decimal places in all parameter regimes presented in Fig. 4. Our error is therefore well below the linewidth. Even though there is no phase transition in the ground state of the model, the polaron becomes extremely heavy in the strong coupling regime. The crossover to a regime of large polaron mass is more rapid in the adiabatic regime, i.e. at small ω_0/t.

The polaron effective mass in higher dimensions is shown as a function of the EP coupling in Fig. 5. The mass increases exponentially for large λ. The crossover to larger effective mass is more rapid, though still continuous, in higher dimensions.

Of course it is of special interest to understand the evolution of the polaronic band structure as the EP coupling strength increases. Figure 6 presents the results for the wavevector dependence of the ground-state energy E_k in 1D at weak (a) and strong (b) EP couplings. As might be expected, for $\lambda = 0.25$ the coherent bandwidth, $\Delta E = \sup_k E_k - \inf_k E_k$, is approximately given by the phonon energy ($\Delta E = 0.782 \sim \omega_0 = 0.8$). If the EP interaction is enhanced a band collapse appears. Note, however, that even in the relatively strong EP coupling regime displayed in Fig. 6 (b) the standard Lang-Firsov formula, $\Delta E_{LF} = 4Dt \exp[-g^2]$ (obtained by performing the Lang-Firsov canonical transformation [30] and taking the expectation value of the kinetic energy over the transformed phonon vacuum), does not give a satisfactory estimate of the bandwidth. So we found $\Delta E_{LF} = 0.0269$ which has to be contrasted with the exact result $\Delta E = 0.0579$. Besides the band narrowing effect, there are several other features worth mentioning for polaronic band states in the crossover region. Most notably, throughout the whole Brillouin zone the band structure differs significantly from that of a rescaled tight-binding (cosine) band containing only nearest-neighbour hopping [65]. Obvi-

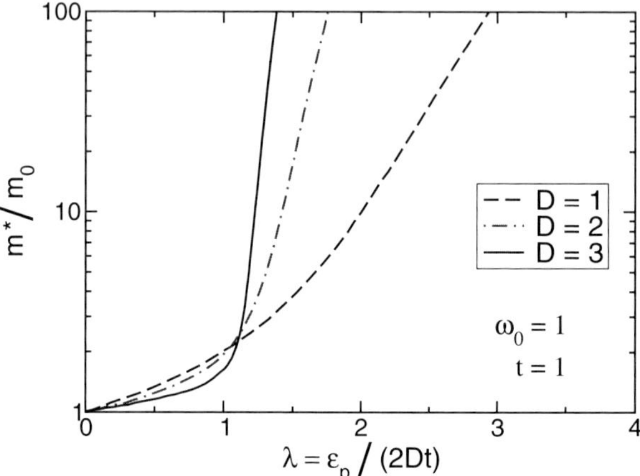

Fig. 5. Effective mass of the Holstein polaron in dimensions 1, 2, and 3.

ously the residual polaron-phonon interaction generates longer-ranged hopping processes [3, 65]. Concomitantly, the mass enhancement due to the EP interaction is weakened at the band minimum. It is important to realize that these effects are most pronounced at intermediate EP couplings and phonon frequencies. In this parameter region even higher-order strong-coupling perturbation theory, with its internal states containing some excited phonons, seems to be intractable because the convergence of the series expansion is

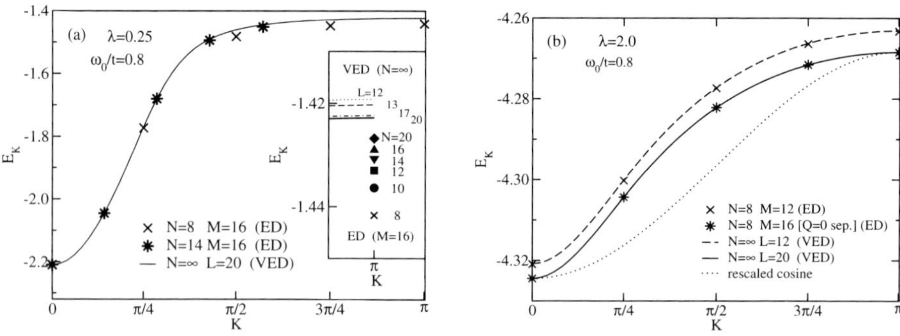

Fig. 6. Band structure of the 1D Holstein model in the weak (a) and strong (b) EP coupling regimes. M and L denote the number of phonons and the depth of the basis in ED and VED calculations, respectively. Within ED the ($Q = 0$) centre of mass motion has been separated if indicated. The inset shows the finite-size dependence being significant for weak EP coupling only. In the strong EP coupling case ED results basically agree with those obtained by VED.

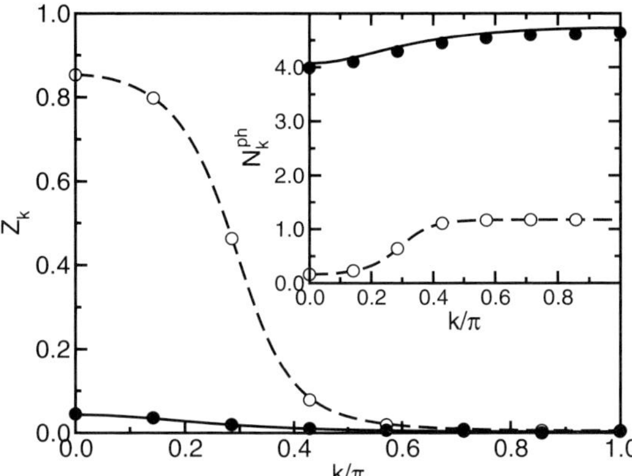

Fig. 7. The quasiparticle weight Z_k and in the inset the total number of phonons N_k^{ph} as a function of the wavevector k for the 1D Holstein model with $(\bar{g}/t)^2 = 0.4$ [dashed line (VED); open circles (ED)], 3.2 [solid line (VED); filled circles (ED)]. $\alpha = \omega_0/t = 0.8$. Data are taken from [60, 65].

poor [73]. Of course the dispersion is barely changed from a rescaled tight-binding band in the very extreme small polaron limit (λ, $g^2 \gg 1$).

Further information about the quasiparticle may be obtained by computing the quasiparticle residue, the overlap squared between a bare electron and a polaron,

$$Z_{\boldsymbol{k}} = |\langle \psi_{\boldsymbol{k}} | c_{\boldsymbol{k}}^\dagger | 0 \rangle|^2, \tag{24}$$

where $|0\rangle$ is the state with no electron and no phonon excitations, and $|\psi_{\boldsymbol{k}}\rangle$ is the polaron wavefunction at momentum $\boldsymbol{k}$. $Z_{\boldsymbol{k}}$ can be measured by angle-resolved photoemission, and gives the bare electronic contribution of the polaronic state. The phonon contribution to the quasiparticle is described by the k-dependent mean phonon number

$$N_{\boldsymbol{k}}^{ph} = \sum_i |\langle \psi_{\boldsymbol{k}} | b_i^\dagger b_i | \psi_{\boldsymbol{k}} \rangle|^2. \tag{25}$$

Figure 7 shows the spectral weight $Z_{\boldsymbol{k}}$ and the mean phonon number $N_{\boldsymbol{k}}^{ph}$ as a function of $\boldsymbol{k}$. The two sets of parameters correspond to the large and small polaron regime respectively. The DMRG cannot straightforwardly compute this quantity. There is excellent agreement between the VED and ED methods in the weak coupling case. In the strong coupling regime there is an approximately 1% disagreement in $N_{\boldsymbol{k}}^{ph}$ due to a lack of phonon degrees of freedom in the variational space of the ED calculation. The results in the weak coupling case show a smooth crossover from the predominantly electronic character of the wavefunction for small k (large $Z_{\boldsymbol{k}}$ and small $N_{\boldsymbol{k}}^{ph} \approx 0$)

to the predominantly phonon character around $\boldsymbol{k} = \pi$ characterised by $Z_{\boldsymbol{k}} \approx 0$ and $N_{\boldsymbol{k}}^{ph} \approx 1$. In the strong coupling regime there is less $\boldsymbol{k}$-dependence. The $Z_{\boldsymbol{k}}$ is already close to zero at small k, indicating strong EP interactions that lead to a large polaron mass. Concomitantly an appropriately defined polaron quasiparticle residue $\widetilde{Z}_0$ tends to one [74, 75]. So we arrive at the conclusion that a well-defined electronic (polaronic) quasiparticle exists for $k = 0$ at weak (strong) EP coupling.

VED is one of the few methods that can also calculate the polaron band dispersion in 3D systems (QMC is another, but is limited to the condition that the polaron bandwidth is much smaller than the phonon frequency, which corresponds to the strong-coupling regime.) Figure 8(a) shows the band dispersion for the 3D polaron along symmetry directions in the Brillouin zone at various EP coupling constants $\bar{g} = g\omega_0$. Starting with weak coupling $\bar{g} = 2.0$

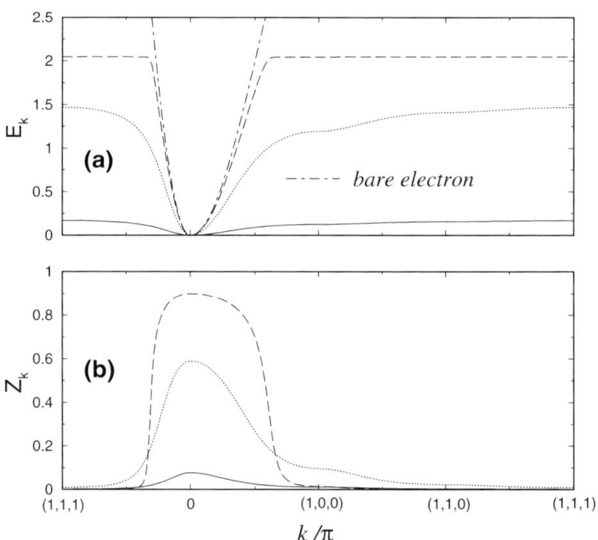

Fig. 8. Ground-state energy $E_{\boldsymbol{k}}$ (shifted by E_0) of the 3D polaron in panel (a) and quasiparticle weight $Z_{\boldsymbol{k}}$ in panel (b) for three different EP coupling constants, $\bar{g} = 4.5$ (solid line), $\bar{g} = 3.5$ (dotted line), and $\bar{g} = 2.0$ (dashed line), for $\omega_0/t = 2$. The dot-dashed line in (a) is the dispersion of a bare electron. The corresponding ground state energies E_0/t are -10.608348, -8.0642850, and -6.588526818 respectively. From [60], ©(1999) by the American Physical Society.

(dashed line), the polaron band is close to the bare electron band at the lower band edge. The deviation between them increases as $\boldsymbol{k}$ increases. When $E_{\boldsymbol{k}} - E_0$ approaches ω_0, we observe a band flattening effect, similar to the 1D and 2D cases, accompanied by a sharp drop of quasiparticle weight $Z_{\boldsymbol{k}}$ (see Fig. 8(b)). The large $\boldsymbol{k}$ lowest energy state can be considered roughly as "a $\boldsymbol{k} = 0$ polaron ground state" plus "an itinerant (or weakly-bound) phonon with momentum $\boldsymbol{k}$". It is the phonon that carries the momentum so as to make $Z_{\boldsymbol{k}}$ essentially vanish and give an approximate bandwidth ω_0. Due to the large spatial extent of the EP correlations in the flattened band, our results are less accurate in this regime. In the case of intermediate coupling $\bar{g} = 3.5$ ($\lambda = 1.0208$), the polaron bandwidth is narrower than the phonon frequency. The upper part of the band has much less dispersion than the lower part but with a substantial $Z_{\boldsymbol{k}}$. This indicates a distinct mechanism for the crossover as a function of $\boldsymbol{k}$. In the case of $\bar{g} = 4.5$, the strong EP interaction leads to a significant suppression of $Z_{\boldsymbol{k}}$ at all $\boldsymbol{k}$. $Z_{\boldsymbol{k}=0}$ approaches the strong-coupling result $Z = \exp(-g^2)$ for $\lambda, g^2 \gg 1$. Note that the inverse effective mass m^*/m_0 and Z differ if the self-energy is strongly $\boldsymbol{k}$-dependent. This discrepancy has its maximum in the intermediate-coupling regime for 1D systems, but vanishes in the limit $\lambda \to \infty$. In the Holstein model with on-site electron-phonon interactions, Z and the inverse effective mass are closely related. However, the two can be made arbitrarily different by increasing the range of the EP interaction [61].

The correlation function between the electron position and the phonon displacement is

$$\chi_{l,j} = \langle \psi_{\boldsymbol{k}} | c_i^\dagger c_i (b_j + b_j^\dagger) | \psi_{\boldsymbol{k}} \rangle . \tag{26}$$

This correlation function can be considered as a measure of the polaron size. It should not be confused with the "polaron radius" in the extreme adiabatic limit, which refers to the spatial extent of a hypothetical symmetry-breaking localised state. The ground-state EP correlation function is plotted for the 1D Holstein polaron in Fig. 9. For parameters close to the (adiabatic) weak

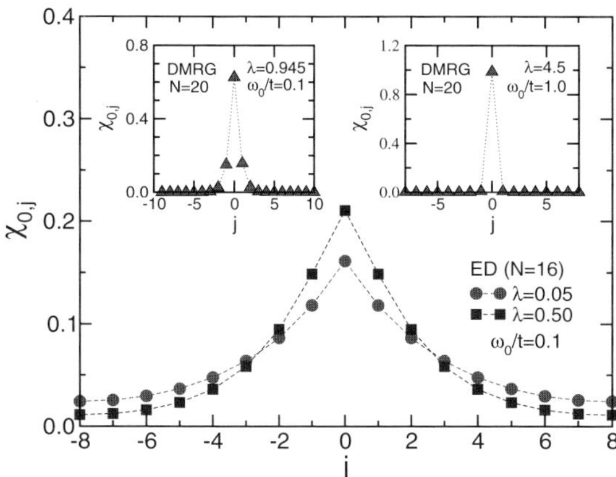

Fig. 9. Renormalised EP correlation function $\chi_{0,j} = \langle n_0 (b_{0+j}^\dagger + b_{0+j}) \rangle / 2g \langle n_0 \rangle$ as a function of the electron-phonon separation j in the $\boldsymbol{k} = 0$ ground state of the 1D Holstein model. Results are taken from [17, 76].

EP coupling regime (main panel), the amplitude of $\chi_{0,j}$ is small and the spatial extent of the electron-induced lattice deformation is spread over the whole (finite) lattice. The DMRG results shown in the left inset indicate a substantial reduction of the polaron's size near the crossover region. Finally a small polaron is formed at large couplings (right inset); now the position of the electron and the phonon displacement is strongly correlated.

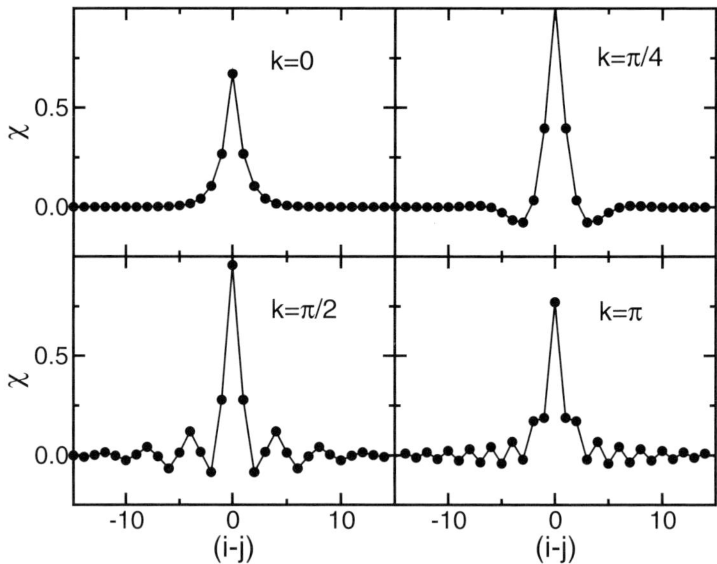

Fig. 10. Lattice deformation χ for 1D as a function of $(i - j)$ for $\omega_0 = 0.8$, $t = 1$, $\bar{g}^2 = 0.4$ and $L = 18$, for four different values of momentum k. The variational Hilbert space for $L = 18$ allows nonzero correlations up to a maximum distance $|i - j|_{max} = 17$. Only distances up to 15 are plotted.

How does the electron–phonon correlation function change as the polaron acquires a nonzero momentum $\boldsymbol{k}$? The answer is shown for 1D in Fig. 10 [60]. The parameters correspond to weak coupling. At $k = 0$, where the group velocity is zero, the deformation is limited to only a few lattice sites around the electron. The correlation is always positive and exponentially decaying. At finite but small $k = \pi/4$, the deformation around the electron increases in amplitude and rings (oscillates in sign) as the polaron acquires a finite group velocity. At $k = \pi/2$ the ringing is strongly enhanced. Note also that the spatial extent of the deformation increases in comparison to $k = 0$. The range of the deformation is maximum at $k = \pi$, where it extends over the entire region shown in the figure. In keeping with the larger extent of the lattice deformation near $k = \pi$, the ground-state energy E_π converges more slowly with the size of the Hilbert space. We have also computed the $k-$dependent χ

for the strong-coupling case $\omega_0 = 0.8$, $\bar{g}^2 = 3.2$ (not shown). We find only weak $k-$dependence, which is a consequence of the crossover to the small polaron regime. The lattice deformation is localised predominantly on the electron site.

Over four decades ago, a simple and intuitive variational approach to the 1D polaron problem was proposed by Toyozawa [28]. This method has been successfully applied to various fields and revisited in a number of guises over the years. It is generally believed to provide a qualitatively correct description of the polaron ground state, aside from predicting a spurious discontinuous change in the mass at intermediate coupling. The Toyozawa variational wavefunction consists of a product of coherent states (displaced oscillators) around the instantaneous electron position. The phonons create a symmetrical cloud around the electron. Numerical studies of the 1D electron-phonon correlation function (two-point function) are in semi-quantitative agreement with the Toyozawa variational wavefunction. The numerically exact three-point function, however, disagrees wildly. Denoting the instantaneous electron position as 0, the Toyozawa variational wavefunction requires that the probability to find phonon excitations, for example, on both sites 3 and 4, should be identical to finding them on sites (-3) and 4. Numerically, however, the latter is many orders of magnitude less probable [61]. This suggests a physical picture in which the $k = 0$ polaron is viewed not as an electron surrounded by a

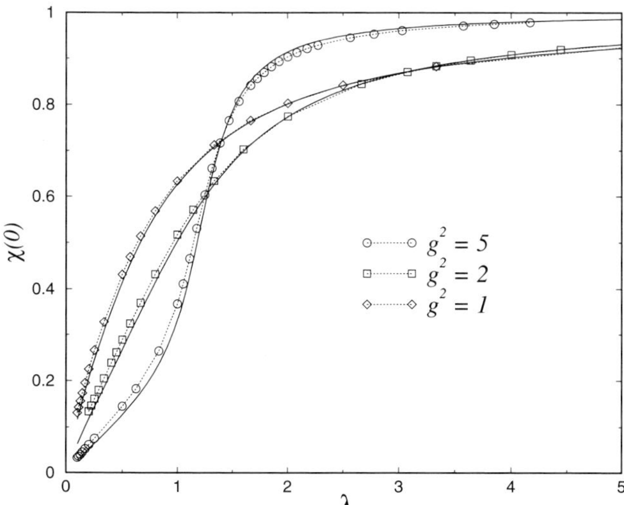

Fig. 11. On-site correlation $\chi(0)$ for the 3D polaron. VED results (solid lines) [61] are compared to DMFT (dotted lines with symbols) [34]. $\chi(0)$ is normalised to 1 when λ is infinite (i.e. $t \to 0$) according to $\lim_{t \to 0} \chi(0) = 2g$. Parameters are $\omega_0 = t = 1$.

symmetrical cloud of phonons, but is instead a coherent superposition of two
"comets," one with a tail extending to the right, and the other to the left.

Studying the properties of Holstein polarons, DMFT is exact in infinite dimensions. An interpolation to 3D lattices is made possible by using a semielliptical free density of states $N(E)$ to mimic the low-energy features. Here DMFT is accurate in the strong-coupling regime, where the surrounding phonons are predominately on the electron site. This is also the regime where strong-coupling perturbation theory works well. However, DMFT fails to compute quantities such as the polaron mass correctly in the weak-coupling regime. The reason is that in DMFT, the lattice problem is mapped onto a self-consistent local impurity model, which preserves the interplay of the electron and the phonons only at the local level. The spatial extent of the EP correlations increases as the EP coupling decreases, which explains the significant discrepancy in the weak-coupling regime. Therefore only the on-site EP correlation has been studied by DMFT, and the results are compared with VED [61] in Fig. 11. There is good qualitative agreement. The curves show a rapid change in slope only for large g, where DMFT is less accurate. It is worth noting that DMFT neglects the k dependence of self-energy, i.e., the inverse effective mass is always equal to the quasiparticle spectral weight. Clearly the difference between m_0/m^* and the spectral weight Z_k is not negligible in the intermediate-to weak-coupling regime.

4 Excited States

In this section we turn our attention from the ground state to the excited states of the Holstein model. Figure 12 plots the energy eigenvalues for a small variational space containing a maximum of 9 phonon excitations. The lowest curve is the polaron ground state at momentum k. Excited states are the polaron with additional bound or unbound (or both) phonon excitations. A ripple can be discerned near the bare electron dispersion. The figure superficially resembles a "band structure," which however encodes ground and excited state information for the many-body (many phonon) polaron problem. The ac conductivity of the polaron, for example, appears as an "interband" transition in this mapping.

What is the nature of the first excited state? We focus on the question of whether an extra phonon excitation forms a bound state with the polaron, or instead remain as two widely separated entities. Using numerical and analytical approaches we show evidence that there is a sharp phase transition (not a crossover) between these two types of states, with a bound excited state appearing only as the EP coupling is increased. Although the ground state energy E_0 is an analytic function of the parameters in the Hamiltonian, there are points at which the energy E_1 of the first excited state is nonanalytic. In previous work, Gogolin has found bound states of the polaron and additional phonon(s), but he does not obtain a phase transition between bound

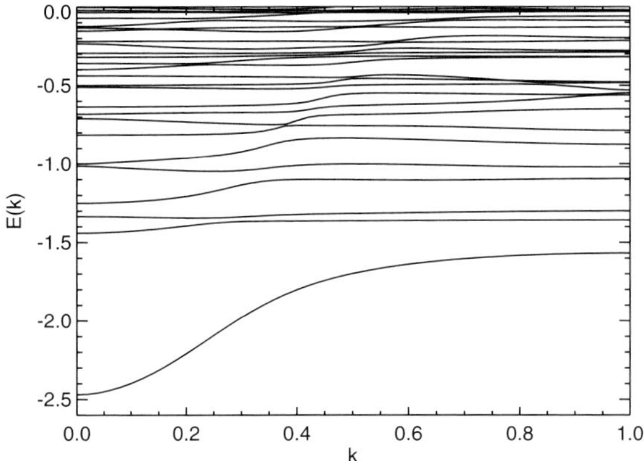

Fig. 12. The ground and excited state 1D polaron energy eigenvalues (those energies below 0) are plotted as a function of k (in units of π) for $\bar{g} = \omega_0 = t = 1$, $L = 9$, $N_{st} = 1185$.

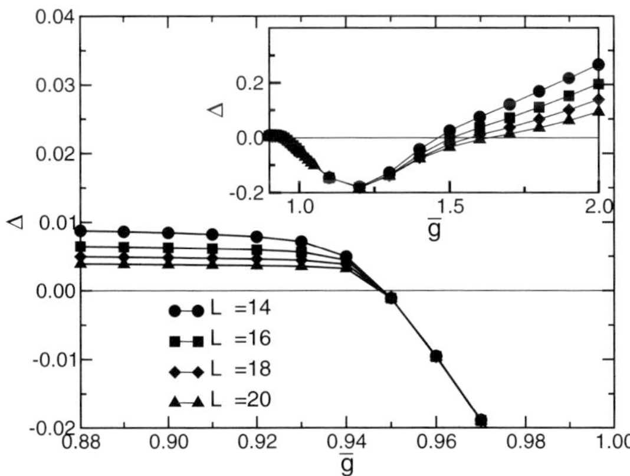

Fig. 13. First excited state binding energy $\Delta = E_1 - E_0 - \omega_0$ as a function of $\bar{g}$ [60]. Results are for $\omega_0 = 0.5$ and various Hilbert space sizes L. Inset: Binding energy over a wider range of $\bar{g}$.

and unbound states because his approximations are limited to strong coupling $g \gg 1$ [77]. A phase transition between a bound and unbound first excited state has been calculated for 3D using a dynamical CPA (coherent potential approximation) [33] and DMFT [34].

We compute the energy difference $\Delta E = E_1 - E_0$, where E_1 and E_0 are the first excited state and the ground state energies at $k = 0$ (the two lowest bands in Fig. 12). In the case where the first excited state of a polaron can be described as a polaron ground state and an unbound extra phonon excitation, this energy difference should in the thermodynamic limit equal the phonon frequency, $\Delta E = \omega_0$. In Fig. 13 we plot the binding energy $\Delta = \Delta E - \omega_0$ for $\omega_0 = 0.5$ as a function of the EP coupling $\bar{g}$ for various sizes of the variational space. We see two distinct regimes. Below $\bar{g}_c \sim 0.95$, Δ varies with the system size but remains positive ($\Delta > 0$). Physically, for $\bar{g} < \bar{g}_c$, the additional phonon excitation would prefer to be infinitely separated from the polaron, but is confined to a distance no greater than $L - 1$ by the variational Hilbert space. As the system size increases, Δ slowly approaches zero from above as the "particle in a box" confinement energy decreases. In the other regime, $\bar{g} > \bar{g}_c$, the data has clearly converged and $\Delta < 0$. This is the regime where the extra phonon excitation is absorbed by the polaron forming an excited polaron bound state. Since the excited polaron forms an exponentially decaying bound state, the method already converges at $L = 14$. In the inset of Fig. 13 we show the binding energy Δ in a larger interval of EP coupling $\bar{g}$. Although the results cease to converge at larger $\bar{g}$, we notice that the binding energy Δ reaches a minimum as a function of $\bar{g}$. As one can demonstrate within the strong coupling approximation, the true binding energy approaches zero exponentially from below with increasing $\bar{g}$.

Figure 14 shows the phase diagram for $k = 0$ separating the two regimes. The phase boundary, given by $\Delta = 0$, was obtained numerically, and compared to strong coupling perturbation theory in t to first and second order. The phase transition where Δ becomes negative at sufficiently large $\bar{g}$ is also seen in ED calculations.

The distribution of the number of phonons in the vicinity of the electron is defined as

$$\gamma(i - j) = \langle \psi_k | c_i^\dagger c_i b_j^\dagger b_j | \psi_k \rangle. \tag{27}$$

In Fig. 15 we compute this distribution for the ground state γ_0 and the first excited state γ_1 slightly below ($\bar{g} = 0.9$), and above ($\bar{g} = 1.0$) the transition for $\omega_0 = 0.5$.

The central peak of the correlation function γ_1 below the transition point is comparable in magnitude to γ_0 (Fig. 15(a,b)). The main difference between the two curves is the long range decay of γ_1 as a function of distance from the electron, onto which the central peak is superimposed. The extra phonon that is represented by this long-range tail extends throughout the whole system and is not bound to the polaron. The existence of an unbound, free phonon is confirmed by computing the difference of total phonon number $N_{0,1}^{ph} = \sum_l \gamma_{0,1}(l)$. This difference should equal one below the transition point. Our numerical values give $N_1^{ph} - N_0^{ph} \sim 1.02$. We attribute the deviation from the exact result to the finite relative separation allowed. Correlation functions above the transition point (Fig. 15(c,d)) are physically different. First, phonon correla-

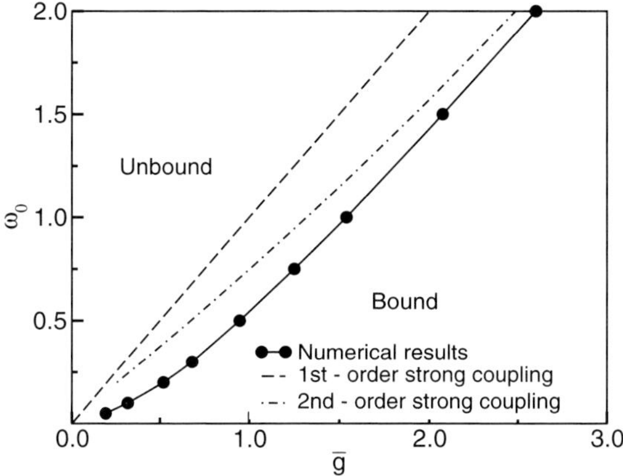

Fig. 14. The phase diagram for the bound to unbound transition of the first excited state, obtained using the condition $\Delta(\omega_0, \bar{g}) = 0$. The corresponding phase diagram for the ground state would be blank: there is no phase transition in the ground state, only a crossover.

tions in γ_1 decay exponentially, which also explains why the convergence in this region is excellent. Second, the size of the central peak in γ_1 is 3 times higher than γ_0. (Note that to match scales in Fig. 15(d) we divided γ_1 by 3). The difference in total phonon number gives $N_1^{ph} - N_0^{ph} \sim 2.33$. We are thus facing a totally different physical picture: The excited state is composed of a polaron which contains several extra phonon excitations (in comparison to the ground-state polaron) and the binding energy of the excited polaron is $\Delta < 0$. The extra phonon excitations are located almost entirely on the electron site. The value of $\gamma_1 - \gamma_0$ at $j = 0$ is 2.16, which almost exhausts the phonon sum.

Next we discuss the role of dimensionality in the excited states. The effect of dimensionality on static properties has been studied previously [61, 78–80]. The eigenvalues of the low-lying $k = 0$ states are shown as functions of $\bar{g}$ in Fig. 16. The energy spectra in D>1 are qualitatively different than in 1D. The 1D polaron ground state becomes heavy gradually as $\bar{g}$ increases. However, in D≥2, the ground state crosses over to a heavy polaron state by a narrow avoided level crossing, which is consistent with the existence of a potential barrier [78]. In the lower panel of Fig. 16, ψ_1 and ψ_4 are nearly free electron states; ψ_2 and ψ_3 are heavy polaron states. The inner product $|\langle \psi_1 | \psi_4 \rangle|$ is equal to 0.99. Just right of the crossing region the effective mass (approximately equal to the inverse of the spectral weight) of the first excited state can be smaller than the ground state by 2 or 3 orders of magnitude, while their energies can differ by much less than ω_0. The narrow avoided crossing description works less well for larger ω_0.

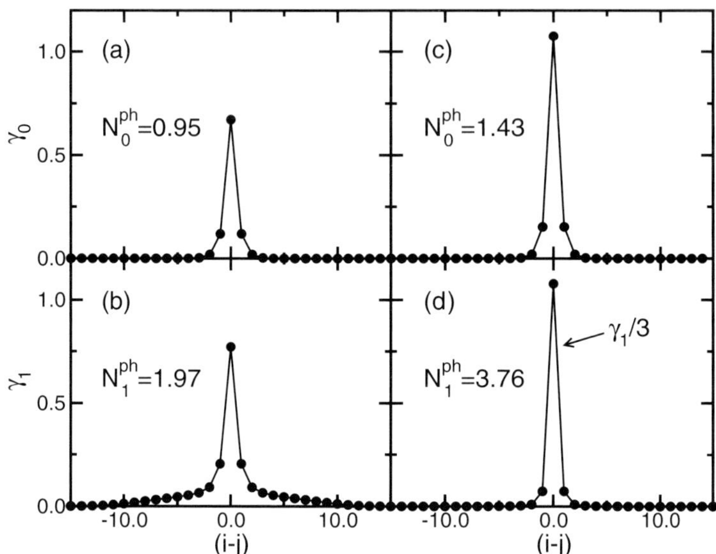

Fig. 15. The phonon number γ as a function of the distance from the electron position for the ground state (a) and the first excited state (b), both computed at $\bar{g} = 0.9$; and the same in (c) and (d) for $\bar{g} = 1.0$. All data are computed at phonon frequency $\omega_0 = 0.5$ and $L = 18$. Note that (d) is a plot of $\gamma_1/3$. In (a,b), $\gamma_1 - \gamma_0$ drops to zero around $|i - j| = 15$. This is a finite-size effect. Computing the same quantity with larger L below the phase transition would lead to a larger extent of the correlation function indicating that the extra phonon excitation is not bound to the polaron.

5 Dynamics of Polaron Formation

How does a bare electron time evolve to become a polaron quasiparticle? The bare electron can be injected by inverse photoemission or tunnelling, or a hole by photoemission, or an exciton (electron-hole bound state) by fast optics.

One approach is to construct a variational many-body Hilbert space including multiple phonon excitations, and to numerically integrate the many-body Schrödinger equation,

$$i\frac{d\psi}{dt} = H\psi \tag{28}$$

in this space [81]. The full many-body wave function is obtained. This method includes the full quantum dynamics of the electrons and phonons. Note that alternative treatments, such as the semiclassical approximation that treats the phonons classically, fail for this problem, particularly in the limit of a wide initial electron wavepacket.

Figure 17 shows snapshots of polaron formation at weak coupling. An initial bare electron wave packet is launched to the right as shown in panel

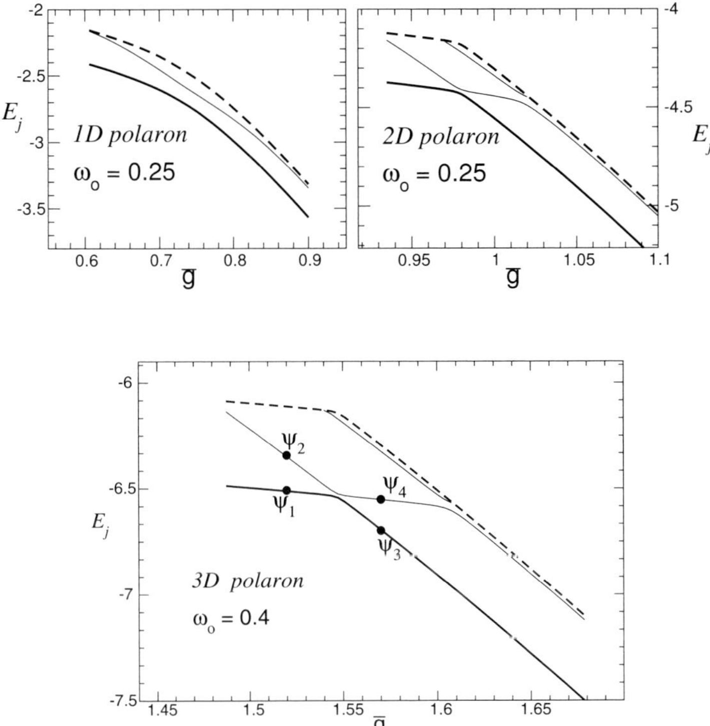

Fig. 16. Eigenvalues of low-lying states as functions of coupling constant in 1D through 3D. Hopping $t = 1$ in all panels. In the adiabatic regime in higher dimensions, the ground state (thick solid lines) shows a fairly abrupt change in slope. In the 3D panel, ψ_1 and ψ_4 are a lightly-dressed electron state; ψ_2 and ψ_3 are a heavy polaron state. The dashed lines are the beginning of the lowest continua.

(a). This initial condition is relevant to the recent experiments [82–86], and to electron injection from a time-resolved STM (scanning tunneling microscope) tip [87]. Although polarons injected optically or by STM can have a range of initial momenta, it would be more realistic to take $k = 0$ for an optically created exciton. In panel (b) the electron is not yet dressed and thus is moving roughly as fast as the free electron (dashed line). In addition, there exists a back-scattered current (which later evolves into a left-moving polaron) on the left side of the wave packet (dot-dashed and thick solid curves). In panel (c) after an elapsed time of one phonon period, the electron density consists of two peaks. The peak on the right (black arrow) is essentially a bare electron. The peak on the left is a polaron wave packet moving more slowly. As time goes on, the bare electron peak decays and the polaron peak grows. Some phonons are left behind (double-dot dashed line), mainly near the injection point. These phonons are of known phase with displacement shown in thin solid grey. Some phonon excitations travel with the polaron (dot double-dashed line). Finally, a coherent polaron wave packet is observed when the polaron separates from the uncorrelated phonon excitations. The velocity operator is defined as

$$\widehat{v}_j \equiv \frac{2\widehat{j}_{j,j+1}}{e\angle c_j^\dagger c_j + c_{j+1}^\dagger c_{j+1}\rangle}, \tag{29}$$

where j is the site index and $\widehat{j}$ is the current operator. $\langle\widehat{v}_j\rangle$ is shown as a dot-dashed line.

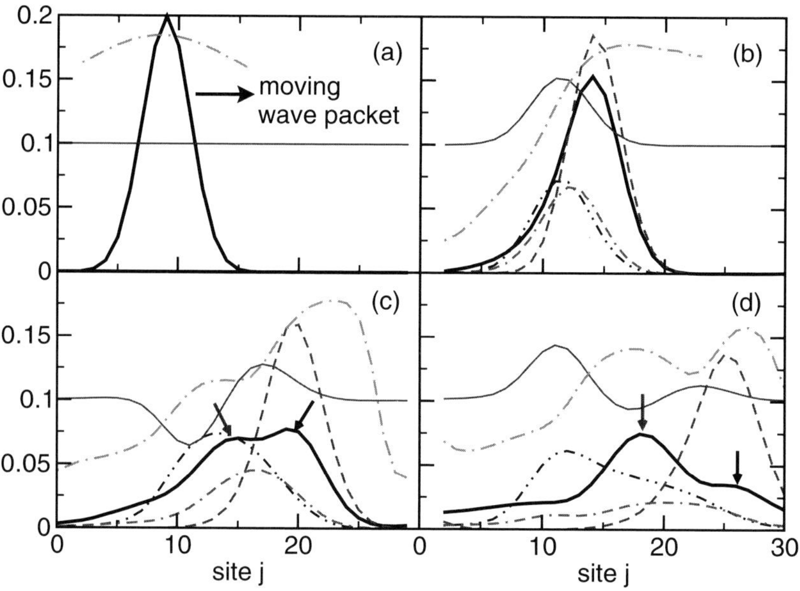

Fig. 17. Snapshots of the polaron-formation process, for $t = \omega_0 = 1$, and $\bar{g} = 0.4$. The calculation is performed on a 30-site periodic lattice. Time is measured in units of the phonon period. Shown are the electron density $\langle c_j^\dagger c_j\rangle$ (thick solid line), phonon density $\langle b_j^\dagger b_j\rangle$ (double-dot dashed line), lattice displacement $\langle\widehat{x}_j\rangle \equiv \langle b_j + b_j^\dagger\rangle$ (thin solid grey line), velocity in units of lattice constant per phonon period (dot dashed line), and EP correlation function $\langle c_j^\dagger c_j b_j^\dagger b_j\rangle$ (dot double-dashed line). The dashed line gives the free-electron wave packet for reference. For clarity, the origins of the thin solid grey and dashed curves are offset by 0.1 and their values are rescaled by a factor of 0.2 and $0.05/(2\pi)$ respectively. The double-dot dashed curve has been rescaled by a factor of 0.5.

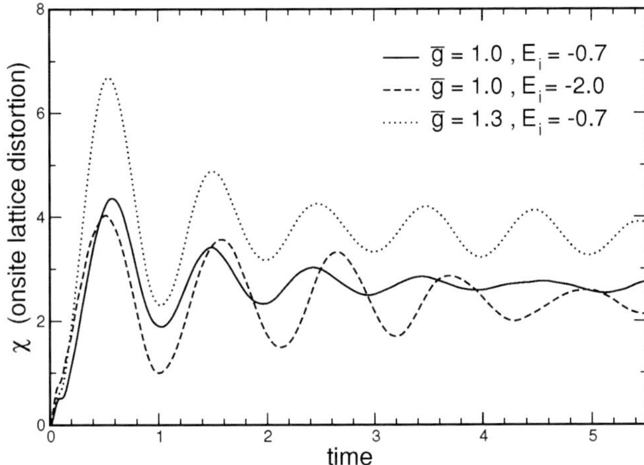

Fig. 18. The on-site electron-phonon correlation function $\chi = \langle c_j^\dagger c_j (b_j + b_j^\dagger) \rangle$ as a function of time measured in phonon periods. For all curves, $\omega_0 = 0.5$ and hopping $t = 1$. The solid line is for a bare electron injected with nonzero initial momentum at energy $E_i = -0.7$, where the bottom of the bare band is at energy -2. The phonon displacement is larger and more weakly damped for larger electron-phonon coupling $\bar{g}$, dotted line. In contrast to a bare electron, an exciton (bound particle-hole pair) is generally created with an initial momentum zero, corresponding to $E_i = -2$, dashed line.

There are regimes where the polaron formation time is a calculable constant of order unity times a phonon period T_0, as seen in some experiments and in Fig. 17, but there are other regimes in which the phonon period is not the relevant timescale. The limit of hopping $t \to 0$ is instructive [88, 89]. After a time $T_0/4$, the expectation of the lattice displacement $\langle \hat{x}_j \rangle$ on the electron site has the same value as a static polaron. It is tempting (but we would argue incorrect) to identify this as the polaron formation time. At later times, $\langle \hat{x}_j \rangle$ overshoots by a factor of two, and after a time T_0, $\langle \hat{x}_j \rangle$ and all other correlations are what they were at time zero when the bare electron was injected. The system oscillates forever. In general an electron emits phonons en route to becoming a polaron, and we propose that the polaron formation time be defined as the time required for the polaron to physically separate from the radiated phonons. The polaron formation time for hopping $t \to 0$ is thus infinite, because the electron is forever stuck on the same site as the radiated phonons.

An electron injected at several times the phonon energy ω_0 above the bottom of the band is another instructive example. The electron radiates successive phonons to reduce its kinetic energy to near the bottom of the band, and then forms a polaron. For very weak EP coupling, the rate for

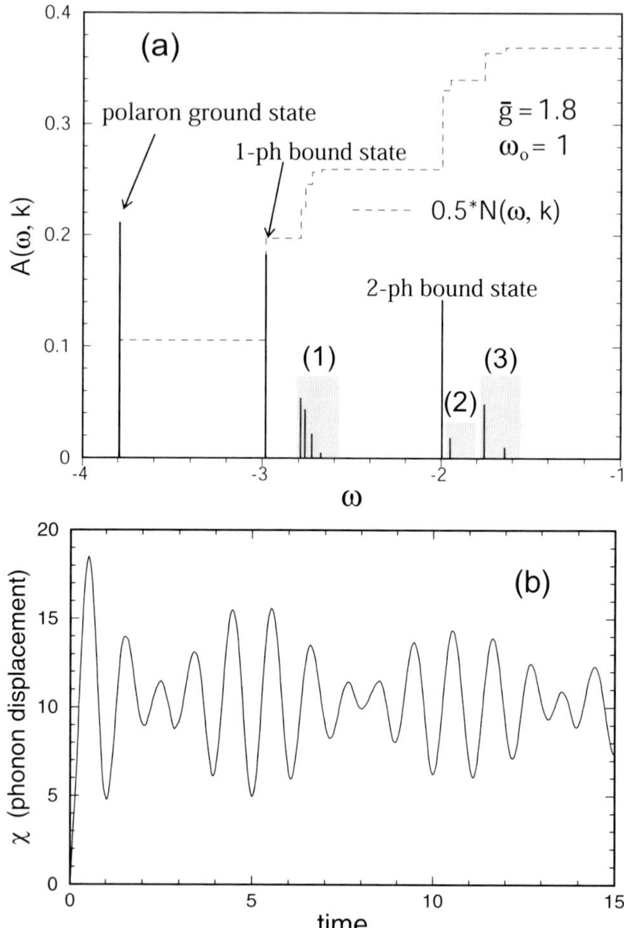

Fig. 19. Panel (a): Spectral function at strong coupling. There are three quasiparticle excited states split off from the continua. Shaded areas (1) and (2) correspond to continuum states. (b): Quantum beat formed by multiple excited states and continua.

radiating the first phonon can be computed by Fermi's golden rule, $\tau_{FGR}^{-1} = \bar{g}^2/[\hbar\, t \sin(k_f)]$, where k_f is the electron momentum after emitting a phonon. The phonon emission time can be arbitrarily long for small $\bar{g}$. For strong coupling, the rate approaches $\tau_{SC}^{-1} = \bar{g}/\hbar$ because the polaron spectral function smoothly spans numerous narrow bands and its standard deviation is equal to $\bar{g}$.

Decaying oscillations in polaron formation (actually the formally equivalent problem of an exciton coupled to phonons [4]) have been observed in a pump-probe experiment that measures reflectivity after a bare exciton is created [83]. The observed oscillatory reflectivity was interpreted as the lattice motion in the phonon-dressed (or "self-trapped") exciton level. Assuming the modulation in the exciton level goes as $\Delta E = -\lambda c_j^\dagger c_j \widehat{x}_j$, where $\widehat{x}_j = \langle b_j + b_j^\dagger \rangle$ is the lattice displacement, the model Hamiltonian applies directly to the experiment. We calculate the corresponding EP correlation function in Fig. 18. In this regime, the polaron formation time (damping time) increases as the electron-phonon coupling $\bar{g}$ increases, and also as the initial electron (exciton) energy approaches the band bottom. We find satisfactory agreement when compared to Fig. 2b of [83]. Both show a damped oscillation with a delayed phase. (Numerical calculations in Figs. 18-19 are performed on an extended system, not a finite cluster.)

Figure 19 shows the spectral function at strong coupling. Three delta functions are visible, corresponding to polaron ground and excited states, along with three continua containing unbound phonons. There is additional structure at higher energy (not shown). The probability to remain in the initial bare particle state $P(\tau) \equiv |\langle \psi(\tau)|c_k^\dagger|0\rangle|^2$ for this spectrum is complicated, and includes oscillating terms that do not decay to zero at zero temperature from the polaron ground and excited states beating against each other. The branching ratios into the various channels are calculated in [90].

6 Spectral Signatures of Holstein Polarons

As already stressed in the introduction the crossover from quasi-free electrons or large polarons to small polarons becomes manifest in the spectral properties above all. Here of particular interest is whether an "electronic" or "polaronic" (quasi-particle) excitation exists in the spectrum. This question has been partially addressed by calculating the wavefunction renormalisation factor [(electronic) quasi-particle weight] Z_k in Sect. 3 (see Fig. 7). More detailed information can be obtained from the one-particle spectral function $A(k,\omega)$. This quantity of great importance can be probed by direct (inverse) photoemission, where a bare electron with momentum k and energy ω is removed (added) from (to) the many-particle system. The intensities (transition amplitudes) of these processes are determined by the imaginary part of the retarded one-particle Green's function, i.e. by

$$A(k,\omega) = -\frac{1}{\pi}\mathrm{Im}\,G(k,\omega) = A^+(k,\omega) + A^-(k,\omega)\,, \tag{30}$$

with

$$A^\pm(k,\omega) = -\frac{1}{\pi}\mathrm{Im}\lim_{\eta \to 0^+}\langle \psi_0|c_k^\mp \frac{1}{\omega + i\eta \mp H}c_k^\pm|\psi_0\rangle$$

$$= \sum_m |\langle \psi_m^\pm|c_k^\pm|\psi_0\rangle|^2\delta[\omega \mp (E_m^\pm - E_0)]\,, \tag{31}$$

where $c_k^+ = c_k^\dagger$ and $c_k^- = c_k$ ($T = 0$; 1D spinless case). These functions test both excitation energies ($E_m^\pm - E_0$) and overlap ($\propto Z_k$) of the N_e-particle ground state $|\psi_0\rangle$ with the exact eigenstates $|\psi_m^\pm\rangle$ of an ($N_e \pm 1$)–particle system. The electron spectral function of the single-particle Holstein model corresponds to $N_e = 0$, i.e., $A^-(k,\omega) \equiv 0$. $A(k,\omega)$ can be determined, e.g., by a combination of KPM and CPT (cf. Sect. 2.2).

Figure 20 (a) shows that at weak EP coupling, the electronic spectrum is nearly unaffected for energies below the phonon emission threshold. Hence, for the case considered with ω_0 lying inside the bare electron band $\varepsilon_k = -2t\cos k$, the signal corresponding to the renormalised dispersion E_k nearly coincides with the tight-binding cosine band (shifted $\propto \varepsilon_p$) up to some k_X, where the phonon intersects the bare electron band. At k_X electron and phonon states "hybridise" and repel each other, forming an avoided-crossing like gap. For $k > k_X$ the lowest absorption signature in each k sector follows the dispersionless phonon mode, leading to the flattening effect [65]. Accordingly the (electronic) spectral weight of these peaks is very low. The high-energy incoherent part of the spectrum is broadened $\propto \varepsilon_p$, with the k-dependent maximum again following the bare cosine dispersion.

Reaching the intermediate EP coupling (polaron crossover) regime a coherent band separates from the rest of the spectrum [$k_X \to \pi$; see panel (b)]. At the same time its spectral weight becomes smaller and will be transferred to the incoherent part, where several sub-bands emerge.

The inverse photoemission spectrum in the strong-coupling case is shown in Fig. 20(c). First, we observe all signatures of the famous polaronic band-collapse. The coherent quasi-particle absorption band becomes extremely narrow. Its bandwidth approaches the strong-coupling result $4Dt\exp(-g^2)$ for $\lambda, g^2 \gg 1$. If we had calculated the polaronic instead of the electronic spectral function, nearly all spectral weight would reside in the coherent part of the spectrum, i.e. in the small-polaron band. This has been demonstrated quite recently [75]. In our case the incoherent part of the spectrum carries most of the spectral weight. It basically consists of a sequence of sub-bands separated in energy by ω_0, which correspond to excitations of an electron and one or more phonons.

Let us emphasise that for all couplings the lowest signature in $A(k,\omega)$ almost perfectly coincides with the coherent polaron band-structure (solid lines) obtained by VED (see Sect. 3), which underlines the high precision of the CPT-KPM approach used here.

Of course, the phonon modes are unaffected by one electron in the solid, i.e. the phonon self-energy is zero. Actually this is true in the thermodynamic limit only. In a finite-cluster calculation there will be an influence of order $\mathcal{O}(1/N)$ and the phonon spectra provide additional useful information concerning the polaron dynamics. For this purpose, we calculate the $T = 0$ phonon spectral function

$$B(q,\omega) = -\frac{1}{\pi}\mathrm{Im}D^R(q,\omega) \tag{32}$$

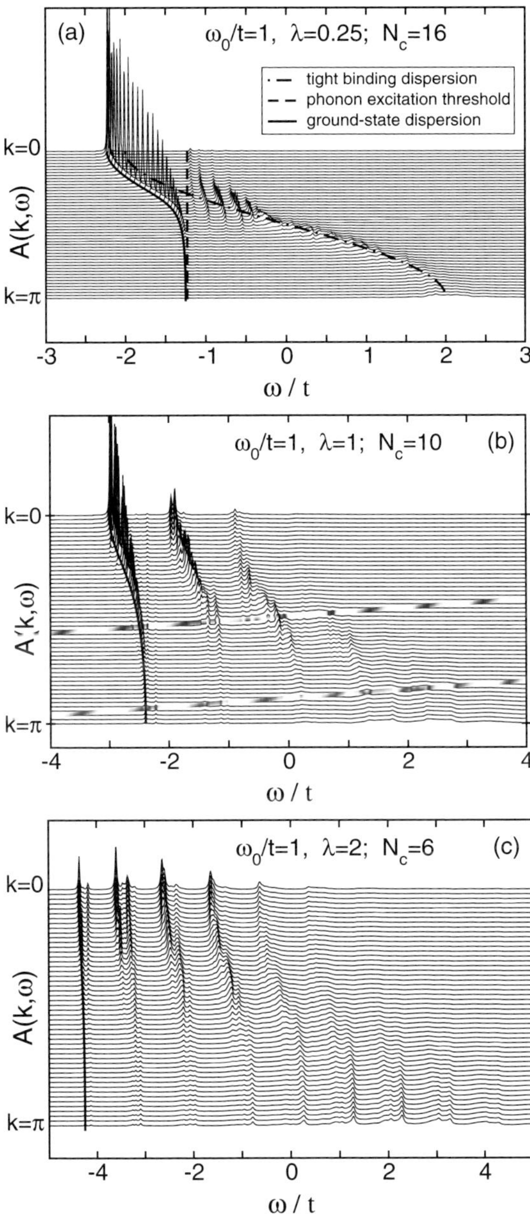

Fig. 20. Spectral function of the 1D Holstein polaron in the weak (a), intermediate (b), and strong (c) EP coupling regimes. CPT data based on finite-cluster ED with N_c sites, and $M = 7$ ($\lambda = 0.25$), $M = 15$ ($\lambda = 1$), $M = 25$ ($\lambda = 2$) phonon quanta. Note that the non-monotonic heights of the lowest energy peaks in (a) are an artifact of the CPT calculation, where some of the wavevectors fit the $N_c = 16$ cluster size, and some don't. Also the dispersionless absorption feature in (c), just above the small-polaron peak, is due to a finite-size effect, but not the double-peak structures of the higher excitation bands. This has been proved by determining the spectral function in the $k = 0$–sector by VED.

which is related to the phonon Green's function

$$D^R(q,\omega) = \lim_{\eta \to 0^+} \langle \psi_0 | \widehat{x}_q \frac{1}{\omega + i\eta - H} \widehat{x}_{-q} | \psi_0 \rangle \,, \tag{33}$$

where $\widehat{x}_q = N^{-1/2} \sum_j \widehat{x}_j e^{-iqj}$ and $\widehat{x}_j = (b_j^\dagger + b_j)/\sqrt{2\omega_0}$.

For the Holstein model (1), $B(q,\omega)$ is symmetric in q. The bare propagator $D_0(q,\omega) = 2\omega_0/(\omega^2 - \omega_0^2)$ is dispersionless. Then, adapting the CPT-KPM approach to the calculation of the phonon spectral function, the cluster expansion is identical to replacing the full real-space Green's function D_{ij} by D_{ij}^c.

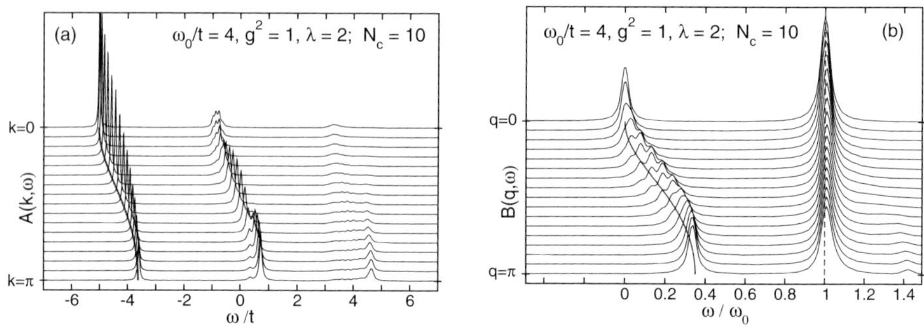

Fig. 21. Electron (a) and phonon (b) spectral functions in the anti-adiabatic intermediate EP coupling regime. Solid (dashed) lines give E_k (ω_0) determined by VED. Note that abscissae are scaled differently.

Figure 21 compares electron (a) and phonon (b) spectra in the high phonon-frequency limit, where the small polaron crossover is determined by g^2. Obviously the phonon spectrum is also renormalised by the EP interaction due to the finite "particle density" $N_e/N_c = 1/10$. So we can detect a clear signature of the polaron band having a width $W \simeq 1.5t$ (cf. Fig. 21 (a)). The dispersionless excitation at $\omega/\omega_0 = 1$ is obtained by adding one phonon with momentum q to the $k = 0$ ground state. Above this pronounced peak, we find an "image" of the lowest polaron band – shifted by ω_0 – with extremely small spectral weight, hardly resolved in Fig. 21(b).

7 Transport and Optical Response

The investigation of transport properties has been playing a central role in condensed matter physics for a long time. Optical measurements, for example, proved the importance of EP interaction in various novel materials such as the cuprates, nickelates or manganites and, in particular, corroborated polaronic scenarios for modelling their electronic transport properties at least at high temperatures [91–93]. Actually, the optical absorption of small polarons is distinguished from that of large (or quasi-free) polarons by the shape and the temperature dependence of the absorption bands which arise from exciting the self-trapped carrier from or within the potential well that binds it [16]. Furthermore, as was the case with the properties of the ground state, the optical spectra of light and heavy electrons, small and large polarons differ significantly as well [4]. In the most simple weak-coupling and anti-adiabatic strong EP coupling limits, the absorption associated with photoionization of Holstein polarons is well understood and the optical conductivity can be analysed analytically ([66, 91, 94–96]; for a detailed discussion of small polaron transport phenomena we refer to [97, 98]). The intermediate coupling and frequency regime, however, is as yet practically inaccessible for a rigorous analysis (here the case of infinite spatial dimensions, where dynamical mean-field theory yields reliable results, is an exception [99, 100]). So far unbiased numerical studies of the optical absorption in the Holstein model were limited to very small 2 to 10-site 1D and 2D clusters [17, 36, 64, 74]. In the following we will exploit the VED and KPM schemes [62, 101], in order to calculate the optical conductivity numerically in the whole parameter range on fairly large systems.

7.1 Optical Conductivity at Zero-Temperature

Applying standard linear-response theory, the real part of the conductivity takes the form

$$\mathrm{Re}\sigma(\omega) = \mathcal{D}\delta(\omega) + \sigma^{reg}(\omega)\,, \tag{34}$$

where $\mathcal{D}$ denotes the so-called Drude weight at $\omega = 0$ and σ^{reg} is the regular part (finite-frequency response) for $\omega > 0$ which can be written in spectral representation at $T = 0$ as [94]

$$\sigma^{reg}(\omega) = \frac{\pi}{\omega N} \sum_{E_m > E_0} |\langle \psi_m |\hat{\jmath}| \psi_0 \rangle|^2 \, \delta[\omega - (E_m - E_0)] \tag{35}$$

with the (paramagnetic) current operator $\hat{\jmath} = -\mathrm{i}et \sum_i (c_i^\dagger c_{i+1} - c_{i+1}^\dagger c_i)$.
Introducing the ω-integrated spectral weight,

$$S^{reg}(\omega) = \int_{0^+}^{\omega} d\omega' \sigma^{reg}(\omega')\,, \tag{36}$$

we arrive at the f-sum rule

$$-E_{kin}/2 = \mathcal{D} + S^{reg}/\pi \qquad \text{(1D case)}\,, \tag{37}$$

where $E_{kin} = -t \sum_i (c_i^\dagger c_{i+1} + c_{i+1}^\dagger c_i)$ is the kinetic energy and $S^{reg} = S^{reg}(\infty)$. Since for the Holstein model the Drude weight can be calculated independently from Kohn's formula or the effective mass,

$$\mathcal{D} = \frac{1}{2N} \frac{\partial^2 E_0(\Phi)}{\partial \Phi^2}\bigg|_{\Phi=0} = \frac{1}{2N} \frac{\partial^2 E_k}{\partial k^2}\bigg|_{k=0} = \frac{1}{2m^*}, \tag{38}$$

the f-sum rule may be used to test the numerics. In (38), the first equality relates $\mathcal{D}$ to the second derivative of the (non-degenerate) ground-state energy with respect to a field-induced phase Φ coupled to the hopping.

We first present $\sigma^{reg}(\omega)$ and its integral $S^{reg}(\omega)$ for the 1D Holstein model in Fig. 22. The upper panel (a) gives the results for intermediate-to-strong EP coupling, i.e. near the polaron crossover, in the adiabatic (light electron) regime. Of course, the optical conductivity threshold is $\omega = \omega_0$ for the infinite system we deal with using VED. In this respect standard ED, defined on finite lattices, suffers from pronounced finite-size effects due to the discreteness in k-space. Knowing that at about $\lambda \simeq 1$ a coherent polaron band with bandwidth much smaller than ω_0 splits off, the first (few) isolated peak(s) at the lower bound of the spectrum can be attributed to one- (few-) ($q = 0-$) phonon emission processes (cf. also Fig. 20(b)). Of course, these transitions have little spectral intensity. At higher energies transitions to the incoherent part of the spectrum take place by "emitting" phonons with finite momentum (to reach the total momentum $k = 0$ ground-state sector). The main signature of $\sigma^{reg}(\omega)$ is that the spectrum is strongly asymmetric, which is characteristic for rather large polarons. The weaker decay at the high-energy side meets the experimental findings for many polaronic materials like TiO_2 [102] even better than standard small-polaron theory.

For $\lambda = 2$ and $\omega_0 = 0.4$, i.e., at larger EP coupling, but not yet in the small-polaron limit, we find a more pronounced and symmetric maximum in the low-temperature optical response (see Fig. 22(b)). The maximum is located below the corresponding one for small polarons at $T = 0$, which on its part lies somewhat below $2\varepsilon_p = 2t\lambda = 2g^2\omega_0$ (being the maximum of the Poisson distribution) because of the $1/\omega$ factor contained in the conductivity. In this case the polaron band structure is more strongly renormalised, but, more importantly, the phonon distribution function in the ground state becomes considerably broadened. Since the current operator connects only different-parity states having substantial overlap as far as the phononic part is concerned, in the optical response multi-phonon emissions/absorptions (i.e., non-diagonal transitions [94]) become increasingly important. Again deviations from the analytical small-polaron result (dashed line in Fig. 22(b)) might be important for relating theory to experiment.

The optical response obtained if the phonon frequency becomes comparable to the electron transfer amplitude is illustrated in Fig. 22(c). Now the lowest one-phonon absorption (threshold) signal is well separated. In contrast to the light electron case Fig. 22(a), the different absorption bands appearing for a heavier electron can be classified according to the number of phonons involved in the optical transition (see inset). Increasing ω_0 at fixed g^2 this becomes even more manifest (at the same time a Poisson distribution of the different sub-bands evolves). The sub-bands are broadened by transitions to different "electronic" levels. For our parameters, a scattering continuum ap-

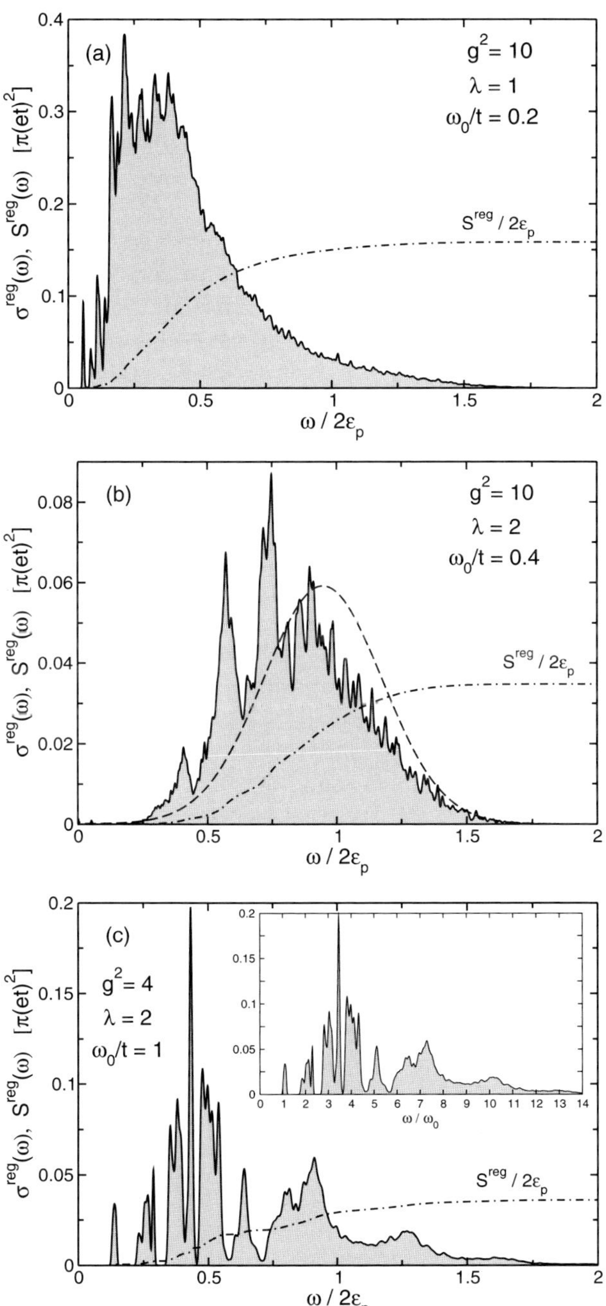

Fig. 22. Optical conductivity of the 1D Holstein polaron at $T = 0$. Results for $\sigma^{reg}(\omega)$ and $S^{reg}(\omega)$ are obtained by VED. In order to reduce finite-M effects, data calculated for $M = 15$ to 20 were averaged. The dashed line in (b) displays the analytical small polaron result, $\sigma^{reg}(\omega) = \frac{\sigma_0}{\sqrt{\varepsilon_p \omega_0}} \frac{1}{\omega} \exp\left[-\frac{(\omega - 2\varepsilon_p)^2}{4\varepsilon_p \omega_0}\right]$ (cf. [91]), where σ_0 was determined to give the same integrated spectral weight as $\sigma^{reg}(\omega > 0)$.

pears above $\omega > 5$ to 6 ω_0. Note that the "fragmentation" of the spectrum appearing at smaller energy transfer is not caused by finite-size effects.

Turning to the sum rules presented in Fig. 23, we notice a monotonic decrease of the total sum rule $S^{tot}/\pi = -E_{kin}/2$, which indicates a suppression of the electronic kinetic energy with increasing EP coupling. In agreement with previous numerical results [17, 43, 63], the kinetic energy clearly shows the crossover from a large polaron, characterised by a E_{kin} that is only weakly reduced from its non-interacting value $[E_{kin}(\lambda = 0) = -2t]$, to a less mobile small polaron in the strong EP-coupling limit, where the strong-coupling perturbation theory result,

$$E_{kin}^{SCPT} = -\frac{4t}{\omega_0}\left\langle\frac{1}{s}\right\rangle_{\kappa=2g^2} - e^{-g^2}\left[2 + \frac{4t}{\omega_0}\left\langle\frac{1}{s}\right\rangle_{\kappa=g^2}\right] \tag{39}$$

($\langle\ldots\rangle_\kappa$ denotes the average with respect to the Poisson distribution with parameter κ), gives a sufficiently accurate description in both the adiabatic and antiadiabatic regimes.

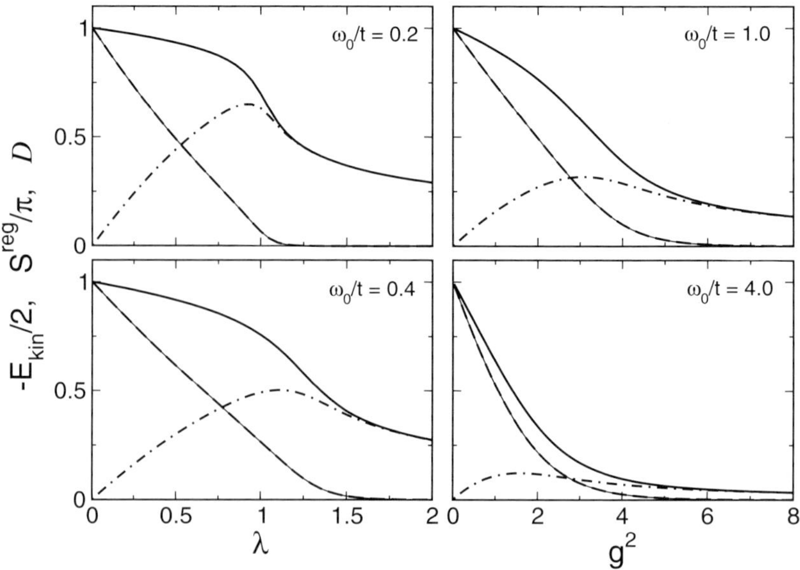

Fig. 23. Renormalised kinetic energy (E_{kin}; solid line) and contribution of σ^{reg} to the f-sum rule (S^{reg}; dot-dashed line) as a function of the EP couplings λ and g in the adiabatic (left panels) and non-to-antiadiabatic (right panels) regimes, respectively. The Drude weights were obtained from the f-sum rule [((37); thin solid line)] and effective mass [((38); dashed line)].

For light electrons (adiabatic regime $\omega_0/t = 0.2$, 0.4; left panels), we found a rather narrow transition region. The drop of S^{tot} in the crossover region $\lambda \simeq 1$ is driven by the sharp fall of the Drude weight, which is a measure of the coherent transport properties of a polaron. By contrast the optical absorption due to inelastic scattering processes, described by the regular (dissipative) part of the optical conductivity, becomes strongly enhanced around $\lambda \simeq 1$ [74] (cf. the behaviour of S^{reg}).

The large to small polaron crossover is considerably broadened for heavy electrons (non-to-antiadiabatic case $\omega_0/t = 1$, 4; right panels). Here E_{kin} decreases more gradually and S^{reg} exhibits a less pronounced maximum at about $g^2 = 1$.

As quoted above, we can calculate the Drude weight independently from the effective mass of the Holstein polaron. Using this data, it is worth mentioning that the f-sum rule (37) is satisfied numerically to at least six digits in the whole parameter regime [62].

7.2 Thermally Activated Transport

If the polaron effects are assumed to be dominant the coherent bandwidth is extremely small. Then the physical picture is that the particle is trapped at a certain lattice site and that hopping occurs infrequently from site to site. There are two kinds of transfer processes [11]. All phonon numbers might remain the same during the hop (diagonal transition) or, alternatively, the number of phonons is changed (non-diagonal transition). In the latter case each hop may be approximated as a statistically independent event and the particle loses its phase coherence by this phonon emission or absorption (inelastic scattering). Diagonal and non-diagonal transitions show a different temperature dependence. While the rate of diagonal (band-type) transitions decreases with increasing temperature, small-polaron theory predicts that the non-diagonal (incoherent hopping) rate is thermally activated and may become the main transport process at higher temperatures (cf., e.g., [94]). Deviations from standard small-polaron theory are expected to occur in the intermediate coupling regime. By means of ED and KPM techniques we are able to study the optical response of Holstein polarons precisely in this regime, at least for small lattices.

AC Conductivity

Our starting point is the Kubo formula for the electrical conductivity at finite temperatures [94],

$$\sigma^{reg}(\omega; T) = \frac{\pi}{\omega N} \frac{1}{Z} \sum_{m,n>0}^{\infty} \left[e^{-\beta E_n} - e^{-\beta E_m} \right] |\langle \psi_n | \hat{\jmath} | \psi_m \rangle|^2 \, \delta(\omega - \omega_{mn}), \quad (40)$$

where $Z = \sum_n^{\infty} e^{-\beta E_n}$ is the partition function and $\beta = T^{-1}$ denotes the inverse temperature ($k_B = 1$). Since in practice the contribution of highly excited phonon states is negligible at the temperatures of relevance, the system is well approximated by a truncated phonon space with at most $M(\lambda, g, \omega_0; T)$ phonons [39]. Then $|\psi_n\rangle$ and $|\psi_m\rangle$ are the eigenstates of H within our truncated Hilbert space. E_n and E_m are the corresponding eigenvalues with $\omega_{mn} = E_m - E_n$.

In order to evaluate temperature-dependent response functions like (40), recently a generalised "two-dimensional" KPM scheme has been proposed [53, 101], which, in our case, can be set up using a current operator density

$$j(x, y) = \sum_{m,n} |\langle \psi_n | \hat{j} | \psi_m \rangle|^2 \, \delta(x - E_n) \, \delta(y - E_m) \,. \tag{41}$$

For the regular part of the conductivity we obtain

$$\sigma^{reg}(\omega) = \frac{2\pi}{\omega N} \frac{1}{Z} \int\limits_{0^+}^{\infty} j(y + \omega, y) \left[e^{-\beta y} - e^{-\beta(y+\omega)} \right] dy \,, \tag{42}$$

where the partition function $Z = 2 \int_{0^+}^{\infty} \rho(E) \exp(-\beta E)$ is easily obtained by integrating over the density of states $\rho(E) = \sum_{n=0}^{D-1} \delta(E - E_n)$, which can be expanded in parallel to $j(x, y)$ (here D is the dimension of the Hilbert space). One advantage of this approach is that the current operator density that enters the conductivity is the same for all temperatures, i.e., it needs to be expanded only once.

Figure 24 gives results for the finite-temperature optical conductivity of a Holstein polaron. Coherent transport related to diagonal transitions within the lowest polaron band is negligible at high temperatures. For instance, the amplitude of the current matrix elements between the degenerate states with momentum $K = \pm\pi/3$ ($K = 0, \pm\pi/3, \pm2\pi/3$, and π are the allowed wave numbers of our 6-site system with periodic boundary conditions) is of the order of 10^{-7} only. In the small polaron limit , where the polaronic sub-bands are roughly separated by the bare phonon frequency, non-diagonal transitions become important for $T > \omega_0$. Let us consider the activated regime in more detail (see Fig. 24 (upper panel)). With increasing temperatures we observe a substantial spectral weight transfer to lower frequencies, and an increase of the zero-energy transition probability in accordance with previous results [103].

In addition, we find a strong resonance in the absorption spectra at about $\omega_0 \sim 2t$, which can be easily understood using a configurational coordinate picture [101]. In order to activate these transitions thermally, the electron has to overcome the "adiabatic" barrier $\Delta = E_{1+} - E_0 = \varepsilon_p/2 - t$, where we have assumed that the first relevant excitation is a state with lattice distortion spread over two neighbouring sites and the particle mainly located at both these sites (in a symmetric $(+)$ or antisymmetric $(-)$ linear combination; $E_{1,\pm} = \mp t - \varepsilon_p/2$). A finite phonon frequency will relax this condition. From Fig. 24, we find the signature to occur above $T > 0.5t$. Obviously this feature is absent in the standard small-polaron transport description which essentially treats the polaron as a quasiparticle without resolving its internal structure.

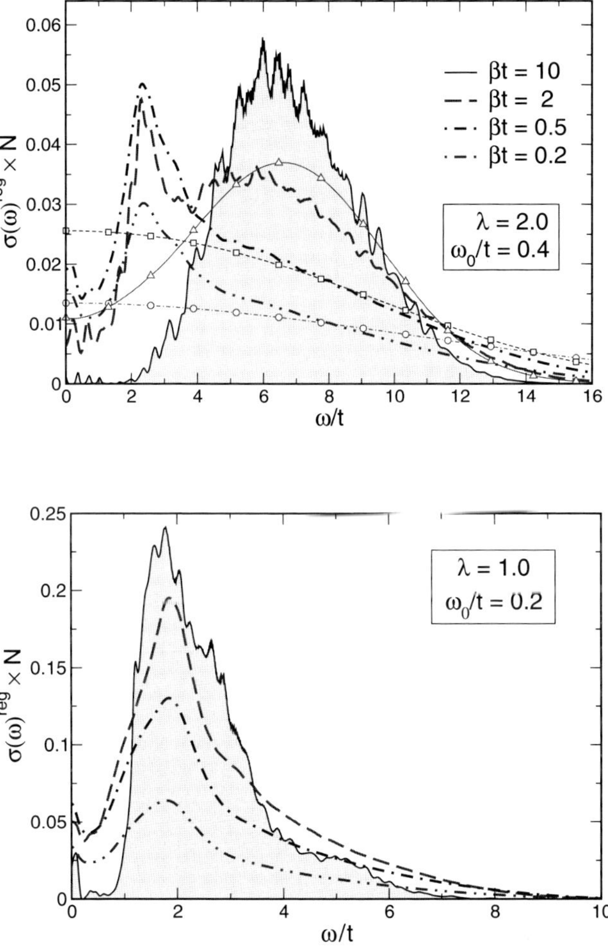

Fig. 24. Optical absorption by Holstein polarons at finite temperatures in the adiabatic strong (upper panel) and intermediate (lower panel) EP coupling regime. Results are obtained by ED for a $N = 6$ site lattice with $M = 45$ phonons. In the upper panel, thin lines with symbols give the analytical results for the small polaron transport [94, 104] at temperatures $\beta t = 2$ (triangles), 0.5 (squares), 0.2 (circles). The deviations observed for high excitation energies at very large temperatures are caused by the necessary truncation of the phonon Hilbert space in ED.

Now let us decrease the EP coupling strength λ keeping $g^2 = 10$ fixed. Results for the optical response in the vicinity of the large to small polaron crossover are depicted in the lower panel of Fig. 24. Here the small polaron maximum has almost disappeared and the $2t$-absorption feature can be activated at very low temperatures ($\Delta \to 0$ for the two-site model with $\lambda = 1$). The gap observed at low frequencies and temperatures is clearly a finite-size effect. The overall behaviour of $\sigma^{reg}(\omega; T)$ resembles that of polarons of intermediate size. At high temperatures these polarons will dissociate readily and the transport properties are equivalent to those of electrons scattered by thermal phonons. Let us emphasise that many-polaron effects become increasingly important in the large-to-small polaron transition region [105] (see also Sect. 8 below). As a result, polaron transport might be changed entirely compared to the one-particle picture discussed so far.

DC Conductivity and Thermopower

We consider dc transport, or $\sigma(\omega)$ in the limit $\omega \to 0$. For simplicity, we consider only a single polaron, or a dilute system of polarons where interactions can be neglected and bipolaron formation is prevented, as by a large repulsive U. We also neglect impurities, which can localise or scatter a polaron.

At zero temperature, the conductivity or mobility of a polaron is infinite. The polaron can be placed in a state of nonzero momentum by a weak electric field acting for a short time. This is an eigenstate, which carries current forever and never decays. At small temperatures $T \ll \omega_0$, an exponentially small number of phonons are thermally excited. The conductivity becomes finite due to scattering of a polaron off thermally excited phonons of density $n_{ph} \sim e^{-\omega_0/T}$. The details depend on the EP scattering process.

In 1D, when a polaron of momentum k encounters a thermally excited phonon, in general part of it is transmitted and part is backscattered. Certain anomalies occur. For example, in the limit of small hopping t, as g approaches 1, the backscattering of the polaron vanishes and the phonon is simultaneously transferred one site in the direction opposite the polaron momentum. The phonon thus recoils opposite to the direction expected, cf. the collision of two balls. This leads to a heat current in the opposite direction as the polaron particle current, which should be observable in the thermopower. A polaron-thermal phonon bound state also exists for sufficiently large g. For this bound state, heat (a phonon excitation) can be transported by an electric field, which again should be observable as a large contribution to the thermopower of the opposite sign as the above. For large g, this bound state or internal polaron excited state can have a much smaller effective mass than the polaron ground state. Perhaps surprisingly, as the temperature increases, the polaron effective mass as measured by the low-frequency ac conductivity can decrease.

We next consider very high temperatures. As T increases, the typical phonon displacement increases as $\widetilde{K}\langle \widehat{x}^2 \rangle = k_B T$, where $\widetilde{K}$ is the phonon spring constant. For quasi-static phonons (large phonon mass), this leads to a disorder potential for the electron that increases without bound as T increases. The disorder Anderson localises the electron, leading to zero dc conductivity. The disorder, however, is not quite static, and rearranges itself on a timescale $\tau \sim 1/\omega_0$. Once every time of order τ, the diagonal energies of the electron site and a neighbouring site become equal, and the electron can hop to a neighbouring site. It is then diffusing with a diffusion constant $\sim a^2\omega_0$, where a is the lattice constant. Using the Einstein relation relating diffusion and mobility, the high temperature resistivity becomes

$$\rho = \frac{\pi k_B T}{n e^2 a^2 \omega_0}. \tag{43}$$

The high temperature resistivity is metallic, i.e. $d\rho/dT > 0$, and can greatly exceed the Ioffe-Regel limit. Numerical studies to confirm or refute this scenario are incomplete.

8 From Few to Many Polarons

Let us now address the important issue of how the character of the (polaronic) quasiparticles may change if we increase the carrier density $n = N_e/N$. Consider first the case of zero electron–electron interaction. Beginning with a noninteracting Fermi gas at $T = 0$, as the Holstein EP interaction g is increased from zero, a singlet superconductor is expected to form. As g increases, the diameter of the Cooper pair decreases. Eventually, the Cooper pair diameter becomes smaller than the distance between Cooper pairs, and the behaviour crosses over from BCS superconductivity to that of Bose condensation, like that of ^{4}He, where the hard core bosons are bipolarons (bound states of two polarons). In this limit, T_c is given approximately by the Bose condensation temperature for ideal bosons of mass m^*, where m^* is the bipolaron mass. The limit of Bose condensation of bipolarons is not given correctly by Eliashberg theory, which describes strong coupling, but not that strong.

8.1 Bipolaron Formation

We investigate how two electrons coupled to phonons may bind together to form a bipolaron, including the bipolaron effective mass, the crossover between two different types of bound states, and the dissociation into two polarons (see also [106, 107]). For problems with more than one electron, the Holstein Hamiltonian is generalised by adding a Hubbard electron-electron interaction term, $U \sum_j n_{j\uparrow} n_{j\downarrow}$. Basis states for the many-body Hilbert space can be written $|b\rangle = |j_1, j_2; \ldots, n_m, n_{m+1}, \ldots, \rangle$, where the up and down electrons are on sites j_1 and j_2, and there are n_m phonons on site m. In a generalisation of the one electron VED method described above, a bipolaron variational space is constructed beginning with an initial state where both electrons are on the same site with no phonons, and operating repeatedly (L-times) with the off-diagonal pieces (t and $\bar{g}$) of the Hamiltonian. All translations of these states are included on an infinite lattice. The method is very efficient in the intermediate coupling regime, where it provides results that are variational in the thermodynamic limit and bipolaron energies that are accurate to 7 digits for the case $L = 18$ and size of the Hilbert space $N_{st} = 2.2 \times 10^6$ phonon and down electron configurations for a given up electron position. In 1D the size of the variational space approximately doubles as L is increased by one, which is the same as for the one electron problem, although the prefactor for two electrons is larger [68].

For large phonon frequency ω_0, the EP interaction leads to a non-retarded attractive on-site interaction of strength $U_0 \equiv 2\omega_0 g^2$. One would expect that as the Hubbard repulsion U becomes larger than this value, the bipolaron would dissociate into two polarons. As can be shown both analytically and numerically, this is not what happens. In the limit of small hopping t, as U exceeds U_0, the bipolaron crosses over from a state S0 with both electrons primarily on the same site, to another bound state S1 with the electrons primarily on nearest neighbour sites. Only for $U > 2U_0$ does the bipolaron dissociate into two polarons. The crossover from $S0$ to $S1$ bipolarons is important in theories of bipolaronic superconductivity applied to real materials, since S1 bipolarons are generally orders of magnitude lighter than S0 bipolarons. Since the superconducting T_c in the dilute limit is inversely proportional to the effective mass, the S1 regime usually provides a more compelling theory.

We now discuss numerical variational results for the singlet bipolaron on an infinite 1D lattice. We have been unable to demonstrate the existence of a bound triplet bipolaron for the Holstein-Hubbard model. Figure 25 shows the ground state electron-electron density correlation function $C(i - j) = \langle \psi_0 | n_i n_j | \psi_0 \rangle$, where $n_i = n_{i\uparrow} + n_{i\downarrow}$ and $|\psi_0\rangle$ denotes the ground state wave function. At $g = 1$, the bipolaron widens with increasing U and transforms into two unbound polarons (which can only move a finite distance apart in the variational space). The value $U = 1.5$ is below the transition to the unbound state at $U_c = 2.17$, calculated by comparing the polaron and bipolaron energies. We see that the probability of electrons occupying the same or neighbouring sites is almost equal. In the unbound regime, the nature of the correlation function changes significantly. At $U = 1.5$, $C(i)$ falls off exponentially, while for $U > U_c$ the typical distance between electrons is the order of the maximum allowed separation L. The electrons can be no farther apart than L in the variational space, although their centre of mass can be anywhere on an infinite lattice. A state of separated polarons is clearly seen for $U = 20$.

Two distinct regimes are seen at $g = 2$ within the bipolaronic region. At $U = 7 < U_0 \equiv 2\omega_0 g^2 = 8$, the correlation function represents the S0 bipolaron, while at $U = 9 > U_0$ we find the largest probability for two electrons to be on neighbouring sites, which is characteristic of the S1 bipolaron. In contrast to previous calculations where phonons were treated classically [108], we find a crossover rather than a phase transition between the two regimes. Moving from strong towards intermediate coupling, the S0 and S1 bipolarons consist of longer exponential tails extending over many lattice sites, and the two regimes can no longer be distinguished. The precision of presented correlation functions in the bipolaron regime is within the size of the plot symbols in the thermodynamic limit.

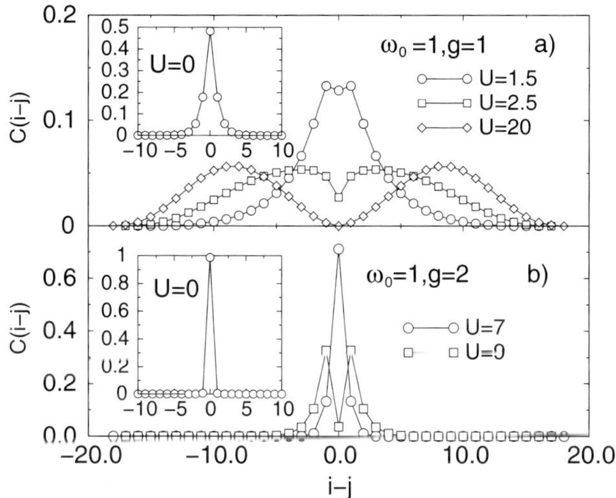

Fig. 25. Electron-electron correlation function $C(i-j)$ calculated at $\omega_0 = 1$, a) $g = 1$ and b) $g = 2$ for different values of U, with $L = 18$. The two ordinate axes have a different range. Insets show results for $U = 0$. All curves are normalised, $\sum_i C(i) = 1$.

Figure 26(a) plots the bipolaron mass ratio $R_m = m_{bi}/2m_{po}$ vs. U for different values of ω_0 and g. In all cases presented in Fig. 26, R_m approaches 1 as U approaches $U = U_c$ in agreement with a state of two free polarons. At fixed $\omega_0 = 1$ the bipolaron mass ratio increases by several orders of magnitude with increasing g at $U = 0$. Increasing U sharply decreases R_m in the S0 regime. Note that the mass scale is logarithmic. In the S1 regime with $U > U_0$, R_m is small, as predicted by the strong coupling result.

In the dilute bipolaron regime, the bipolaron isotope effect is the same as the classic superconductivity isotope effect for T_c. The bipolaron isotope effect, shown in Fig. 26(b), is large in the strong coupling ($\omega_0 = 1, g = 2$) and small U regime, where its value is somewhat below the large g strong coupling prediction $\alpha_{I,S0} \sim 2g^2 - \frac{1}{4} = 7.75$. With increasing U, α_I decreases and in the S1 regime approaches $\alpha_{I,S1} = g^2/2 = 2$. A kink is observed in the crossover regime. With decreasing g or ω_0, α_I also decreases.

The phase diagram $U_c(g)$ is shown in Fig. 27 at fixed $\omega_0 = 1$. Numerical results, shown as circles, indicate the phase boundary between two dissociated

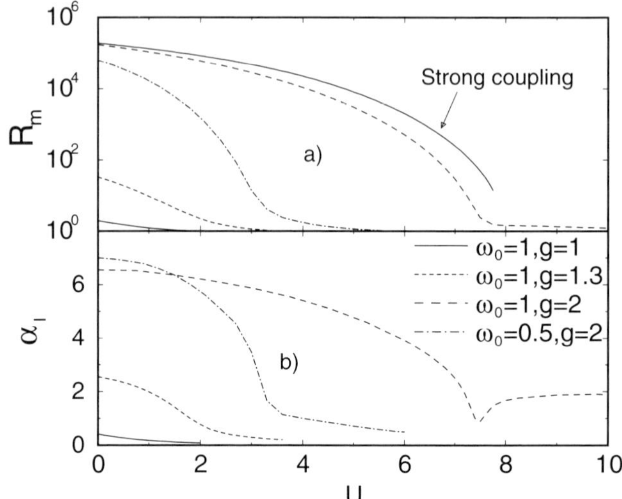

Fig. 26. a) The mass ratio $R_m = m_{bi}/2m_{po}$ vs. U and b) the bipolaron isotope effect α_I vs. U. Numerical results are for $L = 18$. Results for R_m at $\omega_0 = 0.5$ are obtained by extrapolating $L \to \infty$. Precision in all curves is within the linewidth in the thermodynamic limit, except for α_I with $\omega_0 = 0.5$, where the error is estimated to be $\pm 5\%$. The thin line in (a) is the strong coupling expansion result for $\omega_0 = 1$, $g = 2$. Polaron masses in units of the noninteracting electron mass are $m_{po} = 1.35, 1.76, 10.4, 3.06$ from top to bottom respectively.

polarons each having energy E_{po} and a bipolaron bound state with energy E_{bi}. In the inset of Fig. 27 we show the bipolaron binding energy defined as $\Delta = E_{bi} - 2E_{po}$. The phase diagram is obtained from $\Delta = 0$. The dashed line, given by $U_0 = 2\omega_0 g^2$, is a reasonable estimate for the phase boundary at small g. At large g the dashed line roughly represents the crossover between a massive S0 and lighter S1 bipolaron. The S1 region grows with increasing g. The dot-dashed line is the phase boundary between S1 and the unbound polaron phase, as obtained by degenerate strong coupling perturbation theory. Numerical results approach this line at larger g. The dot-dashed line asymptotically approaches $U_c = 4\omega_0 g^2$.

8.2 Many-Polaron Problem

We consider how polarons evolve from the dilute to the concentrated limit, in the regime where spinless or fully spin-polarised fermions prevent bipolaron formation. In 1D with open boundary conditions, the spinless fermion problem is equivalent to infinite Hubbard U. While for very strong EP coupling no significant changes are expected due to the existence of rather independent small (self-trapped) polarons with negligible residual interaction (assuming spinless fermions or strong enough electron-electron repulsion to prevent bipolaron formation), a density-driven crossover from a state with large polarons

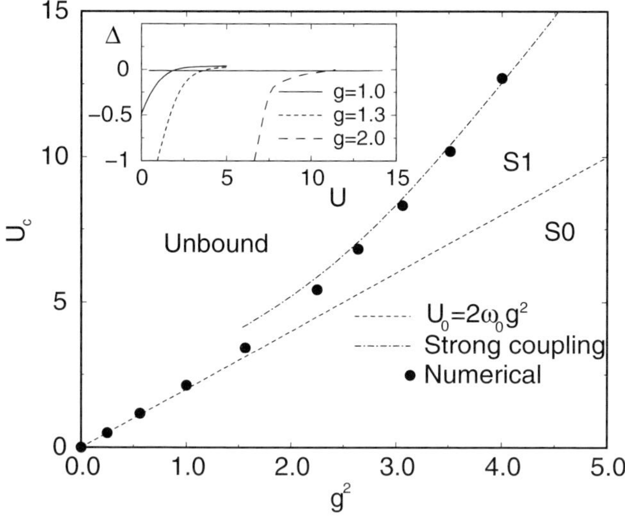

Fig. 27. Phase diagram and binding energy Δ in units of t (inset) calculated at $\omega_0 = 1$. Numerical results are circles. For greater accuracy, results near the weak and strong coupling regime were obtained by extrapolating $L \to \infty$.

to a metal with weakly dressed electrons should occur in the intermediate-coupling regime. This issue has recently been investigated theoretically by ED [109], QMC [105], and variational canonical transformation [75] methods, and is known to be of experimental relevance, e.g., in $La_{2/3}(Sr/Ca)_{1/3}MnO_3$ films [110].

In the spinless fermion Holstein model, the above-mentioned density-driven transition from large polarons to weakly EP-dressed electrons is expected to be possible only in 1D, where large polarons exist at intermediate coupling. The situation is different for Fröhlich-type models [7, 111, 112] with long-range EP interaction, in which large-polaron states exist even for strong coupling and in $D > 1$.

To set the stage, we first comment on the evolution of the one-electron spectral function $A(k, \omega)$ with increasing electron density n in the weak- and strong-coupling limiting cases [105]. In the former the spectra bear a close resemblance to the free-electron case for all n, i.e., there is a strongly dispersive band running from $-2t$ to $2t$, which can be attributed to weakly dressed electrons with an effective mass close to the non-interacting value. As expected, the height (width) of the peaks increases (decreases) significantly in the vicinity of the Fermi momentum. In the opposite strong-coupling limit the spectrum exhibits an almost dispersionless coherent polaron band $\forall\, n < 0.5$. Besides, there are two incoherent features located above and below the Fermi energy, broadened $\propto \varepsilon_p$, which are due to phonon-mediated transitions to high-energy electron states. The most important point, however, is the clear separation of the coherent band from the incoherent parts even at large n, indicating that small polarons are well-defined quasiparticles in the strong-coupling regime, even at high carrier density.

Figure 28 displays the inverse photoemission $[A^+(k,\omega)]$ and photoemission spectra $[A^-(k,\omega)]$ at intermediate EP coupling strength, determined by CPT. At low densities, $n = 0.1$ (upper panel), we can easily identify a (coherent) polaron band crossing the Fermi energy level $E_F = \mu(T \to 0)$, the latter being situated at the point where A^- and A^+ intersect. This large-polaron band has rather small electronic spectral weight especially away from E_F and flattens at large k, as known from single-polaron studies (see Sect. 6, Fig. 20). Below this band, there exist equally spaced phonon satellites, reflecting the Poisson distribution of phonons in the ground state. Above E_F there is a broad dispersive incoherent feature whose maximum closely follows the dispersion relation of free particles.

As the density n increases, a well-separated coherent polaron band can no longer be identified. At about $n \simeq 0.3$ the deformation clouds of the (large) polarons start to overlap leading to a mutual (dynamical) interaction between the particles. Increasing the carrier density further, the polaronic quasiparticles dissociate, stripping their phonon cloud. This is the case shown in the lower panel of Fig. 28. Now diffusive scattering of electrons and phonons seems to be the dominant interaction mechanism. As a result both the phonon peaks in $A^-(k,\omega)$ and the incoherent part of $A^+(k,\omega)$ are washed out, the spectra broaden and ultimately merge into a single wide band. Most notably, the incoherent excitations now lie arbitrarily close to the Fermi level. Obviously the low-energy physics of the system can no longer be described by single-particle small-polaron theory.

9 Polaronic Effects in Strongly Correlated Systems

The interplay of electron-electron and electron-phonon interactions in the formation of dressed quasiparticles is becoming the focus of attention in many contexts, including conducting polymers, ferroelectrics, halide-bridged transition-metal chain complexes, and several important classes of perovskites. Especially research on high-T_c superconductivity (HTSC) and colossal magnetoresistance (CMR) has spurred intense investigations of the competition or, if possible, of the cooperation of these two fundamental interactions (for a recent review see [26], and references therein).

Many experiments have indicated substantial EP interaction in the high-T_c cuprates. The relevance of EP coupling can be seen from the experimental observation of phonon renormalisation [113]. Ion channelling [114, 115], neutron scattering [116] and photo-induced absorption measurements [117] proved the existence of large anharmonic lattice fluctuations, which may be responsible for local phonon-driven charge instabilities in the planar CuO_2 electron system [118, 119]. Photo-induced absorption experiments [120], infrared spectroscopy [121] and reflectivity measurements [122] indicate the for-

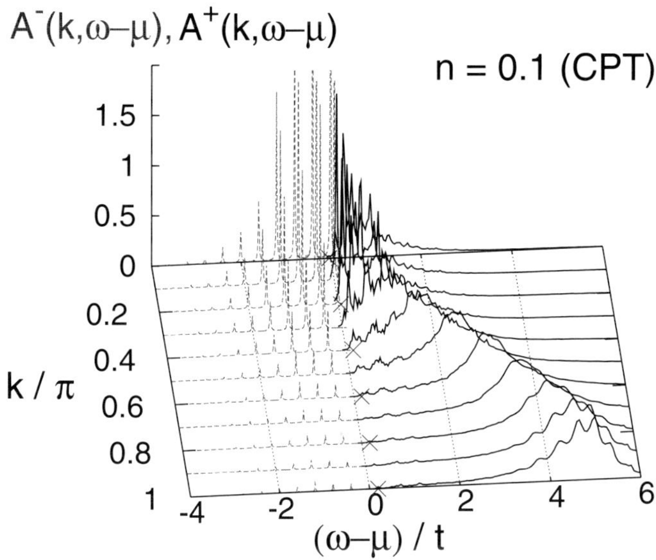

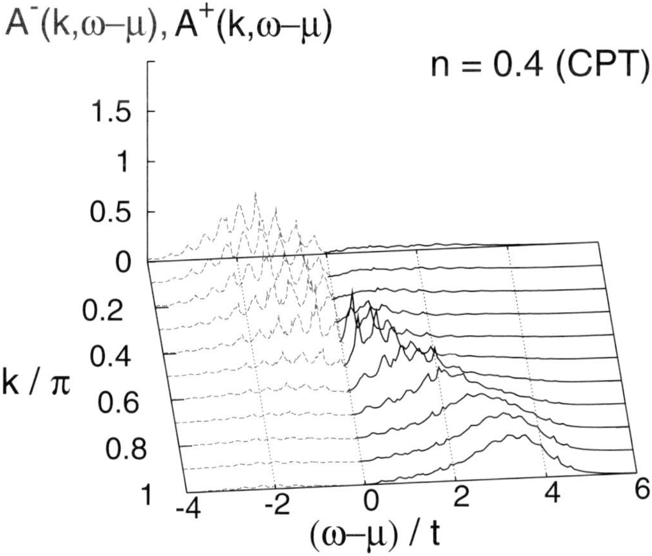

Fig. 28. Single particle spectral functions $A^-(k,\omega)$ (dashed lines) and $A^+(k,\omega)$ (solid lines) for two characteristic band fillings, $n = 0.1$ and $n = 0.4$, at $\omega_0/t = 0.4$ and $\lambda = 1$ ($T = 0$). Results are obtained by CPT using $N_c = 10$). Crosses track the small-polaron band determined by ED.

mation of small polarons in the insulating parent compounds La_2CuO_{4+y} and Nd_2CuO_{4-y} of the hole- and electron-doped superconductors $La_{2-x}Sr_xCuO_{4+y}$ and $Nd_{2-x}Ce_xCuO_{4-y}$, respectively. Recently angle-resolved photoemission spectroscopy data were interpreted in terms of strong EP coupling giving rise to self-localisation of holes (hole polarons) [123]. Based on these experimental findings, several theoretical groups [124–134] promote a (bi)polaronic scenario for HTSC.

Even stronger evidence for polaron formation in doped charge transfer oxides is provided by experiments on the nickelates $La_{2-x}Sr_xNiO_4$ [23, 135]. The isostructural compounds La_2CuO_4 and La_2NiO_4 show a remarkable difference upon the substitution of La by Sr. Both materials become metallic upon doping, but in the nickelates a nearly total substitution of La for Sr is necessary. Also in $La_{2-x}Sr_xNiO_{4+y}$ no superconductivity is found for any x. A resolution of this problem might be given by extended LDA calculations [136], which show that the nickelates are much more susceptible to a breathing polaron instability than the cuprates. The reason is the much stronger magnetic confinement effect of additional holes and nickel spins. These low–spin composite holes are nearly entirely prelocalised and the EP coupling becomes much more effective in forming polarons. For the composition $La_{1.5}Sr_{0.5}NiO_4$ (quarter filling, $x = 0.5$), electron diffraction measurements show a commensurate superstructure spot at the (π, π)-point, which has been interpreted as a sign of truly 2D ordering of breathing-type polarons, i.e., as a polaronic superlattice.

Localised lattice distortions are also suggested to play an important role in determining the electronic and magnetic properties of hole-doped manganese oxides of the form $La_{1-x}[Sr, Ca]_xMnO_3$ [24, 137]. In the region $x_{MI} \sim 0.2 < x < 0.5$ these compounds show a transition from a metallic ferromagnetic low-temperature phase to an insulating paramagnetic high-temperature phase associated with a spectacularly large negative magnetoresistive response to an applied magnetic field [138]. Both breathing-mode collapsed (Mn^{4+}) and (anti) Jahn-Teller distorted (Mn^{3+}) sites are created simultaneously when the holes are localised in passing the metal-insulator transition [139, 140]. The relevance of small polaron transport above T_c is obvious from the activated behaviour of the conductivity [93]. Consequently many theoretical studies focused on polaronic approaches [141–148]. Polaronic features have been established by a variety of experiments. For example, high-temperature thermopower [149, 150] and Hall mobility measurements [25] confirmed the polaronic nature of charge carriers in the paramagnetic phase. More directly the existence of polarons has been demonstrated by atomic pair distribution [151], x-ray and neutron scattering studies [152–154]. Interestingly it seems that the charge carriers partly retain their polaronic character well below T_c, as proved, e.g., by neutron pair-distribution-function analysis [155] and resistivity measurements [156].

Regardless of whether the EP coupling acts as a secondary pairing interaction for HTSC in the cuprates, is responsible for the charge ordering in the nickelates, or triggers the CMR phenomenon in the manganites, EP and particularly polaronic effects need to be reconsidered for the case of strong electronic correlations realised in these materials. For instance, Coulomb or spin exchange interactions may lead to a "prelocalisation" of the charge carriers. Then a rather weak EP coupling can cause polaronic band narrowing and that way might drive the system further into the strongly correlated regime. In the remaining part of this section we will keep track of this problem and present some exact results for composite spin/orbital-lattice polarons.

9.1 Hole Polarons in the Holstein $t - J$ Model

Electronic motion in weakly doped Mott insulators like the HTSC cuprates is determined by the constraint of no double occupancy of sites and anti-ferromagnetic exchange between nearest-neighbour spins. The generic model studied in this context is the 2D $t - J$ Hamiltonian,

$$H_{tJ} = -t \sum_{\langle ij \rangle \sigma} \left(\tilde{c}_{i\sigma}^\dagger \tilde{c}_{j\sigma} + \text{H.c.} \right) + J \sum_{\langle ij \rangle} \left(\boldsymbol{S}_i \boldsymbol{S}_j - \frac{1}{4} \tilde{n}_i \tilde{n}_j \right), \qquad (44)$$

acting in a projected Hilbert space, i.e. $\tilde{c}_{i\sigma}^{(\dagger)} = c_{i\sigma}^{(\dagger)} (1 - \tilde{n}_{i,-\sigma})$, $\tilde{n}_i = \sum_\sigma \tilde{c}_{i\sigma}^\dagger \tilde{c}_{i\sigma}$, and $\boldsymbol{S}_i = \sum_{\sigma,\sigma'} \tilde{c}_{i\sigma}^\dagger \boldsymbol{\tau}_{\sigma\sigma'} \tilde{c}_{i\sigma'}$. Within the $t - J$ model the bare transfer amplitude of electrons (t) sets the energy scale for incoherent transport, while the Heisenberg interaction (J) allows for spin flips leading to coherent hole motion at the bottom of a band with an effective bandwidth determined by J. $J < t$ corresponds to the situation in the cuprates, e.g. $J/t \simeq 0.4$ with $t \simeq 0.3$ eV is commonly used to model the quasi-2D $La_{2-x}Sr_xCuO_4$ system.

In order to study polaronic effects in systems exhibiting besides strong antiferromagnetic exchange a substantial EP coupling the Hamiltionian (44) is often supplemented by a Holstein-type interaction term

$$H = H_{tJ} - \sqrt{\varepsilon_p \omega_0} \sum_i \left(b_i^\dagger + b_i \right) \tilde{h}_i + \omega_0 \sum_i \left(b_i^\dagger b_i + \frac{1}{2} \right) \qquad (45)$$

($\tilde{h}_i = 1 - \tilde{n}_i$ denotes the local density operator of the spinless hole). The resulting Holstein $t - J$ model (HtJM) (45) takes the coupling to the hole as dominant source of the particle-lattice interaction. In the cuprate context an unoccupied site, i.e. a hole, corresponds to a Zhang-Rice singlet (formed by Cu $3d_{x^2-y^2}$ and O $2p_{x,y}$ hole orbitals) for which the coupling should be much stronger than for the occupied (Cu^{2+}) site [157–159]. The hole-phonon coupling constant is denoted by $\varepsilon_p = g^2 \omega_0$, and ω_0 is the bare phonon frequency of an internal vibrational degree of freedom of lattice site i.

The changes of the quasiparticle properties due to the combined effects of hole-phonon/magnon correlations are expected to be very complex and as yet there exist no well-controlled analytical techniques to address this problem.

444   H. Fehske and S. A. Trugman

Naturally such a dressed hole quasiparticle will show the characteristics of both "lattice" and "magnetic" (spin) polarons.

We solved the Holstein $t - J$ Hamiltonian on finite lattices using ED, KPM and a phonon Hilbert space truncation method. To control our truncation procedure we carefully checked the ground-state energy and the weight of the m-phonon states ($|c^m|^2$) in the ground state as a function of the number of phonons retained (M). Convergence is assumed to be achieved if the relative error of both $E_0(M)$ and $|c^m|^2$ is less than 10^{-5}.

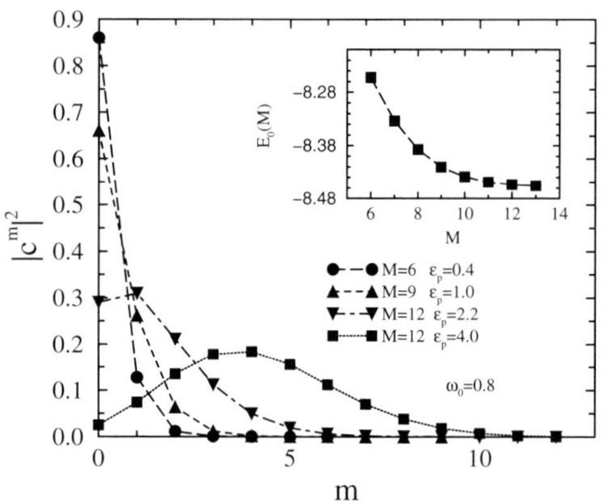

Fig. 29. Phonon-weight function $|c^m|^2$ and ground-state energy $E_0(M)$ for the 2D HtJM with $J = 0.4$ (throughout this section all energies will be measured in units of t).

Figure 29 shows $|c^m|^2$ for the single-hole case at weak, intermediate and strong EP couplings. The curves $|c^m|^2$ are bell-shaped and their maxima correspond to the most probable number of phonon quanta. The importance of multi-phonon states becomes apparent especially in the adiabatic strong-coupling regime $\varepsilon_p \gg t$, ω_0.

Let us start the analysis of the 2D HtJM with a discussion of the single-hole spectral function

$$A_{\boldsymbol{K}}(\omega) = \sum_{n,\sigma} \left| \langle \psi_{n,\boldsymbol{K}}^{(N_e-1)} | \tilde{c}_{\boldsymbol{Q}-\boldsymbol{K},\sigma} | \psi_{0,\boldsymbol{Q}}^{(N_e)} \rangle \right|^2 \delta \left[\omega - (E_{n,\boldsymbol{K}}^{(N_e-1)} - E_{0,\boldsymbol{Q}}^{(N_e)}) \right] . \quad (46)$$

Figure 30 displays $A_{\boldsymbol{K}}(\omega)$ for the allowed (nonequivalent) momenta $\boldsymbol{K}$ of a ten-site square lattice. To visualise the intensities (spectral weights) connected with the various peaks in each $\boldsymbol{K}$-sector we have also shown the integrated density of states

$$N(\omega) = \int_{-\infty}^{\omega} d\omega' \frac{1}{N} \sum_{\boldsymbol{K}} A_{\boldsymbol{K}}(\omega') . \quad (47)$$

In the absence of EP coupling, of course, we reproduced the single-particle spectrum of the pure $t - J$ model [160]. Here one observes a quasiparticle pole corresponding directly to the coherent single-hole ground state (having momentum $(3\pi/5, \pi/5)$ on a ten-site lattice) separated by a pseudogap of about J from the lower edge of a broad incoherent continuum being $\simeq 6t$ wide. In the weak coupling regime the mass renormalisation of the coherent quasiparticle band due to the hole-phonon Holstein coupling is small compared with that arising from hole-spin interactions (magnetic polaron regime). In particular, the integrated density of states is barely changed from that of the pure $t - J$ model. The new structures, nevertheless observed in the $A_{\boldsymbol{K}}$ spectra shown in Fig. 30(a), correspond to predominantly "phononic" side bands separated from the particle-spin excitations by multiples of the bare phonon frequency ω_0. These phonon resonances have less and less "electronic" spectral weight the more phonons are involved. This is because $A_{\boldsymbol{K}}(\omega)$ measures the overlap of these excited states with the state obtained by creating a hole in the zero-phonon Heisenberg ground state.

With increasing ε_p the lowest peaks in each $\boldsymbol{K}$-sector start to separate from the rest of the spectrum. These states become very close in energy and finally a narrow well-separated lattice hole-polaron band evolves in the strong-coupling case (see Fig. 30(b)). Now the hole is heavily dressed by phonons and the quasiparticle pole strength is strongly suppressed (cf. Fig. 31). At the same time spectral weight is transferred to the high-energy part and the whole spectrum becomes incoherently broadened. Therefore we observe an overall smoothing of $N(\omega)$. The gap to the next higher energy band is of the order of ω_0. This excitation will be triggered by an one-phonon absorption process. The crossover to the lattice hole-polaron state is accompanied by a strong increase in the on-site hole-phonon correlations [38], indicating that the lattice polaron quasiparticle comprising a self-trapped hole and the phonon cloud is mainly confined to a single lattice site (small hole polaron). Most notably, compared to the non-interacting single-electron Holstein model [74] or the spinless fermion Holstein-t model [161], the critical EP coupling strength for lattice polaron formation is considerably reduced due to magnetic prelocalisation effects.

To illustrate the formation of the lattice hole-polaron band in some more detail, we first determined the "coherent" band dispersion $E_{\boldsymbol{K}}$ of the 2D HtJM for a 16-site lattice. The band structure is shown in the left panel of Fig. 31 along the principal directions in the Brillouin zone. The minima of the quasiparticle dispersion are found to be located at the momenta $\boldsymbol{K} = (\pm\pi/2, \pm\pi/2)$ (the hidden symmetry of the 4×4 cluster leads to an accidental degeneracy with the $\boldsymbol{K} = (\pm\pi, 0)$, $(0, \pm\pi)$ states). At weak EP couplings the energy dispersion is not significantly changed from that of the standard $t - J$ model, provided that the phonon frequency exceeds the effective bandwidth of the magnetic lattice polaron ($\omega_0 = 0.8 \geq \Delta E_{tJ}$).

Next we evaluated the wave-function renormalisation factor for different band states,

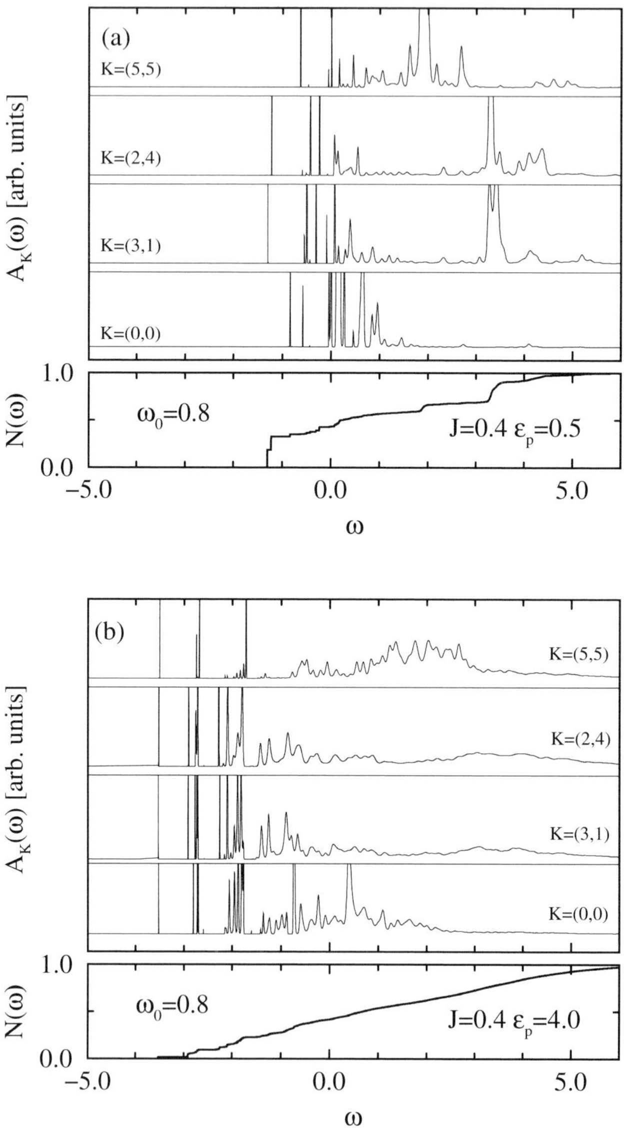

Fig. 30. Wavevector resolved single-hole spectral functions $A_K(\omega)$ and integrated spectral weight $N(\omega)$ for the 2D Holstein $t-J$ model at weak (a) and strong (b) EP coupling. The results, taken from [39], were obtained for a tilted $\sqrt{10} \times \sqrt{10}$ cluster with periodic boundary conditions (K-vectors are given in units of $(\pi/5, \pi/5)$).

$$Z_K = \frac{\left|\langle\psi_{0,K}^{(N_e-1)}|\widetilde{c}_{Q-K,\sigma}|\psi_{0,Q}^{(N_e)}\rangle\right|^2}{\left|\langle\psi_{0,Q}^{(N_e)}|\widetilde{c}_{Q-K,\sigma}^{\dagger}\widetilde{c}_{Q-K,\sigma}|\psi_{0,Q}^{(N_e)}\rangle\right|^2} \, , \tag{48}$$

where $|\psi_{0,Q}^{(N_e-1)}\rangle$ denotes the one-hole state being lowest in energy in the Q-sector. Z_K can be taken as a measure of the "contribution" of the hole (dressed at $\varepsilon_p = 0$ by spin-wave excitations only) to the composite spin/lattice polaron (having total momentum K). The data obtained at weak EP coupling unambiguously confirm the different nature of band states in this regime (see Fig. 31 right panel). We found practically zero-phonon "hole" states at the band minima ($K = (3\pi/5, \pi/5)$, triangles down) and "phonon" states, which are only weakly affected by the hole, around the (flat) band maxima ($K = (\pi, \pi)$, triangles up). With increasing ε_p, a strong "mixing" of holes and phonons takes place, whereby both quantum objects completely lose their own identity. Concomitantly Z_K decreases for the "hole–like" states but increases (first of all) for the "phonon-like" states. At large ε_p, a small lattice hole polaron is formed, which, according to the numerical data, has an extremely small spectral weight. Then the question arises whether the lattice hole polaron is a "good" quasiparticle in the sense that one can construct a quasiparticle operator, $\widetilde{c}_{K\sigma} \to \widetilde{d}_{K\sigma}$, having large spectral weight at the lowest pole in the spectrum. It was demonstrated that such a composite electron/hole-phonon (polaron) operator could be constructed for the Holstein model [74] as well as for the $t - J$ model coupled to buckling/breathing modes [162].

In Fig. 32 we show the optical response in the framework of the single-hole 2D HtJM. In the weak EP coupling regime and for phonon frequencies

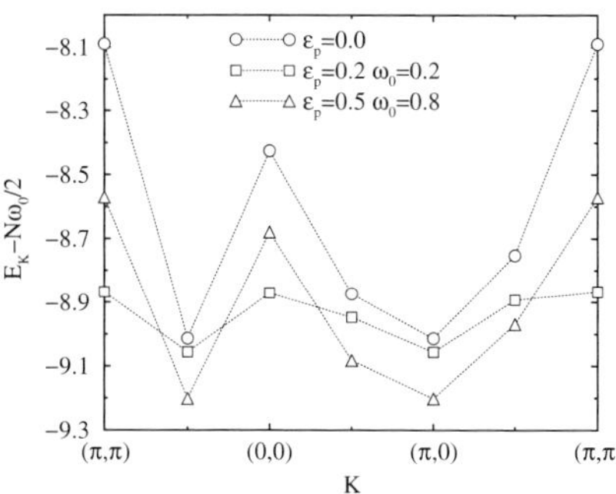
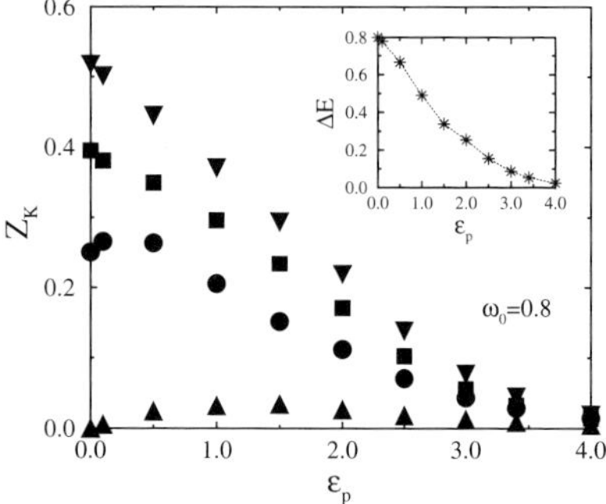

Fig. 31. Left panel: Band dispersion for the 2D HtJM on a 16-site lattice ($J = 0.4$). Right panel: Spectral weight factor Z_K as a function of the EP coupling strength ε_p for the different K vectors of the ten-site lattice: (3,1) (triangles down), (2,4) (squares), (0,0) (circles), and (5,5) (triangles up) [in units of $(\pi/5, \pi/5)$]. The inset shows the narrowing of the coherent bandwidth ΔE with increasing ε_p. From [39], ©(1999) by the American Physical Society.

$\omega_0 > \Delta E_{tJ}$, we recover the main features of the optical absorption spectrum of the 2D $t - J$ model [163], i.e., an "anomalous" broad mid-infrared band ($J < \omega < 2t$), separated from the Drude peak ($\mathcal{D}\delta(\omega)$; not shown) by a pseudo-gap $\simeq J$, and an "incoherent" tail up to $\omega \simeq 7t$ (cf. Fig. 32(a)). At larger EP couplings, the overlap with excited multi-phonon states is enlarged and the optical response is enhanced at higher energies. This redistribution of spectral weight from low to high energies can be seen in Fig. 32(b). As expected the transition to the lattice hole-polaron state, at about $\varepsilon_p^c(J = 0.4, \omega_0 = 0.8) \simeq 2.0$, is accompanied by the development of a broad maximum in $\sigma^{reg}(\omega)$, whereas the Drude weight as well as the low-frequency optical response become strongly suppressed.

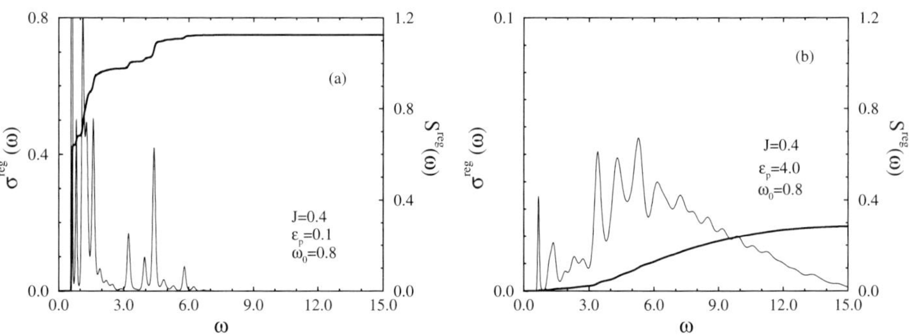

Fig. 32. Optical conductivity in the 2D HtJM with $J = 0.4$. $\sigma^{reg}(\omega)$ and $S^{reg}(\omega)$ were calculated for a ten-site lattice with 15 phonons.

Let us now make contact with the experimentally observed characteristics of the mid-infrared (MIR) spectra in the doped perovskites [22, 160, 164, 165]). The simple 2D HtJM seems to contain the key ingredients to describe, at least qualitatively, the principal features of the optical absorption spectra of these compounds. This can be seen by comparing Figs. 32(a) and (b), corresponding to the weak and strong EP coupling situations realized in the cuprate ($La_{2-x}Sr_xCuO_4$) and nickelate ($La_{2-x}Sr_xNiO_4$) systems, respectively. The EP interaction ratio ε_p/t in the nickelates is estimated to be about one order of magnitude larger than in the cuprates because of the much smaller transfer amplitude ($t \simeq 0.08$ eV [135]). According to the internal structure of the low-spin state, the hopping transport of spin-1/2 composite holes in a spin-1 background is rather complex; implying, within an effective single-band description, a strong reduction of the transition matrix elements [166]. A striking feature of the absorption spectra in the cuprate superconductors is the presence of a MIR band, centred at about 0.5 eV in lightly doped $La_{2-x}Sr_xCuO_4$ (which, using $t \sim 0.3$ eV, means that $\omega \sim 1.5$). Such a strong MIR absorption is clearly visible in Fig. 32(a). Obviously it is quite difficult to distinguish the spectral weight, produced by the dressing of the hole due to the "bag" of reduced antiferromagnetism in its neighbourhood [167], from other (e.g. hole-phonon coupling) processes that may contribute to the MIR band observed experimentally. The results presented for the HtJM in Fig. 32(b) support the claims, however, that the MIR band in the cuprates has a mainly "electronic" origin, i.e., the lattice polaron effects are rather weak [160]. The opposite is true for their isostructural counterpart, the nickelate system, where the MIR absorption band has been ascribed by many investigators to "polaronic" origin [135, 142]. Within the HtJM such a situation can be modelled by the parameter set used in Fig. 32(b). If we fix the energy scale by $t = 0.08$ eV, the maximum in the optical absorption is again located at about 0.5 eV. The whole spectrum clearly shows lattice polaron characteristics, where it seems that the lattice hole polarons are of small-to-intermediate size [16]. Most notably, we are able to reproduce the experimentally observed asymmetry in the shape of the spectrum, in particular the very gradual decay of $\sigma^{reg}(\omega)$ at high energies. It is worth mentioning that this behaviour cannot be obtained from a simple fit to the analytical expressions derived for the small polaron hopping conductivity [135, 168, 169]. Exploiting the f-sum rule we found that there are almost no contributions from band-like carriers in agreement with the experimental findings [135, 142].

Next let us briefly discuss the two-hole problem. In order to study hole-binding effects, we have calculated the hole-hole correlation function

$$C_{ho-ho}(|i-j|) = \langle \psi_0(\varepsilon_p, J)|\widetilde{h}_i\widetilde{h}_j|\psi_0(\varepsilon_p, J)\rangle. \tag{49}$$

Results for $C_{ho-ho}(|i-j|)$ are presented in Fig. 33. At weak EP coupling, $C_{ho-ho}(|i-j|)$ becomes maximum at the largest distance of the ten-site lattice, while in the intermediate EP coupling regime the preference is on next NN pairs. As expected, further increasing ε_p, the maximum in $C_{ho-ho}(|i-j|)$ is shifted to the shortest possible distance, indicating hole-hole attraction. At $\varepsilon_p \gg 1$, the two holes become "self–trapped" sharing a sizeable common lattice distortion, i.e., a nearly immobile hole-bipolaron is formed. The behaviour of C_{ho-ho} is found to be qualitatively similar for higher (lower) phonon frequencies (see inset), except that the crossings of different hole-hole correlation functions occur at larger (smaller) values of ε_p, which again shows the importance of both parameter ratios $\lambda = \varepsilon_p/2Dt$ and $g = \sqrt{\varepsilon_p/\omega_0}$.

Finally let us consider the quarter-filled band case. Here, we have investigated the simpler spinless fermion model (total $S^z = S^z_{max}$). In accordance with previous approximate treatments based on an inhomogeneous variational Lang-Firsov approach [161], we found, as the EP coupling increases, evidence for a transition from a free polaron state to a 2D polaronic superlattice, where the holes are self-trapped on every other site. This crossover is signalled by a pronounced peak in the charge structure factor $S_c(\pi, \pi)$. To visualise the correlations in this state in more detail, in Fig. 34 we have depicted $C_{ho-ho}(|i-j|)$ and the corresponding hole-phonon density correlation function $C_{ho-ph}(|i-j|) = \langle \psi_0|\widetilde{h}_i b_j^\dagger b_j|\psi_0\rangle$ as a function of $|i-j|$. Our exact results

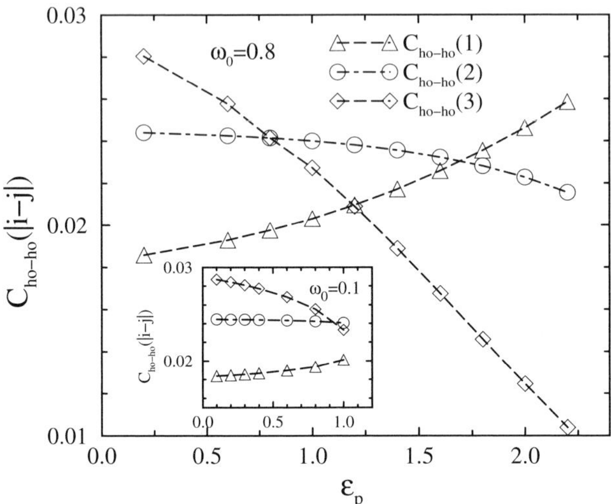

Fig. 33. Non-equivalent hole-hole pair correlation functions $C_{ho-ho}(|i-j|)$ in the two-hole ground state of the HtJM at various ε_p. Here 1, 2, and 3 label NN, next NN, and third NN distances, respectively. Reprinted from [177] with permission from Elsevier.

clearly show the phonon-dressing of the holes and the tendency towards CDW formation as observed, e.g., in $La_{1.5}Sr_{0.5}NiO_4$.

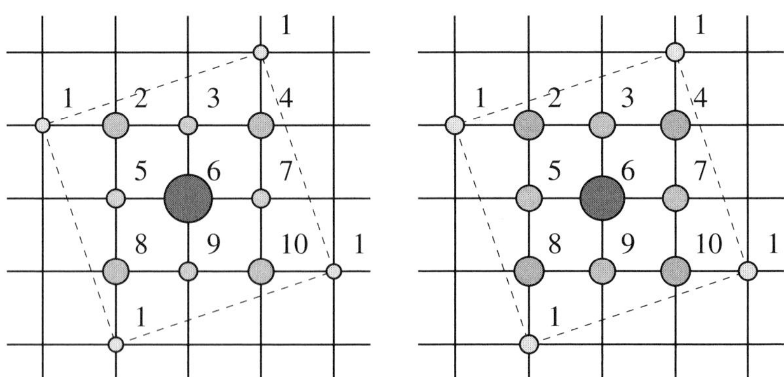

Fig. 34. $C_{ho-ho}(|6-j|)$ (left) and $C_{ho-ph}(|6-j|)$ (right) are displayed at $\varepsilon_p = 3$ and $\omega_0 = 0.8$, where both diameter and gray level of the circles are proportional to the correlation strength. Reprinted from [177] with permission from Elsevier.

9.2 Lattice Polarons in a Generalised Double-Exchange Model

The key elements of the electronic structure of the CMR ($La_{1-x}[Sr, Ca]_x MnO_3$) compounds are the partially filled 3d states. The cubic environment of the Mn sites within the perovskite lattice results in a crystal field splitting of Mn d-orbitals into e_g and t_{2g}. In the case of zero doping ($x = 0$) there are four electrons per Mn site which fill up the three t_{2g} levels and one e_g level, and by Hund's rule coupling, form a $S = 2$ spin state. Doping will remove the electron from the e_g level, and by hopping via bridging oxygen sites the resulting holes acquire mobility.

Due to the specific symmetry of the manganese d and oxygen p orbitals, the transfer of the e_g-electrons shows a pronounced orbital anisotropy. In the limit of large on-site Coulomb (U) and Hund (J_H) interactions the electron transfer is strongly affected by the spin of the core electrons as well. Concentrating on the link between magnetic correlations and transport, early studies on lanthanum manganites attributed the low-T metallic behaviour to Zener's double-exchange (DE) mechanism [170, 171], which maximises the hopping of a strongly Hund's rule coupled electron in a polarised spin background.

As discussed above, it has been realized that physics beyond DE is important not only to explain the phase diagram of the manganites but also the CMR transition itself. The orbital degeneracy in the ground state of Mn^{3+} ions connects the system to the lattice, making it susceptible to Jahn-Teller polaron formation. There are two types of lattice distortions which are important in manganites. First the partially filled e_g states of the Mn^{3+} ion are Jahn-Teller active, i.e., the system can gain energy from a quadrupolar symmetric elongation of the oxygen octahedra which lifts the e_g degeneracy. A second possible deformation is an isotropic shrinking of an MnO_6 octahedron. This "breathing"-type distortion couples to changes in the e_g charge density, i.e., is always associated with the presence of an Mn^{4+} ion.

Restricting the electronic Hilbert space to the large Hund's rule states given by the spin-2 orbital doublet state ^5E $[t_2^3(^4A_2)e]$ for Mn^{3+} (d^4) and the spin-$\frac{3}{2}$ orbital singlet state $^4A_2$ $[t_2^3]$ for Mn^{4+} (d^3), within 2nd order perturbation theory the following Hamiltonian results (for details of the derivation and notation see [172, 173]):

$$H = H_{DE} + H^{2^{nd}\ order}_{spin-orbital} + H_{electron-JT} + H_{electron-breathing} + H_{phonon}$$

$$= \sum_{i,\delta,\alpha,\beta} (a_{i,\uparrow}a^\dagger_{i+\delta,\uparrow} + a_{i,\downarrow}a^\dagger_{i+\delta,\downarrow})\, t^\delta_{\alpha\beta}\, c^\dagger_{i,\alpha} n_{i,\bar\alpha} n_{i+\delta,\bar\beta} c_{i+\delta,\beta}$$

$$+ \sum_{i,\delta,\kappa,\lambda} (J^\delta_{\kappa\lambda}\, \boldsymbol{S}_i \boldsymbol{S}_{i+\delta} + \Delta^\delta_{\kappa\lambda})\, P^\kappa_i P^\lambda_{i+\delta}$$

$$+\bar g \sum_i \left[(n_{i,\varepsilon} - n_{i,\theta})(b^\dagger_{i,\theta} + b_{i,\theta}) + (d^\dagger_{i,\theta}d_{i,\varepsilon} + d^\dagger_{i,\varepsilon}d_{i,\theta})(b^\dagger_{i,\varepsilon} + b_{i,\varepsilon}) \right]$$

$$+\widetilde{g}\sum_i (n_{i,\theta} + n_{i,\varepsilon} - 2n_{i,\theta}n_{i,\varepsilon})(b^\dagger_{i,a_1} + b_{i,a_1})$$

$$+\bar{\omega}_0 \sum_i \left[b^\dagger_{i,\theta}b_{i,\theta} + b^\dagger_{i,\varepsilon}b_{i,\varepsilon} \right] + \widetilde{\omega}_0 \sum_i b^\dagger_{i,a_1} b_{i,a_1} \, . \tag{50}$$

The effective low-energy Hamiltonian H contains Schwinger bosons $a^{(\dagger)}_{i,\mu}$, i.e. $2\boldsymbol{S}_i = a^\dagger_{i,\mu}\tau_{\mu\nu}a_{i,\nu}$ ($\mu,\nu \in \{\uparrow,\downarrow\}$), fermionic holes $c^{(\dagger)}_{i,\alpha}$, phonons $b^{(\dagger)}_{i,\alpha}$ ($\alpha \in \{\theta,\varepsilon\}$), and orbital projectors $P_i^{\kappa(\lambda)}$ ($\kappa, \lambda \in \{\xi,\eta,\zeta\}$). In (50), the first term, being proportional to t, corresponds to the DE interaction [170, 171]. The second term appears to be a bit more involved, since a rather large number of accessible virtual excitations (proportional to t^2 and t_π^2) contribute (see [172, 173]). However, in all cases it is basically the product of a Heisenberg-type spin interaction and two orbital projectors.

The coupling between the orbital degree of freedom of the e_g electrons and the optical phonon modes to lowest order can be modelled by the $E \otimes e$ Jahn-Teller Hamiltonian (third term) and a Holstein-type interaction (fourth term). The energy of the dispersionless optical phonons are given within the harmonic approximation (fifth term).

For analytical methods the above Hamiltonian (50) is far too complex to be understood in full detail, and even its numerical solution on finite lattices is hard. Using high performance computers and the phonon basis optimisation outlined in Sect. 2.2, we were able to calculate the ground-state properties of a small four-site cluster and to address, in particular, short-range correlations between the charge, spin, orbital and lattice degrees of freedom [172, 173] (see also [62]). Figures 35 and 36 give a glimpse of these results. We assumed $t = 0.4$ eV and $t/t_\pi = 3$ for the hopping integrals and characterised the magnetic "order" according to the total spin of the ground state.

Undoped manganites ($\mathrm{LaMnO_3}$, $\mathrm{PrMnO_3}$) usually exhibit A-type antiferromagnetic order and strong Jahn-Teller distortion of the ideal perovskite structure. The origin of the observed magnetic order has been subject to discussions. While different band structure calculations [174] emphasise the importance of lattice distortions for the stability of antiferromagnetism, purely electronic mechanisms were also favoured [175]. In our microscopic model (50), at $x = 0$, only the second and third term will be active, and without EP interaction the competition of the spin-orbital contributions depends sensitively on the values of Coulomb and Hund's rule coupling. Starting from the "ferromagnetic" phase increasing either U or g changes the magnetic order of the ground state to "antiferromagnetism" [172] (see also Fig. 35 (upper left panel)). At larger EP interaction the system tends to develop static Jahn-Teller distortions, which also fixes the orbital pattern and subsequently the spin order. The change of orbital, spin and phonon correlations is illustrated schematically in Fig. 36 (left panels).

Our numerical calculations corroborate the enhancement of ferromagnetic correlations for the weakly doped case (see Fig. 35, upper left panel; $x = 0.25$).

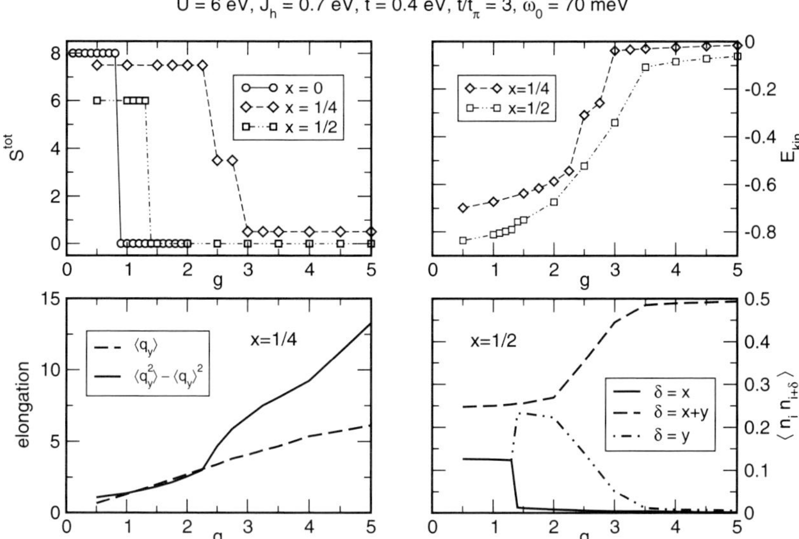

Fig. 35. Upper panels: Total spin S^{tot} and kinetic energy E_{kin} as a function of electron-phonon coupling strength g at various doping levels x. Lower panels: Expectation values $\langle q_y \rangle$ and $\langle q_y^2 \rangle - \langle q_y \rangle^2$ of the bond length in y direction at $x = 1/4$ (left) and density-density correlations at $x = 1/2$ (right). Results obtained by ED for the microscopic model (50) on a four site plaquette, where $g = \bar{g}/\omega_0$, and $\omega_0 = \bar{\omega} = \widetilde{\omega}$ is assumed [173].

However, if strong electron-phonon coupling causes self-trapping of the carriers the spin order switches back to antiferromagnetism. This coincidence can be seen by comparing the total spin of the cluster and the kinetic energy in the ground state, both depicted in the upper panels of Fig. 35. Obviously the change in the magnetic order is accompanied by the appearance of a lattice distortion and a signature in the fluctuation of the bond length ($\propto \langle q_{x/y}^2 \rangle - \langle q_{x/y} \rangle^2$), which reminds us of the data measured close to the critical temperature by Booth et al. [176] (lower left panel). The orbital orientation at the sites surrounding the hole-site is sketched in Fig. 36 (middle panels). Increasing g isolates the lattice sites, each optimising EP interaction individually and uncorrelated with the neighbours.

At half-filling ($x = 0.5$) the picture is more involved. Here strong Coulomb and EP interactions tend to order the charges in the diagonal direction, i.e., in an AB-type structure (compare Fig. 35, lower right panel). This allows for a rather large antiferromagnetic spin exchange $\propto t^2/J_H$. Consequently ferromagnetic order is unstable at much lower values of g. The ferromagnetic to antiferromagnetic transition is not connected to charge localisation and causes only a tiny jump of the kinetic energy. Considering the most relevant eigenstate of the bond orbital density matrix, we observe a symmetric order of

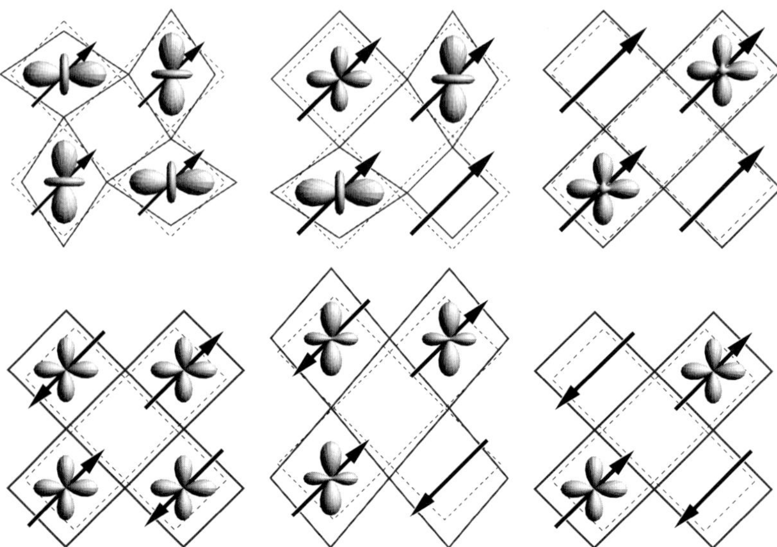

Fig. 36. Schematic evolution of charge, spin, orbital and lattice correlation with increasing doping [$x = 0$, $x = 0.25$ and $x = 0.5$ (from left to right)] at rather weak (upper panels) and strong (lower panels) EP couplings.

complex orbitals along the diagonal [172]. After charge localisation is achieved at large g, neighbouring sites are again uncorrelated with respect to orbital ordering and are in some real mixed-orbital state.

Of course, the results of this section will not provide a quantitative analysis or description of the real 3D HTSC and CMR materials. Nevertheless, ED of even such small systems gives valuable insights into the correlations and driving interactions behind the rich phase diagram of the cuprates and manganites. Moreover the above ED data may serve as a benchmark for approximate theories.

10 Conclusions

In summary, we have performed an extensive numerical analysis of the Holstein model. Combining variational Lanczos diagonalisation, density matrix renormalisation group, kernel polynomial expansion, and cluster perturbation theory techniques we solved for properties of the Holstein polaron and bipolaron problems. Numerical solution of the Holstein model means that we determined the ground-state and low-lying excited states with arbitrary precision in the thermodynamic limit for any dimension. Moreover, we calculated the spectral properties (e.g. photoemission and phonon spectra), optical response and thermal transport, as well as the dynamics of polaron formation. Our approach takes into account the full quantum dynamics of the electrons and phonons and yields unbiased results for all electron-phonon interaction strengths and phonon frequencies, but is of particular value in the intermediate-coupling regime, where perturbation theories and other analytical techniques fail.

Most importantly, polaron formation represents a continuous crossover of the ground state. Nevertheless we found indications for a true phase transition in the first excited state, where a polaron plus phonon system changes from unbound at weak electron–phonon coupling to bound at strong coupling. Obviously electron and phonon excitations become intimately, dynamically connected in the process of polaron formation. Concerning ground-state and spectral properties the (quasi-free) electron (small) polaron self-trapping crossover is related to (i) a significant increase of the particle effective mass, (ii) a substantial narrowing of the polaronic band dispersion, and (iii) a strongly suppressed (electronic) quasiparticle residue. At the same time we observe (iv) an enhancement of the (on-site) electron-phonon correlations and the formation of a phonon drag, (v) a loss of kinetic energy, and (vi) a drop of the Drude weight, accompanied (vii) by a maximum in the spectral weight contained in the regular part of the optical response. All these features are found to be much more pronounced in higher dimensions. The study of bipolarons showed that the two-site bipolaron has a significantly reduced mass and isotope effect compared to the on-site bipolaron, and is bound in the strong-coupling regime up to twice the on-site Coulomb repulsion.

Although we have now achieved a rather complete picture of the single Holstein polaron and bipolaron problem (perhaps dispersive phonons, longer-ranged electron-phonon interaction, finite-temperature and disorder effects deserve closer attention), the situation is disconcerting in the case of a finite carrier density. Here a density-driven crossover from large polarons to quasi-free electrons scattered by unbound phonons might occur in the non-adiabatic intermediate coupling regime. Besides electron-phonon coupling competes with sometimes strong electronic correlations, e.g., in quasi-1D MX chains, quasi-2D high-T_c superconductors, 3D charge-ordered nickelates, or bulk colossal magneto-resistance manganites. The corresponding microscopic models contain (extended) Hubbard, Heisenberg or double-exchange terms, and maybe also a coupling to orbital degrees of freedom, so that they can hardly be solved even numerically with the same precision as the Holstein model. First exact results obtained for the case of a single carrier on finite lattices give strong evidence that the tendency towards lattice (electron-, hole- or Jahn-Teller-) polaron formation is enhanced in strongly correlated electron systems. A more thorough investigation of these systems materials and models will definitely be a great challenge for solid-state theory in the near future.

Acknowledgement. We would like to thank A. Alvermann, I. Batistić, B. Bäuml, A. R. Bishop, J. Bonča, F. X. Bronold, H. Büttner, F. Göhmann, G. Hager, M. Hohenadler, D. Ihle, E. Jeckelmann, T. Katrašnik, L.-C. Ku, J. Loos, G. Schubert, H. Röder, R. N. Silver, A. Weiße and G. Wellein for valuable discussions. This work was supported by Deutsche Forschungsgemeinschaft through SPP1073 FE398/1-3 and Grant No. 436 TSE 113/33/0-3, and the U.S. Department of Energy. Furthermore, we acknowledge generous computer granting by Leibniz-Rechenzentrum München and Norddeutscher Verbund für Hoch- und Höchstleistungsrechnen (HLRN).

References

1. L.D. Landau, Phys. Z. Sowjetunion **3**, 664 (1933)
2. S.I. Pekar, Zh. Eksp. Teor. Fiz. **16**, 341 (1946)
3. Y.A. Firsov, *Polarons* (Izd. Nauka, Moscow, 1975)
4. E.I. Rashba, in *Excitons*, ed. by E.I. Rashba, M.D. Sturge (North-Holland, Amsterdam, 1982), p. 543
5. Y.E. Perlin, M. Wagner (eds.), *The Dynamical Jahn-Teller Effect in Localized Systems.* No. 7 in Modern Problems in Condensed Matter Sciences (North-Holland, Amsterdam, 1984)
6. A.L. Shluger, A.M. Stoneham, J. Phys. Condens. Matter **5**, 3049 (1993)
7. H. Fröhlich, Adv. Phys. **3**, 325 (1954)
8. H. Fröhlich, Fortschr. Phys. **22**, 159 (1974)
9. J.T. Devreese, this volume, and references therein
10. T. Holstein, Ann. Phys. (N.Y.) **8**, 325 (1959)
11. T. Holstein, Ann. Phys. (N.Y.) **8**, 343 (1959)
12. F.M. Peeters, J.T. Devreese, Phys. Status Solidi B **112**, 219 (1982)
13. H. Spohn, Ann. Phys. (N.Y.) **175**, 278 (1987)
14. H. Löwen, Phys. Rev. B **37**, 8661 (1988)
15. B. Gerlach, H. Löwen, Rev. Mod. Phys. **63**, 63 (1991)
16. D. Emin, in *Polarons and bipolarons in High–T_c superconductors and related materials*, ed. by E.K.H. Salje, A.S. Alexandrov, W.Y. Liang (Cambridge University Press, Cambridge, 1995)
17. G. Wellein, H. Fehske, Phys. Rev. B **58**, 6208 (1998)
18. J. Ranninger, in *Polarons in Bulk Materials and Systems With Reduced Dimensionality, International School of Physics Enrico Fermi*, vol. 161, ed. by G. Iadonisi, J. Ranninger, G. De Filippis (IOS Press, Amsterdam, 2006), *International School of Physics Enrico Fermi*, vol. 161, pp. 1–25
19. J.T. Devreese, Bull. Soc. Belge de Phys. (Ser. III) **4**, 259 (1963)
20. N. Tsuda, K. Nasu, A. Yanese, K. Siratori, *Electronic Conduction in Oxides* (Springer-Verlag, Berlin, 1990)
21. Y. Bar-Yam, T. Egami, J.M. de Leon, A.R. Bishop, *Lattice Effects in High–T_c Superconductors* (World Scientific, Singapore, 1992)
22. E.K.H. Salje, A.S. Alexandrov, W.Y. Liang, *Polarons and Bipolarons in High Temperature Superconductors and Related Materials* (Cambridge University Press, Cambridge, 1995)
23. C.H. Chen, S.W. Cheong, A.S. Cooper, Phys. Rev. Lett. **71**, 2461 (1993)
24. A.J. Millis, Nature **392**, 147 (1998)
25. M. Jaime, H.T. Hardner, M.B. Salamon, M. Rubinstein, P. Dorsey, D. Emin, Phys. Rev. Lett. **78**, 951 (1997)
26. T. Egami, in *Polarons in bulk materials and systems with reduced dimensionality, International School of Physics Enrico Fermi*, vol. 161, ed. by G. Iadonisi, J. Ranninger, G.D. Filipis (IOS Press, Amsterdam, 2006), *International School of Physics Enrico Fermi*, vol. 161, pp. 101–117

27. D. Emin, Adv. Phys. **22**, 57 (1973)
28. Y. Toyozawa, Prog. Theor. Phys. **26**, 29 (1961)
29. A.B. Migdal, Sov. Phys. JETP **7**, 999 (1958)
30. I.G. Lang, Y.A. Firsov, Zh. Eksp. Teor. Fiz. **43**, 1843 (1962)
31. F. Marsiglio, Physica C **244**, 21 (1995)
32. A.S. Alexandrov, N.F. Mott, *High Temperature Superconductors and Other Superfluids* (Taylor & Francis, London, 1994)
33. H. Sumi, J. Phys. Soc. Jpn. **36**, 770 (1974)
34. S. Ciuchi, F. de Pasquale, S. Fratini, D. Feinberg, Phys. Rev. B **56**, 4494 (1997)
35. F. Marsiglio, Phys. Lett. A **180**, 280 (1993)
36. A.S. Alexandrov, V.V. Kabanov, D.K. Ray, Phys. Rev. B **49**, 9915 (1994)
37. E.V.L. de Mello, J. Ranninger, Phys. Rev. B **55**, 14872 (1997)
38. G. Wellein, H. Röder, H. Fehske, Phys. Rev. B **53**, 9666 (1996)
39. B. Bäuml, G. Wellein, H. Fehske, Phys. Rev. B **58**, 3663 (1998)
40. H. De Raedt, A. Lagendijk, Phys. Rev. B **27**, 6097 (1983)
41. E. Berger, P. Valášek, W. v. d. Linden, Phys. Rev. B **52**, 4806 (1995)
42. P.E. Kornilovitch, E.R. Pike, Phys. Rev. B **55**, R8634 (1997)
43. M. Hohenadler, H.G. Evertz, W. von der Linden, Phys. Rev. B **69**, 024301 (2004)
44. P. Kornilovitch, this volume
45. M. Hohenadler, W. von der Linden, this volume
46. A.S. Mishchenko, N. Nagaosa, this volume, and references therein
47. A.H. Romero, D.W. Brown, K. Lindenberg, J. Chem. Phys **109**, 6540 (1998)
48. E. Jeckelmann, S.R. White, Phys. Rev. B **57**, 6376 (1998)
49. C. Zhang, E. Jeckelmann, S.R. White, Phys. Rev. Lett. **80**, 2661 (1998)
50. E. Jeckelmann, Phys. Rev. B **66**, 045114 (2002)
51. J.K. Cullum, R.A. Willoughby, *Lanczos Algorithms for Large Symmetric Eigenvalue Computations*, vol. I & II (Birkhäuser, Boston, 1985)
52. R.N. Silver, H. Röder, A.F. Voter, D.J. Kress, J. of Comp. Phys. **124**, 115 (1996)
53. A. Weiße, G. Wellein, A. Alvermann, H. Fehske, Rev. Mod. Phys. **78**, 275 (2006)
54. D. Sénéchal, D. Perez, M. Pioro-Ladrière, Phys. Rev. Lett. **84**, 522 (2000)
55. M. Hohenadler, M. Aichhorn, W. von der Linden, Phys. Rev. B **68**, 184304 (2003)
56. A. Weiße, H. Fehske, G. Wellein, A.R. Bishop, Phys. Rev. B **62**, R747 (2000)
57. V. Cautaudella, G.D. Filippis, G. Iadonisi, C.A. Perroini, in *Polarons in Bulk Materials and Systems With Reduced Dimensionality*, ed. by G. Iadonisi, J. Ranninger, G. De Filippis (IOS Press, Amsterdam, 2006), International School of Physics Enrico Fermi, pp. 119–130
58. V. Cataudella, G.D. Filippis, C.A. Perroini, this volume
59. S.A. Trugman, L.C. Ku, J. Bonča, J. Supercond. **17**, 193 (2004)
60. J. Bonča, S.A. Trugman, I. Batistić, Phys. Rev. B **60**, 1633 (1999)
61. L.C. Ku, S.A. Trugman, J. Bonča, Phys. Rev. B **65**, 174306 (2002)
62. S. El Shawish, J. Bonča, L.C. Ku, S.A. Trugman, Phys. Rev. B **67**, 014301 (2003)
63. H. De Raedt, A. Lagendijk, Phys. Rev. Lett. **49**, 1522 (1982)
64. M. Capone, W. Stephan, M. Grilli, Phys. Rev. B **56**, 4484 (1997)
65. G. Wellein, H. Fehske, Phys. Rev. B **56**, 4513 (1997)

66. H. Fehske, D. Ihle, J. Loos, U. Trapper, H. Büttner, Z. Phys. B **94**, 91 (1994)
67. H. Fehske, G. Wellein, G. Hager, A. Weiße, A.R. Bishop, Phys. Rev. B **69**, 165115 (2004)
68. J. Bonča, T. Katrašnik, S.A. Trugman, Phys. Rev. Lett. **84**, 3153 (2000)
69. S. Sykora, A. Hübsch, K.W. Becker, G. Wellein, H. Fehske, Phys. Rev. B **71**, 045112 (2005)
70. D. Augier, D. Poilblanc, Eur. Phys. J. B **1**, 19 (1998)
71. E.R. Davidson, J. of Comp. Phys. **17**, 87 (1975)
72. R.N. Silver, H. Röder, Phys. Rev. E **56**, 4822 (1997)
73. W. Stephan, Phys. Rev. B **54**, 8981 (1996)
74. H. Fehske, J. Loos, G. Wellein, Z. Phys. B **104**, 619 (1997)
75. J. Loos, M. Hohenadler, H. Fehske, J. Phys. Condens. Matter **18**, 2453 (2006)
76. H. Fehske, A. Alvermann, M. Hohenadler, G. Wellein, in *Polarons in Bulk Materials and Systems With Reduced Dimensionality, International School of Physics Enrico Fermi*, vol. 161, ed. by G. Iadonisi, J. Ranninger, G. De Filippis (IOS Press, Amsterdam, 2006), *International School of Physics Enrico Fermi*, vol. 161, pp. 285–296
77. A.A. Gogolin, Phys. Status Solidi B **109**, 95 (1982)
78. D. Emin, T. Holstein, Phys. Rev. Lett. **36**, 323 (1976)
79. D.R. Jennison, D. Emin, Phys. Rev. Lett. **51**, 1390 (1983)
80. V.V. Kabanov, O.Y. Mashtakov, Phys. Rev. B **47**, 6060 (1993)
81. J.A. Kenrow, T.K. Gustafson, Phys. Rev. Lett. **77**, 3605 (1996)
82. N.H. Ge, et al., Science **279**, 202 (1998)
83. A. Sugita, T. Saito, H. Kano, M. Yamashita, T. Kobayashi, Phys. Rev. Lett. **86**, 2158 (2001)
84. S.L. Dexheimer, A.D.V. Pelt, J.A. Brozik, B.I. Swanson, Phys. Rev. Lett. **84**, 4425 (2000)
85. S. Tomimoto, H. Nansei, S. Saito, T. Suemoto, J. Takeda, , S. Kurita, Phys. Rev. Lett. **81**, 417 (1998)
86. R.D. Averitt, A.J. Taylor, J. Phys. Condens. Matter **14**, R1357 (2002)
87. G.P. Donati, G. Rodriguez, A.J. Taylor, J. Opt. Soc. Am. B **17**, 1077 (2000)
88. B. Krummheuer, V.M. Axt, T. Kuhn, Phys. Rev. B **65**, 195313 (2002)
89. D.W. Brown, K. Lindenberg, B. J.West, J. Chem. Phys **84**, 1574 (1986)
90. L.C. Ku, Ph.D. thesis, UCLA (2003)
91. D. Emin, Phys. Rev. B **48**, 13691 (1993)
92. A.S. Alexandrov, N.F. Mott, *Polarons and Bipolarons* (World Scientic, Singapore, 1995)
93. D.C. Worledge, L. Miéville, T.H. Geballe, Phys. Rev. B **57**, 15267 (1998)
94. G.D. Mahan, *Many-Particle Physics* (Kluwer Academic/Plenum Publishers, New York, 2000)
95. H.G. Reik, D. Heese, J. Phys. Chem. Solids **28**, 581 (1967)
96. J. Loos, Z. Phys. B **71**, 161 (1988)
97. Y.A. Firsov, Semiconductors **29**, 515 (1995)
98. Y.A. Firsov, this volume
99. S. Fratini, S. Ciuchi, Phys. Rev. Lett. **91**, 256403 (2003)
100. S. Fratini, S. Ciuchi, Phys. Rev. B **74**, 075101 (2006)
101. G. Schubert, G. Wellein, A. Weiße, A. Alvermann, H. Fehske, Phys. Rev. B **72**, 104304 (2005)
102. E.K. Kudinov, D.N. Mirlin, Y.A. Firsov, Fiz. Tverd. Tela **11**, 2789 (1969)

103. E.L. Nagaev, Sov. Phys. Solid State **4**, 1611 (1963)
104. H. Böttger, V.V. Bryksin, *Hopping conduction in solids* (Akademie Verlag, Berlin, 1985)
105. M. Hohenadler, D. Neuber, W. von der Linden, G. Wellein, J. Loos, H. Fehske, Phys. Rev. B **71**, 245111 (2005)
106. A.S. Alexandrov, this volume
107. S. Aubry, this volume
108. L. Proville, S. Aubry, Physica (Amsterdam) **133D**, 307 (1998)
109. M. Capone, M. Grilli, W. Stephan, Eur. Phys. J. B **11**, 551 (1999)
110. C. Hartinger, F. Mayr, J. Deisenhofer, A. Loidl, T. Kopp, Phys. Rev. B **69**, 100403(R) (2004)
111. A.S. Alexandrov, P.E. Kornilovitch, Phys. Rev. Lett. **82**, 807 (1999)
112. J.T. Devreese, J. Tempere, Phys. Rev. B **64**, 104504 (2001)
113. H. Rietschel, J. Low Temp. Phys. **95**, 293 (1994)
114. R.P. Sharma, L.E. Rehn, P.M. Baldo, J.Z. Lin, Phys. Rev. Lett. **62**, 2869 (1989)
115. T. Haga, et al., Phys. Rev. B **41**, 826 (1990)
116. T. Egami, et al., in *Electronic Structure and Mechanisms of High Temperature Superconductivity*, ed. by J. Ashkenazi, G. Vezzoli (Plenum Press, New York, 1991)
117. Y.H. Kim, et al., Phys. Rev. B **38**, 6478 (1988)
118. J.D. Jorgenson, Physics Today **44**, 34 (1991)
119. J. Zhong, H.B. Schüttler, Phys. Rev. Lett. **69**, 1600 (1992)
120. D. Mihailović, C.M. Foster, K. Voss, A.J. Heeger, Phys. Rev. B **42**, 7989 (1990)
121. P. Calvani, Solid State Commun. **91**, 113 (1994)
122. J.P. Falck, A. Levy, M.A. Kastner, R.J. Birgenau, Phys. Rev. B **48**, 4043 (1993)
123. O. Rösch, O. Gunnarson, X. Zhou, T. Yoshida, T. Sasagawa, A. Fujimori, Z. Hussain, Z.X. Shen, S. Uchida, Phys. Rev. Lett. **95**, 227002 (2005)
124. G. Verbist, F.M. Peeters, J.T. Devreese, Physica Scripta **T 39**, 66 (1991)
125. D. Emin, Phys. Rev. B **45**, 5525 (1992)
126. D. Emin, Phys. Rev. Lett. **72**, 1052 (1994)
127. A.S. Alexandrov, A.B. Krebs, Usp. Fiz. Nauk **162**, 1 (1992)
128. A.S. Alexandrov, J. Ranninger, Physica C **198**, 360 (1992)
129. A.S. Alexandrov, A.M. Bratkovsky, N.F. Mott, E.K.H. Salje, Physica C **215**, 359 (1993)
130. J.E. Hirsch, Phys. Rev. B **47**, 5351 (1993)
131. N.F. Mott, J. Phys. Condens. Matter **5**, 3487 (1993)
132. A.S. Alexandrov, N.F. Mott, Int. J. Mod. Phys. B **8**, 2075 (1994)
133. J. Ranninger, Acta Phys. Pol. A **85**, 89 (1994)
134. G. Iadonisi, V. Cataudella, D. Ninno, M.L. Chiofalo, Phys. Lett. A **196**, 359 (1995)
135. X.X. Bi, P.C. Eklund, Phys. Rev. Lett. **70**, 2625 (1993)
136. V.I. Anisimov, M.A. Korotin, J. Zaanen, O.K. Andersen, Phys. Rev. Lett. **68**, 345 (1992)
137. G. Zhao, K. Conder, H. Keller, K.A. Müller, Nature **381**, 676 (1996)
138. S. Jin, T.H. Tiefel, M. McCormack, R.A. Fastnach, R. Ramesh, L.H. Chen, Science **264**, 413 (1994)
139. P.B. Allen, V. Perebeinos, Phys. Rev. B **61**, 10747 (1999)
140. S.J.L. Billinge, T. Proffen, V. Petkov, J.L. Sarrao, S. Kycia, Phys. Rev. B **62**, 1203 (2000)

141. A.S. Alexandrov, A.M. Bratkovsky, Phys. Rev. Lett. **82**, 141 (1999)
142. P. Calvani, P. Dore, S. Lupi, A. Paolone, P. Maselli, P. Guira, B. Ruzicka, S.W. Cheong, W. Sadowski, J. Supercond. **10**, 293 (1997)
143. J.D. Lee, B.I. Min, Phys. Rev. B **55**, 12454 (1997)
144. M. Quijada, J. Černe, J.R. Simpson, II.D. Drew, K.H. Ahn, A.J. Millis, R. Shreekala, R. Ramesh, M. Rajeswari, T. Venkatesan, Phys. Rev. B **58**, 16093 (1998)
145. K.H. Kim, J.H. Jung, T.W. Noh, Phys. Rev. Lett. **81**, 1517 (1998)
146. K.A. Müller, J. Supercond. **12**, 3 (1999)
147. D.S. Dessau, T. Saitoh, C.H. Park, Z.X. Shen, P. Villella, N. Hamada, Y. Morimoto, Y. Tokura, J. Supercond. **12**, 273 (1999)
148. A. Weiße, J. Loos, H. Fehske, Phys. Rev. B **68**, 024402 (2003)
149. M. Jaime, M.B. Salamon, M. Rubinstein, R.E. Treece, J.S. Horwitz, D.B. Chrisey, Phys. Rev. B **54**, 11914 (1996)
150. T.T.M. Palstra, A.P. Ramirez, S.W. Cheong, B.R. Zegarski, P. Schiffer, J. Zaanen, Phys. Rev. B **56**, 5104 (1997)
151. S.J.L. Billinge, R.G. DiFrancesco, G.H. Kwei, J.J. Neumeier, J.D. Thompson, Phys. Rev. Lett. **77**, 715 (1996)
152. S. Shimomura, N. Wakabayashi, H. Kuwahara, Y. Tokura, Phys. Rev. Lett. **83**, 4389 (1999)
153. L. Vasiliu-Doloc, S. Rosenkranz, R. Osborn, S.K. Sinha, J.W. Lynn, J. Mesot, O.H. Seeck, G. Preosti, A.J. Fedro, J.F. Mitchell, Phys. Rev. Lett. **83**, 4393 (1999)
154. P. Dai, J.A. Fernandez-Baca, N. Wakabayashi, E.W. Plummer, Y. Tomioka, Y. Tokura, Phys. Rev. Lett. **85**, 2553 (2000)
155. D. Louca, T. Egami, E.L. Brosha, H. Röder, A.R. Bishop, Phys. Rev. B **56**, R8475 (1997)
156. G.M. Zhao, V. Smolyaninova, W. Prellier, H. Keller, Phys. Rev. Lett. **84**, 6086 (2000)
157. H. Fehske, H. Röder, A. Mistriotis, H. Büttner, J. Phys. Condens. Matter **5**, 3635 (1993)
158. R. Fehrenbacher, Phys. Rev. B **49**, 12230 (1994)
159. A. Sherman, M. Schreiber, Phys. Rev. B **52**, 10621 (1995)
160. E. Dagotto, Rev. Mod. Phys. **66**, 763 (1994)
161. H. Fehske, H. Röder, G. Wellein, A. Mistriotis, Phys. Rev. B **51**, 16582 (1995)
162. T. Sakai, D. Poilblanc, D.J. Scalapino, Phys. Rev. B **55**, 8445 (1997)
163. D. Poilblanc, Phys. Rev. B **44**, 9562 (1991)
164. D.B. Tanner, T. Timusk, in *Physical Properties of High–Temparature Superconductors III*, ed. by D.M. Ginsberg (World Scientific, Singapore, 1992), p. 363
165. A.S. Alexandrov, V.V. Kabanov, D.K. Ray, Physica C **224**, 247 (1994)
166. J. Loos, H. Fehske, Phys. Rev. B **56**, 3544 (1997)
167. J.R. Schrieffer, X.G. Wen, S.C. Zhang, Phys. Rev. Lett. **60**, 944 (1988)
168. H.G. Reik, Z. Phys. **203**, 364 (1967)
169. Y.Y. Suzuki, P. Pincus, A.J. Heeger, Phys. Rev. B **44**, 7127 (1991)
170. C. Zener, Phys. Rev. **82**, 403 (1951)
171. K. Kubo, N. Ohata, J. Phys. Soc. Jpn. **33**, 21 (1972)
172. A. Weiße, H. Fehske, Eur. Phys. J. B **30**, 487 (2002)
173. A. Weiße, H. Fehske, New J. Phys. **6**, 158 (2004)

174. D.D. Sarma, N. Shanthi, S.R. Barman, N. Hamada, H. Sawada, K. Terakura, Phys. Rev. Lett. **75**, 1126 (1995)
175. L.F. Feiner, A.M. Oleś, Phys. Rev. B **59**, 3295 (1999)
176. C.H. Booth, F. Bridges, G.H. Kwei, J.M. Lawrence, A.L. Cornelius, J.J. Neumeier, Phys. Rev. Lett. **80**, 853 (1998)
177. H. Fehske, G. Wellein, B. Bäuml, H. Büttner, Physica B **230-232**, 899 (1997)

Lang-Firsov Approaches to Polaron Physics: From Variational Methods to Unbiased Quantum Monte Carlo Simulations

Martin Hohenadler and Wolfgang von der Linden

Institute for Theoretical and Computational Physics, TU Graz, Austria
hohenadler@itp.tu-graz.ac.at, wvl@itp.tu-graz.ac.at

1 Introduction

In the last decades, there has been substantial interest in simple models for electron–phonon (el–ph) interaction in condensed matter. Despite intensive theoretical efforts, it was not before the advent of numerical methods in the 1980's that a thorough understanding on the basis of exact, unbiased results was achieved. Although at the present our knowledge of the rather simple cases of a single carrier (the *polaron problem*) or two carriers (the *bipolaron problem*) in Holstein and Fröhlich models is fairly complete, this is not true for arbitrary band fillings. There is still a major desire to develop more efficient simulation techniques to tackle strongly correlated many-polaron models, which are expected to describe several aspects of real materials currently under investigation, such as quantum dots and quantum wires, high-temperature superconductors or colossal-magnetoresistance manganites.

One of the principle problems in computer simulations of microscopic models is the limitation in both system size and parameter values. Whereas the former can be overcome for the polaron and the bipolaron problem in some cases, it is very difficult to obtain results of similar quality in the many-electron case (see also [1]). Moreover, many approaches still suffer from severe restrictions concerning the parameter regions accessible. For example, interesting materials such as the cuprates and manganites are characterized by small but finite phonon frequencies—as compared to the electronic hopping integral— and intermediate to strong el-ph interaction. Unfortunately, simulations turn out to be most difficult exactly for such parameters, and it is therefore highly desirable to improve existing simulation methods.

In this chapter, we shall mainly review different versions of a recently developed quantum Monte Carlo (QMC) method applicable to Holstein-type models with one, two or many electrons. The appealing advantages of QMC over other numerical methods include the accessibility of rather large systems, the exact treatment of bosonic degrees of freedom (i.e., no truncation is necessary), and the possibility to consider finite temperatures to study phase transitions. The important new aspect here is the use of canonically transformed Hamiltonians, which permits the introduction of exact sampling for the phonon degrees of freedom, enabling us to carry out accurate simulations in practically all interesting parameter regimes.

Additionally, based on a generalization of the Lang-Firsov transformation, we shall present a simple variational approach to the polaron and the bipolaron problem which yields surprisingly accurate results.

This review is organized as follows. In Sect. 2, we present the general model Hamiltonian. Section 3 is devoted to a discussion of the Lang-Firsov transformation, and Sect. 4 contains the derivation of the variational approach. The QMC method is introduced in Sect. 5. Section 6 gives a selection of results for the cases of one, two and many electrons. Finally, we summarize in Sect. 7.

2 Model

In this paper we focus on the *extended Holstein-Hubbard model* defined by

$$H = -t \sum_{\langle ij \rangle \sigma} c_{i\sigma}^{\dagger} c_{j\sigma} + U \sum_{i} \widehat{n}_{i\uparrow} \widehat{n}_{i\downarrow} + V \sum_{\langle ij \rangle} \widehat{n}_i \widehat{n}_j$$
$$+ \frac{\omega_0}{2} \sum_{i} (\widehat{p}_i^2 + \widehat{x}_i^2) - g' \sum_{i} \widehat{n}_i \widehat{x}_i . \tag{1}$$

Here $c_{i\sigma}^{\dagger}$ creates an electron with spin σ at site i, and $\widehat{n}_i = \sum_{\sigma} \widehat{n}_{i\sigma}$ with $\widehat{n}_{i\sigma} = c_{i\sigma}^{\dagger} c_{i\sigma}$. The phonon degrees of freedom at site i are described by the momentum $\widehat{p}_i$ and coordinate (displacement) $\widehat{x}_i$ of a harmonic oscillator. The microscopic parameters are the nearest-neighbour (denoted by $\langle \rangle$) hopping amplitude t, the on-site (Hubbard) repulsion U, the nearest-neighbour Coulomb repulsion V, the Einstein phonon frequency ω_0 and the el-ph coupling g'.

This model neglects both long-range Coulomb and el-ph interaction, which is often a suitable approximation for metallic systems due to screening. Two simple limiting cases of the Hamiltonian (1) are the Holstein model ($U = V = 0$) and the Hubbard model ($g' = V = 0$). In general, the physics of the model (1) is determined by the competition of the various interactions. Depending on the choice of parameters and band filling, it describes fascinating phenomena such as (bi-)polaron formation, Mott– and Peierls quantum phase transitions or superconductivity. As we shall see below, the *adiabaticity ratio*

$$\alpha = \omega_0/t \tag{2}$$

permits us to distinguish two physically different regimes, namely the *adiabatic regime* $\alpha < 1$ and the *non-adiabatic regime* $\alpha > 1$.

We further define the dimensionless el-ph coupling parameter $\lambda = g'^2/(\omega_0 W)$, where $W = 4t\mathrm{D}$ is the bare bandwidth in D dimensions. Alternatively, λ may also be written as $\lambda = 2E_\mathrm{P}/W$, i.e., the ratio of the polaron binding energy in the atomic limit $t = 0$, $E_\mathrm{P} = g'^2/2\omega_0$, and half the bare bandwidth. A useful constant in the non-adiabatic regime is $g^2 = E_\mathrm{P}/\omega_0$. We exclusively consider hypercubic lattices with linear size N and volume N^D, and assume periodic boundary conditions in real space.

3 Lang-Firsov Transformation

The cornerstone of the methods presented here is the canonical *(extended) Lang-Firsov transformation* of the Hamiltonian (1). The original Lang-Firsov (LF) transformation [2] has been used extensively to study Holstein-type models. A well-known, early approximation is due to Holstein [3], who replaced the hopping term by its expectation value in a zero-phonon state, neglecting emission and absorption of phonons during electron transfer. However, this approach yields reliable results only in the non-adiabatic strong-coupling (SC) limit. For $\lambda = \infty$ (or $t = 0$), the LF transformation provides an exact solution of the single-site problem [4].

Whereas transformed Hamiltonians have been treated numerically before [5–7], the first QMC method making use of the LF transformation has been proposed in [8].

We introduce the extended LF transformation by defining the unitary operator

$$\widehat{\varPhi} = \mathrm{e}^S , \quad S = \mathrm{i} \sum_{ij} \gamma_{ij} \widehat{n}_i \widehat{p}_j \tag{3}$$

with real parameters γ_{ij}, $i, j = 1, \ldots, N^\mathrm{D}$. $\widehat{\varPhi}$ as defined in (3) has the form of a translation operator, and fulfills $\widehat{\varPhi}^\dagger = \widehat{\varPhi}^{-1}$. Given an electron at site i, $\widehat{\varPhi}$ mediates displacements γ_{ij} of the harmonic oscillators at all sites j. Hence, the extended transformation is capable of describing an extended phonon cloud, important in the large-polaron or bipolaron regime. We shall use this transformation for the variational approach. However, the standard (local) LF transformation will be expedient as a basis for unbiased QMC simulations, in which the transformed Hamiltonian is treated exactly.

Operators have to be transformed according to $\widetilde{\widehat{A}} = \widehat{\varPhi} \widehat{A} \widehat{\varPhi}^\dagger$. Defining the function $f(\eta) = \mathrm{e}^{\eta \widehat{S}} \widehat{A} \mathrm{e}^{-\eta \widehat{S}}$ we obtain

$$f'(\eta) = \mathrm{e}^{\eta \widehat{S}} [\widehat{S}, \widehat{A}] \mathrm{e}^{-\eta \widehat{S}} , \tag{4}$$

where $f' \equiv \partial f/\partial \eta$. A simple calculation gives

$$[\widehat{S}, c_{i\sigma}] = -\mathrm{i} \sum_l \gamma_{il} \widehat{p}_l \, c_{i\sigma} , \quad [\widehat{S}, c_{i\sigma}^\dagger] = \mathrm{i} \sum_l \gamma_{il} \widehat{p}_l \, c_{i\sigma}^\dagger . \tag{5}$$

Substitution in (4), integration with respect to η and setting $\eta = 1$ results in

$$\tilde{c}^{\dagger}_{i\sigma} = c^{\dagger}_{i\sigma}\, \mathrm{e}^{\mathrm{i}\sum_j \gamma_{ij}\hat{p}_j}\,, \qquad \tilde{c}_{i\sigma} = c_{i\sigma}\, \mathrm{e}^{-\mathrm{i}\sum_j \gamma_{ij}\hat{p}_j}\,. \tag{6}$$

For phonon operators, the relation

$$\tilde{\hat{A}} = \mathrm{e}^{\hat{S}}\,\hat{A}\,\mathrm{e}^{-\hat{S}} = \hat{A} + [\hat{S},\hat{A}] + \frac{1}{2!}[\hat{S},[\hat{S},\hat{A}]] + \cdots, \tag{7}$$

yields

$$\tilde{\hat{x}}_i = \hat{x}_i + \sum_j \gamma_{ij}\hat{n}_j\,, \qquad \tilde{\hat{p}}_i = \hat{p}_i\,. \tag{8}$$

Collecting these results, the transformation of the Hamiltonian (1) leads to

$$\tilde{H} = \underbrace{-t\sum_{\langle ij\rangle\sigma} c^{\dagger}_{i\sigma}c_{j\sigma}\mathrm{e}^{\mathrm{i}\sum_l(\gamma_{il}-\gamma_{jl})\hat{p}_l}}_{\tilde{H}_{\mathrm{kin}}} + \underbrace{\frac{\omega_0}{2}\sum_i(\hat{p}_i^2 + \hat{x}_i^2)}_{\tilde{H}_{\mathrm{ph}}\equiv\tilde{H}^p_{\mathrm{ph}}+\tilde{H}^x_{\mathrm{ph}}} + \underbrace{\sum_{ij}\hat{n}_j\hat{x}_i(\omega_0\gamma_{ij} - g'\delta_{ij})}_{\tilde{H}_{\mathrm{ep}}}$$

$$+ \underbrace{\sum_{ij}\hat{n}_i\hat{n}_j\left(\frac{\omega_0}{2}\sum_l \gamma_{lj}\gamma_{li} - g'\gamma_{ij} + \frac{U}{2}\delta_{ij} + V\delta_{\langle ij\rangle}\right) - \tfrac{1}{2}U\sum_i \hat{n}_i}_{\tilde{H}_{\mathrm{ee}}}. \tag{9}$$

Here the term $\tilde{H}_{\mathrm{ep}}$ describes the coupling between electrons and phonons, whereas $\tilde{H}_{\mathrm{ee}}$ represents both an effective el–el interaction and a single-photon level shift. Hamiltonian (9) will be the starting point for the variational approach in Sect. 4.

For QMC simulations, it is more suitable to require that the el-ph terms in $\tilde{H}_{\mathrm{ep}}$ cancel. This can be achieved by setting $\gamma_{ij} = \gamma\delta_{ij}$ with

$$\gamma = \sqrt{\frac{\lambda W}{\omega_0}}\,. \tag{10}$$

The parameter γ corresponds to the distortion which minimizes the potential energy of the shifted harmonic oscillator $E_{\mathrm{pot}} = \frac{\omega_0}{2}x^2 - g'x$. This leads us to the standard LF transformation

$$\hat{\Phi}_0 = \mathrm{e}^{S_0}\,, \qquad S_0 = \mathrm{i}\gamma\sum_i \hat{n}_i\hat{p}_i\,, \tag{11}$$

and the familiar results for the transformed operators

$$\tilde{c}^{\dagger}_{i\sigma} = c^{\dagger}_{i\sigma}\mathrm{e}^{\mathrm{i}\gamma\hat{p}_i}\,, \qquad \tilde{c}_{i\sigma} = c_{i\sigma}\mathrm{e}^{-\mathrm{i}\gamma\hat{p}_i} \tag{12}$$

and

$$\tilde{\hat{x}}_i = \hat{x}_i + \gamma\hat{n}_i\,, \qquad \tilde{\hat{p}}_i = \hat{p}_i\,. \tag{13}$$

In contrast to the non-local transformation (3), only the oscillator at the site of the electron is displaced. The transformed Hamiltonian reads

$$\widetilde{H} = -t \sum_{\langle ij\rangle\sigma} \underbrace{c_{i\sigma}^{\dagger}c_{j\sigma}\mathrm{e}^{\mathrm{i}\gamma(\widehat{p}_i-\widehat{p}_j)}}_{\widetilde{H}_{\mathrm{kin}}} + \underbrace{\frac{\omega_0}{2}\sum_i(\widehat{p}_i^2+\widehat{x}_i^2)}_{\widetilde{H}_{\mathrm{ph}}}$$

$$+\underbrace{(\tfrac{1}{2}U-E_{\mathrm{P}})\sum_i \widehat{n}_i^2 + V\sum_{\langle ij\rangle}\widehat{n}_i\widehat{n}_j - \tfrac{1}{2}U\sum_i\widehat{n}_i}_{\widetilde{H}_{\mathrm{ee}}} \ . \qquad (14)$$

As we shall discuss in detail in Sect. 5, the difficulties encountered in QMC simulations of the original Hamiltonian (1) are to a certain extent related to (bi-)polaron effects, i.e., to the dynamic formation of spatially rather localized lattice distortions which surround the charge carriers and follow their motion in the lattice.

For a single electron, the aforementioned *Holstein-Lang-Firsov* (HLF) approximation [3] becomes exact in the non-adiabatic SC or small-polaron limit, and agrees qualitatively with exact results also in the intermediate-coupling (IC) regime [9]. Although it overestimates the shift γ of the equilibrium position of the oscillator in the presence of an electron, and does not reproduce the retardation effects when the electron hops onto a previously unoccupied site, the approximation mediates the crucial impact of el-ph interaction on the lattice. Consequently, the transformed Hamiltonian (14) can be expected to be a good starting point for QMC simulations, which then merely need to account for the rather small fluctuations around the (shifted or unshifted) equilibrium positions. In principle, it would also be possible to develop a QMC algorithm based on the Hamiltonian (9)—the basis of our variational approach—with the parameters γ_{ij} determined variationally, but the local LF transformation proves to be sufficient.

The Hamiltonian (14) does no longer contain a term coupling the electron density $\widehat{n}$ and the lattice displacement $\widehat{x}$. By contrast, the extended transformation does not eliminate the interaction term completely. On top of that, the hopping term involves all phonon momenta $\widehat{p}_i$ as well as the parameters γ_{ij}, and the el-el interaction becomes long ranged [cf. (9)].

For electrons with opposite spins with $\widehat{n}_i^2 \neq \widehat{n}_i$, the interaction term $\widetilde{H}_{\mathrm{ee}}$ contains a Hubbard-like attractive interaction. Whereas the latter can be treated exactly in the case of two electrons (Sect. 5.1), the many-electron case requires the introduction of auxiliary fields which complicate simulations. However, no such difficulties arise for the spinless Holstein model considered in Sect. 6.

4 Variational Approach

For simplicity, we shall restrict the following derivation to one dimension; an extension to D > 1 is straightforward. Furthermore, we only consider finite clusters with periodic boundary conditions, although infinite systems may also be treated. The results of this section have originally been presented in [8, 10].

4.1 One Electron

As noted before, the simple variational method presented here is based on the extended transformation (3), leading to the Hamiltonian (9). We treat the γ_{ij} as variational parameters which are determined by minimizing the ground-state energy in a zero-phonon basis in which $\langle \widetilde{H}_{\mathrm{ep}} \rangle = 0$.

For systems with translation invariance the *displacement fields* satisfy the condition $\gamma_{ij} = \gamma_{|i-j|}$. Together with $\sum_i \widehat{n}_i = 1$ for a single electron we get $\widetilde{H}_{\mathrm{ee}} = \frac{\omega_0}{2} \sum_l \gamma_l^2 - g'\gamma_0$.

The eigenvalue problem of the transformed Hamiltonian (9) is solved by making the following ansatz for the one-electron basis states

$$\left\{ |l\rangle = c_{l\sigma}^\dagger |0\rangle \otimes \prod_{\nu=1}^{N} |\phi_0^{(\nu)}\rangle , \quad l = 1, \ldots, N \right\} , \tag{15}$$

where $|\phi_0^{(\nu)}\rangle$ denotes the ground state of the harmonic oscillator at site ν. The non-zero matrix elements of the hopping term are

$$\langle l| \widetilde{H}_{\mathrm{kin}} |l'\rangle = -t\delta_{\langle ll'\rangle} \prod_\nu \langle \phi_0^{(\nu)}| e^{\mathrm{i}(\gamma_{l\nu} - \gamma_{l'\nu})\widehat{p}_\nu} |\phi_0^{(\nu)}\rangle$$

$$= -t\delta_{\langle ll'\rangle} \prod_\nu \int \mathrm{d}x\, \phi(x + \gamma_{l\nu})\phi(x + \gamma_{l'\nu})$$

$$= -t\delta_{\langle ll'\rangle} e^{-\frac{1}{4}\sum_\nu (\gamma_\nu - \gamma_{\nu+l-l'})^2} , \tag{16}$$

where $\phi(x)$ denotes the real-space wavefunction of the harmonic-oscillator ground state. The Kronecker symbol $\delta_{\langle ll'\rangle}$ forces l and l' to represent nearest-neighbor sites. A simple calculation gives for the other terms in (9)

$$\langle l| \widetilde{H}_{\mathrm{ph}} |l'\rangle = \delta_{ll'} \frac{\omega_0}{2} , \quad \langle l| \widetilde{H}_{\mathrm{ep}} |l'\rangle = 0 , \quad \langle l| \widetilde{H}_{\mathrm{ee}} |l'\rangle = \delta_{ll'} \left(\frac{\omega_0}{2} \sum_l \gamma_l^2 - g'\gamma_0 \right) . \tag{17}$$

In the zero-phonon subspace spanned by the basis states (15), the eigenstates of Hamiltonian (9) with momentum k are

$$|\psi_k\rangle = c_{k\sigma}^\dagger |0\rangle \otimes \prod_\nu |\phi_0^{(\nu)}\rangle \tag{18}$$

with eigenvalues

$$E(k) = E_{\mathrm{kin}} + \frac{N\omega_0}{2} + \frac{\omega_0}{2} \sum_l \gamma_l^2 - g'\gamma_0 \tag{19}$$

and the kinetic energy

$$E_{\text{kin}} = -t \sum_{\delta=\pm1} e^{ik\delta} e^{-\frac{1}{4}\sum_\nu (\gamma_\nu - \gamma_{\nu+\delta})^2} . \tag{20}$$

Defining the Fourier transform

$$\overline{\gamma}_q = \frac{1}{\sqrt{N}} \sum_l e^{iql} \gamma_l \tag{21}$$

and using ($\gamma_l \in \mathbb{R}$)

$$\sum_\nu \gamma_\nu \gamma_{\nu+\delta} = \sum_q \overline{\gamma}_q \overline{\gamma}_{-q} e^{iq\delta} = \sum_q \overline{\gamma}_q^2 \cos q\delta , \tag{22}$$

we may write

$$E_{\text{kin}} = -t \sum_\delta e^{ik\delta} e^{-\frac{1}{2}\sum_q (1-\cos q\delta)\overline{\gamma}_q^2} = \varepsilon_0(k) e^{-\frac{1}{2}\sum_q (1-\cos q)\overline{\gamma}_q^2} = \varepsilon(k) \tag{23}$$

with the tight-binding dispersion $\varepsilon_0(k) = 2t\cos k$. Hence the ground-state energy becomes

$$E(k) = \varepsilon(k) + \frac{N\omega_0}{2} + \frac{\omega_0}{2}\sum_q \overline{\gamma}_q^2 - \frac{g'}{\sqrt{N}} \sum_q \overline{\gamma}_q . \tag{24}$$

The variational parameters $\overline{\gamma}_p$ are determined by requiring

$$\frac{\partial E}{\partial \overline{\gamma}_p} = -\overline{\gamma}_p \varepsilon(k)(1 - \cos p) + \omega_0 \overline{\gamma}_p - \frac{g'}{\sqrt{N}} \overset{!}{=} 0 , \tag{25}$$

so that the optimal values $\overline{\gamma}_p$ can be obtained from

$$\overline{\gamma}_p = \frac{g'}{\sqrt{N}} \frac{1}{\omega_0 + \varepsilon(k)(1 - \cos p)} . \tag{26}$$

Since $\varepsilon(k)$ depends implicitly on the $\overline{\gamma}_p$, (26) has to be solved self-consistently. It has the typical form of the random-phase approximation since a variational ansatz for the untransformed Hamiltonian may be written as

$$\widehat{\Phi}^\dagger |\psi_k\rangle = \frac{1}{\sqrt{N}} \sum_j e^{ikj} c_{j\sigma}^\dagger e^{-i\sum_l \gamma_{jl}\widehat{p}_l} |0\rangle \otimes \prod_\nu |\phi_0^{(\nu)}\rangle , \tag{27}$$

with $\widehat{\Phi}$ as defined in (3).

We shall also calculate the quasiparticle spectral weight for momentum $k = 0$, defined as

$$\sqrt{z_0} = \langle 0| \widetilde{c}_{k=0,\sigma} |\psi_0\rangle . \tag{28}$$

Here $|\psi_0\rangle$ denotes the ground state with one electron of momentum $p = 0$ and the oscillators in the ground state $|\phi_0\rangle$. Fourier transformation and the same manipulations as in (16) lead to

$$
\begin{aligned}
\sqrt{z_0} &= \frac{1}{N} \sum_{ij} \langle\phi_0| \langle 0| \tilde{c}_{i\sigma} c_{j\sigma}^\dagger |0\rangle |\phi_0\rangle \\
&= \frac{1}{N} \sum_i \langle\phi_0| e^{-i\sum_k \gamma_{ik}\widehat{p}_k} |\phi_0\rangle \\
&= e^{-\frac{1}{4}\sum_q \tilde{\gamma}_q^2} .
\end{aligned}
\tag{29}
$$

Just as in the HLF approximation, the present variational method becomes exact in the non-interacting limit ($\lambda = 0$) and in the non-adiabatic SC limit. Furthermore, it yields the correct results both for $\alpha = 0$ (classical phonons) and $\alpha = \infty$, and also gives accurate results for large α and finite λ, since the displacements of the oscillators—only local and generally overestimated in the HLF approximation—are determined variationally.

4.2 Two Electrons

As in the one-electron case, the use of a zero-phonon basis leads to $\langle\widetilde{H}_{\mathrm{ep}}\rangle = 0$ and, neglecting the ground-state energy of the oscillators, we also have $\langle\widetilde{H}_{\mathrm{ph}}\rangle = 0$. Hence, $\widetilde{H} = \widetilde{H}_{\mathrm{kin}} + \widetilde{H}_{\mathrm{ee}}$ with the transformed hopping term

$$
\widetilde{H}_{\mathrm{kin}} = -t_{\mathrm{eff}} \sum_{\langle ij\rangle\sigma} c_{i\sigma}^\dagger c_{j\sigma} = \sum_{k\sigma} \varepsilon(k)\, c_{k\sigma}^\dagger c_{k\sigma}
\tag{30}
$$

and $\varepsilon(k) = -2\, t_{\mathrm{eff}} \cos(k)$. Here the effective hopping

$$
t_{\mathrm{eff}} = \frac{1}{2} \sum_{\delta=\pm 1} e^{-\frac{1}{4}\sum_l (\gamma_{l-\delta}-\gamma_l)^2} ,
\tag{31}
$$

where rotational invariance has been exploited. For two electrons of opposite spin (i.e., $\widehat{n}_{i\sigma}\widehat{n}_{j\sigma} = 0$ for $i \neq j$) and $V = 0$, $\widetilde{H}_{\mathrm{ee}}$ in (9) reduces to

$$
\widetilde{H}_{\mathrm{ee}} = 2v_0 - U + 2\sum_{ij} v_{ij}\widehat{n}_{i\uparrow}\widehat{n}_{j\downarrow}, \quad v_{ij} = \frac{\omega_0}{2}\sum_l \gamma_{lj}\gamma_{li} - g'\gamma_{ij} + \tfrac{1}{2}\delta_{ij}U .
\tag{32}
$$

The eigenstates of the two-electron problem have the form

$$
|\psi_k\rangle = \sum_p \bar{d}_p c_{k-p\downarrow}^\dagger c_{p\uparrow}^\dagger |0\rangle ,
\tag{33}
$$

suppressing the phonon component [cf. (18)], and may be written as

$$
|\psi_k\rangle = \frac{1}{\sqrt{N}} \sum_i e^{ikx_i} \sum_l d_l\, c_{i\downarrow}^\dagger c_{i+l\uparrow}^\dagger |0\rangle ,
\tag{34}
$$

with the Fourier transform

$$\boldsymbol{d} = F\overline{\boldsymbol{d}}, \quad (F)_{lp} = \mathrm{e}^{i x_l p}/\sqrt{N}. \tag{35}$$

The normalization of (33) reads

$$\langle \psi_k | \psi_k \rangle = \sum_p |d_p|^2. \tag{36}$$

Using (33), we find for the expectation value of $\widetilde{H}_{\mathrm{kin}}$

$$\langle \psi_k | \widetilde{H}_{\mathrm{kin}} | \psi_k \rangle = \sum_{pp'} \overline{d}_p^* \overline{d}_{p'} \sum_q \varepsilon(q)$$

$$\times \left(\underbrace{\langle 0 | c_{p\uparrow} c_{k-p\downarrow} \widehat{n}_{q\uparrow} c_{k-p'\downarrow}^\dagger c_{p'\uparrow}^\dagger | 0 \rangle}_{\delta_{p,p'} \delta_{q,p}} + \underbrace{\langle 0 | c_{p\uparrow} c_{k-p\downarrow} \widehat{n}_{q\downarrow} c_{k-p'\downarrow}^\dagger c_{p'\uparrow}^\dagger | 0 \rangle}_{\delta_{p,p'} \delta_{q,k-p}} \right)$$

$$= \sum_p |\overline{d}_p|^2 \left[\varepsilon(p) + \varepsilon(k-p) \right]$$

$$= -4\, t_{\mathrm{eff}}\, \boldsymbol{d}^\dagger T_k \boldsymbol{d}. \tag{37}$$

In the last step we have introduced $T_k = \frac{1}{2} F \, \mathrm{diag}[\cos(p) + \cos(k-p)]\, F^\dagger$ and made use of (35).

The expectation value of the interaction term, best computed in the real-space representation (34), takes the form

$$\langle \psi_k | \widetilde{H}_{\mathrm{ee}} | \psi_k \rangle = (2v_0 - U) \sum_l |d_l|^2 + \frac{2}{N} \sum_{ij} v_{ij} \sum_{j'j''} \sum_{ll'} d_l^* d_{l'} \mathrm{e}^{ik(x_l - x_{l'})}$$

$$\times \underbrace{\langle 0 | c_{j'+l\uparrow} c_{j'\downarrow} \widehat{n}_{i\uparrow} \widehat{n}_{j\downarrow} c_{j''\downarrow}^\dagger c_{j''+l'\uparrow}^\dagger | 0 \rangle}_{\delta_{jj'} \delta_{jj''} \delta_{i,j+l} \delta_{l,l'}}$$

$$= (2v_0 - U) \sum_l |d_l|^2 + \frac{2}{N} \sum_{jl} v_{j+l,j} |d_l|^2$$

$$= (2v_0 - U) \boldsymbol{d}^\dagger \boldsymbol{d} + 2 \boldsymbol{d}^\dagger V \boldsymbol{d}, \tag{38}$$

with the diagonal matrix $V_{ij} = \delta_{ij} v_i$.

The minimization of the total energy with respect to $\boldsymbol{d}$ yields the eigenvalue problem

$$(-4t_{\mathrm{eff}}\, T_k + 2V)\, \boldsymbol{d} = (E_0 - 2v_0 + U)\, \boldsymbol{d}. \tag{39}$$

The vector of coefficients $\boldsymbol{d}$ and thereby the ground state are found by minimizing the ground-state energy E_0 through variation of the displacement fields γ_{ij}. Similar to the one-electron case, this procedure takes into account displacements of the oscillators not only at the same but also at surrounding sites of the two electrons, and is therefore capable of describing extended bipolaron states (see Sect. 6.2). Note that the two-electron problem is diagonalized exactly without phonons (i.e., for $\lambda = 0$).

5 Quantum Monte Carlo

In this section, we present an overview of our recently developed QMC algorithms for Holstein-type models [8, 10–12].

As mentioned before, in contrast to the variational approach, the QMC approaches discussed here, based on the local LF transformation (14) which does not contain any free parameters, are unbiased. They yield exact results with only statistical errors that can in principle be made arbitrarily small.

The motivation for the development of improved QMC schemes for Holstein models stems from the fact that calculations with existing methods often suffer from strong autocorrelations, i.e., non-negligible statistical correlations between successive MC configurations [8, 13]. In fact, autocorrelations may render accurate simulations impossible within reasonable computing time. As discussed in [8], the problem becomes particularly noticeable for small phonon frequencies and low temperatures.

Whereas autocorrelations can be avoided to a large extent for one or two electrons by integrating out the phonons analytically, no efficient general schemes exist for finite charge-carrier densities (see discussion in [8]).

In the following, we present a general (i.e., applicable for all densities) solution for this problem in several steps. First, the effects due to el-ph interaction are separated from the free lattice dynamics by means of the LF transformation (14). Since the latter contains the crucial impact of the electronic degrees of freedom on the lattice, simulations may be based only on the purely phononic part of the resulting action. The fermionic degrees of freedom can then be taken into account exactly by *reweighting* of the probability distribution. Consequently, we may completely ignore the electronic weights in the updating process, and thereby dramatically reduce the computational effort. The *principal component representation* of the phonon coordinates allows exact sampling and avoids any autocorrelations.

5.1 Partition Function

We begin by deriving the partition function for the case of a single electron. Then we discuss the differences occurring in the cases of two or more carriers.

One Electron

The partition function is defined as

$$\mathcal{Z} = \mathrm{Tr}\, e^{-\beta \widetilde{H}} \tag{40}$$

with $\widetilde{H}$ given by (14) and the inverse temperature $\beta = (k_\mathrm{B} T)^{-1}$. For a single electron, $\widetilde{H}_{\mathrm{ee}} = -E_\mathrm{P}$ becomes a constant which needs only to be considered in calculating the total energy.

Using the Suzuki-Trotter decomposition [13], we obtain

$$\mathrm{e}^{-\beta\widetilde{H}} \approx (\mathrm{e}^{-\Delta\tau\widetilde{H}_{\mathrm{kin}}}\mathrm{e}^{-\Delta\tau\widetilde{H}^{p}_{\mathrm{ph}}}\mathrm{e}^{-\Delta\tau\widetilde{H}^{x}_{\mathrm{ph}}})^{L} \equiv \mathcal{U}^{L}, \tag{41}$$

where $\Delta\tau = \beta/L \ll 1$. Splitting up the trace into a bosonic and a fermionic part and inserting $L-1$ complete sets of oscillator momentum eigenstates we find the approximation

$$\mathcal{Z}_{L} = \mathrm{Tr}_{\mathrm{f}} \int \mathrm{d}p_{1}\mathrm{d}p_{2}\cdots\mathrm{d}p_{L}\, \langle p_{1}|\mathcal{U}|p_{2}\rangle \cdots \langle p_{L}|\mathcal{U}|p_{1}\rangle \tag{42}$$

with $\mathrm{d}p_{\tau} \equiv \prod_{i}\mathrm{d}p_{i,\tau}$. Each matrix element can be evaluated by inserting a complete set of phonon coordinate eigenstates $\int \mathrm{d}x|x\rangle\langle x|$, since all x-integrals are of Gaussian form and can easily be carried out. The result is

$$\langle p_{\tau}|\mathrm{e}^{-\Delta\tau\widetilde{H}^{x}_{\mathrm{ph}}}|p_{\tau+1}\rangle = C^{N^{\mathrm{D}}}\mathrm{e}^{-\frac{1}{2\omega_{0}\Delta\tau}\sum_{i}(p_{i,\tau}-p_{i,\tau+1})^{2}}, \quad C = \sqrt{\frac{2\pi}{\omega_{0}\Delta\tau}}. \tag{43}$$

The normalization factor in front of the exponential has to be taken into account in the calculation of the total energy, but cancels when we measure other observables. With the abbreviation $\mathcal{D}p = \mathrm{d}p_{1}\mathrm{d}p_{2}\cdots\mathrm{d}p_{L}$ the partition function finally becomes

$$\mathcal{Z}_{L} = C^{N^{\mathrm{D}}L} \int \mathcal{D}p\, w_{\mathrm{b}}\, w_{\mathrm{f}}, \tag{44}$$

where

$$w_{\mathrm{b}} = \mathrm{e}^{-\Delta\tau S_{\mathrm{b}}}, \quad w_{\mathrm{f}} = \mathrm{Tr}_{\mathrm{f}}\,\Omega, \quad \Omega = \prod_{\tau=1}^{L}\mathrm{e}^{-\Delta\tau\widetilde{H}^{(\tau)}_{\mathrm{kin}}}. \tag{45}$$

Here $\widetilde{H}^{(\tau)}_{\mathrm{kin}}$ corresponds to $\widetilde{H}_{\mathrm{kin}}$ with the phonon operators $\widehat{p}_{i}$, $\widehat{p}_{j}$ replaced by the momenta $p_{i,\tau}$, $p_{j,\tau}$ on the τth Trotter slice, and its exponential may be written as

$$\mathrm{e}^{-\Delta\tau\widetilde{H}^{(\tau)}_{\mathrm{kin}}} = D_{\tau}\kappa D^{\dagger}_{\tau}, \quad \kappa_{jj'} = \left(\mathrm{e}^{\Delta\tau t\,h^{\mathrm{tb}}}\right)_{jj'}, \quad (D_{\tau})_{jj'} = \delta_{jj'}\mathrm{e}^{\mathrm{i}\gamma p_{j,\tau}}, \tag{46}$$

where h^{tb} is the $N^{\mathrm{D}} \times N^{\mathrm{D}}$ tight-binding hopping matrix. To save some computer time, we employ the checkerboard breakup [14]

$$\mathrm{e}^{\Delta\tau t\sum_{\langle ij\rangle} c^{\dagger}_{i}c_{j}} \approx \prod_{\langle ij\rangle}\mathrm{e}^{\Delta\tau t c^{\dagger}_{i}c_{j}}. \tag{47}$$

Using (47), the numerical effort scales as $N^{2\mathrm{D}}$ instead of $N^{3\mathrm{D}}$ (see also Sect. 5.6), but the error due to this additional approximation is of the same order $\Delta\tau^{2}$ as the Trotter error in (41).

According to (46), we have the same matrix κ for every time slice, which is transformed by the diagonal unitary matrices D_{τ}. The matrix Ω can be calculated in an efficient way by noting that the transformation matrices $D^{\dagger}_{\tau}$ and $D_{\tau+1}$ at time slice τ may be combined to a diagonal matrix

$$(D_{\tau,\tau+1})_{ij} = \delta_{ij}\mathrm{e}^{\mathrm{i}\gamma(p_{i,\tau+1}-p_{i,\tau})} \, . \tag{48}$$

Due to the cyclic invariance of the fermionic trace, D_1 can be shifted to the end of the product, where it combines with $D_L^\dagger$ to $D_{L,1}$. Hence we can write

$$\Omega = \prod_{\tau=1}^{L} \kappa\, D_{\tau,\tau+1} \, , \tag{49}$$

with periodic boundary conditions in imaginary time. In the one-electron case, the fermionic weight $w_{\mathrm{f}} = \sum_n \langle n| \Omega |n\rangle$ is given by the sum over the diagonal elements of the matrix representation of Ω in the basis of one-electron states (dropping unnecessary spin indices)

$$|n\rangle = c_n^\dagger |0\rangle \, . \tag{50}$$

The bosonic action in (45) contains only classical variables:

$$S_{\mathrm{b}} = \frac{\omega_0}{2} \sum_{i,\tau} p_{i,\tau}^2 + \frac{1}{2\omega_0\Delta\tau^2} \sum_{i,\tau} (p_{i,\tau} - p_{i,\tau+1})^2 \, , \tag{51}$$

where the indices $i = 1, \ldots, N^{\mathrm{D}}$ and $\tau = 1, \ldots, L$ run over all lattice sites and time slices, respectively, and $p_{i,L+1} = p_{i,1}$. It may also be written as

$$S_{\mathrm{b}} = \sum_i \boldsymbol{p}_i^{\mathrm{T}} A \boldsymbol{p}_i \tag{52}$$

with $\boldsymbol{p}_i = (p_{i,1}, \ldots, p_{i,L})$ and a *periodic*, tridiagonal $L \times L$ matrix A with non-zero elements

$$(A)_{l,l} = \frac{\omega_0}{2} + \frac{1}{\omega_0\Delta\tau^2} \, , \quad (A)_{l,l\pm1} = -\frac{1}{2\omega_0\Delta\tau^2} \, . \tag{53}$$

Since $\mathcal{Z}_L$ is a trace, it follows that $(A)_{1,L} = (A)_{L,1} = -(2\omega_0\Delta\tau^2)^{-1}$.

Two Electrons

In contrast to [10], here we also take into account nearest-neighbour Coulomb repulsion V. For any number of electrons, the Hamiltonian (14) simplifies to

$$\widetilde{H} = \widetilde{H}_{\mathrm{kin}} + \widetilde{H}_{\mathrm{ph}} + \widetilde{H}_{\mathrm{ee}} - 2E_{\mathrm{P}} \, , \quad \widetilde{H}_{\mathrm{ee}} = (U - 2E_{\mathrm{P}}) \sum_i \widehat{n}_{i\uparrow}\widehat{n}_{i\downarrow} + V \sum_{\langle ij\rangle} \widehat{n}_i\widehat{n}_j \, . \tag{54}$$

Again, the constant shift can be neglected in the QMC simulation, but in contrast to the single-electron case, we have a non-trivial interaction term. The Suzuki-Trotter decomposition yields

$$\mathrm{e}^{-\beta\widetilde{H}} \approx \left(\mathrm{e}^{-\Delta\tau\widetilde{H}_{\mathrm{kin}}}\mathrm{e}^{-\Delta\tau\widetilde{H}_{\mathrm{ph}}^p}\mathrm{e}^{-\Delta\tau\widetilde{H}_{\mathrm{ph}}^x}\mathrm{e}^{-\Delta\tau\widetilde{H}_{\mathrm{ee}}}\right)^L \equiv \mathcal{U}^L \, . \tag{55}$$

Using the same steps as above we obtain

$$w_{\mathrm{b}} = \mathrm{e}^{-\Delta\tau S_{\mathrm{b}}}\,, \quad w_{\mathrm{f}} = \mathrm{Tr}_{\mathrm{f}}\,\Omega\,, \quad \Omega = \prod_{\tau=1}^{L} \mathrm{e}^{-\Delta\tau \widetilde{H}_{\mathrm{kin}}^{(\tau)}} \mathrm{e}^{-\Delta\tau \widetilde{H}_{\mathrm{ee}}}\,, \tag{56}$$

with S_{b} given by (51).

As pointed out in [10], the numerical effort for two electrons increases substantially in higher dimensions. Therefore, we restrict ourselves to D $= 1$.

Previously, we only considered the case of two electrons of opposite spin (forming a singlet) [10]. Here we shall also present results for the triplet state.

Singlet

In the singlet case we choose the two-electron basis states

$$\left\{ |l\rangle \equiv |i,j\rangle \equiv c_{i\uparrow}^{\dagger} c_{j\downarrow}^{\dagger} |0\rangle\,, \quad i,j = 1,\ldots,N \right\}\,, \tag{57}$$

where we have used a combined index $l = 1,\ldots,N^2$. The tight-binding hopping matrix, denoted as κ, has dimension $N^2 \times N^2$, and the corresponding exponential in (56) can again be written as $\mathrm{e}^{-\Delta\tau \widetilde{H}_{\mathrm{kin}}^{(\tau)}} = D_{\tau}\kappa D_{\tau}^{\dagger}$ [cf. (45)], where

$$(D_{\tau})_{ll'} = \delta_{ll'}\mathrm{e}^{\mathrm{i}\gamma(p_{i,\tau}+p_{j,\tau})} \tag{58}$$

is diagonal in the basis (57).

The remaining contribution to Ω comes from the effective el-el interaction term $\widetilde{H}_{\mathrm{ee}}$ in terms of the sparse matrix

$$(\mathcal{V})_{ll'} = \sum_{k} (\delta_{lk}\,\mathrm{e}^{-\Delta\tau(U-2E_{\mathrm{P}})\delta_{ij}})_{lk}(\mathrm{e}^{-\Delta\tau V\delta_{\langle ij\rangle}})_{kl'}\,. \tag{59}$$

The momenta $\boldsymbol{p}$ merely enter the diagonal matrix D; the $N^2 \times N^2$ matrices $\mathcal{V}$ and κ are fixed throughout the entire MC simulation. Finally, we have

$$\Omega = \prod_{\tau} D_{\tau}\kappa D_{\tau}^{\dagger}\mathcal{V}\,, \tag{60}$$

and the fermionic trace can be calculated as the sum over the diagonal elements of the matrix Ω in the basis (57), i.e.,

$$\mathrm{Tr}_{\mathrm{f}}\,\Omega = \sum_{ij} \langle i,j| \,\Omega\, |i,j\rangle\,. \tag{61}$$

Triplet

For two electrons with parallel spin we use the basis states

$$\left\{ |l\rangle \equiv |i,j\rangle \equiv c_{i}^{\dagger} c_{j}^{\dagger} |0\rangle\,, \quad i = 1,\ldots,N\,, j = i+1,\ldots,N \right\}\,, \tag{62}$$

i.e., double occupation of a site is not possible. Since we can further not distinguish between the states $|i,j\rangle$ and $|j,i\rangle$, the dimension of the electronic Hilbert space is reduced from N^2 (singlet case) to $N(N-1)/2$. Consequently, for the same system size, simulations for the triplet case will be much faster.

Many-Electron Case

The one-electron QMC algorithm can easily be extended to the spinless Holstein model with many electrons. For the latter, assuming $V = 0$, the term in (14) reduces to $\widetilde{H}_{\mathrm{ee}} = -E_{\mathrm{P}} \sum_i \widehat{n}_i$. Therefore, the grand-canonical Hamiltonian becomes

$$\widetilde{\mathcal{H}} = \widetilde{H} - \mu \sum_i \widehat{n}_i = \underbrace{-t \sum_{\langle ij \rangle} c_i^\dagger c_j \mathrm{e}^{\mathrm{i}\gamma(\widehat{p}_i - \widehat{p}_j)} + \widetilde{H}_{\mathrm{ph}}}_{\widetilde{H}_{\mathrm{kin}}} \underbrace{- (E_{\mathrm{P}} + \mu) \sum_i \widehat{n}_i}_{\widetilde{H}'_{\mathrm{ee}}} , \qquad (63)$$

where μ denotes the chemical potential. For half filling $n = 0.5$ [$N/2$ spinless fermions on N sites, cf. (84)], the latter is given by $\mu = -E_{\mathrm{P}}$, whereas for $n \neq 0.5$, it has to be adjusted to yield the carrier density of interest.

The approximation to the partition function may again be cast into the form of (44), with w_{b} as defined by (45) and (51), respectively. The fermionic weight is given by

$$w_{\mathrm{f}} = \mathrm{Tr}_{\mathrm{f}}(\widehat{B}_1 \widehat{B}_2 \cdots \widehat{B}_L) , \quad \widehat{B}_\tau = \mathrm{e}^{-\varDelta\tau \widetilde{H}^{(\tau)}_{\mathrm{kin}}} \mathrm{e}^{-\varDelta\tau \widetilde{H}'_{\mathrm{ee}}} . \qquad (64)$$

Following Blankenbecler *et al.* [15], the fermion degrees of freedom can be integrated out exactly leading to

$$w_{\mathrm{f}} = \det(1 + B_1 \cdots B_L) \equiv \det(1 + \varOmega) , \qquad (65)$$

where the $N^{\mathrm{D}} \times N^{\mathrm{D}}$ matrix B_τ is given by

$$B_\tau = D_\tau \, \kappa \, D_\tau^\dagger \, \mathcal{V} . \qquad (66)$$

Here κ and D_τ are identical to (46), and

$$(\mathcal{V})_{ij} = \delta_{ij} \, \mathrm{e}^{\varDelta\tau(E_{\mathrm{P}} + \mu)} . \qquad (67)$$

There is a close relation to the one-electron Green function

$$G_{ij} = \underbrace{\langle \widetilde{c}_i \widetilde{c}_j^\dagger \rangle}_{G^a_{ij}} + \underbrace{\langle \widetilde{c}_i^\dagger \widetilde{c}_j \rangle}_{G^b_{ij}} . \qquad (68)$$

In real space and imaginary time, we have [15, 16]

$$G^a_{ij} = \langle \widetilde{c}_i \widetilde{c}_j^\dagger \rangle = (1 + \varOmega)^{-1}_{ij} , \quad G^b_{ij} = \delta_{ij} - G^a_{ij} = (\varOmega \, G^a)_{ji} . \qquad (69)$$

At this stage, with the above results for the partition function, a QMC simulation of the transformed Holstein model would proceed as follows. In each MC step, a pair of indices (i_0, τ_0) on the $N^{\mathrm{D}} \times L$ lattice of phonon momenta $p_{i,\tau}$ is chosen at random. At this site, a change $p_{i_0,\tau_0} \mapsto p_{i_0,\tau_0} + \varDelta p$ of the phonon configuration is proposed. To decide upon the acceptance of the new configuration using the Metropolis algorithm [13], the corresponding weights $w_\mathrm{b} w_\mathrm{f}$ and $w'_\mathrm{b} w'_\mathrm{f}$ have to be calculated. Due to the local updating process, the computation of the change of the bosonic weight $\Delta w_\mathrm{b} = w'_\mathrm{b}/w_\mathrm{b}$ is very fast, which is not the case for the fermionic weight $\Delta w_\mathrm{f} = w'_\mathrm{f}/w_\mathrm{f}$. By varying τ_0 sequentially from 1 to L instead of picking random values, the calculation of the ratio of the fermionic weights can be reduced to only two matrix multiplications.

It turns out that a local updating as described above does not permit efficient simulations for small phonon frequencies or low temperatures. Therefore, we shall introduce an alternative global updating in terms of principal components in Sect. 5.4.

5.2 Observables

Using the transformed Hamiltonian (14), the expectation value

$$\langle O \rangle = \mathcal{Z}^{-1} \, \mathrm{Tr}\, \widehat{O}\, \mathrm{e}^{-\beta H} = \mathcal{Z}^{-1} \, \mathrm{Tr}\, \widehat{\widetilde{O}}\, \mathrm{e}^{-\beta \widetilde{H}} \tag{70}$$

of an observable O is computed according to

$$\langle O \rangle = \mathcal{Z}^{-1} \, \mathrm{Tr}_\mathrm{f} \int \mathrm{d}p \, \langle p | \widehat{\widetilde{O}}\, \mathrm{e}^{-\beta \widetilde{H}} | p \rangle \, . \tag{71}$$

As a result of the analytic integration over the phonon coordinates $\widehat{x}$, interesting observables such as the correlation function $\langle \widehat{n}_i \widehat{x}_j \rangle$ are difficult to measure accurately. Other quantities such as the quasiparticle weight, and the closely related effective mass [17], can be determined from the one-electron Green function at long imaginary times [18], but results for one electron or two electrons would not be as accurate as in existing work (e.g., [19–21]).

The situation is strikingly different in the many-electron case, for which many methods fail to produce results of high accuracy for large systems and physically relevant parameters. Moreover, other important observables, such as the one-electron Green function, can be calculated with our approach.

One Electron

The electronic kinetic energy is defined as

$$E_\mathrm{kin} = \langle \widetilde{H}_\mathrm{kin} \rangle = -t \mathcal{Z}^{-1} \sum_{\langle ij \rangle} \mathrm{Tr}\, \left(c_i^\dagger c_j \, \mathrm{e}^{\mathrm{i}\gamma(\widehat{p}_i - \widehat{p}_j)}\, \mathrm{e}^{-\beta \widetilde{H}} \right) . \tag{72}$$

Repeating the steps used to derive the partition function, and noting that the additional phase factors in (72) again lead to the same matrix Ω as in (49), we find

478   Martin Hohenadler and Wolfgang von der Linden

$$E_{\mathrm{kin}} = -t\mathcal{Z}_L^{-1} \sum_{\langle ij\rangle} \int \mathcal{D}p\, w_{\mathrm{b}} \sum_n \langle n|\, \Omega c_i^\dagger c_j\, |n\rangle$$

$$= -t\mathcal{Z}_L^{-1} \sum_{\langle ij\rangle} \int \mathcal{D}p\, w_{\mathrm{b}}\, \langle j|\, \Omega\, |i\rangle \tag{73}$$

with the one-electron states (50). Introducing the matrix elements $(\Omega)_{ij} = \langle i|\, \Omega\, |j\rangle$ and the expectation value with respect to w_{b},

$$\langle O\rangle_{\mathrm{b}} = \frac{\int \mathcal{D}p\, w_{\mathrm{b}}\, O(p)}{\int \mathcal{D}p\, w_{\mathrm{b}}} \tag{74}$$

we obtain

$$E_{\mathrm{kin}} = -t\, \frac{\sum_{\langle ij\rangle} \langle \Omega_{ji}\rangle_{\mathrm{b}}}{\sum_i \langle \Omega_{ii}\rangle_{\mathrm{b}}}\, . \tag{75}$$

Here we have anticipated the reweighting discussed in Sect. 5.3.

The total energy can be obtained from $E = -\partial(\ln \mathcal{Z})/\partial\beta$ as

$$E = E_{\mathrm{kin}} + \frac{\omega_0}{2} \sum_i \langle p_i^2\rangle + E'_{\mathrm{ph}} - E_{\mathrm{P}}\, ,$$

$$E'_{\mathrm{ph}} = \frac{N^{\mathrm{D}}}{2\Delta\tau} - \frac{1}{2\omega_0 \Delta\tau^2 L} \sum_{i,\tau} \left\langle \left(p_{i,\tau} - p_{i,\tau+1}\right)^2\right\rangle\, . \tag{76}$$

To compare with other work we subtract the ground-state energy of the phonons, $E_{0,\mathrm{ph}} = N^{\mathrm{D}}\omega_0/2$.

Two Electrons

For two electrons, exploiting spin symmetry, we have

$$E_{\mathrm{kin}} = -t \sum_{\langle ij\rangle \sigma} \langle \tilde{c}_{i\sigma}^\dagger \tilde{c}_{j\sigma}\rangle = -2t \sum_{\langle ij\rangle} \langle c_{i\uparrow}^\dagger c_{j\uparrow} e^{i\gamma(\widehat{p}_i - \widehat{p}_j)}\rangle\, . \tag{77}$$

A simple calculation gives

$$\langle \tilde{c}_{i\uparrow}^\dagger \tilde{c}_{j\uparrow}\rangle = \mathcal{Z}_L^{-1} \int \mathcal{D}p\, w_{\mathrm{b}} e^{i\gamma(p_{i,1} - p_{j,1})} \mathrm{Tr}_{\mathrm{f}}(\Omega\, c_{i\uparrow}^\dagger c_{j\uparrow})\, . \tag{78}$$

Writing out explicitly the fermionic trace we obtain

$$\mathrm{Tr}_{\mathrm{f}}(\Omega\, c_{i\uparrow}^\dagger c_{j\uparrow}) = \sum_{i'j'} \langle i',j'|\, \Omega c_{i\uparrow}^\dagger c_{j\uparrow}\, |i',j'\rangle$$

$$= \sum_{j'} \langle j,j'|\, \Omega\, |i,j'\rangle\, , \tag{79}$$

and the kinetic energy finally becomes

$$E_{\text{kin}} = -2t\mathcal{Z}_L^{-1} \int \mathcal{D}p\, w_{\text{b}} \sum_{\langle ij \rangle} \sum_{j'} e^{i\gamma(p_{i,1}-p_{j,1})} \langle j, j'| \, \Omega \, |i, j'\rangle \,. \tag{80}$$

In addition to E_{kin}, we shall also consider the correlation function

$$\rho(\delta) = \sum_i \langle \widehat{n}_{i\uparrow}\widehat{n}_{i+\delta\downarrow}\rangle \,, \quad \delta = 0, 1, \ldots, N/2 - 1 \tag{81}$$

depending on the distance δ. We find

$$\rho(\delta) = \mathcal{Z}_L^{-1} \int \mathcal{D}p\, w_{\text{b}} \sum_i \langle i, i+\delta| \, \Omega \, |i, i+\delta\rangle \,. \tag{82}$$

Many-Electron Case

The calculation of observables within the formalism presented here is similar
to the standard determinant QMC method [14–16]. For an equal-time (i.e.,
static) observable O we have

$$\langle O\rangle_{\text{b}} = \frac{\int \mathcal{D}p\, w_{\text{b}} w_{\text{f}} \text{Tr}_{\text{f}}(\widehat{O}\widehat{B}_1 \cdots \widehat{B}_L)}{\int \mathcal{D}p\, w_{\text{b}}} \,. \tag{83}$$

The carrier density

$$n = \frac{1}{N^{\text{D}}} \sum_i \langle \widehat{n}_i\rangle \tag{84}$$

may be calculated from G^b (69) using $\langle \widehat{n}_i\rangle = \langle G^b_{ii}\rangle$.

Similarly, the modulus of the kinetic energy per site is given by

$$\overline{E}_{\text{kin}} = \frac{t}{N^{\text{D}}} \sum_{\langle ij \rangle} \langle G^b_{ji}\rangle \,. \tag{85}$$

Equal–time two–particle correlation functions such as

$$\rho(\delta) = \sum_i \langle \widehat{n}_i \widehat{n}_{i+\delta}\rangle \tag{86}$$

may be calculated in the same way as in [15, 16]. For a given phonon configuration, Wick's Theorem [4] yields

$$\begin{aligned}
\langle \widehat{n}_i \widehat{n}_j\rangle_p &= \langle c_i^\dagger c_i c_j^\dagger c_j\rangle_p \\
&= \langle c_i^\dagger c_i\rangle_p \langle c_j^\dagger c_j\rangle_p + \langle c_i^\dagger c_j\rangle_p \langle c_i c_j^\dagger\rangle_p \\
&= G^b_{ii} G^b_{jj} + G^b_{ij} G^a_{ij} \,,
\end{aligned} \tag{87}$$

and $\langle \widehat{n}_i \widehat{n}_j\rangle$ is then determined by averaging over all phonon configurations.

The time–dependent one–particle Green function

$$G^b(\boldsymbol{k},\tau) = \langle c^\dagger_{\boldsymbol{k}}(\tau)c_{\boldsymbol{k}}\rangle = \langle e^{\tau\mathcal{H}}c^\dagger_{\boldsymbol{k}}e^{-\tau\mathcal{H}}c_{\boldsymbol{k}}\rangle \tag{88}$$

is related to the momentum– and energy–dependent spectral function

$$A(\boldsymbol{k},\omega-\mu) = -\frac{1}{\pi}\mathrm{Im}\,G^b(\boldsymbol{k},\omega-\mu) \tag{89}$$

through

$$G^b(\boldsymbol{k},\tau) = \int\limits_{-\infty}^{\infty} d\omega\,\frac{e^{-\tau(\omega-\mu)}A(\boldsymbol{k},\omega-\mu)}{1+e^{-\beta(\omega-\mu)}}\,. \tag{90}$$

The inversion of the above relation is ill-conditioned and requires the use of the maximum entropy method [12, 13, 22]. Fourier transformation leads to

$$G^b(\boldsymbol{k},\tau) = \frac{1}{N^{\mathrm{D}}}\sum_{ij}e^{\mathrm{i}\boldsymbol{k}\cdot(\boldsymbol{r}_i-\boldsymbol{r}_j)}G^b_{ij}(\tau)\,. \tag{91}$$

The allowed imaginary times are $\tau_l = l\Delta\tau$, with non-negative integers $0 \leq l \leq L$. Within the QMC approach, we have [15, 16]

$$G^b_{ij}(\tau_l) = (G^a B_1 \cdots B_l)_{ji}\,. \tag{92}$$

The one-electron density of states is given by

$$N(\omega-\mu) = -\frac{1}{\pi}\mathrm{Im}\,G(\omega-\mu)\,, \tag{93}$$

where $G(\omega-\mu) = (N^{\mathrm{D}})^{-1}\sum_{\boldsymbol{k}}G(\boldsymbol{k},\omega-\mu)$. It may be obtained numerically via

$$N(\tau) = G^b_{ii}(\tau)\,, \tag{94}$$

and subsequent analytical continuation.

Suzuki-Trotter Error

The error associated with the approximation made in, e.g., (41) can be systematically reduced by using smaller values of $\Delta\tau$. In practice, there are two strategies to handle this so-called *Suzuki-Trotter error*. Owing to the usually large numerical effort for QMC simulations, $\Delta\tau$ is often simply chosen such that the systematic error is smaller than the statistical errors for observables. A second, more satisfactory, but also more costly method is to run simulations at different values of $\Delta\tau$, and to exploit the $\Delta\tau^2$ dependence of the results to extrapolate to $\Delta\tau = 0$.

For the results in Sect. 6, we have used a scaling toward $\Delta\tau = 0$ based on typical values $\Delta\tau = 0.1$, 0.075 and 0.05 to obtain the results for one and two electrons. In contrast, for the numerically more demanding calculations of dynamic properties in the many-electron case, $\Delta\tau = 0.1$ has been chosen. This is justified by the uncertainties in the analytical continuation.

5.3 Reweighting

As pointed out at the end of Sect. 5.1, the calculation of the change of the fermionic weight w_f represents the most time-consuming part of the updating process. Consequently, it would be highly desirable to avoid the evaluation of w_f. This may be achieved by using only the bosonic weight w_b in the updating, and treating w_f as part of the observables. For the expectation value of an observable O, such a reweighting requires calculation of

$$\langle O \rangle = \frac{\langle O\, w_\mathrm{f} \rangle_\mathrm{b}}{\langle w_\mathrm{f} \rangle_\mathrm{b}} \,, \tag{95}$$

where the subscript "b" indicates that the average is computed based on w_b only [cf. (74)].

Reweighting of the probability distribution is frequently used in MC simulations if a minus-sign problem occurs [13]. Here, the splitting into the configuration weight w_b and the observable $O w_\mathrm{f}$ is practicable provided the variance of both w_f and $O w_\mathrm{f}$ is small, which is the case after the LF transformation. Furthermore, we require a significant overlap of the two distributions, which may be quantified using the Kullback-Leibler number [8], in order to avoid prohibitive statistical noise. In fact, our calculations show that, in general, for the *untransformed* model the reweighting method cannot be applied. For a detailed discussion of this point in the one-electron case see [8]. Here we merely note that no problems arise when simulating the transformed model.

Apart from the significant advantage that the fermionic weight w_f only has to be calculated when observables are measured, the reweighting method becomes particularly effective in the present case when combined with the principal component representation introduced in Sect. 5.4. In this case, we will be able to perform an exact sampling of the phonons without any autocorrelations. For a reliable error analysis for observables calculated according to (95) the Jackknife procedure [23] is applied.

5.4 Principal Components

The reweighting method allows us, in principle, to skip enough sweeps between measurements to reduce autocorrelations to a minimum. However, even though a single phonon update requires negligible computer time compared to the evaluation of w_f, for critical parameters, an enormous number of such steps will be necessary between successive measurements [8]. On top of that, reliable results require knowledge of the longest autocorrelation times, which have to be determined in separate simulations for each set of parameters.

Due to the structure of the bosonic action S_b [see (51)], even relatively small (local) changes to the phonon momenta lead to large variations in S_b and hence the weight w_b. As a consequence, only minor changes may be proposed in order to reach a reasonable acceptance rate. Unfortunately, this strategy is the very origin of autocorrelations.

The problem can be overcome by a transformation to the normal modes of the phonons (along the imaginary time axis), so that we can sample completely uncorrelated configurations. As the fermion degrees of freedom are treated exactly, the resulting QMC method is then indeed free of any autocorrelations.

To find such a transformation, let us recall the form of the bosonic action, given by (52), which we write as

$$S_{\mathrm{b}} = \sum_i \boldsymbol{p}_i^{\mathrm{T}} A \boldsymbol{p}_i = \sum_i \boldsymbol{p}_i^{\mathrm{T}} A^{1/2} A^{1/2} \boldsymbol{p}_i \equiv \sum_i \boldsymbol{\xi}_i^{\mathrm{T}} \cdot \boldsymbol{\xi}_i \tag{96}$$

with the *principal components* $\boldsymbol{\xi}_i = A^{1/2} \boldsymbol{p}_i$, in terms of which the bosonic weight takes the simple Gaussian form

$$w_{\mathrm{b}} = \mathrm{e}^{-\Delta\tau \sum_i \boldsymbol{\xi}_i^{\mathrm{T}} \cdot \boldsymbol{\xi}_i} . \tag{97}$$

The sampling can now be performed directly in terms of the new variables $\boldsymbol{\xi}$. To calculate observables we have to transform back to the physical momenta $\boldsymbol{p}$ using $A^{-1/2}$. Comparison with (52) shows that instead of the ill-conditioned matrix A we now have the ideal case that we can easily generate exact samples of a Gaussian distribution. With the new coordinates $\boldsymbol{\xi}$, the probability distribution can be sampled exactly, e.g., by the Box-Müller method [24]. In contrast to a standard Markov chain MC simulation, every new configuration is accepted and measurements can be made at each step, so that simulation times are significantly reduced.

From the definition of the principal components it is obvious that an update of a single variable $\xi_{i,\tau}$, say, actually corresponds to a change of all $p_{i,\tau'}$, $\tau' = 1, \ldots, L$. Thus, in terms of the original phonon momenta $\boldsymbol{p}$, the updating becomes non-local.

The principal component representation can be used for one, two and many electrons, since the bosonic action (97) is identical. This even holds for models including, e.g., spin-spin interactions, as long as the phonon operators enter in the same form as in the Holstein model.

An important point is the combination of the principal components with the reweighting method. Using the latter, the changes to the original momenta $\boldsymbol{p}$, which are made in the simulation, do not depend in any way on the electronic degrees of freedom. Thus we are actually sampling a set of independent harmonic oscillators, as described by S_{b}. The crucial requirement for the success of this method is the use of the LF transformed model, in which the (bi-)polaron effects are separated from the zero-point motion of the oscillators around their current equilibrium positions.

Finally, as there is no need for a warm-up phase, and owing to the statistical independence of the configurations, the present algorithm is perfectly suited for parallelization.

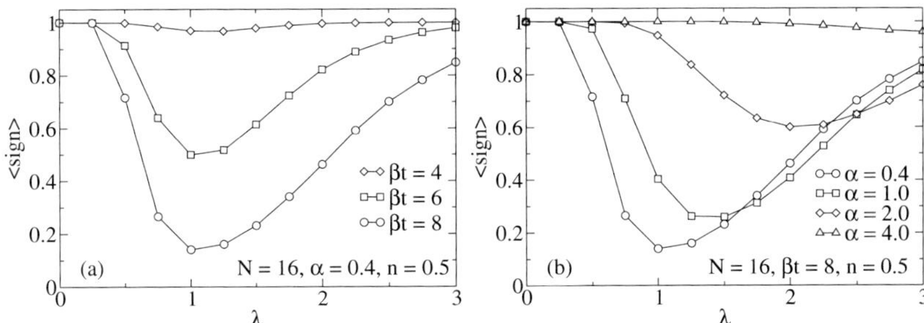

Fig. 1. Average sign $\langle\text{sign}\rangle$ in the many-electron case as a function of el-ph coupling λ in $D = 1$ (a) for different inverse temperatures β, and (b) for different values of the adiabaticity ratio α. Lines are guides to the eye, and errorbars are smaller than the symbols shown. The data presented in Figs. 1 and 2 are for $\Delta\tau = 0.05$. [Taken from [12].]

5.5 Minus-Sign Problem

The motivation for our development of a novel QMC approach to Holstein models was to improve on the performance of existing methods, especially in the many-electron case. As pointed out in [11], the LF transformation causes a sign problem even for the pure Holstein model which, in general, may significantly affect the applicability of the method. Therefore, we briefly discuss the resulting limitations, focussing on the many-electron case.

We shall see that there is a fundamental difference between simulations for one or two electrons—the carrier density being zero in the thermodynamic limit—and grand-canonical calculations at finite density $n > 0$. Whereas for one or two carriers the sign problem turns out to be rather uncritical—the average sign approaches unity upon increasing system size, in contrast to the usual behaviour [13]—restrictions are encountered in simulations of the many-electron case.

Since w_b is strictly positive, we define the average sign as

$$\langle\text{sign}\rangle = \langle w_\text{f}\rangle_\text{b}/\langle|w_\text{f}|\rangle_\text{b} \,. \tag{98}$$

For simplicity, we first show results for $n = 0.5$, while the effect of band filling will be discussed later. The choice $n = 0.5$ is convenient since we know the chemical potential, and we shall see below that the sign problem is most pronounced for a half-filled band. Moreover, most existing QMC results for the spinless Holstein model are for half filling (see references in [8]).

Figure 1(a) shows the dependence of $\langle\text{sign}\rangle$ on the el-ph coupling strength. It takes on a minimum near $\lambda = 1$ (for $\alpha < 1$) that becomes more pronounced with decreasing temperature. At weak coupling (WC) and SC, $\langle\text{sign}\rangle \approx 1$, so that accurate simulations can be carried out. These results are quite similar to the cases of one or two electrons [25].

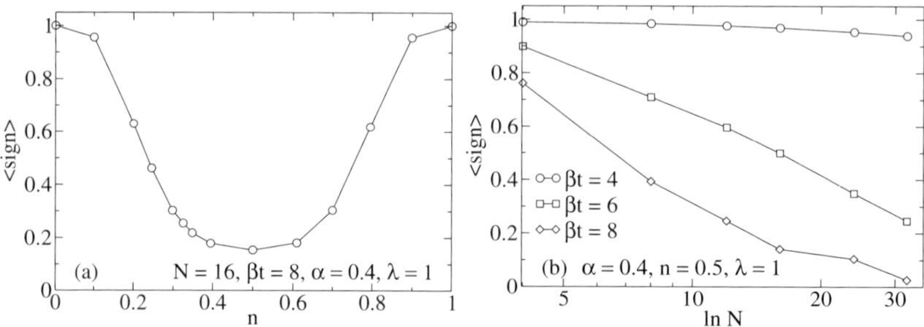

Fig. 2. Average sign in the many-electron case as a function of (a) band filling n, and (b) system size N.

The dependence on phonon frequency (Fig. 1(b)) also bears a close resemblance to the polaron problem [25]. Whereas $\langle\text{sign}\rangle$ becomes very small for $\alpha \ll 1$, it increases noticeably in the non-adiabatic regime $\alpha > 1$, permitting efficient and accurate simulations.

As illustrated in Fig. 2(a), the average sign depends strongly on the band filling n. While it is close to one in the vicinity of $n = 0$ or $n = 1$ (equivalent to one or two electrons), a significant reduction is visible near half filling $n = 0.5$. The minimum occurs at $n = 0.5$, and the results display particle-hole symmetry as expected. Here we have chosen $\beta t = 8$, $\alpha = 0.4$ and $\lambda = 1$, for which the sign problem is most noticeable according to Fig. 1.

In Fig. 2(b), we report the average sign as a function of system size, again for $n = 0.5$. The dependence is strikingly different from the one-electron case. While in the latter $\langle\text{sign}\rangle \to 1$ as $N \to \infty$ [8, 25], here the average sign decreases nearly exponentially with increasing system size, a behaviour well-known from QMC simulations of Hubbard models [13]. Obviously, this limits the applicability of our method. However, we shall see below that we can nevertheless obtain accurate results at low temperatures, small phonon frequencies, and over a large range of the el-ph coupling strength. Moreover, we would like to point out that for such parameters, other methods suffer strongly from autocorrelations, rendering simulations extremely difficult.

The dependence of the sign problem on the dimension of the system is again similar to the single-electron case [25]. The minimum at intermediate λ becomes more pronounced for the same parameters N, α, βt and λ as one increases the dimension of the cluster.

To conclude, we would like to point out that, in principle, the sign problem can be compensated by performing sufficiently long QMC runs, but we have to keep in mind that the statistical errors increase proportional to $\langle\text{sign}\rangle^{-2}$ [13], setting a practical limit to the accuracy.

5.6 Comparison with Other Approaches

The QMC method presented above seems to be most advantageous—as compared to other approaches—in the case of the spinless Holstein model with many electrons. For the latter, other methods are severely restricted by auto-correlations, rendering accurate simulations in the physically important adiabatic, IC regime virtually impossible even at moderately low temperatures. In contrast, the present method enables us to study the single-particle spectrum on rather large clusters and for a wide range of model parameters and band filling (see Sect. 6.3). Unfortunately, the generalization to the spinful Hubbard-Holstein model suffers severely from the sign problem.

For the polaron and the bipolaron problem, our method requires more computer time than other QMC algorithms [26–29]. However, we are able to consider practically all parameter regimes on reasonably large clusters in one (polaron and bipolaron problem) and two dimensions (polaron problem).

Finally, a discussion of the scaling of computer time with the system parameters can be found in [8, 10, 11].

6 Selected Results

We now come to a selection of results obtained with the methods discussed so far, most of which have been published before [8, 10–12]. Note that error bars will be suppressed in the figures if smaller than the symbol size. Moreover, lines connecting data points are guides to the eye only.

6.1 Small-Polaron Cross-Over

The Holstein model with a single electron (for a review see [30]) exhibits a cross-over from a large polaron (D = 1) or a quasi-free electron (D > 1) to a small polaron with increasing el-ph coupling strength.

Quantum Monte Carlo

To investigate the small-polaron crossover, following previous work [5, 17, 19–21, 26, 31–33], we calculate the electronic kinetic energy E_{kin} given by (75). As we shall compare results for different dimensions, we define the normalized quantity

$$\overline{E}_{\mathrm{kin}} = E_{\mathrm{kin}}/(-2t\mathrm{D}) \tag{99}$$

with $\overline{E}_{\mathrm{kin}} = 1$ for $T = 0$ and $\lambda = 0$.

The inverse temperature will be fixed to $\beta t = 10$, low enough to identify the cross-over. Calculations at even lower temperatures can easily be done for $\alpha > 1$, but $\alpha < 1$ requires very large numbers of measurements to ensure satisfactorily small statistical errors. System sizes were 32 sites in 1D, a

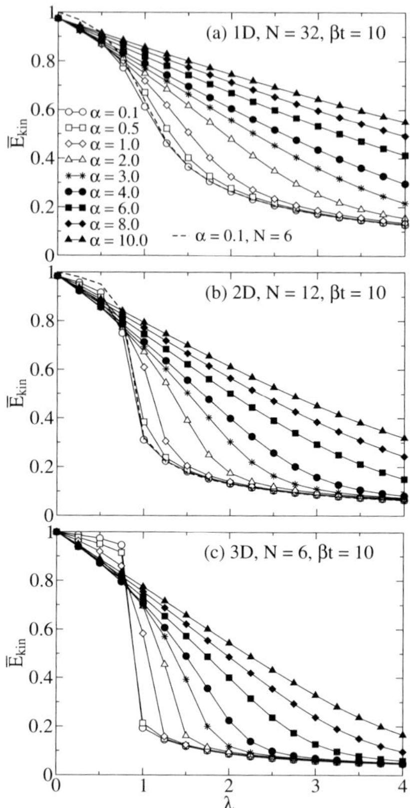

Fig. 3. Normalized kinetic energy $\overline{E}_{\mathrm{kin}}$ (99) of the Holstein model with one electron from QMC as a function of el-ph coupling λ for different adiabaticity ratios α and different dimensions D of the lattice (N denotes the linear cluster size). Here and in subsequent figures, QMC data have been extrapolated to $\Delta\tau = 0$ (see Sect. 5.2). [Taken from [11].]

12×12 cluster in 2D, and a $6 \times 6 \times 6$ lattice in 3D. In contrast to D = 1, 2, where results are well converged with respect to system size, non-negligible finite-size effects (maximal relative changes of up to 20 % between $N = 5$ and $N = 6$ for $\alpha \ll 1$; much smaller changes otherwise) are observed in three dimensions. Moreover, for small N, effects due to thermal population of states with non-zero momentum $\boldsymbol{k}$—absent in ground-state calculations—are visible, as discussed below. Nevertheless, the main characteristics are well visible already for $N = 6$. For a detailed study of finite-size and finite-temperature effects see [25].

Figure 3 shows $\overline{E}_{\mathrm{kin}}$ as a function of the el-ph coupling λ for different phonon frequencies varying over two orders of magnitude, in one to three dimensions. Generally, the kinetic energy is large at WC, where the ground state consists of a weakly dressed electron (D > 1) or a large polaron (D =

1). It reduces more or less strongly—depending on α—in the SC regime, where a small, heavy polaron exists, defined as an electron surrounded by a lattice distortion essentially localized at the same site. The finite values of $\overline{E}_{\mathrm{kin}}$ even for large λ are a result of undirected motion of the electron inside the surrounding phonon cloud. In contrast, the quasiparticle weight is exponentially reduced in the SC regime (see, e.g., [21]), whereas the effective mass becomes exponentially large.

In all dimensions, the phonon frequency has a crucial influence on the behaviour of the kinetic energy. While in the adiabatic regime $\alpha < 1$ the small-polaron cross-over is determined by the condition $\lambda = E_\mathrm{P}/2t\mathrm{D} > 1$, the corresponding criterion for $\alpha > 1$ is $g^2 = E_\mathrm{P}/\omega_0 > 1$. The former condition reflects the fact that the loss in kinetic energy of the electron has to be outweighed by a gain in potential energy in order to make small-polaron formation favourable. The latter condition expresses the increasing importance of the lattice energy for $\alpha > 1$, since the formation of a "localized" state requires a sizable lattice distortion. As a consequence, for large phonon frequencies, the critical coupling shifts to $\lambda_\mathrm{c} > 1$, whereas for $\alpha < 1$ we have $\lambda_\mathrm{c} = 1$. Additionally, the decrease of $\overline{E}_{\mathrm{kin}}$ at λ_c becomes significantly sharper with decreasing phonon frequency.

Concerning the effect of dimensionality, Fig. 3 reveals that, for fixed α, the small-polaron cross-over becomes more abrupt in higher dimensions, with a very sharp decrease in 3D. Nevertheless, there is no real phase transition [34]. Figure 3 also contains results for $N = 6$ in one and two dimensions, i.e., for the same linear cluster size as in 3D (dashed lines). Clearly, for such small clusters, the spacing between the discrete allowed momenta $\boldsymbol{k}$ is too large to permit substantial thermal population, so that results are closer to the ground state [e.g., $\overline{E}_{\mathrm{kin}}(\lambda = 0) \approx 1$], and exhibit a slightly more pronounced decrease near the critical coupling. However, the sharpening of the latter with increasing dimensionality is still well visible.

Variational Approach

To test the validity of the variational approach of Sect. 4 we have calculated the total energy (24) and the quasiparticle weight (29) on a cluster with $N = 4$ for various values of α. A comparison with exact diagonalization results [35] is depicted in Fig. 4. We only consider the regime $\alpha \geq 1$ where the zero-phonon approximation is expected to be justified. The overall agreement is strikingly good. Minor deviations from the exact results increase with decreasing α. For the smallest frequency shown, $\alpha = 1$, the result of the HLF approximation is also reported. Clearly, the variational method represents a significant improvement over the HLF approximation, underlining the importance of taking into account non-local distortions. Similar conclusions can be drawn for larger system sizes (see Fig. 3 in [8]).

In Fig. 5 we present results for the variational displacement fields γ_δ, which provide a measure for the polaron size. For $\alpha = 0.1$ we see an abrupt cross-

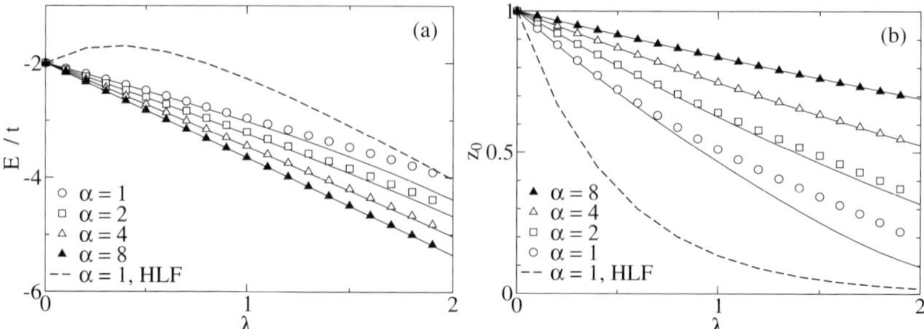

Fig. 4. Total energy E (a) and quasiparticle weight z_0 (b) for $N = 4$ as functions of the el-ph coupling λ for different values of the adiabaticity ratio α. Symbols correspond to variational results and full lines represent exact $T = 0$ data obtained with the Lanczos method [35]. Dashed lines are results of the HLF approximation. [Taken from [8].]

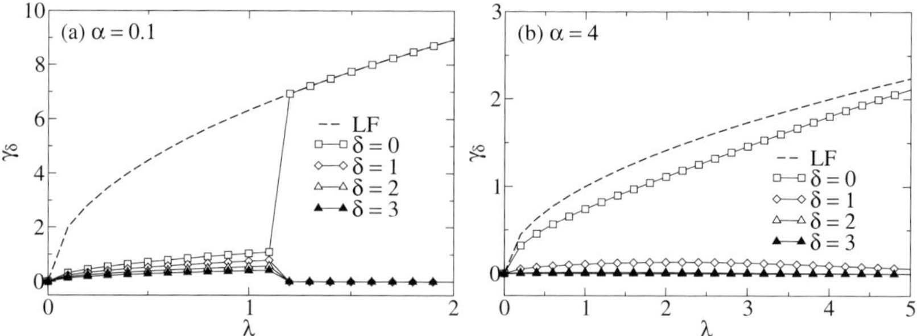

Fig. 5. Polaron-size parameter γ_δ for $N = 16$ as a function of the el-ph coupling λ for various distances δ in the (a) adiabatic and (b) anti-adiabatic regime. Also shown is the LF parameter γ (10). [Taken from [8].]

over from a large to a small polaron at $\lambda \approx 1.2$. For smaller λ, the electron induces lattice distortions at neighboring sites even at a distance of more than three lattice constants. Above $\lambda \approx 1.2$ we have a mobile small polaron extending over a single site only. In contrast, for the anti-adiabatic case $\alpha = 4$, the cross-over is much more gradual, and $\gamma_1 > 0$ even for $\lambda \gg 1$. The same behaviour has been found by Marsiglio [36] who determined the correlation function $\langle \hat{n}_i \hat{x}_{i+\delta} \rangle$ by exact diagonalization; within the variational approach $\langle \hat{n}_i \hat{x}_{i+\delta} \rangle = \gamma_\delta$. Although in Marsiglio's results the cross-over to a small polaron for $\alpha = 0.1$ occurs at a smaller value of the coupling $\lambda \approx 1$, the simple variational approach reproduces the main characteristics.

6.2 Bipolaron Formation in the Extended Holstein-Hubbard Model

In contrast to Cooper pairing of electrons with opposite momentum, two electrons may also form a bound state by travelling sufficiently close in real space.

Table 1. Conditions for the existence of different singlet bipolaron states in the one-dimensional Holstein-Hubbard model [10].

$U = 0$		$U > 0$		
Large bipolaron	Small bipolaron	Two polarons	Inter-site bipolaron	Small bipolaron
$\lambda < 0.5$ or $g < 0.5$	$\lambda > 0.5$ and $g > 0.5$	$U > 2E_\mathrm{P}$ (WC) $U > 4E_\mathrm{P}$ (SC)	$U < 2E_\mathrm{P}$ (WC) $U < 4E_\mathrm{P}$ (SC)	$U \ll 2E_\mathrm{P}$

Bipolaron formation may be studied in the framework of the 1D extended Holstein-Hubbard model, and a brief review of previous work has been given in [10, 37]. Here we merely note that depending on the choice of parameters, the ground state of the model may either consist of two polarons, a large bipolaron, an inter-site bipolaron or a small bipolaron (in the singlet case). A summary of the conditions on the model parameters is given in table 1. Whereas existing work is almost exclusively concerned with the singlet case, here we shall also consider two electrons of the same spin. Triplet bipolarons are expected to play a role, e.g., in the ferromagnetic state of the manganites [38–40]. Furthermore, we are not aware of any previous work for $V > 0$.

Quantum Monte Carlo

Owing to the increased numerical effort compared to the one-electron case, we shall only present results for $N \leq 12$ in one dimension. However, finite-size effects are small even for the most critical parameters [10].

We define the effective kinetic energy of the two electrons as

$$\overline{E}_\mathrm{kin} = E_\mathrm{kin}/(-4t) \,. \tag{100}$$

In Fig. 6(a) we depict $\overline{E}_\mathrm{kin}$ as a function of the el-ph coupling for different values of α and U/t, at $\beta t = 10$, i.e., much closer to the ground state than in some previous work [27].

Figure 6(a) reveals a strong decrease of $\overline{E}_\mathrm{kin}$ near $\lambda = 0.5$ for $\alpha = 0.4$ and $U/t = 0$. With increasing α, the cross-over becomes less pronounced, and shifts to larger values of λ. For the same value of α, the cross-over to a small bipolaron is sharper than the small-polaron cross-over (cf. Fig. 3(a)). For finite on-site repulsion $U/t = 4$, $\overline{E}_\mathrm{kin}$ remains fairly large up to $\lambda \approx 1$ (for $\alpha = 0.4$), in agreement with the SC result $\lambda_\mathrm{c} = 1$ for $U/t = 4$ (see discussion in [37]). At even stronger coupling, the Hubbard repulsion is overcome, and a small bipolaron is formed. Again, the critical coupling increases with phonon frequency. Finally, the kinetic energy in the triplet case (corresponding to $U/t = \infty$) is comparable to the results for $U/t = 4$ up to $\lambda \approx 1$, but significantly larger in the SC regime since on-site bipolaron formation is not possible.

The influence of nearest-neighbour repulsion V is revealed in Fig. 6(b), again for $U/t = 4$. For all values of α shown, the cross-over sharpens noticeably

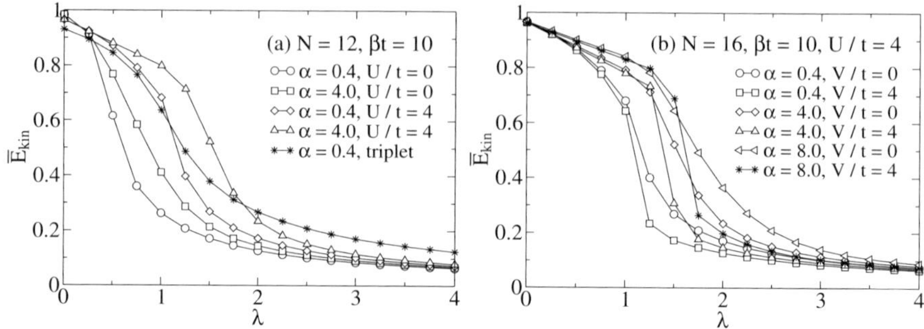

Fig. 6. Normalized kinetic energy $\overline{E}_{\mathrm{kin}}$ (100) from QMC as a function of the el-ph coupling λ for different values of the adiabaticity ratio α, the on-site repulsion U and the nearest-neighbour repulsion V. [(a) taken from [10].]

for $V > 0$. The reason is that $V > 0$ suppresses the (more mobile) inter-site bipolaron state, leading to a direct cross-over from a large to a small bipolaron.

The nature of the bipolaron state is revealed by the correlation function $\rho(\delta)$ (81), which gives the probability for the two electrons to be separated by a distance $\delta \geq 0$, and provides a measure of the bipolaron size. The phonon frequency determines the degree of retardation of the el-ph interaction, and thereby limits the distance between the two electrons in a bound state. In the following, we shall focus on the most interesting case of small phonon frequencies, which has often been avoided in previous work for reasons outlined in Sect. 5.

Starting with $U \ll E_{\mathrm{P}}$, a cross-over from a small to an inter-site bipolaron to two weakly bound polarons takes place upon increasing the Hubbard interaction [41]. Since the latter competes with the retarded el-ph interaction, the phonon frequency is expected to be an important parameter. In Fig. 7, we show the kinetic energy and the correlation function $\rho(\delta)$ as a function of U/t for IC $\lambda = 1$. Starting from a small bipolaron for $U/t = 0$, the kinetic energy increases with increasing Hubbard repulsion, equivalent to a reduction of the effective bipolaron mass [41, 42]. Although the cross-over is slightly washed out by the finite temperature in our simulations, there is a well-conceivable increase in $\overline{E}_{\mathrm{kin}}$ up to $U/t \approx 4$, above which the kinetic energy begins to decrease slowly. The increase of $\overline{E}_{\mathrm{kin}}$ originates from the breakup of the small bipolaron, as indicated by the decrease of $\rho(0)$ in Fig. 7(b). Close to $U/t = 4$, the curves for $\rho(0)$ and $\rho(1)$ cross, and it becomes more favourable for the two electrons to reside on neighboring sites. The inter-site bipolaron only exists below a critical Hubbard repulsion U_{c}. The latter is given by $U_{\mathrm{c}} = 2E_{\mathrm{P}}$ (i.e., here $U_{\mathrm{c}}/t = 4$) at weak el-ph coupling, and by $U_{\mathrm{c}} = 4E_{\mathrm{P}}$ at SC. For an intermediate value $\lambda = 1$ as in Fig. 7, the cross-over from the inter-site state to two weakly bound polarons is expected to occur somewhere in between, but is difficult to locate exactly from the QMC results.

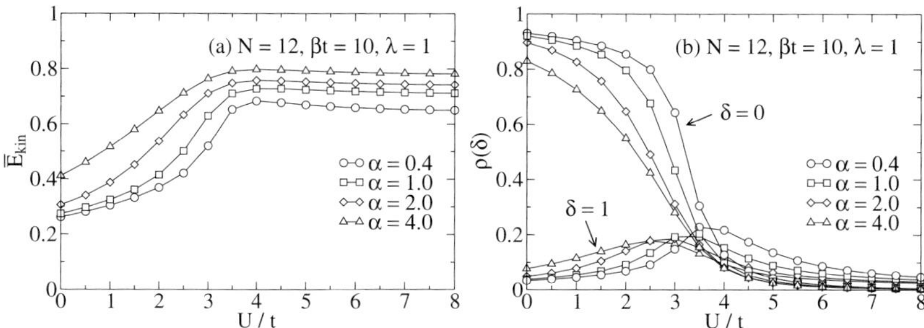

Fig. 7. (a) Normalized kinetic energy $\overline{E}_{\mathrm{kin}}$ and (b) correlation functions $\rho(0)$, $\rho(1)$ from QMC as a function of the Hubbard repulsion U/t for different values of the adiabaticity ratio α. [Taken from [10].]

Figure 7 further illustrates that the cross-over becomes steeper with decreasing phonon frequency. In the adiabatic limit $\alpha = 0$, it has been shown to be a first-order phase transition [43], whereas for $\alpha > 0$ retardation effects suppress any non-analytic behaviour. At the same U/t, $\overline{E}_{\mathrm{kin}}$ increases with α since for a fixed λ, the bipolaron becomes more weakly bound. For the same reason, the cross-over to an inter-site bipolaron—showing up in Fig. 7 as a crossing of $\rho(0)$ and $\rho(1)$—shifts to smaller values of U/t.

Let us now consider the effect of temperature on $\rho(\delta)$. To this end, we plot in Figs. 8(a)–(c) $\rho(\delta)$ at different temperatures, for parameters corresponding to the three regimes of a large, small and inter-site bipolaron, respectively.

For the parameters in Fig. 8(a) ($U/t = 0$, $\lambda = 0.25$), the two electrons are most likely to occupy the same site, but the bipolaron extends over a distance of several lattice constants. Clearly, in this regime, the cluster size $N = 12$ used here is not completely satisfactory, but still provides a fairly accurate description as can be deduced from calculations for $N = 14$ (not shown). Nevertheless, on such a small cluster, no clear distinction between an extended bipolaron and two weakly bound polarons can be made. As the temperature increases from $\beta t = 10$ to $\beta t = 1$, the probability distribution broadens noticeably, i.e., it becomes more likely for the two electrons to be further apart. In particular, for the highest temperature shown, $\rho(0)$ has reduced by about 30 % compared to $\beta t = 10$.

A different behaviour is observed for the small bipolaron, which exist at stronger el-ph coupling $\lambda = 1$. Figure 8(b) reveals that $\rho(\delta)$ peaks strongly at $\delta = 0$, but is very small for $\delta > 0$ at low temperatures. Increasing temperature, $\rho(\delta)$ remains virtually unchanged up to $\beta t = 3$. Only at very high temperatures there occurs a noticeable transfer of probability from $\delta = 0$ to $\delta > 0$. At the highest temperature shown, $\beta t = 0.5$, the two electrons have a non-negligible probability for traveling a finite distance $\delta > 0$ apart, although most of the probability is still contained in the peak located at $\delta = 0$.

Finally, we consider in Fig. 8(c) the inter-site bipolaron, taking $U/t = 4$ and $\lambda = 1$ (cf. Fig. 2 in [10]). At low temperatures, $\rho(\delta)$ takes on a maximum

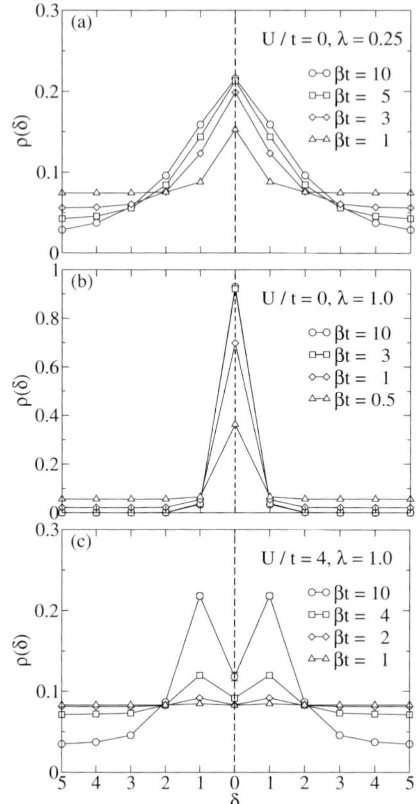

Fig. 8. Correlation function $\rho(\delta)$ from QMC as a function of δ for different inverse temperatures β, $N = 12$ and $\alpha = 0.4$. [Taken from [10].]

for $\delta = 1$. For smaller values of βt, the latter diminishes, until at $\beta t = 1$, the distribution is completely flat, so that all δ are equally likely.

The different sensitivity of the bipolaron states to changes in temperature found above can be explained by their different binding energies. The latter is given by $\Delta E_0 = E_0^{(2)} - 2E_0^{(1)}$, where $E_0^{(1)}$ and $E_0^{(2)}$ denote the ground-state energy of the model with one and two electrons, respectively.

Generally, the thermal dissociation is expected to occur at a temperature such that the thermal energy $k_{\mathrm{B}}T = (\beta T)^{-1}$ becomes comparable to ΔE_0, in accordance with our numerical data. The large and the inter-site bipolarons are relatively weakly bound as a result of the rather small effective interaction $U_{\mathrm{eff}} \approx U - 2E_{\mathrm{P}}$ [37]. The binding energies are $\Delta E_0 \approx -(0.32 \pm 0.08)t$ and $-(0.28 \pm 0.08)t$, respectively, so that we expect a critical inverse temperature $\beta t \approx 2.5\text{–}5$, in agreement with Figs. 8(a) and (c). In contrast, the small bipolaron in Fig. 8(b) has a significantly larger binding energy $\Delta E \approx -(3.43 \pm 0.09)t$, and therefore remains stable up to $\beta t \approx 0.3$. Ther-

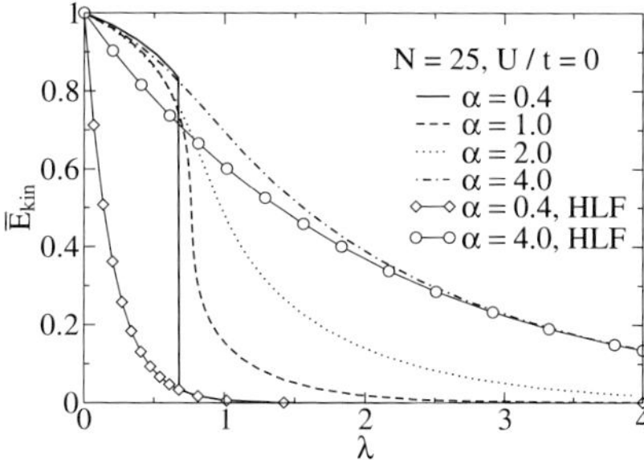

Fig. 9. Variational results for the normalized kinetic energy $\overline{E}_{\mathrm{kin}}$ as a function of the el-ph coupling λ, and for different adiabaticity ratios α. Also shown are results of the HLF approximation. [Taken from [10].]

mal dissociation of bipolarons occurs at even lower temperatures for $V > 0$, especially in the triplet case, owing to the reduced binding energy.

Variational Approach

Whereas the QMC approach is limited to finite temperatures and relatively small clusters, the variational method of Sect. 4 yields ground-state results on much larger systems. To scrutinize the quality of the variational method, we compare the ground-state energy for $U/t = 0$ to the most accurate approach currently available in one dimension, namely the variational diagonalization [41]. We find a good agreement over the whole range of λ. As expected from the nature of the approximation, slight deviations occur for $\alpha \lesssim 1$, similar to the one-electron case.

Despite the success in calculating the total energy—being the quantity that is optimized—one has to be careful not to overestimate the validity of any variational method. To reveal the shortcomings of the current approach, we show in Fig. 9 the normalized kinetic energy $\overline{E}_{\mathrm{kin}} = t_{\mathrm{eff}}$ [see (31) and (100)] as a function of el-ph coupling, and for different α. We have chosen $N = 25$ to ensure negligible finite-size effects. In principle, Fig. 9 displays a behaviour similar to the QMC data in Fig. 6(a). There is a jump-like decrease of $\overline{E}_{\mathrm{kin}}$ near $\lambda = 0.5$ for $\alpha = 0.4$, which becomes washed out and moves to larger λ with increasing phonon frequency. For $\alpha = 0.4$, the cross-over in the variational results is much too steep, regardless of the fact that the latter are for $T = 0$, a common defect of variational methods. Moreover, for $\alpha = 0.4-2$, the variational kinetic energy is too small above the bipolaron cross-over compared to the QMC data, whereas for $\alpha = 4$, the decay of $\overline{E}_{\mathrm{kin}}$ with increasing λ is too slow.

The reason for the failure is the absence of retardation effects, which play a dominant role in the formation of bipolaron states. The increased importance of the phonon dynamics—not included in the variational method—for the two-electron problem leads to a less good agreement with exact results

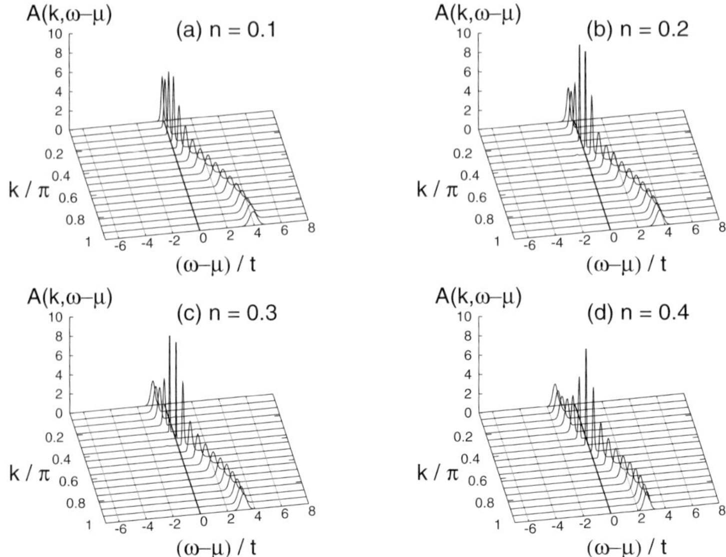

Fig. 10. One-electron spectral function $A(k, \omega - \mu)$ from QMC for different band fillings n, $N = 32$, $\beta t = 8$, $\alpha = 0.4$, and $\lambda = 0.1$. Here and in subsequent figures $\Delta\tau = 0.1$. [Taken from [12].]

than in the one-electron case. In particular, our variational results overestimate the position of the cross-over (Fig. 9) compared to the value $\lambda_c = 0.5$ expected in the adiabatic regime. Nevertheless, the method represents a significant improvement over the simple HLF approximation, due to the variational determination of the parameters γ_{ij}. This is illustrated in Fig. 9, where we also show the HLF result $\overline{E}_{\mathrm{kin}} = \mathrm{e}^{-g^2}$ for $\alpha = 0.4$ and 4.0. In contrast to the variational approach, the HLF approximation yields an exponentially decreasing kinetic energy for all values of the phonon frequency. Whereas such behaviour actually occurs in the anti-adiabatic limit $\alpha \to \infty$, the situation is different for small α [see Figs. 6(a) and 9]. The variational method presented here accounts qualitatively for the influence of the phonon frequency on bipolaron formation.

6.3 Many-Polaron Problem

We review recent results on the carrier-density dependence of photoemission spectra of many-polaron systems in the framework of the spinless Holstein model (63) in one dimension. We shall see that the sensitivity to changes in n strongly depends on the phonon frequency and el-ph coupling strength, with the most interesting physics being observed in the adiabatic, IC regime often realized experimentally. This regime is characterized by the existence of large polarons at low carrier density. At larger densities, a substantial overlap

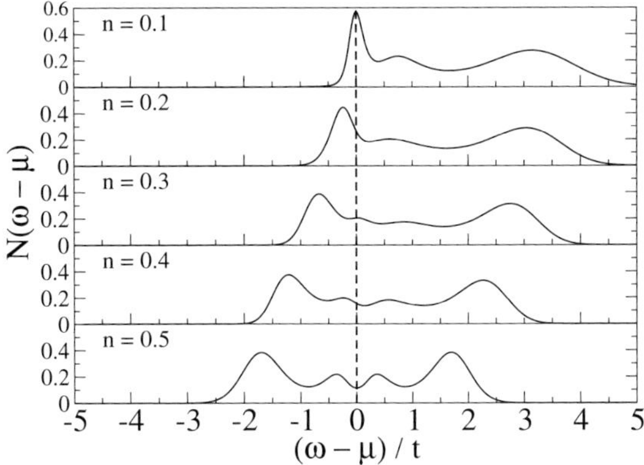

Fig. 11. One-electron density of states $N(\omega - \mu)$ from QMC for different band fillings n, $N = 32$, $\beta t = 8$, $\alpha = 0.4$ and $\lambda = 0.1$. [Taken from [12].]

of the single-particle wavefunctions occurs, leading to a dissociation of the individual polarons and finally to a restructuring of the whole many-particle ground state. Note that the many-polaron problem has since been studied also by means of other methods [44–46], confirming the original findings of [12], and more results along these lines can be found in [1].

Weak Coupling

For WC $\lambda = 0.1$, the sign problem is not severe (Sect. 5.5) so that simulations can easily be performed for large lattices with $N = 32$, making the dispersion of quasiparticle features well visible.

Figure 10 shows the evolution of the one-electron spectral function $A(k, \omega - \mu)$ with increasing electron density n. At first sight, we see that the spectra bear a close resemblance to the free-electron case, i.e., there is a strongly dispersive band running from $-2t$ to $2t$ which can be attributed to weakly dressed electrons. As expected, the height (width) of the peaks increases (decreases) significantly in the vicinity of the Fermi momentum k_F, determined by the crossing of the band with the chemical potential. However, in contrast to the case of a rigid tight-binding band, we shall see below (Fig. 11) that a significant redistribution of spectral weight occurs with increasing n.

We would like to point out that the apparent absence of any phonon signatures in Fig. 10 is not a defect of the maximum entropy method, but results from the large scale of the z-axis chosen. As a consequence, the peaks running close to the bare band dominate the spectra and suppress any small phonon peaks present. At higher resolution, for all densities $n = 0.1 - 0.4$, we observe the band flattening [47–49] at large wavevectors which originates from the intersection of the approximately free-electron dispersion with the bare phonon energy at $\omega - \mu = \omega_0$.

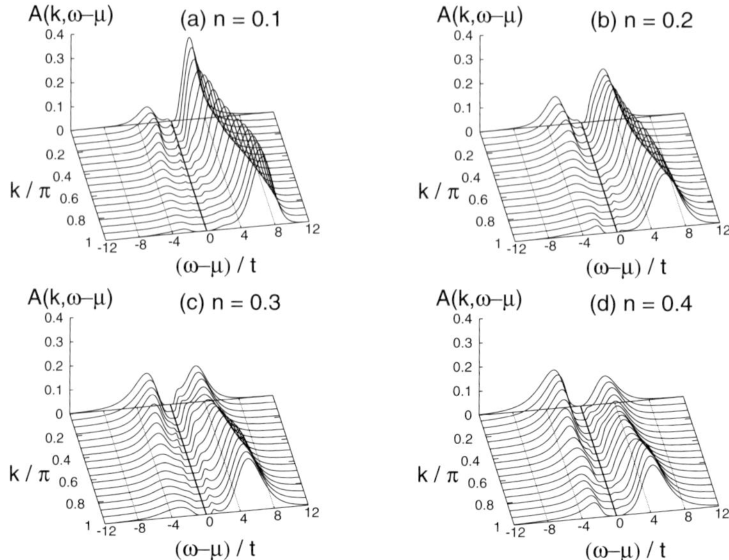

Fig. 12. One-electron spectral function $A(k, \omega - \mu)$ from QMC for different band fillings n, $N = 32$, $\beta t = 8$, $\alpha = 0.4$, and $\lambda = 2$. [Taken from [12].]

To complete our discussion of the WC regime, we show in Fig. 11 the one-electron density of states (DOS) $N(\omega - \mu)$ given by (94). Clearly, for small n, there is a peak with large spectral weight at the Fermi level. In contrast, for large n, the tendency toward formation of a Peierls– (band–) insulating state at $n = 0.5$ suppresses the DOS at the Fermi level, although we are well below the critical value of λ at which the cross-over to the insulating state takes place at $T = 0$ [50, 51]. The additional small features separated from μ by the bare phonon energy ω_0 will be discussed below.

Strong Coupling

We now turn to the SC limit taking $\lambda = 2$. At low density $n = 0.1$ [Fig. 12(a)], we expect the well-known, almost flat polaron band having exponentially reduced spectral weight (given by e^{-g^2} in the single-electron, SC limit) which, nevertheless, can give rise to coherent transport at $T = 0$. As discussed in [12], such weak signatures are difficult to determine accurately using the maximum entropy method. Generally, it is known that the reliability of dynamic properties obtained by means of the maximum entropy method crucially depends on the size of statistical errors and the general structure of the spectra. A detailed discussion of this point has been given in [12].

Besides, the spectrum consists of two incoherent features located above and below the chemical potential, which reflect the phonon-mediated transitions to high-energy electron states. Here, the maximum of the photoemission spectra

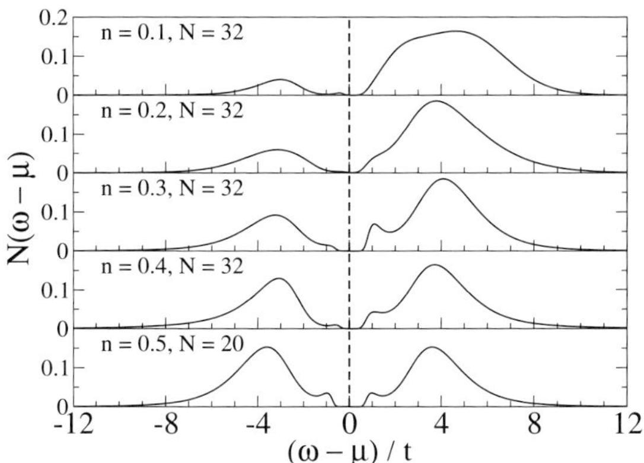

Fig. 13. One-electron density of states $N(\omega - \mu)$ from QMC for different band fillings n and cluster sizes N, $\beta t = 8$, $\alpha = 0.4$ and $\lambda = 2$. [Taken from [12].]

$(\omega - \mu > 0)$ follows a tight-binding cosine dispersion. The incoherent part of the spectra is broadened according to the phonon distribution.

For all band fillings, the chemical potential is expected to be located in a narrow polaron band with little spectral weight. There exists a finite gap to the photoemission (inverse photoemission) parts of the spectrum, so that the system typifies as a polaronic metal. We shall see below that a completely different behaviour is observed at IC. Notice that the incoherent inverse photoemission (photoemission) signatures are more pronounced at small (large) wavevectors.

Finally, for $n = 0.4$ [Fig. 12(d)], the incoherent features lie rather close to the Fermi level, thus being accessible by low-energy excitations. Now, the photoemission spectrum for $k < \pi/2$ is almost symmetric to the inverse photoemission spectrum for $k > \pi/2$ and already reveals the gapped structure which occurs at $n = 0.5$ due to charge-density-wave formation accompanied by a Peierls distortion [51].

As in the WC case discussed above, the properties of the system also manifest itself in the DOS, shown in Fig. 13. Owing to the strong el-ph interaction, the spectral weight at the chemical potential is exponentially small for all fillings n. At half filling, the DOS exhibits particle-hole symmetry, and the system can be described as a Peierls insulator, consisting of a polaronic superlattice. In contrast to the WC case, the ground state is characterized as a polaronic insulator rather than as a band insulator.

Intermediate Coupling

As discussed in the introduction, a crossover from a polaronic state to a system with weakly dressed electrons can be expected in the IC regime. Here we choose $\lambda = 1$, which corresponds to the critical value for the small-polaron

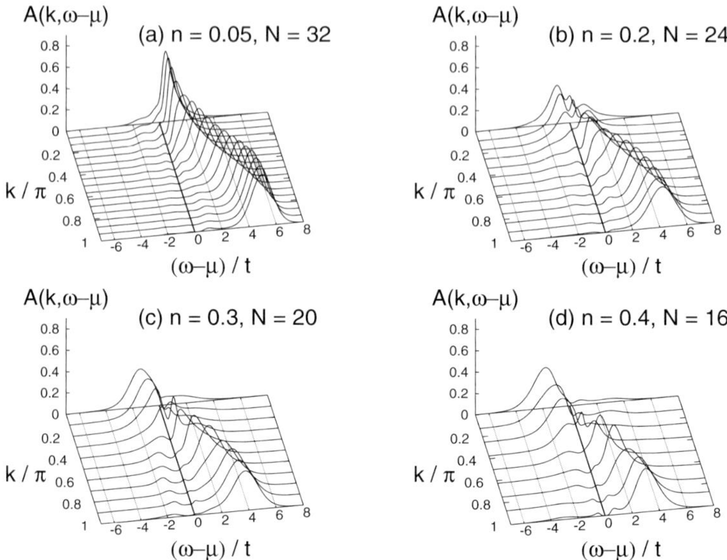

Fig. 14. One-electron spectral function $A(k, \omega - \mu)$ from QMC for different band fillings n and cluster sizes N, $\beta t = 8$, $\alpha = 0.4$, and $\lambda = 1$. [Taken from [12].]

cross-over in the one-electron problem [cf. Fig. 3(a)]. Owing to the sign problem, which is particularly noticeable for $\lambda = 1$ (see Fig. 1), we have to decrease the system size as we increase the electron density n.

We shall see that the cross-over is rather difficult to detect from the QMC results only. However, the data presented here are perfectly consistent with more recent studies employing other methods such as exact diagonalization [12], cluster perturbation theory [45] or self-energy calculations [44].

Figure 14 shows the spectral function for $\lambda = 1$ and increasing band filling. Owing to the overlap of large polarons in the IC regime, we start with a very low density $n = 0.05$ [Fig. 14(a)]. Compared to the behaviour for $\lambda = 2$ [Fig. 12(a)], we notice that the polaron band now lies much closer to the incoherent features, and that there is a mixing of these two parts of the spectrum at small values of k. Nevertheless, the almost flat polaron band is well visible for large k.

With increasing density, the polaron band merges with the incoherent peaks at higher energies, signaling the above-anticipated density-driven cross-over from a polaronic to a (diffusive) metallic state, with the broad main band crossing the Fermi level.

Further information about the density dependence can be obtained from the one-electron DOS. The latter is presented in Fig. 15 for different fillings $n = 0.05 - 0.5$. As in Fig. 14, the cluster size is reduced with increasing n in order to cope with the sign problem. To illustrate the rather small influence of finite-size effects, Fig. 15 also contains results for $N = 10$.

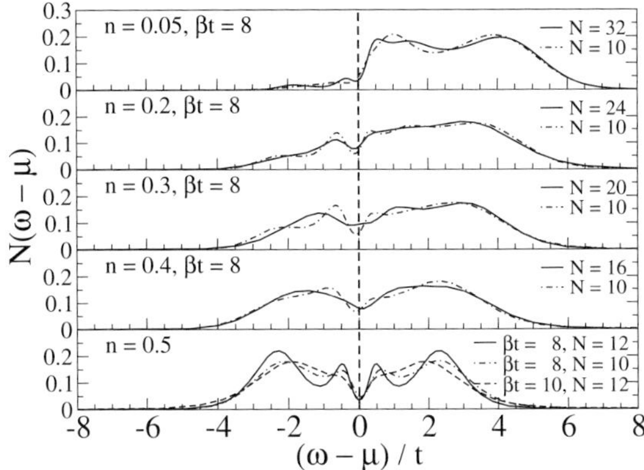

Fig. 15. One-electron density of states $N(\omega - \mu)$ from QMC for different band fillings n, cluster sizes N and inverse temperatures β. Here $\alpha = 0.4$ and $\lambda = 1$. [Taken from [12].]

For low density $n = 0.05$, the DOS in Fig. 15 lies in between the results for WC and SC discussed above. Although the spectral weight at the chemical potential is strongly reduced compared to $\lambda = 0.1$, $N(0)$ is still significantly larger than for $\lambda = 2$.

When the density is increased to $n = 0.2$, the DOS at the chemical potential increases, as a result of the dissociation of polarons. Increasing n further, a pseudogap begins to form at μ, which is a precursor of the charge-density-wave gap at half filling and zero temperature.

In the case of half filling $n = 0.5$, the DOS has become symmetric with respect to μ. There are broad features located either side of the chemical potential, which take on maxima close to $\omega - \mu = \pm E_P$. However, apart from the SC case, where the single-polaron binding energy is still a relevant energy scale, the position of these peaks is rather determined by the energy of the upper and lower bands, split by the formation of a Peierls state. The gap of size $\sim \lambda$ expected for the insulating charge-ordered state at $T = 0$ is partially filled in due to the finite temperature considered here.

Furthermore, we find additional, much smaller features roughly separated from μ by the bare phonon frequency ω_0, whose height decreases with decreasing temperature, as revealed by the results for $\beta t = 10$ (Fig. 15). These peaks—not present at $T = 0$ [51, 52]—arise from thermally activated transitions to states with additional phonons excited, and are also visible in Figs. 11 and 13. While for WC ($\lambda = 0.1$, Fig. 11), the maximum of these features is almost exactly located at $|\omega - \mu| = \omega_0$, it moves to $|\omega - \mu| \approx 1.25\omega_0$ for IC ($\lambda = 1$, Fig. 15), and finally to $|\omega - \mu| \approx 2.5\omega_0$ for SC ($\lambda = 2$, Fig. 13). Although the exact positions of the peaks are subject to uncertainties due to the maximum entropy method, this evolution reflects the shift of the maximum in the phonon distribution function with increasing coupling. The maximum entropy method yields an envelope of the multiple peaks separated by ω_0.

Anti-Adiabatic Regime

The comparison of the spectral functions for $n = 0.1$ and $n = 0.3$ in Fig. 10 of [12] reveals that there is no density-driven cross-over of the system as observed in the adiabatic case even for the critical value $g^2 = 1$. In particular, owing to the large phonon energy, there are no low-energy excitations close to the polaron band, so that the latter remains well separated from the incoherent features even for $n = 0.3$. Furthermore, the spectral weight of the polaron band also remains almost unchanged as we increase the density from $n = 0.1$ to $n = 0.3$. Consequently, almost independent small polarons are formed also at finite electron densities, in accordance with previous findings for small systems [53].

7 Summary

We have reviewed quantum Monte Carlo and variational approaches to Holstein models based on Lang-Firsov transformations of the Hamiltonian. The methods have been applied to investigate single polarons and bipolarons, respectively, as well as a many-polaron system.

The variational methods include displacements of the lattice at all lattice sites, which enables them to quite accurately describe large polaron or bipolaron states.

Using the transformed Hamiltonian, we have shown that quantum Monte Carlo simulation can be based on exact sampling without autocorrelations, which proves to be an enormous advantage for small phonon frequencies or low temperatures. Indeed, we have used a grand-canonical algorithm to obtain dynamical properties of many-polaron systems in all interesting parameter regimes. Such simulations are currently not possible with other Monte Carlo methods.

Acknowledgements

We gratefully acknowledge fruitful collaboration with H Fehske, J Loos, G Wellein, H G Evertz, and D Neuber. We also thank A Alvermann, A S Alexandrov, and P Kornilovitch for useful discussion.

References

1. H. Fehske and S. A. Trugman, in the present volume.

2. I. G. Lang and Y. A. Firsov, Zh. Eksp. Teor. Fiz. **43**, 1843 (1962), [Sov. Phys. JETP **16**, 1301 (1962)], and Yu. A. Firsov, in the present volume.

3. T. Holstein, Ann. Phys. (N.Y.) **8**, 325; **8**, 343 (1959).

4. G. D. Mahan, *Many-Particle Physics*, 2nd ed. (Plenum Press, New York, 1990).

5. E. V. L. de Mello and J. Ranninger, Phys. Rev. B **55**, 14 872 (1997).

6. J. M. Robin, Phys. Rev. B **56**, 13 634 (1997).

7. H. Fehske, J. Loos, and G. Wellein, Phys. Rev. B **61**, 8016 (2000).

8. M. Hohenadler, H. G. Evertz, and W. von der Linden, Phys. Rev. B **69**, 024301 (2004).

9. C. Zhang, E. Jeckelmann, and S. R. White, Phys. Rev. B **60**, 14 092 (1999).

10. M. Hohenadler and W. von der Linden, Phys. Rev. B **71**, 184309 (2005).

11. M. Hohenadler, H. G. Evertz, and W. von der Linden, Phys. Stat. Sol. (b) **242**, 1406 (2005).

12. M. Hohenadler *et al.*, Phys. Rev. B **71**, 245111 (2005).

13. W. von der Linden, Phys. Rep. **220**, 53 (1992).

14. E. Y. Loh and J. E. Gubernatis, in *Electronic Phase Transitions*, edited by W. Hanke and Y. V. Kopaev (Elsevier Science Publishers, North-Holland, Amsterdam, 1992), Chap. 4.

15. R. Blankenbecler, D. J. Scalapino, and R. L. Sugar, Phys. Rev. D **24**, 2278 (1981).

16. J. E. Hirsch, Phys. Rev. B **31**, 4403 (1985).

17. G. Wellein, H. Röder, and H. Fehske, Phys. Rev. B **53**, 9666 (1996).

18. M. Brunner, S. Capponi, F. F. Assaad, and A. Muramatsu, Phys. Rev. B **63**, R180511 (2001).

19. E. Jeckelmann and S. R. White, Phys. Rev. B **57**, 6376 (1998).

20. A. H. Romero, D. W. Brown, and K. Lindenberg, Phys. Rev. B **60**, 14080 (1999).

21. L. C. Ku, S. A. Trugman, and J. Bonča, Phys. Rev. B **65**, 174306 (2002).

22. M. Jarrell and J. E. Gubernatis, Phys. Rep. **269**, 133 (1996).

23. A. C. Davison and D. V. Hinkley, *Bootstrap Methods and their Application* (Cambridge University Press, Cambridge, UK, 1997).

24. W. H. Press, Numerical Recipes in Fortran 77, http://www.numrec.com.

25. M. Hohenadler, Ph.D. thesis, TU Graz, 2004.

26. H. De Raedt and A. Lagendijk, Phys. Rev. Lett. **49**, 1522 (1982).

27. H. De Raedt and A. Lagendijk, Z. Phys. B: Condens. Matter **65**, 43 (1986).

28. P. E. Kornilovitch, Phys. Rev. Lett. **81**, 5382 (1998).

29. A. Macridin, G. A. Sawatzky, and M. Jarrell, Phys. Rev. B **69**, 245111 (2004).

30. H. Fehske, A. Alvermann, M. Hohenadler, and G. Wellein, in *Polarons in Bulk Materials and Systems with Reduced Dimensionality*, Proc. Int. School of Physics *"Enrico Fermi", Course CLXI*, edited by G. Iadonisi, J. Ranninger, and G. De Filippis (IOS Press, Amsterdam, Oxford, Tokio, Washington DC, 2006), pp. 285–296.

31. H. De Raedt and A. Lagendijk, Phys. Rev. B **27**, 6097 (1983).

32. P. E. Kornilovitch, J. Phys.: Condens. Matter **9**, 10675 (1997).

33. A. H. Romero, D. W. Brown, and K. Lindenberg, Phys. Lett. A **254**, 287 (1999).

34. H. Lowen, Phys. Rev. B **37**, 8661 (1988).

35. F. Marsiglio, in *Recent Progress in Many-Body Theories*, edited by E. Schachinger, H. Mitter, and H. Sormann (Plenum Press, New York, 1995), Vol. 4.

36. F. Marsiglio, Physica C **244**, 21 (1995).

37. M. Hohenadler, M. Aichhorn, and W. von der Linden, Phys. Rev. B **71**, 014302 (2005).

38. A. S. Alexandrov and A. M. Bratkovsky, Phys. Rev. Lett. **82**, 141 (1999).
39. A. S. Alexandrov and A. M. Bratkovsky, J. Phys.: Condens. Matter **11**, L531 (1999).
40. D. M. Edwards, Adv. Phys. **51**, 1259 (2002).
41. J. Bonča, T. Katrašnik, and S. A. Trugman, Phys. Rev. Lett. **84**, 3153 (2000).
42. S. El Shawish, J. Bonča, L. C. Ku, and S. A. Trugman, Phys. Rev. B **67**, 014301 (2003).
43. L. Proville and S. Aubry, Eur. Phys. J. B **11**, 41 (1999).
44. J. Loos, M. Hohenadler, and H. Fehske, J. Phys.: Condens. Matter **18**, 2453 (2006).
45. M. Hohenadler, G. Wellein, A. Alvermann, and H. Fehske, Physica B **378-380**, 64 (2006).
46. G. Wellein *et al.*, Physica B **378-380**, 281 (2006).
47. W. Stephan, Phys. Rev. B **54**, 8981 (1996).
48. G. Wellein and H. Fehske, Phys. Rev. B **56**, 4513 (1997).
49. M. Hohenadler, M. Aichhorn, and W. von der Linden, Phys. Rev. B **68**, 184304 (2003).
50. R. J. Bursill, R. H. McKenzie, and C. J. Hamer, Phys. Rev. Lett. **80**, 5607 (1998).
51. M. Hohenadler *et al.*, Phys. Rev. B **73**, 245120 (2006).
52. S. Sykora *et al.*, Phys. Rev. B **71**, 045112 (2005).
53. M. Capone, M. Grilli, and W. Stephan, Eur. Phys. J. B **11**, 551 (1999).

Spectroscopic Properties of Polarons in Strongly Correlated Systems by Exact Diagrammatic Monte Carlo Method

A. S. Mishchenko[1,2] and N. Nagaosa[3]

[1] CREST, Japan Science and Technology Agency (JST), AIST, 1-1-1, Higashi, Tsukuba 305-8562, Japan `andry.mishenko@aist.go.jp`
[2] Russian Research Centre "Kurchatov Institute", 123182 Moscow, Russia `andry@kurm.polyn.kiaae.su`
[3] Department of Applied Physics, The University of Tokyo, 7-3-1 Hongo, Bunkyo-ku, Tokyo 113, Japan `nagaosa@appi.t.u-tokyo.ac.jp`

1 Introduction

The theoretical study of polarons in strongly correlated systems is like an attempt to view the contents of a Pandora's box embedded within another, even more sinister and obscure container of riddles, enigmas and mysteries. This desperate situation occurs because the solution is not known even for the simplest polaron problem, i.e. when a perfectly stable quasiparticle (QP) with momentum as a single quantum number interacts with a well defined bath of bosonic elementary excitations. On the contrary, the definition of the strongly correlated system implies that QPs might be highly unstable and the very notion of QPs, both in electronic and bosonic subsystems, is under question. Thus, one faces the problem of an interplay between ill defined objects and it is crucial to solve the problem without approximations. Further difficulty, pertinent to realistic systems, is an interplay of the momentum and other quantum numbers characterizing the internal states of a QP.

The problem of the polaron originally emerged as that of an electron coupled to phonons (see [1, 2]). In the initial formulation a structureless QP is characterized by the only quantum number, momentum, which changes due to interaction of the QP with phonons [3, 4]. Later, depending on what can be called "particle" and "environment", and how they interact with each other, the polaron concept was related to extreme diversity of physical phenomena. There are many other objects which, having nothing to do with phonons, are isomorphic to the simple polaron [5], as, e.g. an exciton-polaron in the intraband scattering approximation [6–9]. Another example is the problem of a hole in an antiferromagnet which is closely related to polarons since hole movement is accompanied by the spin flips which, in the spin wave approximation, are equivalent to the creation and annihilation of magnons [10, 11].

The concept of the polaron was further generalized to include internal degrees of freedom which, interacting with the environment, change their quantum numbers. An example of a complex QP is the Jahn-Teller polaron, where electron-phonon interaction (EPI) changes the quantum numbers of degenerate electronic states [12–14]. This generalization is important due to its relevance to the colossal magnetoresistance phenomena in the manganese oxides [15, 16]. Another example is the pseudo Jahn-Teller polaron, where EPI is inelastic and leads to transitions between energetically close electronic levels of a QP [17–19]. A further generalization is a system of several QPs which interact both with each other and their environment. For example, effective interaction of two electrons through exchange by phonons can overcome the Coulomb repulsion and form a bound state, a bipolaron [20–24]. On the other hand, the coupling of attracting hole and electron to the lattice vibrations [25–27] can create a lot of qualitatively different objects: a localized exciton, a weakly bound pair of localized hole and localized electron, etc. [7, 28]. Scattering by impurities introduces additional complexity to the polaron problem because the interference of the impurity potential with lattice distortion, which accompanies the polaron movement, can contribute either constructively or destructively to the localization of a QP on the impurity [7, 29, 30].

In addition, a bare QP and bosonic bath can not be considered as well defined in correlated systems. Angle Resolved Photoemission Spectra (ARPES), revealing the Lehmann Function (LF) of quasiparticles, demonstrate broad peaks in many correlated systems: copper oxide high-temperature superconductors [31–33], colossal magnetoresistive manganites [34–36], quasi-one-dimensional Peierls conductors [37, 38], and Verwey magnetites [39]. Besides, phonons are also broadened in many correlated systems, e.g. in high-temperature semiconductors [40] and mixed-valence materials [41, 42]. One of the possible reasons for these broadenings is the interaction of the QPs with the lattice degrees of freedom. However, in many realistic cases other subsystems, not explicitly included into the polaron Hamiltonian, are responsible for the decay of QP and phonons, e.g., other electronic bands, phonon anharmonicity, interaction with nuclear spins, etc. Then, if this auxiliary broadening is known to some approximation, one can formulate the ambitious goal of studying spectral response when the "bare" quasiparticle with known damping interacts with "broadened" bosonic excitations.

None of the traditional numerical methods, to say nothing of analytical ones, can give *approximation free results for measurable quantities* of a polaron, such as optical conductivity or angle resolved photoemission spectra, in a *macroscopic system of arbitrary dimension*. In addition, we are not aware of any numerical method which can incorporate in an approximation free way information on the damping of QP and bosonic bath. Below we describe the basics of a recently developed Diagrammatic Monte Carlo (DMC) method for numerically exact computation of Green functions and correlation functions in imaginary time for few polarons in a macroscopic system [43–51]. Analytic continuation of imaginary time functions to real frequencies is performed by a novel approximation free approach of stochastic optimization (SO) [45, 50, 51], circumventing the difficulties of the popular Maximal Entropy method. Finally we focus on results of application of the DMC-SO machinery to various problems [52–57]

The basic models, related to polaronic objects in correlated systems, which can be solved by DMC-SO methods, are stated in the next section. It is followed in Sect. 1.2 by description of the stumbling blocks encountered by analytic methods. Section 2 concerns the basics of DMC-SO methods. However, those who are not interested in the details of the methods can briefly look through the definitions in the introduction for Sect. 2 and turn to Sect. 3 where LF and optical conductivity of Fröhlich polarons are discussed (see also [58]). Results of studies of the self-trapping phenomenon are presented in Sect. 4 and application of DMC-SO methods to the exciton problem can be found in Sect. 5. The chapter is completed by Sect. 6 devoted to studies of ARPES of high temperature superconductors.

1.1 Formulation of a General Model with Interacting Polarons

In general terms, the simplest problem of a complex polaronic object, where center-of-mass motion does not separate from the rest of the degrees of freedom, is introduced as system of two QPs

$$\widehat{H}_0^{\mathrm{par}} = \sum_{\mathbf{k}} \varepsilon_a(\mathbf{k}) a_{\mathbf{k}}^\dagger a_{\mathbf{k}} + \sum_{\mathbf{k}} \varepsilon_h(\mathbf{k}) h_{\mathbf{k}} h_{\mathbf{k}}^\dagger \tag{1}$$

($a_{\mathbf{k}}$ and $h_{\mathbf{k}}$ are annihilation operators, and $\varepsilon_a(\mathbf{k})$ and $\varepsilon_h(\mathbf{k})$ are dispersions of QPs), which interact with each other

$$\widehat{H}_{\text{a-h}} = -N^{-1} \sum_{\mathbf{pkk'}} \mathcal{U}(\mathbf{p}, \mathbf{k}, \mathbf{k}') a_{\mathbf{p}+\mathbf{k}}^\dagger h_{\mathbf{p}-\mathbf{k}}^\dagger h_{\mathbf{p}-\mathbf{k}'} a_{\mathbf{p}+\mathbf{k}'}. \tag{2}$$

(N is the number of lattice sites) through the instantaneous Coulomb potential and the scattering by bosons

$$\widehat{H}_{\text{par-bos}} = i \sum_{\kappa=1}^{Q} \sum_{\mathbf{k},\mathbf{q}} (b_{\mathbf{q},\kappa}^\dagger - b_{-\mathbf{q},\kappa})$$

$$\left[\gamma_{aa,\kappa}(\mathbf{k},\mathbf{q}) a_{\mathbf{k}-\mathbf{q}}^\dagger a_{\mathbf{k}} + \gamma_{hh,\kappa}(\mathbf{k},\mathbf{q}) h_{\mathbf{k}-\mathbf{q}}^\dagger h_{\mathbf{k}} + \gamma_{ah,\kappa}(\mathbf{k},\mathbf{q}) h_{\mathbf{k}-\mathbf{q}}^\dagger a_{\mathbf{k}} \right] + h.c. \tag{3}$$

($\gamma_{[aa,ah,hh],\kappa}$ are interaction constants) where quanta of Q different branches of bosonic excitations are created or annihilated, which are described by

$$\widehat{H}_{\text{bos}} = \sum_{\kappa=1}^{Q} \sum_{\mathbf{q}} \omega_{\mathbf{q},\kappa} b_{\mathbf{q},\kappa}^\dagger b_{\mathbf{q},\kappa}. \tag{4}$$

In general, each QP can be a composite one with internal degree of freedom represented by T different states

$$\widehat{H}_0^{\mathrm{PJT}} = \sum_{\mathbf{k}} \sum_{i=1}^{T} \epsilon_i(\mathbf{k}) a_{i,\mathbf{k}}^{\dagger} a_{i,\mathbf{k}}, \tag{5}$$

in which quantum numbers can also be changed due to the nondiagonal part of the particle–boson interaction

$$\widehat{H}_{\mathrm{par\text{-}bos}} = i \sum_{\kappa} \sum_{\mathbf{k},\mathbf{q}} \sum_{i,j=1}^{T} \gamma_{ij,\kappa}(\mathbf{k},\mathbf{q}) (b_{\mathbf{q},\kappa}^{\dagger} - b_{-\mathbf{q},\kappa}) a_{i,\mathbf{k}-\mathbf{q}}^{\dagger} a_{j,\mathbf{k}} + h.c. \tag{6}$$

The complicated model (1)-(6) is still too far from the cases encountered in strongly correlated systems. Due to coupling of QPs (1) and (5) and bosonic fields (4) to additional degrees of freedom, these excitations are not well defined from the outset. Namely, the dispersion relation of the QP spectrum $\epsilon(\mathbf{k})$ in a realistic system is ill-defined. One can speak of a Lehmann Function (LF) [59–61] of a QP

$$L_{\mathbf{k}}(\omega) = \sum_{\nu} \delta(\omega - E_{\nu}(\mathbf{k})) \, |\langle \nu | a_{\mathbf{k}}^{\dagger} | \mathrm{vac} \rangle|^2 \tag{7}$$

which is normalized to unity $\int_0^{+\infty} d\omega L_{\mathbf{k}}(\omega) = 1$ and can be interpreted as a probability that a QP has momentum $\mathbf{k}$ and energy ω. (Here $\{|\nu\rangle\}$ is a complete set of eigenstates of the Hamiltonian $\widehat{H}$ in a sector of given momentum $\mathbf{k}$: $H |\nu(\mathbf{k})\rangle = E_{\nu}(\mathbf{k}) |\nu(\mathbf{k})\rangle$.) For noninteracting systems only, the LF reduces to the delta function $L_{\mathbf{k}}^{\mathrm{NONINT}}(\omega) = \delta(\omega - \epsilon(\mathbf{k}))$ and, thus, sets up the dispersion relation $\omega = \epsilon(\mathbf{k})$.

Specific cases of the model (1)-(6) describe an enormous variety of physical problems. Hamiltonians (1) and (2), in the case of attractive potential $\mathcal{U}(\mathbf{p},\mathbf{k},\mathbf{k}') > 0$, describe an exciton with static screening [62,63]. Expressions (1)-(4) describe a bipolaron for repulsive interaction [20–24] $\mathcal{U}(\mathbf{p},\mathbf{k},\mathbf{k}') < 0$ and an exciton–polaron otherwise [25–27]. The simplest model for exciton–phonon interaction, when only the two ($T = 2$) lowest states of relative electron–hole motion are relevant (e.g. in a one-dimensional charge–transfer exciton [64–66]), is defined by Hamiltonians (4)-(6). The same relations (4)-(6) describe the problems of the Jahn-Teller [all ϵ_i in Hamiltonian (5) are the same] and pseudo Jahn-Teller polaron. The problem of a hole in an antiferromagnet in the spin-wave approximation is expressed in terms of Hamiltonians (4)-(6) with $Q = 1$ and $T = 1$. When a hole also interacts with phonons, one has to take into account one more bosonic branch and set $Q = 2$ in (4) and (6). Finally, the simplest nontrivial problem of a polaron, i.e. of a structureless QP interacting with one phonon branch, is described by noninteracting Hamiltonians of QP $\widehat{H}_{\mathrm{par}}$ and phonons $\widehat{H}_{\mathrm{ph}}$

$$\widehat{H}_0 = \sum_{\mathbf{k}} \epsilon(\mathbf{k}) a_{\mathbf{k}}^{\dagger} a_{\mathbf{k}} + \sum_{\mathbf{q}} \omega_{\mathbf{q}} b_{\mathbf{q}}^{\dagger} b_{\mathbf{q}} , \tag{8}$$

and interaction term

$$\widehat{H}_{\mathrm{int}} = \sum_{\mathbf{k},\mathbf{q}} V(\mathbf{k},\mathbf{q})(b_{\mathbf{q}}^{\dagger} - b_{-\mathbf{q}}) a_{\mathbf{k}-\mathbf{q}}^{\dagger} a_{\mathbf{k}} + h.c. . \tag{9}$$

The simplest polaron problem, in turn, can be subdivided into continuous and lattice polaron models.

1.2 Limitations of Analytic Methods in the Problem of Polarons

An analytic solution for the problem of an exciton in a rigid lattice is only available for the small radius Frenkel regime [67] and large radius Wannier regime [68]. However, even the limits of validity for these approximations are not known. Random phase approximation approaches [62, 63], are capable of obtaining some qualitative conclusions for the intermediate radius regime though its quantitative results are not reliable due to uncontrolled errors. The situation is similar with the problem of a structureless polaron, where analytic solutions arc known only in the weak and strong coupling regimes. Besides, reliable results for these regimes are available only for ground state properties.

Although several novel methods, capable of obtaining properties of excited states, were developed recently, variational coherent-states expansion [69] and free propagator momentum average summation [70] as a few examples, all of them, to provide reliable data in a specific regime, need either comparison with exact sum rules [71, 72] or with exact numerical results.

Application of variational methods to the study of excitations is a tricky issue since, strictly speaking, they are valid only for the ground state. As an example of the importance of sum rules in variational treatment, we refer to the problem of the optical conductivity of the Fröchlich polaron. The possibility of the existence of a Relaxed Excited State (RES), which is a metastable state where lattice deformation has adjusted to the electronic excitation rendering stability and narrow linewidth of the spectroscopic response, was briefly mentioned by S. I. Pekar in the early 1950's. Then, the conception of the RES was rigorously formulated by J. T. Devreese with coworkers and has been the subject of extensive investigations for years [5, 48, 57, 73–77]. Calculations of impedance [75] in the framework of technique [78] supported the existence of a narrow stable peak in the optical conductivity. However, even the authors of [75] were skeptical about the fact that the width of the RES in the strong coupling regime appeared to be narrower than the phonon frequency, i.e. inverse time, which, according to the Heisenberg uncertainty principle, is required for the lattice readaptation. In the consequent paper [77] they realized the importance of many-phonon processes and studied two-phonon contribution to optical conductivity. The importance of many phonon processes was confirmed when variational results [75] were compared with exact DMC simulations [48].

The variational result reproduced well the position of the peak in exact data, though it failed in its description of the peak width in the strong coupling regime [48]. Finally, when approach [75] was modified and several sum rules were accurately introduced into the variational model [57], both position and width of the peak were quantitatively reproduced. Studies [57] (see Sect. 3.1), do not address the rather philosophical question of whether the RES exists or not, though they inevitably prove that, in contrast to the foregoing beliefs, there is no stable excited state of the Fröhlich polaron in the strong coupling regime. Note that sometimes excited states can not be handled by analytic methods even for weak couplings: the perturbation theory expression for LF of the Fröhlich polaron model diverges at the phonon energy ω_{ph} [See (34) in Sect. 3.1.] and more elaborate treatment is necessary.

The difficulties of semianalytic methods are enhanced in the intermediate coupling regime where results are sometimes wrong even for ground state properties. For example, the variational approach [79], which has been considered as an intermediate coupling theory, appeared to be valid only in the weak coupling limit [45]. Special interest in the methods giving reliable information on excited states is triggered by the self-trapping phenomenon which occurs just in the intermediate coupling regime. This phenomenon is a dramatic transformation of QP properties when system parameters are slightly changed [3, 7, 9, 80]. In the intermediate coupling regime a "trapped" QP state with strong lattice deformation around it and a "free" state with weakly perturbed lattice may hybridize and resonate because of close energies at some critical value of electron-lattice interaction γ_c. It is clear that, to study self-trapping, one has to apply a method giving reliable information on excited states in the intermediate coupling regime.

2 Diagrammatic Monte Carlo and Stochastic Optimization Methods

In this section we introduce definitions of exciton-polaron properties which can be evaluated by DMC and SO methods. An idea of the DMC approach for numerically exact calculation of Green functions (GFs) in imaginary times is presented in Sect. 2.1, and a short description of the SO method, which is capable of making unbiased analytic continuation from imaginary times to real frequencies, is given in Sect. 2.2. Using a combination of DMC and SO, one can often circumvent the difficulties of analytic and traditional numerical methods. Therefore, a brief comparative analysis of advantages and drawbacks of DMC-SO machinery is given in Sect. 2.3.

To obtain information on QPs it is necessary to calculate Matsubara GF in imaginary time representation and make an analytic continuation to real frequencies [60]. For the two-particle problem (1)-(4) the relevant quantity is the two-particle GF [46, 47]

$$G_{\mathbf{k}}^{\mathbf{pp}'}(\tau) = \langle \text{vac} \mid a_{\mathbf{k}+\mathbf{p}'}(\tau) h_{\mathbf{k}-\mathbf{p}'}(\tau) h_{\mathbf{k}-\mathbf{p}}^{\dagger} a_{\mathbf{k}+\mathbf{p}}^{\dagger} \mid \text{vac} \rangle . \tag{10}$$

(Here $h_{\mathbf{k}-\mathbf{p}}(\tau) = e^{\widehat{H}\tau} h_{\mathbf{k}-\mathbf{p}} e^{-\widehat{H}\tau}$, $\tau > 0$.) In the case of the exciton–polaron, the vacuum state $\mid \text{vac} \rangle$ is the state with filled valence and empty conduction bands. For the bipolaron problem it is a system without particles. In the simpler case of a QP with two-level internal structure described by (4)-(6) the relevant quantity is the one-particle matrix GF [47, 52]

$$G_{\mathbf{k},ij}(\tau) = \langle \text{vac} \mid a_{i,\mathbf{k}}(\tau) a_{j,\mathbf{k}}^{\dagger} \mid \text{vac} \rangle, \quad i,j = 1, 2. \tag{11}$$

For a structureless polaron the matrix (11) reduces to one-particle scalar GF

$$G_{\mathbf{k}}(\tau) = \langle \text{vac} \mid a_{\mathbf{k}}(\tau) a_{\mathbf{k}}^{\dagger} \mid \text{vac} \rangle . \tag{12}$$

Information on the response to an external weak perturbation (e.g. optical absorption) is contained in the current-current correlation function $\langle J_{\beta}(\tau) J_{\delta} \rangle$ (β/δ are Cartesian indexes).

Lehmann spectral representation of $G_{\mathbf{k}}(\tau)$ [60, 61] at zero temperature

$$G_{\mathbf{k}}(\tau) = \int_{0}^{\infty} d\omega\, L_{\mathbf{k}}(\omega)\, e^{-\omega\tau} , \tag{13}$$

with the Lehmann function (LF) $L_{\mathbf{k}}(\omega)$ given in (7), reveals information on the ground and excited states. Here $\{|\nu\rangle\}$ is a complete set of eigenstates of Hamiltonian $\widehat{H}$ in a sector of given momentum $\mathbf{k}$: $H\,|\nu(\mathbf{k})\rangle = E_{\nu}(\mathbf{k})\,|\nu(\mathbf{k})\rangle$. The LF $L_{\mathbf{k}}(\omega)$ has poles (sharp peaks) on the energies of stable (metastable) states of particle. For example, if there is a stable state at energy $E(\mathbf{k})$, the LF reads $L_{\mathbf{k}}(\omega) = Z^{(\mathbf{k})}\,\delta(\omega - E(\mathbf{k})) + \ldots$, and the state with the lowest energy $E_{\text{g.s.}}(\mathbf{k})$ in a sector of a given momentum $\mathbf{k}$ is highlighted by the asymptotic behavior of GF

$$G_{\mathbf{k}}(\tau \gg \max\left[\omega_{\mathbf{q},\kappa}^{-1}\right]) \to Z^{(\mathbf{k})} \exp[-E_{\text{g.s.}}(\mathbf{k})\tau] , \tag{14}$$

where $Z^{(\mathbf{k})}$-factor is the weight of the state. Analyzing the asymptotic behavior of similar n-phonon GFs [45, 52]

$$G_{\mathbf{k}}(n, \tau;\, \mathbf{q}_1, \ldots, \mathbf{q}_n) =$$
$$\langle \text{vac}| b_{\mathbf{q}_n}(\tau) \cdots b_{\mathbf{q}_1}(\tau)\, a_{\mathbf{p}}(\tau) a_{\mathbf{p}}^{\dagger}\, b_{\mathbf{q}_1}^{\dagger} \cdots b_{\mathbf{q}_n}^{\dagger} |\text{vac}\rangle , \quad \mathbf{p} = \mathbf{k} - \textstyle\sum_{j=1}^{n} \mathbf{q}_j . \tag{15}$$

one obtains detailed information about the lowest state. For example, important characteristics of the lowest state wave function

$$\Psi_{\text{g.s.}}(\mathbf{k}) = \sum_{i=1}^{T} \sum_{n=0}^{\infty} \sum_{\mathbf{q}_1 \ldots \mathbf{q}_n} \theta_i(\mathbf{k}; \mathbf{q}_1, \ldots, \mathbf{q}_n) c_{i,\mathbf{k}-\mathbf{q}_1 \ldots -\mathbf{q}_n}^{\dagger}\, b_{\mathbf{q}_1}^{\dagger} \ldots b_{\mathbf{q}_n}^{\dagger} \mid \text{vac} \rangle \tag{16}$$

are partial n-phonon contributions

$$Z^{(\mathbf{k})}(n) \equiv \sum_{i=1}^{T} \sum_{\mathbf{q}_1 \ldots \mathbf{q}_n} \mid \theta_i(\mathbf{k}; \mathbf{q}_1, ..., \mathbf{q}_n) \mid^2 \tag{17}$$

which are normalized to unity $\sum_{n=0}^{\infty} Z^{(\mathbf{k})}(n) \equiv 1$, and the average number of phonons

$$\langle N \rangle \equiv \langle \Psi_{\text{g.s.}}(\mathbf{k}) \mid \sum_{\mathbf{q}} b_{\mathbf{q}}^{\dagger} b_{\mathbf{q}} \mid \Psi_{\text{g.s.}}(\mathbf{k}) \rangle = \sum_{n=1}^{\infty} n Z^{(\mathbf{k})}(n) \tag{18}$$

in the polaronic cloud. Another example is the wave function of relative electron-hole motion of the exciton in the lowest state in the sector of given momentum

$$\Psi_{\text{g.s.}}(\mathbf{k}) = \sum_{\mathbf{p}} \xi_{\mathbf{k}\,\mathbf{p}}(g.s.) a_{\mathbf{k}+\mathbf{p}}^{\dagger} h_{\mathbf{k}-\mathbf{p}}^{\dagger} \mid \text{vac} \rangle . \tag{19}$$

The amplitudes $\xi_{\mathbf{k}\,\mathbf{p}}(g.s.)$ of this wave function can be obtained [46] from asymptotic behavior of the following GF (10)

$$G_{\mathbf{k}}^{\mathbf{p}=\mathbf{p}'}(\tau \to \infty) = \mid \xi_{\mathbf{k}\,\mathbf{p}}(g.s.) \mid^2 e^{-E_{\text{g.s.}}(\mathbf{k})\tau} . \tag{20}$$

Information on the excited states is obtained by the analytic continuation of imaginary time GF to real frequencies which requires us to solve the Fredholm equation $G_{\mathbf{k}}(\tau) = \widehat{\mathcal{F}}\left[L_{\mathbf{k}}(\omega)\right]$ (13)

$$L_{\mathbf{k}}(\omega) = \widehat{\mathcal{F}}_{\omega}^{-1}\left[G_{\mathbf{k}}(\tau)\right] . \tag{21}$$

Equation (13) is a rather general relation between imaginary time GF/correlator and the spectral properties of the system. For example, the absorption coefficient of light by excitons $\mathcal{I}(\omega)$ is obtained as a solution of the same equation [46]

$$\mathcal{I}(\omega) = \widehat{\mathcal{F}}_{\omega}^{-1}\left[\sum_{\mathbf{p}\mathbf{p}'} G_{\mathbf{k}=0}^{\mathbf{p}\mathbf{p}'}(\tau)\right] . \tag{22}$$

Besides, the real part of the optical conductivity $\sigma_{\beta\delta}(\omega)$ is expressed [48] in terms of the current–current correlation function $\langle J_{\beta}(\tau) J_{\delta} \rangle$ by the relation

$$\sigma_{\beta\delta}(\omega) = \pi \widehat{\mathcal{F}}_{\omega}^{-1}\left[\langle J_{\beta}(\tau) J_{\delta} \rangle\right] / \omega . \tag{23}$$

2.1 Diagrammatic Monte Carlo Method

The DMC Method is an algorithm which calculates GF (10)-(12) without any systematic errors. This algorithm is described below for the simplest case of the structureless polaron [45], and generalizations to more complex cases can be found in consequent references[4]. The DMC is based on the Feynman expansion of the Matsubara GF in imaginary time in the interaction representation

$$G_{\mathbf{k}}(\tau) = \left\langle \text{vac} \left| T_\tau \left[a_{\mathbf{k}}(\tau) a_{\mathbf{k}}^\dagger(0) \exp\left\{ -\int_0^\infty \widehat{H}_{\text{int}}(\tau')d\tau' \right\} \right] \right| \text{vac} \right\rangle_{\text{con}} \quad ; \ \tau > 0 \, . \tag{24}$$

Here T_τ is the imaginary time ordering operator, $|\text{vac}\rangle$ is a vacuum state without particle and phonons, and $\widehat{H}_{\text{int}}$ is the interaction Hamiltonian in (9). The symbol of exponent denotes a Taylor expansion which results in multiple integration over internal variables $\{\tau'_1, \tau'_2, \ldots\}$. Operators are in the interaction representation $\widehat{A}(\tau) = \exp[\tau(\widehat{H}_0)]\widehat{A}\exp[-\tau(\widehat{H}_0)]$. The index "con" means that expansion contains only connected terms where no one integral over internal time variables $\{\tau'_1, \tau'_2, \ldots\}$ can be factorized.

The Vick theorem expresses matrix elements of time-ordered operators as a sum of terms, each is a factor of matrix elements of pairs of operators, and expansion (24) becomes an infinite series of integrals with an ever increasing number of integration variables

$$G_{\mathbf{k}}(\tau) - \sum_{m=0,2,4\ldots} \sum_{\xi_m} \int dx'_1 \cdots dx'_m \, \mathcal{D}_m^{(\xi_m)}(\tau; \{x'_1, \ldots, x'_m\}) \, . \tag{25}$$

Here the index ξ_m stands for different Feynman diagrams (FDs) of the same order m. The term with $m = 0$ is the GF of the noninteracting QP $G_{\mathbf{k}}^{(0)}(\tau)$. Function $\mathcal{D}_m^{(\xi_m)}(\tau; \{x'_1, \ldots, x'_m\})$ of any order m can be expressed as a factor of GFs of noninteracting quasiparticle, GFs of phonons, and interaction vortexes $V(\mathbf{k}, \mathbf{q})$. For the simplest case of a Hamiltonian system expressions for GFs of QP $G_{\mathbf{k}}^{(0)}(\tau_2 - \tau_1) = \exp\left[-\epsilon(\mathbf{k})(\tau_2 - \tau_1)\right](\tau_2 > \tau_1)$ and phonons $D_{\mathbf{q}}^{(0)}(\tau_2 - \tau_1) = \exp\left[-\omega_{\mathbf{q}}(\tau_2 - \tau_1)\right](\tau_2 > \tau_1)$ are well known.

An important feature of the DMC method, which is distinct from the row of other exact numerical approaches, is the explicit possibility to include renormalized GFs into an exact expansion without any change of the algorithm. For example, if a damping of QP, caused by some interactions not included in the Hamiltonian, is known, i.e. retarded self-energy of QP $\Sigma_{\text{ret}}(\mathbf{k}, \omega)$ is available, the renormalized GF

$$\widetilde{G}_{\mathbf{k}}^{(0)}(\tau) = \frac{1}{\pi} \int_{-\infty}^\infty d\omega e^{-\omega\tau} \frac{Im\Sigma_{\text{ret}}(\mathbf{k}, \omega)}{[\omega - \epsilon(\mathbf{k}) - Re\Sigma_{\text{ret}}(\mathbf{k}, \omega)]^2 + [Im\Sigma_{\text{ret}}(\mathbf{k}, \omega)]^2} \tag{26}$$

---

[4] Generalization of the technique described below to the case of the exciton (1-2) is given in [46] and its modification for the pseudo-Jahn-Teller polaron (4-6) is developed in [47, 52]. The method for evaluation of the current–current correlation function can be found in [48] and a case of a polaron interacting with two kinds of bosonic fields is considered in [49].

can be introduced instead of the bare GF $G_{\mathbf{k}}^{(0)}(\tau)$. Explicit rules for evaluation of $\mathcal{D}_m^{(\xi_m)}$ do not depend on the order and topology of FD. GFs of noninteracting QPs $G_{\mathbf{k}}^{(0)}(\tau_2-\tau_1)$ (or $\widetilde{G}_{\mathbf{k}}^{(0)}(\tau_2-\tau_1)$) with corresponding times and momenta are ascribed to horizontal lines and noninteracting GFs of phonon $D_{\mathbf{q}}^{(0)}(\tau_2-\tau_1)$ (multiplied by the factor of corresponding vortexes $V(\mathbf{k}',\mathbf{q})V^*(\mathbf{k}'',\mathbf{q})$) are attributed to the phonon propagator arch (see Fig. 1a). Then, $\mathcal{D}_m^{(\xi_m)}$ is the factor of all GSs. For example, the expression for the weight of the second order term (Fig. 1b) is the following

$$\mathcal{D}_2(\tau;\{\tau_2',\tau_1',\mathbf{q}\}) =$$
$$|V(\mathbf{k},\mathbf{q})|^2 \, D_{\mathbf{q}}^{(0)}(\tau_2'-\tau_1')G_{\mathbf{k}}^{(0)}(\tau_1')G_{\mathbf{k}-\mathbf{q}}^{(0)}(\tau_2'-\tau_1')G_{\mathbf{k}}^{(0)}(\tau-\tau_2') \, . \tag{27}$$

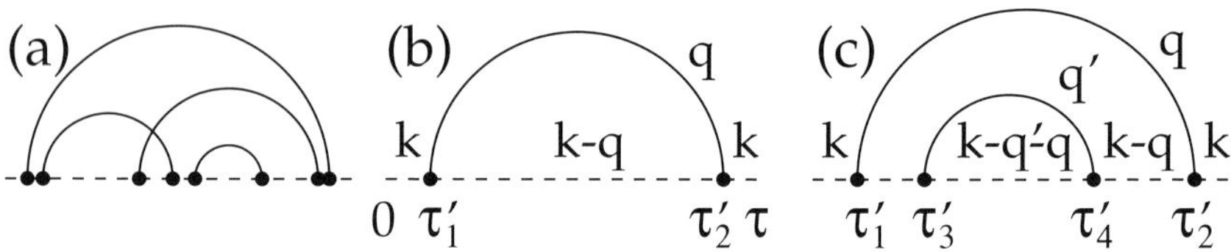

Fig. 1. (a) Typical FD contributing to expansion (25). (b) FD of the second order and (c) fourth order.

The DMC process is a numerical procedure which, basing on the Metropolis principle [81, 82], samples different FDs in the $(\tau, m, \xi_m, \{x_m'\})$ parameter space and collects statistics of the external variable τ in a way that the result of this statistics converges to an exact GF $G_{\mathbf{k}}(\tau)$. Although sampling of the internal parameters of one term in (25) and switching between different orders is performed within the framework of the same numerical process, it is instructive to start with the procedure of evaluation of a specific term $\mathcal{D}_m^{(\xi_m)}(\tau;\{x_1',\ldots,x_m'\})$.

Starting from a set $\{\tau;\{x_1',\ldots,x_m'\}\}$, an update $x_l^{(old)} \to x_l^{(new)}$ of an arbitrary chosen parameter is suggested. This update is accepted or rejected according to the Metropolis principle. After many steps, altering all variables, statistics of the external variable converges to exact dependence of the term on τ. A suggestion for a new value of the parameter $x_l^{(new)} = \widehat{S}^{-1}(R)$ is generated by random number $R \in [0,1]$ with a normalized distribution function $W(x_l)$ in the range $x_l^{(min)} < x_l < x_l^{(max)}$. There are only two restrictions for this otherwise arbitrary function. First, the new parameters $x_l^{(new)}$ must not violate FD topology, i.e., the internal time τ_1' in Fig. 1c must be in the range $[x^{(min)} = 0, x^{(max)} = \tau_3']$. Second, the distribution must be nonzero for the whole domain allowed by FD topology. This ergodicity property is crucial since it is necessary to sample the whole domain for convergence to an exact answer. At each step, update $x_l^{(old)} \to x_l^{(new)}$ is accepted with probability $P_{acc} = M$ (if $M < 1$) and always otherwise. The ratio M has the following form

$$M = \frac{\mathcal{D}_m^{(\xi_m)}(\tau; \{x_1', \ldots, x_l^{(new)}, \ldots, x_m'\})/W(x_l^{(new)})}{\mathcal{D}_m^{(\xi_m)}(\tau; \{x_1', \ldots, x_l^{(old)}, \ldots, x_m'\})/W(x_l^{(old)})} \ . \tag{28}$$

For uniform distribution $W = \text{const} = \left(x_l^{(max)} - x_l^{(max)}\right)^{-1}$, the probability of any combination of parameters is proportional to the weight function $\mathcal{D}$. However, for better convergence the distribution $W(x_l^{new})$ must be as close as possible to the actual distribution given by function $\mathcal{D}_m^{(\xi_m)}(\{\ldots, x_l^{(new)}, \ldots, \})$.

For sampling over FDs of all orders and topologies it is enough to introduce two complimentary updates. Update $\mathcal{A}$ transforms FD $\mathcal{D}_m^{(\xi_m)}(\tau; \{x_1', \ldots, x_m'\})$ into higher order FD $\mathcal{D}_{m+2}^{(\xi_{m+2})}(\tau; \{x_1', \ldots, x_m'; \ \mathbf{q}', \tau_3', \tau_4'\})$ with extra phonon arch, connecting some time points τ_3' and τ_4' by phonon propagator with momentum $\mathbf{q}'$ (Fig. 1c). Note that the ratio of weights $\mathcal{D}_{m+2}^{(\xi_{m+2})}/\mathcal{D}_m^{(\xi_m)}$ is not dimensionless. The dimensionless Metropolis ratio

$$M = \frac{p_\mathcal{A}}{p_\mathcal{B}} \frac{\mathcal{D}_{m+2}^{(\xi_{m+2})}(\tau; \{x_1', \ldots, x_m'; \ \mathbf{q}', \tau', \tau''\})}{\mathcal{D}_m^{(\xi_m)}(\tau; \{x_1', \ldots, x_m'\})W(\mathbf{q}', \tau', \tau'')} \ . \tag{29}$$

contains a normalized probability function $W(\mathbf{q}', \tau', \tau'')$, which is used for generation of new parameters[5]. The complementary update $\mathcal{B}$, removing the phonon propagator, uses ratio M^{-1} [45].

Note that all updates are local, i.e. they do not depend on the structure of the whole FD. Neither rules nor CPU time, needed for updates, depend on the FD order. The DMC method does not imply any explicit truncation of the FD order due to the finite size of computer memory. Even for strong coupling, where the typical number of phonon propagators N_{ph}, contributing to the result, is large, the influence of finite memory size is not significant. Really, according to the Central Limit Theorem, the number of phonon propagators obeys a Gaussian distribution centered at $\bar{N}_{ph}$ with half width of the order of $\sqrt{\bar{N}_{ph}}$ [83]. Hence, if a memory for at least $2\bar{N}_{ph}$ propagators is reserved, the diagram order hardly surpasses this limit.

2.2 Stochastic Optimization Method

The problem of inverting integral equation (13) is an ill posed problem. Due to incomplete noisy information about GF $G_\mathbf{k}(\tau)$, which is known with statistical errors on a finite number of imaginary times in a finite range $[0, \tau_{\max}]$, there is infinite number of approximate solutions which reproduce GF within some

---

[5] The factor $p_\mathcal{A}/p_\mathcal{B}$ depends on the probability to address additional/removal processes.

range of deviations and the problem is to choose the "best one". Another problem, which has been a stumbling block for decades, is the saw tooth noise instability. It occurs when a solution is obtained by a naive method, e.g. by using a least-squares approach for minimizing deviation measure

$$D[\widetilde{L}_{\mathbf{k}}(\omega)] = \int_0^{\tau_{\max}} \left| G_{\mathbf{k}}(\tau) - \widetilde{G}_{\mathbf{k}}(\tau) \right| G_{\mathbf{k}}^{-1}(\tau) d\tau . \tag{30}$$

Here $\widetilde{G}_{\mathbf{k}}(\tau)$ is obtained from approximate LF $\widetilde{L}_{\mathbf{k}}(\omega)$ by applying integral operator $\widetilde{G}_{\mathbf{k}}(\tau) = \mathcal{F}\left[\widetilde{L}_{\mathbf{k}}(\omega)\right]$ in (13). Saw tooth instability corrupts LF in the ranges where the actual LF is smooth. Fast fluctuations of the solution $\widetilde{L}_{\mathbf{k}}(\omega)$ often have much larger amplitude than the value of the actual LF $L_{\mathbf{k}}(\omega)$. Standard tools for saw tooth noise suppression are based on the early 1960's idea of the Fillips-Tikhonov regularization method [84–87]. A nonlinear functional, which suppresses large derivatives of approximate solution $\widetilde{L}_{\mathbf{k}}(\omega)$, is added to the linear deviation measure (30). The most popular variant of regularization methods is the Maximal Entropy Method [61].

However, a typical LF of a QP in a boson field consists of δ-functional peaks and a smooth incoherent continuum with a sharp border [45, 54]. Hence, suppression of high derivatives, as a general strategy of the regularization method, fails. Moreover, any specific implementation of the regularization method uses predefined mesh in the ω space, which could be absolutely unacceptable for the case of sharp peaks. If the actual location of a sharp peak is between predefined discrete points, the rest of spectral density can be distorted beyond recognition. Finally, the regularization Maximal Entropy approach requires assumption of Gaussian distribution of statistical errors in $G_{\mathbf{k}}(\tau)$, which might be invalid in some cases [61].

Recently, a Stochastic Optimization (SO) method, which circumvents the difficulties mentioned above, was developed [45]. The idea of the SO method is to generate a large enough number M of statistically independent nonregularized solutions $\{\widetilde{L}_{\mathbf{k}}^{(s)}(\omega)\}, s = 1, ..., M$, for which the deviation measures $D^{(s)}$ are smaller than some upper limit D_u, depending on the statistic noise of the GF $G_{\mathbf{k}}(\tau)$. Then, using the linearity of expressions (13), (30), the final solution is found as the average of particular solutions $\{\widetilde{L}_{\mathbf{k}}^{(s)}(\omega)\}$

$$L_{\mathbf{k}}(\omega) = M^{-1} \sum_{s=1}^{M} \widetilde{L}_{\mathbf{k}}^{(s)}(\omega) . \tag{31}$$

The particular solution $\widetilde{L}_{\mathbf{k}}^{(s)}(\omega)$ is parameterized in terms of the sum

$$\widetilde{L}_{\mathbf{k}}^{(s)}(\omega) = \sum_{t=1}^{K} \chi_{\{P_t\}}(\omega) \tag{32}$$

of rectangles $\{P_t\} = \{h_t, w_t, c_t\}$ with height $h_t > 0$, width $w_t > 0$, and center c_t. Configuration

$$\mathcal{C} = \{\{P_t\},\, t = 1, ..., K\}\,, \tag{33}$$

which satisfies normalization condition $\sum_{t=1}^{K} h_t w_t = 1$, defines function $\widetilde{G}_{\mathbf{k}}(\tau)$. The procedure of generating a particular solution starts from the stochastic choice of the initial configuration $\mathcal{C}_s^{\text{init}}$. Then, the deviation measure is optimized by a randomly chosen consequence of updates until the deviation is less than D_u. In addition to updates, which do not change the number of terms in the sum (32), there are updates which increase or decrease the number K. Hence, since the number of elements K is not fixed, any spectral function can be reproduced with desired accuracy.

Although each particular solution $\widetilde{L}_{\mathbf{k}}^{(s)}(\omega)$ suffers from saw tooth noise in the area of smooth LF, statistical independence of each solution leads to self averaging of this noise in the sum (32). Note that suppression of noise happens without suppression of high derivatives and, hence, sharp peaks and edges are not smeared out in contrast to regularization approaches. Therefore, saw tooth noise instability is defeated without corruption of sharp peaks and edges. Moreover, continuous parameterization (32) does not need predefined mesh in ω-space. Besides, since the Hilbert space of the solution is sampled directly, any assumption about distribution of statistical errors is not necessary.

The SO method was successfully applied to restore LF of the Fröhlich polaron [45], the Rashba-Pekar exciton-polaron [54], the hole-polaron in the t-J-model [49, 53], and many-particle spin system [88]. Calculation of the optical conductivity of a polaron by the SO method can be found in [48]. The SO method appeared to be helpful in cases when GF's asymptotic limit, giving information about the ground state, can not be reached. For example, sign fluctuations of the terms in expansion (25) for a hole in the t-J-model lead to poor statistics at large times [53], though the SO method is capable of recovering the energy and Z-factor even from GF known only at small imaginary times [53].

2.3 Advantages and Drawbacks of DMC-SO Machinery

Among numerical methods, capable of obtaining quantitative results in the problem of excitons (1) and (2), one can list time-dependent density functional theory [89], the Hanke-Sham technique of correcting particle-hole excitation energy [90, 91], and approaches directly solving the Bethe-Salpeter equation [92–94]. The latter ones provide rather accurate information on the two-particle GF. However, usage of finite mesh in direct/reciprocal space, which is avoided in the DMC method, leads to its failure in the Wannier regime [93].

In contrast to the DMC method, none of the traditional numerical methods can give reliable results for measurable properties of excited states of polaron at arbitrary range of electron-phonon interaction for the macroscopic system in the thermodynamic limit. The exact diagonalization method [95–98] can study excited states though only on rather small finite size systems and results of this method are not even justified in the variational sense in the thermodynamic limit [99]. There is a batch of rather effective variational "exact translation" methods [99–103] where the basis is chosen in momentum space and, hence, the variational principle is applied in the thermodynamic limit. Although these methods can reveal few discrete excited states, they fail for long-range interaction and for dispersive, especially acoustic phonons due to catastrophic growth of variational basis. A non perturbative theory, which is able to give information about spectral properties in the thermodynamic limit at least for one electron, is Dynamical Mean Field Theory [104–107]. However it gives an exact solution only in the case of infinite dimensions which does not correspond to a realistic system and can be considered only as a guide for extrapolation to finite dimensions [108].

Recently developed cluster perturbation theory, where exact diagonalization of a cluster is further improved by taking into account inter-cluster interaction [109–113], is applicable for study of the excited states, but limited to one-dimensional lattices or two-dimensional systems with short-range interaction. The traditional density-matrix renormalization group method [114–118] is very effective though mostly limited to one-dimensional systems and ladders. Finally, the recently developed path integral quantum Monte Carlo algorithm [119–122] is valid for any dimension and properly takes into account quasi long-range interactions [123]. The path integral method is capable of obtaining the density of states [119, 120] and isotope exponents [121, 124]. However, calculations of the measurable characteristics of excited states, such as ARPES or optical conductivity, by this method have never been reported.

In conclusion, none of the methods, except the DMC-SO combination, can obtain at the moment approximation-free results for *measurable physical quantities* for a few QPs interacting with a macroscopic bosonic bath *in the thermodynamic limit*. Indeed, there are limitations of the DMC and SO methods. The DMC method does not work in many-fermion systems due to the sign problem and the SO method fails at high temperatures, comparable to energies of dominant spectral peaks, because even the very small statistical noise of GFs turns the Fredholm equation (13) into essentially "ill defined" problem [84].

3 Spectral Properties of the Fröhlich Polaron

Before the development of DMC-SO methods, information on the excited states of polaron models, especially the Fröhlich one, was very limited. Knowledge of LF was based on results of infinite-dimensions approximation [125], exact diagonalization [96, 97, 126], or strong coupling expansion [127]. None of the above techniques was capable of obtaining the LF of a polaron without approximations, especially for long-range interaction where the difficulties of traditional numerical methods dramatically increase. In a similar way, optical conductivity (OC) of Fröhlich model was known only in the strong coupling expansion approximation [128], within the framework of the perturbation theory [129], or was based on the variational Feynman path integral technique [75]. In this section we consider exact DMC-SO results on LF [45] and OC [48, 57] of the Fröhlich polaron model.

3.1 Lehmann Function of the Fröhlich Polaron

The perturbation theory expression for the high-energy part ($\omega > 0$) of the LF for arbitrary interaction potential $V(|\,\mathbf{q}\,|)$ reads [45] (frequency of the optical phonon ω_{ph} is set to unity)

$$L_{\mathbf{k}=0}(\omega > 0) = \frac{1}{\sqrt{2}\pi^2} \frac{\sqrt{\omega - 1}}{\omega^2}\, |\, V(\sqrt{2(\omega - 1)})\,|^2\, \theta(\omega - 1)\,. \qquad (34)$$

Low-energy part of the LF for the short-range interaction $V(|\,\mathbf{q}\,|)\;=$

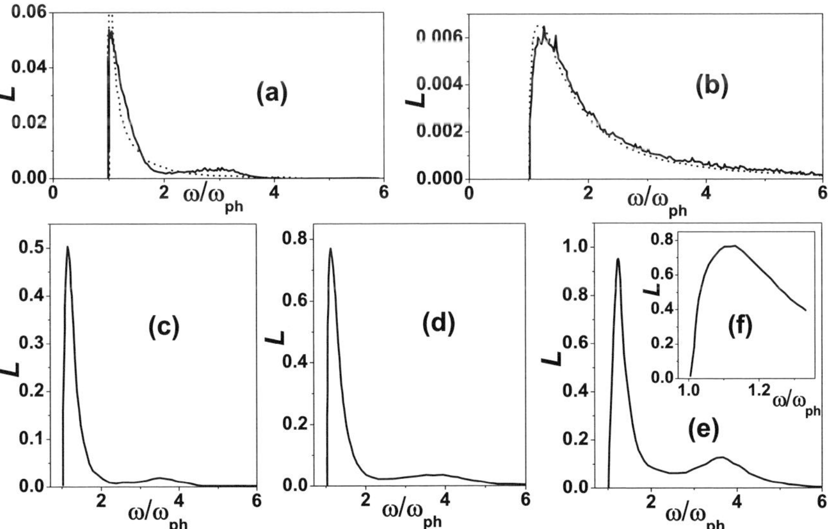

Fig. 2. T Comparison of the numerical results (solid lines) and the perturbation theory (dashed lines) for the LFs of the Fröhlich model with $\alpha = 0.05$ (a) and the short-range interaction model with $\alpha = 0.05$ and $\kappa = 1$ (b). LFs of the Fröhlich polaron for $\alpha = 0.5$ (c), $\alpha = 1$ (d) and $\alpha = 2$ (e). Energy is measured from that of the ground state of the polaron. The initial fragment of the LF for $\alpha = 1$ is shown in the inset (f).

$i\left(2\sqrt{2}\alpha\pi\right)^{1/2}\left(q^2 + \kappa^2\right)^{-1/2}$, reducing to the Fröhlich one when $\kappa \to 0$, is

$$L_{\mathbf{k}=0}(\omega < 0) = \frac{\alpha}{(\kappa + \sqrt{2})^2}\delta\left(\omega + \alpha\frac{\sqrt{2}}{\kappa + \sqrt{2}}\right)\,. \qquad (35)$$

Comparison of the low-energy parts of the LF of the Fröhlich model, obtained by DMC-SO and taken from (35), shows perfect agreement for $\alpha = 0.05$: the accuracy for the polaron energy and Z-factor is about 10^{-4}. On the other hand, the high-energy part of the numerical result (Fig. 2) significantly deviates from that of the analytic expression (34). This is not surprising since for Fröhlich polarons the perturbation theory expression diverges as $\omega \to \omega_{ph}$ and, therefore the perturbation theory breaks down. When perturbation theory is obviously valid, e.g. for the case of finite $\kappa = 1$, there is a perfect agreement between analytic expression and DMC-SO results (Fig. 2b). Note that the high-energy part of $L_{\mathbf{k}=0}(\omega)$ is successfully restored by the SO method despite the fact that the total weight of the feature for $\alpha = 0.05$ is less than 10^{-2}.

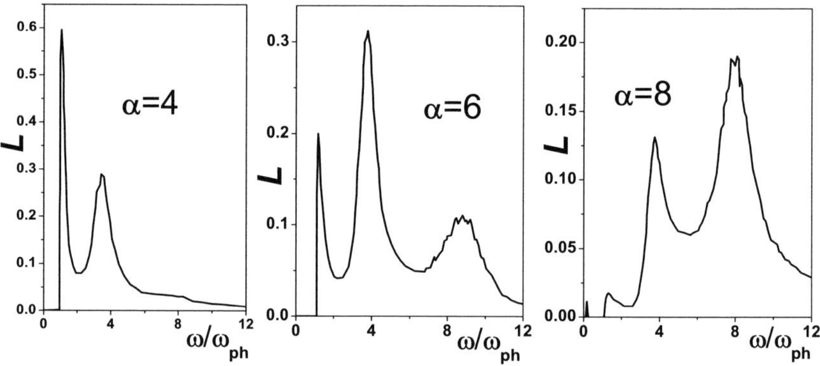

Fig. 3. Evolution of spectral density with α in the cross-over region from intermediate to strong couplings. The polaron ground state peak is shown only for $\alpha = 8$. Note that the spectral analysis still resolves it, despite its very small weight $< 10^{-3}$.

The main deviation of the actual LF from the perturbation theory result is the extra broad peak in the actual LF at $\omega \sim 3.5$. To study this feature $L_{\mathbf{k}=0}(\omega)$ was calculated for $\alpha = 0.5$, $\alpha = 1$, and $\alpha = 2$ (Fig. 2c-e). The peak is seen for higher values of the interaction constant and its weight grows with α. Near the threshold, $\omega = 1$, LF demonstrates the square-root dependence $\sim \sqrt{\omega - 1}$ (Fig. 2f).

To trace the evolution of the peak at higher values of α the LF was calculated [45] for $\alpha = 4$, $\alpha = 6$, and $\alpha = 8$ (Fig. 3). At $\alpha = 4$ the peak at $\omega \sim 4$ already dominates. Moreover, a distinct high-energy shoulder appears at $\alpha = 4$, which transforms into a broad peak at $\omega \sim 8.5$ in the LF for $\alpha = 6$. The LF for $\alpha = 8$ demonstrates further redistribution of the spectral weight between different maxima without significant shift of the peak positions.

3.2 Optical Conductivity of the Fröhlich Polaron: Validity of the Franck-Condon Principle in Optical Spectroscopy

The FC principle [130, 131] and its validity have been widely discussed in studies of optical transitions in atoms, molecules [132, 133], and solids [9, 134]. Generally, the FC principle means that if only one of two coupled subsystems, e.g. an electronic subsystem, is affected by an external perturbation, the second subsystem, e.g., the lattice, is not fast enough to follow the reconstruction of the electronic configuration. It is clear that the justification for the FC principle is the short characteristic time of the measurement process $\tau_{mp} \ll \tau_{ic}$, where τ_{mp} is related to the energy gap between the initial and final states, ΔE, through the uncertainty principle: $\tau_{mp} \simeq \hbar/(\Delta E)$ and τ_{ic} is the time necessary to adjust the lattice when the electronic component is perturbed. Then, the spectroscopic response depends considerably on the value of the ratio τ_{mp}/τ_{ic} For example, in mixed valence systems, where the ionic valence fluctuates between configurations f^5 and f^6 with characteristic time $\tau_{ic} \approx 10^{-13}$s, spectra of fast and slow experiments are dramatically different [135, 136]. Photoemission experiments with short characteristic times $\tau_{mp} \approx 10^{-16}$s (FC regime), reveal two lines, corresponding to f^5 and f^6 states. On the other hand slow Mössbauer isomer shift measurements with $\tau_{mp} \approx 10^{-9}$s show a single broad peak with mean frequency lying between signals from pure f^5 and f^6 shells. Finally, according to the paradigm of measurement process time, magnetic neutron scattering with $\tau_{mp} \approx \tau_{ic}$ revealed both coherent lines with all subsystems dynamically adjusted and broad incoherent remnants of strongly damped excitation of f^5 and f^6 shells [137, 138]. Actually, the meaning of the times τ_{ic} and τ_{mp} varies with the system and with the measurement process.

To study the interplay between measurement process time τ_{mp} and adjustment time τ_{ic}, the OC of the Fröhlich polaron was studied in [57] from the weak to the strong coupling regime by three methods. The DMC method gives a numerically exact answer which is compared with memory function formalism (MFF), which is able to take dynamical lattice relaxation into account, and strong coupling expansion (SCE) which assumes an FC approach. It was found that near critical coupling $\alpha_c \approx 8.5$ a dramatic change of the OC spectrum occurs: the dominating peak of OC splits into two satellites. In this critical regime the upper (lower) one quickly decreases (increases) its spectral weight as the value of the coupling constant increases. Besides, while the OC follows the prediction of MFF at $\alpha < \alpha_c$, its dependence switches to that predicted by SCE for larger couplings. It was concluded that, for the OC measurement of polarons, the adjustment time $\tau_{ic} \approx \hbar/\mathcal{D}$ is set by typical nonadiabatic energy $\mathcal{D}$. Nonadiabaticity destroys FC classification at $\alpha < \alpha_c$ while the FC principle rapidly regains its validity at large couplings due to fast growth of energy separation between initial and final states of optical transitions.

Comparison of exact DMC-SO data for OC with existing results of approximate methods showed [48] that the Feynman path integral technique [75] of Devreese, De Sitter, and Goovaerts, where OC is calculated starting from the Feynman variational model [139], is the only successful description of the evolution of the energy of the main peak in OC with coupling constant α (see [58]). However, starting from the intermediate coupling regime this approach fails to reproduce the peak width. Subsequently, the path integral approach was rewritten in terms of MFF [140]. Then, in [57] the extended MFF formalism, which introduces dissipation processes fixed by exact sum rules, was developed [141].

As shown in Fig. 4a, in the weak coupling regime, the MFF, with or without dissipation, is in very good agreement with DMC data, showing significant improvement with respect to the weak coupling perturbation approach [129] which provides a good description of OC spectra only for very small values of α. For $1 \leq \alpha \leq 8$, where standard MFF fails to reproduce peak width (Fig. 4b-d) and even the peak position (Fig. 4c), the damping, introduced to extended MFF scheme, becomes crucial. Results of extended MFF are accurate for the peak energy and quite satisfactory for the peak width (Fig. 4b-e). Note that the broadening of the peak in DMC data is not a consequence of the poor quality of the analytic continuation procedure since DMC-SO methods are capable of revealing such fine features as 2- and 3-phonon thresholds of emission (Fig. 4b).

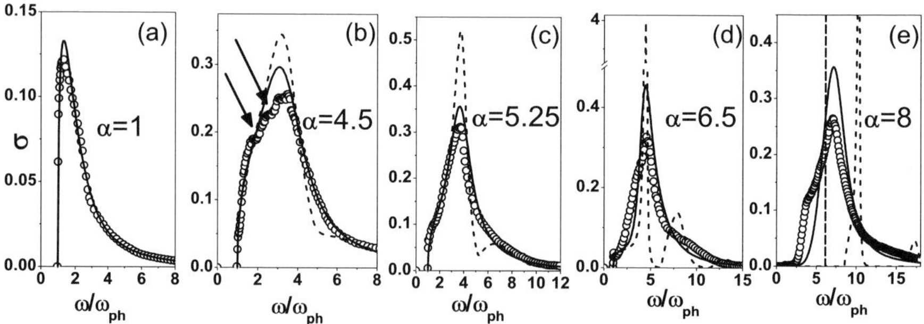

Fig. 4. Comparison of the optical conductivity calculated by DMC method (circles), extended MFF (solid line), and DSG [75, 140] (dashed line) for different values of α. The slanted arrows indicate 2- and 3-phonon thresholds of absorption.

However, a dramatic change of OC occurs around the critical coupling strength $\alpha_c \approx 8.5$. The dominating peak of OC splits into two, the energy of the lower one corresponding to the predictions of SCR expansion and that of the upper one obeying extended MFF value (Fig. 5a). The shoulder, corresponding to dynamical extended MFF contribution, rapidly decreases its intensity with increase of α and at large α (Fig. 5b-c) the OC is in good agreement with strong coupling expansion, assuming FC scheme. Finally, comparing energies of the peaks, obtained by DMC, extended MFF and FC strong

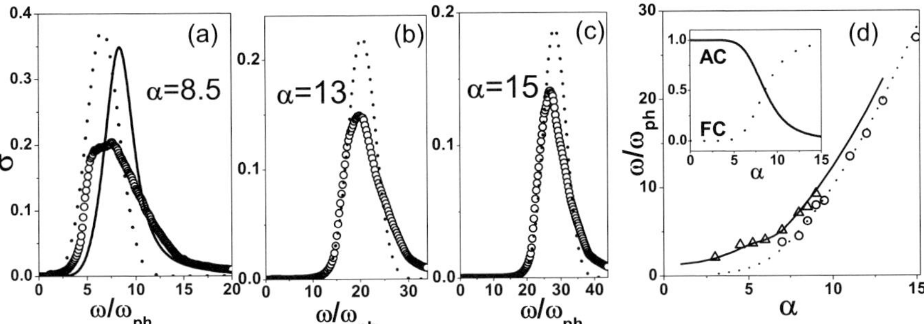

Fig. 5. (a)-(c) Comparison of the optical conductivity calculated within the DMC method (circles), the extended MFF (solid line), and SCE (dotted line) for different values of α. (d) The energy of lower- and higher-frequency features (circles and triangles, respectively) compared with the FC transition energy with the SCE (dashed line) and with the energy of the peak obtained from the extended MFF (solid line). In the inset, the weights of FC and adiabatically connected transitions are shown as a function of α (for $\eta = 1.3$.)

coupling expansion (Fig. 5d), we conclude that at critical coupling $\alpha_c \approx 8.5$ the spectral properties rapidly switch from dynamic, when the lattice relaxes at transition, to the FC regime, where nuclei are frozen in the initial configuration.

In order to get an idea of the FC breakdown authors of [57] consider the following arguments. The approximate adiabatic states are not exact eigenstates of the system. These states are mixed by nondiagonal matrix elements of the nonadiabatic operator $\mathcal{D}$ and exact eigenstates are linear combinations of the adiabatic wavefunctions. Being interested in the properties of transition from ground (g) to an excited (ex) state, whose energy corresponds to that of the OC peak, it is necessary to consider mixing of only these states and express exact wavefunctions as linear combinations [142, 143] of ground and excited adiabatic states. The coefficients of superposition are determined from standard techniques [142, 143] where nondiagonal matrix elements of the nonadiabatic operator [142] are expressed in terms of matrix elements of the kinetic energy operator M, the gap between excited and ground states $\Delta E = E_{ex} - E_g$ and the number n_β of phonons in the adiabatic state:

$$\mathcal{D}^{\pm} = M(\Delta E)^{-1}\sqrt{n_\beta + 1/2 \pm 1/2 + M^2(\Delta E)^{-2}}. \tag{36}$$

The extent to which the lattice can follow the transition between electronic states, depends on the degree of mixing between initial and final exact eigenstates through the nonadiabatic interaction. If initial and ground states are strongly mixed, the adiabatic classification has no sense and, therefore, the FC processes have no place and the lattice is adjusted to the change of electronic states during the transition. In the opposite limit adiabatic approximation is valid and FC processes dominate. The estimation for the weight of the FC component I_{FC} [57] is equal to unity in the case of zero mixing and zero in the case of maximal mixing. The weight of the adiabatically connected (AC) transition $I_{AC} = 1 - I_{FC}$ is defined accordingly. The non-diagonal matrix element M is proportional to the root square of α with a coefficient η of the order of unity. In the strong coupling regime, assuming that $\Delta E \approx \gamma \alpha^2$ ($\gamma \approx 0.1$ from MC data), and $n_\beta \approx \Delta E$ ($n_\beta \gg 1$), one gets

$$I_{FC} = \left[1 + 4(\tau_{mp}/\tau_{ic})^2\right]^{-1} , \tag{37}$$

where $\tau_{mp} = 1/\Delta E$ and $\tau_{ic} = 1/D$. For η of the order of unity one obtains qualitative description of a rather fast transition from AC- to FC-dominated transition, when I_{FC} and I_{AC} exchange half of their weights in the range of α from 7 to 9. The physical reason for such a quick change is the faster growth of energy separation $\Delta E \sim \alpha^2$ compared to that of the matrix element $M \sim \alpha^{1/2}$. Finally, for large couplings, initial and final states become adiabatically disconnected. The rapid AC-FC switch has nothing to do with the self-trapping phenomenon where crossing and hybridization of the ground and an excited states occurs. This phenomenon is a property of the transition between different states and is related to the choice of whether the lattice can or can not adiabatically follow the change of electronic state at the transition.

4 Self-Trapping

In this section we consider the self-trapping (ST) phenomenon which, due to essential importance of many-particle interaction of QP with bosonic bath of *macroscopic system*, has not been addressed by exact methods before. We start with a basic definition of the ST phenomenon and introduce the adopted criterion for its existence. Then, the generic features of ST are demonstrated on a simple model of the Rashba-Pekar exciton-polaron in Sect. 4.1. It is shown in Sect. 4.2 that the criterion is not a dogma since even in a one-dimensional system, where ST is forbidden by criterion of existence, one can observe all of the main features of ST due to the peculiar nature of the electronic states.

In general terms [7, 80], ST is a dramatic transformation of QP properties when system parameters are changed slightly. The physical reason of ST is a quantum resonance, which happens at some critical interaction constant γ_c, between a "trapped" (T) state of QP with strong lattice deformation around it and a "free" (F) state. Naturally, the ST transition is not abrupt because of nonadiabatic interaction between T and F states and all properties of the QP are analytic in γ [144]. At small $\gamma < \gamma_c$, the ground state is an F state which is weakly coupled to phonons while excited states are T states and have a large lattice deformation. At critical couplings $\gamma \approx \gamma_c$ a crossover and hybridization of these states occurs. Then, for $\gamma > \gamma_c$ the roles of the states exchange. The lowest state is a T state while the upper one is an F state.

First, and up to now the only quantitative criterion for ST existence was given in terms of the ground state properties in the adiabatic approximation.

This criterion considers stability of the delocalized state in undistorted lattice $\Delta = 0$ with respect to the energy gain due to lattice distortion $\Delta' \neq 0$. ST phenomenon occurs when a completely delocalized state with $\Delta = 0$ is separated from a distorted state with $\Delta' \neq 0$ by a barrier of adiabatic potential. One of these states is stable while the other one is meta-stable. The criterion of barrier existence is defined in terms of the stability index

$$s = d - 2(1 + l) , \tag{38}$$

where d is the system dimensionality. Index l determines the range of the force $\lim_{q \to 0} \psi(q) \sim q^{-l}$, where $\psi(\mathbf{R})$ is the kernel of interaction $U(\mathbf{R}_n) = \psi(\mathbf{R}_n - \mathbf{R}_{n'})\nu(\mathbf{R}_{n'})$ connecting potential $U(\mathbf{R}_n)$ with generalized lattice distortion $\nu(\mathbf{R}_{n'})$ [7]. The barrier exists for $s > 0$ and does not exist for $s < 0$. The discontinuous change of the polaron state, i.e. ST, occurs in the former case but not in the latter case. When $s = 0$, this scaling argument alone can not determine the presence or absence of ST, and more detailed discussion for each model is needed.

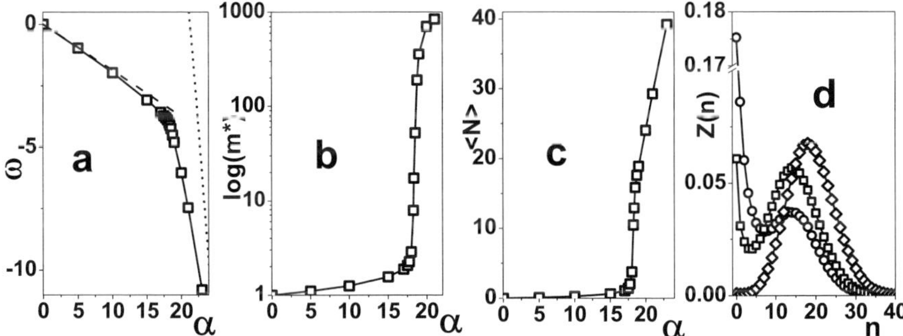

Fig. 6. The ground-state energy (a), effective mass (b), and average number of phonons as function of coupling constant (c). Partial weights of n-phonon states (d) in the polaron ground state ($\mathbf{k} = 0$) at $\gamma = 18$ (circles), $\gamma = 18.35$ (squares), and $\gamma = 19$ (diamonds). The dotted line in panel (a) is the result of the strong coupling limit and the dashed line is the result of perturbation theory.

4.1 Typical Example of the Self-Trapping: Rashba-Pekar Exciton–Polaron

A classical example of a system with ST phenomenon is the three dimensional continuous Rasba-Pekar exciton-polaron in the approximation of intraband scattering, i.e. when polar electron-phonon interaction (EPI) with dispersionless optical phonons $\omega_{\mathrm{ph}} = 1$ does not change the wave function of internal electron-hole motion. The system is defined as a structureless QP with dispersion $\epsilon(\mathbf{k}) = k^2/2$ and short range coupling to phonons [7,54]. The general criterion of the existence of ST is satisfied for the three dimensional system with short range interaction [7, 50, 54] and, thus, one expects to observe typical features of the phenomenon.

It is shown [54] that in the vicinity of the critical coupling $\gamma_c \approx 18$ the average number of phonons $\langle N \rangle$ in (18) and effective mass m^* quickly increase in the ground state by several orders of magnitude (Fig. 6b-c). Besides, a quantum resonance between polaronic phonon clouds of F and T state is demonstrated. Distribution of partial n-phonon contributions $Z^{(\mathbf{k}=0)}(n)$ in (17) has one maximum at $n = 0$ in the weak coupling regime, which corresponds to weak deformation, and one maximum at $n \gg 1$ in the strong coupling regime, which is the consequence of a strong lattice distortion. However, due to F-T resonance there are two distinct peaks at $n = 0$ and $n \gg 1$ for $\gamma \approx \gamma_c$ (Fig. 6d).

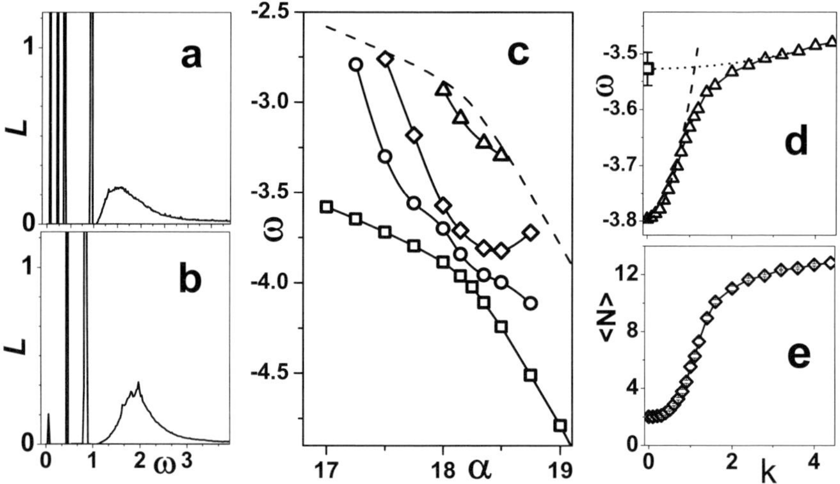

Fig. 7. LF $L_{(\mathbf{k}=0)}(\omega)$ at critical coupling $\gamma = \gamma_c$ (a) and for $\gamma > \gamma_c$ (b). Energy is counted from the polaron ground state. (c) Dependence of energy of ground state (squares) and stable excited states (circles, diamonds, and triangles) on the coupling constant. The dashed line is the threshold of the incoherent continuum. Dependence of energy (d) and average number of phonons (e) on the wave vector at $\gamma < \gamma_c$ (circles and rectangles). The dashed line is the effective mass approximation $E^{(\mathbf{k})} = E_{gs} + \mathbf{k}^2/2m^*$ for parameters $E_{gs} = -3.7946$ and $m^* = 2.258$, obtained by DMC estimators for given value of γ. The dotted line is a parabolic dispersion law which is fitted to the last 4 points of the energy dispersion curve with parameters $E_1 = -3.5273$ and $m_1^* = 195$. The empty squares represent the energy of the first excited stable state at zero momentum obtained by the SO method.

Near the critical coupling γ_c the LF of the polaron has several stable states (Fig. 7a-b) below the threshold of the incoherent continuum $E_{gs} + \omega_{ph}$. Any state above the threshold is unstable because emission of a phonon with transition to the ground state at $\mathbf{k} = 0$ with energy E_{gs} is allowed. On the other hand, decay is forbidden by conservation laws for states below the threshold. Dependence of the energies of ground and excited resonances on the interaction constant resembles a picture of crossing of several states interacting with each other (Fig. 7c).

According to the general picture of the ST phenomenon, the lowest F state in the weak coupling regime at $\mathbf{k} = 0$ has small effective mass $m^* \approx m$ of the order of the bare QP mass m. In contrast, the effective mass of excited state $m^* \gg m$ is large. Hence, below the critical coupling the energy of the F state, which is lowest at $\mathbf{k} = 0$, has to reach a flat band of T state at some momentum. Then, F and T states have to hybridize and exchange in energy. DMC data visualize this picture (Fig. 7 d-e). After the F state crosses the flat band of the excited T state, the average number of phonons increases and dispersion becomes flat.

It is natural to assume that above the critical coupling the situation is opposite: ground state is the T state with large effective mass while excited F state has small, nearly bare, effective mass. Indeed, this assumption was confirmed in the framework of another model which is considered in Sect. 6.1. Moreover, it was shown that in the strong coupling regime excited resonance inherits not only bare effective mass around $\mathbf{k} = 0$ but the whole dispersion law of the bare QP [49].

4.2 Degeneracy Driven Self-Trapping

According to the criterion (38), ST phenomenon in a one-dimensional system does not occur. Although this statement is probably valid for the case of a single band in the relevant energy range, it is not the case for the generic multiband cases. This fact has been unnoticed for many years, which prevented a proper explanation of the puzzling physics of the quasi-one-dimensional compound Anthracene-PMDA, although its optical properties [65, 66, 145–148] directly suggested resonance of T and F states. The reason is that in Anthracene-PMDA, in contrast to conditions at which criterion (38) is obtained, there are two, nearly degenerate exciton bands. Then one can consider a quasi-degenerate self-trapping mechanism when the ST phenomenon is driven by nondiagonal interaction of phonons with quasidegenerate exciton levels [52]. Such a mechanism was already suggested for the explanation of the properties of mixed valence systems [143] though its relevance was never proved by an exact approach.

The minimal model to demonstrate the mechanism of quasi-degenerate self-trapping involves one optical phonon branch with frequency $\omega_{ph} = 0.1$ and two exciton branches with energies $\epsilon_{1,2}(q) = \Delta_{1,2} + 2[1 - \cos(q)]$, where $\Delta_1 = 0$ and $\Delta_2 = 1$. The presence of short range diagonal γ_{22} and nondiagonal γ_{12} interactions (with corresponding dimensionless constants $\lambda_{22} = \gamma_{22}^2/(2\omega)$ and $\lambda_{12} = \gamma_{12}^2/(2\omega)$) leads to classical self-trapping behavior even in a one-dimensional system [52] (see Fig. 8).

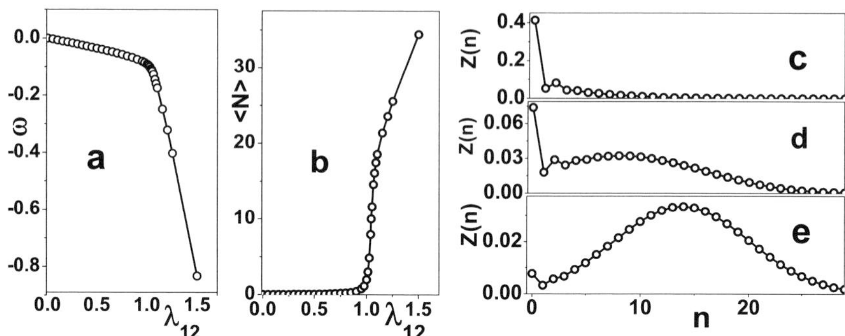

Fig. 8. Dependence of energy (a) and average number of phonons (b) on the non-diagonal coupling constant λ_{12} at $\lambda_{11} = 0$ and $\lambda_{22} = 0.25$. Phonon distributions in polaron cloud below ST point at $\lambda_{12} = 1.0125$ (c), at ST point at $\lambda_{12} = 1.0435$ (d), and above ST coupling at $\lambda_{12} = 1.0625$ (e).

5 Exciton

Despite numerous efforts over the years, there has been no rigorous technique to solve for exciton properties even for the simplest model (1)-(2) which treats electron-electron interactions as a static renormalized Coulomb potential with averaged dynamical screening. The only solvable cases are the Frenkel small-radius limit [67] and the Wannier large-radius limit [68] which describe molecular crystals and wide gap insulators with large dielectric constant, respectively. Meanwhile, even accurate data for the limits of validity of the Wannier and Frenkel approximations have not been available. As discussed in Sects. 1.2 and 2.3, semianalytic approaches have little to add to the problem when quantitative results are needed whereas traditional numerical methods fail to reproduce them even in the Wannier regime. In contrast, DMC results do not contain any approximation.

To study the conditions of validity of limiting regimes by the DMC method, the electron–hole spectrum of a three dimensional system was chosen in the form of symmetric valence and conduction bands with width E_c and direct gap E_g at zero momentum [46]. For large ratio $W = E_c/E_g$, when $W > 30$, exciton binding energy is in good agreement with Wannier approximation results (Fig. 9a) and probability density of relative electron-hole motion corresponds (Fig. 9c) to a hydrogen-like result. The striking result is the requirement of rather large valence and conduction bandwidths ($W > 20$) for applicability of the Wannier approximation. For smaller values of W the binding energy and wave function of relative motion (Fig. 9d) deviate from large radius results. In a similar way, the conditions of validity of the Frenkel approach are also rather restricted. Moreover, even strong localization of wave function does not guarantee good agreement between exact and Frenkel approximation results for binding energy. At $1 < W < 10$ the wave function is already strongly localized though binding energy differ considerably from the Frenkel

approximation result. For example, at $W = 0.4$ relative motion is well localized (Fig. 9e) whereas the binding energy of the Frenkel approximation is two
times larger than the exact result (Inset in Fig. 9a).

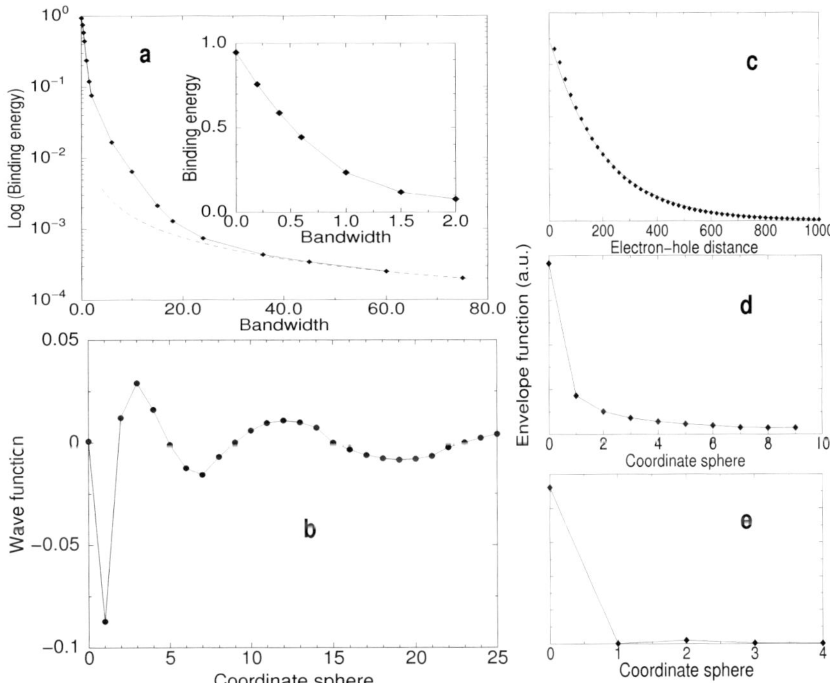

Fig. 9. Panel (a): dependence of the exciton binding energy on the bandwidth
$E_c = E_v$ for conduction and valence bands. The dashed line corresponds to the
Wannier model. The solid line is the cubic spline, the derivatives at the right and left
ends being fixed by the Wannier limit and perturbation theory, respectively. Inset
in panel (a): the initial part of the plot. Panel (b): the wave function of internal
motion in real space for the optically forbidden monopolar exciton. Panels (c)-(e):
the wave function of internal motion in real space: (c) Wannier $[E_c = E_v = 60]$; (d)
intermediate $[E_c = E_v = 10]$; (e) near-Frenkel $[E_c = E_v = 0.4]$ regimes. The solid
line in the panel (c) is the Wannier model result while the solid lines in other panels
are to guide the eye only.

A study of the conditions necessary for formation of a charge transfer
exciton in three dimensional systems is crucial to finalize the protracted discussion of numerous models concerning properties of mixed valence semiconductors [149]. A decade ago the unusual properties of SmS and SmB_6 were explained by invoking the excitonic instability mechanism assuming the charge-
transfer nature of the optically forbidden exciton [150, 151]. Although this
model explained quantitatively the phonon spectra [152, 153], optical proper-

ties [154, 155], and magnetic neutron scattering data [138], its basic assumption has been criticized as being groundless [156, 157]. To study the excitonic wavefunction, dispersions of valence and conduction bands were chosen as it is typical for mixed valence materials: an almost flat valence band is separated from a broad conduction band, having a maximum in the centre and minimum at the border of the Brillouin zone [46]. Results presented in Fig. 9b support the assumption of [150, 151] since the wave function of relative motion has almost zero on-site component and maximal charge density at near neighbors.

6 Polarons in Undoped High Temperature Superconductors

It is now well established that the physics of high temperature superconductors is that of hole doping a Mott insulator [158–160]. Even a single hole in a Mott insulator, i.e. a hole in an antiferromagnet in the case of infinite Hubbard repulsion U, is substantially influenced by many-body effects [10] because its jump to a neighboring site disturbs the antiferromagnetic arrangement of spins. Hence a thorough understanding of the dynamics of doped holes in Mott insulators has attracted a great deal of recent interest. The two major interactions relevant to electrons in solids are electron–electron interactions (EEI) and electron–phonon interactions (EPI). The importance of the former at low doping is no doubt essential since the Mott insulator is driven by strong Hubbard repulsion, while the latter was considered to be largely irrelevant to superconductivity based on the observations of a small isotope effect on the optimal T_c [161] and an absence of a phonon contribution to the resistivity (for a review see [162]).

On the other hand, there is now accumulating evidence that the EPI plays an important role in the physics of cuprates such as (i) an isotope effect on superfluid density ρ_s and T_c away from optimal doping [163], (ii) neutron and Raman scattering [164–166] experiments showing strong phonon softening with both temperature and hole doping, indicating that EPI is strong [167, 168]. Furthermore, recent studies of cuprates by angle resolved photoemission spectroscopy (ARPES), which spectra are proportional to the LF (7) [32], resulted in the discovery of the dispersion "kinks" at around 40-70meV measured from the Fermi energy, in the correct range of the relevant oxygen related phonons [169–171]. These particular phonon–oxygen buckling and half-breathing modes are known to soften with doping [164, 172] and with temperature [164–166, 170–172] indicating strong coupling. The quick change of the velocity can be predicted by any interaction of a quasiparticle with a bosonic mode, either with a phonon [170, 171] or with a collective magnetic resonance mode [173–175]. However, the recently discovered "universality" of the kink energy for LSCO over the entire doping range [176] casts doubts on the validity of the latter scenario as the energy scale of the magnetic excitation changes strongly with doping.

Besides, measured in undoped high T_c materials ARPES revealed apparent contradiction between momentum dependence of the energy and linewidth of the QP peak. On the one hand the experimental energy dispersion of the broad peak in many underdoped compounds [31, 177] obeys the theoretical predictions [178, 179], whereas the experimental peak width is comparable with the bandwidth and orders of magnitude larger than that obtained from the theory of Mott insulators [53]. Early attempts to interpret this anomalously short lifetime of a hole by an interaction with additional nonmagnetic bosonic excitations, e.g. phonons [180], faced the generic question: is it possible that interaction with media leaves the energy dispersion absolutely unrenormalized, while inducing a decay whose inverse life-time is comparable or even larger than the QP energy dispersion? The possibility of an extrinsic origin of this width can be ruled out since the doping induces further disorder, while a sharper peak is observed in the overdoped region.

In order to understand whether phonons can be responsible for the peculiar shape of the ARPES in the undoped cuprates, the LF of a hole interacting with phonons in a Mott insulator was studied by DMC-SO [49]. The case of the LF of a single hole corresponds to the ARPES in an undoped compound. For a system with large Hubbard repulsion U, when U is much larger than the typical bandwidth W of noninteracting QP, the problem reduces to the t-J model [11, 158, 181, 182]

$$\widehat{H}_{\text{t-J}} = -t \sum_{\langle ij \rangle s} c_{is}^{\dagger} c_{js} + J \sum_{\langle ij \rangle} (\mathbf{S}_i \mathbf{S}_j - n_i n_j / 4) \ . \tag{39}$$

Here $c_{j\sigma}$ is the projected (to avoid double occupancy) fermion annihilation operator, n_i (< 2) is the occupation number, $\mathbf{S}_i$ is spin $1/2$ operator, J is an exchange integral, and $\langle ij \rangle$ denotes nearest-neighbor sites in a two dimensional square lattice. Different theoretical approaches revealed [53, 158, 183] basic properties of the LF. The LF has a sharp peak in the low energy part of the spectrum which disperses with a bandwidth $W_{J/t} \sim 2J$ and, therefore, the large QP width in experiment can not be explained. The more complicated $tt't''$-J model takes into account hoppings to the second t' and third t'' nearest neighbors and, hence, dispersion of the hole changes [32, 178, 179, 184–186]. However, for parameters which are necessary for the description of dispersion in realistic high T_c superconductors [31, 178], the peak in the low energy part remains sharp and well defined for all momenta [187].

After expressing spin operators in terms of Holstein-Primakoff spin wave operators and diagonalizing the spin part of Hamiltonian (39) by Fourier and Bogoliubov transformations [10, 188–190], the $tt't''$-J Hamiltonian is reduced to the boson-holon model, where the hole (annihilation operator is $h_{\mathbf{k}}$) with dispersion $\varepsilon(\mathbf{k}) = 4t' \cos(k_x) \cos(k_y) + 2t''[\cos(2k_x) + \cos(2k_y)]$ propagates in the magnon (annihilation operator is $\alpha_{\mathbf{k}}$) bath

$$\widehat{H}^0_{\text{t-J}} = \sum_{\mathbf{k}} \varepsilon(\mathbf{k}) h^\dagger_{\mathbf{k}} h_{\mathbf{k}} + \sum_{\mathbf{k}} \omega_{\mathbf{k}} \alpha^\dagger_{\mathbf{k}} \alpha_{\mathbf{k}} \tag{40}$$

with magnon dispersion $\omega_{\mathbf{k}} = 2J\sqrt{1 - \gamma^2_{\mathbf{k}}}$, where $\gamma_{\mathbf{k}} = (\cos k_x + \cos k_y)/2$. The hole is scattered by magnons as described by

$$\widehat{H}^{\text{h-m}}_{\text{t-J}} = N^{-1/2} \sum_{\mathbf{k},\mathbf{q}} M_{\mathbf{k},\mathbf{q}} \left[h^\dagger_{\mathbf{k}} h_{\mathbf{k}-\mathbf{q}} \alpha_{\mathbf{k}} + h.c. \right] \tag{41}$$

with the scattering vertex $M_{\mathbf{k},\mathbf{q}}$. Parameters t, t' and t'' are hopping amplitudes to the first, second and third nearest neighbors, respectively. If hopping integrals t' and t'' are set to zero and the bare hole has no dispersion, the problem (40-41) corresponds to the *t-J* model.

Short range interaction of a hole with dispersionless optical phonons $\widehat{H}^{\text{e-ph}} = \Omega_0 \sum_{\mathbf{k}} b^\dagger_{\mathbf{k}} b_{\mathbf{k}}$ of frequency Ω_0 is introduced by the Holstein Hamiltonian

$$\widehat{H}^{\text{e-ph}} = N^{-1/2} \sum_{\mathbf{k},\mathbf{q}} \frac{\sigma}{\sqrt{2M\Omega_0}} \left[h^\dagger_{\mathbf{k}} h_{\mathbf{k}-\mathbf{q}} b_{\mathbf{q}} + h.c. \right] , \tag{42}$$

where σ is the momentum and isotope independent coupling constant, M is the mass of the vibrating lattice ions, and Ω_0 is the frequency of the dispersionless phonon. The coefficient in front of the square brackets is the standard Holstein interaction constant $\gamma = \sigma/\sqrt{(2M\Omega_0)}$. In the following we characterize the strength of EPI in terms of the dimensionless coupling constant $\lambda = \gamma^2/4t\Omega_0$. Note, if interaction with magnetic subsystem (41) is neglected and hole dispersion $\varepsilon(\mathbf{k})$ is chosen in the form $\varepsilon(\mathbf{k}) = 2t[\cos(k_x) + \cos(k_y)]$, the problem (40), (42) corresponds to the standard Holstein model where the hole with near neighbor hopping amplitude t interacts with dispersionless phonons.

We consider the evolution of ARPES of a single hole in the *t-J*-Holstein model (40)-(42) from the weak to the strong coupling regime and dispersion of the LF in the strong coupling regime in Sect. 6.1. It occurs that properties of the LF in the strong coupling regime of the EPI explain the puzzle of broad lineshape in ARPES in underdoped high T_c superconductors. Therefore, in order to suggest a crucial test for the mechanism of phonon-induced broadening, we present calculations of the effect of the isotope substitution on ARPES in Sect. 6.2.

6.1 Spectral Function of a Hole Interacting with Phonons in the *t-J* Model: Self-Trapping and Momentum Dependence

Previously, the LF of the *t-J*-Holstein model was studied by exact diagonalization method on small clusters [191] and in the non-crossing approximation (NCA)[6] for both phonons and magnons [192, 193]. However, the small system size in the exact diagonalization method implies a discrete spectrum and,

---

[6] NCA is equivalent to self-consistent Born approximation (SCBA)

therefore, the problem of lineshape could not be addressed. The latter method omits the FDs with mutual crossing of phonon propagators and, hence, is an invalid approximation for phonons in strong and intermediate couplings of EPI. This statement was demonstrated by DMC, which can sum all FDs for the Holstein model both exactly and in the NCA [49]. Exact results and those of the NCA are in good agreement for small values $\lambda \leq 0.4$ and drastically different for $\lambda > 1$. For example, for $\Omega_0/t = 0.1$ the exact result shows a sharp crossover to strong coupling regime for $\lambda > \lambda_H^c \approx 1.2$ whereas the NCA result does not undergo such crossover even at $\lambda = 100$. On the other hand, the NCA is valid for the interaction of a hole with magnons since spin S=1/2 can not flip more than once and the number of magnons in the polaronic cloud can not be large. Note that the t-J-Holstein model is reduced to the problem of a polaron which interacts with several bosonic fields (3)-(4).

DMC expansion in [49] takes into account mutual crossing of phonon propagators and, in the framework of partial NCA, neglects mutual crossing of magnon propagators, to avoid the sign problem. NCA for magnons is justified for $J/t \leq 0.4$ by good agreement of results of NCA and exact diagonalization on small clusters [10, 188, 190, 194, 195]. Recently results of exact diagonalization were compared in the limit of small EPI for t-J-Holstein model, boson-holon model (40-42) without NCA, and boson-holon model with NCA [196]. Although agreement is not so good as for the pure t-J model, it was concluded that NCA for magnons is still good enough to suggest that one can use NCA for a qualitative description of the t-J-Holstein model.

Figures 10a-e show the low energy part of the LF in the ground state at $\mathbf{k} = (\pi/2, \pi/2)$ in the weak, intermediate, and strong coupling regimes of interaction with phonons. Dependence on the coupling constant of energies of resonances (Fig. 10f), $Z^{\mathbf{k}=(\pi/2,\pi/2)}$-factor of lowest peak (Fig. 10g), and average number of phonons in the polaronic cloud $\langle N \rangle$ (Fig. 10h) demonstrates a picture which is typical for ST (see [54, 80] and Sect. 4). Two states cross and hybridize in the vicinity of critical coupling constant $\lambda_{t\text{-}J}^c \approx 0.38$, $Z^{\mathbf{k}=(\pi/2,\pi/2)}$-factor of lowest resonance sharply drops and the average number of phonons in the polaronic cloud quickly rises. According to the general understanding of the ST phenomenon, above the critical couplings $\lambda > \lambda_{t\text{-}J}^c$ one expects that the lowest state is dispersionless while the upper one has small effective mass. This assumption is supported by the momentum dependence of the LF in the strong coupling regime (Fig. 11a-e). Dispersion of the upper broad shake-off Franck-Condon peak nearly perfectly obeys the relation

$$\varepsilon_{\mathbf{k}} = \varepsilon_{min} + W_{J/t}/5\{[\cos k_x + \cos k_y]^2 + [\cos(k_x + k_y) + \cos(k_x - k_y)]^2/4\}, \quad (43)$$

which describes dispersion of the pure t-J model in the broad range of exchange constant $0.1 < J/t < 0.9$ [194] (Fig. 11f). Note that this property of the shake-off peak is general for the whole strong coupling regime (Fig. 11f).

Momentum dependence of the shake-off peak, reproducing that of the free particle, is the direct consequence of the adiabatic regime. Actually, the phonon frequency Ω_0 is much smaller than the coherent bandwidth $2J$ of the

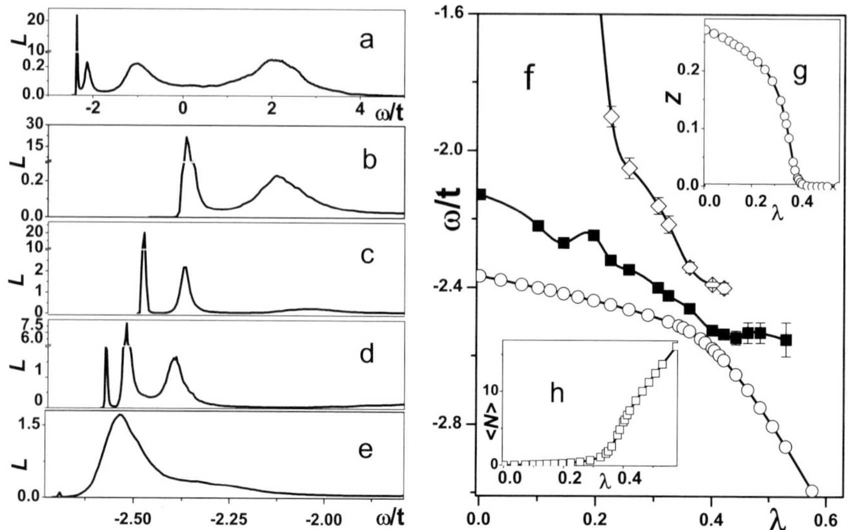

Fig. 10. (a) The LF of a hole in the ground state $\mathbf{k} = (\pi/2, \pi, 2)$ at $J/t = 0.3$ and $\lambda = 0$. The low energy part of the LF of a hole in the ground state $\mathbf{k} = (\pi/2, \pi, 2)$ at $J/t = 0.3$: (b) $\lambda = 0$; (c) $\lambda = 0.3$; (d) $\lambda = 0.4$; (e) $\lambda = 0.46$. Dependence on coupling strength λ at $J/t = 0.3$: (f) energies of lowest LF resonances; (g) Z-factor of lowest peak; (h) average number of phonons $\langle N \rangle$.

t-J model, giving the adiabatic ratio $\Omega/2J = 1/6 \ll 1$. Besides, as experience with the OC of the Fröhlich polaron (Sect. 3.2) shows, there is one more important parameter in the strong coupling limit. Namely, the ratio between measurement process time $\tau_{mp} = \hbar/\Delta E$ where ΔE is the energy separation of shake-off hump from the ground state pole, and that of the characteristic lattice time $\tau \approx 1/\Omega_0$ is much less than unity. Hence, the fast photoemission probe sees the ions frozen in one of many possible configurations [197]. The LF in the FC limit is a sum of transitions between a lower $E_{\mathrm{low}}(Q)$ and an upper $E_{\mathrm{up}}(Q)$ sheet of adiabatic potential, weighted by the adiabatic wave function of the lower sheet $\mid \psi_{\mathrm{low}}(Q) \mid^2$ [198]. If EPI is absent both in initial $E_{\mathrm{low}}(Q) = Q^2/2$ and final $E_{\mathrm{up}}(Q) = \mathcal{D} + Q^2/2$ states, the LF is peaked at the energy $\mathcal{D}$. Then, if there is EPI $\Delta E_{\mathrm{up}}(Q) = -\lambda Q$ only in the final state, i.e. when the hole is created in the Mott insulator, the upper sheet of adiabatic potential $E_{\mathrm{up}}(Q) = \mathcal{D} - \lambda^2/2 + (Q - \lambda)^2/2$ has the same energy $\mathcal{D}$ at $Q = 0$. Since the probability function $\mid \psi_{\mathrm{low}}(Q) \mid^2$ has maximum at $Q = 0$, the peak of the LF broadens but its energy does not shift [198] (Fig. 11g).

The behavior of the LF is the same as that observed in the ARPES of undoped cuprates. The LF consists of a broad peak and a high energy incoherent continuum (see Fig. 11a). Besides, dispersion of the broad peak "c" in Fig. 11 reproduces that of the sharp peak in the pure t-J model (Fig. 11b-f). The lowest dispersionless peak, corresponding to the small radius polaron, has very small weight and, hence, can not be seen experimentally. On the other hand,

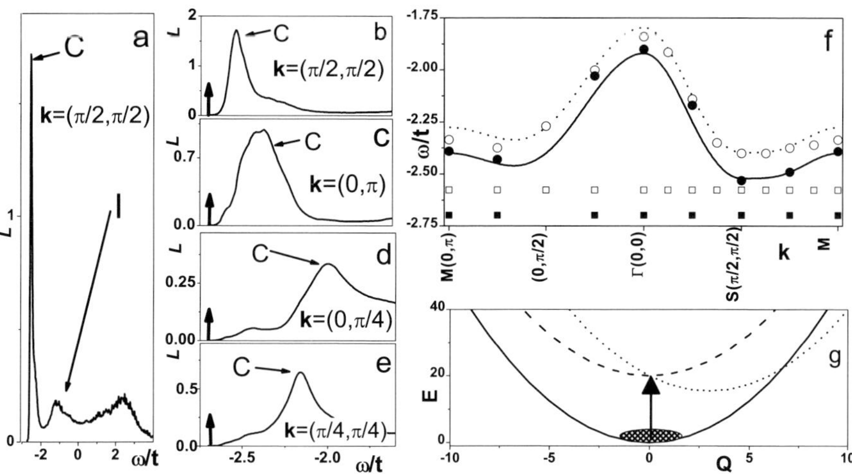

Fig. 11. The LF of a hole at $J/t = 0.3$ and $\lambda = 0.46$: (a) full energy range for $\mathbf{k} = (\pi/2, \pi/2)$; (b–e) low energy part for different momenta. Slanted arrows show broad peaks which can be interpreted in ARPES spectra as coherent (C) and incoherent (I) part. Vertical arrows in panels (b) (c) indicate the position of "invisible" lowest resonance. (f) Dispersion of resonance energies at $J/t = 0.3$: broad resonance (filled circles) and lowest polaron pole (filled squares) at $\lambda = 0.46$; broad resonance (open circles) and lowest polaron pole (open squares) at $\lambda = 0.4$. The solid curves are dispersions (43) of a hole in pure t-J model at $J/t = 0.3$ ($W_{J/t=0.3} = 0.6$): $\varepsilon_{min} = -2.396$ ($\varepsilon_{min} = -2.52$) for dotted (solid) line. Panel (g) shows ground state potential $Q^2/2$ (solid line), excited state potential without relaxation $D + Q^2/2$ (dashed line), and relaxed excited state potential $D + (Q - \lambda)^2/2 - \lambda^2/2$ (dotted line).

according to experiment, the momentum dependence of the spectral weight $Z^{(\mathbf{k})'}$ of broad resonance exactly reproduces the dispersion of the $Z^{(\mathbf{k})}$-factor of the pure t-J model. The reason for such perfect mapping is that in the adiabatic case $\Omega_0/2J \ll 1$ all the weight of the sharp resonance in t-J model without EPI is transformed at strong EPI into the broad peak. This picture implies that the chemical potential in the heavily underdoped cuprates is not connected with the broad resonance but pinned to the real quasiparticle pole with small Z-factor. This conclusion was recently confirmed experimentally [177].

Comparing the critical EPI for a hole in the t-J-Holstein model (40-42) $\lambda^c_{t\text{-}J} \approx 0.38$ and that for Holstein model $\lambda^c_H \approx 1.2$ with the same value of hopping t, we conclude that *spin-hole interaction accelerates the transition into the strong coupling regime*. The reason for enhancement of the role of EPI is found in [196]. Comparison of the EPI driven renormalization of the effective mass in the t-J-Holstein and Holstein models shows that large effective mass in the t-J model is responsible for this effect. The enhancement of the role of EPI by EEI takes place at least for a single hole at the bottom of the t-J

band. Had the comparison been made with the half-filled model, the result would have been smaller enhancement or no enhancement at all [199]. On the other hand, the coupling constant of the half-breathing phonon is increased by correlations [200]. Finally, we conclude that the effect of enhancement of the effective EPI by EEI is not unambiguous and depends on the details of interaction and filling. However, this effect is present for small filling in the t-J-Holstein model.

6.2 Isotope Effect on ARPES in Underdoped High-Temperature Superconductors

The magnetic resonance mode and the phonon modes are the two major candidates to explain the "kink" structure of the electron energy dispersion around 40-70 meV below the Fermi energy, and the isotope effect (IE) on ARPES should be the smoking-gun experiment to distinguish between these two. Gweon et al. [201] performed the ARPES experiment on O^{18}-replaced Bi2212 at optimal doping and found an appreciable IE, which however can not be explained within the conventional weak-coupling Migdal-Eliashberg theory. Namely the change of the spectral function due to O^{18}-replacement has been observed at higher energy region beyond the phonon energy (~ 60meV). This is in sharp contrast to the weak coupling theory prediction, i.e., that the IE should occur only near the phonon energy. Hence the IE in optimal Bi2212 remains still a puzzle. On the other hand, the ARPES in undoped materials, as described in Sect. 6.1, has recently been understood in terms of the small polaron formation [49, 198, 202]. Therefore, it is essential to compare experiments in undoped systems with the DMC-SO data presented in this section, where theory can offer quantitative results.

In addition to the high-T_c problem, strong EPI mechanism of ARPES spectra broadening was considered as one of alternative scenarios for diatomic molecules [203], colossal magnetoresistive manganites [34], quasi-one-dimensional Peierls conductors [37, 38], and Verwey magnetites [39]. Therefore, exact analysis of the IE on ARPES at strong EPI is of general interest for conclusive experiments in a broad variety of compound classes.

Dimensionless coupling constant $\lambda = \gamma^2/4t\Omega$ in (42) is an invariant quantity for the simplest case of IE. Indeed, assuming the natural relation $\Omega \sim 1/\sqrt{M}$ between phonon frequency and mass, we find that λ does not depend on the isotope factor $\kappa_{\rm iso} = \Omega/\Omega_0 = \sqrt{M_0/M}$, which is defined as the ratio of phonon frequency in isotope substituted (Ω) and normal (Ω_0) systems. We chose adopted parameters of the $tt't''$-J model which reproduce the experimental dispersion of ARPES [178]: $J/t = 0.4$, $t'/t = -0.34$, and $t''/t = 0.23$. The frequency of the relevant phonon [32] is set to $\Omega_0/t = 0.2$ and the isotope factor $\kappa_{\rm iso} = \sqrt{16/18}$ corresponds to substitution of O^{18} isotope for O^{16}.

To sweep aside any doubts of possible instabilities of analytic continuation, we calculate the LF for normal compound ($\kappa_{\rm nor} = 1$), isotope substituted

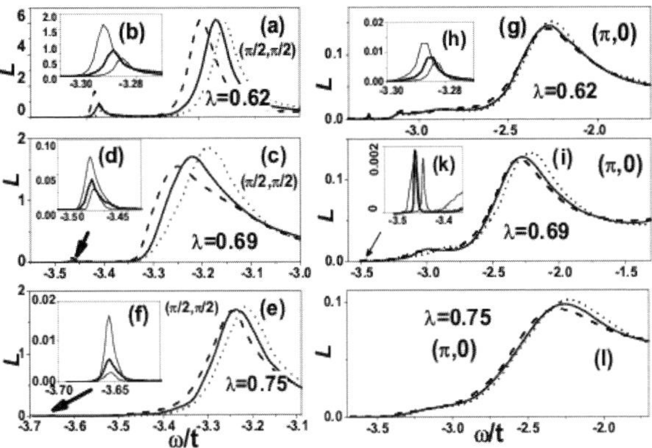

Fig. 12. Low energy part of hole LFs: normal compound (solid line), isotope substituted compound (dotted line) and "antiisotope" substituted compound (dashed line). LFs at different couplings in the nodal (a, c, e) and antinodal (g, i, l) points. Insets (b, d, f, h, k) show low energy peak of real QP.

($\kappa_{\mathrm{iso}} = \sqrt{16/18}$) and "anti-isotope" substituted ($\kappa_{\mathrm{ant}} = \sqrt{18/16}$) compounds. Monotonic dependence of LF on κ ensures stability of analytic continuation and gives us the possibility to evaluate the error-bars of a quantity $\mathcal{A}$ using quantities $\mathcal{A}_{\mathrm{iso}} - \mathcal{A}_{\mathrm{nor}}$, $\mathcal{A}_{\mathrm{nor}} - \mathcal{A}_{\mathrm{ant}}$, and $(\mathcal{A}_{\mathrm{iso}} - \mathcal{A}_{\mathrm{ant}})/2$.

Since LF is sensitive to strengths of EPI only for low frequencies [55], we concentrate on the low energy part of the spectrum. Figure 12 shows IE on the hole LF for different couplings in nodal and antinodal points, respectively. The general trend is a shift of all spectral features to larger energies with increase of the isotope mass ($\kappa < 1$). One can also note that the shift of broad FCP is much larger than that of the narrow real-QP peak. Moreover, for large couplings λ the shift of QP energy approaches zero and only the decrease of QP spectral weight Z is observed for larger isotope mass. On the other hand, the shift of FCP is not suppressed for larger couplings. Except for the LF in nodal point at $\lambda = 0.62$ (Fig. 12a,b), where LF still has significant weight of QP δ-functional peak, there is one more notable feature of the IE. With increase of the isotope mass the height of FCP increases. Taking into account the conservation law for LF $\int_{-\infty}^{+\infty} L_{\mathbf{k}}(\omega) = 1$ and the insensitivity of the high energy part of LF to EPI strength [55], the narrowing of the FCP for larger isotope mass can be concluded. To understand the trends of the IE in the strong coupling regime we analyze the exactly solvable independent oscillators model (IOM) [60]. The LF in IOM is the Poisson distribution

$$L(\omega) = \exp[-\xi_0/\kappa] \sum_{l=0}^{\infty} \frac{[\xi_0/\kappa]^l}{l!} \mathcal{G}_{\kappa,l}(\omega) \,, \tag{44}$$

where $\xi_0 = \gamma_0^2/\Omega_0^2 = 4t\lambda/\Omega_0$ is the dimensionless coupling constant for the normal system and $\mathcal{G}_{\kappa,l}(\omega) = \delta[\omega+4t\lambda-\Omega_0\kappa l]$ is the δ-function. The properties of the Poisson distribution quantitatively explain many features of the IE on LF[7].

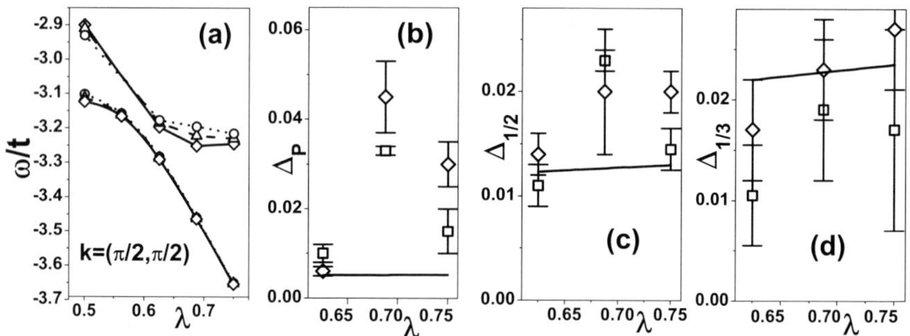

Fig. 13. (a) Energies of ground state and broad peaks for normal (triangles), isotope substituted (circles) and "antiisotope" substituted (diamonds) compounds. Comparison of IOM estimates (lines) with DMC data in the nodal (squares) and antinodal (diamonds) points: (b) shift of the FCP top, (c) FCP leading edge at $1/2$ of height, and (d) FCP leading edge at $1/3$ of height.

The energy $\omega_{\mathrm{QP}} = -4t\lambda$ of the zero-phonon line $l = 0$ in (44) depends only on isotope independent quantities which explains the very weak isotope dependence of QP peak energy in the insets of Fig. 12. Besides, change of the zero-phonon line weight $Z^{(0)}$ obeys the relation $Z_{\mathrm{iso}}^{(0)}/Z_{\mathrm{nor}}^{(0)} = \exp\left[-\xi_0(1-\kappa)/\kappa\right]$ in IOM. These IOM estimates agree with DMC data within 15% in the nodal point and within 25% in the antinodal one. IE on FCP in the strong coupling regime follows from the properties of zero $M_0 = \int_{-\infty}^{+\infty} L(\omega)d\omega = 1$, first $M_1 = \int_{-\infty}^{+\infty} \omega L(\omega)d\omega = 0$, and second $M_2 = \int_{-\infty}^{+\infty} \omega^2 L(\omega)d\omega = \kappa\xi_0\Omega_0^2$ moments of shifted Poisson distribution (44). Moments M_0 and M_2 establish the relation $\mathcal{D} = h_{\mathrm{iso}}^{\mathrm{FCP}}/h_{\mathrm{nor}}^{\mathrm{FCP}} = 1/\sqrt{\kappa} \approx 1.03$ between heights of FCP in normal and substituted compounds. DMC data in the antinodal point perfectly agree with the above estimate for all couplings. This is consistent with the idea that the anti-nodal region remains in the strong coupling regime even though the nodal region is in the crossover region. In the nodal point DMC

---

[7] Cautions should be made about the approximate form of EPI (42). Strictly speaking, the actual momentum dependence of the interaction constant σ [204, 205] can slightly change the obtained differences between nodal and antinodal points though the general trends have to be left intact because ST is caused solely by the short range part of EPI [80].

data agree well with the IOM estimate for $\lambda = 0.75$ ($\mathcal{D} \approx 1.025$) whereas at $\lambda = 0.69$ and $\lambda = 0.62$ influence of the ST point leads to anomalous values of $\mathcal{D}$: $\mathcal{D} \approx 1.07$ and $\mathcal{D} \approx 0.98$, respectively. Shift of the low energy edge at half maximum $\Delta_{1/2}$ must be proportional to change of the root square of second moment $\Delta_{\sqrt{M_2}} = \sqrt{\xi_0}\,\Omega_0[1 - \sqrt{\kappa}]$. As we found in numeric simulations of (44) with Gaussian functions[8] $\mathcal{G}_{\kappa,l}(\omega)$, relation $\Delta_{1/2} \approx \Delta_{\sqrt{M_2}}/2$ is accurate to 10% for $0.62 < \lambda < 0.75$. Also, simulations show that the shift of the edge at one third of maximum $\Delta_{1/3}$ obeys relation $\Delta_{1/3} \approx \Delta_{\sqrt{M_2}}$. DMC data with IOM estimates are in good agreement for strong EPI $\lambda = 0.75$ (Fig. 13). However, shift of the FCP top Δ_p and $\Delta_{1/2}$ are considerably enhanced in the self-trapping (ST) transition region. The physical reason for enhancement of IE in this region is a general property regardless of the QP dispersion, range of EPI, etc. The influence of nonadiabatic matrix elements, mixing excited and ground states, on the energies of resonances essentially depends on the phonon frequency. While in the adiabatic approximation the ST transition is sudden and nonanalytic in λ [80], nonadiabatic matrix elements turn it to a smooth crossover [144]. Thus, as illustrated in Fig. 13a, the smaller the frequency the sharper the kink in the dependence of excited state energy on the interaction constant.

In the undoped case the present results can be directly compared with experiments. It is found that the IE on the ARPES lineshape of a single hole is anomalously enhanced in the intermediate coupling regime while it can be described by the simple independent oscillators model in the strong coupling regime. The shift of FCP top and change of the FCP height are relevant quantities to pursue experimentally in the intermediate coupling regime since IE on these characteristics is enhanced near the self trapping point. In contrast, shift of the leading edge of the broad peak is the relevant quantity in the strong coupling regime since this value increases with coupling as $\sqrt{\lambda}$. These conclusions, depending on whether the self-trapping phenomenon is encountered in specific case, can be applied fully or partially to other compounds with strong EPI [34, 37–39].

6.3 Conclusions and Perspectives

In this chapter, we have focused mainly on the polaron problem in strongly correlated systems. This offers an approach from the limit of low carrier concentration doped into the (Mott) insulator, which is complementary to the conventional Eliashberg-Migdal approach for the EPI in metals. In the latter case, we have the Fermi energy ε_F as a relevant energy scale, which is usually much larger than the phonon frequency Ω_0. In this case, the adiabatic Migdal approximation is valid and the vertex corrections, which correspond to the multi-phonon cloud and are essential to the self-trapping phenomenon, are

---

[8] Results are almost independent of the parameter η of the Gaussian distribution $\mathcal{G}_{\kappa,l}(\omega) = 1/(\eta\sqrt{2\pi})\exp(-[\omega + 4t\lambda - \Omega_0\kappa l]/(2\eta^2))$ in the range $[0.12, 0.2]$.

suppressed by the ratio Ω_0/ε_F. Therefore an important issue is the crossover from the strong coupling polaronic picture to the weak coupling Eliashberg-Migdal picture. This occurs as one increases the carrier doping into the insulator. As is observed by ARPES experiments in high temperature supercon-ductors, the polaronic states continue to survive even at finite doping [177]. This suggests a novel polaronic metallic state in underdoped cuprates, which is common also in CMR manganites [36] and is most probably universal in transition metal oxides. In the optimal and overdoped region, the Eliashberg-Migdal picture becomes appropriate [170, 171], but still a nontrivial feature of the EPI is its strong momentum dependence leading to the dichotomy between the nodal and anti-nodal regions. It is an interesting observation that the high-est superconducting transition temperature is attained at the crossover region between the two pictures above, which suggests that both the itinerancy and strong coupling to the phonons are essential to the quantum coherence. It should be noted that this crossover occurs in a nontrivial way also in the mo-mentum space, i.e., the nodal and anti-nodal regions behave quite differently as discussed in Sect. 6.2. However, the relevance of the EPI to the high T_c superconductivity is still left for future investigations.

We hope that this chapter convinces the reader of the vital role of ARPES experiments and numerically exact solutions to the EPI problem. The com-bination of the two offers a powerful tool for the momentum-energy resolved analysis of these rather complicated strongly correlated electronic systems. This will pave a new path to a deeper understanding of many-body electronic systems.

Acknowledgement. We thank Y. Toyozawa, Z. X. Shen, T. Cuk, T. Devereaux, J. Zaanen, S. Ishihara, A. Sakamoto, N. V. Prokof'ev, B. V. Svistunov, E. A. Burovski, J. T. Devreese, G. de Filippis, V. Cataudella, P. E. Kornilovitch, O. Gunnarsson, N. M. Plakida, and K. A. Kikoin, for collaborations and discussions.

References

1. J. Appel: *Solid State Physics*, Vol. 21, ed by H. Ehrenreich, F. Seitz and D. Turnbull (Academic, New York 1968).
2. S. I. Pekar: *Untersuchungen über die Elektronentheorie der Kristalle,* (Akademie Verlag, Berlin 1954)
3. L. D. Landau: Sow. Phys. **3**, 664 (1933).
4. H. Fröhlich, H. Pelzer, S. Zienau: Philos. Mag. **41**, 221 (1950)
5. J. T. Devreese: *Encyclopedia of Applied Physics* Vol. 14, ed by G. L. Trigg (VCH, New York 1996), p. 383
6. A. I. Anselm, Yu. A. Firsov: J. Exp. Theor. Phys. **28**, 151 (1955); ibid. **30**, 719 (1956)
7. M. Ueta, H. Kanzaki, K. Kobayashi, Y. Toyozawa, E. Hanamura: *Excitonic Processes in Solids*, (Springer-Verlag, Berlin 1986)
8. Y. Toyozawa: Progr. Theor. Phys. **20** 53 (1958).

9. Y. Toyozawa: *Optical Processes in Solids*, (University Press, Cambridge 2003)
10. C. L. Kane, P. A. Lee, N. Read: Phys. Rev. B **39**, 6880 (1989)
11. Yu. A. Izymov: Usp. Fiz. Nauk **167**, 465 (1997) [Physics-Uspekhi **40**, 445 (1997)]
12. J. Kanamori: Appl. Phys. **31**, S14 (1960).
13. A. Abragam, B. Bleaney: *Electron Paramagnetic Resonance of Transition Ions*, (Clarendon Press, Oxford 1970)
14. K. I. Kugel, D. I. Khomskii: Sov. Phys. Usp. **25**, 231 (1982)
15. A. J. Millis, P. B. Littlewood, B. I. Shraiman: Phys. Rev. Lett. **74**, 5144 (1995)
16. A. S. Alexandrov, A. M. Bratkovsky: Phys. Rev. Lett. **82**, 141 (1999)
17. E. I. Rashba: Sov. Phys. JETP **23**, 708 (1966)
18. Y. Toyozawa, J. Hermanson: Phys. Rev. Lett. **21**, 1637 (1968)
19. I. B. Bersuker: *The Jahn-Teller Effect*, (IFI/Plenum, New York 1983)
20. V. L. Vinetskii: Zh. Exp. Teor. Fiz **40**, 1459 (1961) [Sov. Phys. - JETP **13**, 1023 (1961)]
21. P. W. Anderson: Phys. Rev. Lett. **34** 953 (1975)
22. H. Hiramoto, Y. Toyozawa: J. Phys. Soc. Jpn. **54**, 245 (1985)
23. A. Alexandrov, and J. Ranninger: Phys. Rev. B **23** 1796 (1981)
24. A. Alexandrov, and J. Ranninger: Phys. Rev. B **24** 1164 (1981)
25. H. Haken: Il Nuovo Cimento **3**, 1230 (1956)
26. F. Bassani, G. Pastori Parravicini: *Electronic States and Optical Transitions in Solids*, (Pergamon, Oxford 1975)
27. J. Pollman, H. Büttner: Phys. Rev. B **16**, 4480 (1977)
28. A. Sumi: J. Phys. Soc. Jpn. **43**, 1286 (1977)
29. Y. Shinozuka, Y. Toyozawa: J. Phys. Soc. Jpn. **46**, 505 (1979)
30. Y. Toyozawa: Physica **116B**, 7 (1983)
31. B. O. Wells, Z.-X. Shen, A. Matsuura et al: Phys. Rev. Lett. **74**, 964 (1995)
32. A. Danmascelli, Z.-X. Shen, and Z. Hussain: Rev. Mod. Phys. **75**, 473 (2003)
33. X. J. Zhou, T. Yoshida, D.-H. Lee et al: Phys. Rev. Lett. **92**, 187001 (2004)
34. D. S. Dessau, T. Saitoh, C.-H. Park et al: Phys. Rev. Lett. **81**, 192 (1998);
35. N. Mannella, A. Rosenhahn, C. H. Booth et al: Phys. Rev. Lett. **92**, 166401 (2004)
36. N. Mannella, W. L. Yang, X. J. Zhou et al: Nature **438**, 474 (2005)
37. L. Perfetti, H. Berger, A. Reginelli et al: Phys. Rev. Lett. **87**, 216404 (2001)
38. L. Perfetti, S. Mitrovic, G. Margaritondo et al: Phys. Rev. B **66**, 075107 (2002)
39. D. Schrupp, M. Sing, M. Tsunekawa et al: Eur. Phys. Lett. **70**, 789 (2005)
40. R. J. Mc Queeney, T. Egami, G. Shirane and Y. Endoh: Phys. Rev. B **54** R9689 (1996)
41. H. A. Mook, R. M. Nicklow: Phys. Rev. B **20** 1656 (1979)
42. H. A. Mook, D. B. McWhan, F. Holtzberg: Phys. Rev. B **25** 4321 (1982)
43. N. V. Prokof'ev, B. V. Svistunov, I. S. Tupitsyn: J. Exp. Theor. Phys. **114**, 570 (1998) [Sov. Phys. - JETP **87**, 310 (1998)]
44. N. V. Prokof'ev, B. V. Svistunov: Phys. Rev. Lett. **81**, 2514 (1998)
45. A. S. Mishchenko, N. V. Prokof'ev, A. Sakamoto, B. V. Svistunov: Phys. Rev. B **62**, 6317 (2000)
46. E. A. Burovski, A. S. Mishchenko, N. V. Prokof'ev, B. V. Svistunov: Phys. Rev. Lett. **87**, 186402 (2001)
47. A. S. Mishchenko, N. Nagaosa, N. V. Prokof'ev, B. V. Svistunov, E. A. Burovski: Nonlinear Optics **29**, 257 (2002)

48. A. S. Mishchenko, N. Nagaosa, N. V. Prokof'ev, A. Sakamoto, B. V. Svistunov: Phys. Rev. Lett. **91**, 236401 (2003)

49. A. S. Mishchenko, N. Nagaosa: Phys. Rev. Lett. **93**, 036402 (2004)

50. A. S. Mishchenko: Usp. Phys. Nauk **175**, 925 (2005) [Physics-Uspekhi **48**, 887 (2005)]

51. A. S. Mishchenko, N. Nagaosa: J. Phys. Soc. J. **75**, 011003 (2006)

52. A. S. Mishchenko, N. Nagaosa: Phys. Rev. Lett. **86**, 4624 (2001)

53. A. S. Mishchenko, N. V. Prokof'ev, B. V. Svistunov: Phys. Rev. B **64**, 033101 (2001)

54. A. S. Mishchenko, N. Nagaosa, N. V. Prokof'ev, A. Sakamoto, B. V. Svistunov: Phys. Rev. B **66** 020301 (2002)

55. A. S. Mishchenko, N. Nagaosa: Phys. Rev. B **73**, 092502 (2006)

56. A. S. Mishchenko, N. Nagaosa: J. Phys. Chem. Solids **67**, 259 (2006)

57. G. De Fillipis, V. Cataudella, A. S. Mishchenko, J. T. Devreese, C. A. Perroni: Phys. Rev. Lett. **96**, 136405 (2006)

58. J. T. Devreese, in the present volume.

59. A. A. Abrikosov, L. P. Gor'kov, I. E. Dzyaloshinskii: *Quantum field theoretical method in statistical physics* (Pergamon Press, Oxford 1965)

60. G. D. Mahan: *Many particle physics* (Plenum Press, Plenum Press 2000)

61. M. Jarrell, J. Gubernatis: Phys. Rep. **269**, 133 (1996)

62. R. Knox: *Theory of Excitons*, (Academic Press, New York 1963)

63. I. Egri: Phys. Rep **119**, 364 (1985)

64. D. Haarer: Chem. Phys. Lett **31**, 192 (1975)

65. D. Haarer, M. R. Philpot, H. Morawitz: J. Chem. Phys **63**, 5238 (1975)

66. A. Elscner, G. Weiser: Chem. Phys **98** 465 (1985)

67. J. I. Frenkel: Phys. Rev. **17**, 17 (1931)

68. J. H. Wannier: Phys. Rev. **52**, 191 (1937)

69. G. De Filippis, V. Cataudella, V. Marigliano Ramaglia, C. A. Perroni: Phys. Rev. B **72**, 014307 (2005)

70. M. Berciu: cond-mat/0602195

71. J. T. Devreese, L. F. Lemmens, J. Van Royen: Phys. Rev. B **15**, 1212 (1977)

72. P. E. Kornilovitch: Europhys. Lett. **59**, 735 (2002)

73. J. Devreese, R. Evrard: Phys. Lett. **11**, 278 (1966)

74. E. Kartheuzer, R. Evrard, J. Devreese: Phys. Rev. Lett. **22**, 94 (1969)

75. J. Devreese, J. De Sitter, M. Goovaerts: Phys. Rev. B **5**, 2367 (1972)

76. J. T. Devreese: Internal structure of free Fröhlich polarons, optical absorption and cyclotron resonance. In *Polarons in Ionic crystals and Polar Semiconductors* (North Holland, Amsterdam 1972) pp 83–159

77. M. J. Goovaerts, J. M. De Sitter, J. T. Devreese: Phys. Rev. B **7**, 2639 (1973)

78. R. Feynman, R. Hellwarth, C. Iddings, and P. Platzman: Phys. Rev. **127**, 1004 (1962)

79. T. D. Lee, F. E. Low, D. Pines: Phys. Rev. **90**, 297 (1953)

80. E. I. Rashba: Self-trapping of excitons. In *Modern Problems in Condensed Matter Sciences*, vol. 2, ed by V. M. Agranovich and A. A. Maradudin (Notrh Holland, Amsterdam 1982) pp 543–602

81. N. Metropolis, A. W. Rosenbluth, M. N. Rosenbluth, A. M. Teller and E. Teller: J. Chem. Phys. **21**, 1087 (1953)

82. D. P. Landau, K. Binder: *A Guide to Monte Carlo Simulations in Statistical Physics*, (University Press, Cambridge 2000)

83. A. W. Sandvik, J. Kurkijärvi: Phys. Rev. B **43**, 5950 (1991)
84. A. N. Tikhonov, V. Y. Arsenin: *Solutions of Ill-Posed Problems*, (Winston, Washington 1977)
85. E. Perchik: math-ph/0302045
86. D. L. Phillips: J. Assoc. Comut. Mach. **9** 84 (1962)
87. A. N. Tikhonov: DAN USSR **151** 501 (1963)
88. S. S. Aplesnin: J. Exp. Theor. Phys **97** 969 (2003)
89. G. Onida, L. Reining, A. Rubio: Rev. Mod. Phys. **74**, 601 (2002)
90. L. J. Sham, T. M. Rice: Phys. Rev. **144**, 708 (1965).
91. L. X. Benedict, E. L. Shirley, R. B. Bohn: Phys. Rev. Lett. **80**, 4514 (1998)
92. S. Albrecht, L. Reining, R. Del Sole, G. Onida: Phys. Rev. Lett. **80**, 4510 (1998)
93. M. Rohlfing, S. G. Louie: Phys. Rev. Lett. **81**, 2312 (1998)
94. A. Marini, R. Del Sole: Phys. Rev. Lett. **91**, 176402 (2003)
95. W. Stephan: Phys. Rev. B **54**, 8981 (1996)
96. G. Wellein, H. Fehske: Phys. Rev. B **56**, 4513 (1997)
97. H. Fehske, J. Loos, G. Wellein: Z. Phys. B **104**, 619 (1997)
98. H. Fehske, J. Loos, G. Wellein: Phys. Rev. B **61**, 8016 (2000)
99. J. Bonča, S. A. Trugman, I. Batistić: Phys. Rev. B **60**, 1633 (1999)
100. L.-C. Ku, S. A. Trugman, J. Bonča: Phys. Rev. B **65**, 174306 (2002)
101. S. E. Shawish, J. Bonča, L.-C. Ku, S. A. Trugman: Phys. Rev. B **67**, 014301 (2003)
102. O. S. Barisic: Phys. Rev. B **65**, 144301 (2002)
103. O. S. Barisic: Phys. Rev. B **69**, 064302 (2004)
104. A. Georges, G. Kotliar: Phys. Rev. B **45**, 647 (1992)
105. M. Jarrel: Phys. Rev. Lett. **69**, 168 (1992)
106. P. G. J. van Dongen, D. Vollhardt: Phys. Rev. Lett. **65**, 1663 (1990)
107. A. Georges, G. Kotliar, W. Krauth, M. J. Rozenberg: Rev. Mod. Phys. **68**, 13 (1996)
108. S. Ciuchi, F. de Pasquale, S. Fratini, D. Feinberg: Phys. Rev. B **56** 4494 (1997)
109. D. Sénéchal, D. Perez, M. Pioro-Landriére: Phys. Rev. Lett. **84**, 522 (2000)
110. D. Sénéchal, D. Perez, M. Plouffe: Phys. Rev. B **66**, 075129 (2002)
111. M. Hohenadler, M. Aichhorn, W. von der Linden: Phys. Rev. B **68**, 18430 (2003)
112. M. Hohenadler, M. Aichhorn, W. von der Linden: Phys. Rev. B **71**, 014302 (2005)
113. M. Hohenadler, D. Neuber, W. von der Linden, G. Wellein, J. Loos, H. Fehske: *ibid.* **71**, 245111 (2005)
114. S. R. White: Phys. Rev. Lett. **69**, 2863 (1992)
115. S. R. White: Phys. Rev. B **48**, 10345 (1993)
116. S. R. White: Phys. Rev. Lett. **77**, 363 (1996)
117. E. Jeckelmann, S. R. White: Phys. Rev. B **57**, 6376 (1998)
118. G. Hager, G. Wellein, E.Jeckelmann, H. Fehske: Phys. Rev. B **71**, 075108 (2005)
119. P. E. Kornilovitch: Phys. Rev. Lett. **81**, 5382 (1998)
120. P. E. Kornilovitch: Phys. Rev. B **60**, 3237 (1999)
121. P. E. Spenser, J. H. Samson, P. E. Kornilovitch, A. S. Alexandrov: Phys. Rev. B **71**, 184310 (2005)
122. J. P. Hague, P. E. Kornilovitch, A. S. Alexandrov, J. H. Samson: Phys. Rev. B **73**, 054303 (2006)

123. A. S. Alexandrov, P. E. Kornilovitch: Phys. Rev. Lett. **82**, 807 (1999)
124. A. S. Alexandrov, P. E. Kornilovitch: Phys. Rev. B **70**, 224511 (2004)
125. S. Ciuchi, F. de Pasquale, D. Feinberg: Europhys. Lett. **30**, 151 (1995)
126. A. S. Alexandrov, V. V. Kabanov, D. K. Ray: Phys. Rev. B **49**, 9915 (1994)
127. A. S. Alexandrov, J. Ranninger: Phys. Rev. B **45**, 13109 (1992)
128. L. D. Landau, S. I. Pekar: Zh. Eksp. Teor. Fiz. **18**, 419 (1948) [Sov. Phys. JETP **18**, 341 (1948)]
129. V. L. Gurevich, I. G. Lang, Yu. A. Firsov: Fiz. Tverd. Tela (Leningrad) **4**, 1252 (1962) [Sov. Phys. Solid State **4**, 918 (1962)]
130. J. Franck, E. G. Dymond: Trans. Faraday Soc. **21**, 536 (1926)
131. E. U. Condon: Phys. Rev. **32**, 858 (1928)
132. D. N. Bertran, J. J. Hopfield: J. Chem. Phys. **81**, 5753 (1984)
133. X. Urbain, B. Fabre, E. M. Staice-Casagrande et al: Phys. Rev. Lett. **92**, 163004 (2004)
134. M. Lax: J. Chem. Phys. **20**, 1752 (1952)
135. D. I. Khomskii: Usp. Fiz. Nauk **129**, 443 (1979) [Sov. Phys. Usp. **22**, 879 (1979)]
136. C. E. T. Goncalves da Silva, L. M. Falicov: Phys. Rev. B **13**, 3948 (1976).
137. P. A. Alekseev, J. M. Mignot, J. Rossat-Mignot: J. Phys.: Condens. Matter **7**, 289 (1995)
138. K. A. Kikoin, A. S. Mishchenko: J. Phys.: Condens. Matter **7**, 307 (1995)
139. R. Feynman: Phys. Rev. **97**, 660 (1955)
140. F. M. Peeters, J. T. Devreese: Phys. Rev. B **28**, 6051 (1983)
141. V. Cataudella, G. De Filippis, and C.A. Perroni, in the present volume.
142. E. G. Brovman, Yu. Kagan: Zh. Eksp. Teor. Fiz. **52**, 557 (1967) [Sov. Phys. JETP **25**, 365 (1967)]
143. K. A. Kikoin, A. S. Mishchenko: Zh. Eksp. Teor. Fiz. **104**, 3810 (1993) [Sov. Phys. JETP **77**, 828 (1993)]
144. B. Gerlach, H. Löwen: Rev. Mod. Phys. **63**, 63 (1991)
145. A. Brillante, M. R. Philpott: J. Chem. Phys. **72**, 4019 (1980)
146. D. Haarer: Chem. Phys. Lett. **27**, 91 (1974)
147. D. Haarer: J. Chem. Phys. **67**, 4076 (1977)
148. M. Kuwata-Gonokami, N. Peyghambarian, K. Meissner et al: Nature **367**, 47 (1994)
149. S. Curnoe, K. A. Kikoin: Phys. Rev. B **61**, 15714 (2000)
150. K. A. Kikoin, A. S. Mishchenko: Zh. Eksp. Teor. Fiz. **94**, 237 (1988) [Sov. Phys. JETP **67**, 2309 (1988)]
151. K. A. Kikoin, A. S. Mishchenko: J. Phys.: Condens. Matter **2**, 6491 (1990)
152. P. A. Alekseev, A. S. Ivanov, B. Dorner et al: Europhys. Lett. **10**, (1989) 457.
153. A. S. Mishchenko, K. A. Kikoin: J. Phys.: Condens. Matter **3**, 5937 (1991).
154. G. Trawaglini P. Wachter: Phys. Rev. B **29**, 893 (1984)
155. P. Lemmens, A. Hoffman, A. S. Mishchenko et al: Physica B **206&207**, 371 (1995)
156. T. Kasuya: Europhys. Lett. **26**, 277 (1994)
157. T. Kasuya: Europhys. Lett. **26**, 283 (1994)
158. E. Manousakis: Rev. Mod. Phys. **63**, 1 (1991)
159. E. Dagotto: Rev. Mod. Phys. **66**, 763 (1994)
160. P. A. Lee, N. Nagaosa, X. G. Wen: Rev. Mod. Phys. **78**, 17 (2006)
161. B. Batlogg, R. J. Cava, A. Jayaraman et al: Phys. Rev. Lett. **58**, 2333 (1987)

162. O. Gunnarsson, M. Calandra, J. E. Han: Rev. Mod. Phys. **75**, 1085 (2003)
163. R. Khasanov, D. G. Eshchenko, H. Luetkens et.al: Phys. Rev. Lett. **92**, 057602 (2004)
164. L. Pintschovius, M. Braden: Phys. Rev. B, **60**, R15039 (1999).
165. C. Thomsen, M. Cardona, B. Gegenheimer et. al: Phys. Rev. B **37**, 9860 (1988)
166. V. G. Hadjiev, X. Zhou, T. Strohm, et. al: Phys. Rev. B **58**, 1043 (1998)
167. G. Khaliullin, P. Horsch: Physica C **282-287**, 1751 (1997)
168. O. Rösch, O. Gunnarsson: Phys. Rev. Lett. **93**, 237001 (2004)
169. A. Lanzara, P. V. Bogdanov, X. J. Zhou et al: Nature **412**, 510 (2001)
170. T. Cuk, F. Baumberger, D. H. Lu et al: Phys. Rev. Lett. **93**, 117003 (2004)
171. T. P. Devereaux, T. Cuk, Z.-X. Shen, N. Nagaosa: Phys. Rev. Lett. **93**, 117004 (2004)
172. R. J. McQueeney, Y. Petrov, T. Egami et al: Phys. Rev. Lett. **82**, 628 (1999)
173. A. V. Chubukov, M. R. Norman: Phys. Rev. B **70**, 174505 (2004)
174. M. Eschrig, M. R. Norman: Phys. Rev. Lett. **85**, 3261 (2000)
175. M. Eschrig, M. R. Norman: Phys. Rev. B **67**, 144503 (2003)
176. X. J. Zhou, T. Yoshida, A. Lanzara et al: Nature **423**, 398 (2003)
177. K. M. Shen, F. Ronnig, D. H. Lu et al: Phys. Rev. Lett. **93**, 267002 (2004)
178. T. Xiang, J. M. Wheatley: Phys. Rev. B **54**, R12653 (1996)
179. B. Kyung, R. A. Ferrell: Phys. Rev. B **54**, 10125 (1996)
180. J. J. M. Pothuizen, R. Eder, N. T. Hien et al: Phys. Rev. Lett. **78**, 717 (1997)
181. K. A. Chao, J. Spalek, A. M. Oles: J. Phys. C **10**, L271 (1977)
182. C. Gross, R. Joynt, T. M. Rice: Phys. Rev. B **36**, 381 (1987)
183. M. Brunner, F. F. Assaad, A. Muramatsu: Phys. Rev. B **62**, 15480 (2000).
184. V. I. Belinicher, A. L. Chernyshev, V. A. Shubin: Phys. Rev. B **53**, 335 (1996)
185. V. I. Belinicher, A. L. Chernyshev, V. A. Shubin: Phys. Rev. B **54**, 14914 (1996)
186. T. Tohyama, S. Maekawa: Superconductors Science and Technology **13**, R17 (2000)
187. J. Bała, A. M. Oleś, J. Zaanen: Phys. Rev. B **52** 4597 (1995)
188. S. Schmitt-Rink, C. M. Varma, A. E. Ruckenstein: Phys. Rev. Lett. **60**, 2793 (1988)
189. Z. Liu, E. Manousakis: Phys. Rev. B **44**, 2414 (1991)
190. Z. Liu, E. Manousakis: Phys. Rev. B **45**, 2425 (1992)
191. B. Bauml, G. Wellein, H. Fehske: Phys. Rev. B **58**, 3663 (1998)
192. A. Ramšak, P. Horsch, P. Fulde: Phys. Rev. B **46**, 14305 (1992)
193. B. Kyung, S. I. Mukhin, V. N. Kostur, R. A. Ferrell: Phys. Rev. B **54**, 13167 (1996)
194. F. Marsiglio F, A. E. Ruckenstein, S. Schmitt-Rink, C. Varma: Phys. Rev. B **43**, 10882 (1991)
195. G. Martinez, P. Horsch: Phys. Rev. B **44**, 317 (1991)
196. O. Rösch, O. Gunnarsson: Phys. Rev. B **73**, 174521 (2006)
197. A. S. Mishchenko: Pis'ma Zh. Eksp. Teor. Fiz. **66**, 460 (1997) [JETP Lett. **66**, 487 (1997)]
198. O. Rösch, O. Gunnarsson: Europhys. Phys. J. B **43**, 11 (2005)
199. G. Sangiovanni, O. Gunnarsson, E. Koch, C. Castellani, M. Capone: cond-mat/0602606.
200. O. Rösch, O. Gunnarsson: Phys. Rev. B **70**, 224518 (2004).
201. G.-H. Gweon, T. Sasagawa, S. Y. Zhou et al: Nature **430**, 187 (2004)

202. O. Rösch, O. Gunnarsson, X. J. Zhou et al: Phys. Rev. Lett. **95**, 227002 (2005)
203. G. A. Sawatzky: Nature (London) **342B**, 480 (1989)
204. O. Rösch, O. Gunnarsson: Phys. Rev. Lett. **92**, 146403 (2004)
205. S. Ishihara , N. Nagaosa: Phys. Rev. B **69**, 144520 (2004)

Part IV

Polarons in Contemporary Materials

Photoinduced Polaron Signatures in Infrared Spectroscopy

Dragan Mihailovic

Jozef Stefan Institute and International Postgraduate School, SI-1000, Ljubljana, Slovenia `dragan.mihailovic@ijs.si`

Summary. We present a brief review of photoinduced infrared absorption (PIA) measurements applied to polaronic systems. The experimental technique and theoretical models are discussed, together with a few examples where the technique has been succesfully applied, particularly conducting polymers and oxides such as cuprates, manganites and tungstates. The symmetry breaking effects of carriers and the appearance of local modes is shown to give useful information on carrier localisation. The very small experimentally observed isotope effect is also shown to be consistent with theory.

1 Introduction

When a photon is absorbed in matter, it excites an electron into an unoccupied state at an energy $\hbar\nu$ above the occupied state from which it is excited. This can cause many different effects on different energy scales and on vastly different timescales. Commonly, the material emits secondary light in the form of luminescence at frequencies which are lower than $\hbar\nu$ due to direct radiative recombination of the electron and hole. Alternatively, the electron and hole can recombine non-radiatively by scattering with other electrons and losing energy by scattering from phonons. These processes take place immediately after photon absorption and are very fast, occuring on the timescale of tens of femtoseconds to tens of picoseconds.

In many functional materials such as conducting polymers, cuprates, manganites and tungstates, the photoexcited carriers become self-trapped to form polarons. These polarons can cause a change in the infrared spectrum which is detectable by standard infrared spectrometry techniques. The measurement of photoinduced changes in infrared absorbance is experimentally reasonably straightforward with current technology and many different materials have been investigated in this way up till now.

The measurement of photoinduced absorption in polaronic materials can give information on polaron binding energies, their lifetimes, and also give microscopic information on phonon coupling to the photoinduced carriers. The photoinduced vibrational spectrum may also reveal information on the carrier localisation site.

The spectra observed in infrared spectroscopy involve dipolar optical transitions, which means that only charge excitations can be probed. This has been a crucial point in the assignment of the mid-infrared spectral features to polarons in high-temperature superconducting cuprates for example [1] as will be discussed later in this chapter.

The observation of a characteristic polaron signature is relatively unambiguous and can be used as proof for the existence of polarons, but the opposite is not necessarily true: The non-observation of the characteristic polaronic spectrum does not rule out the existence of polarons, but merely implies the non-existence of long-lived metastable polaronic states. Complementary information to photoinduced absorption on carrier mobility can be obtained by measurement of photoconductivity, and the two techniques have often been used together [2].

The technique of photoinduced absorption has been extended in recent years to time-resolved spectroscopy, enabling short-lived photoinduced states to be probed with resolution of a few tens of femtoseconds. These can be performed either one wavelength at a time, or over a wide range of wavelengths, giving spectral information on the evolution of photoinduced states with time. Other related techniques for detection of nonequilibrium states are electroabsorption or electroreflectance and photoinduced dichroism. These do not directly give information on polaronic states and will not be discussed here.

In this review we concentrate on photoinduced infrared absorption studies performed on oxides with long-lived localised and self-trapped states, presenting a brief overview of the current status of experiment and theory, discussing a few reperesentative examples of polaronic features in functional materials. The non-photoinduced optical response of polarons has been reviewed recently by Calvani [3], which gives a complementary overview to the one presented here.

2 Experimental Technique

Photoinduced infrared spectroscopy is conceptually rather simple. Typically, transmission spectra are recorded with and without laser illumination of the sample using a stabilized Fourier transform infrared spectrometer. The crucial experimental consideration is drift stability, which determines the experimental parameters such as scan repetition rate and scan time. The experimental setup is shown in Fig. 1.

The sample is usually kept in a cryostat, and the laser excitation wavelength can vary, depending on the absorption spectrum of the material under investigation. The samples are typically dispersed in KBr or Nujol in the forms

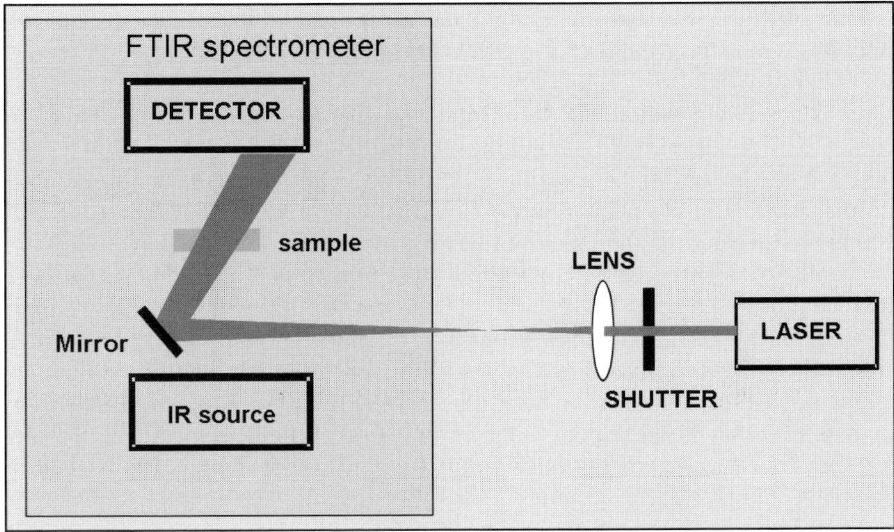

Fig. 1. The typical experimental setup for measuring photoinduced absorption. The sample is usually placed in a cryostat. The spectra with laser on and off are repeated over many cycles to eliminate the drift of the spectrometer.

of powders, where the concentration is adjusted to obtain the correct optical density D. D should be adjusted to allow small changes in the IR spectrum of the order of 10^{-6} to be observed without saturation of the spectra (which can produce spurious spectral features). The size of the grains of the material in the KBr dispersion also needs to be carefully adjusted, because scattering from micron-sized grains can strongly interfere with the photoinduced spectrum, particularly at frequencies above 4000 cm^{-1} (0.5 eV). The problem can be reduced by performing systematic measurements with different grain sizes and identifying the scattering contribution to the spectrum. The photoinduced transmission is given by: $\Delta T = (T_{\text{on}} - T_{\text{off}})$ where T_{off} and T_{on} are the transmission of the sample with laser on and off respectively. The photoinduced absorption A is related to the photoinduced transmission via $T = e^{-A}$, where A is the absorbance, so $\Delta T = -\Delta A = -(A_{\text{on}} - A_{\text{off}})$.

The incident photons can in principle create bound electron-hole pairs (self-trapped excitons), or the carriers may remain independently trapped as electron and hole polarons. The two cases can be distinguished experimentally by measurements of the intensity of the photoinduced response as a function of excitation laser intensity. We may model the process by assuming that the photoexcited carrier density n_{pe} is governed by the relaxation rate equation:

$$\frac{dn_{pe}}{dt} = -R + G \tag{1}$$

where R is the recombination rate, and G is the laser generation rate. In terms of the laser fluence F, $G = \alpha F$. In the steady state ($R = G$) (relevant for continuous laser excitation) n_{pe} is time-independent, so we can distinguish between single particle and biparticle recombination. Since $R \propto F \propto n_{pe}$ for single-particle recombination (e.g. for excitons) and $R \propto F \propto n_{pe}^2$ for biparticle recombination, the laser intensity dependences will be different. In the case of single-particle recombination, we can write $\frac{dn_{pe}}{dt} = -n_{pe}/\tau + G$, which has a steady state solution $n_{pe} = G\tau$, so the measured photoinduced absorption (PIA) is linearly proportional to the product of the lifetime and the laser intensity. In the case of biparticle recombination (where the electron and hole are uncorrelated), the PIA shows a square-root dependence on laser intensity.

Laser heating is a common problem with the PIA technique, particularly when photoinduced optical modes are being investigated. The shift of anharmonic optical phonons due to laser heating can be quite substantial, resulting in derivative-like lineshapes in the phonon region of the PIA spectrum. By performing PIA measurements at different temperatures, the heating effect can be easily identified and accounted for by comparing the spectra at different temperatures to the photoinduced spectra.

3 Theoretical Models of the Infrared Response due to Polarons

The photoinduced absorption measures the response of a dilute photoexcited polaron system, and the appropriate theory which describes the situation is for non-interacting polarons. In normal IR spectroscopy, where doping levels may be quite high, this limit may no longer apply, and the spectra may show deviations from simple polaron theory described in the following paragraphs. The first calculations of the polaronic response in the infrared spectrum were given by Reik [4] and Eagles [5] and were later discussed by Devreese [6], Alexandrov et al. [7], Mischenko et al. [8] and generalised by Emin [9] to include both small and large polarons. The processes leading to the infrared absorption of a small polaron are shown in Fig. 2.

The spectrum reflects the hopping of trapped charges, which emit and re-absorb phonons in the hopping process. A carrier is excited from its self-localised state to an adjacent site. For equivalent sites, the difference in energy is just equal to the energy associated with small polaron formation $-2E_p$. The predicted spectrum contains information on which phonons are involved in the hopping process, and in principle could reveal which phonons are involved as well as the symmetry of the electron-phonon interaction. However, typically very little structure is observed in the photoinduced spectrum, indicating that the charges usually do not couple only to a single dispersionless mode, but to many modes, and at different wavevectors q, resulting in a smeared-out spectrum.

The frequency-dependent conductivity in the low temperature limit $\sigma_{PT}(\omega, T \to 0)$ is given by Reik [4] as:

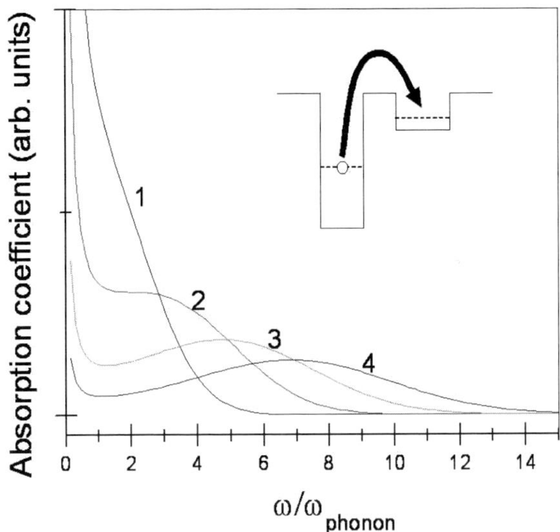

Fig. 2. The small polaron absorption spectrum for different values of $E_\mathrm{p}/\omega_\mathrm{ph}$. The inset shows the process responsible for the infrared absorption

$$\sigma_{PT}(\omega, T \to 0) = \frac{\sqrt{2\pi}t^2 e^2 N}{\hbar^3 \omega_0^2} e^{-\eta} \left(\frac{\omega_0}{\omega}\right)^{\omega/\omega_0} e^{(\omega/\omega_0)} \eta^{\omega/\omega_0} \tag{2}$$

Where η is the number of phonons in the polaron cloud, ω_0 is the average phonon frequency, t is the electronic overlap integral and N is the number density of the carriers of charge e, η and ω_0 are parameters which describe the electron–phonon interaction, and are related to the the electron–phonon coupling constant $\alpha_j(\mathbf{q})$ by:

$$\omega_0 = \frac{\sum\limits_{j,\mathbf{q}} \alpha_j(\mathbf{q}) \sin^2(\tfrac{1}{2}q)\omega_j^2(\mathbf{q})}{\sum\limits_{j,q} \alpha_j(\mathbf{q}) \sin^2(\tfrac{1}{2}q)\omega_j(\mathbf{q})} \tag{3}$$

and

$$\eta = \frac{1}{\omega_0} \sum\limits_{j,\mathbf{q}} 2\alpha_j(\mathbf{q}) \sin^2(\tfrac{1}{2}q)\omega_j(\mathbf{q}).$$

The sum is over $-\pi < \mathbf{q} < \pi$ for the jth optical branch. The characteristic phonon frequency ω_0 represents an averaged frequency of the phonons involved in the hopping, weighted by the electron-phonon interaction. If the dispersion of the phonon is small, η can be simplified to:

$$\eta \approx \sum\limits_{j,\mathbf{q}} 2\alpha_j(\mathbf{q}) \sin^2(\tfrac{1}{2}q),$$

which represents a weighted zone average of $\alpha_j(\mathbf{q})$ over $\mathbf{q}$. Reik's formulation explicitly gives the spectrum in terms of the $\mathbf{q}$-dependent electron–phonon coupling.

A more compact form has been suggested by Emin, where the photoinduced change in the infrared transmission spectrum (which is assumed to be proportional to the photoinduced absorption) is given by:

$$-\frac{\Delta T_{\mathrm{Pl}}}{T} \propto \alpha \propto \frac{1}{\hbar\omega} \exp\left(-\frac{(2E_{\mathrm{p}} - \hbar\omega)^2}{4E_{\mathrm{p}}\hbar\omega_0}\right). \tag{4}$$

The predicted small polaron spectrum (the spectral envelope) is the same as predicted by Reik. Emin's derivation also includes a $1/\omega$ divergence which is usually ignored as unphysical, since the model does not apply for low freqencies $\omega < \omega_0$. The MIR spectrum is quite characteristic, exhibiting a broad feature which peaks at a frequency $\hbar\omega = 2E_p$. This is true for strong coupling (small polaron limit). For weaker coupling, the peak shifts to lower frequency, and becomes less distinct, eventually disappearing towards zero frequency [9] as shown in Fig. 2.

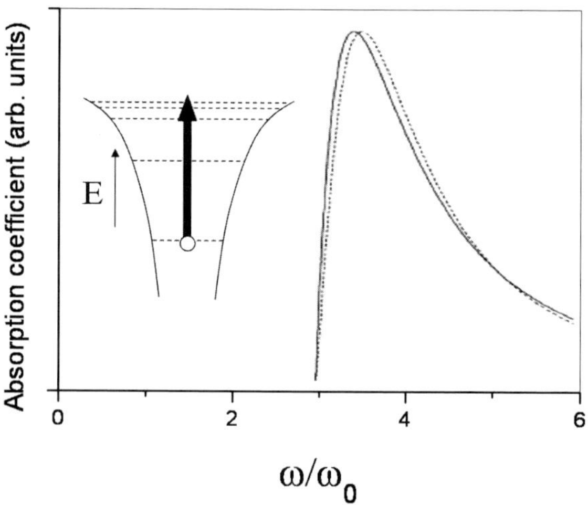

Fig. 3. The absorption coefficient for a large polaron shows an onset at $3E_p$, where E_p is the large polaron binding energy. The inset shows the electronic transition from the bound polaronic state to the continuum of unbound states.

Typically, in photoinduced absorption measurements, the small polaron feature is observed as a symmetric peak, suggesting that photoexcited polarons which are observed by the PIA technique are strongly coupled to the lattice and hence are localised.

The infrared signature for large polarons is qualitatively different, which reflects the different processes involved (see inset to Fig. 3). In this case the self-trapped carrier is excited from the ground states of the potential well in which it is self-trapped into the continuum of unbound states.

As we can see from the spectrum in Fig. 3, the large polaron spectrum is expected to be quite different than in the case of small polarons and is shown in Fig. 2. The spectrum shows an onset at $3E_\mathrm{p}$, where E_p is the polaron binding energy. Spectra of large polarons have been discussed very often in infrared reflectance spectroscopy [10], but not in photoinduced spectroscopy, so we shall not discuss it further here.

The predictions of Reik [4] were also confirmed by Alexandrov, Kabanov and Ray [7] on the basis of cluster calculations, particularly for the strong coupling regime. The authors also conclude that in the intermediate coupling regime the results differ regarding the position of the peak which is more asymmetric than in the strong coupling case. Spectral structure is expected to be observed corresponding to the different spectral weight of states with a different number of phonons involved.

They also predict that the small polaron spectrum should be more or less temperature independent. This prediction is particularly relevant, because experimentally, the photoinduced spectra appear to be T-independent. The drop in intensity at high temperatures is usually attributed to the change of the lifetime of the photoinduced carriers due to thermal excitation, but systematic studies of the T-dependence of the lifetime do not exist at present.

Devreese and co-workers developed an all-coupling theory of optical absorption for Fröhlich polarons [6]. The key predictions of the theory have been confirmed by the recent numerical and analytical calculations as shown [11]. In particular, the main peak positions are in good agreement with the results of Monte-Carlo calculations [12].

3.1 Photoinduced Local Modes (PILMs)

In addition to the MIR electronic spectrum produced by the polaron, the lattice vibrations in the immediate vicinity of the self-trapped charge may be modified as a result of the lattice deformation surrounding the polaron. The photoinduced spectrum therefore contains information on the frequency shifts and other signatures of the lattice deformations. The polaron may also cause the lattice to change its symmetry. For example if a polaron is formed on a site which is not at the centre of inversion of a centrosymmetric lattice, locally it will cause the loss of the inversion lattice symmetry, which means that new modes may become visible by infrared spectroscopy. Such an example has been observed in the case of cuprates and will be discussed in some detail below. Generally, the new local modes associated with polaronic distortions are difficult to assign very accurately because PIA spectroscopy is always performed in powder samples and it is difficult to obtain any detailed information on selection rules. Nevertheless, to some extent inelastic neutron diffraction can provide complementary information on local modes, and again cuprates are a good example of this. Another problem associated with the identification of local modes in photoinduced spectroscopy arises in assigning the trapped site for both the electron *and* the hole (both should in princinple be observable).

4 Some Examples

4.1 Conducting Polymers

Materials in which polarons have been discussed in greatest detail are conducting polymers. There is a general concensus that carriers introduced into conducting polymers can form polarons, and photoinduced spectroscopy has been a very useful tool in their study. The polaronic features are generally very universal, as is shown in a set of spectra for different derivatives (DOO-, DHO-, MEH- and DMO-) of poly-phenyline-vinyline (PPV) [13] (Fig. 4). The measurements were used to determine the polaronic character of the doped charges, while the red shift of the IRAV mode frequencies in the alkoxy derivatives of PPV indicated that the binding energy of bipolarons decreases as the length of the side chain increases. An interesting feature of the spectra of the PPV derivatives is the fact that overtones are observable at frequency multiples of the vibrational frequencies, which is clear confirmation that the spectra are of polaronic origin, and that the electron–phonon (vibronic) interaction is strong in these systems. Many similar systems have been investigated by PIA by numerous groups, showing similar qualitative features to the example shown here.

4.2 Cuprates

The first photoinduced experiments on cuprates were performed by Kim et al [14] and Ruani et al [15]. They observed infrared photoinduced vibrational modes, which they attributed to self-trapped states. Subsequent experiments revealed the presence of a polaronic mid-infrared feature which was ubiquitously present in undoped cuprates [1, 16] (Fig. 5). The authors concluded that the photoinduced infrared conductivity of dilute carriers photoinjected into $Tl_2Ba_2CaCu_2O_8$, $YBa_2Cu_3O_{7-\delta}$ and La_2CuO_4 is well described by nonadiabatic polaron hopping (polaron transport theory). The similar spectral shape and systematic trends in both photoinduced $\sigma_P(\omega)$ and chemically doped $\sigma(\omega)$ indicate that carriers in the concentrated (metallic) regime retain much of the character of carriers in the dilute (photoexcited) regime, implying that the charge carriers in the normal state of high-T_c cuprates are polarons or bipolarons. Furthermore, they concluded that the absence of a shift in the mid-infrared spectrum at the superconducting transition suggests that "if bipolaronic superconductivity is present in the cuprates, bipolarons are formed above T_c".

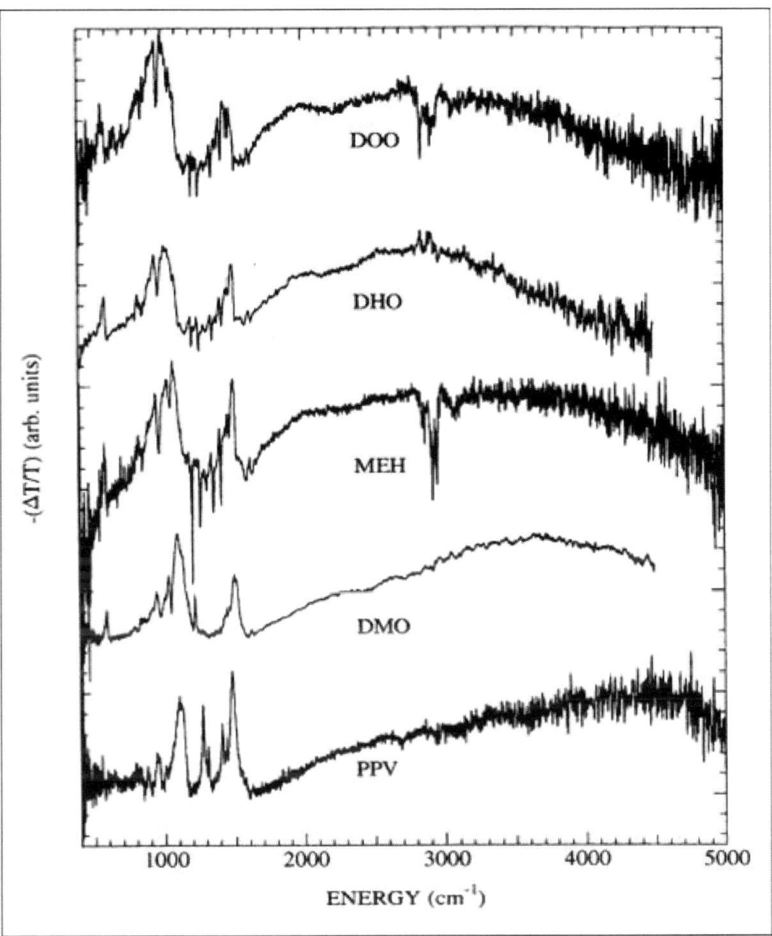

Fig. 4. Infrared photoinduced absorption spectra of PPV and its alkoxy derivatives. The spectra have been normalised in intensity. From [13].

An investigation of the recombination dynamics of photogenerated charge carriers in La_2CuO_4 and Nd_2CuO_4 by Kim, Cheong and Fisk [17] revealed that the spectrum is actually a superposition of two peaks, which show different dynamics. The low-energy (LE) feature, (with a peak energy at around 0.12 eV in LCO and 0.16 eV in NCO) shows a square-root dependence on photoexcitation intensity I above 45 K, $\Delta T/T \propto I^{1/2}$, indicative of biparticle recombination. At 4.2 K on the other hand a fourth-root dependence $\Delta T/T \propto I^{1/4}$ is observed. Giving one possible explanation for this important observation, the authors have suggested that the LE peak might originate from the recombination of momentum-space-paired holes in the CuO_2 planes with trapped electrons. An alternative, real-space pairing model [18] involves

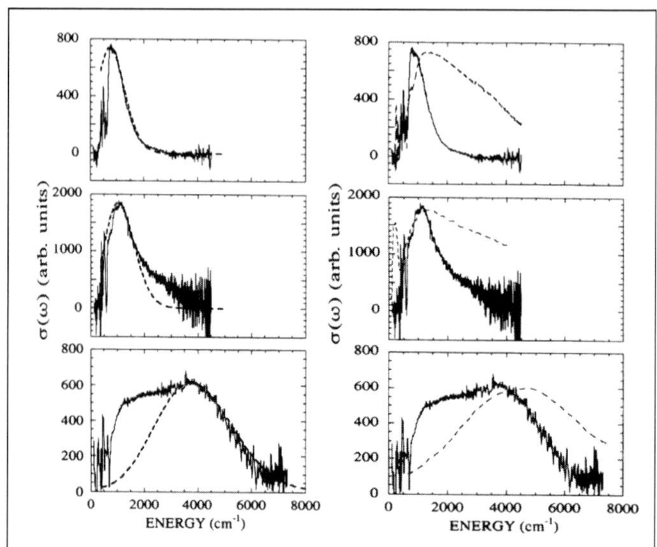

Fig. 5. Infrared photoinduced absorption spectra of three undoped cuprates $Tl_2Ba_2CaCu_2O_8$ (top), $YBa_2Cu_3O_{7-\delta}$ (middle) and La_2CuO_4 (bottom). The left panel shows the data and a fit using Reik's model. The right set of panels shows a comparison of the mid-IR feature in doped cuprates compared to the PIA spectra in the pristine materials. From [1].

the recombination of polarons and bipolarons with their charge compensating electrons outside the CuO_2 planes. In this model, it is not necessary to have momentum-paired states. It is sufficient that the photoexcited electrons and holes form polarons and bipolarons at high density. A cross-over from single polarons to bipolaron clustering with increasing carrier density is indeed predicted by Monte-Carlo calculations of the JT mode [23]. This behaviour can explain the cross-over from bi-particle recombination to four-particle recombination and the change of intensity dependence exponent from $1/2$ to $1/4$. Experimentally, the cross-over from 2-particle to 4-particle recombination dynamics (corresponding to e–h recombination and 2e–2h recombination respectively) has so far been observed in La_2CuO_4, $YBa_2Cu_3O_6$ and Nd_2CuO_4 [19].

The temperature dependence of the amplitude of the LE peak can then be explained well by the two level system, such as the intersite JT polaron model [26]. The T-dependence of the amplitude of the LE feature is shown in Fig. 6 [26].

A systematic investigation of photoinduced infrared excitations performed by Kim and co-workers [19] also showed that the LE peak shifts slightly with temperature, with a similar temperature dependence as the intensity. The shift occurs at a temperature which is close to the critical temperature $T_c = 38$ K of the *doped* $La_{2-x}Sr_xCuO_4$ system. Similar behaviour was observed in

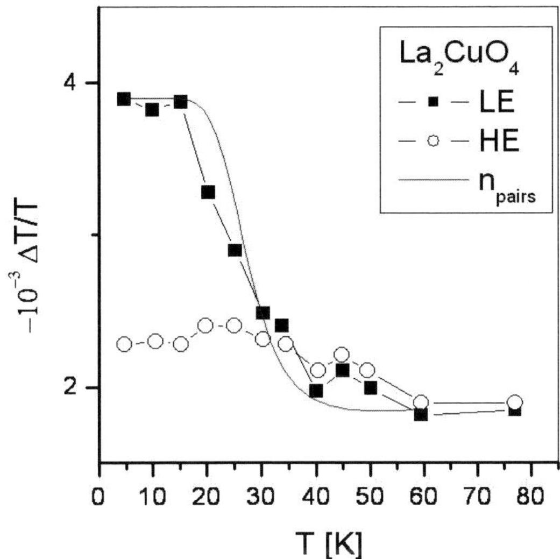

Fig. 6. The peak PIA transmission intensity as a function of temperature in La$_2$CuO$_4$ for the LE and HE peaks shown by square and circles respectively. The LE peak shows a characteristic, bipolaronic two-level system temperature dependence (line is a fit to the data using (4) from the paper by Kabanov and Mihailovic [24] with a gap E$_g$ = 0.06 eV obtained from the position of the LE PIA peak in Fig. 5. In contrast, the HE peak shows very weak temperature dependence, suggesting that it is associated with polarons localised on impurities, or on the surface.

YBCO, where the shift occurs near 90 K, which is near the superconducting T_c of optimally doped YBCO. This is quite remarkable, since the system is completely *undoped* in both cases. The implication is that the photoexcited carriers are forming bound states with the same characteristic energy scale, as they are upon chemical doping. The shift in frequency implies that there is a slight difference in binding energy between the polaron state and bipolaron state ($\sim$ 120 cm^{-1} in Nd$_2$CuO$_4$ and $\sim$ 220 cm^{-1} in La$_2$CuO$_4$).

More recent PIA measurements [20] on different sized nano-grained samples have shown that the two peaks in the MIR spectrum of La$_2$CuO$_4$ can be assigned to surface and bulk polarons respectively. The intensity of the 0.5 eV peak in the MIR was found to scale with the surface area, while the 0.12 eV peak did not show any such scaling. In fact, the intensity of the 0.5 eV peak was found to scale in the same way as the ESR signal from surface states of Cu^{2+}, which confirms the assignment of the 0.5 eV band to surface states in La$_2$CuO$_4$.

Further confirmation of the assignment of the MIR peak to polarons has recently come from measurements of the MIR reflectivity in a magnetic field by Dordevic et al [21], who found that the infrared reflectance of the MIR does not show any change up to 33 T to a very high degree of accuracy, effectively ruling out magnetic excitations as the origin of the MIR spectrum in cuprates.

The spectra of electron-doped cuprates such as Nd_2CuO_4 were studied by Calvani et al, showing similar features as the hole-doped cuprates [22], who also studied local modes and overtones which were attributed to local modes strongly coupled to the photoinduced carriers.

The photoinduced local modes in $YBa_2Cu_3O_7$, $Tl_2Ba_2Gd_2Cu_2O_8$ and La_2CuO_4 have been analysed in detail by Mihailovic et al. [23]. By comparison with Raman spectra of the pristine (undoped) materials and the doped (superconducting) phase of the same materials it was found that the Raman frequencies of the *undoped* material did not correspond to the PI mode frequencies, but a number of high-frequency modes appeared which were close in frequency to the A_g-symmetry Raman modes observed in the *metallic* (superconducting) phase (Fig. 7). The local symmetry and structure surrounding the photoexcited charges thus resemble the doped phase of the material much better than the parent insulator.

The conclusion that could be drawn was that the photoexcited holes localised in the CuO_2 planes, while the electrons were trapped in the intemediate dielectric layers. This creates a dipole moment along the c axis and caused a loss of inversion symmetry of the local site. The consequence is that previously symmetric Raman vibrations acquire a dipole moment which was observed in the photoinduced infrared spectra.

More recent PIA measurements on La_2CuO_4 as a function of annealing and isotope effect clearly showed an isotope effect whose magnitude was in agreement with the expected shift on the basis of the mass difference, with the conclusion that the photoinduced spectral features were indeed due to local vibrations and not of some other origin (see last section).

In their original work [23] the authors also reported that the photoinduced data local modes show a clear correspondence with a phonon anomaly observed in inelastic neutron scattering [24]. Neutron scattering data reveal that a new high-frequency in-plane "half-breathing" O mode appears upon doping. The mode appears approximately in the middle of the Brillouin zone, and extends to the zone boundary [24, 25]. This in-plane O half-breathing mode has later been interpreted as the signature of polaron formation [26]. Thus the neutron data on $La_{2-x}Sr_xCuO_4$ are particularly useful for comparison with the photoinduced modes. A clear correspondence is seen between the photoinduced bleaching (black arrow in Fig. 8) and the zone-center mode which appears upon doping in neutron scattering. The bleaching of this mode is indeed expected, since the intensity of the high-frequency zone-center branch of the mode is depleted upon doping, so photodoping should also deplete the intensity of the corresponding IR vibration. Similarly, an induced absorption is observed where the observed effect of the doped carriers is to increase IR mode intensity (white arrow in Figs. 7 and 8). However, these are not as clear

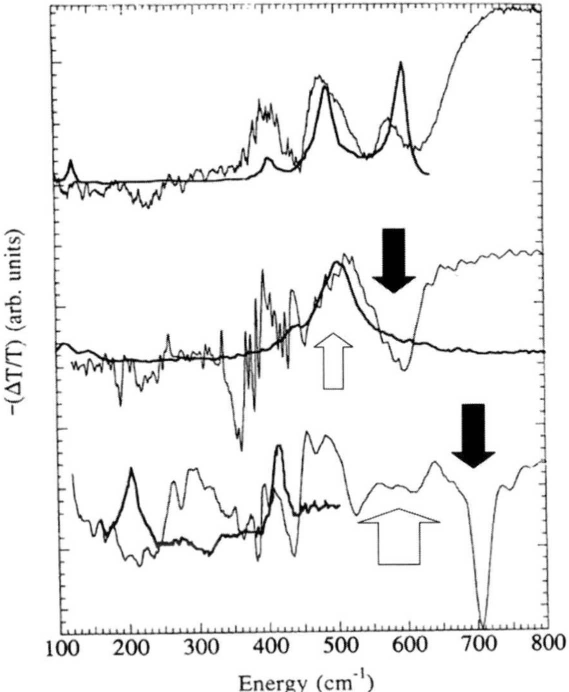

Fig. 7. The photoinduced local mode spectra for $Tl_2Ba_2Gd_2Cu_2O_8$ (top), $YBa_2Cu_3O_7$ (middle) and La_2CuO_4 (bottom). The bold lines are the Raman spectra of the complementary doped (superconducting) materials. The bold arrow indicates the position of the anomalous neutron half-breathing mode (HBM) at the zone centre. The white arrow shows the position of the HBM at the zone boundary.

as the bleachings, presumably because the new local modes are not as well defined, but form a more continuous set of IR bands.

4.3 Manganites

Manganites are systems in which polarons have been extensively studied by various means, and is an ideal system for photoinduced absorption studies. In these materials there appears to be no controversy as to the origin of the PIA spectra, as there has been in the cuprates. Thus, the photoinduced spectrum (Fig. 9) in weakly hole-doped $(La_{1-x}Sr_xMn)_{1-d}O_3$ (LSMO) is unambiguously attributed to photon-assisted hopping of anti-Jahn-Teller polarons [27].

The small polaron model of Emin [9] is seen to give a very good fit to the data. The polaron binding energy E_p was found to be in agreement with other measurements and from the change in the peak position with doping, it was concluded that the binding energy of the polarons decreases with increasing doping. No shifts of the spectra were observed with temperature, in agreement

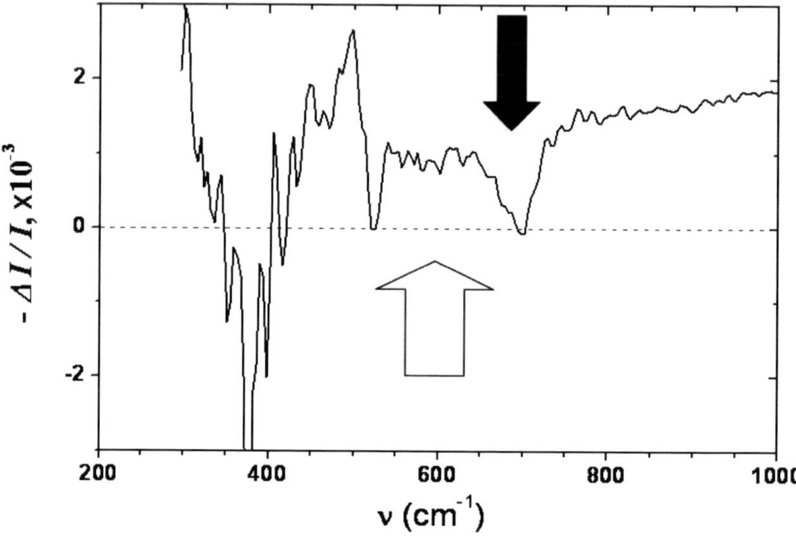

Fig. 8. The photoinduced local mode spectra for La_2CuO_4. The bold arrow indicates the position of the anomalous neutron half-breathin mode (HBM) at the zone centre which is seen as a bleached line, while the white arrow shows the position of the HBM at the zone boundary, which appears as an absorption. The PILMs at lower energy have not been unambiguously assigned.

with the prediction of small polaron theory. In addition, no changes in the photoexited spectra were observed around the Neel temperature. We note here that the normal IR spectra do not show such agreement between E_p from the infrared and activation energy measurements from transport. The MIR spectra were interpreted in terms of small polarons [28] for $La_{2/3}Ca_{1/3}MnO_3$, but large polarons in $La_{2/3}Sr_{1/3}MnO_3$.

From the square-root dependence of the PI peak amplitude on the laser photoexcitation intensity, Fig. 10, the process of recombination was determined to be biparticular, which means that two independent particles interact during recombination. Hence the PI absorption corresponds to excitations of independent electron-like and hole-like carriers created in the photoexcitation process, with no evidence for pairing and clustering in this range of intensities, as was observed in the case of the cuprates.

4.4 Tungstates

Recent indications of possible surface superconductivity at 91 K in sodium tungsten bronze Na_xWO [29] has enhanced experimental interest in the idea

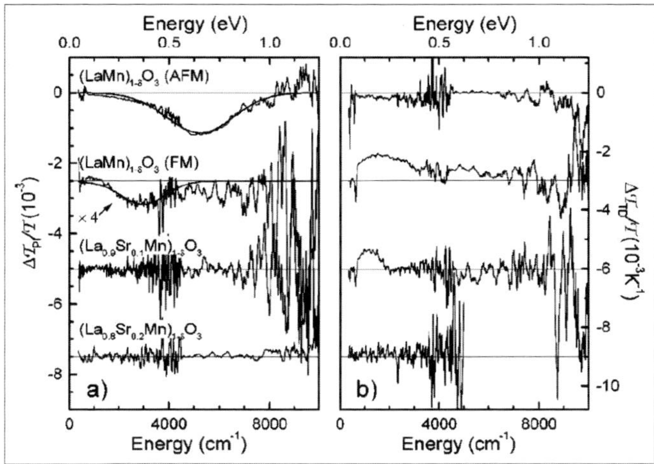

Fig. 9. The low-temperature (25K) photoinduced spectra for (La,Sr,Mn)O$_3$ for different doping levels x and δ (left). The spectra on the right are thermal difference spectra for comparison. The lines represent the theoretical small polaron absorption spectra fit to the data.

of high-temperature bipolaronic superconductivity, which had led to discovery of cuprate high temperature superconductors [30]. Measurements of polaron properties in this material may therefore be important for reaching a deeper understanding of the phenomenon. Polarons and bipolarons in tungsten bronzes were studied in the past with electron spin resonance and optical spectroscopy [31]. An optical absorption peak found at 0.71 eV in tungsten

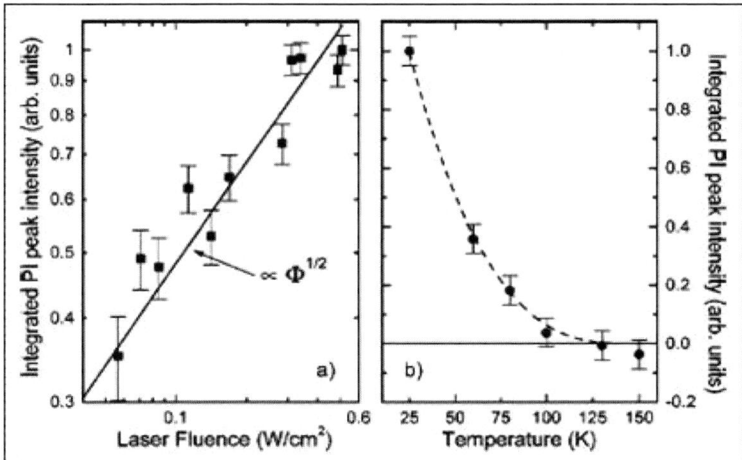

Fig. 10. The photoinduced polaron peak intensity as a function of laser fluence Φ (a) and temperature (b) in LSMO The solid line in (a) is $\sqrt{\Phi}$ fit to the data.

trioxide WO was attributed to photon assisted hopping of small polarons [31] (Fig. 11). It was also shown that bipolarons present in oxygen deficient WO appear to dissociate under light illumination forming single polarons.

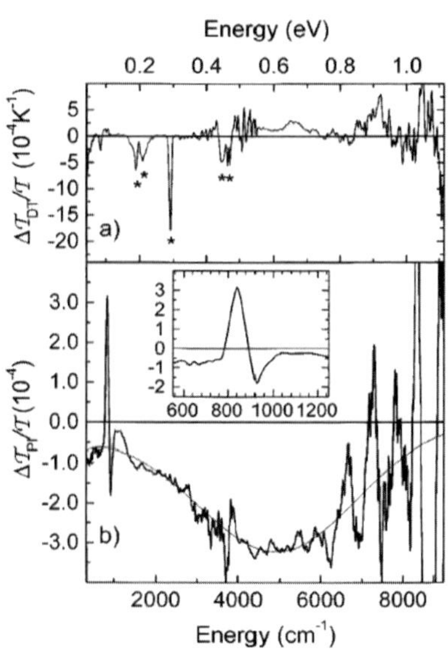

Fig. 11. The photoinduced and thermal difference spectrum of WO_3 at 25 K. The line is a theory fit to the small polaron model. The insert shows the PILM spectrum. Stars indicate water vapour and CO_2 artifacts.

A photoinduced mid-infrared small polaron peak centered at 4800 cm^{-1} (0.59 eV) was observed in photoinduced IR transmittance spectra measured in WO_{3-x} powder. The position of the PI feature is similar to the peak position previously reported in IR absorption measurements of WO_{3-x} single crystals The shape of the PI spectrum in WO_{3-x} is consistent with photon-assisted small-polaron hopping and the polaron binding energy of $E_p = 345 \pm 20$ meV is extracted from the theoretical fit. The binding energy is similar to that extracted from analyses of photoinduced absorption measurements in different undoped high T_c cuprates, implying some common physics in these materials. In comparison with the PI spectra of the parent insulating compounds of high T_c cuprates, the PI spectra in WO_{3-x} show a simpler structure, showing a single-component spectrum without the low frequency peak around 1000 cm^{-1}. Nevertheless, the high frequency part of PI spectra of high T_c cuprates have similar peak energies, although the PI absorption is broader in some cuprates. A similar peak energy in WO_3 implies a similar polaron binding energy as for the surface polaronic states in La_2CuO_4.

4.5 The ^{18}O Isotope Effect on the PIA in La$_2$CuO$_4$

The origin of the mid-infrared spectrum in the case of cuprates has been controversial ever since their discovery, and although the undoped cuprates showed unambiguous polaronic behaviur, the possibility of magnetic origin could be dismissed if the polaronic spectrum could be shown to have an isotope effect. The crucial confirmation of a phonon-mediated mechanism for superconductivity in low-T_c metallic superconductors was given by the existence of an isotope effect on T_c. Similar experiments on the isotope effect in cuprate superconductors have shown behaviour which is more complicated than in the metallic superconductors. One clear conclusion has been that the BCS mechanism is not operative, since superconducting T_c was found to be doping dependent. Moreover, a superconducting gap was not observed in HTS materials, but rather a suppression of low-frequency spectral weight starting well above T_c, and associated with a characteristic pseudogap temperature T^*. Isotope effect measurements revealed features on superconducting cuprates in the infrared which showed rather strong changes in the low-frequency spectra and very little effect in the mid-infrared. One possible method to elucidate the origin of the infrared spectrum in cuprates might be to measure the photoinduced polaronic spectrum with different isotopes. Unfortunately, both theory and experiment show that the polaron spectrum in the infrared is nearly independent of O isotope, and so the measurements of PIA cannot be used to prove the existence of polarons.

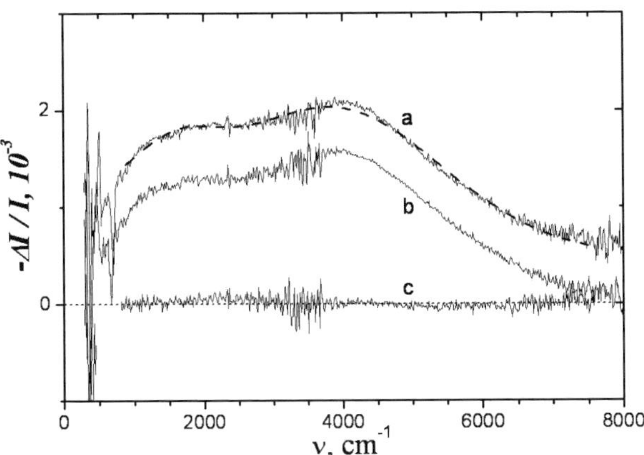

Fig. 12. The normalized PIA spectra of La$_2$CuO$_4$ samples; a) La$_2$Cu$_{16}$O$_4$ (shifted), b) La$_2$Cu$_{18}$O$_4$, c) the difference between a and b; the fit of the La$_2$Cu$_{16}$O$_4$ PIA spectrum with (1) is shown by the dashed line (the parameters are shown in the caption of Fig. 3).

The oxygen isotope effect (OIE) in La_2CuO_4 has been studied by Yusupov et al [32]. Theory predicts that there is no isotope effect on the peak position, only on the linewidth of the polaronic peak, and that this is relatively small. Fig. 12 shows a measurement of the PIA spectra for isotope substituted $La_2Cu^{16}O_4$ and $La_2Cu^{18}O_4$. We also plot the difference spectrum. We observe that the differences are too small to be measurable with PIA. Nevertheless, the observations are entirely consistent with the predictions of polaron theory. The expected polaronic PIA spectrum for different isotopic mass using two small-polaron absorption bands is of the form [23]:

$$\frac{\alpha}{n_p} = \frac{2\pi\sigma}{cn_p} = \frac{2\pi^{3/2}e^2}{m'\omega c}\frac{t'}{\Delta}\exp\left[-\frac{(2E_p - \hbar\omega)^2}{\Delta^2}\right], \tag{5}$$

where α/n_p is an absorption coefficient per unit polaron density, σ is an ac conductivity, E_p is a polaron binding energy, Δ is expressed as $\Delta = \sqrt{4E_pE_{vib}}$, E_{vib} in the low-temperature limit is the zero-point vibrational energy $\hbar\omega_{ph}/2$, t' is the intersite electronic transfer energy and m' is the electronic effective mass. A comparison of the fit to the $La_2Cu^{16}O_4$ PIA spectrum with the predicted $La_2Cu^{18}O_4$ PIA spectrum shown in Fig. 13, suggests that the oxygen isotope effect will be too small to be observed in the PIA.

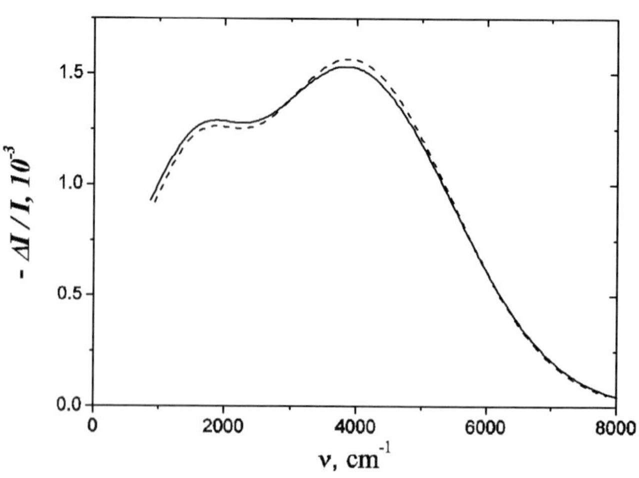

Fig. 13. The fit of the PIA spectrum of $La_2Cu_{16}O_4$ sample with (5) (solid line) with the parameter values $\hbar\omega_{ph,1} = 504 \pm 7$ cm^{-1}, $E_{p,1} = 2214 \pm 4$ cm^{-1} and $\hbar\omega_{ph,2} = 251 \pm 12$ cm^{-1} and $E_{p,2} = 848 \pm 8$ cm^{-1} and the predicted spectrum change due to the isotope substitution (dotted line) $\hbar\omega_{ph,1} = 480 \pm 7$ cm^{-1}, $E_{p,1} = 2214 \pm 4$ cm^{-1}.

It was assumed that the effective phonon frequencies $\omega_{ph,i}$ are totally defined by the oxygen atom mass and thus decrease by 4.87% in respect to that for ^{16}O-sample. In order to conserve the integral intensity of the spectrum the intensities of the MIR features have also been corrected according to (5), such that $\alpha/n_p \propto \omega_{ph}^{-1/2}$. We note that the difference in the low frequency region might be overestimated because of the effect of the remaining tail of the physically-inconsistent $1/\omega$ divergence appearing within the theory of Emin [9].

In conclusion, the mid-infrared PIA spectra of stoichiometric $La_2Cu^{18}O_4$ and $La_2Cu^{16}O_4$ do not show any significant difference within experimental error, which is found to be fully consistent with small-polaron theory.

5 Discussion

The observation of a photoinduced polaronic spectrum and associated photoinduced local modes in many oxides and polymers has given unambiguous proof for the existence of self-trapped long-lived states. The spectra give a direct readout of the polaron binding energy, and it has been shown that systematic variations can be monitored, as a function of doping or ionic substitution for example. The data are in quantitative agreement with the data available from other systems. On the other hand, the absence of a characteristic polaronic feature is not proof that polarons are not present. One possible reason for not observing a polaron signature in the PIA spectra occurs when the recombination lifetime of the carriers is shorter than 10^{-4} s, so the magnitude of the time-integrated signal is small and thus limited by noise. For such cases, it is necessary to use time-resolved techniques to detect the presence of polarons. The theories describing the photoinduced small polaron response in the infrared are well tested by experiments and it can be said that the agreement is very good. The main limitations of the technique arise from the fact that single crystals cannot be measured, and hence it is difficult to obtain information on the symmetry properties of photoinduced local modes.

In the case of cuprates, PIA spectroscopy has given a wealth of information on polaron formation not available from other techniques, particularly on local symmetry breaking and the polaron pairing (clustering) dynamics of photoexcited carriers. Interestingly, the PIA spectra reveal similar polaronic binding energies around $E_p = 0.15$–0.25 eV in cuprates, manganites and cuprates, but the cuprates appear to be unique in that they also have a low-energy (LE) polaronic feature ($E_p = 0.05$–0.06 meV) which is not present in other materials. Finally, we wish to remark on the possibility that the mid-infrared spectrum in cuprates might be due to spin fluctuations, rather than lattice polarons. The polaron systematics observed in the cuprates is very similar to the spectra observed in many different materials (including those which do not include magnetic ions), and the fact that magnetic transitions are not allowed by symmetry in the dipolar infrared spectrum means that the

PIA spectra would be undetectable in time-integrated spectra. Furthermore, the observation that the mid-infrared spectrum is independent of magnetic field up to 33 T to a high degree of accuracy conclusively shows that the mid-infrared is polaronic, and not magnetic in origin. Indeed, to our knowledge, there has so far been no report of photoinduced MIR spectra arising from magnetic excitations.

References

1. D.Mihailovic, C.M.Foster, K.Voss and A.J.Heeger, Phys. Rev.B **42**, 7989 (1990).
2. G.Yu et al., Phys.Rev. B **48**, 7545 (1993)
3. P.Calvani, La Rivista del Nuovo Cimento **24**, 1 (2001)
4. H.G.Reik, Solid State Comm **1**, 67 (1963)
5. D.M.Eagles, Phys. Rev. **145,** 645 (1966)
6. J.T.Devreese, Z.Phys. B **104**, 601 (1997), J.T.Devreese, J.DeSitter, M.Goovaerts, Phys. Rev.B **5**, 2367 (1972), J.Tempere and J.T.Devreese Phys.Rev.B **64**, 104504 (2001)
7. A.S.Alexandrov, V.V.Kabanov and D.K.Ray, Physica C**224**, 247 (1994), V.V.Kabanov and D.K.Ray, Phys. Rev.B **52**, 13021 (1995)
8. Mishchenko et al., Phys.Rev.Lett **83**, 4852 (1999)
9. D.Emin, Phys. Rev. B **48**, 13691 (1993)
10. A recent example for manganites is given by Ch. Hartinger et al, Phys.Rev B **69** 100403 (2004)
11. G. De Filippis, V. Cataudella, A. S. Mishchenko, C. A. Perroni, and J.T. Devreese, Phys. Rev. Lett. **96**, 136405, 1-4 (2006).
12. J. T. Devreese, in the present volume.
13. K.F.Voss et al., Phys. Rev.B **43**, 5109 (1991)
14. Y.H.Kim et al., Phys. Rev. B **36**, 7252 (1987)
15. G.Ruani, C.Taliani and R.Zamboni, Physica C **153-155**, 645 (1988)
16. P.Calvani et al., Phys.Rev.B **53**, 2756 (1996)
17. Y.H.Kim, S-W.Cheong and Z.Fisk, Phys.Rev.Lett. **67**, 227 (1991)
18. T.Mertelj, V.V.Kabanov and D.Mihailovic, Phys. Rev.Lett. **94**, 147003 (2005)
19. F.Li, Y.H.Kim and S-W.Cheong, Physica C **257**, 167 (1996)
20. R.Yusupov et al., unpub. data
21. S.V.Dordevic et al, Phys.Rev.B **73**, 132501 (2006), and unpublished data (for fields up to 33 T).
22. P.Calvani et al., Phys.Rev.B **53**, 2756 (1996), S.Lupi et al., Phys. Rev.Lett., **83**, 4852 (1999)
23. D.Mihailovic et al, Phys.Rev.B **44** p237 (1991)
24. R.J.McQueebey et al, Phys.Rev.Lett. **82**, 628 (1999)
25. Y.Petrov et al, cond-mat/0003414
26. D.Mihailovic and V.V. Kabanov, Phys. Rev.B **63**, 054505 (2001), ibid., **65**, 212508 (2002)
27. T.Mertelj et al., **61**, 15102 (2000)
28. Ch.Hartinger et al., Phys.Rev.B **69**, 100403 (2004)
29. S. Reich, Y. Tsabba, Eur.Phys. J. B **9**, 1 (1999), A. Shengelaya, S. Reich, Y. Tsabba, K.A. Müller, Eur.Phys. J. B **12**, 13 (1999).

30. J.G. Bednorz, K.A. Müller, Z. Physik B **64**,189 (1986).

31. O.F. Schirmer, E. Salje, Sol. Stat. Comm. **33**, 333 (1980), O.F. Schirmer, E. Salje, J. Phys C **13**, L1067 (1980), R.G. Egdell, G.B. Jones, J. Sol. Stat. Chem. 81, 137 (1989).

32. R.Yusupov, K.Conder, H.Keller, K.A.Mueller, T.Mertelj and D.Mihailovic, Eur.Phys.J.B (2007)

Polarons in Colossal Magnetoresistive and High-Temperature Superconducting Materials

Guo-meng Zhao

Department of Physics and Astronomy, California State University, Los Angeles, CA 90032, U.S.A. gzhao2@calstatela.edu

Summary. We review some unconventional oxygen-isotope effects in colossal magnetoresistive manganites and high temperature superconducting cuprates to assess the role of electron-phonon coupling in the basic physics of these materials. These include the unconventional oxygen-isotope effects on the Curie temperature and electrical transport in doped ferromagnetic manganites, on the supercarrier mass of superconducting cuprates, and on the antiferromagnetic ordering temperature of undoped parent cuprates. These unconventional isotope effects clearly demonstrate that the formation of polarons/bipolarons due to strong electron-phonon coupling is relevant to the basic physics of these materials and may be important for the occurrence of colossal magnetoresistance and high-temperature superconductivity. We also identify the phonon modes that are strongly coupled to conduction electrons from the angle-resolved photoemission spectroscopy, tunneling spectra, and optical data of doped cuprates. We consistently show a strong electron-phonon coupling feature at about 20 meV along the antinodal direction, which should also be important to the pairing mechanism of high-temperature superconductivity.

1 Introduction

Over the last decades, very interesting phenomena such as high-temperature superconductivity and colossal magnetoresistance have been discovered in several doped perovskite oxides (e.g., cuprates [1] and manganites [2]). These doped oxides are characterized by strong electron-phonon interactions, significant carrier densities ($\geq 10^{21}$ cm^{-3}), and low mobility of the order or even less than the Mott-Ioffe-Regel limit ($ea^2/\hbar \sim 1$ cm^2/Vs) (where a is the lattice constant). The very nature of the low-temperature "metallic" state of these materials cannot be understood within the framework of the canonical theory of metals where electron-phonon coupling is rather weak and the Fermi energy is much larger than a characteristic phonon energy. Since the electron-phonon interactions in these perovskite oxides are much stronger than in normal metals, new theoretical approaches based on the formation of polarons/bipolarons are required to understand the basic physics of these materials.

The concept of polarons was first introduced by Landau in 1933 [3]. If an electron is placed into the conduction band of an ionic crystal, the electron is "trapped by digging its own hole" due to a strong Coulombic interaction of the electron with its surrounding positive ions. The electron together with the lattice distortions induced by itself is called a polaron (lattice polaron). Lattice polarons are not "bare" charge carriers, but are carriers which are dressed by lattice distortions. Later on, the polaron problem was treated in great detail. One of the examples is Holstein's treatment where an electron is trapped by self induced deformation of two-atomic molecules (the Holstein polaron) [4, 5]. In this case, the polaron moves by thermally activated hopping at high temperatures with a diffusion coefficient $\omega a^2 \exp[-(E_p/2 - t)/k_B T)]$, where ω is the vibration frequency, E_p is the polaron binding energy, and t is the bare hopping integral. Further extensive theoretical studies (for reviews see [6, 7]) have shown that polarons behave like heavy particles, and can be mobile with metallic conduction at sufficiently low temperatures. Under certain conditions [8] they form a polaronic Fermi liquid with some properties being very different from ordinary metals.

It is well known that electrons can change their mass in solids due to the interactions with ions, spins, and themselves. The renormalized (effective) mass of electrons is independent of the ion mass M in ordinary metals where the Migdal adiabatic approximation is believed to be valid. However, the effective mass of polarons m^* will depend on M. This is because the polaron mass $m^* = m \exp(\gamma E_p/\hbar\omega)$ [6, 9], where m is the band mass in the absence of the electron-phonon interaction, γ is a constant, and ω is a characteristic phonon frequency which depends on the masses of ions. Hence, there is a large isotope effect on the polaronic carrier mass, in contrast to the zero isotope effect on the effective carrier mass in ordinary metals.

The total exponent of the isotope effect on the effective carrier mass is defined as $\beta = \sum -d\ln m^*/d\ln M_i$ (M_i is the mass of the ith atom in a unit cell). For polaronic carriers, this definition leads to

$$\beta = -\frac{1}{2}\ln(m^*/m). \tag{1}$$

It is interesting that the simple relation $m^* = m \exp(\gamma E_p/\hbar\omega)$ is even valid in the weak coupling region in the case of the long-range Fröhlich electron-phonon interaction [9]. Then the polaron mass enhancement factor m^*/m in this case is simply equal to $\exp(-2\beta)$.

Therefore, if electron-phonon coupling in a solid is strong enough to form polarons and/or bipolarons, one will expect a substantial isotope effect on the effective mass of carriers. In this chapter, we will review some unconventional oxygen-isotope effects in both manganites and cuprates. These include oxygen-isotope effects on the Curie temperature and electrical transport in doped ferromagnetic manganites, on the supercarrier mass of superconducting cuprates, and on the antiferromagnetic ordering temperature of undoped parent cuprates. The observed large unconventional isotope effects clearly demonstrate that the formation of polarons/bipolarons due to strong electron-phonon coupling is relevant to the basic physics of these materials and may be important for the occurrence of colossal magnetoresistance and high-temperature superconductivity.

2 Polarons in Manganites

Doped manganites $Ln_{1-x}A_xMnO_3$ (where Ln is a trivalent rare earth ion and A is a divalent ion) have been found to exhibit some remarkable features. The undoped parent compound $LaMnO_3$ (with Mn^{3+}) is an insulating antiferromagnet [10]. When Mn^{4+} ions are introduced by substituting a divalent ion (e.g., Ca) for La^{3+}, the materials undergo a transition from a high-temperature paramagnetic and insulating state to a ferromagnetic and metallic ground state for $0.2 \leq x \leq 0.5$ [11]. For $0.5 \leq x \leq 0.8$, the materials exhibit an insulating, charge-ordered and antiferromagetic ground state. The temperature at which the insulator-metal transition occurs can be increased by applying a magnetic field. As a result, the electrical resistance of the material can be decreased by a factor of 1000 or more [2], if the temperature is held in the region of the transition. This phenomenon is now known as colossal magnetoresistance (CMR).

The physics of manganites has primarily been described by the double-exchange model [12, 13]. Crystal fields split the Mn 3d orbitals into three localized t_{2g} orbitals, and two higher energy e_g orbitals which are hybridized with the oxygen p orbitals. Each manganese ion has a core spin of $S = 3/2$, and a fraction $(1 - x)$ have extra electrons in the e_g orbitals with spin parallel to the core spin due to strong Hund's exchange. The electron can hop to an adjacent Mn site with unoccupied e_g orbitals without loss of spin polarization, but with an energy penalty that varies with the angle between the core spins. This double-exchange model accounts qualitatively for ferromagnetic ordering and carrier mobility that depends on the relative orientation of Mn moments which near T_C will therefore be strongly dependent on the applied field. However, Millis, Littlewood and Shraiman [14] have pointed out that double-exchange alone cannot fully explain the data of $La_{1-x}Sr_xMnO_3$. They suggest that lattice-polaronic effects due to strong electron-phonon coupling (arising from a strong Jahn-Teller effect) should be involved. The basic argument [15] is that in the high-temperature paramagnetic state the electron-phonon coupling constant λ is large and the carriers are polarons, while the growing ferromagnetic order increases the bandwidth and thus decreases λ sufficiently for metallic behavior to occur below the Curie temperature T_C. On the other hand, Alexandrov and Bratkovsky [16] show that, in order to explain CMR quantitatively, one needs to consider the formation of small bipolarons (pairs of small polarons) in the paramagnetic state.

Although it is generally accepted that a strong electron-phonon interaction plays an important role in the basic physics of manganites [17], a quantitative explanation of the CMR is lacking. The very nature of charge carriers in the ferromagnetic metallic state and in the paramagnetic state is still under intensive debate [14–16]. Moreover, electron-energy-loss (EELS) [18], and O $1s$ x-ray absorption spectroscopy [19] consistently show that doped holes in manganites are of oxygen p character as expected for doped charge-transfer insulators. This would raise a question of whether the double-exchange mechanism is still valid in doped charge-transfer insulators [20]. Thus, the present understanding of the physics in manganites is far from complete, and further theoretical and experimental studies are essential.

2.1 Giant Oxygen-Isotope Shift of the Curie Temperature

A first observation of the oxygen isotope effect on the Curie temperature was made in $La_{1-x}Sr_xMnO_{3+y}$ system by Zhao and Morris in 1995 [21]. In Fig. 1, the normalized magnetizations for the ^{16}O and ^{18}O samples of $La_{0.9}Sr_{0.1}MnO_{3+y}$ are plotted as a function of temperature. The oxygen isotope shifts of T_C were determined from the differences between the midpoint temperatures on the transition curves of the ^{16}O and ^{18}O samples. It is clear that the ^{18}O sample has a lower T_C than the ^{16}O sample by ~ 6.7 K. It should be noted that since the value of y is substantial (> 0.05) when samples are prepared below 1100 °C, the Curie temperatures in these samples are much higher than those for the corresponding single crystal samples where y is close to zero. Actually the extra oxygen in the above chemical formula is caused by the existence of cation vacancies.

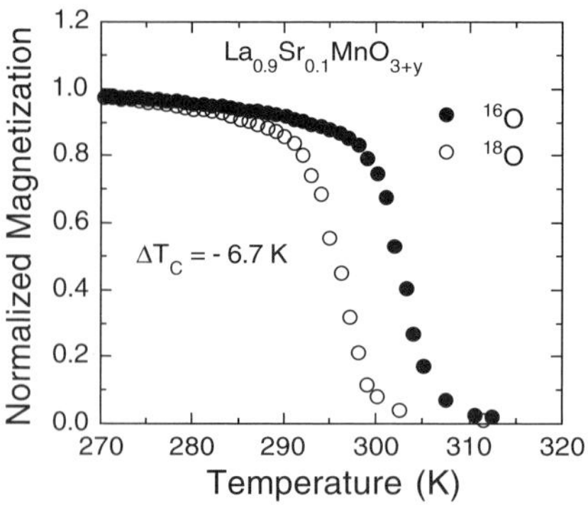

Fig. 1. Oxygen isotope effect on the Curie temperature of $La_{0.9}Sr_{0.1}MnO_{3+y}$. Reproduced from [21].

On the other hand, the oxygen-isotope shift of T_C in $La_{0.8}Ca_{0.2}MnO_{3+y}$ is very large [22], as seen from Fig. 2. The samples with the same isotope mass have the same T_C, while the samples with a heavier oxygen isotope mass (about 95% of ^{18}O) have a much lower T_C. The relative isotope shift of T_C is as large as 10%. Such a large oxygen isotope shift of the ferromagnetic transition is very unusual since lattice vibrations were believed to play no role in the magnetic interactions of most magnetic materials. It is a clear-cut experiment to establish what many have suspected, that atomic motion must be included in any viable description of the manganites [14]. It was also the first experiment in condensed matter physics to demonstrate that there can be a giant isotope shift of a magnetic transition temperature. The oxygen

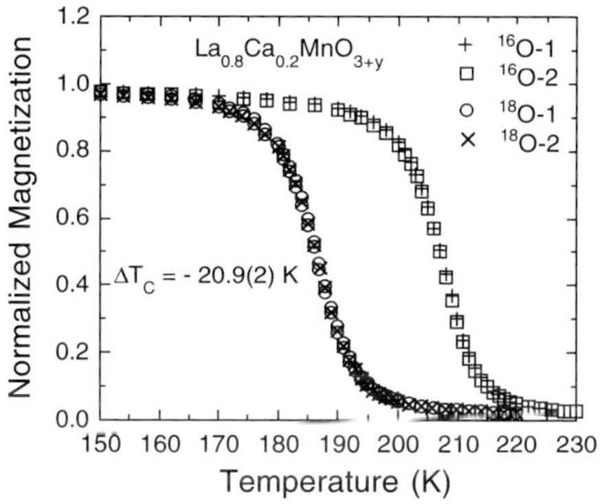

Fig. 2. Oxygen isotope effect on the Curie temperature of $La_{0.8}Ca_{0.2}MnO_{3+y}$. The figure is reproduced from [22].

isotope exponent is defined as usual, $\alpha_O = -\,(\Delta T_C/T_C)/(\Delta M_O/M_O)$, where both T_C and the oxygen isotope mass M_O are referred to a ^{16}O sample. With this definition, we obtain $\alpha_O = 0.85$ for $La_{0.8}Ca_{0.2}MnO_{3+y}$ and 0.142 for $La_{0.9}Sr_{0.1}MnO_{3+y}$.

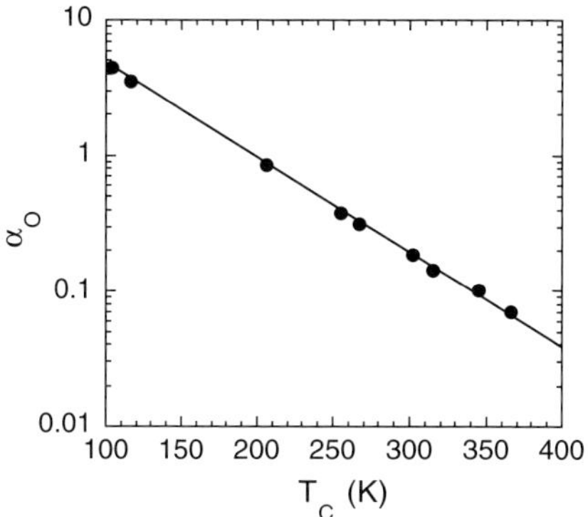

Fig. 3. α_O vs the Curie temperature of ^{16}O samples. The data are from [25].

The oxygen isotope effect on the Curie temperature has also been studied in other manganite systems [23–25]. It is found that [25] the α_O value increases rapidly as the Curie temperature decreases. In Fig. 3, we plot α_O vs the Curie temperature of ^{16}O samples. The data can be well fit by an equation

$$\alpha_O = 21.9 \exp(-0.016 T_C). \tag{2}$$

The above simple empirical relation between α_O and T_C is quite unexpected. From a simple argument based on the double-exchange model and the small polaron theory discussed above, one would expect that [22] $T_C \propto \exp(-2\alpha_O)$. At least this scenario cannot quantitatively explain (2) possibly because it is too simple.

It is also interesting that (2) is very similar to the relation between the pressure effect $(d \ln T_C/dP)$ and T_C. Plotted in Fig. 4 is $d \ln T_C/dP$ (for fixed $x = 0.3$) as a function of T_C. The data can also be well fit by the equation

$$d \ln T_C/dP = 4.4 \exp(-0.016 T_C). \tag{3}$$

Combining (2) and (3), one has $\alpha_O = 5.0 d \ln T_C/dP$. Such a simple relation between the oxygen isotope exponent and the pressure effect implies that the oxygen isotope and pressure effects have the same origin, and that the observed oxygen isotope effect must be intrinsic.

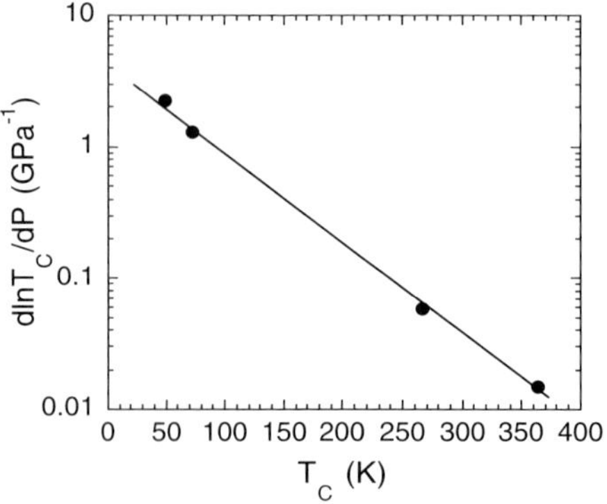

Fig. 4. $d \ln T_C/dP$ vs Curie temperature. The quantity $d \ln T_C/dP$ is referred to the pressure effect in the low pressure region. Reproduced from [23], ©(1997) by the American Physical Society.

2.2 Oxygen-Isotope Effects on Electrical Transport Below the Curie Temperature

The very nature of charge carriers and the electrical transport mechanism in the low-temperature metallic state of doped manganites have not been fully understood. At low temperatures, a dominant T^2 contribution in resistivity is generally observed, and has been ascribed to electron-electron scattering [26]. In contrast, Jaime et al. [27] have shown that the resistivity is essentially temperature independent below 20 K and exhibits a strong T^2 dependence above 50 K. They proposed single magnon scattering with a cutoff at long wavelengths to explain their data. In their scenario [27], they considered a case where the manganese e_g minority (spin-up) band lies slightly above the Fermi level (in the majority spin-down band) with a small energy gap of about 1 meV. This is in contradiction with the optical data [28] which show that the minimum band gap between the e_g minority and majority bands is even larger than 0.5 eV and that the size of the gap strongly depends on the chemical pressure. Alternatively, Zhao et al. [29] have recently shown that the temperature dependent part of the resistivity at low temperatures is mainly due to scattering by a soft optical phonon mode.

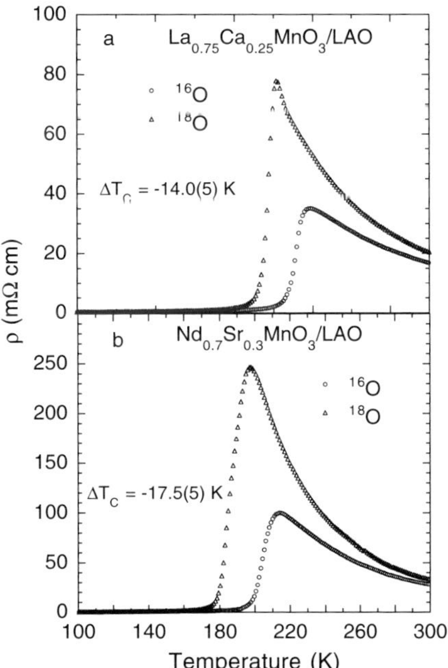

Fig. 5. The resistivity of the oxygen-isotope exchanged films of (a) La$_{0.75}$Ca$_{0.25}$MnO$_3$; (b) Nd$_{0.7}$Sr$_{0.3}$MnO$_3$. From [31], ©(2000) by the American Physical Society.

Various CMR theories [14, 16, 30] predict different natures of charge carriers in the ferromagnetic state. In a theory by Alexandrov and Bratkovsky [16], polarons are considered as the carriers even in the low-temperature metallic state, while others [14, 30] believe that polaronic effects are not important at low temperatures. The clarification of the nature of charge carriers in the ferromagnetic state can discriminate these theoretical models. One of the effective and direct ways to prove or disprove the existence of polaronic carriers is to study the isotope effect on the effective carrier mass. This is because the polaron mass depends strongly on the isotope mass.

The oxygen-isotope effect on the intrinsic low-temperature resistivity has been studied [31] in high-quality epitaxial thin films of $La_{0.75}Ca_{0.25}MnO_3$ and $Nd_{0.7}Sr_{0.3}MnO_3$. The residual resistivity of these compounds shows a strong dependence on the oxygen-isotope mass. The quantitative data analyses suggest that the nature of the charge carriers in the ferromagnetic state of doped manganites are intermediate-size polarons.

Figure 5 shows the resistivity of the oxygen-isotope exchanged films of $La_{0.75}Ca_{0.25}MnO_3$ (LCMO) and $Nd_{0.7}Sr_{0.3}MnO_3$ (NSMO) over 100-300 K. It is apparent that the ^{18}O samples have lower metal-insulator crossover temperatures and much sharper resistivity drop. The Curie temperature T_C normally coincides with a temperature where $d\ln\rho/dT$ exhibits a maximum. We find that the oxygen-isotope shift of T_C is 14.0(6) K for LCMO, and 17.5(6) K for NSMO, in excellent agreement with the results for the bulk samples [25].

In Fig. 6 we plot the low-temperature resistivity of the oxygen-isotope exchanged films of (a) LCMO; (b) NSMO. In both cases, the residual resistivity ρ_o for the ^{18}O samples is larger than for the ^{16}O samples by about 15%. Repeating the van der Pauw measurements at 5 K several times with different contact configurations indicates that the uncertainty of the difference in ρ_o of the two isotope samples is less than 3%. We should mention that the intrinsic resistivity cannot be obtained from ceramic samples where the boundary resistivity is dominant. Thus one cannot use ceramic samples to study the isotope effect on the intrinsic resistivity. Moreover, the van der Pauw technique is particularly good to precisely measure the resistivity difference between the oxygen-isotope exchanged films which have the same thickness. Thus the data shown in Figs. 5 and 6 represent the first precise measurements on the intrinsic resistivity of the isotope substituted samples.

It is known that [33] the residual resistivity $\rho_o \propto m^*/n\tau_o$. Here $\hbar/\tau_o$ is the scattering rate which is associated with the random potential produced by randomly distributed trivalent and divalent cations [33], and/or with impurities; m^* is the effective mass of carriers at low temperatures, and n is the mobile carrier concentration. If the chemical potential is far above the mobility edge, $\rho_o \propto (m^*)^2$, that is, $\hbar/\tau_o \propto m^*$. This is what one expects from the simple Born approximation. On the other hand, one can show [34] that $\rho_o \propto m^*$ if the chemical potential is slightly above the mobility edge. Therefore, the large oxygen-isotope effect on ρ_o implies that the effective mass of the carriers strongly depends on the oxygen mass, as expected for polaronic charge carriers.

If the charge carriers at low temperatures are of small or intermediate-size polarons, the temperature dependence of the resistivity should agree with

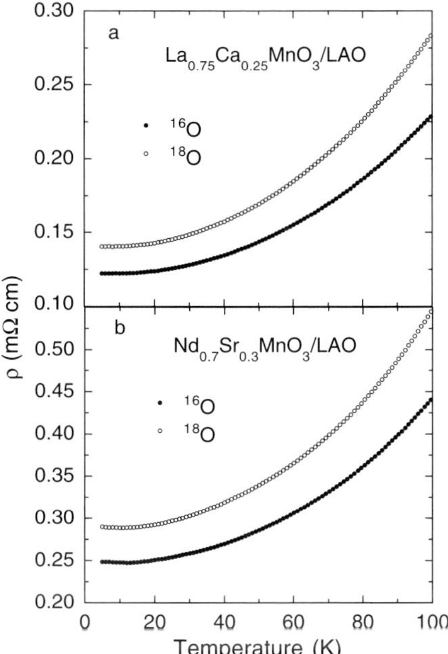

Fig. 6. The low-temperature resistivity of the oxygen-isotope exchanged films of (a) $La_{0.75}Ca_{0.25}MnO_3$; (b) $Nd_{0.7}Sr_{0.3}MnO_3$. After [31], ©(2000) by the American Physical Society.

polaron metallic conduction. This is indeed the case as recently demonstrated by Zhao et al. [29]. There are three contributions to the resistivity: the residual resistivity ρ_o, the term $AT^{4.5}$ contributed from 2-magnon scattering [35], and the term $B\omega_s/\sinh^2(\hbar\omega_s/2k_BT)$, which arises from polaron coherent motion involving a relaxation due to a soft optical phonon mode that is strongly coupled to the carriers [29]. Here ω_s is the frequency of a soft optical mode. The temperature dependent part of the resistivity is then given by

$$\rho(T) - \rho_o = AT^{4.5} + B\omega_s/\sinh^2(\hbar\omega_s/2k_BT). \tag{4}$$

It was shown that the parameter B is proportional to m^*/n [29]. This implies that B and ρ_o should have a similar relative change upon the isotope substitution if the chemical potential is near the mobility edge. The coefficient A has an analytical expression in the case of a simple parabolic polaron band (occupied by single-spin holes) [35]. In terms of the hole density per cell n, the average spin stiffness D, and the effective hopping integral t^*, the coefficient A can be written as [35]

$$A = (\frac{3a\hbar}{32\pi e^2})(2 - n/2)^{-2}(6\pi^2 n)^{5/3}(2.52 + 0.0017\frac{D}{a^2 t^*})$$

$$\{\frac{a^2 k_B}{D(6\pi^2)^{2/3}(0.5^{2/3} - n^{2/3})}\}^{9/2}. \tag{5}$$

Here we have used the relations: $ak_F = (6\pi^2 n)^{1/3}$ (where $\hbar k_F$ is the Fermi momentum); $E_F = t^*(6\pi^2)^{2/3}(0.5^{2/3} - n^{2/3})$ (where the Fermi energy E_F is measured from the band center); the effective spin $S^* = 2 - n/2$. The value of t^* can be estimated to be about 40 meV from the measured effective plasma frequency $\hbar\Omega_p^* = 1.1$ eV and $n \sim 0.3$ in $La_{0.7}Ca_{0.3}MnO_3$ [36]. Since D is about 100 meV $\mathring{A}^2$ (see below), one expects that the term $0.0017D/a^2 t^* <<$ 2.52, and thus can be dropped out in (5).

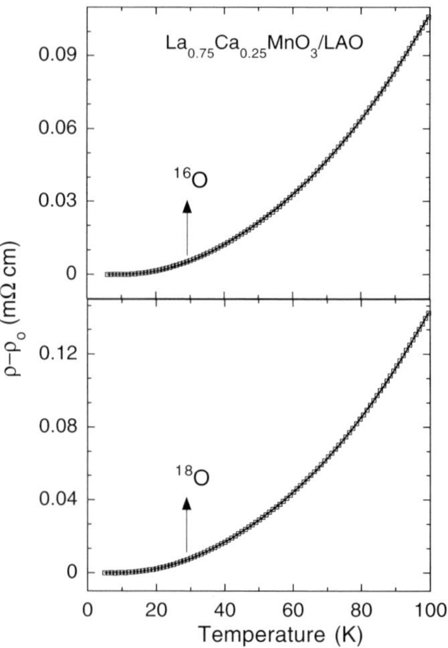

Fig. 7. $\rho(T) - \rho_o$ for the ^{16}O and ^{18}O films of $La_{0.75}Ca_{0.25}MnO_3$. The solid lines are fitted curves by (4). After [31], ©(2000) by the American Physical Society.

From (5), one sees that there are two parameters n and D that determine the magnitude of A. In doped manganites, n should be approximately equal to the doping level x. The average spin stiffness D should be close to the long-wave spin stiffness $D(0)$ if there is negligible magnon softening near the zone boundary. On the other hand, $D < D(0)$ if there is a magnon softening near the zone boundary as in the case of low T_C materials [37]. In any case, one might expect that the average D should be proportional to T_C so that D/k_BT_C is a universal constant in the manganite system. Since the magnon softening becomes unimportant when the T_C is higher than 350 K [37], then $D \simeq D(0)$ in the compound $La_{0.7}Sr_{0.3}MnO_3$ with the highest $T_C = 378$ K [38]. Thus, the universal value of D/k_BT_C should be close to the value of $D(0)/k_BT_C$ in $La_{0.7}Sr_{0.3}MnO_3$, which was found to 5.8 ± 0.2 Å^2 [38].

The $\rho(T) - \rho_o$ data below 100 K are fitted by (4) for the LCMO ^{16}O and ^{18}O samples, as shown in Fig. 7. It is striking that the fits to the data of both isotope samples are very good. The fitting parameters A, B and $\hbar\omega_s$ are summarized in Table 1. Since the fits are excellent, the uncertainties in the fitting parameters are very small (see Table 1). From Table 1, one can see that ρ_o increases by 15(3)%, and B by 17(3)%. This provides additional evidence that the scattering rate $\hbar/\tau_o$ is nearly isotope-mass independent, in agreement with the above argument. Thus the observed large oxygen-isotope effects on both ρ_o and B suggest that the effective mass of carriers depends strongly on the oxygen-isotope mass. This is consistent with the presence of small or intermediate-size polarons in the ferromagnetic state of manganites.

In addition, ω_s decreases by about 10(1)% upon replacing ^{16}O with ^{18}O. This may imply that the soft mode might be associated with the motion of the oxygen atoms and has a large anharmonicity. It was shown that the tilt/rotation mode of the oxygen octahedra in cuprates has a strong electron-phonon coupling (quadratic coupling) and a large anharmonicity [39, 40]. The large anharmonicity of the mode can possibly lead to a decrease of ω_s by 12.5% upon replacing ^{16}O with ^{18}O [39]. In a similar perovskite superconductor $Ba(Pb_{0.75}Bi_{0.25})O_3$, both neutron and tunneling experiments [41] show that a soft mode with $\hbar\omega_s/k_B = 70$ K is related to rotational vibrations of the oxygen octahedra, and has a strong electron-phonon coupling. Moreover, the frequency of the rotational mode in $Ba(Pb_{0.75}Bi_{0.25})O_3$ is nearly the same as that of the soft mode ($\hbar\omega_s/k_B = 74$ K) in the $La_{0.75}Ca_{0.25}MnO_3$ ^{16}O film. Both the frequency of the soft mode and its isotope dependence can be quantitatively explained if the soft mode in the manganites is also associated with the rotational vibrations of the oxygen octahedra.

Table 1. The summary of the fitting parameters A, B and $\hbar\omega_s/k_B$ for the ^{16}O and ^{18}O films of $La_{0.75}Ca_{0.25}MnO_3$ (LCMO), and the summary of the T_C and ρ_o values for the LCMO films and the $Nd_{0.7}Sr_{0.3}MnO_3$ films (NSMO). The uncertainty in T_C is ±0.3 K. The uncertainty in ρ_o is discussed in the text.

Compounds	T_C (K)	ρ_0 ($\mu\Omega$cm)	A (mΩcm/K$^{4.5}$)	B ($\mu\Omega$cm/K)	$\hbar\omega_s/k_B$ (K)
LCMO(^{16}O)	231.3	122.4	$1.20(2)\times10^{-11}$	0.370(3)	74.4(2)
LCMO(^{18}O)	217.3	140.5	$1.89(2)\times10^{-11}$	0.434(3)	66.8(3)
NSMO(^{16}O)	203.9	248.2			
NSMO(^{18}O)	186.4	289.2			

Now we turn to the discussion on the magnitude of the parameter A and its isotope dependence. From (5), one can see that n and/or D should be isotope dependent in order to explain the large isotope effect on the parameter A. As discussed above, the D in (5) should be proportional to T_C. Then one can obtain D values from the T_C values and the value of $D/k_B T_C = 5.8$ Å^2 (see the above discussion). Substituting the D and A values (see Table 1) into (5), we find $n = 0.235$ for the La$_{0.75}$Ca$_{0.25}$MnO$_3$ ^{16}O film, and $n = 0.240$ for the La$_{0.75}$Ca$_{0.25}$MnO$_3$ ^{18}O film. The fact that $n \simeq x$ for both isotope samples is consistent with our previous interpretation of the isotope dependence of ρ_o being caused only by m^*. The $AT^{4.5}$ term in our resistivity data is in quantitative agreement with the 2-magnon scattering theory [35].

From the oxygen-isotope effects on the residual resistivity and the thermoelectric power at low temperatures, one can deduce the exponent of the oxygen-isotope effect on m^* based on a polaronic Fermi liquid model [34]. The exponent β_O of the oxygen-isotope effect on m^* was estimated [34] to be about -0.7 for La$_{0.75}$Ca$_{0.25}$MnO$_3$ and about -1.1 for Nd$_{0.7}$Sr$_{0.3}$MnO$_3$. From 1, one finds that the polaronic mass enhancement factor $m^*/m = \exp(-2\beta)$ $\simeq \exp(-2\beta_O) = 4$-$9$. This mass enhancement factor is consistent with the existence of intermediate size polarons.

Very recent angle-resolved photoemission spectroscopic data [42] of a layered manganite La$_{1.2}$Sr$_{1.8}$Mn$_2$O$_7$ ($T_C = 126$ K) suggest the coexistence of bipolaronic carriers (pseudo-gapped state) along the Mn-O binding direction and polaronic charge carriers along the diagonal direction. The polaron mass is enhanced by a factor of 5.6 [42], which is in good agreement with (4-9) obtained from our isotope experiments.

The coexistence of polaronic and bipolaronic carriers in the ferromagnetic state of the layered manganite [42] is also consistent with our earlier results [43, 44] for a 3-dimensional manganite (La$_{0.5}$Nd$_{0.5}$)$_{0.67}$Ca$_{0.33}$MnO$_3$ with a similar T_C (~ 100 K). The ground state of these low T_C manganites is phase-separated into ferromagnetic metallic regions where mobile polarons reside and charge-ordered insulating regions where localized bipolarons sit.

2.3 Oxygen-Isotope Effects on Electrical Transport Above the Curie Temperature

The first experimental evidence for small polaronic charge carriers in the paramagnetic state was provided by transport measurements [45]. It was found that the activation energy E_ρ deduced from the conductivity data is one order of magnitude larger than the activation energy E_s obtained from the thermoelectric power data. Such a large difference in the activation energies is the hallmark of the small-polaron hopping conduction. The giant oxygen-isotope shifts of the ferromagnetic transition temperature T_C give clear evidence for the presence of polaronic charge carriers in this system [22, 23]. Moreover, the fast and local techniques have directly shown that the doped charge carriers are accompanied by local Jahn-Teller distortions [46–49]. However, all these experiments cannot make a distinction between small polarons and small bipolarons since both are dressed by local lattice distortions. Small bipolarons are normally much heavier than small polarons, and should be localized in the presence of small random potentials. In order to discriminate between polarons and bipolarons and to place constraints on the CMR theories, it is essential to study the oxygen-isotope effects on the intrinsic electrical properties. This is because the activation energies in resistivity and thermoelectric power will depend on the oxygen isotope mass if there coexist localized bipolaronic charge carriers and mobile polaronic carriers.

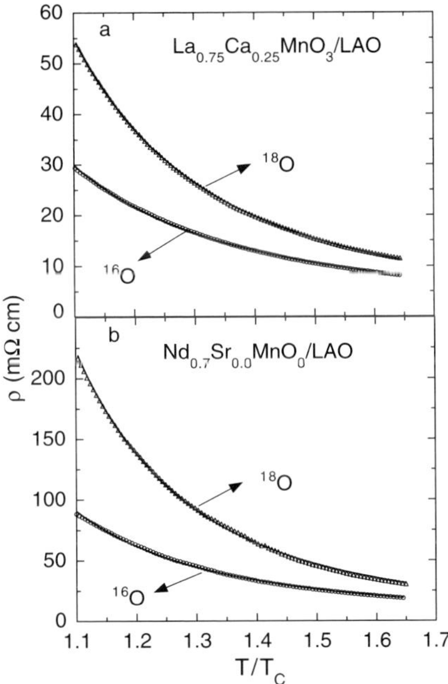

Fig. 8. The resistivity of the oxygen-isotope exchanged films of La$_{0.75}$Ca$_{0.25}$MnO$_3$ and Nd$_{0.7}$Sr$_{0.3}$MnO$_3$. The maximum temperature of the data points for the ^{16}O film of La$_{0.75}$Ca$_{0.25}$MnO$_3$ is 380 K. The solid lines are the fitted curves by (7). As in [45], we excluded the data points below $1.1T_C$ for the fitting. After [50], ©(2000) by the American Physical Society.

The resistivity for the oxygen-isotope exchanged high-quality epitaxial thin films of La$_{0.75}$Ca$_{0.25}$MnO$_3$ and Nd$_{0.7}$Sr$_{0.3}$MnO$_3$, and the thermoelectric power for the oxygen-isotope exchanged ceramic samples of La$_{0.75}$Ca$_{0.25}$MnO$_3$ were precisely measured by Zhao et al. [50]. The data cannot be explained by a simple small-polaron model, but are in quantitative agreement with a model where the formation of localized small bipolarons is essential.

Figure 8 shows the resistivity of the oxygen-isotope exchanged films of LCMO and NSMO above $1.1\,T_C$. From the figure, one can see that there is a large difference in the intrinsic resistivity between the two isotope samples. Such a large isotope effect is reversible upon the oxygen isotope back-exchange.

It is known that the resistivity can be generally expressed as $\rho = 1/\sigma = 1/ne\mu$, where n is the mobile carrier concentration and μ is the mobility of the carriers. For adiabatic small-polaron hopping, the mobility is given by [5]

$$\mu = \frac{ed^2}{h}\frac{\hbar\omega_o}{k_BT}\exp(-E_a/k_BT). \qquad (6)$$

Here d is the site to site hopping distance, which is equal to $a/\sqrt{2}$ in manganites since the doped holes in this system mainly reside on the oxygen sites [18]; ω_o is the characteristic optical phonon frequency; $E_a = (\eta E_p/2)f(T) - t$; $f(T) = [\tanh(\hbar\omega_o/4k_BT)]/(\hbar\omega_o/4k_BT)$ for $T > \hbar\omega_o/4k_B \simeq 200$ K [51]; E_p is the polaron binding energy; t is the "bare" hopping integral; $\eta \leq 1$ [5, 52]. In the harmonic approximation, E_p is independent of the isotope mass M.

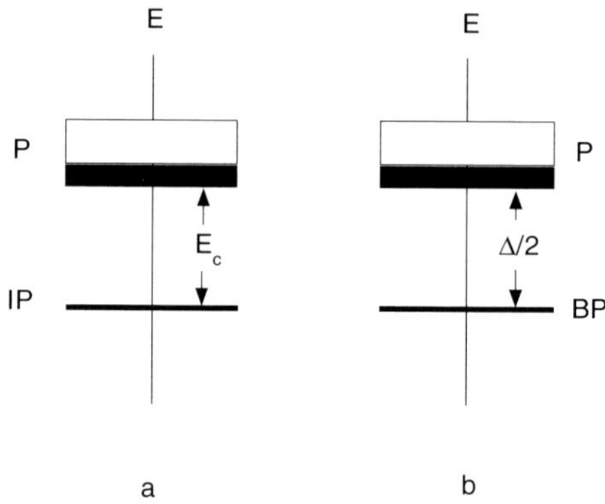

Fig. 9. A schematic diagram of the polaron band, and polaron trapping into impurity (IP) states (a) [51], or into localized bipolaron (BP) states (b) [16]. The bipolaron binding energy Δ is isotope-mass dependent [52], while E_c is independent of the isotope-mass M [51]. After [50], ©(2000) by the American Physical Society.

The mobile polaron density n can be calculated for the two possible cases shown in Fig. 9. If we assume a simple parabolic band, then $n = 2(k_BT/1.05a^2W_p)^{3/2}\exp(-E_s/k_BT)$ for $T \ll W_p/k_B$ [16, 51]. Here W_p is the polaron bandwidth; $E_s = E_c$ if polarons are trapped into impurity (IP) states [51], while $E_s = \Delta/2$ if polarons are bound into localized bipolaron (BP) states [16]. The bipolaron binding energy $\Delta = 2(1 - \gamma)E_p - V_c - W_p$, where V_c is the Coulombic repulsion between bound polarons [52]. In fact, the above $n(T)$ expression is the same as that for semiconductors when the chemical potential is pinned to the impurity levels. Using the above $n(T)$ expression and (6), we finally have

$$\rho = \frac{C}{\sqrt{T}} \exp(E_\rho/k_B T), \tag{7}$$

where $E_\rho = E_a + E_s$, and $C = (ah/e^2\sqrt{k_B})(1.05W_p)^{1.5}/\hbar\omega_o$. The quantity C should strongly depend on the isotope mass M and decrease with increasing M. This is because W_p decreases strongly with increasing M according to $W_p \propto \exp(-\gamma E_p/\hbar\omega_o) = \exp(-g^2)$ [9, 52]. It is worth noting that (7) is valid only if $T<<W_p/k_B$. For $T>>W_p/k_B$, the prefactor in (7) should be proportional to T/ω_o [45].

The thermoelectric power is given by [51]

$$S = E_s/eT + S_o, \tag{8}$$

where S_o is a constant depending on the kinetic energy of the polarons and on the polaron density [51]. One should note that (8) is valid only if there is one type of carrier (e.g., holes).

One can make a distinction between the two cases shown in Fig. 9. If small polarons are bound to impurity centers, there will be no isotope effect on E_s since E_c is independent of M [51]. On the other hand, if small polarons are bound to localized bipolaron states, both E_ρ and E_s in (7) and (8) will depend on M due to the fact that Δ is M dependent. In general, the isotope shift of E_ρ will be larger than the isotope shift of E_s. This is because $E_u = (\eta E_p/2)f(T) - t$, and $f(T) = [\tanh(\hbar\omega_o/4k_B T)]/(\hbar\omega_o/4k_B T)$, which may depend on M if the temperature is not so high compared with $\hbar\omega_o/k_B$.

The data are now fitted by (7) (see solid lines in Fig. 8). It is apparent that the fits are quite good for both isotope samples. The fitting parameters are summarized in Table 2. From Table 2, one can see that, upon replacing ^{16}O with ^{18}O, the parameter C for LCMO/NSMO decreases by 35(5)/40(7)%, while E_ρ increases by 13.2(3)/14.2(8) meV. The huge oxygen-isotope effect on the parameter C is consistent with (7).

One can obtain the isotope shift of E_s by measuring the thermoelectric power for two isotope samples according to (8). In Fig. 10, the thermoelectric power S as a function of $1/T$ is plotted for the ^{16}O and ^{18}O samples of La$_{0.75}$Ca$_{0.25}$MnO$_3$. Both T_C's and the isotope shift of the ceramic samples [25] are the same as those in the corresponding thin films. Since the grain-boundary effect on S is negligible, the thermoelectric power obtained in ceramic samples should be intrinsic. From the slopes of the straight lines in Fig. 10, one finds $E_s = 13.2\pm0.3$ meV for the ^{16}O sample and 18.7 ± 0.3 meV for the ^{18}O. The isotope shift is $\delta E_s = 5.5\pm0.6$ meV, which is about half the isotope shift of E_ρ. The observed oxygen-isotope effects on both E_ρ and E_s do not support a simple small-polaron model [51], but provide evidence that small polarons are bound into localized bipolaron states.

One can use the values of the parameter C to calculate the polaron bandwidth W_p according to the relation: $C = (ah/e^2\sqrt{k_B})(1.05W_p)^{1.5}/\hbar\omega_o$. The calculated W_p values are listed in Table 2. In the calculation, $\hbar\omega_o = 74$ meV

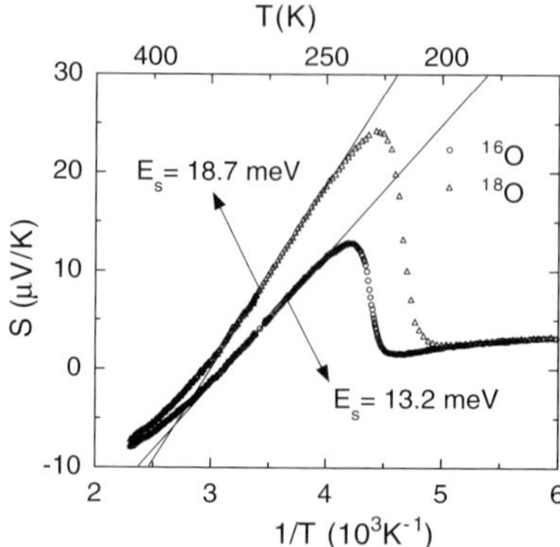

Fig. 10. The thermoelectric power $S(T)$ of the oxygen-isotope exchanged ceramic samples of La$_{0.75}$Ca$_{0.25}$MnO$_3$. After [50], ©(2000) by the American Physical Society.

is used for the ^{16}O samples [52], and $\hbar\omega_o$ for the ^{18}O samples is 5.7% lower than for the ^{16}O samples. From the W_p values (see Table 2), one can see that the data satisfy $T < W_p/k_B$, which justifies the use of (7).

Furthermore, one can quantitatively explain the isotope dependence of E_s if small polarons form localized bipolarons. In this scenario [52], $\delta\Delta = -\delta W_p$. From Table 2, δW_p= -11.2±1.6 meV for La$_{0.75}$Ca$_{0.25}$MnO$_3$. So $\delta E_s = \delta\Delta/2 =$ 5.6±0.8 meV for La$_{0.75}$Ca$_{0.25}$MnO$_3$, in quantitative agreement with the value (5.5±0.6 meV) deduced from the thermoelectric power data.

Table 2. Summary of the fitting and measured parameters for the ^{16}O and ^{18}O films of La$_{0.75}$Ca$_{0.25}$MnO$_3$ (LCMO) and Nd$_{0.7}$Sr$_{0.3}$MnO$_3$ (NSMO). The errors of the parameters come from the fitting and from the van der Pauw measurement. The absolute uncertainty of the thickness of the films was not included in the error calculations since it only influences the absolute values of the resistivity.

Compounds	LCMO(^{16}O)	LCMO(^{18}O)	NSMO(^{16}O)	NSMO(^{18}O)
T_C (K)	231.5(3)	216.5(3)	204(1)	186(1)
C (mΩcmK$^{0.5}$)	17.3(5)	12.9(3)	23.2(8)	16.2(7)
E_ρ (meV)	72.8(2)	86.0(1)	78.8(4)	92.9(4)
E_s (meV)	13.2(3)	18.7(3)		
W_p (meV)	49.0(9)	38.8(7)	60(2)	45(2)

Therefore, the oxygen isotope effects on the intrinsic electrical transport in the paramagnetic state of doped manganites can be quantitatively explained by a scenario [16] where the small polarons form localized bound pairs (bipolarons) in the paramagnetic state. The coexistence of small polarons and bipolarons in the paramagnetic state may lead to a dynamic phase separation into the insulating antiferromagnetically coupled region where the bipolarons reside, and into the ferromagnetically coupled region where the polarons sit. This simple picture can naturally explain the observation of the ferromagnetic clusters (intrinsic electronic inhomogeneity) in the paramagnetic state [53].

3 Polarons in Cuprates

Developing the microscopic theory for high-T_c superconductivity is one of the most challenging problems in condensed matter physics. Eighteen years after the discovery of the high-T_c cuprate superconductors by Bednorz and Müller [1], there have been no microscopic theories that can describe the physics of high-T_c superconductors completely and unambiguously. Due to the high T_c values and the observation of a small oxygen-isotope effect in a 90 K cuprate superconductor $YBa_2Cu_3O_{7-y}$ (YBCO) [54, 55], many theorists believe that the electron-phonon interaction is not important in bringing about high-T_c superconductivity. Most physicists have thus turned their minds towards alternative pairing interactions of purely electronic origin.

On the other hand, there is overwhelming evidence that electron-phonon coupling is very strong in cuprates [56–71]. In particular, the studies of various unconventional oxygen-isotope effects Zhao and his coworkers have initiated since 1992 clearly indicate that the electron-phonon interactions are so strong that polarons/bipolarons are formed in doped cuprates and manganites [22, 56–60, 62–67, 69], in agreement with a theory of high-temperature superconductivity [6]. However, such clear experimental evidence for strong electron-phonon interactions from the unconventional isotope effects has been generally ignored. In the 2001 Nature paper [68], Lanzara et al. appear to provide evidence for a strong coupling between doped holes and the 70 meV half-breathing phonon mode from angle-resolved photoemission spectroscopy (ARPES). They further show that this 70 meV phonon mode can lead to d-wave pairing symmetry and is mainly responsible for high-temperature superconductivity [72]. Recently, Devereaux et al. [73] have proposed that the 40 meV B_{1g} phonon mode rather than 70 meV half-breathing phonon mode is responsible for d-wave high-temperature superconductivity. This pairing mechanism contradicts the very recent ARPES data from their own group, which show that multiple phonon modes at 27 meV, 45 meV, 61 meV, and 75 meV are strongly coupled to doped holes in deeply underdoped $La_{2-x}Sr_xCuO_4$ [70]. The strong coupling to the multiple phonon modes is not in favor of d-wave gap symmetry but may support a general s-wave gap symmetry [74, 75].

3.1 Oxygen-Isotope Effect on T_N in La_2CuO_4

The antiferromagnetic order (AF) observed in the parent insulating compounds like La_2CuO_4 signals a strong electron-electron Coulomb correlation. On the other hand, if there is a very strong electron-phonon coupling such that the Migdal adiabatic approximation breaks down, one might expect that the antiferromagnetic exchange energy should depend on the isotope mass. Following this simple argument, Zhao and his co-workers initiated studies of the oxygen isotope effect on the AF ordering temperature in several parent compounds in 1992. A noticeable oxygen-isotope shift of T_N was consistently observed in La_2CuO_4 [56].

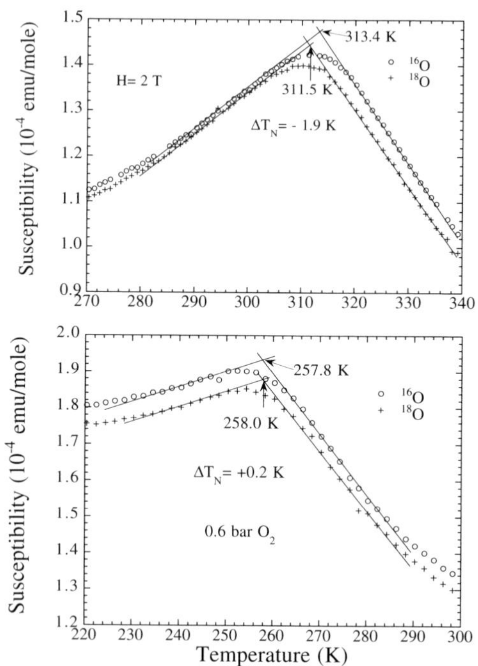

Fig. 11. The temperature dependence of the susceptibility for the ^{16}O and ^{18}O samples of undoped La_2CuO_4 (upper panel), and of the oxygen-doped La_2CuO_{4+y} (lower panel). After [56], ©(1994) by the American Physical Society.

Figure 11 shows the temperature dependence of the susceptibility for the ^{16}O and ^{18}O samples of undoped La_2CuO_4 (upper panel), and of oxygen doped La_2CuO_{4+y} (lower panel). One can see that the AF ordering temperature T_N for the ^{18}O sample is lower than the ^{16}O sample by about 1.9 K in the case of the undoped samples. For the oxygen-doped samples, there is a negligible isotope effect.

It is known that the antiferromagnetic properties of La_2CuO_{4+y} can be well understood within mean-field theory which leads to a T_N formula [76]:

$$k_B T_N = J'[\xi(T_N)/a]^2, \tag{9}$$

where J' is the interlayer coupling energy, $\xi(T_N)$ is the in-plane AF correlation length at T_N with $\xi(T_N) \propto \exp(J/T_N)$ for $y = 0$ (J is the in-plane exchange energy). When T_N is reduced to about 250 K by oxygen doping, a mesoscopic phase separation has taken place so that $\xi(T_N) = L$ ([77]), where L is the size of the antiferromagnetically correlated clusters, and depends only on the extra oxygen content y. In this case, we have $T_N = J'(L/a)^2$. Since L is independent of the isotope mass, a negligible isotope shift of T_N in the oxygen-doped La_2CuO_{4+y} suggests that J' is independent of the isotope mass. Then we easily find for undoped compounds

$$\Delta T_N/T_N = (\Delta J/J)\frac{B}{1+B}, \tag{10}$$

where $B = 2J/T_N \simeq 10$. From the measured isotope shift of T_N for the undoped samples, we obtain $\Delta J/J \simeq -0.6\%$.

Recently, Eremin et al. [78] have considered strong electron-phonon coupling within a three-band Hubbard model. They showed that the antiferromagnetic exchange energy J depends on the polaron binding energy E_p^O due to oxygen vibrations, on the polaron binding energy E_p^{Cu} due to copper vibrations, and on their respective vibration frequencies ω_O and ω_{Cu}. At low temperatures, J is given by [78]

$$J = J_\circ(1 + \frac{3E_p^O\hbar\omega_O}{\Delta_{pd}^2} + \frac{3E_p^{Cu}\hbar\omega_{Cu}}{\Delta_{pd}^2}). \tag{11}$$

Here Δ_{pd} is the charge-transfer gap, which is measured to be about 1.5 eV in undoped cuprates. The oxygen-isotope effect on J can be readily deduced from (11):

$$\frac{\Delta J}{J} = (\frac{3E_p^O\hbar\omega_O}{\Delta_{pd}^2})(\frac{\Delta\omega_O}{\omega_O}). \tag{12}$$

Substituting the unbiased parameters $\hbar\omega_O = 0.075$ eV, $\Delta J/J \simeq -0.6\%$, $\Delta_{pd} = 1.5$ eV, and $\Delta\omega_O/\omega_O = 6.0\%$ into (12), we find that $E_p^O = 1.0$ eV. The total polaron binding energy should be larger than 1.0 eV since E_p^{Cu} is not zero.

The polaron binding energy can be also estimated from optical data where the energy of the mid-infrared peak in the optical conductivity is equal to $2\gamma E_p$, where γ is between 0.2-0.3 [52]. The peak position ($2\gamma E_p$) was found to be about 0.6 eV for $La_{1.97}Sr_{0.03}CuO_4$ [79], implying that $E_p = 1.0$-1.5 eV. This is in quantitative agreement with the value estimated from the isotope effect. These results thus consistently suggest that the polaron binding energy of undoped La_2CuO_4 is over 1 eV. Doping will reduce the value of E_p due to screening. For the optimal doping ($x = 0.15$), the optical data suggest that $2\gamma E_p = 0.12$ eV, which is a factor of 5 smaller than that for $x = 0.03$.

Very recently, angle-resolved photoemission spectroscopy (ARPES) data of undoped La_2CuO_4 have been explained in terms of polaronic coupling between phonons and charge carriers [80]. From the width of the phonon side band in the ARPES spectra, the authors find the polaron binding energy to be about 1.9 eV, in good agreement with their theoretical calculation based on a shell model [80]. On the other hand, the observed binding energy of the side band should be consistent with a polaron binding energy of about 1.0 eV [80]. This should be the lower limit because the binding energy of the side band decreases rapidly with doping and because the sample may be lightly doped [80]. Therefore, the ARPES data suggest that 1.0 eV $< E_p < 1.9$ eV, which is in quantitative agreement with the value deduced from our earlier isotope experiments.

3.2 Oxygen-Isotope Effect on the In-Plane Supercarrier Mass

One of the most remarkable oxygen-isotope effects we have observed is the oxygen-isotope effect on the penetration depth [57–60, 64–67, 69]. Zhao et al. made the first observation of this effect in optimally doped $YBa_2Cu_3O_{6.93}$ in 1994 [57]. By precisely measuring the diamagnetic signals for the ^{16}O and ^{18}O samples, it was possible to deduce the oxygen-isotope effects on the penetration depth $\lambda(0)$ and on the supercarrier density n_s. It turns out that $\Delta n_s \simeq 0$, and $\Delta\lambda(0)/\lambda(0) = 3.2$ % [57]. These isotope effects thus suggest that the effective supercarrier mass depends on the oxygen-isotope mass.

In fact, for highly anisotropic materials, the observed isotope effect on the angle-averaged $\lambda(0)$ is the same as the isotope effect on the in-plane penetration depth $\lambda_{ab}(0)$. From the magnetic data for $YBa_2Cu_3O_{6.93}$, $La_{1.85}Sr_{0.15}CuO_4$, and $Bi_{1.6}Pb_{0.4}Sr_2Ca_2Cu_3O_{10+y}$, one finds that $\Delta\lambda_{ab}(0)/\lambda_{ab}(0) = 3.2\pm0.7\%$ for the three optimally doped cuprates [67]. Several independent experiments have consistently shown that the carrier densities of the two isotope samples are the same within 0.0004 per unit cell [59, 60, 65]. Therefore, the observed oxygen-isotope effect on the in-plane penetration depth is caused only by the isotope dependence of the in-plane supercarrier mass. Recently, direct measurements of the in-plane penetration depth by low energy muon-spin-relaxation (LEμSR) technique [69] have confirmed the earlier isotope-effect results. It is found that [69] $\Delta\lambda_{ab}(0)/\lambda_{ab}(0) = 2.8\pm1.0\%$. It is remarkable that the isotope effect obtained from the most advanced and expensive technology (LEμSR) is the same as that deduced from simple magnetic measurements.

Figure 12 shows the doping dependence of the exponent (β_O) of the oxygen-isotope effect on the in-plane supercarrier mass in $La_{2-x}Sr_xCuO_4$. Here the exponent is defined as $\beta_O = -d\ln m_{ab}^{**}/d\ln M_O$. It is apparent that the exponent increases with decreasing doping, in agreement with the fact that doping reduces electron-phonon coupling due to screening. The large oxygen-isotope effect on the in-plane supercarrier mass cannot be explained within the conventional phonon-mediated pairing mechanism where the effective mass of

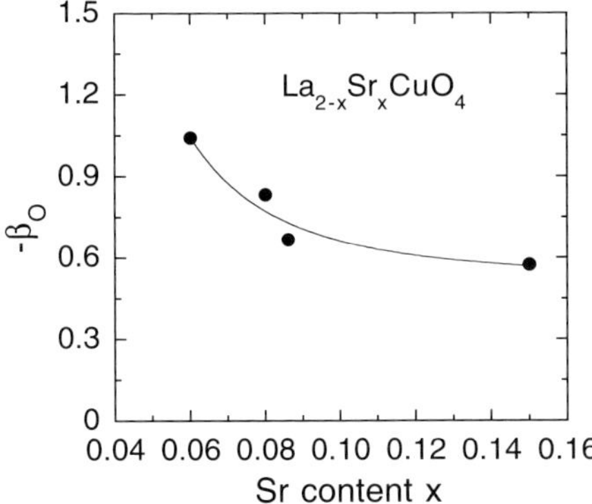

Fig. 12. The doping dependence of the exponent β_O of the oxygen-isotope effect on the in-plane supercarrier mass in La$_{2-x}$Sr$_x$CuO$_4$. The exponent is defined as $\beta_O = -d\ln m_{ab}^{**}/d\ln M_O$. The data are from [64, 66, 67].

supercarriers is independent of the isotope mass [81]. In particular, the substantial oxygen-isotope effect on m_{ab}^{**} in optimally doped cuprates indicates that the polaronic effect does not vanish in the optimal doping regime where the BCS-like superconducting transition occurs. This suggests that polaronic carriers may be bounded into the Cooper pairs in optimally doped and over-doped cuprates.

3.3 Strong Electron–Phonon Coupling Features Along the Diagonal Direction

In conventional superconductors, strong electron-phonon coupling features can be identified from single-particle tunneling spectra. For high-T_c cuprates, high-quality tunneling spectra are difficult to obtain because of a short coherence length. Moreover, due to a strong gap anisotropy, the energies of the strong coupling features will depend on the tunneling directions. Only if one can make a directional tunneling, one may be able to identify the electron-phonon coupling features from the tunneling spectrum. On the other hand, the observation of the electron self-energy renormalization effect in the form of a "kink" in the band dispersion may reveal coupling of electrons with phonon modes. The "kink" feature at an energy of about 65 meV has been seen in the band dispersion of various cuprate superconductors along the diagonal ("nodal") direction [68]. From the measured dispersion, one can extract the real part of electron self-energy that contains information about coupling of electrons with collective boson modes. The energy of a broad peak in the electron self-energy should correspond to the averaged energy of all the bosonic modes that couple to electrons if the superconducting gap is zero along the diagonal direction. The fine coupling structures can be only revealed in the high-resolution ARPES data. Very recently, such fine coupling structures have been clearly seen in the raw data of electron self-energy of deeply underdoped $La_{2-x}Sr_xCuO_4$ along the diagonal direction [70]. Using the maximum entropy method (MEM) procedure, they are able to extract the electron-phonon spectral density $\alpha^2 F(\omega)$ that contains coupling features at 27 meV, 45 meV, 61 meV and 75 meV. These ARPES data and exclusive data analysis [70] clearly indicate that the phonons rather than the magnetic collective mode are responsible for the electron self-energy effect.

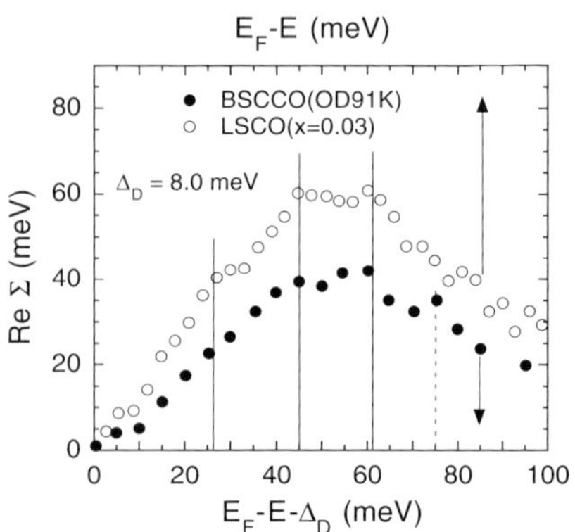

Fig. 13. The real part of electron self-energy along the diagonal direction for a slightly overdoped BSCCO with $T_c = 91$ K (OD91K) [82] and for $La_{2-x}Sr_xCuO_4$ with $x = 0.03$ [70]. The energy scale for BSCCO is shifted down by $\Delta_D = 8.0$ meV. The solid vertical lines (27 meV, 45 meV, 61 meV) mark the energies of the pronounced phonon peaks in the electron-phonon spectral density $\alpha^2 F(\omega)$ that is determined from the MEM procedure [70]. The dashed vertical line indicates the energy of a pronounced phonon peak (75 meV) in the superconducting LSCO with $x = 0.07$.

In fact, the fine coupling structures also appear in the earlier high-resolution ARPES data of a slightly overdoped BSCCO with $T_c = 91$ K (OD91K) [82]. Fig. 13 shows the real part of electron self-energy along the diagonal direction for the OD91K sample. Plotted in the same figure is the real part of electron self-energy along the diagonal direction for $La_{2-x}Sr_xCuO_4$ (LSCO) with $x = 0.03$. One can clearly see the fine structures in the raw data of both LSCO and BSCCO. The solid vertical lines mark the energies of the pronounced phonon peaks (27 meV, 45 meV, 61 meV) in the electron-phonon spectral density $\alpha^2 F(\omega)$ that is determined from the MEM procedure [70]. The dashed vertical line indicates the energy of a pronounced phonon peak (75 meV) in the superconducting LSCO with $x = 0.07$. In order for the fine structures in the self-energy of BSCCO to be aligned with those for LSCO, the energy scale for BSCCO has to be shifted down by $\Delta_D = 8.0$ meV. This suggests that the superconducting gap along the diagonal direction is not zero. This is consistent with an extended s-wave gap symmetry that has a finite gap of about 10 meV along the diagonal direction [74]. The magnitude of Δ_D surely depends on impurities and disorder, so it may vary from sample to sample.

It is interesting to note that the coupling feature at 75 meV is invisible in the deeply underdoped LSCO ($x = 0.03$), but becomes pronounced in the superconducting LSCO ($x = 0.07$) and in BSCCO (OD91K). This is consistent with the neutron experiments that clearly demonstrate that the coupling to the 75 meV half-breathing mode increases with increasing doping [61]. Nevertheless, the electron-phonon coupling constant for this mode is still small (< 0.2) due to a high phonon energy. This implies that this mode alone cannot cause high-temperature superconductivity, in contrast with the claim by Lanzara et al. [68]. In fact, the low energy phonon modes below 30 meV along the diagonal direction in LSCO ($x = 0.03$) contribute a large electron-phonon coupling constant (about 0.8).

3.4 Strong Electron–Phonon Coupling Features Along the Antinodal Direction

The electron self-energy effect along the antinodal direction has been studied for several BSCCO crystals (OD91K, OD71K,and OD58K) [83]. The kink feature in the band dispersion in the antinodal direction is much stronger than that along the diagonal direction. This indicates a much stronger electron-boson coupling. One of the puzzling issues is that the averaged energy of the boson modes shifts to a much lower energy (about 20 meV) [83]. Figure 14 shows the boson energy as a function of the antinodal gap Δ_M for several overdoped BSCCO. The boson energy is calculated according to $E_{boson} = E_{kink} - \Delta_M$, where E_{kink} is the kink energy in the band dispersion. Since the antinodal gap Δ_M was found to be very close to the peak energy in the energy distribution curve (EDC) [84], one can simply take Δ_M being equal to the EDC peak energy. One can see that the boson energy is about 20 meV for heavily overdoped BSCCO and about 16 meV for nearly optimally doped BSCCO. The strong coupling feature at about 20 meV agrees with the electron-boson spectral density $\alpha^2 F(\omega)$ deduced from a break-junction spectrum, as shown in Figure 15. The spectral density clearly shows strong coupling features at about 20 meV, 35 meV, 60 meV, and 72 meV. In particular, the coupling to phonon modes near 20 meV is strongly enhanced (the coupling constant for the 20 meV phonon peak is about 2.6). This unusual enhancement is possible if the extended van Hover singularity is about 20 meV below the Fermi level and the electron-phonon matrix element for the 20 meV phonon modes has a maximum around $q = 0$, where q is the phonon wavevector. The large density of states at the van Hover singularity (20 meV below the Fermi level) and strong Fermi surface nesting along the antinodal direction greatly enhance the phase space available for 20 meV small-q phonons to scatter quasiparticles from the states near the antinodal regime to the extended saddle points. The first principle calculation indeed shows that unusual long-range Madelung-like interactions lead to very large matrix elements especially for zone center modes that are related to vibrations of cations (e.g., La, Sr, Ba, Ca) [86]. The phonon energies for the vibrations of the cations are between 15 meV to 25 meV.

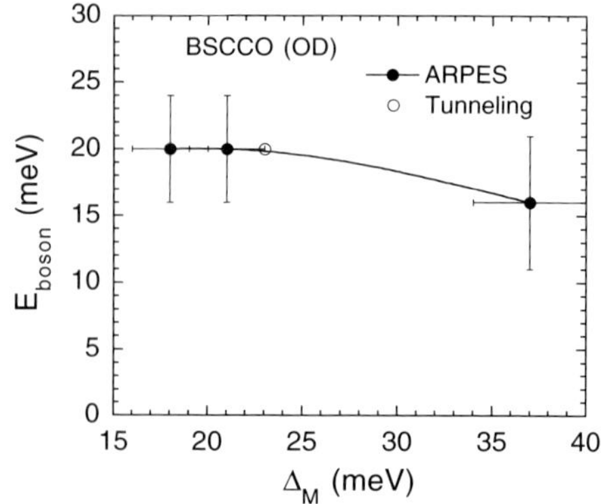

Fig. 14. The boson energy as a function of the antinodal gap Δ_M for several overdoped BSCCO. The boson energy extracted from ARPES [83] is calculated according to $E_{boson} = E_{kink} - \Delta_M$, where E_{kink} is the kink energy in the band dispersion. One data point (open circle) is from the tunneling data (see Fig. 15).

Further, a strong coupling feature in optical data, which was previously explained as due to a strong coupling between electrons and the magnetic resonance mode [87, 88], is actually consistent with a strong electron-phonon coupling at a phonon energy of about 20 meV. It is known that the electron-phonon spectral density $\alpha^2 F(\omega)$ can be obtained through inversion of optical data. Marsiglio et al. [89] introduced a dimensionless function $W(\omega)$ which is defined as the second derivative of the normal state optical scattering rate $\tau^{-1}(\omega) = (\Omega_p^2/4\pi)\text{Re}\,\sigma^{-1}(\omega)$ multiplied by frequency ω. Here Ω_p is the bare plasma frequency and $\sigma(\omega)$ the normal state optical conductivity. Specifically,

$$W(\omega) = \frac{1}{2\pi} \frac{d^2}{d\omega^2} \frac{\omega}{\tau(\omega)} \tag{13}$$

which follows directly from experiment. Marsiglio et al. [89] made the very important observation that within the phonon range $W(\omega) \simeq \alpha^2 F(\omega)$.

In the superconducting state, a phonon mode that is strongly coupled to electrons will appear at an energy of $2\Delta(\boldsymbol{k}) + \omega_{ph}$ (where ω_{ph} is the phonon energy), that is, the energies of the phonon structures shift up by the pair-breaking energy $2\Delta(\boldsymbol{k})$ [90]. Because the 20 meV phonon modes are much more strongly coupled to the states near the antinodal regime and because there is a large quasiparticle density of states at the maximum gap edge, there must be a maximum at $2\Delta_M + \omega_{ph}$ in $W(\omega)$. For slightly overdoped BSCCO

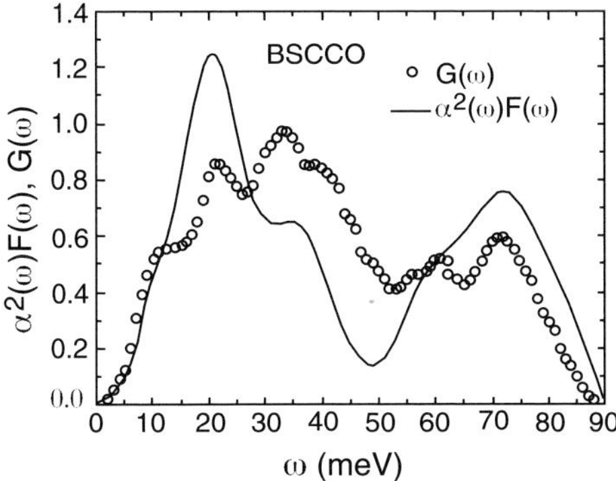

Fig. 15. The electron-phonon spectral density $\alpha^2 F(\omega)$ for a slightly overdoped $\text{Bi}_2\text{Sr}_2\text{CaCu}_2\text{O}_{8+y}$ (BSCCO) crystal, which was deduced from an SIS break-junction spectrum [85].

with $T_c = 90$ K, $\Delta_M = 26.0(5)$ meV [74, 91], so we should expect a maximum in $W(\omega)$ to be at about 72 meV. This is in quantitative agreement with the result shown in Fig. 16.

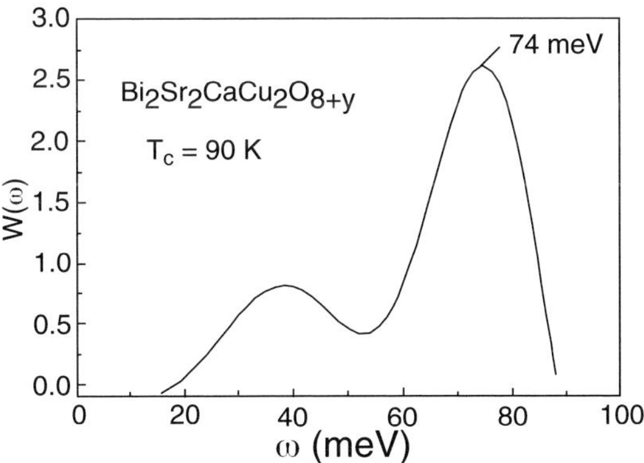

Fig. 16. The optically determined electron-boson spectral density $W(\omega)$ for a slightly overdoped BSCCO crystal with $T_c = 90$ K. After [88], ©(2000) by the American Physical Society.

Recently, Devereaux et al. have calculated the electron-phonon interactions for the oxygen buckling mode (B_{1g}) and the in-plane half-breathing mode [73]. They find that the 40 meV B_{1g} mode couples strongly to electronic states near the antinodal regime. They use an electron-phonon matrix element that is suitable only for $\text{YBa}_2\text{Cu}_3\text{O}_{7-y}$ where a large buckling distortion occurs. For other cuprates, the CuO_2 plane is flat and the buckling effect is negligible. Raman data have indeed shown that the coupling constant of the B_{1g} mode in

BSCCO is more than one order of magnitude smaller than that in YBCO [92]. Even for YBCO, the coupling constant of this mode was deduced to be about 0.05 from the Raman data [92], in agreement with the earlier first principle calculation [93]. Moreover, if this 40 meV phonon were strongly coupled to the electronic states near the antinodal regime, one would expect a maximum in $W(\omega)$ to occur at about 92 meV in slightly overdoped BSCCO with $T_c = 90$ K. This is in total disagreement with experiment (see Fig. 16).

Previously, the energy of the maximum in $W(\omega)$ was claimed to be in quantitative agreement with the theoretical prediction based on the strong coupling between electrons and the magnetic resonance mode [87, 88]. Later on, a more rigorous theoretical approach [94] shows that the maximum in $W(\omega)$ should occur at about $2\Delta_M + E_r$ rather than at $\Delta_M + E_r$, where E_r is the magnetic resonance energy. For BSCCO with $T_c = 90$ K, $E_r = 43$ meV, so we should expect a maximum in $W(\omega)$ to occur at about 95 meV, in disagreement with experiment (see Fig. 16)

The strong electron-phonon coupling feature at about 20 meV is also consistent with the energy distribution curves obtained from ARPES. For a conventional strong-coupling superconductor (e.g., Pb), a dip feature in the single-particle density of states can be clearly seen in the superconducting state. The dip feature appears near $\Delta + \omega_c$, where ω_c is the phonon cut-off energy [81]. From Fig. 15, one can see that the cut-off energy for the 20 meV phonon peak is about 40 meV, i.e., $\omega_c \simeq 40$ meV. Since the intensity of EDC is proportional to the single-particle spectral weight, the dip feature should appear in EDC at $\Delta_M + \omega_c$. Because Δ_M is close to the EDC peak energy along the antinodal direction, the separation energy between the peak and dip in the EDC should be close to $\omega_c \simeq 40$ meV, in agreement with experiment [66].

4 Conclusion

The various unconventional oxygen-isotope effects observed in both manganites and cuprates clearly demonstrate the existence of polaronic/bipolaronic charge carriers due to strong electron-phonon coupling. The very recent angle-resolved photoemission spectra for both manganites and cuprates [42, 80] have confirmed our earlier conclusions from the unconventional isotope effects. The formation of polarons/bipolarons in both manganites and cuprates plays an important role in the basic physics of these materials and is essential for the occurrence of colossal magnetoresistance and high-temperature superconductivity. We further show that strong electron-phonon coupling with the 20 meV phonon modes in cuprates are also important to the pairing mechanism of high-temperature superconductivity.

Acknowledgement. This research is partly supported by a Cottrell Science Award from Research Corporation.

References

1. J. G. Bednorz and K. A. Müller, Z. Phys. B **64**, 189 (1986). **332**, 814 (1988).
2. R. von Helmolt, J. Wecker, B. Holzapfel, L. Schultz, and K. Samwer, Phys. Rev. Lett. **71**, 2331 (1993); S. Jin, T. H. Tiefel, M. McCormack, R. A. Fastnacht, R. Ramesh, and L. H. Chen, Science **264**, 413 (1994).
3. L. D. Landau, Phys. Z. Sowjetunion **3**, 664 (1933).
4. T. Holstein, Ann. Phys. (N. Y.), **8**, 325 (1959).
5. D. Emin and T. Holstein, Ann. Phys. (N. Y.), **53**, 439 (1969).
6. A. S. Alexandrov and N. F. Mott, *Polarons and Bipolarons*, 67-106 (World Scientific, Singapore, 1995).
7. J. T. Devreese, in Encyclopedia of Applied Physics, vol. 14, p. 383, VCH Publishers (1996); J. Tempere and J. T. Devreese, Phys. Rev. B **64**, 104504 (2001); J. T. Devreese (in the present volume).
8. A. S. Alexandrov, Phys. Rev. B **61**, 12315 (2000).
9. A. S. Alexandrov and P. E. Kornilovitch, Phys. Rev. Lett. **82**, 807 (1999).
10. E. O. Wollan, and W. C. Koeler, Phys. Rev. **100**, 545 (1955).
11. G. H. Jonker, and J. H. van Santen, Physica **16**, 337 (1950).
12. C. Zener, Phys. Rev. **82**, 403 (1951).
13. P. W. Anderson, and H. Hasegawa, Phys. Rev. **100**, 675 (1955).
14. A. J. Millis, P. B. Littlewood, and B. I. Shraiman, Phys. Rev. Lett. **74**, 5144 (1995).
15. A. J. Millis, B. I. Shraiman, and R. Mueller, Phys. Rev. Lett. **77**, 175 (1996); H. Röder, J. Zang, and A. R. Bishop, Phys. Rev. Lett. **76**, 1356 (1996).
16. A. S. Alexandrov and A. M. Bratkovsky, Phys. Rev. Lett. **82**,141 (1999); A. S. Alexandrov and A. M. Bratkovsky, J. Phys.:Condens. Matter, **11**, 1989 (1999).
17. A. J. Millis, Nature **392**, 147 (1998).
18. H. L. Ju *et al.*, Phys. Rev. Lett. **79**, 3230 (1997).
19. T. Saitoh *et al.*, Phys. Rev. B **51**, 13942 (1995).
20. G. M. Zhao, Phys. Rev. B **62**, 11 639 (2001).
21. G.-M. Zhao and D. E. Morris, 1995 (unpublished).
22. G.-M. Zhao, K. Conder, H. Keller, and K. A. Müller, Nature **381**, 676 (1996).
23. G. M. Zhao, M. B. Hunt and H. Keller, Phys. Rev. Lett. **78**, 955 (1997).
24. J.-S. Zhou and J. B. Goodenough, Phys. Rev. Lett. **80**, 2665 (1998).
25. G. M. Zhao *et al.*, Phys. Rev. B **60**, 11 914 (1999).
26. A. Urushibara *et al.*, Phys. Rev. B **51**, 14103 (1995).
27. M. Jaime *et al.*, Phys. Rev. B **58**, R5901 (1998).
28. A. Machida, Y. Moritomo, and A. Nakamura, Phys. Rev. B **58**, R4281 (1998).
29. G. M. Zhao *et al.*, Phys. Rev. Lett. **84**, 6086 (2000).
30. A. Moreo, S. Yunoki, and E. Dagotto, Science **283**, 2034 (1994).
31. G. M. Zhao *et al.*, Phys. Rev. B **63**, 060402R (2000).
32. W. Prellier *et al.*, Appl. Phys. Lett. **75**, 1446 (1999).
33. W. E. Pickett and D. J. Singh, Phys. Rev. B **55**, R8642 (1997); D. A. Papaconstantopoulos and W. E. Pickett, Phys. Rev. B **57**, 12751 (1998).
34. A. S. Alexandrov, G. M. Zhao, H. Keller, B. Lorenz, Y. S. Wang, and C. W. Chu, Phys. Rev. B **64**, R140404 (2001).
35. K. Kubo and N. A. Ohata, J. Phys. Soc. Jpn. **33**, 21 (1972).
36. J. R. Simpson *et al.*, Phys. Rev. B **60**, R16 263 (1999).
37. H. Y. Hwang *et al.*, Phys. Rev. Lett. **80**, 1316 (1998).

38. M. C. Martin *et al.*, Phys. Rev. B **53**, 14 285 (1996).
39. V. H. Crespi and M. L. Cohen, Phys. Rev. B **48**, 398 (1993).
40. W. E. Pickett, R. E. Cohen, and H. Krakauer, Phys. Rev. Lett. **67**, 228 (1991).
41. W. Reichardt, B. Batlogg, and J. P. Remeika, Physica B **135**, 501 (1985).
42. N. Mannella *et al.*, Nature **438**, 474 (2005).
43. G. M. Zhao et al., Solid State Commun. **104**, 57 (1997)
44. M. R. Ibarra, G. M. Zhao, J. M. De Teresa, B. Garcia-Landa, Z. Arnold, C. Marquina, P. A. Algarabel, and H. Keller, Phys. Rev. B **57**, 7446 (1998).
45. M. Jaime, M. B. Salamon, M. Rubinstein, R. E. Treece, J. S. Horwitz, and D. B. Chrisey, Phys. Rev. B **54**, 11914 (1996).
46. S. J. L. Billinge, R. G. DiFrancesco, G. H. Kwei, J. J. Neumeier, and J. D. Thompson, Phys. Rev. Lett. **77**, 715 (1996).
47. C. H. Booth, F. Bridges, G. H. Kwei, J. M. Lawrence, A. L. Cornelius, and J. J. Neumeier, Phys. Rev. Lett. **80**, 853 (1998).
48. D. Louca, T. Egami, E. L. Brosha, H. Röder, and A. R. Bishop, Phys. Rev. B **56**, R8475 (1997).
49. A. Lanzara, N. L. Saini, M. Brunelli, F. Natali, A. Bianconi, P. G. Radelli, and S. W. Cheong, Phys. Rev. Lett. **81**, 878 (1998).
50. G. M. Zhao *et al.*, Phys. Rev. B **63**, 11949R (2000).
51. I. G. Austin and N. F. Mott, Adv. Phys. **18**, 41 (1969).
52. A. S. Alexandrov and A. M. Bratkovsky, J. Phys.: Condens. Matter, **11**, L531 (1999).
53. J. M. De Teresa, M. R. Ibarra, P. A. Algarabel, C. Ritter, C. Marquina, J. Blasco, J. Garcia, A. del Moral, and Z. Arnold, Nature (London), **386**, 256 (1997).
54. B. Batlogg, R. J. Cava, A. Jayaraman, R. B. van Dover, G. A. Kourouklis, S. Sunshine, D. W. Murphy, L. W. Rupp, H. S. Chen, A. White, K. T. Short, A. M. Mujsce, and E. A. Rietman, Phys. Rev. Lett. **58**, 2333 (1987).
55. L. C. Bourne, M. F. Crommie, A. Zettl, H. zur Loye, S. W. Keller, K. L. Leary, A. M. Stacy, K. J. Chang, M. L. Cohen, and D. E. Morris, Phys. Rev. Lett., **58**, 2337 (1987).
56. G. M. Zhao, K. K. Singh, and D. E. Morris, Phys. Rev. B **50**, 4112 (1994).
57. G. M. Zhao and D. E. Morris, Phys. Rev. B **51**, 16487R (1995).
58. G. M. Zhao, K. K. Singh, A. P. B. Sinha, and D. E. Morris,, Phys. Rev. B **52**, 6840 (1995).
59. G. M. Zhao, M. B. Hunt, H. Keller, and K. A. Müller, Nature (London) **385**, 236 (1997).
60. G. M. Zhao, K. Conder, H. Keller, and K. A. Müller, J. Phys.: Condens. Matter, **10**, 9055 (1998).
61. R. J. McQueeney, Y. Petrov, T. Egami, M. Yethiraj, G. Shirane, and Y. Endoh, Phys. Rev. Lett. **82**, 628 (1999).
62. A. Lanzara, G. M. Zhao, N. L. Saini, A. Bianconi, K. Conder, H. Keller, and K. A. Müller, J. Phys.: Condens. Matter **11**, L541 (1999).
63. A. Shengelaya, G. M. Zhao, C. M. Aegerter, K. Conder, I. M. Savic, and H. Keller, Phys. Rev. Lett. **83**, 5142 (1999).
64. J. Hofer, K. Conder, T. Sasagawa, G. M. Zhao, M. Willemin, H. Keller, and K. Kishio, Phys. Rev. Lett. **84**, 4192 (2000).
65. G. M. Zhao, H. Keller, and K. Conder, J. Phys.: Condens. Matter, **13**, R569 (2001).

66. G. M. Zhao, Phil. Mag. B **81**, 1335 (2001).

67. G. M. Zhao, V. Kirtikar, and D. E. Morris, Phys. Rev. B **63**, 220506R (2001).

68. A. Lanzara *et al.*, Nature (London) 412, 510(2001).

69. R. Khasanov *et al.*, Phys. Rev. Lett. **92**, 057602 (2004).

70. X. J. Zhou *et al.*, Phys. Rev. Lett. **95**, 117001 (2005).

71. G. M. Zhao, Phys. Rev. B **71**, 104517 (2005).

72. Z.-X. Shen, A. Lanzara, S. Ishihara, and N. Nagaosa, Phil. Mag. B **82**, 1349 (2002).

73. T. P. Devereaux, T. Cuk, Z.-X. Shen, and N. Nagaosa,Phys. Rev. Lett. **93**, 117004 (2004).

74. G. M. Zhao, Phys. Rev. B **64**, 024503 (2001); G. M. Zhao, Phil. Mag. B **84**, 3861 (2004); G. M. Zhao, Phil. Mag. B **84**, 3869 (2004).

75. B. H. Brandow, Phys. Rev. B **65**, 054503 (2002).

76. T. Thio et al., Phys. Rev. B **38**, 905 (1988).

77. J. H. Cho, F. C. Chou, and D. C. Johnston, Phys. Rev. Lett. **70**, 222 (1993).

78. I. Eremin, O. Kamaev, and M. V. Eremin, Phys. Rev. B **69**, 094517 (2004).

79. X. X. Bi and P. C. Eklund, Phys. Rev. Lett. **70** 2625 (1993).

80. O. Rösch *et al.*, Phys. Rev. Lett. **95** 227002 (2005).

81. J. P. Carbotte, Rev. Mod. Phys. **62**, 1027 (1990).

82. P. D. Johnson, T. Valla, A.V. Fedorov, Z. Yusof, B. O. Wells, Q. Li, A. R. Moodenbaugh, G. D. Gu, N. Koshizuka, C. Kendziora, Sha Jian, and D. G. Hinks, Phys. Rev. Lett. **87**, 177007 (2001).

83. A. D. Gromko, A. V. Fedorov, Y.-D. Chuang, J. D. Koralek, Y. Aiura, Y. Yamaguchi, K. Oka, Yoichi Ando, and D. S. Dessau, Phys. Rev. B **68**, 174520 (2003).

84. H. Ding *et al.*, Phys. Rev. Lett. **74**, 2784 (1995).

85. R. S. Gonnelli, G. A. Ummarino, and V. A. Stepanov, Physica C **275**, 162 (1997).

86. W. E. Pickett, R. E. Cohen, and H. Krakauer, Phys. Rev. Lett. **67**, 228 (1991).

87. J. P. Carbotte, E. Schachinger, and D. N. Basov, Nature (London) **401**, 354 (1999).

88. E. Schachinger and J. P. Carbotte, Phys. Rev. B, **62**, 9054 (2000).

89. F. Marsiglio, T. Startseva, and J.P. Carbotte, Physics Lett. A **245**, 172 (1998).

90. J. Orenstein, Nature (London) **401**, 333 (1999).

91. V.M. Krasnov, A. Yurgens, D. Winkler, P. Delsing, and T. Claeson, Phys. Rev. Lett. **84**, 5860 (2000).

92. M. Opel *et al.*, Phys. Rev. B **60**, 9836 (1999).

93. C. O. Rodriguez, A. I. Liechtenstein, I. I. Mazin, O. Jepen, O. K. Anderson, M. methfessel, Phys. Rev. B **42**, 2692 (1990).

94. A. Abanov, A. V. Chubukov, and J. Schmalian, Phys. Rev. B **63**, 180510R (2001).

Polaron Effects in High-Temperature Cuprate Superconductors

Annette Bussmann-Holder[1] and Hugo Keller[2]

[1] Max-Planck-Institut für Festkörperforschung, Heisenbergstr. 1, D-70569
Stuttgart, Germany `A.Bussmann-Holder@fkf.mpg.de`
[2] Physik-Institut der Universität Zürich, Winterthurerstr. 190, CH-8057 Zürich,
Switzerland `keller@physik.unizh.ch`

Summary. High-temperature cuprate superconductors (HTSC) are modelled within
a two-component approach where strong inter-component interactions are driven by
the lattice. It is shown that a polaronic renormalization of the single-particle energies
together with the inter-component interactions lead to a substantial enhancement
of the superconducting transition temperature T_c, various unconventional isotope
effects, and strain-induced enhancements of T_c. Further consequences of polaronic
renormalization effects are discussed.

1 Introduction

The discovery of high-temperature cuprate superconductors [1] was motivated
by the knowledge that the copper ion is one of the strongest Jahn-Teller
ions known [2]. The solid-state Jahn-Teller effect is a consequence of strong
electron-lattice coupling and the formation of polaronic states, i.e., states
where the electronic and lattice wave functions are not separable, have been
attributed to it [3]. In HTSC the role played by electron-lattice interactions
has frequently been questioned due to the antiferromagnetic properties of
the undoped or slightly doped parent compounds. Since these are half-filled
insulating band systems, a strong correlation energy U at the copper lattice
site is present which prevents double occupancy. The energy scale set by U
is clearly the largest in these systems, and it has been argued [4] that – as
a consequence – the physical properties of cuprates are dominated by it. In
support of this argument is the observation of an almost vanishing isotope
effect on T_c at optimum doping [5] and the likely d-wave symmetry of the
superconducting order parameter. In addition, T_c is too large to be explained
within conventional phonon mediated BCS theory. Also, there is evidence that
spin fluctuations persist in the superconducting phase [6], and the conventional
Fermi liquid picture seems to be inadequate to describe the physics of the
underdoped phase.

However, the total neglect of the lattice and its interactions with the electronic systems is certainly an incomplete approach in achieving a physically realistic approach to HTSC, since many experiments clearly reveal striking effects stemming from the lattice and correlated with superconductivity. Extended x-ray absorption fine structure (EXAFS) [7, 8] as well as ion channeling results [9] show that the local crystallographic structure deviates pronouncedly from the average one in the superconducting regime. The application of strain enhances T_c substantially [10] and induces changes in the Fermi surface topology [11]. In addition, local structural responses are observed which depend on the substrate and change sign for strained samples as compared to compressed ones, accompanied by giant changes in T_c [12, 13]. The isotope effect on T_c increases with decreasing T_c in the underdoped regime and even exceeds the BCS value in the very proximity to the antiferromagnetic state [5, 14–19]. The London penetration depth has an isotope effect [16–22], not present within conventional BCS theory, which finds an interpretation only if unconventional lattice effects are considered. The electronic dispersion shows a universal kink at energies compatible with phonon frequencies [23], and isotope effects on the dispersion have been reported [24]. Atomic pair distribution function (PDF) data [25] have demonstrated the existence of short range correlated atomic displacements. The anisotropic thermal conductivity data of single crystal $YBa_2Cu_3O_{6.92}$ has been interpreted in terms of unusually strong electron phonon coupling [26]. This has also been concluded from Raman [27] and infrared data [28], where strong phonon responses appear at the onset of superconductivity. Another important issue in HTSC is the observation of substantial heterogeneity which shows up in pronounced time-dependent electronic properties [29] and gives rise to spatial fluctuations in the superconducting energy gap [30] but also modifies the infrared response [31], the Josephson plasma resonance [32], is manifested in two distinctly different relaxation times in the time resolved optical response [33], modifies the EPR [34] and NMR data [35], and is present in x-ray [36] and neutron scattering data [37]. An interpretation of this inhomogeneous state in terms of dynamic stripes has been given early [38], where the onset temperature T* for this pseudo ordered state (pseudogap state) occurs at much higher temperatures than T_c for small doping, gradually decreases with increasing T_c, and finally merges with it around optimum doping. Within models based on strong correlations, T* has been interpreted as the opening of a spin gap [39]. However, various experiments on different HTSC compounds have shown that T* carries an unusually large isotope effect which is sign reversed [40–42] and absent within purely electronic models. This observation is in contrast to results obtained by NQR techniques [43] where a small isotope effect with the same sign as the one on T_c has been reported. Heterogeneity seems to appear also in the symmetry of the superconducting order parameter: While Andreev reflection measurements have detected a combined s+d order parameter [44], Raman scattering, microwave and temperature dependent surface impedance measurements have been interpreted in terms of multiple coupled order parameters [45].

2 Intrinsic Heterogeneity and Complexity of Cuprates

How can this complexity of cuprates be understood? Since it is evident from the experimental findings that models based on strong correlations only are incomplete, the role played by the lattice has to be reinvestigated. It is important to mention that soon after the discovery of HTSC it was argued that unconventional electron-lattice interactions, i.e. different from those considered by the BCS theory, have to be taken into account [46]. Thus, the high static dielectric constants of cuprates have motivated models based on strong electron-lattice anharmonicity related to incipient ferroelectric instabilities [47]. Multiphonon processes are a consequence of such a scenario and facilitate interband interactions and multicomponent superconductivity [48]. Related to these ideas are models based on polaron formation [49–51] and bipolaron superconductivity [49, 52], where it was, however, argued that small polarons give rise to an activated mobility with semiconducting temperature dependence of the resistivity, which is not observed in cuprates [53]. On the other hand, for conventional large polarons a sign change in the Seebeck coefficient near room temperature has been predicted [53] which has not been observed experimentally for all cuprates [54]. With respect to bipolaronic superconductivity it has been argued that Bose condensation of small bipolarons takes place and that these objects are already present in the normal state [49]. In conflict with this assumption is the observation of a well defined Fermi surface and the formation of Cooper pairs below T_c with a clear gap opening. A way out of this dilemma has been proposed by Hirsch [55] and Goodenough and coworkers [56] who proposed intermediate size polarons with extensions of up to four lattice constants. They point out that the electron-phonon coupling could overcome the Coulomb repulsion within the spatial extent of the intermediate size polaron and lead to pairing within these "correlation bags" [57]. Importantly, the antiferromagnetic matrix is partially conserved outside the bags and fluctuations are still possible even for large doping concentrations. A clear shortcoming of both approaches is that quantitative comparison with experimental data is not made and that unconventional isotope effects, such as those mentioned above, are not considered. On the other hand, however, heterogeneity is common to such a scenario and self-organization from a polaron liquid into a polaron glass signals the onset of the pseudogap phase [58]. Characteristic of this scenario is that charge rich and charge poor areas are formed in real space evidencing the onset of coexisting ground states. A similar result has been achieved by looking for exact solutions of a nonlinear electron-lattice interaction model, where spatial modulations of the charge density, incommensurate with the underlying lattice, have been obtained [59]. Extensions of these solutions have been made later by taking into account the coupling of these local solutions to the "unperturbed" background [60]. The model thus corresponds to a two-component scenario with coupling between locally distorted charge rich (spin poor) regions to undistorted charge poor (spin rich) regimes. Such a scenario resembles closely the "correlation bag" approach [57] but extended to account for selfconsistent coupling to the host "matrix".

3 Two-Component Scenario

Doping the antiferromagnetic parent compounds of HTSC introduces a distribution of holes into a stoichiometric system where strong electronic correlations are present at the copper site. Obviously, all energy scales are destabilized by the "extra" charge, since charge and lattice mismatch are a consequence, and a rapid destruction of the antiferromagnetic order around the doped holes sets in. In accordance with the considerations of [53–57] we postulate that a stabilization of this energetically unstable situation can be achieved through local lattice displacements around the doped holes, i.e., polaron formation (in the sense of coupled spin-charge-lattice deformation) which remains, however, dynamic (Fig. 1). In the following, we consider two subsystems: the antiferromagnetic background and the polaron distorted regimes around the doped holes. The interactions between these two systems happen via the lattice which modulates the hopping integrals. For both components it is assumed that the electronic energies are p-d hybridized states [61, 62]. In order to differentiate between them, the electronic states within the antiferromagnetic background are denoted by creation and annihilation operators d^+, d, with density $n_d = d^+ d$, those within the lattice distorted areas are labelled c^+, c, and $n_c = c^+ c$.

The above considerations are taken into account by using a Hamiltonian which explicitly incorporates effects stemming from the lattice, where spin– and charge–lattice coupling are included. Correspondingly, the Hamiltonian is given by [61]:

$$H = H_d + H_c + H_{cd} + H_L + H_{L-d} + H_{L-c} \qquad (1)$$

H_d and H_c refer to the pure electronic energies in the c and d channels, the third part corresponds to the interaction between both channels resembling a hybridization term. Further, the non-renormalized lattice part H_L is considered and the two final terms, H_{L-d} and H_{L-c}, are those where interactions between c and d electrons (holes) with the lattice are taken into account. The terms in (1) are then defined as:

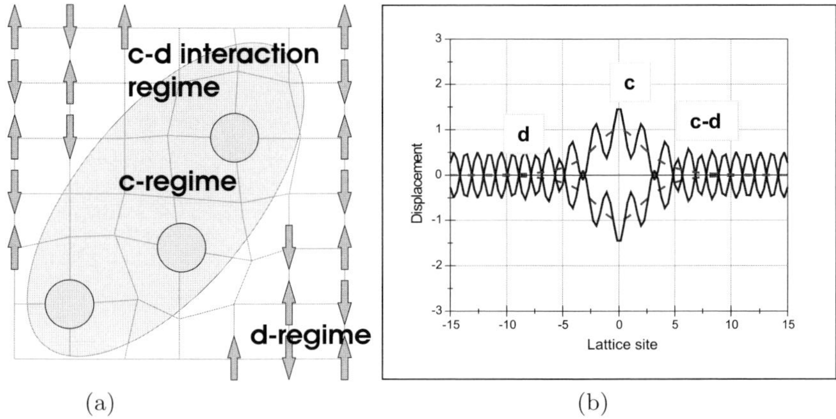

(a)   (b)

Fig. 1. a) Schematic representation of c and d dominated regimes as discussed in the text; b) Calculated displacement pattern based on an electron-phonon interaction model where the notations of the different regimes are the same as in (a).

$$H_d = \sum_{i,\sigma} \varepsilon_d d^+_{i,\sigma} d_{i,\sigma} + \sum_{i,j,\sigma,\sigma'} t_{ij}(d^+_{i,\sigma} d_{j,\sigma'} + c.c.) + U \sum_i n_{d,i\uparrow} n_{d,i\downarrow}$$

$$H_c = \sum_{i,\sigma} \varepsilon_c c^+_{i,\sigma} c_{i,\sigma} + \sum_{i,j,\sigma,\sigma'} t_{ij}(c^+_{i,\sigma} c_{j,\sigma'} + h.c.)$$

$$H_{cd} = \sum_{i,j,\sigma,\sigma'} t_{cd}(c^+_{i,\sigma} d_{j,\sigma'} + c.c.) \tag{2}$$

$$H_L = \sum_i \frac{p_i^2}{2M_i} + \frac{1}{2}M\omega^2 Q_i^2$$

$$H_{L-d} = \sum_{i,j,\sigma,\sigma'} [g n_{d,i} Q_i + \widetilde{g}(c^+_{i,\sigma} d_{j,\sigma'} + d^+_{i,\sigma} c_{j,\sigma'})Q_j]$$

$$H_{L-c} = \sum_{i,j,\sigma,\sigma'} [g n_{c,i} Q_i]$$

Here ε is the site i dependent energy, t the hopping integral, U the on-site Coulomb repulsion within the antiferromagnetic matrix. Note, that U is strongly reduced by the electron–lattice interactions in the hole rich distorted regions where metallic droplets are formed which appear as in-gap states in the vicinity of the Fermi surface. In these regions U loses its meaning, whereas the electron–lattice coupling sets the important energy scale. For simplicity the lattice Hamiltonian is taken to be purely harmonic with p and Q being momentum and conjugate displacement coordinates at site i with frequency ω and M is the ionic mass. The coupling between the lattice and the electronic system consists of an on-site coupling proportional to g which is also present within the d-regime due to the long range strain fields created by the distortions around the doped holes. The intersite coupling $\widetilde{g}$ which provides the interaction between the c and d dominated regions originates from the co-existence regime of c and d-states where charge transfer processes are possible through the local lattice distortions. Principally, the spin- and charge-lattice coupling constants are not identical, but for simplicity these are taken to be the same throughout this chapter.

A decoupling of lattice and electronic degrees of freedom can be obtained by a Lang-Firsov transformation [63] which corresponds to the following replacements:

$$\widetilde{c}_i = c_i \exp[\sum_q g_q[b_q^+ - b_q]], \quad \widetilde{c}_i^+ = c_i^+ \exp[-\sum_q g_q[b_q^+ - b_q]]$$

$$\widetilde{d}_i = d_i \exp[\sum_q g_q[b_q^+ - b_q]], \quad \widetilde{d}_i^+ = d_i^+ \exp[-\sum_q g_q[b_q^+ - b_q]] \qquad (3)$$

$$\widetilde{b}_q = b_q + \sum_q g_q d_i^+ d_i, \quad \widetilde{b}_q^+ = b_q^+ + \sum_q g_q d_i^+ d_i$$

and a corresponding transformation for the momentum q dependent phonon creation and annihilation operators b^+, b which couple to the c-channel. The consequences of the transformation are that the site energies experience a rigid band shift, i.e., $\varepsilon \to \widetilde{\varepsilon} = \varepsilon - \Delta^*$, with $\Delta^* = \frac{1}{2N} \sum_q \frac{g_q^2}{\hbar \omega_q}$, whereas the hopping integrals are exponentially renormalized as $t \to \widetilde{t} = t \exp[-g^2 \coth \frac{\hbar \omega}{2kT}]$ [64]. In addition, another consequence of the transformation is that lattice induced $n_c n_d$ density-density interactions between c and d bands appear, which give rise to interband interactions between c and d electronic states [61]. These interactions are of crucial importance since they facilitate – as will be shown below – multiband superconductivity. It is important to note that the quasi-particles are now dressed and carry a local distortion with them which reduces their mobility. This means that the band states related to the doped holes within the metallic droplets lose dispersion, whereas those related to the d holes, which are nearly localized in the undoped case, gain dispersion due to their interaction with the c states. A strange metal phase is the consequence.

At high temperatures the polaronic distorted areas are randomly distributed over the lattice, forming a polaron liquid (Fig. 2) [58]. A transition to an ordered state (analogous to the polaron glass state as postulated in [58]) takes place with decreasing temperature and ordering into patterns (stripe segments) becomes energetically favourable in order to minimize the interaction energy attributed to strain fields.

It is, however, important to mention that the patterning remains dynamic in order to ensure the interplay between c and d dominated regions. Static stripes have been observed at around $1/8$ doping in $LaSrCuO_4$ [65] where superconductivity is suppressed. The transition from the polaron "liquid" to the polaron "glass" is attributed to the opening of the pseudogap at T^*. Theoretically this transition can be modeled as a transition from a disordered phase to an ordered one [66], which carries a close analogy to transitions ob-

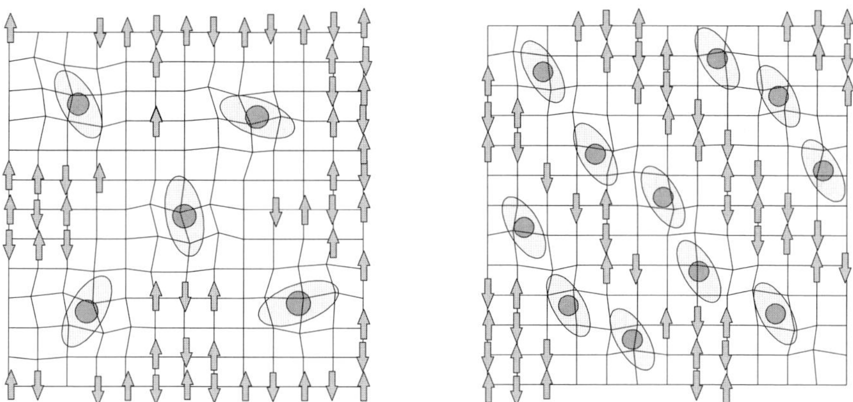

Fig. 2. Disordered polaron gas (left panel), ordered polaron liquid (right panel).

served in superionic conductors. Here a coupling of the disordered polaron state (liquid) to the strain fields takes place, and the transition temperature becomes a function of the coupling g of the polarons through the strain fields, the number of polarons, i.e. the doping level δ, and the energy ε_p to remove a single polaron from its ordered arrangement, i.e. $k_B T^* = (\varepsilon_p - g/2)/\ln\delta$ [66]. Clearly T^* decreases with doping as observed experimentally, but since also ε_p and g are doping dependent, the T^* versus doping dependence cannot easily be determined within this phenomenological approach. On the other hand, the ordering of polarons into dynamical patterns resembles closely charge ordering in the fashion of charge density wave formation which has been used previously to calculate the doping dependence of T^* [61]. The dynamical patterning is accompanied by the formation of in-gap electronic states close to the Fermi energy which mark the onset of the insulator metal transition. The dynamically broken translational symmetry of this new phase restricts these states to special k-vectors where intensity is gained with increasing doping. This is expected to be of relevance to ARPES data where doping induced new states have been observed along the nodal direction [67–69].

4 Isotope and Strain Effects

In what follows, the superconducting properties of the two-component system are addressed together with isotope and strain effects. For this purpose the Hamiltonian is cast into an effective BCS scheme extended to account for multiband superconductivity [70]. The terms which are of major interest are the band energies, where we use the simplified bands suggested by LDA calculations [71]:

$$E_{c,d} = -2t_1(\cos k_x a + \cos k_y b) + 4t_2 \cos k_x a \cos k_y b$$
$$+ 2\,t_3(\cos 2k_x a + \cos 2k_y b) \mp t_4(\cos k_x a - \cos k_y b)^2/4 - \mu. \quad (4)$$

Here t_1, t_2, t_3 account for nearest, second nearest and third nearest neighbor hopping integrals in the CuO_2 planes, and t_4 is the interlayer hopping, which is particularly relevant to bi- and multi-layer materials but also to single layer ones via apical oxygen ions. We use $t_1 = 0.1$ eV, $t_2/t_1 = 0.3$, $t_3/t_2 = 0.2$ and $t_4 = t_2$ throughout the paper in accordance with band structure calculations [71]. μ is the chemical potential which controls the number of holes. According to recent angle resolved photoemission spectroscopy (ARPES) results [72], the chemical potential scales directly with doping. Due to the polaronic renormalization all hopping integrals are exponentially renormalized, where, in this study, it is assumed for simplicity that the renormalizations are identical for each hopping term. This means that all t have to be replaced by $\tilde{t} = t \exp(-g^2 \coth \frac{\hbar\omega}{2kT})$ and an additional level shift $\Delta^* = \frac{1}{2N} \sum_q |g(q)|^2/(\hbar\omega)$ has to be added. At the mean-field level the reduced Hamiltonian corresponding to (1) and (2) then reads [70]:

$$H = H_0 + H_1 + H_2 + H_{12}$$

$$H_0 = \sum_{k_1\sigma} \xi_{k_1} c^+_{k_1\sigma} c_{k_1\sigma} + \sum_{k_2\sigma} \xi_{k_2} d^+_{k_2\sigma} d_{k_2\sigma}$$

$$H_1 = -\sum_{k_1 k'_1 q} V_1(k_1, k'_1) c^+_{k_1+q/2\uparrow} c^+_{-k_1+q/2\downarrow} c_{-k'_1+q/2\downarrow} c_{k'_1+q/2\uparrow} \tag{5}$$

$$H_2 = -\sum_{k_2 k'_2 q} V_2(k_2, k'_2) d^+_{k_2+q/2\uparrow} d^+_{-k_2+q/2\downarrow} d_{-k'_2+q/2\downarrow} d_{k'_2+q/2\uparrow}$$

$$H_{12} = -\sum_{k_1 k_2 q} V_{12}(k_1, k_2)\{c^+_{k_1+q/2\uparrow} c^+_{-k_1+q/2\downarrow} d_{-k_2+q/2\downarrow} d_{k_2+q/2\uparrow} + h.c.\}$$

Here H_0 is the kinetic energy of the bands $i = 1, 2$ with band energy $\xi_k = \tilde{\varepsilon} - \varepsilon_k - \mu$, and $\tilde{\varepsilon}_i$ denotes the position of the c and d band. Note, that the d-band refers to the lower Hubbard band whereas the c-band forms in-gap states close to the Fermi energy with strongly limited k-space weight. The pairing potentials $V_i(k_i, k'_i)$ are assumed to be represented in factorized form as $V_i(k_i, k'_i) = V_i \varphi_{k_i} \psi_{k'_i}$, where $\varphi_{k_i}, \psi_{k_i}$ are cubic harmonics for anisotropic pairing which yields for dimension D=2 and on-site pairing: $\varphi_{k_i} = 1$, $\psi_{k_i} = 1$, extended s-wave: $\varphi_{k_i} = \cos k_x a + \cos k_y b = \gamma_{k_i}$, and d-wave: $\varphi_{k_i} = \cos k_x a - \cos k_y b = \eta_{k_i}$ where a, b are the lattice constants along x and y directions, with $a \neq b$ to account for the orthorhombicity. It is important to note that the pairing interactions in the c and d channel can have different origins: while in the c channel they can result from effectively attractive electron-lattice interactions, they can be of magnetic origin in the d channel. However, more important than the intraband potentials are those interactions between the bands, i.e., the interband interactions, where pairwise exchange between c and d channels takes place. These interactions are clearly driven by the lattice. By performing a BCS mean-field analysis of the above equations [70, 73], the gap equations can be derived and the superconducting transition temperature evaluated. The details of this procedure are presented in the Appendix.

If the intraband interactions V are constants, the resulting gaps are momentum independent. A more interesting case is obtained by assuming the following general momentum dependence of the intraband interactions [70, 73]: $V_i = V_0(1 + \gamma_k \gamma_{k'} + \eta_k \eta_{k'})$, where the first term yields on-site pairing, the second extended s-wave pairing and the last term d-wave pairing. In the following calculation it is assumed that the symmetry of V_1 corresponds to on-site s-wave pairing while V_2 has either extended s-wave pairing or d-wave pairing symmetry. For both bands the dispersion relation (4) is assumed to apply, however, as already outlined above, the k-space weight of the respective bands differs [74]. In addition, the polaron induced exponential renormalizations are taken into account, which have important consequences for isotope effects [64, 74] as will be shown below. Importantly, they also lead to an increase in the effective mass which can be substantial. Within this scenario the self-consistent set of equations is solved numerically as a function of $V_{12} = \widetilde{V}_{12}\sqrt{N_c N_d}$, where N_c, N_d are the densities of states of the c and d band, respectively. The results are shown in Fig. 3 where both cases $V_1 \sim V_0, V_2 \sim V_0$ and $V_1 \sim V_0, V_2 \sim V_0(\cos k_x a - \cos k_y b)$ are considered. Obviously small values of V_{12} are sufficient to induce superconductivity in both channels. With increasing V_{12} substantial enhancements of T_c are obtained such that T_c easily exceeds 100K [70].

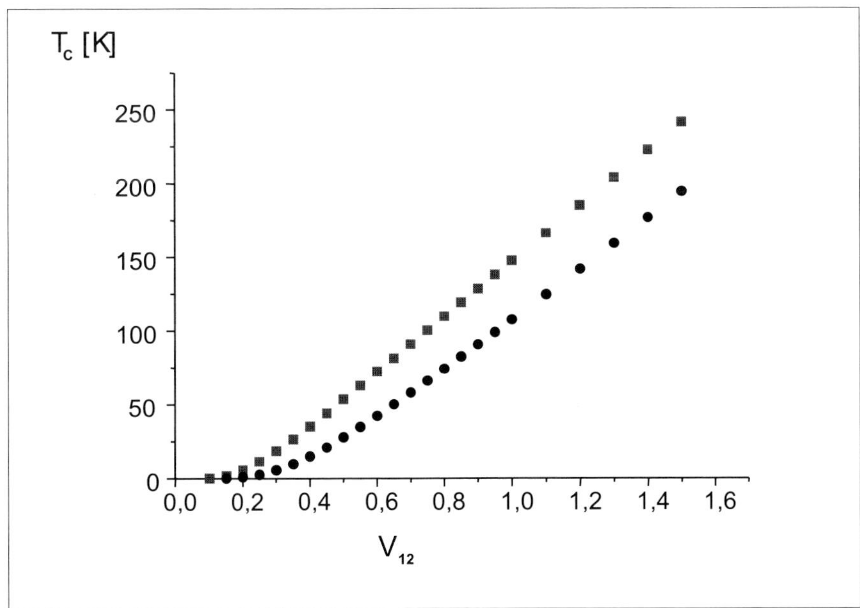

Fig. 3. The dependence of T_c on the interband pairing V_{12}. Squares refer to coupled s+d order parameters, circles to s+s combinations.

Interestingly, the combination of s+d wave symmetry in the two component systems has an additional T_c-increasing factor which increases with increasing interband coupling (Fig. 3). This finding clearly shows that a mixed order parameter symmetry favors superconductivity, as compared to two identical pairing interactions, and from now on only this combination is considered in the following.

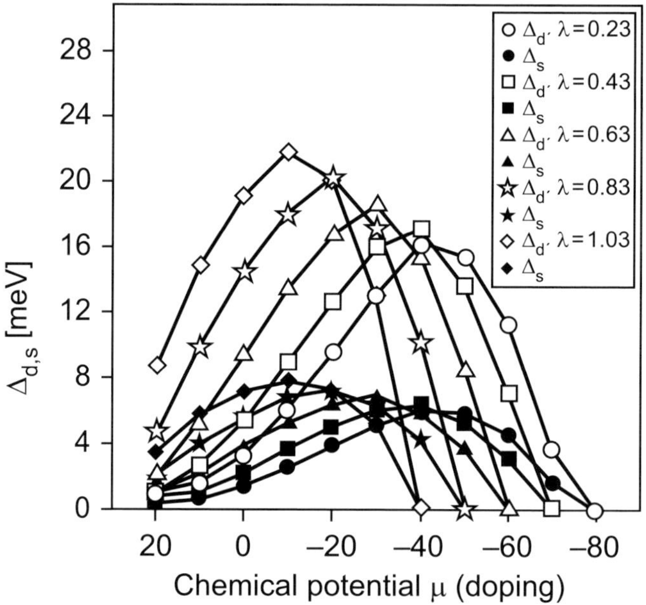

Fig. 4. Doping dependence of the superconducting gaps Δ_s, Δ_d for various coupling constants as indicated in the figure.

The corresponding superconducting gaps Δ_s, Δ_d have been calculated as functions of the chemical potential and for the case of combined s and d-wave symmetry and are shown in Fig. 4. This combination admits direct comparison of our results to those obtained via Andreev reflection spectroscopy [44], where unambiguous evidence for coupled order parameters with different symmetry has been obtained. From our calculation we observe that the ratio of s to d-wave gaps is nearly independent of doping in agreement with the experimental results. The comparison between theory and experiment is made in Fig. 5, where the average gap $E_g = \sqrt{\Delta_s^2 + \Delta_d^2}$ is shown as a function of T_c and good agreement between both is observed.

The transition temperature has been calculated as a function of doping for various values of the polaronic coupling g (Fig. 6). The typical dome shaped dependence, as observed experimentally, is found. In addition, it is seen that a substantial enhancement of T_c takes place with increasing polaronic coupling,

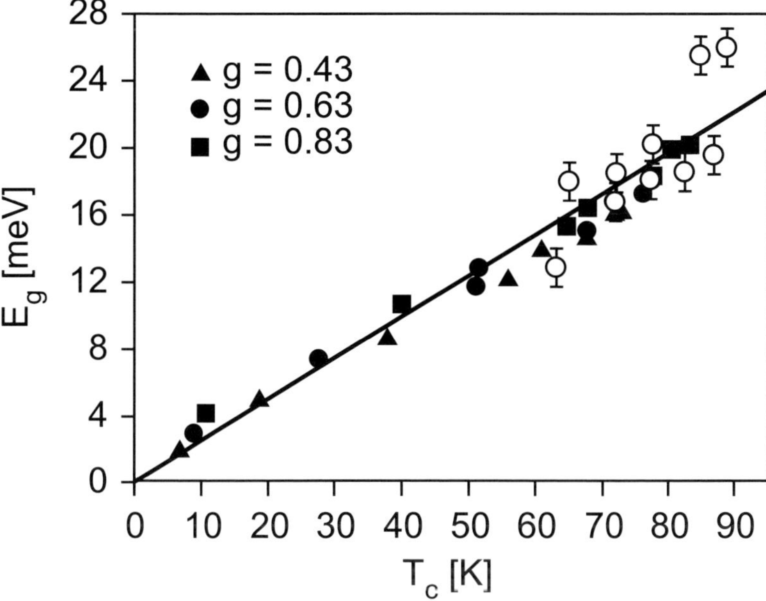

Fig. 5. Average superconducting gap $E_g = \sqrt{\Delta_s^2 + \Delta_d^2}$ as a function of T_c. Full symbols with coupling constants g as indicated in the figure are calculated values. Open circles are experimental results from [44].

even though the main enhancement effects are a consequence of interband interactions. Note that the decrease in T_c on the overdoped side is more rapid than on the underdoped side. It is important to mention that with too strong polaronic coupling, T_c collapses since localization sets in and the doped carriers become immobile. This is expected to happen in the underdoped regime where T^* is large, reflecting the strong polaronic coupling, and a metal insulator transition sets in. It is proposed here to test this new scenario which differs substantially from the Mott–Hubbard approach for the metal insulator transition, by performing isotope experiments at the transition line. It is also important to mention that the above results have been obtained at a mean-field level, where the spatial heterogeneity has been averaged out and is only accounted for by attributing the two components to special k-space-regions on the Fermi surface.

In order to explain the effects on T_c which arise through strain, the phase diagram has been calculated as a function of doping and the ratio of the second-nearest neighbour hopping integral with respect to the nearest neigh-bour one, i.e., t_2/t_1. This ratio has been shown to be crucial to T_c experimentally [11] as well as from trends seen in band structure calculations [71]. Experimentally, ARPES data have provided evidence that highly dispersive bands at the Fermi energy change their dispersion by applying strain as com-pared to unstrained samples [11]. This feature has been explained as being due

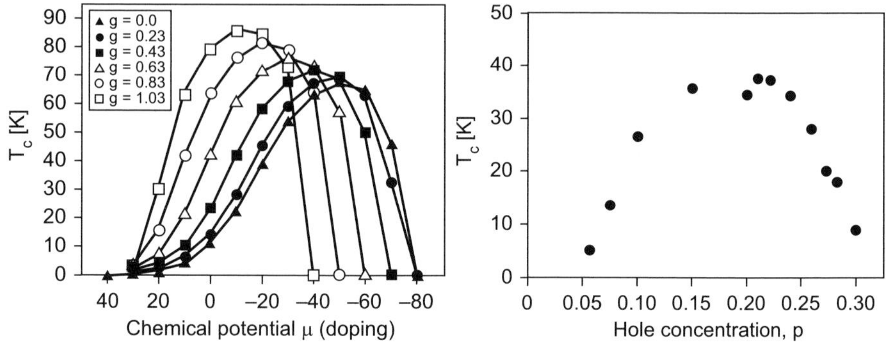

Fig. 6. (Left panel) T_c as a function of doping (chemical potential μ) for various coupling constants g as indicated in the figure. (Right panel) For comparison the experimental doping dependence of T_c for the system $La_{2-x}Sr_xCuO_4$ [75].

to a change in t_2/t_1, since the Fermi surface changes from hole- to electron-like. From band structure calculations a phenomenological relation between T_c and t_2/t_1 has been derived [71], where increases in T_c with increasing hopping integral ratio have been observed. Here we have calculated the phase diagram within the above outlined two-component approach and using s+d order parameters. The calculation can be carried through for different cuprate

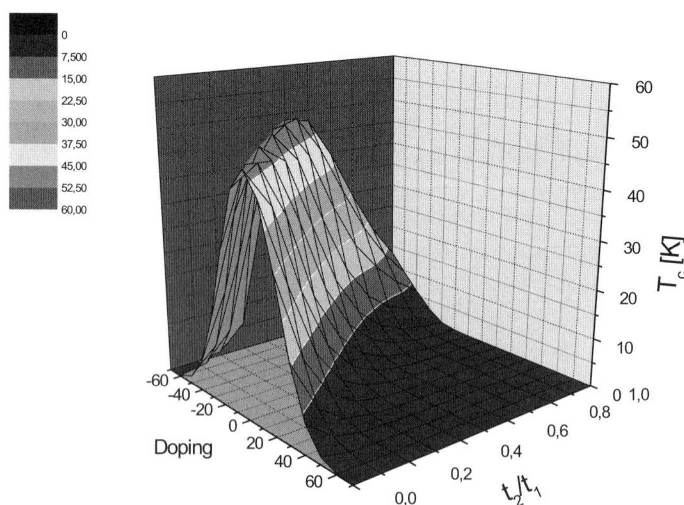

Fig. 7. The calculated phase diagram as a function of doping (chemical potential) and t_2/t_1.

families by varying the inter-component coupling constant V_{12} which greatly determines the optimum value of T_c within a given family. This is adjusted so as to yield the correct maximum T_c for the respective cuprate family. As a representative system we have chosen $LaSrCuO_4$ and the results are shown in Fig. 7. For t_2 being zero, the phase diagram is symmetric with respect to the chemical potential. With finite and increasing t_2 it becomes increasingly asymmetric and simultaneously T_c increases to reach a maximum around $t_2/t_1{=}0.35$ [71]. The increase in T_c is then approximately 15K, a value which is close to the one observed in strained samples [10, 11]. However, for larger increases in the hopping ratio T_c decreases rapidly, even though it does not vanish even at values as high as $t_2/t_1 = 1$. This trend is approximately the same in all other cuprate families, thereby defining the optimum ratio for the maximum T_c to be within $0.3 < t_2/t_1 < 0.4$. These results are not in full accordance with the band structure derived trends [71], since those find increases in t_2/t_1 for each cuprate family with increasing maximum T_c. This discrepancy can be attributed to the fact that the band structure derived ratio has been related to the hybridization between Cu 4s states and the apical oxygen p_z states which are not considered here. However, an approximate estimate of these hybridization effects can be included in our approach by investigating the dependence of T_c on the interplanar hopping integral t_4 alone. This is shown in Fig. 8 where t_4 is varied between values relevant for the bonding band and those relevant to the antibonding one.

As can be seen in the figure, reducing the magnitude of t_4 from its band structure derived value induces substantial enhancements in T_c in the underdoped to optimum doped regime. Only on the strongly overdoped side these increases are marginal. Since t_4 can be related to the Cu apical oxygen distance, increases in this distance will reduce the interplanar hopping integral and enhance T_c. This is precisely what has been observed [13] and derived within LDA [71].

Experimental results for an isotope effect on T_c have been obtained soon after the discovery of cuprate superconductors [5, 14–19]. Universal trends for various families have been obtained, where typically at optimum doping the oxygen isotope exponent $\alpha = -d\ln T_c/d\ln M$ vanishes or becomes small, whereas with decreasing doping α systematically increases, being substantially enhanced as compared to the BCS value in the region of the insulating antiferromagnetic state. BCS theory predicts a doping independent value of $\alpha = 0.5$ which might be reduced by strong coupling effects. It is thus not possible to obtain the observed enormous doping dependent variations in α by a standard electron-phonon mediated superconducting pairing mechanism. In the above scenario an isotope effect on T_c can only arise through the exponential band narrowing effect proportional to g^2 since the level shift Δ^* is independent of the ionic mass [64, 74]. Since with this assumption all hopping integrals experience the same exponential renormalization through the polaron transformation, it is possible to investigate the isotope effect arising from the individual hopping terms separately. It is found that no isotope effect

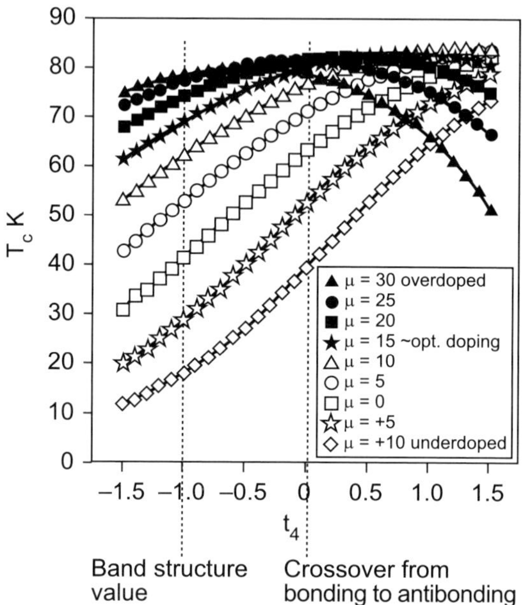

Fig. 8. T_c as a function of the interplanar hopping integral t_4 for different doping levels. The dashed lines refer to the band structure derived value [71] (left line) and to the crossover from bonding to antibonding character (right line).

arises from the third-nearest neighbour integral t_3. Thus its effect will not be addressed in the following. The most important effects can be attributed to the renormalizations of the nearest, second-nearest neighbour and interplanar hopping terms t_1, t_2, t_4, respectively. Obviously t_1 alone yields the wrong tendency for the isotope effects (Fig. 9), since it decreases in the underdoped regime and even adopts negative values there. In contrast to this observation is the combined effect arising from t_2, t_4 on the isotope exponent. Here the correct trend is followed, i.e., with decreasing T_c, α increases and exceeds the BCS value for small doping. It has, however, to be pointed out that within our calculations the experimentally observed huge enhancements of α in the underdoped regime [5, 14–19] are not present. We attribute this discrepancy to the fact that the calculations are obtained at a meanfield level and that a doping dependence in the polaronic coupling strength is not taken into account. This should, however, be quite strong, as already outlined previously, i.e. increase from optimum doping to underdoping.

Finally, we have investigated whether an isotope effect on the London penetration depth exists within the present scenario. Our analysis closely follows the one of [76, 77], however, concentrating on the consequences of polaronic coupling. While previous results already yielded the correct trend using a tight binding dispersion with nearest neighbour hopping only [76], we use here the dispersion relation (4). The penetration depth λ_L is obtained from the superfluid stiffness ρ_S via the relation:

$$\lambda_L^{-2} = \mu_0 e^2 n_s / m^* \propto \rho_s, \tag{6}$$

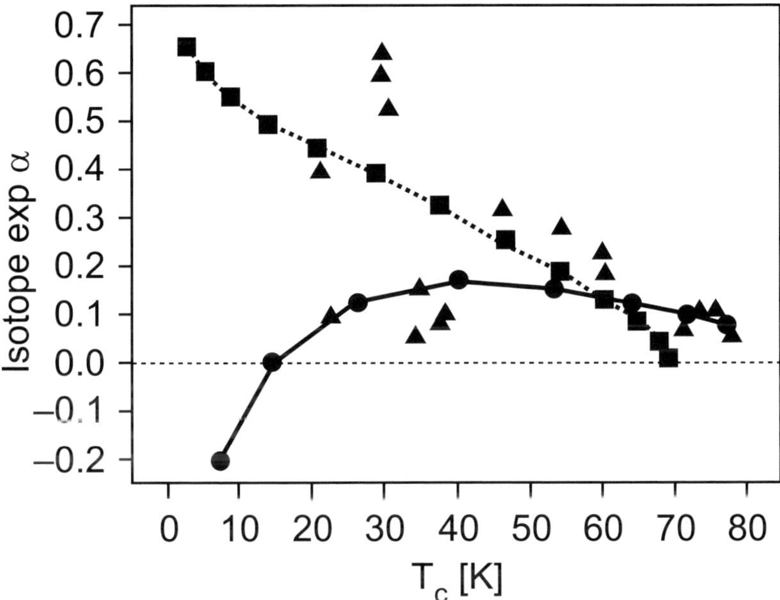

Fig. 9. The calculated isotope exponent α as a function of the corresponding T_c. The circles are obtained by renormalizing t_1 only, whereas for the squares the combined effect stemming from renormalizations of t_2 and t_4 is shown. Triangles are taken from the experiments [5] and [14, 18].

where m^* is the effective mass, n_s the superconducting carrier density and μ_0 the vacuum permeability. The superfluid stiffness is calculated within linear response theory in analogy with the previously published analysis [76]. The results are shown in Fig. 10.

Even though the isotope effect on λ_L is small, it is clearly present and shows systematic depressions for the ^{18}O system as compared to the ^{16}O compound, in agreement with the experimental data [22]. There has been some discussion about the origin of this isotope effect since mainly two sources could cause it: either an isotope induced change in the effective carrier mass m^* [78], or an isotope dependence of the carrier density [77]. Experimentally this question cannot be uniquely clarified. However, back-exchange experiments have clearly evidenced that the carrier density is not changed upon isotopic substitution. One could conclude from this fact that m^* is isotope dependent whereas the carrier density remains the same. However, the enhanced mass of ^{18}O as compared to ^{16}O may affect the carrier density since this is coupled to

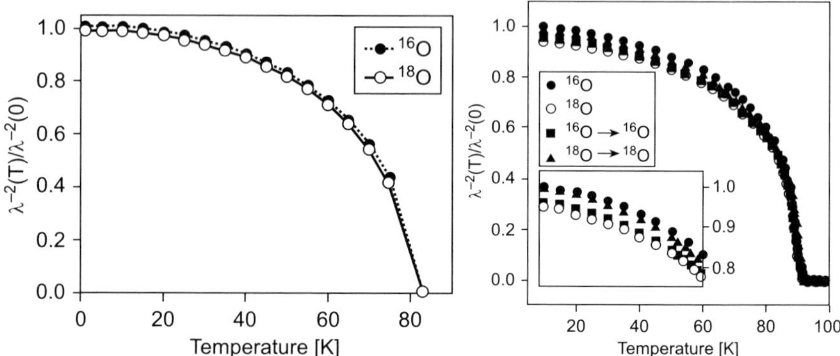

Fig. 10. The calculated isotope effect on the penetration depth (left panel) compared to experimental results (right panel) [22].

the lattice through polaron formation. The results are shown in Fig. 11 with T_c being a function of ρ_s. Besides observing a clear isotope effect on ρ_s, there is a linear relation between T_c and ρ_s reproducing the Uemura plot [79]. It is important to note that the equality in (6) is a consequence of approximating the Fermi surface by a sphere, i.e. using a parabolic band dispersion. In the present approach this is not true since explicitly further nearest neighbour hopping integrals are included, whereby the effective carrier mass does not correspond to the band mass [80]. Therefore the observed isotope effect is most likely due to an isotope effect on the effective carrier density [16–19] and [81]. A similar conclusion has recently been reached from field cooled dc magnetization measurements [82].

In agreement with experiments [18, 19] and [81] an interesting correlation of the isotope exponent β on λ_L and α on T_c is obtained by comparing them to each other for various couplings [64]. This is shown in Fig. 12. Obviously there is a linear relation between them and a sign reversal of β as compared to α. This finding explains our recent results of a linear correlation between the isotope effect on the superconducting gap and the one on the penetration depth [64, 74]. The sign reversal of β with respect to α has also been predicted in [76] and observed experimentally [18, 19] and [81, 82]. The linear relation obtained here has also been emphasized in [83] in the limit of the quantum critical end point in the underdoped regime. It is important to note that the above predicted linear relation between the two isotope exponents has been verified experimentally [18, 19] and [81, 82].

Finally, we address the question of what kind of lattice distortion is driving the above discussed isotope effects. Obviously the nearest neighbour hopping integral yields the wrong effect. From this finding we can directly exclude three lattice modes as being involved in the isotope effect, viz. the longitudinal half breathing mode, the full breathing mode and the buckling mode. All three modes couple to the nearest neighbour hopping integral only and are thus

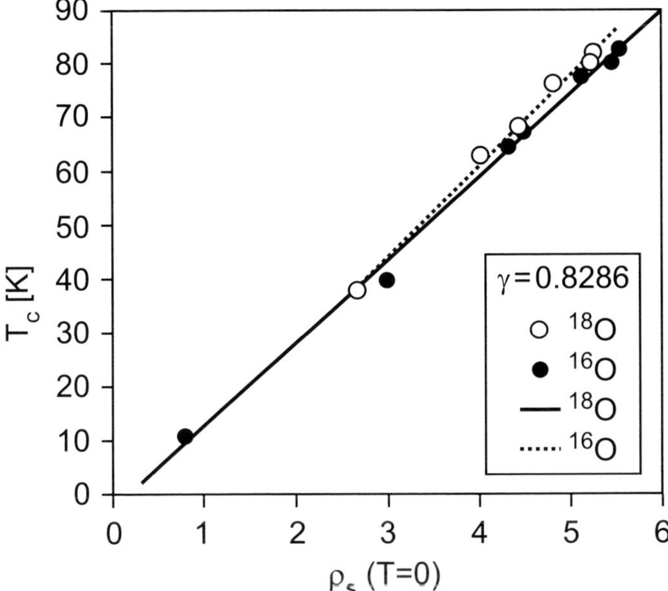

Fig. 11. The dependence of T_c on the superfluid density ρ_s for two oxygen isotopes.

irrelevant for the isotope effects. However, the Q_2 type pseudo-Jahn–Teller mode (Fig. 13) couples to the second nearest neighbour hopping and is thus the most likely candidate to cause the isotope effects [64, 74]. In addition, as has been shown above, a c-axis component must also be present [13] since the interplanar hopping element causes a substantial enhancement of α in the underdoped regime.

5 Conclusions

In conclusion, we have shown that various properties of cuprate superconductors can be understood within a two-component scenario, where polaron distorted charge–rich areas couple through the lattice to the antiferromagnetic matrix. Within the polaronic regimes the on-site Coulomb repulsion is largely depressed and the dispersion of the mobile doped carriers reduced. Simultaneously, the localization of the spin carrying carriers is lifted and dispersion is gained by them through the interaction with the extra charges. These intermediate sized polarons start to self-organize into patterns of stripe segments with decreasing temperature but always remain dynamic. The important consequences of polaron formation are that the electronic hopping integrals experience an exponential band narrowing which carries an isotope effect. Experimentally observed isotope effects together with the doping dependent phase diagram are correctly reproduced. Our analysis shows that a

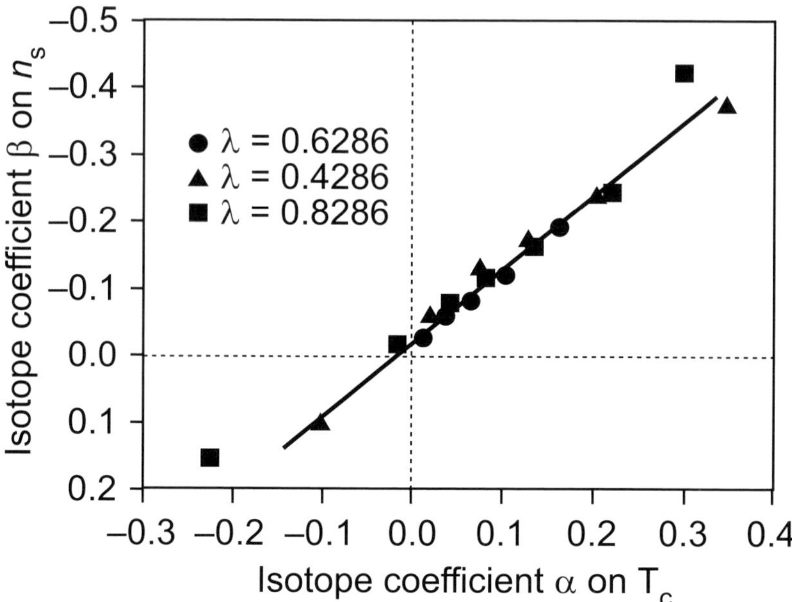

Fig. 12. Correlation between the isotope exponent β on the penetration depth and isotope exponent α on T_c.

special role is played by the second nearest neighbour hopping integral. Its variations as compared to the nearest neighbour one explain strain induced T_c enhancements as well as the isotope effect on T_c. Symmetry arguments have consequently led us to conclude that the underlying lattice distortion which couples to t_2 is the Q_2 type Jahn–Teller mode.

Acknowledgement. It is a pleasure to acknowledge stimulating discussions with K. A. Müller, A. R. Bishop, D. Pavuna and A. Simon. This work was partially supported by the Swiss National Science Foundation and MaNEP, as well as by the EU project CoMePhS.

Appendix

The gap equations are derived from (5) in the following way:

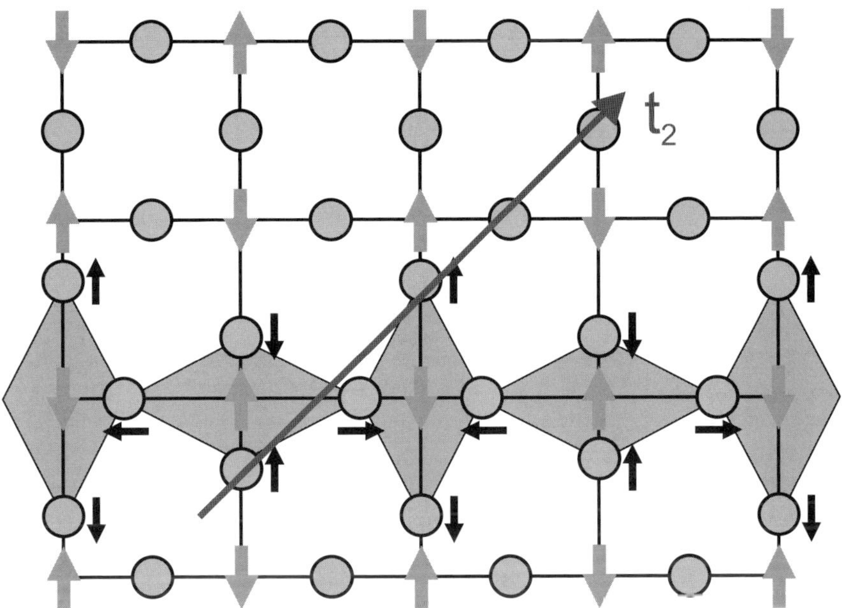

Fig. 13. The relevant ionic displacements which are governed by t_2 only. Here only displacements in the CuO_2 plane are considered. The circles represent the oxygen ions, the large arrows are copper ions with the antiferromagnetic order. The rhombohedra show the displacements (small arrows) of the Q_2 type mode which is dominated by t_2

$$
\begin{aligned}
H_{red} &= \sum_{k_1\sigma} \xi_{k_1} c^+_{k_1\sigma} c_{k_1\sigma} + \sum_{k_2\sigma} \xi_{k_2} d^+_{k_2\sigma} d_{k_2\sigma} + \bar{H}_1 + \bar{H}_2 + \bar{H}_{12} \\
\bar{H}_i &= -\sum_{k'_i} [\Delta_{k'_i} c^+_{k'_i\uparrow} c^+_{-k_i\downarrow} + \Delta^*_{k'_i} c_{-k'_i\downarrow} c_{k_i\uparrow}] \\
&\qquad + \sum_{k_i,k'_i} V_i(k_i,k'_i) < c^+_{k_i\uparrow} c^+_{-k_i\downarrow} >< c_{-k'_i\downarrow} c_{k'_i\uparrow} > \\
\bar{H}_{12} &= -\sum_{k_1,k_2} [V_{12}(k_1,k_2) < c^+_{k_1\uparrow} c^+_{-k_1\downarrow} > d_{-k_2\downarrow} d_{k_2\uparrow} \\
&\qquad + V_{12}(k_1,k_2) < d_{-k_2\downarrow} d_{k_2\uparrow} > c^+_{k_1\uparrow} c^+_{-k_1\downarrow} \\
&\qquad + V^*_{12}(k_1,k_2) d^+_{k_2\uparrow} d^+_{-k_2\downarrow} < c_{-k_1\downarrow} c_{k_1\uparrow} > \\
&\qquad + V^*_{12}(k_1,k_2) c_{-k_1\downarrow} c_{k_1\uparrow} < d^+_{k_2\uparrow} d^+_{-k_2\downarrow} > \\
&\qquad - V_{12}(k_1,k_2) < c^+_{k_1\uparrow} c^+_{-k_1\downarrow} >< d_{-k_2\downarrow} d_{k_2\uparrow} > \\
&\qquad - V^*_{12}(k_1,k_2) < c_{-k_1\downarrow} c_{k_1\uparrow} >< d^+_{k_2\uparrow} d^+_{-k_2\downarrow} >]
\end{aligned}
\tag{A.1}
$$

$$(i = 1, 2)$$

and for $i = 2$, c is replaced by d. $\xi_{k_i} = \varepsilon_i + \varepsilon_{k_i} - \mu$. Here it is assumed that $< c^+_{k_1+q/2\uparrow} c^+_{-k_1+q/2\downarrow} >=< c^+_{k_1\uparrow} c^+_{-k_1\downarrow} > \delta_{q,0}$ and equivalently for the d operators. In addition the following definitions are introduced:

$$\Delta^*_{k'_i} = \sum_{k_i} V_i(k_i, k'_i) < c^+_{k_i\uparrow} c^+_{-k_i\downarrow} >$$

together with:

$$A^*_{k_1} = \sum_{k_2} V_{12}(k_1, k_2) < d^+_{k_2\uparrow} d^+_{-k_2\downarrow} > \, , \; B^*_{k_2} = \sum_{k_2} V_{12}(k_1, k_2) < c^+_{k_2\uparrow} c^+_{-k_2\downarrow} >,$$

and $V^*_{12} = V_{12}$. Applying standard techniques we obtain:

$$< c^+_{k_1\uparrow} c^+_{-k_1\downarrow} >= \frac{\bar{\Delta}^*_{k_1}}{2E_{k_1}} \tanh \frac{E_{k_1}}{2k_B T} = \bar{\Delta}^*_{k_1} \Phi_{k_1}$$

$$< d^+_{k_2\uparrow} d^+_{-k_2\downarrow} >= \frac{\bar{\Delta}^*_{k_2}}{2E_{k_2}} \tanh \frac{E_{k_2}}{2k_B T} = \bar{\Delta}^*_{k_2} \Phi_{k_2} \tag{A.2}$$

with $E^2_{k_2} = \xi^2_{k_2} + |\bar{\Delta}_{k_2}|^2$, $\bar{\Delta}_{k_2} = \Delta_{k_2} + B_{k_2}$ and $E^2_{k_1} = \xi^2_{k_1} + |\bar{\Delta}_{k_1}|^2$, $\bar{\Delta}_{k_1} = \Delta_{k_1} + A_{k_1}$, which results in the self-consistent set of coupled equations:

$$\bar{\Delta}_{k_1} = \sum_{k'_1} V_1(k_1, k'_1) \bar{\Delta}_{k'_1} \Phi_{k'_1} + \sum_{k_2} V_{1,2}(k_1, k_2) \bar{\Delta}_{k_2} \Phi_{k_2}$$

$$\bar{\Delta}_{k_2} = \sum_{k'_2} V_2(k_2, k'_2) \bar{\Delta}_{k'_2} \Phi_{k'_2} + \sum_{k_1} V_{1,2}(k_1, k_2) \bar{\Delta}_{k_1} \Phi_{k_1} \tag{A.3}$$

from which the temperature dependencies of the gaps and the superconducting transition temperature have to be determined.

References

1. Bednorz J G, Müller K A, (1986) Z Phys: Cond Mat 64: 189
2. Müller K A, in Magnetic resonance and relaxation (ed. Blinc, R.) 192-208 (North Holland Pbl Inc, 1966)
3. Höck K-H, Nickisch H, Thomas H (1983) Helv Phys Acta 56: 237
4. Anderson P W, (1987) Science 235: 1196; Schrieffer J R, Wen X G, Zhang S C, (1989) Phys Rev B 39: 11663; Kane C L, Lee P A, Read N (1989) Phys Rev B 39: 6880
5. Franck J P, in Physical Properties of high temperature superconductors IV (ed. Ginsberg, D M) 189-293 (World Scientific, Singapore, 1994)
6. Shirane G, Birgeneau R J, Endoh Y, Gehring P, Kastner M A, Kitazawa K, Kojima H, Tanaka I, Thurston T R, Yamada K (1989) Phys Rev Lett 63: 330
7. Bianconi A, Saini N L, Lanzara A, Missori M, Rossetti T, Oyanagi H, Yamaguchi H, Oka K, Ito T (1996) Phys Rev Lett 76: 3412

8. Saini N L, Lanzara A, Oyanagi H, Yamaguchi H, Oka K, Ito T, Bianconi A (1997) Phys Rev B 55: 12759
9. Haga T, Yamaya K, Abe Y, Tajima Y, Hidaka Y (1990) Phys Rev B 41: R826
10. Locquet J-P, Perret J, Fompeyrine J, Mächler E, Seo J W, van Tendeloo G (1998) Nature 394: 453
11. Abrecht M, Ariosa D, Cloetta D, Mitrovic S, Onellion M, Xi X X, Margaritondo G, Pavuna D (2003) Phys Rev Lett 91: 057002
12. Oyanagi H, Saini N L, Tsukuda A, Naito M, J Phys Chem Sol, to be published

13. Abrecht M, Photoemission studies of thin films grown by pulsed laser deposition; epitaxial strain effects on the electronic structure of high temperature superconductors, (2005) Thesis, EPFL
14. Crawford M K, Kunchur M N, Farneth W E, Carron III E M, Poon S J, (1990) Phys Rev B 41: 282
15. Franck J P, Jung J J, Mohamed M A K, Gygax S, Sproule G I (1991) Phys Rev B 44
16. Zhao G M, Conder K, Keller H, Müller K A (1998) J Phys Cond Mat 10:9055 and refs therein
17. Zhao G M, Keller H, Conder K (2001) J Phys Cond Mat 13: R569 and refs therein
18. Khasanov R, Shengelaya A, Morenzoni E,Conder K, Savić I M, Keller H (2004) J Phys Cond Mat 16: S4439 and refs therein
19. Keller H (2005) Structure and Bonding 114: 143 and refs. therein
20. Zhao G M, Hunt M B, Keller H, Müller K A, (1997) Nature (London) 385: 236
21. Hofer J, Conder K, Sasagawa T, Zhao G M, Willemin M, Keller H, Kishio K, (2000) Phys Rev Lett 84: 4192
22. Khasanov R, Eshchenko D G, Luethens H, Morenzoni E, Prokscha T, Suter A,Garifianov N, Mali M, Roos J, Conder K, Keller H (2004) Phys Rev Lett 92: 057602
23. Lanzara A, Bogdanov P V, Zhou X J, Kellar S A, Feng D L, Lu E D, Yoshida T, Eisaki H, Fujimori A, Kishio K, Shimoyama J-I, Noda T, Uchida S, Hussain Z, Shen Z-X (2001) Nature (London) 412: 510
24. Gweon G H, Sasagawa T, Zhou S Y, Graf J, Takagi H, Lee D-H, Lanzara A (2004) Nature 430: 187
25. Bozin E S, Billinge S J L, Kwei G H, Takagi H (1999) Phys Rev B 59: 4445
26. Ando Y, Takeya J, Abe Y, Sun X F, Lavrov A F (2002) Phys Rev Lett 88: 147004
27. Sacuto A, Combescot R, Bontemps N, Müller C A, Viallet V, Colson D (1998) Phys Rev B 58: 11721
28. Genzel L, Wittlin A, Bauer M, Cardona M, Schönherr E, Simon A (1989) Phys Rev B 40: 2170
29. Kremer R K, Simon A, Sigmund E, Hizhnyakov V, in Phase Separation in Cuprate Superconductors, Eds Sigmund E, Müller K A (1994) Springer-Verlag Berlin pp 66
30. Pan S H, O'Neal J P, Badzey R L, Chamon C, Ding H, Engelbrecht J R, Wang Z, Eisaki H, Uchida S, Gupta A K, Ng K-W, Hudson E W, Lang K M, Davis J C (2001) Nature 413: 282
31. Dumm M, Komiya S, Ando Y, Basov D N (2003) Phys Rev Lett 91: 077004
32. Dordevic S V, Komiya S, Ando Y, Basov D N (2003) Phys Rev Lett 91: 167401

33. Mihailovic D, Kabanov V V (2005) Structure and Bonding 114: 331 and refs therein
34. Kochelaev B I, Sichelschmidt J, Elschner B, Lemor W, Loidl A (1997) Phys Rev Lett 79: 4274
35. Haase J, Slichter C P (2003) J Supercond 1: 473
36. Islam Z, Liu X, Sinha S K, Lang J C, Moss S C, Haskel D, Srajer G, Wochner P, Lee D R, Haeffner D R, Welp U (2004) Phys Rev Lett 93: 157008
37. McQueeney R J, Egami T, Shirane G, Endoh Y (1996) Phys Rev B 54: R9689
38. Zaanen J, Gunnarsson O (1989) Phys Rev B 40: 7391
39. Monthoux P, Pines D (1994) Phys Rev B 50: 16015
40. Lanzara A, Zhao G, Saini N L, Bianconi A, Conder K, Keller H, Müller K A (1999) J Phys Cond Mat 11: L545
41. RubioTemprano D, Mesot J, Janssen S, Conder K, Furrer A, Mutka H, Müller K A (2000) Phys Rev Lett 84: 1999
42. RubioTemprano D, Mesot J, Janssen S, Conder K, Furrer A, Sokolov A, Trounov V, Kazakov S M, Karpinski J, Müller K A (2001) Eur Phys J B 19: R5
43. Raffa R, Ohno T, Mali M, Roos J, Brinkmann D, Conder K, Eremin M (1998) Phys Rev Lett 81: 5912
44. Kohen A, Leibowitch G, Deutscher G (2003) Phys Rev Lett 90: 207005
45. Srikanth H, Zhai Z, Sridhar S, Erb A, Walzer E (1998) Phys Rev B 57: 7986 and refs therein
46. Bussmann-Holder A, Bishop A R (1991) Phys Rev B 44: 2853
47. Bussmann-Holder A, Simon A, Büttner H (1989) Phys Rev B 39: 207
48. Bussmann-Holder A, Genzel L, Simon A, Bishop A R (1993) Z Phys B 92: 149
49. Alexandrov A S, Ranninger J, Robaszkiewicz S (1986) Phys Rev B 33: 4526
50. Bill A, Kresin V Z, Wolf S A (1998) Phys Rev B 57: 10814
51. Emin D (1992) Phys Rev B 45: 5525
52. Mihailovic D, Kabanov V V (2001) Phys Rev B 63: 054505
53. Goodenough J B, Zhou J-S (1994) Phys Rev B 49: 4251
54. Takagi H, Ido T, Ishibashi S, Vota M, Uchida S, Tokura Y (1989) Phys Rev B 40: 2254
55. Hirsch J E (1993) Phys Rev B 47: 5351
56. Goodenough J B, Zhou J-S, Chan J, in Lattice effects in High-T_c superconductors, editors: Bar-Yam Y, Egami T, Mustre-de Leon J, Bishop A R (World Scientific, Singapore, New Jersey, London, Hong Kong, 1992) p 486
57. Goodenough J B, Zhou J-S (1990) Phys Rev B 42: 4276
58. Goodenough J B, Zhou J-S, Chan J (1993) Phys Rev B 47: 5275
59. Bussmann-Holder A, Simon A, Buttner H, Bishop A R (2000) Phil Mag B 80: 1955
60. Bussmann-Holder A, Bishop A R (2004) Phys Rev B 70: 184303
61. Bussmann-Holder A, Müller K A, Micnas R, Büttner H, Simon A, Bishop A R, Egami T (2001) J Phys Cond Mat 13: L168
62. Bussmann-Holder A, Bishop A R, Büttner H, Egami T, Micnas R, Müller K A (2001) J Phys Cond Mat 13: L545
63. Lang S G, Firsov Yu A (1963) Sov Phys JETP 16: 1302-1312
64. Bussmann-Holder A, Keller H (2005) Europ Phys J 44: 487
65. Tranquada J M, Axe J D, Ichikawa N, Nakamura Y, Uchida S, Nachumi B (1996) Phys Rev B 54: 7489
66. Rice M J, Strässler S, Toombs G A (1974) Phys Rev Lett 32: 596

67. Zhou X J et al (2001) Phys Rev Lett 86: 5578
68. Ino A, Kim C, Nakamura M, Yoshida T, Mizokawa T, Fujimori A,. Shen Z-X, Kakeshita T, Eisaki H, Uchida S (2002) Phys Rev B 65: 094504
69. Yoshida T, Zhou X J, Sasagawa T, Yang W L, Bogdanov P V, Lanzara A, Hussain Z, Mizokawa T, Fujimori A, Eisaki H, Shen Z-X, Kakeshita T, Uchida S (2003) Phys Rev Lett 91: 027001
70. Bussmann-Holder A, Micnas R, Bishop A R (2003) Europ Phys J 37: 345
71. Pavarini E, Dasgupta I, Saha-Dasgupta T, Jepsen O, Andersen O K (2001) Phys Rev Lett 87: 047003
72. Shen K M, Ronning F, Lu D H, Lee W S, Ingle N J C, Meevasana W, Baumberger F, Damascelli A, Armitage N P, Miller L L, Kohsaka Y, Azuma M, Takano M, Takagi H, Shen Z-X (2004) Phys Rev Lett 93: 267002
73. Micnas R, Robaszkiewicz S, Bussmann-Holder A (2005) Structure and Bonding 114: 13
74. Bussmann-Holder A, Keller H, Müller K A (2005) Structure and Bonding 114: 367
75. Torrance J B, Tokura Y, Nazzal A I, Bezinge A, Huang T C, Parkin S S P (1988) Phys Rev Lett 61: 1127 and refs therein
76. Bussmann-Holder A, Micnas R, Bishop A R (2004) Phil Mag 84: 1257
77. Bill A, Kresin V Z, Wolf S A (1998) Phys Rev B 57: 10814
78. Alexandrov A S (1992) Phys Rev B 46: 149332
79. Uemura Y J, Luke G M, Sternlieb B J, Brewer J H, Carolan J F, Hardy W N, Kadono R, Kempton J R, Kiefl R F, Kreitzman S R, Mulhern P, Riseman T M, Ll D, Williams R, Yang B X, Uchida S, Takagi H, Gopalakrishnan J, Sleight A W, Subramanian M A, Chien C L, Cieplak M Z, Xiao Gang, Lee V Y, Statt B W, Stronach C E, Kossler W J, Yu X H (1989) Phys Rev Lett 62: 2317
80. Chandrasekhar B S, Enzel D (1993) Ann Phys 2: 353
81. Khasanov R, Shengelaya A, Conder K, Morenzoni E, Savić I M, Keller H (2003) J Phys Cond Mat 15: L17
82. Tallon J L, Islam R S, Storey J, Williams G V M, Cooper J R (2005) Phys Rev Lett 94: 237002
83. Schneider T, Keller H (2001) Phys Rev Lett 86: 4899

Current Rectification, Switching, Polarons, and Defects in Molecular Electronic Devices

A.M. Bratkovsky

Hewlett-Packard Laboratories, 1501 Page Mill Road, Palo Alto, California 94304, U.S.A. `alex.bratkovski@hp.com`

Summary. Devices for nano- and molecular-size electronics are currently a focus of research aimed at an efficient current rectification and switching. A few generic molecular scale devices are reviewed here on the basis of first-principles and model approaches. Current rectification by (ballistic) molecular quantum dots can produce the rectification ratio ≤ 100. Current switching due to conformational changes in the molecules is slow, on the order of a few kHz. Fast switching (~ 1 THz) may be achieved, at least in principle, in a degenerate molecular quantum dot with strong coupling of electrons with vibrational excitations. We show that the mean-field approach fails to properly describe *intrinsic* molecular switching and present an exact solution to the problem. Defects in molecular films result in spurious peaks in conductance, apparent negative differential resistance, and may also lead to unusual temperature and bias dependence of current. The observed switching in many cases is *extrinsic*, caused by changes in molecule-electrode geometry, molecule reconfiguration, metallic filament formation through, and/or changing amount of disorder in a molecular film. We give experimental examples of telegraph "switching" and "hot spot" formation in molecular films.

1 Introduction

Current interest in molecular electronics is largely driven by expectations that molecules can be used as nanoelectronics components able to complement/replace standard silicon CMOS technology [1, 2] on the way down to $\leq$10nm circuit components. The first speculations about molecular electronic devices (diodes, rectifiers) were apparently made in the mid-1970s [3]. That original suggestion of a molecular rectifier generated a large interest in the field and a flurry of suggestions of various molecular electronics components, especially coupled with premature estimates that silicon-based technology cannot scale to below 1μm feature size. The Aviram-Ratner's Donor-insulator-Acceptor construct $\mathrm{TTF}-\sigma-\mathrm{TCNQ}$ ($D^+ - \sigma - A^-$, see details below), where carriers were supposed to tunnel asymmetrically in two directions through an insulating saturated molecular $\sigma - $ 'bridge', has never materialized, in spite of extensive experimental effort over a few decades [4]. The end result in some cases appears to be a slightly electrically anisotropic *insulator*, rather than a diode, unsuitable as a replacement for silicon devices. This comes about because in order to assemble a reasonable quality monolayer of these molecules in a Langmuir-Blodgett trough (avoiding defects that will short the device after electrode deposition) one needs to attach a long 'tail' molecule C18 [$\equiv (CH_2)_{18}$] that can produce enough of a Van-der-Waals force to keep molecules together, but C18 is a wide-band insulator with a bandgap $E_g \approx 9 - 10eV$. The outcome of these studies may have been anticipated, but if one were able to assemble the Aviram-Ratner molecules without the tail, they could not rectify anyway. Indeed, a recent ab-initio study [5] of a $D^+\sigma A^-$ prospective molecule showed no appreciable asymmetry of its I-V curve. The molecule was envisaged by Ellenbogen and Love [6] as a 4-phenyl ring Tour wire with a dimethylene insulating bridge in the middle directly connected to Au electrodes via thiol groups. Donor-acceptor asymmetry was produced by side NH_2^+ and NO_2^- moieties, which is a frequent motif in molecular devices using the Tour wires. The reason for poor rectification is simple: the bridge is too short, it is a transparent piece of one-dimensional insulator, whereas the applied field is three dimensional and it cannot be screened efficiently with an appreciable voltage drop on the insulating group in this geometry. Although there is only 0.7eV energy separation between levels on the D and A groups, one needs about 4eV bias to align them and get a relatively small current because total resonant transparency is practically impossible to achieve. Remember, that the model calculation implied an ideal coupling to electrodes, which is impossible in reality and which is known to dramatically change the current through the molecule (see below). We shall discuss below some possible alternatives to this approach.

It is worth noting that studies of energy and electron transport in molecular crystals [7] started already in the early 1960s. It was established in the mid-1960s in what circumstances charge transport in biological molecules involves electron tunneling [8]. It was realized in the mid-1970s that since the organic molecules are 'soft', energy transport along linear biological molecules, proteins, etc. may proceed by low energy nonlinear collective excitations, like Davydov solitons [9] (see review [10]).

To take over from current silicon CMOS technology, molecular electronics should provide smaller, more reliable, functional components that can be produced and assembled concurrently and are compatible with CMOS for integration. The small size of units that molecules may hopefully provide is quite obvious. However, meeting other requirements seems to be a very long shot. To beat alternative technologies for e.g. dense (and cheap) memories, one should aim at a few TB/in^2 ($> 10^{12} - 10^{13}$ bit/cm^2), which corresponds to linear bit (footprint) sizes of $3 - 10$ nanometers, and an operation lifetime of ~ 10 years. The latter requirement is very difficult to meet with organic molecules that tend to oxidize and decompose, especially under conditions of very high applied electric field (given the operational voltage bias of $\sim 1V$ for molecules integrated with CMOS and their small sizes on the order of a few nanometers). In terms of areal density, one should compare this with rapidly developing technologies like ferroelectric random access (FERAM)[11] or phase-change memories (PCM)[12]. The current smallest commercial nano-ferroelectrics are about $400{\times}400$ nm^2 and 20-150 nm thick [13], and 128×128 arrays of switching ferroelectric pixels bits have already been demonstrated with a bit size ≤ 50nm (with density $\sim$TB/in^2)[14]. The phase-change memories based on chalcolgenides GeSbTe (GST) seem to scale even better than the ferroelectrics. As we see, the mainstream technology for random-access memory approaches molecular size very rapidly. For instance, the so-called "nanopore" molecular devices [34] have comparable sizes and are yet to demonstrate a repeatable behavior (see below).

In terms of parallel fabrication of molecular devices, one is looking at *self-assembly* techniques (see, e.g. [15, 16], and references therein). Frequently, the Langmuir-Blodgett technique is used for self-assembly of molecules on water, where molecules are prepared to have a hydrophilic "head" and a hydrophobic "tail" to make the assembly possible, see e.g. [17, 18]. The allowances for a corresponding assembly, especially of hybrid structures (molecules integrated on silicon CMOS), are on the order of a *fraction* of an Angstrom, so actually a *picotechnology* is required [2]. Since it is problematic to reach such a precision any time soon, the all-in-one molecule approach was advocated, meaning that a fully functional computing unit should be synthesized as a single supermolecular unit [2]. The hope is that perhaps directed self-assembly will help to accomplish building such a unit, but self-assembly on a large scale is impossible without defects [15, 16], since the entropic factors work against it. Above some small defect concentration ("percolation") threshold the mapping of even a simple algorithm on such a self-assembled network becomes impossible [19].

There is also a big question about electron transport in such a device consisting of large organic molecules. Even in high-quality pentacene crystals, perhaps the best materials for thin film transistors, the mobility is a mere 1-2 cm^2/V$\cdot$s (see e.g. [20]), as a result of carrier trapping by interaction with a lattice. The situation with carrier transport through long molecules ($> 2 - 3$nm) is, of course, substantially different from the transport through short rigid molecules that have been envisaged as possible electronics components. Indeed, in *short* molecules the dominant mode of electron transport would be resonant tunneling through *electrically active* molecular orbital(s) [21], which, depending on the work function of the electrode, affinity of the molecule, and symmetry of coupling between molecule and electrode may be one of the lowest unoccupied molecular orbitals (LUMO) or highest occupied molecular orbitals (HOMO)[22, 23]. Indeed, it is well known that in longer wires containing more than about 40 atomic sites, the tunneling time is comparable to or larger than the characteristic phonon times, so that the polaron (and/or bipolaron) can be formed inside the molecular wire [24]. There is a wide range of molecular bulk conductors with (bi)polaronic carriers. The formation of polarons (and charged solitons) in polyacetylene (PA) was discussed a long time ago theoretically in [25] and formation of bipolarons (bound states of two polarons) in [26]. Polarons in PA were detected optically in [27] and since then studied in great detail. There is an exceeding amount of evidence for polaron and bipolaron formation in conjugated polymers such as polyphenylene, polypyrrole, polythiophene, polyphenylene sulfide [28], Cs-doped biphenyl [29], n-doped bithiophene [30], polyphenylenevinylene(PPV)-based light emitting diodes [31], and other molecular systems. Given the above problems with electron transport through large molecules one should look at the short- to medium-size molecules first.

The latest wave of interest in molecular electronics is mostly related to recent studies of carrier transport in synthesized linear conjugated molecular wires (Tour wires [1]) with apparent non-linear I-V characteristics [negative differential resistance (NDR)] and "memory" effects [32–34], various molecules with a mobile *microcycle* that is able to move back and forth between metastable conformations in solution (molecular shuttles) [35] and demonstrate some sort of "switching" between relatively stable resistive states when sandwiched between electrodes in a solid state device [36] (see also [37]). There are also various photochromic molecules that may change conformation ("switch") upon absorption of light [38], which may be of interest to some photonics applications but not for general purpose electronics. Non-adiabatic exciton-phonon effects in quantum dots [39] could be of direct relevance for such devices.

One of the most serious problems with using this kind of molecules is *power dissipation*. Indeed, the studied organic molecules are, as a rule, very resistive (in the range of $\sim 1 M\Omega - 1 G\Omega$, or more). Since usually the switching bias exceeds 0.5V the dissipated power density would be in excess of 10 kW/cm^2, which is orders of magnitude higher than the presently manageable level. One can drop the density of switching devices, but this would undermine a main advantage of using molecular size elements. This is a common problem that CMOS faces too, but organic molecules do not seem to offer a tangible advantage yet. There are other outstanding problems, like understanding an actual switching mechanism, which seems to be rather molecule–independent [37], stability, scaling, etc. It is not likely, therefore, that molecules will displace silicon technology, or become a large part of a hybrid technology in the foreseeable future.

The first major moletronic applications would most likely come in the area of chemical and biological sensors. One of the current solutions in this area is to use functionalized nanowires. When a target analyte molecule attaches from the environment to such a nanowire, it changes the electrostatic potential "seen" by the carriers in the nanowire. Since the conductance of the nanowire device is small, even one chemisorbed molecule could make a detectable change in conductance [40]. Semiconducting nanowires can be grown from seed metal nanoparticles [15], or they can be carbon nanotubes (CNT), which are studied extensively due to their relatively simple structure and some unique properties like very high conductance [41].

In this chapter we shall address various generic problems related to electron transport through molecular devices, and describe some specific molecular systems that may be interesting for applications as rectifiers and switches, and some pertaining physical problems. We shall first consider systems where an elastic tunneling is dominant, and interaction with vibrational excitations on the molecules only renormalizes some parameters describing tunneling. We shall also describe a situation where the coupling of carriers to molecular vibrons is strong. In this case the tunneling is substantially inelastic and, moreover, it may result in current hysteresis when the electron-vibron interaction is so strong that it overcomes Coulomb repulsion of carriers on a central narrow-band/conjugated unit of the molecule separated from electrodes by wide band gap saturated molecular groups like $(CH_2)_n$, which we shall call a molecular quantum dot (molQD). Another very important problem is to understand the nature and the role of imperfections in organic thin films. This is addressed in the last section of the chapter.

2 Role of Molecule–Electrode Contact: Extrinsic Molecular Switching due to Molecule Tilting

We predicted some time ago that there should be a strong dependence of the current through conjugated molecules (like the Tour wires [1]) on the geometry of molecule-electrode contact [22, 23]. The apparent "telegraph" switching observed in STM single-molecule probes of the three-ring Tour molecules, inserted into a SAM of non-conducting shorter alkanes, has been attributed to this effect [33]. The theory predicts very strong dependence of the current through the molecule on the tilting angle between the backbone of a molecule and a normal to the electrode surface. Other explanations, like rotation of the middle ring, charging of the molecule, or effects of the moieties on the middle ring, do not hold. In particular, switching of the molecules *without* any NO_2 or NH_2 moieties have been practically the same as with them.

The simple argument in favor of the "tilting" mechanism of the conductance lies in a large anisotropy of the molecule-electrode coupling through π-conjugated molecular orbitals (MOs). In general, we expect the overlap and the full conductance to be maximal when the lobes of the p-orbital of the end atom at the molecule are oriented perpendicular to the surface, and smaller otherwise, as dictated by the symmetry. The overlap integrals of a p-orbital with orbitals of other types differ by a factor of about 3 to 4 for the two orientations. Since the conductance is proportional to the square of the matrix element, which contains a product of two metal-molecule hopping integrals, the total conductance variation with overall geometry may therefore reach two orders of magnitude, and in special cases be even larger.

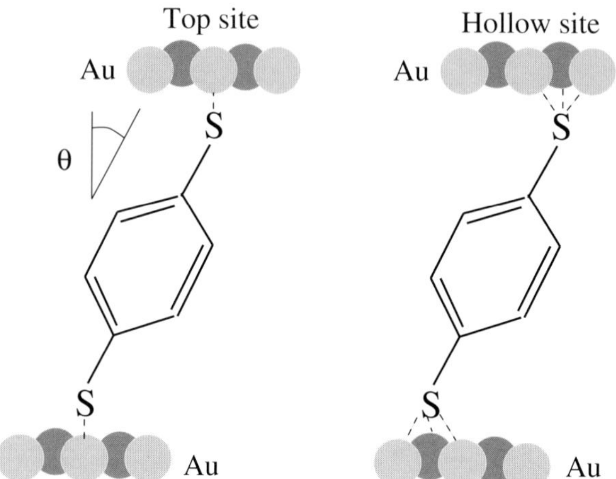

Fig. 1. Schematic representation of the benzene-dithiolate molecule on top and hollow sites. End sulfur atoms are bonded to one and three surface gold atoms, respectively, θ is the tilting angle.

In order to illustrate the geometric effect on current we have considered a simple two-site model with $p-$orbitals on both sites, coupled to electrodes with $s-$orbitals [23]. For non-zero bias the transmission probability has the resonant form with line widths for hopping to the left (right) lead Γ_L (Γ_R). The current has the approximate form (with $\Gamma = \Gamma_L + \Gamma_R$)

$$
I \approx \begin{cases} \frac{q^2}{h} \frac{\Gamma^2}{t^2} V \propto \sin^4 \theta, & qV \ll E_{\text{LUMO}} - E_{\text{HOMO}}, \\ \frac{8\pi q}{h} \frac{\Gamma_L \Gamma_R}{\Gamma_L + \Gamma_R} \propto \sin^2 \theta, & qV > E_{\text{LUMO}} - E_{\text{HOMO}}, \end{cases} \tag{1}
$$

where θ is the tilting angle [22, 23], Fig. 1.

The tilting angle has a large effect on the I-V curves of benzene-dithiolate (BDT) molecules, especially when the molecule is anchored to the Au electrode in the top position, Fig. 2. By changing θ from 5° to just 15°, one drives the I-V characteristic from one with a gap of about 2V to an ohmic one with a large relative change of conductance. Even changing θ from 10° to 15° changes the conductance by about an order of magnitude. The I-V curve for the hollow site remains ohmic for tilting angles up to 75° with moderate changes of conductance. Therefore, if the molecule in measurements snaps from the top to the hollow position and back, it will lead to an apparent switching [33]. It has recently been realized that the geometry of a contact strongly affects coherent spin transfer between molecularly bridged quantum dots [42]. It is worth noting that another frequently observed *extrinsic* mechanism of "switching" in organic layers is due to electrode material diffusing into the layer and forming metallic "filaments" (see below).

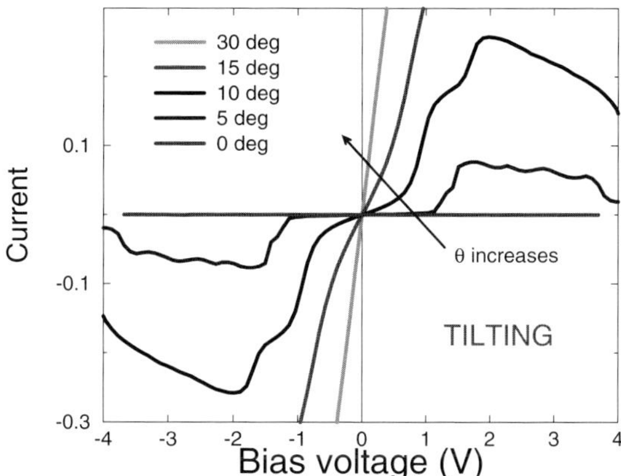

Fig. 2. Effect of tilting on I-V curve of the BDT molecule, Fig. 1. Current is in units of $I_0 = 77.5\mu A$, θ is the tilting angle.

3 Molecular Quantum Dot Rectifiers

Aviram and Ratner speculated about a rectifying molecule containing donor (D) and acceptor (A) groups separated by a saturated σ-bridge (insulator) group, where the (inelastic) electron transfer will be more favorable from A to D [3]. The molecular rectifiers actually synthesized, $C_{16}H_{33} - \gamma Q3CNQ$, were of somewhat different $D - \pi - A$ type, i.e. the "bridge" group was conjugated [4]. Although the molecule did show rectification (with considerable hysteresis), it performed rather like an anisotropic insulator with tiny currents on the order of $10^{-17}A$/molecule, because of the large alkane "tail" needed for LB assembly. It was recently realized that in this molecule the resonance does not come from the alignment of the HOMO and LUMO, since they cannot be decoupled through the conjugated $\pi-$bridge, but rather due to an asymmetric voltage drop across the molecule [43]. Rectifying behavior in other classes of molecules is likely due to asymmetric contact with the electrodes [44, 45], or an asymmetry of the molecule itself [46]. To make rectifiers, one should avoid using molecules with long insulating groups, and we have suggested using relatively short molecules with "anchor" end groups for their self-assembly on a metallic electrodes, with a phenyl ring as a central conjugated part [47]. This idea has been tested in [48] with a phenyl and thiophene rings attached to a $(CH_2)_{15}$ tail by a CO group. The observed rectification ratio was ≤ 10, with some samples showing the ratio of about 37.

We have recently studied a more promising rectifier like $-$S-$(CH_2)_2$-Naph$-$ $(CH_2)_{10}$$-S-$ with a theoretical rectification ≤ 100 [49], Fig. 3. This system has been synthesized and studied experimentally [50]. To obtain an accurate description of transport in this case, we employ an ab-initio non-equilibrium Green's function method [51–53]. The present calculation takes into account only elastic tunneling processes. Inelastic processes may substantially modify the results in the case of strong interaction of the electrons with molecular vibrations, see [54] and below. There are indications in the literature that the carrier might be trapped in a polaron state in saturated molecules somewhat longer than those we consider in the present paper [55]. One of the barriers in the present rectifiers is short and relatively transparent, so there will be no appreciable Coulomb blockade effects. The structure of the present molecular rectifier is shown in Fig. 3. The molecule consists of a central conjugated part (naphthalene) isolated from the electrodes by two insulating aliphatic chains $(CH_2)_n$ with lengths $L_1(L_2)$ for the left (right) chain.

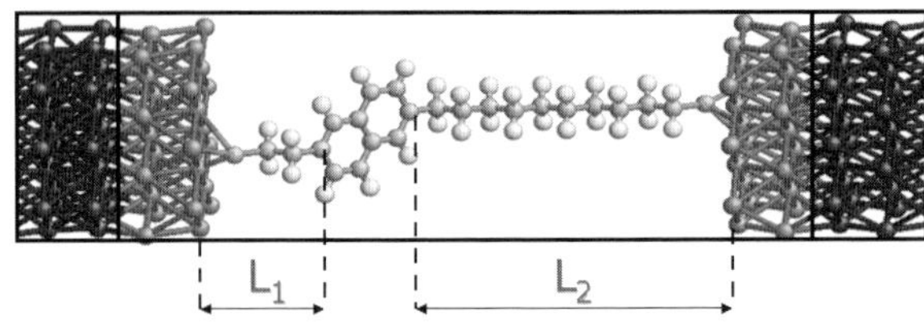

Fig. 3. Stick figure representing the naphthalene conjugated central unit separated from the left (right) electrode by saturated (wide band gap) alkane groups with length $L_{1(2)}$.

The principle of molecular rectification by a molecular quantum dot is illustrated in Fig. 4, where the electrically "active" molecular orbital, localized on the middle conjugated part, is the LUMO, which lies at an energy Δ above the electrode Fermi level at zero bias. The position of the LUMO is determined by the work function of the metal $q\phi$ and the affinity of the molecule $q\chi$, $\Delta = \Delta_{\mathrm{LUMO}} = q\,(\phi - \chi)$. The position of the HOMO is given by $\Delta_{\mathrm{HOMO}} = \Delta_{\mathrm{LUMO}} - E_g$, where E_g is the HOMO-LUMO gap. If this orbital is considerably closer to the electrode Fermi level E_F, then it will be brought into resonance with E_F prior to other orbitals. It is easy to estimate the forward and reverse bias voltages, assuming that the voltage mainly drops on the insulating parts of the molecule,

$$V_F = \tfrac{\Delta}{q}(1 + \xi), \qquad V_R = \tfrac{\Delta}{q}\left(1 + \tfrac{1}{\xi}\right), \tag{2}$$

$$V_F/V_R = \xi \equiv L_1/L_2, \tag{3}$$

where q is the elementary charge. A significant difference between forward and reverse currents should be observed in the voltage range $V_F < |V| < V_R$. The current is obtained from the Landauer formula

$$I = \frac{2q^2}{h} \int dE \left[f(E) - f(E + qV) \right] g(E, V). \tag{4}$$

We can make qualitative estimates in the resonant tunneling model, with the conductance $g(E, V) \equiv T(E, V)/q$, where $T(E, V)$ is the transmission given by the Breit-Wigner formula

$$T(E, V) = \frac{\Gamma_L \Gamma_R}{(E - E_{MO})^2 + (\Gamma_L + \Gamma_R)^2/4}, \tag{5}$$

E_{MO} is the energy of the molecular orbital. The width $\Gamma_{L(R)} \sim t^2/D = \Gamma_0 e^{-2\kappa L_{1(2)}}$, where t is the overlap integral between the MO and the electrode, D is the electron band width in the electrodes, κ the inverse decay length of the resonant MO into the barrier. The current above the resonant threshold is

$$I \approx \frac{2q}{\hbar} \Gamma_0 e^{-2\kappa L_2}. \tag{6}$$

We see that increasing the spatial asymmetry of the molecule (L_2/L_1) changes the operating voltage range linearly, but it also brings about an *exponential* decrease in current [47]. This severely limits the ability to optimize the rectification ratio while simultaneously keeping the resistance at a reasonable value. To calculate the I-V curves, we use an ab-initio approach that combines the Keldysh non-equilibrium Green's function (NEGF) with pseudopotential-based real space density functional theory (DFT) [51–53]. The main advantages of our approach are (i) a proper treatment of the open boundary condition; (ii) a fully atomistic treatment of the electrodes and (iii) a self-consistent calculation of the non-equilibrium charge density using NEGF. The transport Green's function is found from the Dyson equation

$$\left(G^R \right)^{-1} = \left(G_0^R \right)^{-1} - V, \tag{7}$$

where the unperturbed retarded Green's function is defined in operator form as $\left(G_0^R \right)^{-1} = (E + i0)\widehat{S} - \widehat{H}$, H is the Hamiltonian matrix for the scatterer (molecule plus screening part of the electrodes). S is the *overlap* matrix, $S_{i,j} = \langle \chi_i | \chi_j \rangle$ for non-orthogonal basis set orbitals χ_i, and the coupling of the scatterer to the leads is given by the Hamiltonian matrix $V = \text{diag}[\Sigma_{l,l}, \, 0, \, \Sigma_{r,r}]$, where l (r) stands for left (right) electrode. The self-energy part $\Sigma^<$, which is used to construct the non-equilibrium electron density in the scattering region, is found from $\Sigma^< = -2i\text{Im}\left[f(E)\Sigma_{l,l} + f(E + qV)\Sigma_{r,r} \right]$, where $\Sigma_{l,l(r,r)}$ is the self-energy of the left (right) electrode, calculated for the semi-infinite leads using an iterative technique [51–53]. $\Sigma^<$ accounts for the steady charge "flowing in" from the electrodes. The transmission probability is given by

632   A.M. Bratkovsky

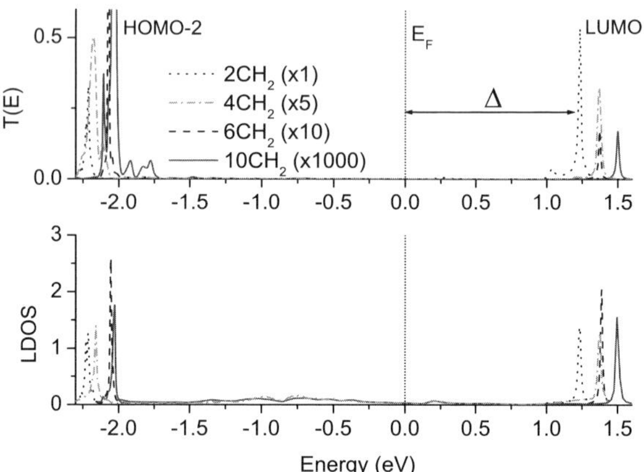

Fig. 4. Transmission coefficient versus energy E for rectifiers $-$S-(CH$_2$)$_2$-C$_{10}$H$_6$$-$(CH$_2$)$_n$$-S-$, $n = 2, 4, 6, 10$. Δ indicates the distance of the closest MO to the electrode Fermi energy (E$_F = 0$).

$$T(E, V) = 4\mathrm{Tr}\left[(\mathrm{Im}\Sigma_{l,l})\, G^R_{l,r}\,(\mathrm{Im}\Sigma_{r,r})\, G^A_{r,l}\right], \tag{8}$$

where $G^{R(A)}$ are the retarded (advanced) Green's function, and Σ the self-energy part connecting left (l) and right (r) electrodes [51–53], and the current is obtained from (4). The calculated transmission coefficient $T(E)$ is shown for a series of rectifiers $-$S-(CH$_2$)$_m$-C$_{10}$H$_6$$-$(CH$_2$)$_n$$-S-$ for $m = 2$ and $n = 2, 4, 6, 10$ at zero bias voltage in Fig. 4. We see that the LUMO is the molecular orbital transparent to electron transport, which lies above E_F by an amount $\Delta = 1.2 - 1.5$eV. The transmission through the HOMO and HOMO-1 states, localized on the terminating sulfur atoms, is negligible, but the HOMO-2 state conducts very well. The HOMO-2 defines the threshold reverse voltage V_R, thus limiting the operating voltage range. Our assumption, that the voltage drop is proportional to the lengths of the alkane groups on both sides, is quantified by the calculated potential ramp. It is close to a linear slope along the (CH$_2$)$_n$ chains [49]. The forward voltage corresponds to the crossing of the LUMO(V) and $\mu_R(V)$, which happens at about 2V. Although the LUMO defines the forward threshold voltages in all molecules studied here, the reverse voltage is defined by the HOMO-2 for "right" barriers (CH$_2$)$_n$ with $n = 6, 10$. The I-V curves are plotted in Fig. 5. We see that the rectification ratio for current in the operation window I_+/I_- reaches a maximum value of 35 for the "2-10" molecule ($m = 2$, $n = 10$). Series of molecules with a central *single phenyl* ring [47] do not show any significant rectification. One can manipulate the system in order to increase the energy asymmetry of the conducting orbitals (reduce Δ). To shift the LUMO towards E_F, one can attach an electron withdrawing group, like $-$C$\equiv$N [49]. The molecular

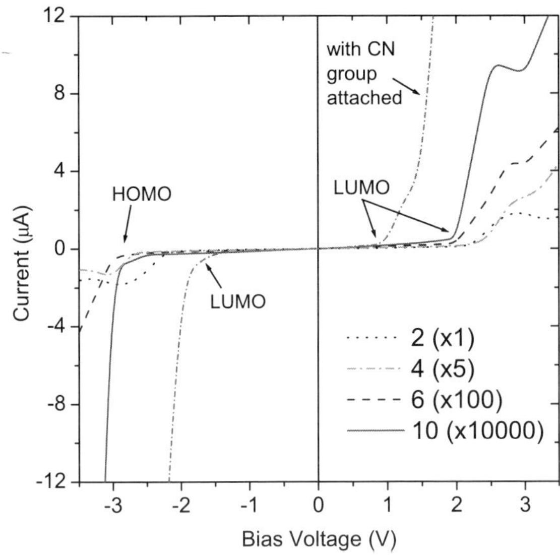

Fig. 5. I-V curves for naphthalene rectifiers $-S-(CH_2)_2-C_{10}H_6-(CH_2)_n-S-$, $n = 2, 4, 6, 10$. The short-dash-dot curve corresponds to a cyano-doped (added group -C≡N) $n = 10$ rectifier.

rectification ratio is not great by any means, but one should bear in mind that this is a device necessarily operating in a ballistic quantum-mechanical regime because of the small size. This is very different from present Si devices with carriers diffusing through the system. As silicon devices become smaller, however, the same effects will eventually take over, and tend to diminish the rectification ratio, in addition to effects of finite temperature and disorder in the system.

4 Molecular Switches

There are various molecular systems that exhibit some kind of current "switching" behavior [33, 35–37], "negative differential resistance"[32], and "memory"[34]. The switching systems are basically driven between two states with considerably different resistances. This behavior is not really sensitive to a particular molecular structure, since this type of bistability is observed in complex rotaxane-like molecules as well as in very simple alkane chains $(CH_2)_n$ assembled into LB films [37], and is not even exclusive to organic films. The data strongly indicate that the switching has an extrinsic origin, and is related either to bistability of molecule-electrode orientation [22, 23, 33], or transport assisted by defects in the film [56, 57].

4.1 Extrinsic Switching in Organic Molecular Films: Role of Defects and Molecular Reconfigurations

Evidently, large defects can be formed in organic thin films as a result of electromigration in a very strong field, as was observed long ago [58, 59]. It was concluded some decades ago that conduction through absorbed [56] and Langmuir-Blodgett [57] monolayers of fatty acids $(CH_2)_n$, which we denote as Cn, is associated with *defects*. In particular, Polymeropoulos and Sagiv studied a variety of absorbed monolayers from C7 to C23 on Al/Al_2O_3 substrates and found that the exponential dependence on the length of the molecular chains is only observed below the liquid nitrogen temperature of 77K, and no discernible length dependence was observed at higher temperatures [56]. The temperature dependence of current was strong, and was attributed to transport assisted by some defects. The current also varied strongly with the temperature in [57] for LB films on Al/Al_2O_3 substrates in He atmosphere, which is not compatible with elastic tunneling. Since the He atmosphere was believed to hinder the Al_2O_3 growth, and yet the resistance of the films increased about 100-fold over 45 days, the conclusion was made that the "defects" somehow anneal out with time. Two types of switching have been observed in 3-30 μm thick films of polydimethylsiloxane (PDMS), one as a standard dielectric breakdown with electrode material "jet evaporation" into the film with subsequent Joule melting of the metallic filament under bias of about 100V, and a low-voltage (<1 V) "ultraswitching" that has a clear "telegraph" character and resulted in intermittent switching into a much more conductive state [60]. The exact nature of this switching also remains unclear, but there is a strong expectations that the formation of metallic filaments that may even be in a ballistic regime of transport, may be relevant to the phenomenon.

Recently, direct evidence was obtained of the formation of "hot spots" in the LB films that may be related to the filament growth through the film imaged with the use of AFM current mapping [61]. The system investigated in this work has been Pt/stearic acid (C18)/Ti (Pt/C18/Ti) crossbar molecular structure, consisting of planar Pt and Ti electrodes sandwiching a monolayer of 2.6-nm-long stearic acid (C18H36OH) molecules with typical zero-bias resistance in excess of $10^5 \Omega$. The devices has been switched reversibly and repeatedly to higher ("on") or lower ("off") conductance states by applying sufficiently large bias voltage V_b to the top Ti electrode with regards to the Pt counterelectrode, Fig. 6.

Interestingly, reversible switching was not observed in symmetric Pt/C18/Pt devices. The local conductance maps of the Pt/C18/Ti structure have been constructed by using an AFM tip and simultaneously measuring the current through the molecular junction biased to $V_b = 0.1$V (the AFM tip was not used as an electrode, only to apply local pressure at the surface). The study

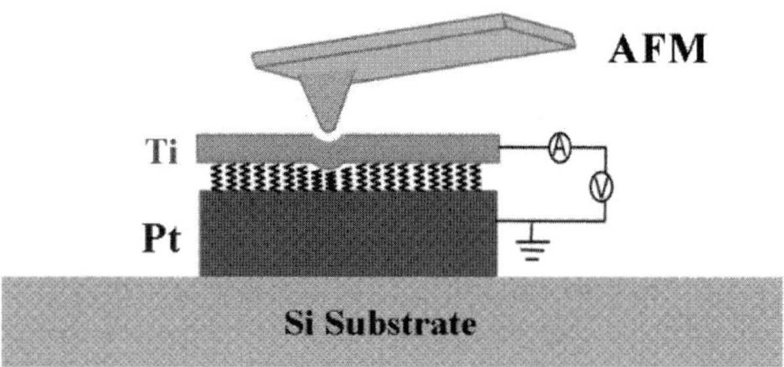

Fig. 6. Experimental setup for mapping local conductance. AFM produces local deformation of top electrode and underlying organic film. The *total* conductance of the device is measured and mapped. (Courtesy C.N. Lau).

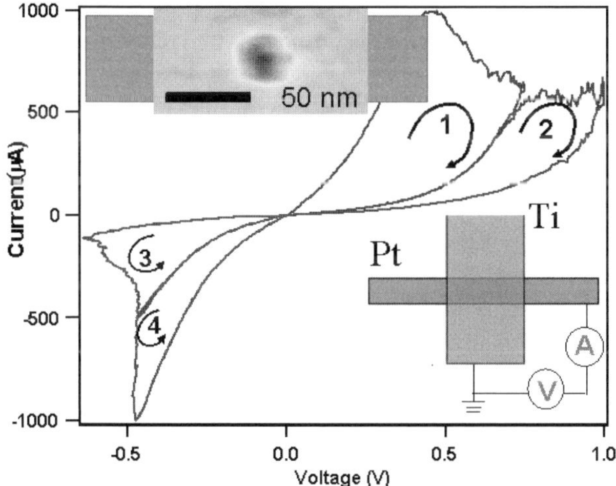

Fig. 7. I-V characteristic showing the reversible switching cycle of the device (bottom inset) with organic film. The arrows indicate sweep direction. A negative bias switches the device to a high conductance state, while the positive one switches it to low conductance state. The mapping according to the schematic in Fig. 6 shows the appearance of "hot spots" after switching (top inset). (Courtesy C.N. Lau).

revealed that the film showed pronounced switching between electrically very distinct states, with zero-bias conductances $0.17\mu S$ ("off" state) and $1.45\mu S$ ("on" state), Fig. 7.

At every switching "on" there appeared a local conductance peak on the map with a typical diameter ~ 40nm, which then disappeared upon switching "off", Fig. 7 (top inset). The switching has been attributed to local conducting filament formation due to electromigration processes. It remains unclear how exactly the filaments dissolve under opposite bias voltage, why they tend to appear in new places after each switching, and why conductance in some cases strongly depends on temperature. It is clear, however, that switching in such a simple molecule without any redox centers, mobile groups, or charge reception centers should be *extrinsic*. Interestingly, very similar "switching" between two resistive states has also been observed for tunneling through thin *inorganic* perovskite oxide films [62].

There have been plenty of reports on non-linear I-V characteristics like negative differential resistance (NDR) and random switching recently for molecules assembled on metal electrodes (gold) and silicon. Reports on NDR for molecules with metal contacts (Au, Hg) have been made in [32, 45, 63]. It became very clear though that most of these observations are related to molecular reconfigurations and bond breaking and making, rather to any intrinsic mechanism, like redox states, speculated about in the original [32]. Thus, the NDR in Tour wires was related to molecular reconfiguration with respect to metallic electrodes [22, 33], NDR in ferrocene-tethered alkyl monolayers [63] was found to be related to oxygen damage at high voltage [64]. Structural changes and bond breaking have been found to result in NDR in experiments with STM [65–67] and mercury droplet contacts [68].

Several molecules, like styrene, have been studied on degenerate Si surfaces and showed an NDR behavior [69]. However, those results have been carefully checked later and it was found that the styrene molecules do not exhibit NDR, but rather sporadically switch between states with different current while held at the same bias voltage (the blinking effect) [70].

The STM map of the styrene molecules (indicated by arrows) on the Silicon (100) surface shows that the molecules are blinking, see Fig. 8. The blinking is absent at clean Si areas, dark (D) and bright (C) defects. This may indicate a dynamic process occurring during the imaging. Comparing the panel (a) and (b) one may see that some molecules are actually decomposing. The height versus voltage spectra over particular points are shown in Fig. 8c. The featureless curve 1 was taken over a clean silicon dimer. The other spectra were recorded over individual molecules. Each of these spectra have many sudden decreases and increases in current as if the molecules are changing between different states during the measurement causing a change in current and a response of the feedback control, resulting in a change in height so there exists one or more configurations that lead to measurement of a different height. Evidently, these changes have the same origin as the blinking of molecules in STM images. Figure 8b reveals clear structural changes associated with those particular spectroscopic changes. In each case where a dramatic change in spectroscopy occurred, the molecules in the image have changed from a bright feature to a dark spot. This is interpreted as a decomposition of the molecules. A detailed look at each decomposed styrene molecule, at locations 3, 4, 5, and 6, shows that the dark spot is not in precise registry with the original bright feature, indicating that the decomposition product involves reaction with an adjacent dimer [70].

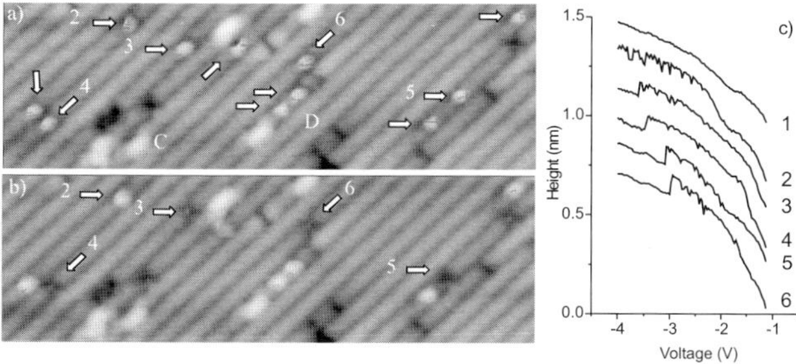

Fig. 8. STM images or styrene molecules on clean Si(100) before and after spectroscopy over the area 75 ×240 Å. (Top left) Bias -2 V, current 0.7 nA. Only the styrene molecules (indicated by arrows) are blinking during imaging. The clean Si surface, bright defects (marked C), and dark spots (marked D) do not experience blinking. (Bottom left) bias -2 V, current 0.3 nA. STM image of individual styrene molecules indicated with numbers 2-6. Styrene molecules 3-6 have decomposed. Decomposition involves the changing of the styrene molecule from a bright feature to a dark depression and also involves the reaction with an adjacent dimer. Styrene molecule 2 does not decompose and images as usual with no change of position. (Right) Height-voltage spectra taken over clean silicon (1) and styrene molecules (2-6). The spectra taken over molecules show several spikes in height related to blinking in the images. In spectra 3-6, an abrupt and permanent change in height is recorded and is correlated with decomposition, as seen in the bottom left image. Spectrum 2 has no permanent height change, and the molecule does not decompose. (Courtesy J. Pitters and R. Wolkow).

The fact that the structural changes and related NDR behavior are not associated with any resonant tunneling through the molecular levels or redox processes, but are perhaps related to inelastic electron scattering or other extrinsic processes, becomes evident from current versus time records shown in Fig. 9 [70]. The records show either no change of the current with time (1), or one or a few random jumps between certain current states (telegraph noise). The observed changes in current at a fixed voltage obviously cannot be explained by shifting and aligning of molecular levels, as was suggested in [71], they must be related to an adsorbate molecule's structural changes with time. Therefore, the explanation by Datta et al. that the resonant level alignment is responsible for NDR does not apply [71]. As mentioned above, similar telegraph switching and NDR has been observed in Tour wires [33] and other molecules. Therefore, the observed negative differential resistance apparently has similar origin in disparate molecules adsorbed on different substrates, and has to do with molecular reconformation/reconfiguration on the surface.

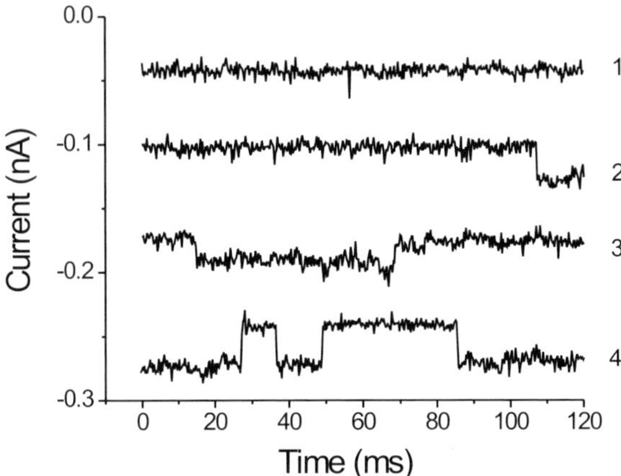

Fig. 9. Variation of current through styrene molecules on Si(100) with time. Tunneling conditions were set at -3 V and 0.05 nA. Abrupt increases and decreases in current relate to changes of the molecule during the spectroscopy. Some experiments show no changes in current (curve 1), others show various kinds of telegraph switching. (Courtesy J. Pitters and R. Wolkow).

4.2 Intrinsic Polarization and Extrinsic Conductance Switching in Molecular Ferroelectric PVDF

The only well established, to the best of our knowledge, intrinsic molecular switching (of polarization, not current) under bias voltage was observed in molecular ferroelectric block co-polymers polyvinylidene [72]. Ferroelectric polymer films have been prepared with the 70% vinylidene fluoride copolymer, P(VDF-TrFE 70:30), formed by horizontal LB deposition on aluminum-coated glass substrates with evaporated aluminum top electrodes. The polymer chains contain random sequences of $(CH_2)_n(CF_2)_m$ blocks, the fluorine sites carry a strong negative charge, and in the ferroelectric phase most of the carbon–fluorine bonds point in one direction. The fluorine groups can be rotated and aligned in very strong electric field, $\sim 5MV/cm$. As a result, the whole molecular chain orders, and in this way the macroscopic polarization can be switched between the opposite states. The switching process is extremely slow, however, and takes 1-10 seconds (!) [73, 74]. This is not surprising, given the strong Coulomb interaction between charged groups and the metal electrodes, pinning by surface roughness, and steric hindrance to rotation. This behavior should be suggestive of other switching systems based on one of a few monolayers of molecules, and other nontrivial behavior involved [74].

The switching of current was also observed in films of PVDF 30 monolayers thick. The conductance of the film was following the observed hysteresis loop for the polarization, ranging from $\sim 1 \times 10^{-9} - 2 \times 10^{-6}\Omega^{-1}$[75]. The phenomenon of conductance switching has these important features: (i) It is connected with the bulk polarization switching; (ii) there is a large $\sim 1000 : 1$ contrast between the ON and OFF states; (iii) the ON state is obtained only when the bulk polarization is switched in the positive direction; (iv) the conductance switching is much faster than the bulk polarization switching. The conductance switches ON only after the 6s delay, after the bulk polarization switching is nearly complete, presumably when the last layer switches into alignment with the others, while the conductance switches OFF without a noticeable delay after the application of reverse bias as even one layer reverses (this may create a barrier to charge transfer). The slow ($\sim$ 2s) time constant for polarization switching is probably nucleation limited as has been observed in high-quality bulk films with low nucleation site densities [76]. The duration of the conductance switching transition ($\sim$ 2ms) may be limited only by the much faster switching time of individual layers.

The origin of conductance switching by 3 orders of magnitude is not clear. It may indeed be related to a changing amount of disorder for tunneling/hopping electrons. It is conceivable that the carriers are strongly trapped in polaron states inside PVDF and find the optimal path for hopping in the material, which is incompletely switched. This is an interesting topic that is certainly in need of further experimental and theoretical study.

5 Molecular Quantum Dot Switching

5.1 Electrically Addressable Molecules

For many applications one needs an *intrinsic* molecular "switch", i.e. a bistable voltage-addressable molecular system with very different resistances in the two states that can be accessed very quickly. There is a trade-off between the stability of a molecular state and the ability to switch the molecule between two states with an external perturbation (we discuss an electric field, switching involving absorbed photons is impractical at a nanoscale). Indeed, the applied electric field, on the order of a typical breakdown field $E_b \leq 10^7$V/cm, is much smaller than a typical atomic field $\sim 10^9$V/cm, characteristic of the energy barriers. A small barrier would be a subject for sporadic thermal switching, whereas a larger barrier $\sim 1 - 2$eV would be impossible to overcome with the applied field. One may only change the relative energy of the minima by an external field and, therefore, redistribute the molecules statistically slightly inequivalently between the two states. An intrinsic disadvantage of the conformational mechanism, involving motion of ionic groups, exceeding the electron mass by many orders of magnitude, is a slow switching speed ($\sim$kHz). In case of supramolecular complexes like rotaxanes and catenanes [35] there are two entangled parts which can change mutual positions as a result of redox reactions (in solution). Thus, for the

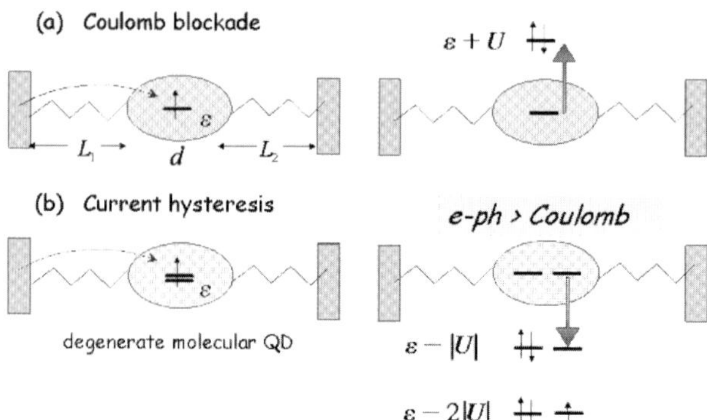

Fig. 10. Schematic of the molecular quantum dot with central conjugated unit separated from the electrodes by wide-band insulating molecular groups. First electron tunnels into the dot and occupies an empty (degenerate) state there. If the interaction between the first and second incoming electron is repulsive, $U > 0$, then the dot will be in a Coulomb blockade regime (a). If the electrons on the dot effectively attract each other, $U < 0$, the system will show current hysteresis (b).

rotaxane-based memory devices a slow switching speed of $\sim 10^{-2}$ seconds was reported.

As a possible conformational E-field addressable molecular switch, we have considered a bistable molecule with -CONH$_2$ dipole group [77]. The barrier height is $E_b = 0.18$eV. Interaction with an external electric field changes the energy of the minima, but estimated switching field is huge, ~ 0.5V/A. At non-zero temperatures, temperature fluctuations might result in statistical dipole flipping at lower fields. The I-V curve shows hysteresis in the 3 to 4 Volts window for two possible conformations. One can estimate the thermal stability of the state as 58 ps at room temperature, and 33 ms at 77 K.

We explored the possibility for a fast molecular switching where switching is due to strong correlation effects on the molecule itself, the so-called molecular quantum dot (MQD). The molecular quantum dot consists of a central *conjugated* unit (containing half-occupied, and, therefore, extended $\pi-$orbitals), Fig. 10. Frequently, those are formed from the p-states on carbon atoms, which are not *saturated* (i.e. they do not share electrons with other atoms forming strong $\sigma-$bonds, with typical bonding-antibonding energy difference about 1Ry). Since the $\pi-$orbitals are half-occupied, they form the HOMO-LUMO states. The size of the HOMO-LUMO gap is then directly related to the size of the conjugated region d, Fig. 10, by a standard estimate $E_{\mathrm{HOMO-LUMO}} \sim \hbar^2/md^2 \sim 2-5$ eV. It is worth noting that in the conjugated linear polymers like polyacetylene $(-\overset{|}{C} = \overset{|}{C})_n$ the spread of the $\pi-$electron would be $d = \infty$ and the expected $E_{\mathrm{HOMO-LUMO}} = 0$. However, such a one-dimensional metal is impossible, Peierls distortion (C=C bond length dimerization) sets in and opens up a gap of about $\sim 1.5\text{eV}$ at the Fermi level [17]. In a molecular quantum dot the central conjugated part is separated from the electrodes by insulating groups with saturated $\sigma-$bonds, like e.g. the alkane chains, Fig. 3. Now, there are two main possibilities for carrier transport through the MQD. If the length of at least one of the insulating groups $L_{1(2)}$ is not very large (a conductance $G_{1(2)}$ is not much smaller than the conductance quantum $G_0 = 2e^2/h$), then the transport through the MQD will proceed by resonant tunneling processes. If, on the other hand, both groups are such that the tunnel conductance $G_{1(2)} \ll G_0$, the charge on the dot will be quantized. Then we will have another two possibilities: (i) the interaction of the extra carriers on the dot is *repulsive* $U > 0$, and we have a Coulomb blockade [78], or (ii) the effective interaction is *attractive*, $U < 0$, then we would obtain the current *hysteresis* and switching (see below). Coulomb blockade in molecular quantum dots has been demonstrated in [79]. In these works, and in [80], the three-terminal active molecular devices have been fabricated and successfully tested.

Much faster switching, compared to the conformational one described in the previous section, may be caused by coupling to the vibrational degrees of freedom, if the vibron-mediated attraction between two carriers on the molecule is stronger than their direct Coulomb repulsion [54], Fig. 10b. The attractive energy is the difference of two large interactions, the Coulomb repulsion and the phonon mediated attraction, on the order of 1eV each, hence $|U| \sim 0.1\text{eV}$.

5.2 Polaron Mechanism of Carrier Attraction and Switching

Although the correlated electron transport through mesoscopic systems with repulsive electron-electron interactions has received considerable attention in the past, and continues to be the focus of current studies, much less was known about the role of electron-phonon correlations in "molecular quantum dots" (MQD). Some while ago we proposed a negative$-U$ Hubbard model of a d-fold degenerate quantum dot [81] and a polaron model of resonant tunneling through a molecule with degenerate level [54]. We found that the *attractive* electron correlations caused by any interaction within the molecule could lead to a molecular *switching* effect where I-V characteristics have two branches with high and low current at the same bias voltage. This prediction has been confirmed and extended further in our theory of *correlated* transport through degenerate MQDs with a full account of both the Coulomb repulsion and realistic electron-phonon (e-ph) interactions [54]. We have shown that while the phonon side-bands significantly modify the shape of hysteretic I-V curves in comparison with the negative-U Hubbard model, switching remains robust. It shows up when the effective interaction of polarons is attractive and the state of the dot is multiply degenerate, $d > 2$.

First, we shall describe the simplest model of a single atomic level coupled with a single one-dimensional oscillator using the first quantization represen-

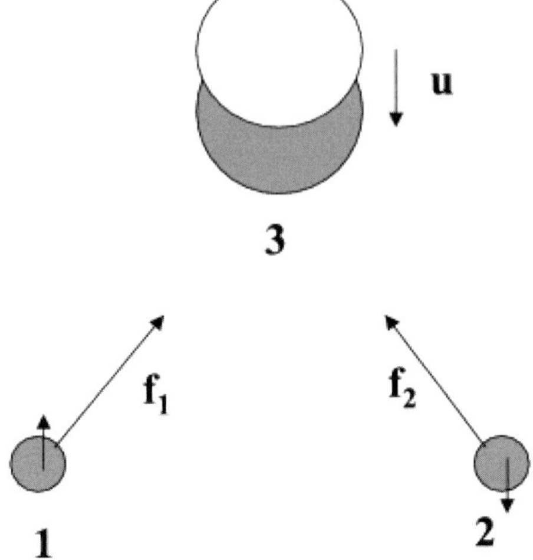

Fig. 11. Two localized electrons at sites **1** and **2** shift the equilibrium position of the ion at site **3**. As a result, the two electrons *attract* each other.

tation for its displacement x [82],

$$H = \varepsilon_0 \widehat{n} + f x \widehat{n} - \frac{1}{2M} \frac{\partial^2}{\partial x^2} + \frac{kx^2}{2}. \tag{9}$$

Here M and k are the oscillator mass and the spring constant, f is the interaction force, and $\hbar = c = k_B = 1$. This Hamiltonian is readily diagonalized with the *exact* displacement transformation of the vibration coordinate x,

$$x = y - \widehat{n} f/k, \tag{10}$$

to the transformed Hamiltonian without electron-phonon coupling,

$$\widetilde{H} = \varepsilon \widehat{n} - \frac{1}{2M} \frac{\partial^2}{\partial y^2} + \frac{ky^2}{2}, \tag{11}$$

$$\varepsilon = \varepsilon_0 - E_p, \tag{12}$$

where we used $\widehat{n}^2 = \widehat{n}$ because of the Fermi-Dirac statistics. It describes a small polaron at the atomic level ε_0 shifted down by the polaron level shift $E_p = f^2/2k$, and entirely decoupled from ion vibrations. The ion vibrates near a new equilibrium position, shifted by f/k, with the "old" frequency $(k/M)^{1/2}$. As a result of the local ion deformation, the total energy of the whole system decreases by E_p since a decrease of the electron energy by $-2E_p$ overruns an increase of the deformation energy E_p. Note that the authors of [83] made an illegitimate replacement of the square of the occupation number operator $\widehat{n} = c_0^\dagger c_0$ in (11) by its "mean-field" expression $\widehat{n}^2 = n_0 \widehat{n}$ which contains the average population of a single molecular level, n_0, in disagreement with the exact identity, $\widehat{n}^2 = \widehat{n}$. This leads to a spurious *self-interaction* of a single polaron with itself [i.e. the term $\varepsilon = \varepsilon_0 - n_0 E_p$ instead of (12)], and a resulting non-existent nonlinearity in the rate equation.

Lattice deformation also strongly affects the interaction between electrons. When a short-range deformation potential and molecular e-ph interactions are taken into account together with the long-range Fröhlich interaction, they can overcome the Coulomb repulsion. The resulting interaction becomes attractive at a short distance comparable to a lattice constant. The origin of the attractive force between two small polarons can be readily understood from a similar Holstein-like toy model as above [84], but with two electrons on neighboring sites **1,2** interacting with an ion **3** between them, Fig. 11. For generality, we now assume that the ion is a three-dimensional oscillator described by a displacement vector $\mathbf{u}$, rather than by a single-component displacement x as in (1).

The vibration part of the Hamiltonian in the model is

$$H_{ph} = -\frac{1}{2M}\frac{\partial^2}{\partial \mathbf{u}^2} + \frac{k\mathbf{u}^2}{2}, \tag{13}$$

Electron potential energies due to the Coulomb interaction with the ion are about

$$V_{1,2} = V_0 - \mathbf{u}\nabla_{R_{1,2}}V_0(R_{1,2}), \tag{14}$$

where $\mathbf{R_{1(2)}}$ is the vector connecting ion site **3** with electron **1(2)**. Hence, the Hamiltonian of the model is given by

$$H = E_a(\widehat{n}_1 + \widehat{n}_2) + \mathbf{u}\cdot(\mathbf{f_1}\widehat{n}_1 + \mathbf{f_2}\widehat{n}_2) - \frac{1}{2M}\frac{\partial^2}{\partial \mathbf{u}^2} + \frac{k\mathbf{u}^2}{2}, \tag{15}$$

where $\mathbf{f_{1,2}} = Ze^2\mathbf{e_{1,2}}/a^2$ is the Coulomb force, and $\widehat{n}_{1,2}$ are occupation number operators at every site. This Hamiltonian is also readily diagonalized by the same displacement transformation of the vibronic coordinate $\mathbf{u}$ as above,

$$\mathbf{u} = \mathbf{v} - (\mathbf{f_1}\widehat{n}_1 + \mathbf{f_2}\widehat{n}_2)/k. \tag{16}$$

The transformed Hamiltonian has no electron-phonon coupling,

$$\widetilde{H} = (\varepsilon_0 - E_p)(\widehat{n}_1 + \widehat{n}_2) + V_{ph}\widehat{n}_1\widehat{n}_2 - \frac{1}{2M}\frac{\partial^2}{\partial \mathbf{v}^2} + \frac{k\mathbf{v}^2}{2}, \tag{17}$$

and it describes two small polarons at their atomic levels shifted by the polaron level shift $E_p = f_{1,2}^2/2k$, which are entirely decoupled from ion vibrations. As a result, the lattice deformation caused by two electrons leads to an effective interaction between them, V_{ph}, which should be added to their Coulomb repulsion, V_c,

$$V_{ph} = -\mathbf{f_1}\cdot\mathbf{f_2}/k. \tag{18}$$

When V_{ph} is negative and larger by magnitude than the positive V_c, the resulting interaction becomes attractive. That is V_{ph} rather than E_p, which is responsible for the hysteretic behavior of MQDs, as discussed below.

5.3 Exact Solution

The procedure, which fully accounts for all correlations in MQD is as follows, see [54]. The molecular Hamiltonian includes the Coulomb repulsion, U^C, and the electron-vibron interaction as

$$H = \sum_{\mu} \varepsilon_{\mu}\widehat{n}_{\mu} + \frac{1}{2}\sum_{\mu\neq\mu'} U^C_{\mu\mu'}\widehat{n}_{\mu}\widehat{n}_{\mu'}$$
$$+ \sum_{\mu,q}\widehat{n}_{\mu}\omega_q(\gamma_{\mu q}d_q + H.c.) + \sum_{q}\omega_q(d^{\dagger}_q d_q + 1/2). \tag{19}$$

Here d_q annihilates phonons, ω_q is the phonon (vibron) frequency, and $\gamma_{\mu q}$ are the e-ph coupling constants (q enumerates the vibron modes). This Hamiltonian conserves the occupation numbers of molecular states $\widehat{n}_{\mu}$.

One can apply the canonical unitary transformation e^S, with

$$S = -\sum_{q,\mu}\widehat{n}_{\mu}\left(\gamma_{\mu q}d_q - H.c.\right)$$

integrating phonons out. The electron and phonon operators are transformed as

$$\widetilde{c}_{\mu} = c_{\mu}X_{\mu}, \qquad X_{\mu} = \exp\left(\sum_{q}\gamma_{\mu q}d_q - H.c.\right) \tag{20}$$

and

$$\widetilde{d}_q = d_q - \sum_{\mu}\widehat{n}_{\mu}\gamma^*_{\mu q}, \tag{21}$$

respectively. This Lang-Firsov transformation shifts ions to new equilibrium positions with no effect on the phonon frequencies. The diagonalization is *exact*:

$$\widetilde{H} = \sum_{i}\widetilde{\varepsilon}_{\mu}\widehat{n}_{\mu} + \sum_{q}\omega_q(d^{\dagger}_q d_q + 1/2) + \frac{1}{2}\sum_{\mu\neq\mu'} U_{\mu\mu'}\widehat{n}_{\mu}\widehat{n}_{\mu'}, \tag{22}$$

where

$$U_{\mu\mu'} \equiv U^C_{\mu\mu'} - 2\sum_{q}\gamma^*_{\mu q}\gamma_{\mu' q}\omega_q, \tag{23}$$

is the renormalized interaction of polarons comprising their interaction via molecular deformations (vibrons) and the original Coulomb repulsion, $U^C_{\mu\mu'}$. The molecular energy levels are shifted by the polaron level-shift due to the deformation created by the polaron,

$$\widetilde{\varepsilon}_{\mu} = \varepsilon_{\mu} - \sum_{q}|\gamma_{\mu q}|^2\omega_q. \tag{24}$$

If we assume that the coupling to the leads is weak, so that the level width $\Gamma \ll |U|$, we can find the current from [85]

$$I(V) = I_0 \int\limits_{-\infty}^{\infty} d\omega \left[f_1(\omega) - f_2(\omega) \right] \rho(\omega), \tag{25}$$

$$\rho(\omega) = -\frac{1}{\pi} \sum_{\mu} \mathrm{Im} \widehat{G}^R_\mu(\omega), \tag{26}$$

where $|\mu\rangle$ is a complete set of one-particle molecular states, $f_{1(2)}(\omega) = \left(\exp \frac{\omega + \Delta \mp eV/2}{T} + 1 \right)^{-1}$ the Fermi function. Here $I_0 = q\Gamma$, $\rho(\omega)$ is the molecular DOS, $\widehat{G}^R_\mu(\omega)$ is the Fourier transform of the Green's function $\widehat{G}^R_\mu(t) = -i\theta(t) \left\langle \{ c_\mu(t), c_\mu^\dagger \} \right\rangle$, $\{ \cdots, \cdots \}$ is the anticommutator, $c_\mu(t) = e^{iHt} c_\mu e^{-iHt}$, $\theta(t) = 1$ for $t > 0$ and zero otherwise. We calculate $\rho(\omega)$ *exactly* for the Hamiltonian (22), which includes both the Coulomb U^C and e-ph interactions.

The retarded GF becomes

$$\begin{aligned} G^R_\mu(t) = -i\theta(t)[&\langle c_\mu(t) c_\mu^\dagger \rangle \langle X_\mu(t) X_\mu^\dagger \rangle \\ &+ \langle c_\mu^\dagger c_\mu(t) \rangle \langle X_\mu^\dagger X_\mu(t) \rangle]. \end{aligned} \tag{27}$$

The phonon correlator is simply

$$\begin{aligned} \langle X_\mu(t) X_\mu^\dagger \rangle = \exp \sum_q &\frac{|\gamma_{\mu q}|^2}{\sinh \frac{\beta \omega_q}{2}} \\ &\times \left[\cos\left(\omega t + i\frac{\beta \omega_q}{2} \right) - \cosh \frac{\beta \omega_q}{2} \right], \end{aligned} \tag{28}$$

where the inverse temperature $\beta = 1/T$, and $\langle X_\mu^\dagger X_\mu(t) \rangle = \langle X_\mu(t) X_\mu^\dagger \rangle^*$. The remaining GFs $\langle c_\mu(t) c_\mu^\dagger \rangle$, are found from the equations of motion *exactly*. Finally, for the simplest case of a coupling to a single mode with the characteristic frequency ω_0 and $\gamma_q \equiv \gamma$ we find [54]

$$\begin{aligned} G^R_\mu(\omega) = \mathcal{Z} \sum_{r=0}^{d-1} C_r(n) \sum_{l=0}^{\infty} I_l(\xi) \\ \left[e^{\frac{\beta \omega_0 l}{2}} \left(\frac{1-n}{\omega - rU - l\omega_0 + i\delta} + \frac{n}{\omega - rU + l\omega_0 + i\delta} \right) \right. \\ + (1 - \delta_{l0}) e^{-\frac{\beta \omega_0 l}{2}} \\ \left. \times \left(\frac{1-n}{\omega - rU + l\omega_0 + i\delta} + \frac{n}{\omega - rU - l\omega_0 + i\delta} \right) \right], \end{aligned} \tag{29}$$

where

$$\mathcal{Z} = \exp\left(-\sum_{\mathbf{q}} |\gamma_q|^2 \coth \frac{\beta \omega_q}{2} \right) \tag{30}$$

is the familiar *polaron narrowing* factor, the degeneracy factor

646   A.M. Bratkovsky

$$C_r(n) = \frac{(d-1)!}{r!(d-1-r)!}n^r(1-n)^{d-1-r},\tag{31}$$

$\xi = |\gamma|^2/\sinh\frac{\beta\omega_0}{2}$, $I_l(\xi)$ the modified Bessel function, and δ_{lk} the Kroneker symbol.

To simplify our discussion, we assume that the Coulomb integrals do not depend on the orbital index, i.e. $U_{\mu\mu'} = U$, and consider a coupling to a single vibrational mode, $\omega_q = \omega_0$. Applying the same transformation to the retarded Green's function, one obtains the exact spectral function [54] for a $d-$ fold degenerate MQD (i.e. the density of molecular states, DOS) as

$$\rho(\omega) = \mathcal{Z}d\sum_{r=0}^{d-1} C_r(n) \sum_{l=0}^{\infty} I_l(\xi)$$

$$\times \left[e^{\beta\omega_0 l/2}\left[(1-n)\delta(\omega - rU - l\omega_0) + n\delta(\omega - rU + l\omega_0)\right]\right.$$

$$+ (1-\delta_{l0})e^{-\beta\omega_0 l/2}[n\delta(\omega - rU - l\omega_0)$$

$$\left. + (1-n)\delta(\omega - rU + l\omega_0)]\right],\tag{32}$$

where $\mathcal{Z} = \exp\left[-|\gamma|^2\coth\frac{\beta\omega_0}{2}\right]$, is the above polaron narrowing factor, $\xi = |\gamma|^2/\sinh(\beta\omega_0/2)$, $\beta = 1/T$, and δ_{lk} the Kroneker symbol.

The important feature of DOS, (32), is its nonlinear dependence on the average electronic population $n = \langle c_\mu^\dagger c_\mu \rangle$, which leads to the switching, hysteresis, and other nonlinear effects in I-V characteristics for $d > 2$ [54]. It appears due to *correlations* between *different* electronic states via the correlation coefficients $C_r(n)$. There is no nonlinearity if the dot is nondegenerate, $d = 1$, since $C_0(n) = 1$. In this simple case the DOS, (32), is a *linear* function of the average population that can be found as a textbook example of an exactly solvable problem [86].

In the present case of MQD weakly coupled with leads, one can apply the Fermi-Dirac golden rule to obtain an equation for n. Equating incoming and outgoing numbers of electrons in MQD per unit time we obtain the self-consistent equation for the level occupation n as

$$(1-n)\int_{-\infty}^{\infty} d\omega\,\{\Gamma_1 f_1(\omega) + \Gamma_2 f_2(\omega)\}\,\rho(\omega)$$

$$= n\int_{-\infty}^{\infty} d\omega\,\{\Gamma_1[1 - f_1(\omega)] + \Gamma_2[1 - f_2(\omega)]\}\,\rho(\omega),\tag{33}$$

where $\Gamma_{1(2)}$ are the transition rates from left (right) leads to MQD, and $\rho(\omega)$ is found from (29) and (25). For $d = 1, 2$ the kinetic equation for n is linear, and the switching is *absent*. Switching appears for $d \geq 3$, when the kinetic equation becomes non-linear. Taking into account that $\int_{-\infty}^{\infty} \rho(\omega) = d$, (33) for the symmetric leads, $\Gamma_1 = \Gamma_2$, reduces to

$$2nd = \int d\omega \rho(\omega)(f_1 + f_2),\tag{34}$$

which automatically satisfies $0 \leq n \leq 1$. Explicitly, the self-consistent equation for the occupation number is

$$n = \frac{1}{2}\sum_{r=0}^{d-1} C_r(n)[na_r^+ + (1-n)b_r^+],\tag{35}$$

where

$$\begin{aligned}a_r^+ = \mathcal{Z}\sum_{l=0}^{\infty} I_l(\xi)\Big(& e^{\frac{\beta\omega_0 l}{2}}[f_1(rU - l\omega_0) + f_2(rU - l\omega_0)] \\ & +(1-\delta_{l0})e^{-\frac{\beta\omega_0 l}{2}}[f_1(rU + l\omega_0) + f_2(rU + l\omega_0)]\Big),\end{aligned}\tag{36}$$

$$\begin{aligned}b_r^+ = \mathcal{Z}\sum_{l=0}^{\infty} I_l(\xi)\Big(& e^{\frac{\beta\omega_0 l}{2}}[f_1(rU + l\omega_0) + f_2(rU + l\omega_0)] \\ & +(1-\delta_{l0})e^{-\frac{\beta\omega_0 l}{2}}[f_1(rU - l\omega_0) + f_2(rU - l\omega_0)]\Big).\end{aligned}\tag{37}$$

The current is expressed as

$$j \equiv \frac{I(V)}{dI_0} = \sum_{r=0}^{d-1} Z_r(n)[na_r^- + (1-n)b_r^-],\tag{38}$$

where

$$\begin{aligned}a_r^- = \mathcal{Z}\sum_{l=0}^{\infty} I_l(\xi)\Big(& e^{\frac{\beta\omega_0 l}{2}}[f_1(rU - l\omega_0) - f_2(rU - l\omega_0)] \\ & +(1-\delta_{l0})e^{-\frac{\beta\omega_0 l}{2}}[f_1(rU + l\omega_0) - f_2(rU + l\omega_0)]\Big),\end{aligned}\tag{39}$$

$$\begin{aligned}b_r^- = \mathcal{Z}\sum_{l=0}^{\infty} I_l(\xi)\Big(& e^{\frac{\beta\omega_0 l}{2}}[f_1(rU + l\omega_0) - f_2(rU + l\omega_0)] \\ & (1-\delta_{l0})e^{-\frac{\beta\omega_0 l}{2}}[f_1(rU - l\omega_0) - f_2(rU - l\omega_0)]\Big).\end{aligned}\tag{40}$$

5.4 Absence of Switching of Single- or Double-Degenerate MQD

If the transition rates from electrodes to MQD are small, $\Gamma \ll \omega_0$, the rate equations for n and the current, $I(V)$ are readily obtained by using the exact

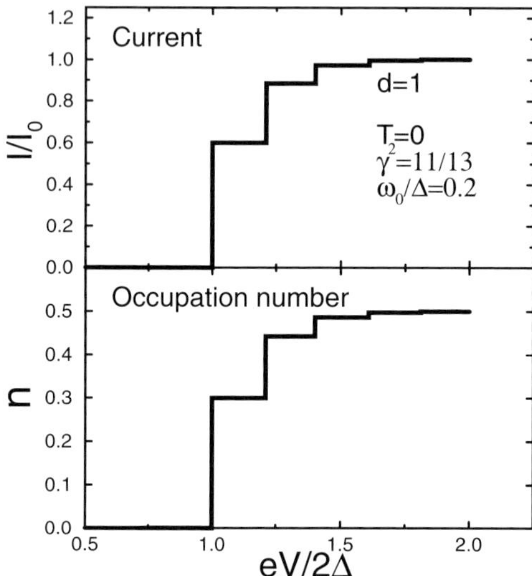

Fig. 12. Current-voltage characteristic of the nondegenerate ($d = 1$) MQD at $T = 0$, $\omega_0/\Delta = 0.2$, and $\gamma^2 = 11/13$. There is the phonon ladder in I-V, but no hysteresis.

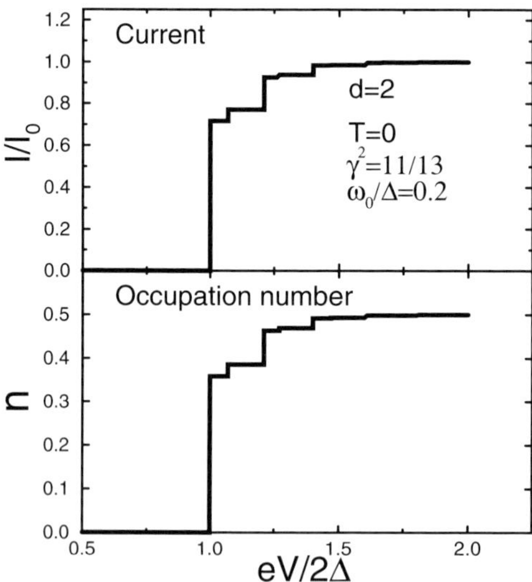

Fig. 13. Current-voltage characteristic of two-fold degenerate MQDs ($d = 2$) does not show hysteretic behavior. The parameters are the same as in Fig. 12. Larger number of elementary processes for conductance compared to the nondegenerate case of $d = 1$ generates more steps in the phonon ladder in comparison with Fig. 12.

molecular DOS, (32) and the Fermi-Dirac Golden rule. In particular, for the nondegenerate MQD and $T = 0$K the result is

$$n = \frac{b_0^+}{2 + b_0^+ - a_0^+},$$ (41)

and

$$j = \frac{2b_0^- + a_0^- b_0^+ - a_0^+ b_0^-}{2 + b_0^+ - a_0^+}.$$ (42)

The general expressions for the coefficients are given at arbitrary temperatures in [54], and they are especially simple in the low-temperature limit:

$$a_0^\pm = \mathcal{Z} \sum_{l=0}^{\infty} \frac{|\gamma|^{2l}}{l!} [\theta(l\omega_0 - \Delta + eV/2)$$
$$\pm \theta(l\omega_0 - \Delta - eV/2)],$$ (43)

$$b_0^\pm = \mathcal{Z} \sum_{l=0}^{\infty} \frac{|\gamma|^{2l}}{l!} [\theta(-l\omega_0 - \Delta + eV/2)$$
$$\pm \theta(-l\omega_0 - \Delta - eV/2)],$$ (44)

$j = I/I_0$, $I_0 = ed\Gamma$, Δ is the position of the MQD level with respect to the Fermi level at $V = 0$, and $\theta(x) = 1$ if $x > 0$ and zero otherwise. The current is single valued, Fig. 12, with the familiar steps due to phonon-side bands. The double-degenerate level provides more elementary processes for conductance reflected in a larger number of steps on the phonon ladder compared to $d = 2$ case, Fig. 13.

On the contrary, the mean-field approximation (MFA) leads to the opposite conclusion. Galperin et $al.$ [83] have replaced the occupation number operator $\hat{n}$ in the e-ph interaction by the average population n_0 [(2) of [83]] and found the average steady-state vibronic displacement $\langle d + d^\dagger \rangle$ proportional to n_0 (this is an explicit $neglect$ of all quantum fluctuations on the dot accounted for in the exact solution). Then, replacing the displacement operator $d + d^\dagger$ in the bare Hamiltonian, (11), by its average, [83], they obtained a new molecular level, $\tilde{\varepsilon}_0 = \varepsilon_0 - 2\varepsilon_{reorg}n_0$ shifted linearly with the average population of the level. This is in stark disagreement with the conventional constant polaronic level shift, (12,24) (ε_{reorg} is $|\gamma|^2\omega_0$ in our notations). The MFA spectral function turned out to be highly nonlinear as a function of the population, e.g. for the weak-coupling with the leads $\rho(\omega) = \delta(\omega - \varepsilon_0 - 2\varepsilon_{reorg}n_0)$, see (17) in [83]. As a result, the authors of [83] have found multiple solutions for the steady-state population, (15) and Fig. 1, and switching, Fig. 4 of [83], which actually do not exist being an artefact of the model.

In the case of a double-degenerate MQD, $d = 2$, there are two terms, which contribute to the sum over r, with $C_0(n) = 1 - n$ and $C_1(n) = n$. The rate equation becomes a quadratic one [54]. Nevertheless there is only one physical root for any temperature and voltage, and the current is also single-valued, Fig. 13.

Note that the mean-field solution by Galperin et al. [83] applies at any ratio Γ/ω_0, including the limit of interest to us, $\Gamma \ll \omega_0$. where their transition between the states with $n_0 = 0$ and 1 only sharpens, but none of the results change. Therefore, MFA predicts a current bistability in the system where it does not exist at $d = 1$. [83] plots the results for $\Gamma \geq \omega_0$, $\Gamma \approx 0.1 - 0.3$ eV, which corresponds to molecular bridges with a resistance of about a few 100 kΩ. Such model "molecules" are rather "metallic" in their conductance and could hardly show any bistability at all because the carriers do not have time to interact with vibrons on the molecule. Indeed, taking into account the coupling with the leads beyond the second order and the coupling between the molecular and bath phonons could hardly provide any non-linearity because these couplings do not depend on the electron population. This rather obvious conclusion for molecules strongly coupled to the electrodes can be reached in many ways, see e.g. a derivation in [87, 88]. While [87, 88] do talk about telegraph current noise in the model, there is no hysteresis in the adiabatic regime, $\Gamma \gg \omega_0$ either. This result certainly has nothing to do with our mechanism of switching [54] that applies to molecular quantum dots ($\Gamma \ll \omega_0$) with $d > 2$. Such a regime has not been studied in [87–89], which have applied the adiabatic approximation, as being a "too challenging problem". Nevertheless, Mitra et al. [89] have misrepresented our formalism [54] claiming that it "lacks of renormalization of the dot-lead coupling" (due to electron–vibron interaction), or "treats it in an average manner". In fact, the formalism [54] is exact, fully taking into account the polaronic renormalization, phonon-side bands and polaron-polaron correlations in the exact molecular DOS, (32).

As a matter of fact, most of the molecules are very resistive, so the actual molecular quantum dots are in the regime we study, see [90]. For example, the resistance of fully conjugated three-phenyl ring Tour-Reed molecules chemically bonded to metallic Au electrodes [32] exceeds 1GΩ. Therefore, most of the molecules of interest to us are in the regime that we discussed, not that of [87, 88].

5.5 Nonlinear Rate Equation and Switching

Now, consider the case $d = 4$, where the rate equation is non-linear, to see if it produces multiple physical solutions. For instance, in the limit $|\gamma| \ll 1$, and $T = 0$, we have $b_r = a_r$, $\mathcal{Z} = 1$, and, if the remaining interaction is $U = U^C < 0$, we recover the negative-U model [81], and the kinetic equation for $d = 4$ is

$$2n = 1 - (1 - n)^3 \tag{45}$$

in the voltage range $\Delta - |U| < eV/2 < \Delta$. This equation has an additional nontrivial physical solution $n = (3 - \sqrt{5})/2 = 0.38$. The current is simplified as $I/I_0 = 2n$. The current-voltage characteristics will show a *hysteretic* behavior

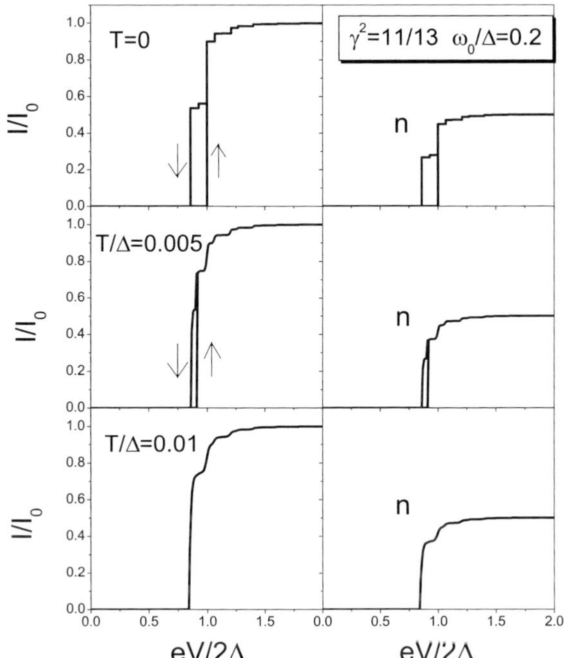

Fig. 14. The I-V curves for tunneling through the molecular quantum dot, Fig. 10b with the electron–vibron coupling constant $\gamma^2 = 11/13$.

in this case for $d = 4$. When the voltage increases from zero, 4-fold degenerate MQD remains in a low-current state until the threshold $eV_2/2 = \Delta$ is reached. Remarkably, when the voltage *decreases* from the value above the threshold V_2, the molecule remains in the high-current state down to the voltage $eV_1/2 = \Delta - |U|$ well below the threshold V_2.

In fact, there is a hysteresis of current at all values of the electron-phonon constant γ, e.g. $\gamma^2 = 11/13$ (selected in order to avoid an accidental commensurability of ω_0 and U), Fig. 14. Indeed, the exact equation for average occupation of the dot reads

$$
\begin{aligned}
2n = {} & (1 - n)^3[na_0^+ + (1 - n)b_0^+] \\
& + 3n(1 - n)^2[na_1^+ + (1 - n)b_1^+] \\
& + 3n^2(1 - n)[na_2^+ + (1 - n)b_2^+] \\
& + n^3[na_3^+ + (1 - n)b_3^+].
\end{aligned}
\tag{46}
$$

We solved this nonlinear equation for the case $\omega_0/\Delta = 0.2$, $U^C = 0$, so that the attraction between electrons is $U = -2\gamma^2\omega_0 = -0.4$ (all energies are in units of Δ), and found an additional stable solution for an average occupation number (and current) that results in a hysteresis curve, Fig. 14. The bistability region shrinks down with temperature, and the hysteresis loop practically closes at $T/\Delta = 0.01$. As we see from (36),(37), the electron levels with phonon sidebands $\Delta \pm l\omega_0$, $\Delta + U \pm l\omega_0$, $\Delta + 2U \pm l\omega_0$, $\Delta + 3U \pm l\omega_0$ with $l = 0, 1, ...$ contribute to electron transport with different weights, and this creates a complex picture of steps on the I-V curve, Fig. 14.

Note that switching required a degenerate MQD ($d > 2$) and the weak coupling to the electrodes, $\Gamma \ll \omega_0$. Different from the non-degenerate dot, the rate equation for a multi-degenerate dot, $d > 2$, weakly coupled to the leads has multiple physical roots in a certain voltage range and a hysteretic behavior due to *correlations* between different electronic states of MQD [54].

Summarizing this section, we have calculated the I-V characteristics of the nondegenerate and two-fold degenerate MQDs showing no hysteretic behavior, and conclude that mean field approximation [83] leads to a non-existent hysteresis in a model that was solved exactly in [54]. Different from the non-degenerate and two-fold degenerate dots, the rate equation for a multi-degenerate dot, $d > 2$, weakly coupled to the leads, has multiple physical roots in a certain voltage range showing hysteretic behavior due to *correlations* between different electronic states of MQD [54]. Our conclusions are important for searching of the current-controlled polaronic molecular switches. Incidentally, C_{60} molecules have the degeneracy $d = 6$ of the lowest unoccupied level, which makes them one of the most promising candidate systems, if the weak-coupling with leads is secured.

6 Role of Defects in Molecular Transport

The interesting behavior of electron transport in molecular systems, as described above, refers to ideal systems without imperfection in ordering and composition. In reality, one expects that there will be a considerable disorder and defects in organic molecular films. As mentioned above, the conduction through absorbed [56] and Langmuir-Blodgett [57] monolayers of fatty acids $(CH_2)_n$ was associated with *defects.* An absence of tunneling through self-assembled monolayers of C12-C18 (inferred from an absence of thickness dependence at room temperature) has been reported by Boulas et al. [55]. On the other hand, the tunneling in alkanethiol SAMs was reported in [91, 92], with an exponential dependence of monolayer resistance on the chain length L, $R_\sigma \propto \exp(\beta_\sigma L)$, and no temperature dependence of the conductance in C8-C16 molecules was observed over the temperatures $T = 80 - 300K$ [92].

The electrons in alkane molecules are tightly bound to the C atoms by $\sigma-$bonds, and the band gap (between the highest occupied molecular orbital, HOMO, and lowest unoccupied molecular orbital, LUMO) is large, $\sim 9 - 10eV$ [55]. In conjugated systems with $\pi-$electrons the molecular orbitals are extended, and the HOMO-LUMO gap is correspondingly smaller, as in e.g.

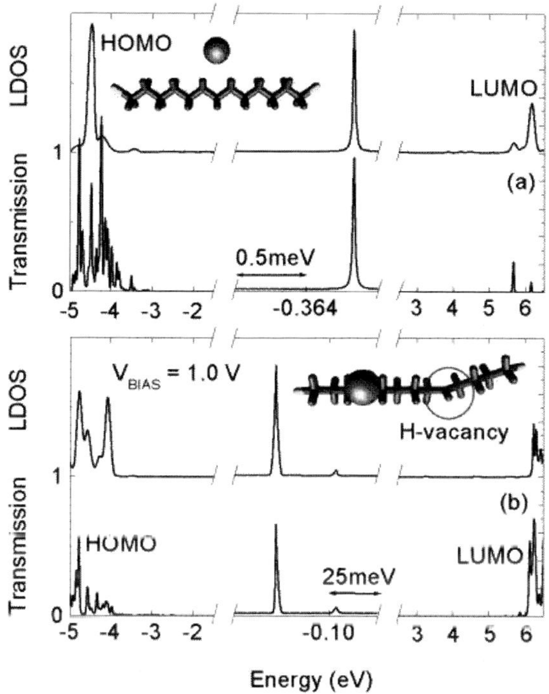

Fig. 15. Local density of states and transmission as a function of energy for (a) C13 with Au impurity and (b) C13 with Au impurity and H vacancy (dangling bond). Middle sections show closeups of the resonant peaks due to deep defect levels with respect to the HOMO and LUMO molecular states. The HOMO-LUMO gap is about 10 eV.

polythiophenes, where the resistance was also found to scale exponentially with the length of the chain, $R_\pi \propto \exp(\beta_\pi L)$, with $\beta_\pi = 0.35$ Å^{-1} instead of $\beta_\sigma = 1.08$ Å^{-1}[91]. In stark contrast with the temperature-independent tunneling results for SAMs [92], recent extensive studies of electron transport through 2.8 nm thick eicosanoic acid (C20) LB monolayers at temperatures 2K-300K have established that the current is practically temperature independent below $T < 60$K, but very strongly temperature dependent at higher temperatures $T = 60 - 300$K [93].

A large amount of effort went into characterizing the organic thin films and possible defects there [18, 94, 95]. It has been found that the electrode material, like gold, gets into the body of the film, leading to the possibility of metal ions existing in the film as single impurities and clusters. Electronic states on these impurity ions are available for the resonant tunneling of carriers in very thin films (or hopping in thicker films, a crossover between the regimes depending on the thickness). Depending on the density of the impurity states, with increasing film thickness the tunneling will be assisted by impurity "chains", with an increasing number of equidistant impurities [96]. One-impurity channels produce steps on the I-V curve but no temperature dependence, whereas the inelastic tunneling through pairs of impurities at low temperatures defines the temperature dependence of the film conductance, $G(T) \propto T^{4/3}$, and the voltage dependence of current $I(V) \propto V^{7/3}$ [97]. This behavior has been predicted theoretically and observed experimentally for tunneling through amorphous Si [98] and Al_2O_3 [99]. Due to the inevitable disorder in a "soft" matrix, the resonant states on different impurities within a "channel" will be randomly moving in and out of resonance, creating mesoscopic fluctuations of the I-V curve. The tunneling may be accompanied by interaction with vibrons on the molecule, causing step-like features on the I-V curve [54, 80].

During processing, especially top electrode deposition, small clusters of the electrode material may form in the organic film, causing Coulomb blockade, which also can show up as steps on the I-V curve. It has long been known that a strong applied field can cause localized damage to thin films, presumably due to electromigration and the formation of conducting filaments [58, 59]. The damaged area was about 30nm in diameter in 40-160 monolayer thick LB films [58] and 5-10μm in diameter in films 500-5000Å thick, and showed switching behavior under external bias voltage cycling [59]. As discussed above, recent spatial mapping of a conductance in LB monolayers of fatic acids with the use of conducting AFM has revealed damage areas 30-100nm in diameter, frequently appearing in samples after a "soft" electrical breakdown, which is sometimes accompanied by a strong temperature dependence of the conductance through the film [61].

A crossover from tunneling at low temperatures to an activation-like dependence at higher temperatures is expected for electron transport through organic molecular films. There are recent reports about such a crossover in individual molecules like the 2nm long Tour wire with a small activation energy $E_a \approx 130$meV [100]. Very small activation energies on the order of 10-100 meV have been observed in polythiophene monolayers [101]. Our present results suggest that this may be a result of interplay between the drastic renormalization of the electronic structure of the molecule in contact with electrodes, and disorder in the film (Fig. 16, right inset). We report the ab-initio calculations of point-defect assisted tunneling through alkanedithiols $S(CH_2)_nS$ and thiophene T3 (three rings SC_4) self-assembled on gold electrodes. The length of the alkane chain was in the range $n = 9 - 15$.

We have studied single and double defects in the film: (i) single Au impurity, Figs. 15a,16a, (ii) Au impurity and H vacancy (dangling bond) on the chain, Figs. 15b,16c, (iii) a pair of Au impurities, Fig. 16b, (iv) Au and a "kink" on the chain (one C=C bond instead of a C-C bond). Single defect states result in steps on the associated I-V curve, whereas molecules in the presence of two defects generally exhibit a negative differential resistance (NDR). Both types of behavior are generic and may be relevant to some observed unusual transport characteristics of SAMs and LB films [56, 57, 61, 93, 100, 101]. We have used an ab-initio approach that combines the Keldysh non-equilibrium Green's function (NEGF) method with self-consistent pseudopotential-based real space density functional theory (DFT) for systems with *open* boundary conditions provided by semi-infinite electrodes under external bias voltage [49, 51–53]. All present structures have been relaxed with the Gaussian98 code prior to transport calculations [102]. The conductance of the system at a given energy is found from (8) and the current from (4).

The equilibrium position of an Au impurity is about 3Å away from the alkane chain, which is a typical Van-der-Waals distance. As the density maps show (Fig. 16), there is an appreciable hybridization between the s- and d-states of Au and the sp- states of the carbohydrate chain. Furthermore, the Au^+ ion produces a Coulomb center trapping a 6s electron state at an energy $\epsilon_i = -0.35$eV with respect to the Fermi level, almost in the middle of the HOMO-LUMO ~ 10eV gap in Cn. The tunneling evanescent resonant state is a superposition of the HOMO and LUMO molecular orbitals. Those orbitals have a very complex spatial structure, reflected in an asymmetric line shape for the transmission. Since the impurity levels are very deep, they may be understood within the model of "short-range impurity potential" [103]. Indeed, the impurity wave function *outside* of the narrow well can be fairly approximated as

$$\varphi(r) = \sqrt{\frac{2\pi}{\kappa}} \frac{e^{-\kappa r}}{r}, \tag{47}$$

where κ is the inverse radius of the state, $\hbar^2\kappa^2/2m^* = E_i$, where $E_i = \Delta - \epsilon_i$ is the depth of the impurity level with respect to the LUMO, and Δ=LUMO$-F$ is the distance between the LUMO and the Fermi level F of gold and, consequently, the radius of the impurity state $1/\kappa$ is small. The energy distance $\Delta \approx 4.8$eV in alkane chains $(CH_2)_n$[55] (≈ 5eV from DFT calculations), and $m^* \sim 0.4$ the effective tunneling mass in alkanes [92]. For one impurity in a rectangular tunnel barrier [103] we obtain the Breit-Wigner form of transmission $T(E, V)$, as before, (5). Using the model with the impurity state wave function (47), we may estimate for an Au impurity in C13 ($L = 10.9$Å) the width $\Gamma_L = \Gamma_R = 1.2 \times 10^{-6}$, which is within an order of magnitude compared with the calculated value 1.85×10^{-5}eV. The transmission is maximal and equals unity when $E = \epsilon_i$ and $\Gamma_L = \Gamma_R$, which corresponds to a symmetrical position for the impurity with respect to the electrodes.

The electronic structure of the alkane backbone, through which the electron tunnels to an electrode, shows up in the asymmetric lineshape, which is substantially non-Lorentzian, Fig. 16. The current remains small until the bias has aligned the impurity level with the Fermi level of the electrodes, resulting in a step in the current, $I_1 \approx \frac{2g}{\hbar}\Gamma_0 e^{-\kappa L}$ (Fig. 16a). This step can be observed only when the impurity level is not very far from the Fermi level F, such that

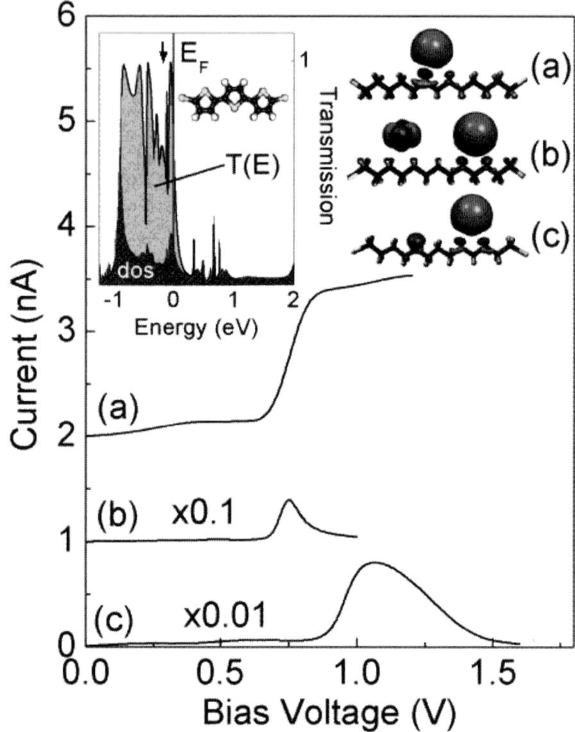

Fig. 16. Current-voltage characteristics of an alkane chain C13 with (a) single Au impurity (6s-state), (b) two Au impurities (5d and 6s-states on left and right ions, respectively), and (c) Au impurity and H vacancy (dangling bond). Double defects produce the negative differential resistance peaks (b) and (c). Inset shows the density of states, transmission, and stick model for polythiophene T3. There is significant transmission at the Fermi level, suggesting an ohmic I-V characteristic for T3 connected to gold electrodes. Disorder in the film may localize states close to the Fermi level (schematically marked by arrow), which may assist in hole hopping transport with an apparently very low activation energy (0.01-0.1eV), as is observed.

biasing the contact can produce alignment before a breakdown of the device may occur. The most interesting situations that we have found relate to the *pairs* of point defects in the film. If the concentration of defects is $c \ll 1$, the relative number of configurations with pairs of impurities will be very small, $\propto c^2$. However, they give an exponentially larger contribution to the current. Indeed, the optimal position of two impurities is symmetrical, a distance $L/2$ apart, with current $I_2 \propto e^{-\kappa L/2}$. The conductance of a two-impurity chain is [103]

$$g_{12}(E) = \frac{4q^2}{\pi\hbar} \frac{\Gamma_L \Gamma_R t_{12}^2}{|(E - \epsilon_1 + i\Gamma_L)(E - \epsilon_2 + i\Gamma_R) - t_{12}^2|^2}. \tag{48}$$

For a pair of impurities with slightly differing energies separated by a distance r_{12} we have $t_{12} = 2(E_1 + E_2)e^{-\kappa r_{12}}/\kappa r_{12}$. The interpretation of the two-impurity channel conductance (48) is fairly straightforward: if there were no coupling to the electrodes, i.e. $\Gamma_L = \Gamma_R = 0$, the poles of g_{12} would coincide with the bonding and antibonding levels of the two-impurity "molecule". The coupling to the electrodes gives them a finite width and produces, generally, two peaks in conductance, whose relative positions in energy change with the bias. The same consideration is valid for longer chains too, and gives an intuitive picture of the formation of the impurity "band" of states. The maximal conductivity $g_{12} = q^2/\pi\hbar$ occurs when $\epsilon_1 = \epsilon_2$, $\Gamma_L\Gamma_R = t_{12}^2 = \Gamma_2^2$, where Γ_2 is the width of the two-impurity resonance, and it corresponds to the symmetrical position of the impurities along the normal to the contacts separated by a distance equal to half of the molecule length, $r_{12} = L/2$. The important property of the two-impurity case is that it produces negative differential resistance (NDR). Indeed, under external bias voltage the impurity levels shift as

$$\epsilon_i = \epsilon_{i0} + qV z_i/L, \tag{49}$$

where z_i are the positions of the impurity atoms counted from the center of the molecule. Due to disorder in the film, under bias voltage the levels will be moving in and out of resonance, thus producing NDR peaks on the I-V curve. The most pronounced negative differential resistance is presented by a gold impurity next to a Cn chain with an H-vacancy on one site, Fig. 15b (the defect corresponds to a dangling bond). The defects result in two resonant peaks in transmission. Surprisingly, the H vacancy (dangling bond) has an energy very close to the electrode Fermi level F, with $\epsilon_i = -0.1\text{eV}$ (Fig. 15b, right peak). The relative positions of the resonant peaks move with an external bias and cross at 1.2V, producing a pronounced NDR peak in the I-V curve, Fig. 16c. No NDR peak is seen in the case of an Au impurity and a kink C=C on the chain because the energy of the kink level is far from that of the Au 6s impurity level. The calculated values of the peak current through the molecules were large: $I_p \approx 90$ nA/molecule for an Au impurity with H vacancy, and ≈ 5 nA/molecule for double Au impurities.

We have observed a new mechanism for the NDR peak in a situation with two Au impurities in the film. Namely, Au ions produce two sets of deep impurity levels in Cn films, one stemming from the 6s orbital, another from the 5d shell, as clearly seen in Fig. 16b (inset). The 5d-states are separated in energy from 6s, so that now the tunneling through s-d pairs of states is allowed in addition to s-s tunneling. Since the 5d-states are at a lower energy than the s-state, the d- and s-states on different Au ions will be aligned at a certain bias. Due to the different angular character of those orbitals, the tunneling between the s-state on the first impurity and a d-state on another impurity will be described by the hopping integral analogous to the Slater-

Koster $sd\sigma$ integral. The peak current in that case is smaller than for the pair Au-H vacancy, where the overlap is of $ss\sigma$ type (cf. Figs. 16b,c).

Thiophene molecules behave very differently since the $\pi-$states there are conjugated and, consequently, the HOMO-LUMO gap is much narrower, just below 2 eV. The tail of the HOMO state in the T3 molecule (with three rings) has a significant presence near the electrode Fermi level, resulting in a practically "metallic" density of states and hence ohmic I-V characteristic. This behavior is quite robust and is in apparent disagreement with experiment, where tunneling has been observed [91]. However, in actual thiophene devices the contact between the molecule and electrodes is obviously very poor, and it may lead to unusual current paths and temperature dependence [101].

We have presented the first parameter-free DFT calculations of a class of organic molecular chains incorporating single or double point defects. The results suggest that the present generic defects produce deep impurity levels in the film and cause a resonant tunneling of electrons through the film, strongly dependent on the type of defects. Thus, a missing hydrogen produces a level (dangling bond) with an energy very close to the Fermi level of the gold electrodes F. In the case of a single impurity, it produces steps on the I-V curve when one electrode's Fermi level aligns with the impurity level under a certain bias voltage. The two-defect case is much richer, since in this case we generally see a formation of the negative-differential resistance peaks. We found that the Au atom together with the hydrogen vacancy (dangling bond) produces the most pronounced NDR peak at a bias of 1.2V in C13. Other pairs of defects do not produce such spectacular NDR peaks. A short range impurity potential model reproduces the data very well, although the actual lineshape is different.

There is a remaining question of what may cause the strong temperature dependence of conductance in "simple" organic films like $[\mathrm{CH}_2]_n$. The activation-like conductance $\propto \exp(-E_a/T)$ has been reported with a small activation energy $E_a \sim 100 - 200\mathrm{meV}$ in alkanes [93, 100] and even smaller, 10-100 meV, in polythiophenes [101]. This is much smaller than the value calculated here for alkanes and expected from electrical and optical measurements on C_n molecules, $E_a \sim \Delta \sim 4\mathrm{eV}$ [55], which correspond nicely to the present results. In conjugated systems, however, there may be a rather natural explanation of small activation energies. Indeed, the HOMO in T3 polythiophene on gold is dramatically broadened, shifted to higher energies and has a considerable weight at the Fermi level. The upward shift of the HOMO is just a consequence of the work function difference between gold and the molecule. In the presence of (inevitable) disorder in the film some of the electronic states on the molecules will be localized in the vicinity of E_F. Those states will assist the thermally activated hopping of holes within a range of small activation energies ≤ 0.1 eV. Similar behavior is expected for Tour wires [100], where $E_F - \mathrm{HOMO} \sim 1$ eV [53], if the electrode-molecule contact is poor, as is usually the case.

With regards to carrier hopping in monolayers of saturated molecules, one may reasonably expect that in many studied cases the organic films are riddled with metallic protrusions (filaments), emerging due to electromigration in a very strong electric field, and/or metallic, hydroxyl, etc. inclusions[58, 59, 61]. It may result in a much smaller tunneling distance d for the carriers and the image charge lowering of the barrier. The image charge lowering of the barrier in a gap of width d is $\Delta U = q^2 \ln 4/(\epsilon d)$, meaning that a decrease of about 3.5eV may only happen in an unrealistically narrow gap $d = 2 - 3\mathring{A}$ in a film with dielectric constant $\epsilon = 2.5$, but it will add to the barrier lowering. More detailed characterization and theoretical studies along these lines may help to resolve this very unusual behavior. We note that such a mechanism cannot explain the crossover with temperature from tunneling to hopping reported for single molecular measurements, which has to be a property of the device, but not a single molecule [100].

7 Conclusions

Studying molecules as possible building blocks for ultradense electronic circuits is a fascinating quest that spans over 30 years. It was inspired decades ago by the notion that silicon technology is approaching its limiting feature size, estimated at around 1985 to be about $1\mu m$ [104]. More than thirty years later and with FET gate lengths getting below 10nm [105], the same notion that silicon needs to be replaced at some point by other technologies arises again. We do not know whether alternatives will continue to be steamrollered by silicon technology, which is a leading nanotechnology at the moment, but the mounting resistance to the famed Moore's law requires us to look hard at other solutions for power dissipation, leakage current, crosstalk, speed, and other very serious problems. There are very interesting developments in studying electronic transport through molecular films but the mechanisms of some observed conductance "switching" and/or nonlinear electric behavior remain elusive, and this interesting behavior remains intermittent and not very reproducible. Most of the currently observed switching is extrinsic in nature. For instance, we have discussed the effect of molecule-electrode contact: the tilting of the angle at which the conjugated molecule attaches to the electrode may dramatically change its conductance, and that probably explains extrinsic "telegraph" switching observed in Tour wires [23, 33] and molecule reconfigurations may lead to similar phenomena in other systems [70]. Defects in molecular films have also been discussed and may result in spurious peaks in I-V curves. We have outlined some designs of the molecules that may demonstrate rectifying behavior, which we call molecular "quantum dots". We have shown that at least in some special cases molecular quantum dots may exhibit fast ($\sim$THz) intrinsic switching [54].

Acknowledgement. The author is grateful to Jeanie Lau, Jason Pitters and Robert Wolkow for kind permission to use their data and figures. The work has been partly supported by DARPA.

References

1. J.M. Tour, Acc. Chem. Res. **33**, 791 (2000).
2. C. Joachim, Nanotechnology **13**, R1 (2002).
3. A. Aviram and M.A. Ratner, Chem. Phys. Lett. **29**, 277 (1974).
4. A.S. Martin, J.R. Sambles, and G.J. Ashwell, Phys. Rev. Lett. **70**, 218 (1993); R. M. Metzger, B. Chen, U. Höpfner, M. V. Lakshmikantham, D. Vuillaume, T. Kawai, X. Wu, H. Tachibana, T. V. Hughes, H.Sakurai, J. W. Baldwin, C. Hosh, M. P. Cava, L. Brehmer, and C. J. Ashwell, J. Am. Chem. Soc. **119**, 10455 (1997).
5. K. Stokbro, J. Taylor, and M. Brandbyge, J. Am. Chem. Soc. **125**, 3674 (2003).
6. J.C. Ellenbogen and J. Love, IEEE Proc. **70**, 218 (1993).
7. A.S. Davydov, *Theory of Molecular Excitons* (McGraw-Hill, New York, 1962).
8. F. Gutmann, Nature **219**, 1359 (1968).
9. A.S. Davydov, J. Theor. Biol. **66**, 379 (1977); A.S. Davydov and N.I. Kislukha, Phys. Status Solidi (b) **75**, 735 (1976).
10. A. Scott, Phys. Reports **217**, 1 (1992).
11. J.F. Scott, *Ferroelectric Memories* (Springer, Berlin, 2000).
12. G. Atwood and R. Bez, presentation at ISIF11 (Honolulu, 21-27 April, 2006); W.Y. Cho et al., ISSCC Dig. Tech. papers (2004); G. Wicker, SPIE **3891**, 2 (1999); S.R. Ovshinsky, Phys. Rev. Lett. **21**, 1450 (1968).
13. D. J. Jung, K. Kim, and J. F. Scott, J. Phys. Condens. Mat.**17**, 4843 (2005).
14. Y. Hiranaga and Y. Cho, 11th Intl. Mtg. Ferroel. (IMF 11, Iguassu Falls, Brazil, Sep 5-9, 2005).
15. T. I. Kamins, Interface **14**, 46 (2005); Self-assembled semiconductor nanowires, in *The Nano-Micro Interface*, edited by H. J. Fecht and M. Werner (Wiley-VCH, 2004), p. 195.
16. E. Rabani, D.R. Reichman, P.L. Geissler, and L.E. Brus, Nature **426**, 271 (2003).
17. M.C. Petty, *Langmuir-Blodgett Films* (Cambridge University Press, Cambridge, 1996).
18. A. Ulman, *Characterization of Organic Thin Films* (Butterworth-Heinemann, Boston, 1995).
19. G. Snider, P. Kuekes, and R.S. Williams, Nanotechnology **15**, 881 (2004).
20. B. Stadlober, M. Zirkl, M. Beutl, G. Leising, S. Bauer-Gogonea, and S. Bauer, Appl. Phys. Lett. **86**, 242902 (2005); J.M. Shaw and P.F. Seidler, IBM J. Res. Dev. **45**, 3 (2001).
21. S. Datta *et al.*, Phys. Rev. Lett. **79**, 2530 (1997).
22. A.M. Bratkovsky and P.E. Kornilovitch, Phys. Rev. B **67**, 115307 (2003).
23. P.E. Kornilovitch and A.M. Bratkovsky, Phys. Rev. B **64**, 195413 (2001).
24. N. Ness, S. A. Shevlin, and A. J. Fisher, Phys. Rev. B **63**, 125422 (2001).
25. W.P. Su and J.R. Schrieffer, Proc. Natl. Acad. Sci. **77**, 5626 (1980); S.A. Brazovskii, Sov. Phys. - JETP **51**, 342 (1980).

26. S. A. Brazovskii and N. N. Kirova, Zh. Eksp. Teor. Fiz. Pis'ma Red. **33**, 6 (1981) [JETP Lett. **33**, 4 (1981)].
27. A. Feldblum *et al.*, Phys. Rev. B **26**, 815 (1982).
28. R.R. Chance, J.L. Bredas, and R. Silbey, Phys. Rev. B **29**, 4491 (1984).
29. M.G. Ramsey *et al.*, Phys. Rev. B **42**, 5902 (1990).
30. D. Steinmuller, M.G. Ramsey, and F.P. Netzer, Phys. Rev. B **47**, 13323 (1993).
31. L.S. Swanson *et al.*, Synth. Metals **55**, 241 (1993).
32. J. Chen, M. A. Reed, A. M. Rawlett, and J. M. Tour, Science **286**, 1550 (1999).
33. Z.J. Donhauser, B.A. Mantooth, K.F. Kelly, L.A. Bumm, J.D. Monnell, J.J. Stapleton, D.W. Price, Jr., A. M. Rawlett, D. L. Allara, J. M. Tour, and P. S. Weiss, Science **292,** 2303 (2001); Z. J. Donhauser, B.A. Mantooth, T.P. Pearl, K.F. Kelly, S.U. Nanayakkara, and P.S. Weiss, Jpn. J. Appl. Phys. **41**, 4871 (2002).
34. C. Li, D. Zhang, X. Liu, S. Han, T. Tang, C. Zhou, W. Fan, J. Koehne, J. Han, M. Meyyappan, A.M. Rawlett, D.W. Price, and J.M. Tour, Appl. Phys. Lett. **82**, 645 (2003); M.A. Reed, J. Chen, A.M. Rawlett, D.W. Price, and J. M. Tour, Appl. Phys. Lett. **78**, 3735 (2001).
35. C.P. Collier, E. W. Wong, M. Belohradský, F. M. Raymo, J. F. Stoddart, P. J. Kuekes, R. S. Williams, and J. R. Heath, Science **285**, 391 (1999); C.P. Collier, G. Mattersteig; E.W. Wong, Y. Luo, K. Beverly, J. Sampaio, F.M. Raymo, J.F. Stoddart, J.R. Heath, Science **289**, 1172 (2000).
36. Y. Chen, D.A.A. Ohlberg, X. Li, D.R. Stewart, and R.S. Williams, J.O. Jeppesen, K.A. Nielsen, and J.F. Stoddart, D.L. Olynick, and E. Anderson, Appl. Phys. Lett. **82**, 1610 (2003); Y. Chen, G.Y. Jung, D.A A Ohlberg, X. Li, D.R Stewart, J.O Jeppesen, K.A Nielsen, J F. Stoddart, and R S. Williams, Nanotechnology **14** 462 (2003).
37. D. R. Stewart, D. A. A. Ohlberg, P. A. Beck, Y. Chen, R. S. Williams, J. O. Jeppesen, K. A. Nielsen, and J. F. Stoddart, Nanoletters **4**, 133 (2004).
38. *Molecular Switches*, edited by Ben L. Feringa (Wiley-VCH, 2001).
39. V.M. Fomin, V.N. Gladilin, J.T. Devreese, E.P. Pokatilov, S.N. Balaban, and S.N. Klimin, Phys. Rev. B 57, 2415 (1998).
40. G. Zheng, F. Patolsky, Yi Cui, W.U. Wang, C.M. Lieber, Nature Biotechnology **23**, 1294 (2005).
41. A. Bachtold, P. Hadley, T. Nakanishi, and C. Dekker, Science **294**: 1317 (2001); P.G. Collins, M.S. Arnold, and P.Avouris, Science **292**, 706 (2001); T. Rueckes, K. Kim, E. Joselevich, G.Y. Tseng, C.-L. Cheung, and C.M. Lieber, Science **289**, 94 (2000).
42. M. Ouyang and D.D. Awschalom, Science **301**, 1074 (2003).
43. C. Krzeminski, C. Delerue, G. Allan, D. Vuillaume, and R.M. Metzger, Phys. Rev. B **64**, 085405 (2001).
44. C. Zhou, M.R. Deshpande, M.A. Reed, L. Jones II, and J.M. Tour, Appl. Phys. Lett. **71**, 611 (1997).
45. Y. Xue, S.Datta, S. Hong, R. Reifenberger, J.I. Henderson, C.P. Kubiak, Phys. Rev. B **59**, 7852(R) (1999).
46. J. Reichert, R. Ochs, D. Beckmann, H. B. Weber, M. Mayor, and H. v. Löhneysen, Phys. Rev. Lett. **88**, 176804 (2002).
47. P.E. Kornilovitch, A.M. Bratkovsky, and R.S. Williams, Phys. Rev. B **66**, 165436 (2002).
48. S. Lenfant, C. Krzeminski,C. Delerue, G. Allan, D. Vuillaume, Nanoletters **3**, 741 (2003).

49. B. Larade and A.M. Bratkovsky, Phys. Rev. B **68**, 235305 (2003).

50. S. Chang, Z. Li, C.N. Lau, B. Larade, and R.S. Williams, Appl. Phys. Lett. **83**, 3198 (2003).

51. J. Taylor, H. Guo and J. Wang, Phys. Rev. B **63**, R121104 (2001)

52. J. Taylor, H. Guo and J. Wang, Phys. Rev. B **63**, 245407 (2001)

53. A.P. Jauho, N.S. Wingreen and Y. Meir, Phys. Rev. B **50**, 5528 (1994).

54. A.S. Alexandrov and A.M. Bratkovsky, Phys. Rev. B **67**, 235312 (2003).

55. C. Boulas, J.V. Davidovits, F. Rondelez, and D. Vuillaume, Phys. Rev. Lett. **76**, 4797 (1996).

56. E.E. Polymeropoulos and J. Sagiv, J. Chem. Phys. **69**, 1836 (1978).

57. R.H. Tredgold and C.S. Winter, J. Phys. D **14**, L185 (1981).

58. H. Carchano, R. Lacoste, and Y. Segui, Appl. Phys. Lett. **19**, 414 (1971)

59. N.R. Couch, B. Movaghar, and I.R. Girling, Sol. St. Commun. **59**, 7 (1986).

60. I. Shlimak and V. Martchenkov, Sol. State Commun. **107**, 443 (1998).

61. C.N. Lau, D. Stewart, R.S. Williams, and D. Bockrath, Nano Lett. **4**, 569 (2004).

62. J. Rodriguez Contreras, H. Kohlstedt, U. Poppe, R. Waser, C. Buchal, and N. A. Pertsev, Appl. Phys. Lett. **83**, 4595 (2003).

63. R.A. Wassel, G.M. Credo, R.R. Fuierer, D.L. Feldheim, C.B. Gorman, J. Am. Chem. Soc. **126**, 295 (2004).

64. J. He and S.M. Lindsay, J. Am. Chem. Soc. **127**, 11932 (2005).

65. J. Gaudioso, L.J. Lauhon, and W. Ho, Phys. Rev. Lett. **85**, 1918 (2000).

66. S.-W. Hla, G. Meryer, and K.-H. Rieder, Chem. Phys. Lett. **370**, 431 (2003).

67. G. Yang and G. Liu, J. Phys. Chem. B **107**, 8746 (2003).

68. A. Salomon, R. Arad-Yellin, A. Shanzer, A. Karton, and D.J. Cahen, Am. Chem. Soc. **126**, 11648 (2004).

69. N.P. Guisinger, M.E. Greene, R. Basu, A.S. Baluch, M.C. Hersam, Nano Lett. **4**, 55 (2004).

70. J.L. Pitters and R.A. Wolkow, Nano Lett. **6**, 390 (2006).

71. T. Rakshit, G.C. Liang, A.W. Ghosh, and S. Datta, Nano Lett. **4**, 1803 (2004).

72. A.V. Bune, V.M. Fridkin, S. Ducharme, L.M. Blinov, S.P. Palto, A. Sorokin, S.G. Yudin, and A. Zlatkin, Nature **391**, 874 (1998).

73. S. Ducharme, V. M. Fridkin, A. V. Bune, S. P. Palto, L. M. Blinov, N. N. Petukhova, and S. G. Yudin, Phys. Rev. Lett. **84**, 175 (2000).

74. A.M. Bratkovsky and A.P. Levanyuk, Phys. Rev. Lett. **87**, 019701 (2001).

75. A. Bune, S. Ducharme, V. Fridkin, L.Blinov, S. Palto, N. Petukhova, and S. Yudin, Appl. Phys. Lett. **67** 3975 (1995).

76. T. Furukawa, M. Date, M. Ohuchi, and A. Chiba, J. Appl. Phys. **56**, 1481 (1984).

77. P.E. Kornilovitch, A.M. Bratkovsky, and R.S. Williams, Phys. Rev. B **66**, 245413 (2002).

78. D.V. Averin and K.K. Likharev, in: *Mesocopic Phenomena in Solids*, edited by B.L. Altshuler *et al.* (North-Holland, Amsterdam, 1991).

79. H. Park, J. Park, A.K.L. Lim, E.H. Anderson, A.P. Alivisatos, and P.L. McEuen, Nature **407**, 57 (2000); J. Park, A.N. Pasupathy, J.I. Goldsmith, C. Chang, Y. Yaish, J.R. Petta, M. Rinkoski, J.P. Sethna, H.D. Abruña, P.L. McEuen, and D.C. Ralph, *ibid.* **417,** 722 (2002); W. Liang, M.P. Shores, M. Bockrath, J.R. Long, and H. Park, *ibid.* **417**, 725 (2002).

80. N.B. Zhitenev, H. Meng, and Z. Bao, Phys. Rev. Lett. **88**, 226801 (2002).

81. A.S. Alexandrov, A.M. Bratkovsky, and R.S. Williams, Phys. Rev. B **67**, 075301 (2003).
82. A.S. Alexandrov and A.M. Bratkovsky, cond-mat/0606366.
83. M. Galperin, M.A. Ratner, and A. Nitzan, Nano Lett. **5**, 125 (2005).
84. A.S. Alexandrov and N.F. Mott, *Polarons and Bipolarons* (World Scientific, Singapore, 1996).
85. Y. Meir and N.S. Wingreen, Phys. Rev. Lett. **68**, 2512 (1992).
86. G.D. Mahan, *Many-Particle Physics*, 2nd ed. (Plenum, New York, 1993).
87. A. Mitra, I. Aleiner, and A. Millis, Phys. Rev. Lett. **94**, 076404 (2006).
88. D. Mozyrsky, M. B. Hastings, and I. Martin, Phys. Rev. B **73**, 035104 (2006).
89. A. Mitra, I. Aleiner, and A. Millis, Phys. Rev. B **69**, 245302 (2004).
90. H. Park, J. Park, A.K.L. Lim, E.H. Anderson, A.P. Alivisatos, and P.L. McEuen, Nature **407**, 57 (2000); J. Park, A.N. Pasupathy, J.I. Goldsmith, C. Chang, Y.Yaish, J.R. Retta, M. Rinkoski, J.P. Sethna, H.D. Abruna, P.L. McEuen, and D.C. Ralph, Nature (London) **417**, 722 (2000); W. Liang *et al.*, Nature (London) **417**, 725 (2002).
91. H. Sakaguchi, A. Hirai, F. Iwata, A. Sasaki, and T. Nagamura, Appl. Phys. Lett. **79**, 3708 (2001); X.D. Cui, X. Zarate, J. Tomfohr, O.F. Sankey, A. Primak, A.L. Moore, T.A. Moore, D. Gust, G. Harris, and S. M. Lindsay, Nanotechnology **13**, 5 (2002).
92. W. Wang, T. Lee, and M.A. Reed, Phys. Rev. B **68**, 035416 (2003).
93. D.R. Stewart, D.A.A. Ohlberg, P.A. Beck, C.N. Lau, and R.S. Williams, Appl. Phys. A **80**, 1379 (2005).
94. G.L. Fisher, A.E. Hooper, R.L. Opila, D.L. Allara, and N. Winograd, J. Phys. Chem. B **104**, 3267 (2000); K. Seshadri, A.M. Wilson, A. Guiseppi-Elie, D.L. Allara, Langmuir **15**, 742 (1999).
95. A.V. Walker, T.B. Tighe, O. Cabarcos, M.D. Reinard, S. Uppili, B.C. Haynie, N. Winograd, and D. L. Allara, J. Am. Chem. Soc. **126**, 3954 (2004).
96. M. Pollak and J.J. Hauser, Phys. Rev. Lett. **31**, 1304 (1973); I.M. Lifshitz and V.Ya. Kirpichenkov, Zh. Eksp. Teor. Fiz. **77**, 989 (1979).
97. L.I. Glazman and K.A. Matveev, Sov. Phys. JETP **67**, 1276 (1988).
98. Y. Xu, D. Ephron, and M.R. Beasley, Phys. Rev. B **52**, 2843 (1995).
99. C.H. Shang, J. Nowak, R. Jansen, and J.S. Moodera, Phys. Rev. B **58**, 2917 (1998).
100. Y. Selzer, M.A. Cabassi, T.S. Mayer, D.L.Allara, J. Am. Chem. Soc. **126**, 4052 (2004).
101. N.B. Zhitenev, A. Erbe, and Z. Bao, Phys. Rev. Lett. **92**, 186805 (2004).
102. M. J. Frisch, GAUSSIAN98, Revision A.9, Gaussian, Inc., Pittsburgh, PA, 1998.
103. A.I. Larkin and K.A. Matveev, Zh. Eksp. Teor. Fiz. **93**, 1030 (1987).
104. N.G. Rambidi and V.M.Zamalin, *Molecular Microelectronics: Origin and Outlook* (Znanie, Moscow, 1985).
105. M. Ieong, B. Doris, J. Kedzierski, K. Rim, M.Yang, Science **306**, 2057 (2004).

Index

Springer Series in
MATERIALS SCIENCE

Editors: R. Hull   R. M. Osgood, Jr.   J. Parisi   H. Warlimont

Springer Series in
MATERIALS SCIENCE

Editors: R. Hull   R. M. Osgood, Jr.   J. Parisi   H. Warlimont